Wound Healing Research

Prasun Kumar • Vijay Kothari
Editors

Wound Healing Research

Current Trends and Future Directions

 Springer

Editors
Prasun Kumar
Department of Chemical Engineering
Chungbuk National University
Cheongju, Korea (Republic of)

Vijay Kothari
Institute of Science
Nirma University
Ahmedabad, India

ISBN 978-981-16-2679-1   ISBN 978-981-16-2677-7   (eBook)
https://doi.org/10.1007/978-981-16-2677-7

This Springer imprint is published by the registered company Springer Nature Singapore Pte Ltd.
The registered company address is: 152 Beach Road, #21-01/04 Gateway East, Singapore 189721, Singapore

Contents

About the Editors

Prasun Kumar is presently working as an Assistant Professor at the Department of Chemical Engineering, Yeungnam University, Republic of Korea. He holds a Ph.D. in Biotechnology from CSIR-Institute of Genomics and Integrative Biology, Delhi, India. His main areas of research are biopolymers, microbial biodiversity, bioenergy, microbial biofilms, quorum sensing, quorum quenching, and genomics. He has over seven years of experience in applied microbiological research including over 2 years postdoctoral research experience as BK-21 plus fellow at Chungbuk National University, Republic of Korea. His research work is oriented towards valorization of lignocellulosic biowastes into value-added products and antibiofilm compounds. To his credits, there are over 29 articles in various peer-reviewed SCI journals including Trends in Microbiology, Biotechnology Advances, and Bioresource Technology. To date, his work has fetched decent citations with an h index of 24 and an i10 index of 33. He has been contributing to scientific society by actively reviewing articles for more than 32 SCI journals and was awarded the peer review award by Publons in the year 2018. He serves as the editorial board member of few international journals and also worked as the guest editor for the journals, namely 'Polymers' and 'Frontiers in Bioengineering and Biotechnology'.

Vijay Kothari is a microbiologist. His primary research interest is AMR (antimicrobial resistance). His group is actively involved in investigating antimicrobial/anti-virulence potential of natural products as well as synthetic compounds. In recent past, his lab has extensively investigated the anti-pathogenic activity of various traditional medicine formulations, e.g. *Panchvalkal*, *Panchagavya*, *Triphala*, etc., against different antibiotic-resistant bacterial strains including the wound-infective species. His lab was awarded SRISTI-DBT-BIRAC Appreciation Award for validation of anti-infective potential of a traditional polyherbal formulation—Herboheal. He has also been awarded two AIMS (Artificial Intelligence Molecular Screen) award projects by Atomwise Inc., USA for identifying novel anti-infective leads. Dr. Kothari has also contributed substantially to the field as an active editor and reviewer, and has been conferred the *Sentinel of Science* award by Publons in 2016 recognizing his contribution as a peer reviewer.

Part I
Cellular and Physiological Aspects of Wound Healing

Classification of Wounds and the Physiology of Wound Healing

Ankit Gupta

Abbreviations

Ang	Angiopoietin
BM-MSC	Bone marrow mesenchymal stem cell
CTGF	Connective tissue growth factor
DAMPs	Damage-associated molecular patterns
DETC	Dendritic epidermal T-cell
ECM	Extracellular matrix
EGF	Epidermal growth factor
FGF-2	Fibroblast growth factor-2
GPCR	G protein-coupled receptor
HGF	Hepatocyte growth factor
IGF	Insulin-like growth factor
IL	Interleukin
LC	Langerhans cells
MC	Mast cell
MMP	Matrix metalloproteinases
MMP-2	Matrix metalloproteinase-2
NETs	Neutrophil extracellular traps
PA	Plasminogen activator
PAI	Plasminogen activator inhibitor
PARs	Pattern recognition receptors
PDGF	Platelet-derived growth factor
PGE2	Prostaglandin E2
TGF-α	Transforming growth factor-α
TGF-β	Transforming growth factor-β

A. Gupta (✉)
Chittorgarh, Rajasthan, India

© The Author(s), under exclusive license to Springer Nature Singapore Pte Ltd. 2021
P. Kumar, V. Kothari (eds.), *Wound Healing Research*,
https://doi.org/10.1007/978-981-16-2677-7_1

TNF-α	Tumor necrosis factor-α
tPA	Tissue plasminogen activator
TRM	CD+ resident memory T-cells
TRM	Resident memory T-cells
uPA	Urokinase plasminogen activator
VEGF	Vascular endothelial growth factor.

1   Introduction

Skin is the outermost covering of the body that is frequently exposed to external stress, pathogens, etc. and acts as a barrier against the outer environment (Rinnerthaler et al. 2015; Wong et al. 2016; Gallo 2017; Losquadro 2017; Gravitz 2018). It shields internal tissues and organs of the body and provides protection against mechanical stress, microbial infection, fluid imbalance, maintains thermal dysregulation, and also permits the sensations of touch, heat, and cold (Richmond and Harris 2014; Belkaid and Tamoutounour 2016; Chen et al. 2018b; Choi and Di Nardo 2018; Kwiecien et al. 2019; Kabashima et al. 2019). As the skin is subjected to a range of external and internal pressures, it is susceptible to various types of injuries or damage. When the integrity of the multiple layers of skin, mucosal surfaces, or organ tissue is lost due to any mechanical force (such as accidental or intentional etiology), disease, or microbial infection, etc., it leads to cellular damage and the occurrence of wound (Kujath and Michelsen 2008; Wilkins and Unverdorben 2013; Putnam et al. 2015; Gonzalez et al. 2016; Obagi et al. 2019; Herman and Bordoni 2020). In other words, loss in the skin, mucosal membrane, or tissue integrity due to internal or external factors is called as wound (Kujath and Michelsen 2008; Sarabahi and Tiwari 2012; Wilkins and Unverdorben 2013; Putnam et al. 2015; Herman and Bordoni 2020). As the skin is constantly subjected to a variety of stress factors, several kinds of immune cells, including Langerhans cells (LCs), $\gamma\delta$ T-cells, regulatory T-cells (T_{reg}), and resident memory T-cells (T_{RM}) are recruited into the skin, which plays a vital role in sustaining the physiological homeostasis (Hikosaka and Wurtz 1989; Liu et al. 2016; Ono and Kabashima 2016; Sorg et al. 2017; Kabashima et al. 2019). The basic underlying architecture of the skin and various immunological barriers present in the skin has been highlighted in Fig. 1.

The occurrence of wounds allows the entry of bacteria, viruses, or external chemicals into the body, which in turn can reason inflammation and can reason local infection (wound infection) or systemic infection (septicemia) (Percival 2002). This is a potential threat to the human organs, body, and sometimes can also lead to life-threatening conditions. Recent reports suggest that every year worldwide, scores of people are susceptible to irregular wound healing that in turn leads to long-term recovery, due to improper treatment, and moderately effective wound healing therapies (Fife and Carter 2012; Leavitt et al. 2016; Sen 2019; Rodrigues et al.

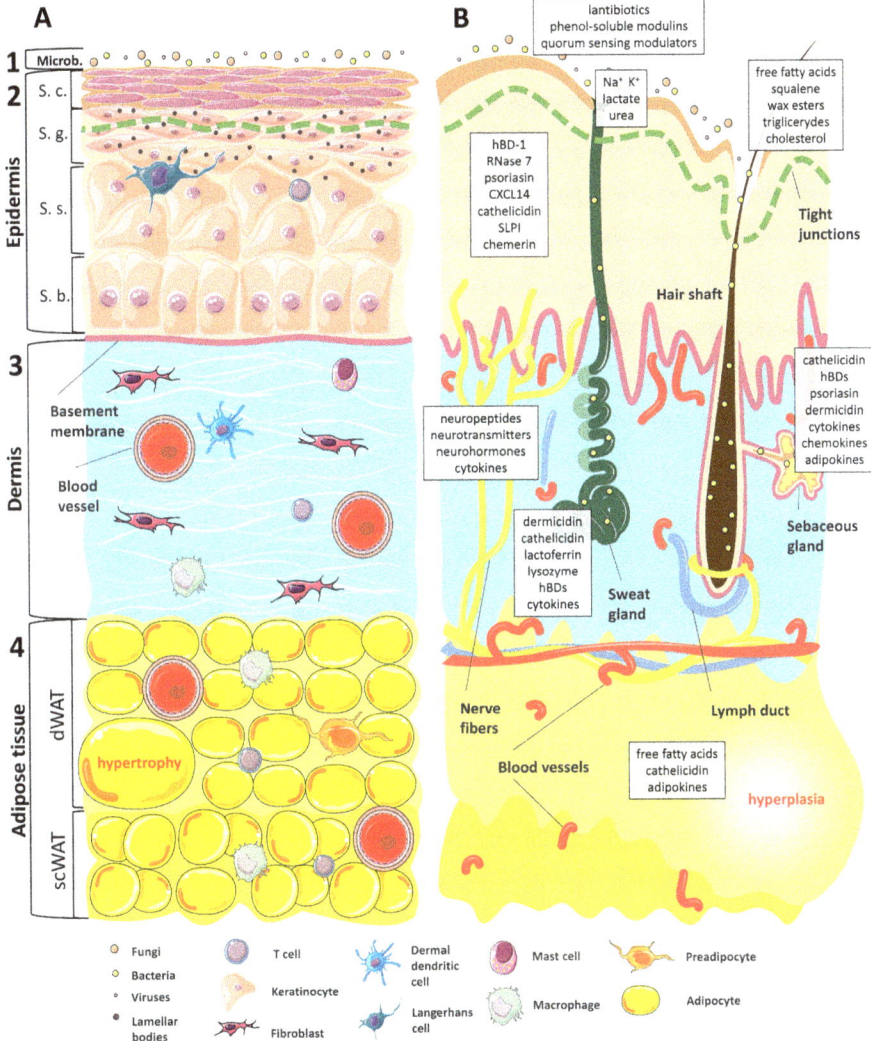

Fig. 1 The basic architecture and immunological barrier of the skin. (**a**) Structural outline of the varying layers of the skin and its components. The first layer consists of microbes such as bacteria, fungi, and viruses located at the top of the skin surface, followed by hair follicles and sweat glands. This forms the first line of defense by producing antimicrobial factors. The second layer is the epidermis which comprises four distinct strata and possesses the keratinocytes and other immune cells that are triggered upon infection. The layer below this, known as the dermis, is enriched in leukocytes and also serves as a reservoir for immune cells recruited through the bloodstream. The lowermost layer of the skin is called the adipose tissue and subdivided into white and subcutaneous white adipose tissue. This layer predominantly contains lipid-dependent immune cells and also provides defense by secreting antimicrobial peptides and factors. (**b**) Functional overview of each layer of skin corresponding to the reference mentioned in (**a**). The figure is adapted from Kwiecien, K. et al. 2019, Cytokine Growth Factor Rev. (Kwiecien et al. 2019)

2019). Therefore, wound care and rapid wound healing are critical and have clinical significance, and therefore more effective therapies of wound healing are needed.

From the early period of human life, the human body is continuously exposed to various types of injuries and diseases. Therefore, humans have always tried to find easy and effective ways of wound care and wound healing (to stop bleeding, minimize microbial infection on wounds, and in accelerating the healing process), for the successful treatment of different types of wounds (Gottrup and Leaper 2004; Broughton et al. 2006; Shah 2011; Sarabahi and Tiwari 2012; Jones 2015). Wound care has evolved over thousands of years and its treatment is an ancient area of specialization in medical science. Through evolution, the wound healing ability by regeneration of organs was replaced by repair through inflammation and subsequent deposits of the matrix proteins at the wound site (Sarabahi and Tiwari 2012). Wound care has evolved from ancient Greek medical practice (460–136 BC) through the middle ages (476 AD–1453) to the modern era of wound care (fifteenth century to twenty-first century) (Sarabahi and Tiwari 2012). During this period, we have always tried to achieve rapid wound healing to prevent wound infection and to avoid other clinical complications (Sarabahi and Tiwari 2012). The recent developments in the branches of cellular pathology, human physiology, molecular biology, microbiology, surgery, polymer chemistry, and allied fields have helped us substantially in understanding the basic mechanism of the wound healing process. With this knowledge, clinicians are now trying to find new protocols/methods of wound care and wound healing (Sarabahi and Tiwari 2012).

Wound healing is a multifaceted biological process that requires the intricate collaboration of various cell types and their products in sequential steps (Kujath and Michelsen 2008; Sorg et al. 2017; Kabashima et al. 2019; Rodrigues et al. 2019). During this process, several types of immune cells are recruited at the wound site in response to micro-environmental conditions caused by inflammatory challenges (Singh et al. 2017). These cells and chemical factors help in maintaining homeostasis upon the inflammatory challenges (Singer and Clark 1999; Gurtner et al. 2008; Eming et al. 2014; Martin and Nunan 2015; Oishi and Manabe 2018; Lim et al. 2019; Rodrigues et al. 2019). At the same time, the local environment at the site of infection changes with the improving health status of the individual. It is critical that physicians understand the fundamental physiological processes involved in the proper treatment of any cellular damage or any wound. Understanding the physiology of a typical trajectory of infection and healing through different phases guides the way for comprehending the basic principles of wound healing. With the help of this knowledge, health care professionals can develop the necessary skills to care for a wound, and the body can accomplish the complex task of tissue repair. Additionally, successful wound healing should be prioritized to reduce morbidity arising out of improper wound management (Kujath and Michelsen 2008; Gurtner et al. 2008; Singh et al. 2017).

In this chapter, we talk about two key aspects of wound healing, which are (i) classification of various types of wounds, and (ii) the cellular physiology of wound healing. Here, we first discuss the classification of these wound types in detail, which is fundamental to identifying a particular wound. Proper identification of the wound is required for a planned and successful treatment of any patient without spreading any further contamination/infection and damage to the wound or the patient's body. Next, we discuss the cellular basis of wound healing and provide detailed information on different stages of wound healing. We also provide a detailed role of the several cell types and chemical factors involved in this process. Finally, we summarize our current understanding of the fundamental mechanism and the physiology of wound healing.

2   Classification of Wounds

Various types of trauma can damage the organ tissues and lead to the occurrence of the wound. Once the body tissues are damaged, proper wound care is required to achieve complete wound healing. In wound care, we ensure that wounds are appropriately cleaned, dressed, and treated to stop further infection (Kujath and Michelsen 2008; Wilkins and Unverdorben 2013; Onyekwelu et al. 2017; Herman and Bordoni 2020). In wound care, the first and crucial step is to identify the wound and wound type. For efficient wound management, one requires detailed knowledge of the reason of tissue damage and occurrence of a wound, likelihood of surgical site infection, characteristics of the wound, and wound type (Vu et al. 2009; Levy et al. 2013; Mioton et al. 2013; Herman and Bordoni 2020). This is possible only when we have a demarcation between various wound types or injuries, and the wounds are properly classified into different classes. Wound classification is also required because the wounds of a healthy person and a diseased person (a person having diabetes, malnutrition, patients receiving grafts, or any other disorder) shows a drastic variation in bacterial contamination, infection, as well as they may also show a differential ability to achieve wound healing (Percival 2002; Vu et al. 2009; Jones 2015; Onyekwelu et al. 2017; Herman and Bordoni 2020).

Proper classification of wounds can be achieved, when one can accurately predict the probability of postoperative complications, surgical site infections, and reoperation (Mioton et al. 2013; Herman and Bordoni 2020). Several medical practitioners have attempted wound classification, but the disparities observed across various injuries and associated wounds make it a challenging and complex process (Belkaid and Tamoutounour 2016; Choi and Di Nardo 2018; Rodrigues et al. 2019). The main problem with wound classification is the low inter-rater reliability of wounds among medical professionals (Levy et al. 2013; Onyekwelu et al. 2017; Herman and Bordoni 2020). Further, wound classification does not effectively work in neonatal surgical wounds and chronically ill children, and it requires a different classification scheme for this demography (Baharestani 2007; Vu et al. 2009; Herman and Bordoni 2020). Once the details of wound characteristics (based on their etiology, morphology, skin integrity, stage of infection, etc.) have been documented properly, we have to plan for an optimized treatment to avoid further contamination and damage to the wound or patient's body. This is required to select a suitable treatment protocol at the time of diagnosis. Thus, it is necessary to classify wounds in several groups or subgroups for their proper identification, as described below.

2.1 Classification of the Wounds According to Etiology

Based on the homeostatic response of the wound, it becomes important to classify them. The severity of the wound depends on the mechanism of injury and any understated comorbidities of the patient. The first mode of classification discussed herein is the classification of wounds based on etiology. Etiology is nothing but the characterization of the wounds based on the comprehensive assessment of the cause of the injury (Gamelli and He 2003; Spanholtz et al. 2009; Kuhajda et al. 2014). Some wounds appear easy for the clinicians to classify, whereas some may represent more complicated tasks to identify. Assessing the root cause of the problem and rectifying it is also critical in preventing a recurrence of the wound. Thus, intentional or unintentional wounds depending on the severity, are classified into several different types, as described below (Table 1). The four main types of wounds, when classified by etiology, are trauma wounds, burn injuries, penetrating injuries, and surgical wounds.

2.2 Classification of the Wounds According to the Rank-Wakefield System

To accurately assess wounds, different management strategies have to be developed. This depends on the type and extremity of injury, which is of great importance. Sometimes, injuries of high extremity require amputation. Therefore, it becomes essential to classify wounds based on the treatment strategy that has to be employed. One such classification system is the Rank and Wakefield classification (Mackay 1995; Vidyarthi and Gupta 2003; Purcell 2016). According to the Rank and Wakefield point of view, wounds are divided into two broad categories: tidy and untidy

Table 1 Classification of the wounds according to etiology

	Type of wounds	Description	References
1.	Blunt trauma wounds	When a blunt object directly comes in contact with the body.	Simon et al. (2020)
2.	Burn injuries	When a small area of the body or other tissue is affected by burns.	Spanholtz et al. (2009)
3.	Penetrating injuries	When the skin is pierced due to the penetration of a foreign object in the body and leads to an open wound.	Kuhajda et al. (2014)
4.	Incisional wounds	When an incision/cut is made through muscle causing damage and disruption to the tissues.	Gamelli and He (2003)

Table 2 Classification of the wounds according to the Rank-Wakefield system

	Type of wounds	Description	References
1.	Tidy wounds	These are inflicted by sharp objects causing minimal contamination and tissue remains well-vascularized on the edge of the skin. Primary healing occurs with time.	Mackay (1995), Vidyarthi and Gupta (2003), Purcell (2016)
2.	Untidy wounds	Results from tearing, crushing, avulsion, vascular injury, or burns. These contain devitalized tissue and require debridement of all the devitalized tissue to create a tidy wound for the initial assessment. Healing occurs through secondary approaches.	Mackay (1995), Vidyarthi and Gupta (2003)

wounds (Table 2). This was one of the earliest and simplest modes of classification and served the purpose of guiding wound management strategies.

2.3 Classification of the Wounds According to the Duration of Wound Healing

Time plays a vital role in injury management and healing or repair of wounds. Thus, based on the time frame of the healing process, wounds can be surgically categorized as acute and chronic. These two terms are considered as references to the cause as well as the time frame of the wound recovery (Bowler et al. 2001; Moreo 2005; Morton and Phillips 2016; Powers et al. 2016; Maqsood 2018). The timeframe for this criterion to be applied and to consider a wound as chronic is approximately 6 weeks. Any wound which can be expected to heal in a shorter period can be classified as acute (Bowler et al. 2001; Moreo 2005; Korting et al. 2011; Jacobsona et al. 2017; Maqsood 2018). Chronic wounds are ones that do not proceed through the standard stages of wound healing in an appropriate and timely fashion. Chronic wounds are frequently found to stall in the inflammation phase of healing (Quirinia 2000; Alexiadou and Doupis 2012). One of the critical examinations to be performed on chronic wound patients is the vascular examination since the main driving factor for potential healing of these wounds is the proper recruitment of immune cells through the bloodstream. Chronic wounds are also characterized by an absence of balance in the production and degradation of cells in the wound healing process (Alexiadou and Doupis 2012; Morton and Phillips 2016; Powers et al. 2016) (Table 3). The differing types and subtypes of wounds, when classified according to the duration of wound healing, are described below.

Table 3 Classification of the wounds according to the duration of wound healing

	Type of wounds	Description	References
1.	**Acute wounds**	These arise out of damage to bony structures and soft tissues, which involves a trauma injury caused by environmental factors such as knife cuts, insect bites, burns, etc.	Moreo (2005), Maqsood (2018)
	Scrapes or abrasion	Skin causes friction when rubbed against rough surfaces. E.g., Skinned knees and rope burns.	Korting et al. (2011)
	Missile or velocity wounds	Deep body tissue damage caused by a high-speed gunshot.	Hamdan (2006), Lone et al. (2009)
	Contusions or avulsion	Wounds caused due to breaking of bones by a forcible strike on the body or getting pulled away from the rest of the bone. E.g., Hit by a ball or loss of nail, tooth, etc.	Broder (2011)
	Cut or crush wounds	When a sharp or heavy object falls on the body, causing a slice or cut. Road injuries are also included in this category, where damage occurs to the dermis and parts of the hypoderm.	Bowler et al. (2001), Montella et al. (2014)
	Lacerations	Tearing of soft body tissues, which can either be internal or external. E.g., Punching body and childbirth.	Pergialiotis et al. (2014)
	Radiation wounds or ulcers	Injuries caused to the underlying soft tissues by ionizing radiation. E.g., Chemotherapy.	Jacobsona et al. (2017)
2.	**Chronic wounds**	These are caused by metabolic perturbations or tissues injuries which heal slowly.	Morton and Phillips (2016), Powers et al. (2016)
	Venous/vascular ulcers	Occurs in the lower extremities in the legs also known as stasis ulcer or dermatitis. Usually seen in old age groups.	Vasudevan (2014)
	Diabetic wounds/ ulcers	Due to neuropathic conditions and compromised immune system, the body is unable to prevent infection and turns small wounds into chronic.	Alexiadou and Doupis (2012)
	Pressure ulcers	Also known as bedsores and normally found in the paralytic condition. Due to the immobility of the body, blood flow is restricted in muscles and tissues.	Bhattacharya and Mishra (2015)
	Ischemic wounds	It occurs as a result of a clinical blockage of blood supply to vascular beds, resulting in glucose and oxygen shortage for cellular metabolism.	Quirinia (2000)

2.4   Classification of the Wounds According to the Integrity of the Skin

Another method of classification, that is based on the degree of damage to the integrity of the skin. This method heavily dictates wound management and healing

Table 4 Classification of the wounds according to the integrity of the skin

	Type of wounds	Description	References
1.	Open wound	These injuries occur due to skin laceration with or without tissue loss.	Bauer and Aiken (1989), John et al. (2014)
2.	Closed wound	These injuries occur without any disruption in skin integrity and the skin remains intact.	John et al. (2014)

strategy since the integrity of the skin is the key component required for proper repair and regeneration (Bauer and Aiken 1989; John et al. 2014). It is a broad and straightforward mode of classification since it does not entail many categories or subtypes. A different version of the same classification emerges when we classify wounds into superficial and deep wounds (Bauer and Aiken 1989; John et al. 2014). The major criterion is the capacity of the skin to heal itself, which depends upon the extent of damage to the skin, which can then leads us to the classification as superficial, deep dermal, or full-thickness (Table 4). When classified according to the integrity of the skin, the wounds can be divided into two groups, as described below.

2.5   Classification of the Wounds According to the Bacterial Contamination or the Degree of Contamination

It is essential to restrict the spread of injury and infection caused by the wounds, and for that, it is essential to confirm that the wounds are properly cleaned and appropriately dressed. One can surmise that a surgical wound is considered a contaminated wound when an external object has come into contact with the skin leading to a high risk of infection (Devaney and Rowell 2004; Sarabahi and Tiwari 2012; Onyekwelu et al. 2017; Gorvetzian et al. 2018; Herman and Bordoni 2020)'. The extent to which the contamination has occurred dictates the wound management strategy and the urgency to clean up the wound (Table 5). When there is a presence of wound contamination, the optimal method of management is to proceed with wound closure after an initial period of delay. Thus, to classify the condition and cleanliness of wounds, the Centers for Diseases Control and Prevention (CDC) has established four classes of the degree of contamination (Devaney and Rowell 2004; Sarabahi and Tiwari 2012; Onyekwelu et al. 2017; Gorvetzian et al. 2018; Herman and Bordoni 2020)' as described below.

Table 5 Classification of the wounds based on the degree of contamination

	Type of wounds	Description	References
1.	Clean wounds	Uninfected surgical wounds where no internal damage is seen and only skin microflora has been contaminated such as hernia repair, exploratory laparotomy.	Sarabahi and Tiwari (2012), Herman and Bordoni (2020)
2.	Clean-contaminated wounds	These wounds occur when surgical wounds with microbial flora, under an uncontrolled condition, penetrate the genitourinary tract, respiratory, and alimentary tract. E.g., Hysterectomy, lobectomy.	Margenthaler et al. (2003), Devaney and Rowell (2004)
3.	Contaminated wounds	These wounds are marked by the introduction of microflora in a previously uncontaminated part of the body due to a major break in aseptic technique or gross spillage from the intestinal tract (cholecystectomy with bile spillage or acute inflammation).	Devaney and Rowell (2004), Gorvetzian et al. (2018)
4.	Heavily contaminated wounds	These are typically old traumatic wounds in which necrotic tissues are present that involve clinical infection, such as infection including repair of a perforated bowel.	Onyekwelu et al. (2017), Gorvetzian et al. (2018)

2.6 Classification of the Accidental Wounds Based on Their Origin

A clue to the cause for any skin lesions or wound can be determined from the physical assessment of the injury. Molecular mechanisms governing skin wound healing are still not completely understood. This puts the onus on the primary level of wound assessment and classification to guide wound management strategies. Clinicians find it difficult to have a unanimous denomination of wounds, since each tissue that has to heal, or the reason that initiates the wound will dictate a distinct approach (Ikpeme et al. 2010; Abrahamian and Goldstein 2011; Iyer and Balasubramanian 2012; Okonkwo and DiPietro 2017; VanHoy et al. 2020; Schaefer and Tannan 2020; Sveen et al. 2020). The final mode of classification discussed here involves assessing the root cause or origin of the wound. The major benefit of this approach is that it expedites the initial step towards wound healing and thereby improves clinical outcomes. For this purpose, Table 6 briefly describes the classification of wounds based on the morphology and pathophysiological process for the type of damage.

Table 6 Classification of the wounds according to the origin

	Type of wounds	Description	References
1.	**Mechanical wounds**	Disruption of the skin integrity including the mucous layer, caused by any mechanical force on the body. Mechanical wounds can be divided into abraded/abrasion wounds, puncture wounds, incised wounds, cut wounds, crush wounds, torn wound, bite wound, shot wound, etc.	Sveen et al. (2020)
2.	**Chemical wounds**		
	Acid wounds	These occur due to exposure to acids causing chest pain and vomiting.	VanHoy et al. (2020)
	Base wounds	Similar to acid wounds but more toxic. Necrotic tissue becomes liquified, and lysis of cell and protein occurs.	VanHoy et al. (2020)
3.	**Wounds caused by radiation**	Tissue damage caused by radiation exposure.	Iyer and Balasubramanian (2012)
4.	**Wounds caused by thermal stress**		
	Burning wounds	These are skin injuries caused by contact with hot surfaces, flame, hot liquids, or steam.	Schaefer and Tannan (2020)
	Freezing wounds	Injuries caused by the cold temperature that leads to contraction of the blood vessels and results in thrombosis.	Sarabahi and Tiwari (2012)
5.	**Wounds caused by diseases**		
	Bone infection	Inflammation of bones by bacteria or fungus.	Ikpeme et al. (2010)
	Diabetes	Due to neuropathic conditions and compromised immune system body is unable to prevent infection and turns small wounds into chronic. This is a type of chronic wound.	Boniakowski et al. (2017), Okonkwo and DiPietro (2017)
	Gangrene	Necrosis caused by bacterial infection leading to the death of body tissue.	Tan et al. (2018)
	Immunosuppressive disorder wounds	Surgical wounds associated with the compromised immune system.	Raje and Dinakar (2015)
6.	**Surgical wounds**	An incision/cut usually made during surgery using a scalpel.	Onyekwelu et al. (2017)
7.	**Microbial infection**	Penetration of pathogens or microbes through cuts on the body, causing diseases.	Aly (1996)
8.	**Wounds caused by animals**	Wounds caused by an animal bite.	Bjornstig et al. (1991), Abrahamian and Goldstein (2011)

3 Physiology of Wound Healing

A consecutive loss of function in any anatomical structure, which leads to tissue disruption can be described as a wound. As soon as any damage is done to the tissue, organ, or body, multiple parallel and interrelated pathways are activated (Singer and Clark 1999; Gurtner et al. 2008; Eming et al. 2014; Martin and Nunan 2015; Oishi and Manabe 2018; Lim et al. 2019; Rodrigues et al. 2019). These cellular and extracellular pathways work in a coordinated manner, and their corresponding functions must be carried out in the proper sequence, at the appropriate time to achieve wound healing (Gosain and DiPietro 2004; Guo and Dipietro 2010; Bielefeld et al. 2013; Sgonc and Gruber 2013; Eming et al. 2014; Bonifant and Holloway 2019; Rodrigues et al. 2019). Once the healing is completed, these pathways are stopped in a precise order to avoid extreme reactions or delayed responses (Bayat et al. 2003; Diegelmann and Evans 2004; Rodrigues et al. 2019). Despite the intricate nature of wound healing pathways, it is noteworthy that these mechanisms that regularly take place in the human body are precisely programmed and work without complication (Guo and Dipietro 2010). Interruptions in these processes can lead to delayed wound healing and a high risk of patient mortality.

Wound healing is a common phenomenon of repair, growth, and tissue regeneration (Hunt et al. 2000; Diegelmann and Evans 2004; Broughton et al. 2006; Boateng et al. 2008; Velnar et al. 2009; Wang et al. 2018). In other words, this is a complex and dynamic biological process that requires intricate spatial and temporal synchronization of various cell types and mediators, interacting in an extremely sophisticated cascade of cellular events (Hunt et al. 2000; Gonzalez et al. 2016; Sorg et al. 2017; Kabashima et al. 2019; Rodrigues et al. 2019). During the healing process, various immune cells, such as neutrophils, monocytes, Langerhans cells (LCs), $\gamma\delta$ T-cells, CD+ resident memory T-cells (TRM), and others are recruited at the wound site. In response to micro-environmental conditions and different cytokines, these cells maintain homeostasis upon inflammatory challenges (Hunt et al. 2000; Gurtner et al. 2008; Singh et al. 2017; Rodrigues et al. 2019). Details of these cells and chemical factors are provided in Tables 7 and 8, respectively, and also in succeeding sections. Since the process of wound repair is continuous, at the time of injury, various cellular and biological events are activated to restore the integrity of skin (Hunt et al. 2000; Gonzalez et al. 2016; Sorg et al. 2017; Kabashima et al. 2019; Rodrigues et al. 2019). Therefore, to better understand this physiological process that is happening at the wound site and nearby tissues, molecular events in wound repair are categorized into the following stages: hemostasis, inflammatory phase, angiogenesis, growth phase, re-epithelialization, and tissue maturation and remodeling (Hunt et al. 2000; Guo and Dipietro 2010; Sgonc and Gruber 2013; Eming et al. 2014; Gonzalez et al. 2016; Sorg et al. 2017; Bonifant and Holloway 2019; Rodrigues et al. 2019). These stages of wound healing proceed sequentially but also overlap with each other (Table 9 and Fig. 2).

The primary response in the process of wound healing is: contraction of the wounded blood vessels as well as platelet activation to form a fibrin clot, ultimately

Table 7 Cells involved in wound healing. The table has been reproduced with permission from Singh S. et al., 2017, Surgery (Singh et al. 2017) and Greaves N. S. et al., J. 2013, Dermatol. Sci. (Greaves et al. 2013)

Cell type	Time of action	Growth factors released	Function	References
Platelets	Immediately after injury	TGF-α, TGF-β, IL-1, lipoxins, leukotrienes, thromboxane-A$_2$, TNF-α, serotonin IGF-1, CTGF, VEGF, PDGF, EGF, FGF-2	• Release of inflammatory mediators (EGF, TGF-β, PDGF, FGF, serotonin, histamine, thromboxane, prostaglandins, bradykinin) • Activation of the coagulation cascade • Thrombus formation	Abe et al. (2001), Egozi et al. (2003), Workalemahu et al. (2003), Anitua et al. (2004), Sun et al. (2009), Johnston et al. (2011), Greaves et al. (2013), Singh et al. (2017)
Neutrophils	1–48 h	TGF-β, CTGF, TNF-α, IL-1	• Release of proteolytic enzymes and ROS generation • Increase vascular permeability • Phagocytosis of bacteria • Wound debridement	Egozi et al. (2003), Werner and Grose (2003), Workalemahu et al. (2003), Strid et al. (2004), Hu et al. (2010), Greaves et al. (2013), Singh et al. (2017)
Keratinocytes	8 h	TGF-β, EGF, FGF-2, VEGF, IGF-1, TNF-α, IL-1, uPA, tPA, PAI-1	• Releases inflammatory mediators • Stimulate adjacent keratinocytes • Neovascularization	Sahni and Francis (2000), Egozi et al. (2003), Workalemahu et al. (2003), Sun et al. (2009), Hu et al. (2010), Greaves et al. (2013), Singh et al. (2017)
Lymphocytes	72–120 h	IL-1, interferon-γ	• Regulates proliferative phase of wound healing • Collagen deposition	Werner and Grose (2003), Strid et al. (2004), Hu et al. (2010), Mahdavian Delavary et al. (2011), Greaves et al. (2013), Singh et al. (2017)
Fibroblasts	120 h	TGF-β, HGF, FGF-2, PDGF, VEGF, CTGF,	• Synthesis of granulation tissue • Produces ECM components	Abe et al. (2001), Yang et al. (2005), Conway et al. (2006), Giannouli

(continued)

Table 7 (continued)

Cell type	Time of action	Growth factors released	Function	References
		IGF-1, Ang-1, Ang-2, uPA	• Collagen synthesis • Release of inflammatory mediators and various proteases	and Kletsas (2006), Sun et al. (2009), Greaves et al. (2013), Singh et al. (2017)

CTGF connective tissue growth factor; *ECM* extracellular matrix; EGF: Epidermal growth factor; *FGF-2* Fibroblast growth factor-2; *HGF* hepatocyte growth factor; *IGF-1* insulin-like growth factor; *IL* Interleukin; *PAI* plasminogen activator inhibitor; *PDGF* platelet-derived growth factor; *TGF-α* transforming growth factor-α; *TGF-β* transforming growth factor-β; *TNF-α* tumor necrosis factor-α; *tPA* tissue plasminogen activator; *ROS* reactive oxygen species; *uPA* urokinase plasminogen activator; *VEGF* vascular endothelial growth factor

stopping the bleeding (Clark 2003; Geer and Andreadis 2003; Cogle et al. 2004; Rodrigues et al. 2019). Immediately after clot formation, the injured tissues release growth factors and pro-inflammatory cytokines (see Table 8 for detail). As soon as the bleeding is controlled, various inflammatory cells, for instance, neutrophils, monocytes, and macrophages, are recruited at the wound site to promote the inflammatory phase. Additionally, the adaptive immune system also gets activated to fight against a variety of self and foreign antigens (Park and Barbul 2004; Brown and Watson 2011; Davies et al. 2013a, b; Hoeffel and Ginhoux 2018; Larouche et al. 2018; Rodrigues et al. 2019). In the next phase, closely aligned with the inflammation phase, new blood vessel formation ensues through angiogenesis (Rodrigues et al. 2019). This involves the activity of neuromas cell types within the perivascular space. Here, along with the proliferation of the endothelial cells, circulating progenitor cells from the bone marrow and pericytes within the basal lamina also participate in new blood vessel formation within the perivascular space (Asahara et al. 1997; Ceradini et al. 2004; Armulik et al. 2011; Ansell and Izeta 2015; Kosaraju et al. 2016; Zhan et al. 2018; Rodrigues et al. 2019). It is then followed by fibroblast migration and proliferation, synthesis of the matrix proteins, ECM formation, keratinocyte proliferation, and differentiation, regeneration of hair follicles, etc. during the growth and proliferative phase of wound healing (Martin 1997; Werner et al. 2007; Donati et al. 2017; Rodrigues et al. 2019). Finally, the reorganization and remodeling of ECM, as well as the rearrangement of granulation tissue to scar tissue, completes the wound healing process, and synthesis as well as cross-linking of collagen provide stability to the healing tissue. In this section, we discuss each of these processes in detail to understand the physiology of wound healing (Table 9).

Table 8 Growth factors involved in wound healing. The table has been reproduced with permission from Singh S. et al., 2017, Surgery. (Singh et al. 2017) and Greaves N. S. et al., J. 2013, Dermatol. Sci. (Greaves et al. 2013)

Factor	Released from	Phase of wound healing	Action	Physiological effects during wound healing	References
TGF-α	Platelets, macrophages	Angiogenesis	• Granulation tissue formation • Stimulates proliferation of fibroblasts and epithelial cell	• Leucocyte recruitment	Abe et al. (2001), Egozi et al. (2003), Workalemahu et al. (2003), Mahdavian Delavary et al. (2011), Greaves et al. (2013), Singh et al. (2017)
TGF-β	Platelets, neutrophils, macrophages, fibroblasts, keratinocyte, mast cell	Fibroplasia, angiogenesis	• Chemotaxis • Trans-differentiation of fibroblasts • Stimulation of angiogenesis through collagen matrix construction • Wound contraction • MMP stimulation and release of growth factors	• Leucocyte recruitment	Abe et al. (2001), Workalemahu et al. (2003), Yang et al. (2005), Sun et al. (2009), Mahdavian Delavary et al. (2011), Greaves et al. (2013), Singh et al. (2017)
EGF	Platelets Macrophages	Fibroplasia	• Stimulates proliferation of different types of cells	• Re-epithelialization	Schultz et al. (1991), Johnston et al. (2011), Greaves et al. (2013), Singh et al. (2017)
HGF	Fibroblasts	Angiogenesis	• Accelerates healing and prevents fibrosis • Higher expression of HGF-NK2	• Re-epithelialization • Leucocyte recruitment	Kankuri et al. (2005), Conway et al. (2006), Greaves et al. (2013), Singh et al. (2017)
FGF-2	Platelet, macrophage Keratinocyte, fibroblast, endothelial cell, Fibrocyte	Fibroplasia, angiogenesis	• Helps in cell proliferation and migration • Formation of granulation tissue and skin repair	• Endothelial cell proliferation • ECM regulation via MMP-1 up-regulation	Anitua et al. (2004), Giannouli and Kletsas (2006), Xie et al. (2008), Chrissouli et al. (2010), Greaves et al. (2013), Singh et al. (2017)

(continued)

Table 8 (continued)

Factor	Released from	Phase of wound healing	Action	Physiological effects during wound healing	References
PDGF	Platelets, fibroblasts Endothelial cells, macrophages	Fibroplasia	• Chemotaxis • Fibroblast proliferation • Collagen deposition	• Leucocyte recruitment	Anitua et al. (2004), Chrissouli et al. (2010), Mahdavian Delavary et al. (2011), Greaves et al. (2013), Singh et al. (2017)
VEGF	Platelets, fibroblasts Endothelial cells, macrophages, keratinocyte, Myofibroblast, Fibrocyte, mast cell	Angiogenesis	• Chemotaxis • Fibroblast proliferation • Collagen deposition	• Vascular permeability • Endothelial cell migration and proliferation	Gaudry et al. (1997), Sahni and Francis (2000), Ohtani et al. (2007), Greaves et al. (2013), Singh et al. (2017)
CTGF	Platelet, neutrophil Monocyte, $\gamma\delta$ + −T-cell, fibroblast	Fibroplasia, angiogenesis	• Inhibit advanced granulation tissue formation	• Scarring	Werner and Grose (2003), Workalemahu et al. (2003), Mahdavian Delavary et al. (2011), Greaves et al. (2013), Singh et al. (2017)
IGF-1	Platelet, fibroblast, keratinocyte, DETC, BM-MSC	Fibroplasia	• Re-epithelialization and granulation tissue formation of epidermal tissue • Stimulate GH secretion	• Keratinocyte pro-survival	Gillitzer and Goebeler (2001), Anitua et al. (2004), Sharp et al. (2005), Mahdavian Delavary et al. (2011), Greaves et al. (2013), Singh et al. (2017)
Ang-1	Fibroblast, Myofibroblast	Angiogenesis	• Determines the fate of blood vessel formation • Express mRNA	• Endothelial cell proliferation	Suri et al. (1998), Papapetropoulos et al. (2000), Kumar et al. (2009), Greaves et al. (2013), Singh et al. (2017)
Ang-2	Fibroblast, Myofibroblast	Angiogenesis	• Reduce VEGF in wounds	• Antagonist to Ang-1	Suri et al. (1998), Papapetropoulos et al. (2000), Kumar et al. (2009), Staton et al.

					(2010), Greaves et al. (2013), Singh et al. (2017)
Serotonin	Platelets	Fibroplasia, apoptosis	• Vasoconstriction • Platelet aggregation • Chemotaxis • Increase vascular permeability	• Cellular proliferation	Greaves et al. (2013), Singh et al. (2017), Sadiq et al. (2018)
TNF-α	Platelets, neutrophils, monocyte, macrophage, keratinocyte, mast cell	Fibroplasia	• Chemotaxis • Nitric oxide release • Activation of other growth factors	• Expression of growth factors • Leucocyte chemotaxis • ECM degradation • Keratinocyte migration	Werner and Grose (2003), Strid et al. (2004), Strid et al. (2009), Mahdavian Delavary et al. (2011), Greaves et al. (2013), Singh et al. (2017)
PGE$_2$	Keratinocytes, macrophages, endothelial cells	Fibroplasia	• Vasodilation • Platelet disaggregation • Increased vascular permeability • Pain • Fever	• Inhibit fibroblast migration • Inhibit FPCL contraction	Sandulache et al. (2006), Greaves et al. (2013), Singh et al. (2017)
Thromboxane-A$_2$	Platelets	Fibroplasia	• Platelet aggregation • Vasoconstriction	• Enhances aggregation of blood thrombocytes	Greaves et al. (2013), Singh et al. (2017), Rodrigues et al. (2019)
Leukotrienes	Platelets, leukocytes	Fibroplasia, angiogenesis	• Amplified vascular permeability • Chemotaxis • Leukocyte adhesion • Chemotaxis (neutrophils)	• Keratinocyte migration.	Greaves et al. (2013), Singh et al. (2017), Luo et al. (2017)
IL–1	Platelets, neutrophil Macrophage, monocyte, keratinocyte, endothelial cells, lymphocytes	Fibroplasia	• Chemotaxis	• Leucocyte chemotaxis • Expression of growth factors • ECM degradation	Werner and Grose (2003), Strid et al. (2004), Hu et al. (2010), Greaves et al. (2013), Singh et al. (2017)

(continued)

Table 8 (continued)

Factor	Released from	Phase of wound healing	Action	Physiological effects during wound healing	References
Lipoxins	Platelets, leukocytes	Fibroplasia	• Dampen inflammatory response • Inhibit chemotaxis (neutrophils)	• 12/15-lipoxygenase are expressed in epithelial cells	Gronert (2005), Greaves et al. (2013), Singh et al. (2017)
Interferon-γ	Fibroblasts, lymphocytes	Angiogenesis	• Macrophage maturation • Nitric oxide release	• Enhance vascular endothelial growth factor mRNA expression	Ishida et al. (2004), Greaves et al. (2013), Singh et al. (2017)
MMP-2	Dermal fibroblast	Fibroplasia, angiogenesis	• Inhibitors help in regulating extracellular matrix degradation	• ECM/collagen degradation enables endothelial cell migration	Toriseva and Kahari (2009), Greaves et al.(2013), Singh et al. (2017), Yen et al. (2018)
uPA	Macrophage, keratinocyte, endothelial cell, monocyte, fibroblast	Fibroplasia, angiogenesis	• Induces matrix proteolysis	• Fibrin dissolution and ECM degradation activation of growth factors and MMPs	Madhyastha et al. (2008), Toriseva and Kahari (2009), Greaves et al. (2013), Singh et al. (2017), Yen et al. (2018)
tPA	Keratinocyte	Fibroplasia, angiogenesis	• Synaptic plasticity and remodeling	• Dissolution of fibrin • Degradation of ECM • Activation of MMPs and various growth factors	Lund et al. (2006), Toriseva and Kahari (2009), Greaves et al. (2013), Singh et al. (2017), Yen et al. (2018)
PAI-1	Migrating epidermal keratinocyte, fibroblast	Fibroplasia, angiogenesis	• Soluble inhibitor of proteolysis • Matrix-bound regulator for cell migration.	• PA (uPA, tPA) activity regulator • Keratinocyte migration	Schafer et al. (1994), Ghosh and Vaughan (2012), Greaves et al. (2013), Singh et al. (2017)

Ang Angiopoietin; *BM-MSC* bone marrow mesenchymal stem cell; *CTGF* connective tissue growth factor; *DETC* dendritic epidermal T-cell; *ECM* extracellular matrix; *EGF* epidermal growth factor; *FGF-2* Fibroblast growth factor-2; *GH* growth hormone; *HGF* hepatocyte growth factor; *IGF-1* insulin-like growth factor; *IL* Interleukin; *MMP* matrix metalloproteinases; *MMP-2* matrix metalloproteinase-2; *PA* plasminogen activator; *PAI* plasminogen activator inhibitor; *PDGF* platelet-derived growth factor; *PGE2* Prostaglandin E2; *TGF-α* transforming growth factor-α; *TGF-β* transforming growth factor-β; *TNF-α* tumor necrosis factor-α; *tPA* tissue plasminogen activator; *uPA* urokinase plasminogen activator; *VEGF* vascular endothelial growth factor

Table 9 Major stages in the physiological process of wound healing

Phase	Main process	Events during the process	Time	References
Hemostasis	• Vasoconstriction • Platelet plug formation (primary hemostasis) • Coagulation and reinforcement of the platelet plug	• Vascular contraction • Platelet aggregation and degranulation • Thrombus formation	Immediately after injury	Mathieu et al. (2006), Guo and Dipietro (2010), Mahdavian Delavary et al. (2011), Bielefeld et al. (2013), Sun et al. (2014), Rodrigues et al. (2019)
Inflammatory phase	• Release of growth factors and cytokines by platelets, immune cells, and disrupted matrix • Invasion of inflammatory cells (neutrophils, monocytes, macrophages)	• Vascular exudation • Neutrophil infiltration • Monocyte conversion to macrophage • Matrix enrichment in proteoglycans • Lymphocyte infiltration	Day 0–3	Mathieu et al. (2006), Guo and Dipietro (2010), Mahdavian Delavary et al. (2011), Bielefeld et al. (2013), Sun et al. (2014), Rodrigues et al. (2019)
Growth and proliferative phase	• Granulation tissue formation and neovascularization • Formation of endothelial cells and new vessel • Pericytes in neovascularization and wound healing • Circulating progenitor cells in neovascularization and wound healing • Regeneration of hair follicles	• Re-epithelialization • Angiogenesis • Fibroblast infiltration and proliferation • Collagen formation • ECM formation	Day 3–15	Mathieu et al. (2006), Guo and Dipietro (2010), Mahdavian Delavary et al. (2011), Bielefeld et al. (2013), Sun et al. (2014), Rodrigues et al. (2019)
Tissue remodeling	• ECM reorganization and remodeling • Myofibroblast formation • Contraction of the wound • Cell apoptosis	• Vascular maturation and regression • Conversion of fibroblast to fibrocyte • Collagen degradation and formation	Day 15– Month to years	Mathieu et al. (2006), Guo and Dipietro (2010), Mahdavian Delavary et al. (2011), Bielefeld et al. (2013), Sun et al. (2014), Rodrigues et al. (2019)

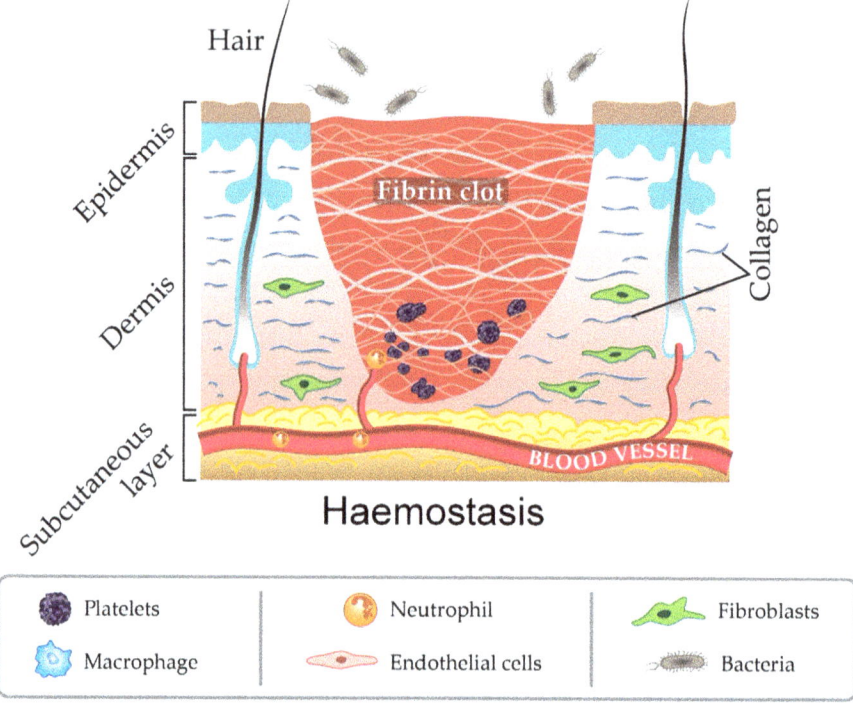

Fig. 2 Hemostasis in wound healing. This is the first phase of wound healing, which occurs as soon as any damage or injury is done on any tissue or organ. In response to the injury, the outermost layer of the skin works towards preventing excess loss of blood in a process known as vasoconstriction. Thereafter, the platelets are forced into action, where they release essential factors and drive the platelet plug formation. The final part of hemostasis stage involves the reinforcement of the platelet plug through the coagulation process. The figure is adapted with permission from Negut, I. et al. 2018, Molecules (Negut et al. 2018)

3.1 Coagulation and Hemostasis

This is the first phase in the process of wound healing, which occurs as soon as any damage or injury is done on any tissue or organ (Fig. 2, Table 9). This event occurs on a micro- or macro-vascular scale where various cellular responses promote blood clotting and prevent exsanguination or blood loss at the site of injury (Mathieu et al. 2006; Guo and Dipietro 2010; Mahdavian Delavary et al. 2011; Bielefeld et al. 2013; Singh et al. 2017; Rodrigues et al. 2019). The key point of this mechanism is (i) to prevent exsanguination in order to protect the vascular system and keep the organs unharmed during the time of injury, and (ii) to provide a scaffold for migrating cells essential for complete healing (Robson et al. 2001; Velnar et al. 2009; Guo and Dipietro 2010; Bielefeld et al. 2013; Smith et al. 2015; Singh et al. 2017; Negut et al. 2018; Rodrigues et al. 2019).

Hemostasis and reparative wound healing are strongly controlled by the balance between coagulation, endothelial cells, thrombocytes, and fibrinolysis (Falanga 2005; Velnar et al. 2009; Smith et al. 2015). This process is completed in three steps: (i) Vasoconstriction—contraction of the muscular wall of the blood vessels, in order to stop the bleeding; (ii) Primary hemostasis—platelet aggregation and platelet plug formation; and (iii) Secondary hemostasis—activation of the coagulation cascade and formation of fibrin mesh. These processes in combination form the thrombus (platelet plug and the fibrin mesh), which then allows the entry of various immune cells and growth factors to achieve wound healing (Martin 1997; Robson et al. 2001; Werner and Grose 2003; Falanga 2005; Bielefeld et al. 2013; Rodrigues et al. 2019).

3.1.1 Vasoconstriction

During this process, the blood vessels constrict rapidly to minimize or stop bleeding at the injury site. This reflexive contraction of the vascular smooth muscle is triggered by the molecule endothelin, which is released in response to the damaged endothelium (Martin 1997; Velnar et al. 2009; Godo and Shimokawa 2017; Rodrigues et al. 2019). In addition to this, some other molecules such as catecholamines, epinephrine, norepinephrine, prostaglandins, and PDGF (platelet-derived growth factor) also regulate the process of vasoconstriction. Due to increased acidosis and hypoxia at the wound site, the initial contraction of the vessels temporarily reduces bleeding and requires further activation of the coagulation cascade for the complete wound healing (Martin 1997; Robson et al. 2001; Werner and Grose 2003; Falanga 2005; Velnar et al. 2009; Godo and Shimokawa 2017; Rodrigues et al. 2019).

3.1.2 Primary Hemostasis (Formation of the Platelet Plug)

Blood platelets play a crucial role in this healing phase. Under normal conditions, anti-thrombotic properties of the endothelial cell layer do not allow platelet activation, attachment, and aggregation (Nording and Langer 2018; Suzuki-Inoue et al. 2018; Rodrigues et al. 2019). As a primary response to an injury or blood vessel rupture, the inside-out signaling pathway gets activated, leading to integrin activation and increased platelet binding to other platelets as well as ECM (Goto 2008; Golebiewska and Poole 2015). Subsequently, activation of the outside-in signaling cascade further triggers the plate activation and modifies the actin cytoskeleton or change in platelet shape. The self-association of activated platelets and their binding to ECM seals the ruptured blood vessel by forming a "platelet plug" (Goto 2008; Rumbaut and Thiagarajan 2010; Golebiewska and Poole 2015; Rodrigues et al. 2019). Several molecular factors such as integrins ($\alpha IIb\beta3$, $\alpha2\beta1$, etc.), active molecules (ADP, serotonin, calcium, histamine, etc.), glycoproteins (platelet glycoprotein Ib-IX-V and glycoprotein VI), thromboxane A2, fibronectin, vitronectin, thrombospondins, etc. regulate platelet activation and help in the "platelet plug" formation (Goto 2008; Rumbaut and Thiagarajan 2010; Golebiewska and Poole 2015; Rodrigues et al. 2019).

Platelets also secrete growth factors and cytokines, including PDGF (platelet-derived growth factor), TGF-β (transforming growth factor-β), IGF (insulin-like growth factors), and EGF (epidermal growth factor) within the "platelet plug." These molecules act as promoters in wound healing by providing the scaffold for migration of various immune cells, including keratinocytes, leukocytes, fibroblasts, and endothelial cells, as a reservoir for the subsequent phases of healing (Robson et al. 2001; Falanga 2005).

3.1.3 Coagulation and Platelet Plug Fortification

This process is called secondary hemostasis, where the platelet plug is reinforced by fibrin mesh to promote the platelet plug or the thrombus formation (Monagle and Massicotte 2011; Xu et al. 2016). The "platelet plug" provides the surface for the coagulation process. Here, the intrinsic and extrinsic coagulation pathways get stimulated by the exposure of the sub-endothelial matrix and lead to the activation of factor X. The activation of factor X induces conversion of the precursor molecule prothrombin into thrombin, which finally leads to cleavage of fibrinogen into fibrin (Goto 2008; Vojacek 2017; Rodrigues et al. 2019). The factor XIII crosslinks with fibrin, which then binds to the plate plug and forms a thrombus or secondary hemostasis plug (Monagle and Massicotte 2011; Xu et al. 2016; Tomaiuolo et al. 2017). The thrombus offers a framework for the migration of various immune cells at the wound site for the subsequent phases of healing.

3.2 Inflammatory Stage

Inflammation is a primary nonspecific immune response of the body to harmful environmental signals, such as any pathogen or bacterial infection, any damage or injury to the cells, or due to any toxic substances (Audial and Bonnotte 2015; Kuprash and Nedospasov 2016; Watson et al. 2017; Chen et al. 2018a; Negut et al. 2018). Any cellular and vascular response to injury after an immediate hemorrhage is characterized as the inflammatory phase (Table 9). The inflammation can be characterized by swelling, pain, redness of the tissue, and loss of cellular function (Fig. 3). This is a defense mechanism of the immune cells, where the principal focus is initiating the wound healing process and suppressing the infection (Ferrero-Miliani et al. 2007; Medzhitov 2010; Nathan and Ding 2010; Chen et al. 2018a). A variety of mediators are released after injury, like activation of platelets and their cytokines, and the byproducts of hemostasis, which triggers the inflammatory response from the injured tissue cells and capillaries (Ferrero-Miliani et al.

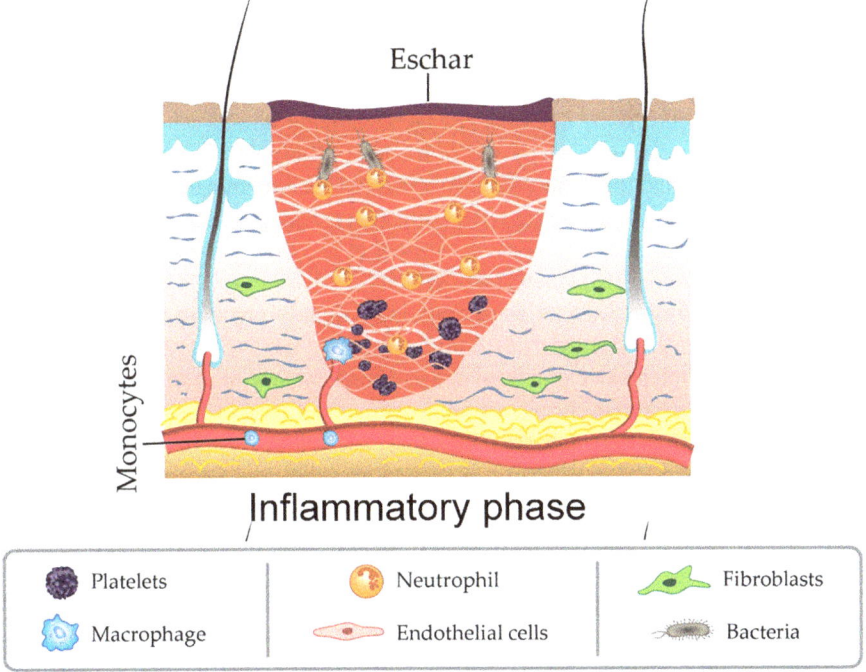

Fig. 3 Inflammatory phase in wound healing. Once hemostasis phase is complete, the blood vessels dilate and recruit the white blood cells, antibodies, growth factors, and enzymes at the site of injury. The neutrophils begin to mount the typical host response and endeavor to destroy toxic agents, foreign antigens, and microbes. The other kind of phagocytic cells, the macrophages, then proceed to autolyze any necrotic tissue. Finally, the keratinocytes, fibroblasts, and endothelial cells are activated for further stages of wound healing. The figure is adapted with permission from Negut, I. et al. 2018, Molecules (Negut et al. 2018)

2007; Medzhitov 2010; Nathan and Ding 2010; Audial and Bonnotte 2015; Chen et al. 2018a).

The inflammatory cells are recruited at the wound site through a complex process. Notably, the injury leads to an immediate increase in intracellular Ca^{+2}, and a calcium wave (Ca^{+2}) appears during the initial few seconds of the injury. The increased amount of intracellular Ca^{+2} modifies gene transcription by activating protein kinase-C and Ca^{+2}/CaMK (calmodulin-dependent protein kinase) (Tran et al. 1999; Cordeiro and Jacinto 2013; Audial and Bonnotte 2015). Apart from the increased amount of intracellular Ca^{+2}, other cellular responses or damage signals including reactive oxygen species (ROS) gradient (such as H_2O_2), damage-associated molecular patterns (DAMPs) (purines, peptides, uric acid, and ECM components), nitric oxide (NO), lipid mediators, and various chemokines (CC cytokines and CXC cytokines) activates the transcription-independent pathways for the localization of the inflammatory cells (neutrophils, macrophages, mast cells, etc.) (Jacinto et al. 2001; Lansdown 2002; McNeil and Steinhardt 2003;

Halliwell 2006; Kufareva et al. 2015; Audial and Bonnotte 2015; Hughes et al. 2017; Rodrigues et al. 2019). In combination, these molecules diffuse through epithelial tissues and trigger instantaneous gene transcription-independent responses through regulating Tyr kinases receptor and ERK or JNK signaling (Tran et al. 1999; Cordeiro and Jacinto 2013). These chemotactic damage stimuli, which are produced by epithelial cells, are sensed by the immune cells present in the nearby tissues situated in and around the wound site. This triggers the inflammatory response at the site of injury. As a result, the immune cells are recruited wound sites to clear the foreign antigens and microbes (Ferrero-Miliani et al. 2007; Nathan and Ding 2010; Audial and Bonnotte 2015; Chen et al. 2018a).

The inflammatory phase can be divided into (i) early phase and (ii) late inflammatory phase. The early phase, which initiates during the late phase of coagulation, activates the complement cascade of the immune system to stop infection at the wound site. At this stage, the neutrophils are co-localized at the site of injury in response to growth factors and cytokines such as TGF-β, complement components (C3a and C5a), and the molecules released by platelets and bacteria (Velnar et al. 2009; Cerqueira et al. 2016). Here, the neutrophil cells remove foreign antigens or bacteria through phagocytosis by releasing the ROS (reactive oxygen species) and proteolytic enzymes (Velnar et al. 2009; Roberts et al. 2018). Once all the foreign bodies are removed through neutrophils, these cells are eliminated from the wound site before the initiation of the late phase. During the late phase of inflammation, various molecules such as cytokines (PDGF, TGF-β) clotting factors, complement components (C3a and C5a), leukotriene B4, and platelet factor IV, and other breakdown products attract and recruit another phagocytotic cell macrophage at the site of injury. Once migrated, these macrophages continue the phagocytosis process for the clearance of toxic substances. Additionally, these cells provide a scaffold for various growth factors (TGF-α, TGF-β, EGF, etc.), and activate keratinocytes, fibroblasts, and endothelial cells for the later stages of wound healing (Hunt et al. 2000; Diegelmann and Evans 2004; Velnar et al. 2009; Wang et al. 2018).

In response to chemotactic damage signals, a variety of immune cells like macrophages, neutrophils, dendritic cells, mast cells, and T-cells are recruited at the site of infection to initiate wound healing and stop further infection (Hunt et al. 2000; Diegelmann and Evans 2004; Broughton et al. 2006; Velnar et al. 2009; Rodrigues et al. 2019). Each cell has a defined architecture and function in wound healing, which are as follows:

3.2.1 Neutrophils in Wound Healing

Neutrophils are the most abundant and highly motile type of granulocytes. These cells are generally found in blood and are recruited as "first responders" at the wound site. At the time of injury, various surface receptors (GPCRs (G protein-coupled receptors), integrins Fc receptors, and pattern recognition receptors (PARs), etc.) present on neutrophils detects stress signals (such as ROS gradient, DAMPs, lipid mediators, and chemokines) secreted from the endothelial cells (Robson et al. 2001; Velnar et al. 2009; Lammermann et al. 2013; Cerqueira et al. 2016; Rodrigues et al. 2019). In response to these signals, neutrophils migrate from the bloodstream towards the site of injury. Once recruited at the injury site, activated neutrophils initiate phagocytosis and snuff out toxic agents, foreign antigen, and microbes by producing toxic granules, proteolytic enzymes, oxidative burst, and neutrophil extracellular traps (NETs) (Velnar et al. 2009; Thieblemont et al. 2016; Roberts et al. 2018; Oliveira et al. 2018).

The toxic granules secreted by the neutrophils have specific functions. The primary granules or Azurophilic granules secreted from neutrophils contain azurocidin, myeloperoxidase, serine proteases (elastase, cathepsin-G, and protease-3), and lysozyme; while the secondary granules contain matrix metalloprotease 8 (MMP-8), lactoferrin, human cationic antimicrobial protein (hCAP-18), and collagenase-2, etc. (Reeves et al. 2002; Faurschou and Borregaard 2003; Wilgus et al. 2013; Rodrigues et al. 2019). The proteases present in these toxic granules degrade the basement membrane as well as the ECM and aid the neutrophils in escaping from the bloodstream and entering the injured tissue or the wound site (Velnar et al. 2009; McCarty and Percival 2013). These proteases also cleave laminin, elastin, vitronectin, fibronectin, collagen-IV, and possess antimicrobial activity. At times, an increase in the neutrophil-derived proteolytic enzyme production can result in cleavage of growth factor receptors, growth factors, and ECM causing tissue damage (Sedmak and Orosz 1991; Rodrigues et al. 2019; Westby et al. 2020).

The activated neutrophils also secrete NETs into the extracellular space. NETs are chromatin filaments lined with cytosolic proteins and histones. This is one of the most important molecules critical for neutrophil-mediated phagocytosis, which is required to capture the pathogens, foreign antigens, and cell debris present in the extracellular space through the process of NETosis (Jorch and Kubes 2017; Rodrigues et al. 2019). The process of vital NETosis helps these cells stay alive and perform phagocytosis to eliminate bacteria, viruses, and cell debris at the site of injury (Jorch and Kubes 2017; Rodrigues et al. 2019).

In the initial step of phagocytosis, neutrophils get recruited at the wound site in reaction to various stress signals such as ROS gradient, DAMPs, lipid mediators, and chemokines, etc. In response to these signals, neutrophils mobilize towards the injury site along with the bloodstream. The neutrophils present in the bloodstream secrets toxic granules that degrade the basement membrane as well as the ECM and allow neutrophils to enter the site of injury (Velnar et al. 2009; McCarty and Percival 2013). Once recruited at the wound site, neutrophils spread the chromatin filaments lined with proteases outside of the cell through NETosis (Jorch and Kubes 2017; Rodrigues et al. 2019). Thereafter, the neutrophils detect antigens using GPCRs, Fc receptors, integrins, and PARs, and a phagocytic cup is formed to engulf the foreign antigens and microbes (Jorch and Kubes 2017; Rodrigues et al. 2019). As soon as the neutrophil engulfs the foreign molecules, the toxic granules, proteolytic enzymes, oxidative bursts secreted from the cell degrade the internalized molecule and eliminate bacteria, viruses, and cell debris at the wound site (Velnar et al. 2009; Braem et al. 2015; Jorch and Kubes 2017; Roberts et al. 2018; Rodrigues et al. 2019). Once the neutrophil-mediated phagocytosis is completed and when neutrophils have achieved their charge, they undergo necrosis/apoptosis and are thrown away from the wound surface by macrophages, which is termed as efferocytosis (Bratton and Henson 2011; Jun et al. 2015; Jorch and Kubes 2017; Rodrigues et al. 2019).

3.2.2 Macrophages in Wound Healing

Macrophages can be defined as monocytes-derived cells that proliferate due to the differentiation of the monocytes present within the wound (Gordon and Taylor 2005; Malissen et al. 2014; Gordon et al. 2014; Das et al. 2015; Rodrigues et al. 2019). Macrophages are critical in the inflammatory response of wound healing. The M-CSFR is also recognized as CSF1R (colony-stimulating factor 1 receptor), that is, the receptor for IL-34, and CSF1 or M-CSF (cytokines colony-stimulating factor-1) controls/regulates the development of macrophages (Greter et al. 2012; Stanley and Chitu 2014; Easley-Neal et al. 2019). These cells situated in the skin from the time of development and are recruited at the wound site in response to platelet and mast cells degranulation which then leads to an increase in stromal-derived factor 1 (SDF1/ CXCL12) and hypoxia-inducible factors (Gordon and Taylor 2005; Gordon et al. 2014; Das et al. 2015; Gordon and Martinez-Pomares 2017). These chemical factors also raise additional monocytes and amplify the macrophage inflammatory response. After recruitment at the wound site, the tissue-resident macrophage removes toxic metabolites, dead cells, and necrotic tissues through phagocytosis (Gordon and Taylor 2005; Gordon et al. 2014; Das et al. 2015; Gordon and Martinez-Pomares 2017).

Macrophage-mediated phagocytosis is a stepwise process wherein during the early stages of healing, the microbicidal and pro-inflammatory macrophages (M1-macrophages) release various chemical molecules like IL-6 and IL-1β, and TNF-β to fight infection (Gordon and Taylor 2005; Das et al. 2015; Gordon and Martinez-Pomares 2017; Rodrigues et al. 2019). The early macrophages recruited at the wound site release several chemoattractants, including MCP-1 (monocyte chemoattractant protein-1), within the damaged tissue, which can raise additional monocytes and intensify the macrophage inflammatory response (DiPietro et al. 1995; Evans et al. 2013). These M1-macrophages identify and engulf the pathogens into the phagosomes and kill them through ROS response and oxidative burst (Slauch 2011; Das et al. 2015; Rodrigues et al. 2019). The M1-macrophages also secrete MMPs that digest the ECM and thrombus and also activates the conventional inflammation pathway (Takeuchi and Akira 2010; Malissen et al. 2014; Rodrigues et al. 2019).

By and large, neutrophils and macrophages both show similar stages of the phagocytosis process. However, the maturation of phagosomes (early endosomes, late endosomes, and lysosomes) shown by the macrophage is absent in the neutrophil and ensues through fusion with numerous granules comprising proteases and antimicrobial agents (Jorch and Kubes 2017; Rodrigues et al. 2019). Both the cells also show differences in the phagocytic receptors present on the cell surface.

The macrophages also demonstrate a phenotypic conversion from M1-macrophage (pro-inflammatory macrophage) to alternatively activated macrophage (M2-macrophage) (Galli et al. 2011; Rodrigues et al. 2019). These M2 macrophages participate in the process of angiogenesis for the production of new blood vessels and also in the proliferation stage and re-epithelialization phase of the wound healing (Rodrigues et al. 2019). The involvement of the macrophages in the physiology of wound healing has been addressed significantly; however, the knowledge of macrophage origin and their function in wound healing is still not completely understood (Gordon and Taylor 2005; Slauch 2011).

3.2.3 Mast Cells in Wound Healing

Mast cells (MCs) are ubiquitous resident cells and important for the innate immune system, and these cells are present in abundance in the barrier organs such as skin (Kalesnikoff and Galli 2008; Wulff and Wilgus 2013; Komi et al. 2020). MCs originate in the bone marrow and move to skin and mucosa, a perivascular region of the connective tissue (Sonoda et al. 1983; Trabucchi et al. 1988; Kaur et al. 2017). Mast cells, expressed in various tissues, show heterogeneity in phenotype and functions, which in turn is influenced by the change in the microenvironment (Nakano et al. 1985; Rao and Brown 2008). MCs primarily function as effector molecules in initiating allergic reactions by activating acute inflammation, mediating immunoglobulin E (IgE) reactions, and also help to combat helminth infestations (Akimoto et al. 1998; Wulff and Wilgus 2013; Rodrigues et al. 2019). These cells also participate in the re-epithelialization process as well as in the angiogenesis and help regulate mechanoresponsive mechanisms and promote scarring (Kischer et al. 1978; Akimoto et al. 1998; Foley and Ehrlich 2013).

Mast cells can prevent skin infection by releasing antimicrobial peptides in the early stages of wound healing (Siebenhaar et al. 2007; Di Nardo et al. 2008; Wang et al. 2012). These cells generate chymase and tryptase enzymes, histamine, VEGF, etc., which in turn help in ECM degradation, allow neutrophil flux, and promote keratinocyte proliferation and re-epithelialization, and enhance collagen synthesis and fibroblast proliferation (Caughey 2007; Rodrigues et al. 2019), which in conjunction contributes to wound healing process.

3.2.4 Role of Dendritic Cells (DCs) in Wound Healing

DCs are accessory cells or antigen-presenting cells (APC) that work as key modulators between innate and adaptive immunity and are involved in priming the T-cell response (Steinman et al. 1983; Said and Weindl 2015; Dixon et al. 2017; Rodrigues et al. 2019). Within the epidermis, DCs are present as Langerhans cells (LCs) (Steinman et al. 1983; Romani et al. 2003). Epidermal LCs are considered as the first immunological barrier. In humans, these cells possess unique surface receptors and atypical Birbeck granule (a unique cytoplasmic organelle) (Romani et al. 2003; Taylor et al. 2005; Said and Weindl 2015). At the time of infection, both Langerhans and Dendritic cells are found in the skin-draining lymph nodes (Steinman and Cohn 1973; Steinman et al. 1983; Romani et al. 2003; Rodrigues et al. 2019). The dendritic cells have a stronger antigen-presenting ability as compared to the macrophages (Hume 2008; Rodrigues et al. 2019). In humans, we observe the presence of CD1C+ DC, CD14+ DC, CD141+ DC, and CD207+ DC at the wound site (Romani et al. 2003; Taylor et al. 2005; Said and Weindl 2015). When the DCs encounter the antigens, the LCs downregulate e-Catherin expression, which in turn allows their migration into the draining lymph nodes and activates T-cell mediated adaptive response (Tamoutounour et al. 2013; Rodrigues et al. 2019). In humans, specific plasmacytoid dendritic cells (pDCs) are also recruited at the site of injury. This cell type senses the purines released by the wounded cells and triggers acute inflammatory response and secrets interferon-α and interferon-β (INF-α and INF-β) (Hume 2008; Rodrigues et al. 2019).

3.2.5 T-Cells in Wound Healing

T-cells or T-lymphocytes constitute a significant component of the immune system. T-cells interact with APCs and proliferate and differentiate into a particular type of effector T-cell (Skapenko et al. 2005; Wilson and Brooks 2013; Heath and Carbone 2013). Among the two variants ($\gamma\delta$ + T-cells and $\alpha\beta$ + T-cells), the $\alpha\beta$ + T-cells are found abundantly in the human skin (Mestas and Hughes 2004; Povoleri et al. 2018; Rodrigues et al. 2019). The $\gamma\delta$ + T-cells are of two types: Vγ5- negative or Vγ5-positive, and allow increased localization of neutrophils and DCs into the skin in reaction to IL-7 and TNF-α (Rodrigues et al. 2019). DETCs or $\alpha\beta$ + T-cells in the epidermis are the only subtypes of T-cells that produce chemical factors such as growth factors and cytokines (Jameson et al. 2002, 2004; Rodrigues et al. 2019). These cells act on the keratinocytes and are crucial in the healing process, which facilitates wound re-epithelialization and wound closure (Jameson et al. 2002, 2004; Rodrigues et al. 2019). The DETCs keep examining the epidermis for the infection, and as soon as it gets any damage signals, the DETCs get activated and change its morphology (Jameson et al. 2002, 2004; Rodrigues et al. 2019). The activated DETCs release several growth factors including KGF-1, KGF2, and IF-1, regulating keratinocyte proliferation at the site of wound and leads to re-epithelialization (Jameson et al. 2002, 2004; Rodrigues et al. 2019).

The $\alpha\beta$ + T-cells presenting the CD8+ receptor become cytotoxic T-cells (CD8+ killer T-cells), while the T-cells presenting CD4+ receptor become helper T-cells (CD4+ helper T-cells) (Skapenko et al. 2005; Wilson and Brooks 2013; Heath and Carbone 2013; Rodrigues et al. 2019). Upon inflammation, populations of antigen-specific T-cells become tissue-resident memory T-cells that play a critical role in the inflammatory response as well as attract antigen-specific memory T-cells towards the site of inflammation (Mijnheer and van Wijk 2019). Additionally, the CD8+ killer

T-cells and CD4+ helper T-cells are also found at the site of chronic inflammation (Skapenko et al. 2005; Wilson and Brooks 2013; Mijnheer and van Wijk 2019). The specific helper cells (CD4+ helper T-cells) play a key role in the immune-pathogenesis of autoimmune diseases and can be categorized into two major subsets: Th1 and Th2 cells (Skapenko et al. 2005; Mijnheer and van Wijk 2019). Here, the Th1 cells secrete IFN-γ and fight the intracellular bacteria, while the Th2 cells release IL-4 and IL-13 and fight against helminth infection (Skapenko et al. 2005; Wilson and Brooks 2013; Zhang et al. 2014; Mijnheer and van Wijk 2019). Conversely, the CD8+ killer T-cells possess the cytolytic activity and produce pro-inflammatory cytokines. These killer T-cells secrete TNF and INF-γ and protect the body against viral infection and tumors (Wilson and Brooks 2013; Zhang et al. 2014; Mijnheer and van Wijk 2019).

3.3 Growth and Proliferative Phase of Wound Healing

Once the initial immune responses are completed and the inflammatory phase is over, the tissue repair process is initiated (Falanga 2005; Velnar et al. 2009; Negut et al. 2018; Rodrigues et al. 2019). The main aim of this stage is to ensure the closure of the wound area through the process of angiogenesis, fibroplasia, and re-epithelialization (Falanga 2005; Velnar et al. 2009; Gonzalez et al. 2016; Han and Ceilley 2017; Rodrigues et al. 2019). This phase can be identified by the initiation of proliferation, phenotypic alteration, and migration of fibroblast and endothelial cells, followed by ECM deposition as well as granulation tissue formation (Bauer et al. 2005; Greaves et al. 2013; Gonzalez et al. 2016; Han and Ceilley 2017). Overall, the proliferative phase can be explained by two significant events (i) granulation tissue formation, angiogenesis, and neovascularization, and (ii) re-epithelization (Fig. 4, Table 9).

3.3.1 Granulation Tissue Formation, Angiogenesis, and Neovascularization

In this process, activated fibroblast and endothelial cells support the synthesis of granulation tissues, ECM, and collagen; it provides support for the further cellular influx, new blood vessels, and various inflammatory cells (Gurtner et al. 2008; Velnar et al. 2009; Eming et al. 2014; Rodrigues et al. 2019). The freshly synthesized matrix works as a replacement for the interim network of fibrin and fibronectin (Velnar et al. 2009). Fibroblasts are present in the body's connective tissues and maintain the structural integrity of these tissues (Lynch and Watt 2018; Rodrigues et al. 2019). These cells have high plasticity and have heterogeneity based on their activation status (Rodrigues et al. 2019). Fibroblast cells synthesize and remodel the extracellular matrix, secrete cytokines as well as growth factors, and help in immunomodulation during the healing process (Lynch and Watt 2018; Rodrigues

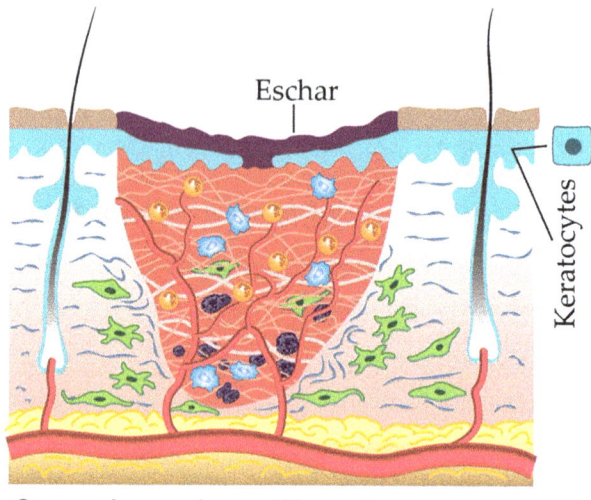

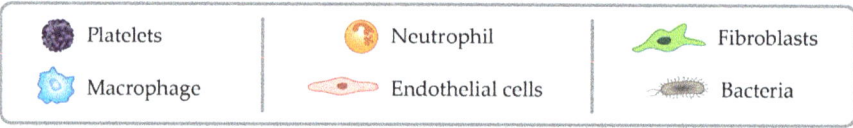

Fig. 4 Growth and proliferative phase in wound healing. During the proliferative phase, tissue repair and neovascularization occur with the help of keratinocytes and endothelial cells. Initially, the fibroblasts, with the help of collagen and other ECM factors, drive the granulation tissue formation. Thereafter, epithelial cells resurface the wound under the command of a cascade of growth factors in a process known as re-epithelialization. The figure is adapted with permission from Negut, I. et al. 2018, Molecules (Negut et al. 2018)

et al. 2019). The lower lineage fibroblast expresses myofibroblast markers such as α-SMA, which forms a large amount of ECM required for scar formation, and the collagen fibrils provide higher mechanical strength to the tissue (Rodrigues et al. 2019). The dermal papilla fibroblast contributes to hair follicle development and regeneration, as well as differentiates into myofibroblasts or smooth muscle cells (Fujiwara et al. 2011; Rodrigues et al. 2019). The myofibroblast cells dedifferentiate and release molecular factors such as fibronectin, hyaluronan, periostin, osteopontin, endothelin, angiotensin, vitronectin, CCN2, and Cx43 and further assist in wound healing (Velnar et al. 2009; Rodrigues et al. 2019).

The microvascular endothelial cells (ECs) are primarily associated with the formation of new blood cells; in response to activation signals from adjacent cells, the ECs migrate within the fibrin/fibronectin-rich clot, whereas the activated ECs establish and strengthen the interaction with the neighboring perivascular cells (Rodrigues et al. 2019). Here, the endothelial cells sense the proangiogenic activation signals such as PDGF-B, FGF, VEGF, TGF-alpha, and angiopoietins, secreted from the adjacent cells, and start proliferating to initiate the process of angiogenesis

(Tonnesen et al. 2000; DiPietro 2016; Rodrigues et al. 2019). During the process of angiogenesis, the ECs exhibits heterogeneity and are present as tip cells and stalk cells (Gerhardt et al. 2003; Rodrigues et al. 2019). Heterogeneity of ECs is regulated by Notch pathways involving effector molecules (*Eg*. VEGF), which are generated by proliferating keratinocytes, subcutaneous adipose stromal cells, and macrophages in the wound environment (Hellstrom et al. 2007; Suchting et al. 2007; Rodrigues et al. 2019), which regulates the heterogenicity observed for ECs. During angiogenesis, the ECs present at the tip sense VEGF and other signaling molecules secreted from macrophages, epidermal cells, and the subcutaneous adipose tissues (Rodrigues et al. 2019). In response to these chemicals, the tip ECs extend their filopodia and control and organize the formation of new capillaries, and the stalk ECs trail the tip cells and preserve the integrity of the vascular system (Rodrigues et al. 2019).

Pericytes and circulating progenitor cells are crucial for the process of neovascularization during wound healing. The pericyte cells are rooted within the basement membrane of the vascular system and play a vital role in(de)stabilization of the microvasculature system and in the generation of a vascular barrier to bacteria and other external agents (Hall et al. 2014; Rodrigues et al. 2019). Pericytes can wrap around numerous endothelial cells as they have a large body and extended cell membrane (Crisan et al. 2008; Hall et al. 2014; Rodrigues et al. 2019). The pericytes show multipotent regeneration capacity and contribute to wound healing by interacting with ECs and hematopoietic cells (HCs) and depositing the ECM to the injury site (Rodrigues et al. 2019). The hematopoietic stem cells and endothelial progenitor cells (HSCs and EPCs) participate in new blood vessel formation by binding to endothelial cells at sites of hypoxia/ischemic tissue (Rajantie et al. 2004; Cogle et al. 2004; Rodrigues et al. 2019). Here, the progenitor cells generated at bone marrow sense chemokines and are attracted towards increasing chemokine gradients, and once they are incorporated into sprouting endothelium, they differentiate into endothelial cells (Takahashi et al. 1999; Rajantie et al. 2004; Rodrigues et al. 2019).

3.3.2   Re-Epithelialization

The term re-epithelialization means restoring an intact epidermis after a cutaneous injury (Falanga 2005; Velnar et al. 2009; Clark 2014; Rousselle et al. 2019; Rodrigues et al. 2019). Re-epithelialization can also be defined as the migration of keratinocytes over the vascularized tissue by producing specific matrix molecules (Falanga 2005; Velnar et al. 2009; Rousselle et al. 2019; Rodrigues et al. 2019). The movement of epithelial cells from the epidermis begins within hours after injury over the denuded surface (Falanga 2005; Clark 2014; Rodrigues et al. 2019). The epithelial cells form a barrier across new tissue between the wound and the environment. The epidermis plays a crucial role in conferring protection against microorganisms, extreme temperature, UV radiation, and excessive water loss (Watt 2014; Rodrigues et al. 2019). The primary cells responsible for epithelialization are the basal keratinocytes released from the wound edge, and they become activated for migration as soon as the injury occurs (Seifert and Maden 2014; Xu et al. 2015; Rodrigues et al. 2019). Various modulators are involved in successful wound closure, including cytokines, growth factors, extracellular matrix, cellular receptors, matrix metalloproteinase, etc. (Rousselle et al. 2019; Rodrigues et al. 2019).

In general, re-epithelialization includes several stages like the movement of keratinocytes from the wound edge, proliferation of keratinocytes, and its differentiation. Migration of keratinocytes is the early phase of wound re-epithelialization, wherein the epithelial sheet migrates towards the center of the wound (Odland and Ross 1968; Rodrigues et al. 2019). This event is observed for a few hours to 1 day from the time of the injury. This involves flattening and extension of keratinocytes, the formation of lamellipodia, and ruffling of cytoplasmic projections. Also, events like the loss of cell-cell and cell-matrix interaction, recantation of intracellular tonofilament at the end of cytoplasm, and the formation of actin filament are observed (Odland and Ross 1968; Stenn and Depalma 1988; Takeo et al. 2015; Galliot et al. 2017). While in the proliferative stage of keratinocytes, cells are restricted from the foremost edge of the wound (Krawczyk 1971; Clark and Henson 1996; Rodrigues et al. 2019). Therefore, to increase the strength of cells, epidermal stem cells from interfollicular epidermis present near the hair follicle bulge, or sebaceous gland proliferate towards the edge of the wound nearly 2 to 3 days after the injury (Tumbar et al. 2004; Ito et al. 2005; Levy et al. 2007; Gurtner et al. 2008; Lau et al. 2009; Snippert et al. 2010; Mascre et al. 2012). As the suprabasal layer of differentiated keratinocytes loses its ability to proliferate, only the basal keratinocytes can move (Morasso and Tomic-Canic 2005). The proliferation of keratinocytes is regulated by growth factors name a few EGF, ECM, and integrins (Michopoulou and Rousselle 2015). There are various other growth factors, including FGFs (fibroblast growth factors), HGF (hepatocyte growth factor), HB-EGF (heparin-binding epidermal growth factor), IGF-1 (insulin growth factor-1), TGF-α, TGF-β, GM-CSF (epidermal granulocyte-macrophage colony-stimulating factor), NGF (nerve growth factor) which are secreted by multiple cells during injury. Simultaneously, wound-related signals are also released, e.g., nitric oxide, which is synthesized by macrophages (Witte and Barbul 2002; Werner and Grose 2003; Santoro and Gaudino 2005; Raja et al. 2007; Barrientos et al. 2008; Pastar et al. 2014). Finally, the epidermis is renewed by the proliferation of stem cells, undergoing terminal differentiation of their progeny, leaving the basal layer and moving towards the surface (Koster and Roop 2007).

In addition to re-epithelialization, keratinocytes have mainly three sites of origination: the intrafollicular epidermis (IFE) during wound healing; hair follicles and sebaceous glands; and melanocytes. At the time of injury, interfollicular-derived stem cells give rise to new progenitors with active keratinocytes, which expand and repair the wound (Aragona et al. 2017). IFE has been found to produce high levels of integrins, which are more adhesive as compared to amplifying cells (Jones et al. 1995; Watt 2002b; Nieuwenhuis et al. 2007). Furthermore, there are three main integrins attached to the basal cells: $\alpha3\beta1$ and $\alpha6\beta4$ that bind to laminin, $\alpha2\beta1$ that binds to collagen, and $\alpha v\beta5$ that binds to vitronectin; several other integrins are also present over the basal, lateral, and apical surface (Watt 2002a; Raymond et al. 2007).

The second most important site from where the keratinocytes can originate in the hair follicle and sebaceous gland. The role of the hair follicle and sebaceous gland is to give rise to the hair shaft, whereas the sebaceous gland lubricates the surface of the skin by releasing lipids. The bulge is present at the hair follicle base and is further categorized into (a) lower bulge, (b) mid bulge, and (c) upper bulge stem cells. A secondary hair germ is present below the bulge consisting of progenitor cells derived from the lower bulge stem cells. To this, the junctional zone and the sebaceous gland are attached vertically above the hair bulge. The sebaceous glands have four sub-divisions of stem cells, which include the LRIG1+ stem cells, bulge stem cells, and LGR6+ stem cells, as well as BLIMP1 expressing cells (Oshima et al. 2001; Barker et al. 2008; Fuchs and Horsley 2011; Brownell et al. 2011). Also present are melanocytes, which are dendritic, neural crest-derived cells, producing melanin. The role of melanin is crucial in the pigmentation of the skin, occurring due to burns and protects the skin from UV radiation and ROS (Nishimura et al. 2002; Costin and Hearing 2007; Solanas and Benitah 2013).

3.4   Tissue Maturation and Remodeling in Wound Healing

This is the last stage of wound healing, where the closure of an acute and chronic wound is achieved, which in turn can go on for several months or sometimes more than a year (Binstock 1991; Greenhalgh 1998; Velnar et al. 2009; Gurtner et al. 2011; Gonzalez et al. 2016; Negut et al. 2018; Rodrigues et al. 2019). This stage is also termed as the maturation phase, wherein the development of fresh epithelium and scar tissue formation occurs once the wound is closed, and the body attempts to regain normal tissue structure to achieve complete healing (Witte and Barbul 1997; Greenhalgh 1998; Ramasastry 2005; Gurtner et al. 2011; Rodrigues et al. 2019). In other words, this stage determines whether the wound will recur or not. The primary aim of this stage is to reorganize wound tissue, degradation, and regeneration of the extracellular matrix, leading to normal healing (Hunt et al. 2000; Baum and Arpey 2005). The variety of growth factors such as TGF-β, PDGF, and FGF, are activated during injury and tissue repair, regulate the remodeling phase (Steenfos 1994; Amjad et al. 2007). The remodeling phase can also be characterized by the loss of multiple cell types, i.e., decrease in tissue cellularity occurs (Fig. 5, Table 9). Also, the disappearance of myofibroblast, fibroblast, pericytes, endothelial cells, and deposition of ECM (particularly interstitial collagen) takes place (Gurtner and Evans 2000; Li et al. 2007; Petreaca and Green 2014; Rodrigues et al. 2019).

   In the remodeling phase, the reorganization of tissue occurs mainly through collagen. During this, the wounded tissue has less strength as compared to the unwounded tissue. To increase the tensile strength, collagen bundles increase in diameter and recover approximately 80% of the original strength (Robson et al. 2001; Clark 2014; Qin 2016). At this stage, subsequent variations in the composition and deposition of the ECM matrix take place. Collagen type III, formed during the proliferative stage, undergoes degradation, and production of robust collagen type I

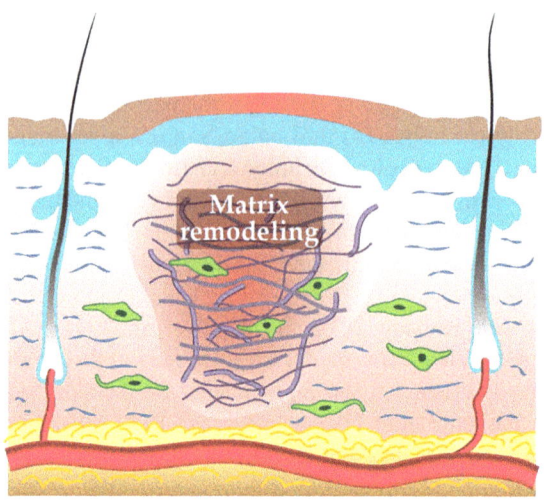

Maturation and tissue remodelling

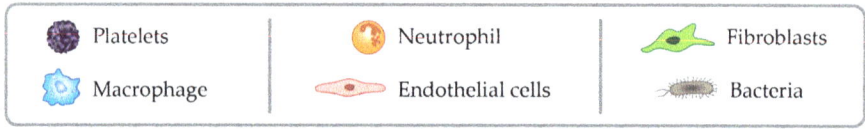

Fig. 5 Maturation and tissue remodeling phase in wound healing. This is the final phase of healing that occurs after the wound closure has been achieved. It involves remodeling of the initially formed type III collagen into the more robust interwoven type I collagen. This phase is also driven by a variety of growth factors, but the density of blood vessels and the amount of cellular activity is drastically reduced. The maturation phase typically ends with the process of scar formation. The figure is adapted with permission from Negut, I. et al. 2018, Molecules (Negut et al. 2018)

increases (Clark and Henson 1996; Greenhalgh 1998; Singer and Clark 1999; Hunt et al. 2000; Cheung and Anseth 2006; Gurtner et al. 2008; Guo and Dipietro 2010). The newly synthesized collagen type I matrix is highly organized, in contrast to the collagen type I bundle. Finally, the wound margins are brought closer to each other, and connective tissues start to shrink in size. During the remodeling and maturation process, the density of macrophages, blood vessels, fibroblasts, and inflammatory cells reduces due to emigration and apoptosis from the wound area (Greenhalgh 1998; Medrado et al. 2003; Hinz and Gabbiani 2003; Rodrigues et al. 2019). Macrophages are crucial during wound remodeling, wherein a fibrolytic phenotype is taken up, which helps in breaking the excessive ECM and engulfment of ECM debris (Rodrigues et al. 2019). At the same time, myofibroblasts move to the border of the wound and help in the contraction of smooth muscle cells during healing. However, they also express the "don't eat me" signal by overexpressing CD47 and stop phagocytosis of the fibroblast by macrophages (Chawla et al. 2016; Wernig et al. 2017; Kohale et al. 2018; Rodrigues et al. 2019). Thus, the interaction between macrophages and fibroblast cells is crucial for the deposition of excessive matrix and collagen, which results in hypertrophic scars (HTS) formation (Greenhalgh 1998; Rodrigues et al. 2019). Also, the punctured blood vessels formed during angiogenesis are repaired, and the newly formed blood vessels are subjected to pruning and give rise to a stable and well-perfused blood vessel, thereby completing the process of maturation and remodeling (Rodrigues et al. 2019).

4 Discussion

In this chapter, we have highlighted two critical aspects of wound care and wound healing, which are the following: (i) Identification of wounds and their classification, and (ii) Physiological mechanism of wound healing. The occurrence of wounds and poor wound care is a potential threat to the body tissues and can ultimately cause damage to the human body organs and result in life-threatening conditions. We show that identifying wound type such as the reason for tissue damage, the likelihood of surgical site infection, and characteristics of the wound represent the first and crucial step in proper diagnosis and treatment of wounds. We also demonstrate that despite several available classification schemes such as etiology, morphology, the integrity of skin, stage of infection, etc., we still lack the proper identification of wounds, and scores of people die every year due to medical complications. Thus, we must find more effective and promising wound healing therapies with optimum clinical significance.

Next, we discussed the cellular and molecular mechanisms entailing the wound healing process. Here, we have first explained the cellular events that happen after the body encounters any wound, and we list out the involvement of various immune cells (Table 7) as well as cytokines and growth factors (Table 8) in the physiology of wound healing. We have presented that the different stages of wound healing are well defined, are precisely programmed, and work in a synchronized manner without any complication to achieve tissue repair and restore tissue function. In the past several years, the process of wound healing has been addressed by various researchers, but there is still a debate over the recruitment and clearance of several immune cells in wound healing, and this needs to be investigated. Additionally, most of our knowledge of wound healing and skin repair is derived from murine models. As we know that there are several differences in the immune cells recruited at the wound site, skin architecture, and elasticity, adherence of skin to the underlying structure, etc. between humans and rodents, we need more studies on human systems to get a more accurate understanding of this process. Furthermore, the differences in the time and the actual process of wound healing observed due to variation in nutrition, age, sex, hormone level, oxygenation, stress level, and disease such as alcoholism, diabetes, skin fibrosis, and immunosuppressive disorders also demand a further understanding of wound healing mechanism.

To overcome all these limitations, which lead to impaired wound healing, several advanced strategies and technologies such as the use of skin grafts, use of ECM and cell-based therapies, use of the biodegradable scaffold, etc. are being employed to facilitate wound healing and to reduce the healing time. However, these advanced methodologies are moderately effective and need to be standardized for patients with chronic wounds or other comorbidities for improved and effective wound healing. Overall, we believe that an improved classification system for various wounds, the development of newer and advanced strategies for the discovery of cellular alteration and cellular diversity, as well as other technological advances, will help in achieving promising, effective, and clinically significant wound healing therapies.

Acknowledgments I would like to thank Dr. Vijay Kothari for providing me this opportunity to write this chapter and for his valuable suggestions. I would also like to thank Dr. Bharat Iyer and Dr. Himanshu Joshi for their valuable inputs.

Conflict of Interest The author has no conflict of interest to declare.

References

Abe R, Donnelly SC, Peng T, Bucala R, Metz CN (2001) Peripheral blood fibrocytes: differentiation pathway and migration to wound sites. J Immunol 166(12):7556–7562. https://doi.org/10.4049/jimmunol.166.12.7556

Abrahamian FM, Goldstein EJ (2011) Microbiology of animal bite wound infections. Clin Microbiol Rev 24(2):231–246. https://doi.org/10.1128/CMR.00041-10

Akimoto S, Ishikawa O, Igarashi Y, Kurosawa M, Miyachi Y (1998) Dermal mast cells in scleroderma: their skin density, tryptase/chymase phenotypes and degranulation. Br J Dermatol 138(3):399–406. https://doi.org/10.1046/j.1365-2133.1998.02114.x

Alexiadou K, Doupis J (2012) Management of diabetic foot ulcers. Diabetes Ther 3(1):4. https://doi.org/10.1007/s13300-012-0004-9

Aly R (1996) Microbial infections of skin and nails. In: Baron S (ed) Medical microbiology. Galveston, TX, University of Texas Medical Branch at Galveston

Amjad SB, Carachi R, Edward M (2007) Keratinocyte regulation of TGF-beta and connective tissue growth factor expression: a role in suppression of scar tissue formation. Wound Repair Regen 15(5):748–755. https://doi.org/10.1111/j.1524-475X.2007.00281.x

Anitua E, Andia I, Ardanza B, Nurden P, Nurden AT (2004) Autologous platelets as a source of proteins for healing and tissue regeneration. Thromb Haemost 91(1):4–15. https://doi.org/10.1160/TH03-07-0440

Ansell DM, Izeta A (2015) Pericytes in wound healing: friend or foe? Exp Dermatol 24 (11):833–834. https://doi.org/10.1111/exd.12782

Aragona M, Dekoninck S, Rulands S, Lenglez S, Mascre G, Simons BD, Blanpain C (2017) Defining stem cell dynamics and migration during wound healing in mouse skin epidermis. Nat Commun 8:14684. https://doi.org/10.1038/ncomms14684

Armulik A, Genove G, Betsholtz C (2011) Pericytes: developmental, physiological, and pathological perspectives, problems, and promises. Dev Cell 21(2):193–215. https://doi.org/10.1016/j.devcel.2011.07.001

Asahara T, Murohara T, Sullivan A, Silver M, van der Zee R, Li T, Witzenbichler B, Schatteman G, Isner JM (1997) Isolation of putative progenitor endothelial cells for angiogenesis. Science 275 (5302):964–967. https://doi.org/10.1126/science.275.5302.964

Audial S, Bonnotte B (2015) Inflammation. Rev Prat 65(3):403–408

Baharestani MM (2007) An overview of neonatal and pediatric wound care knowledge and considerations. Ostomy Wound Manag 53(6):34–36

Barker N, van Es JH, Jaks V, Kasper M, Snippert H, Toftgard R, Clevers H (2008) Very long-term self-renewal of small intestine, colon, and hair follicles from cycling Lgr5+ve stem cells. Cold Spring Harb Symp Quant Biol 73:351–356. https://doi.org/10.1101/sqb.2008.72.003

Barrientos S, Stojadinovic O, Golinko MS, Brem H, Tomic-Canic M (2008) Growth factors and cytokines in wound healing. Wound Repair Regen 16(5):585–601. https://doi.org/10.1111/j.1524-475X.2008.00410.x

Bauer MS, Aiken S (1989) The healing of open wounds. Semin Vet Med Surg 4(4):268–273

Bauer SM, Bauer RJ, Velazquez OC (2005) Angiogenesis, vasculogenesis, and induction of healing in chronic wounds. Vasc Endovasc Surg 39(4):293–306. https://doi.org/10.1177/153857440503900401

Baum CL, Arpey CJ (2005) Normal cutaneous wound healing: clinical correlation with cellular and molecular events. Dermatol Surg 31(6):674–686.; discussion 686. https://doi.org/10.1111/j.1524-4725.2005.31612

Bayat A, McGrouther DA, Ferguson MW (2003) Skin scarring. BMJ 326(7380):88–92. https://doi.org/10.1136/bmj.326.7380.88

Belkaid Y, Tamoutounour S (2016) The influence of skin microorganisms on cutaneous immunity. Nat Rev Immunol 16(6):353–366. https://doi.org/10.1038/nri.2016.48

Bhattacharya S, Mishra RK (2015) Pressure ulcers: current understanding and newer modalities of treatment. Indian J Plast Surg 48(1):4–16. https://doi.org/10.4103/0970-0358.155260

Bielefeld KA, Amini-Nik S, Alman BA (2013) Cutaneous wound healing: recruiting developmental pathways for regeneration. Cell Mol Life Sci 70(12):2059–2081. https://doi.org/10.1007/s00018-012-1152-9

Binstock J (1991) Health objective series. Stress management consulting for workplace mental health and wellness. AAOHN J 39(2):62–63

Bjornstig U, Eriksson A, Ornehult L (1991) Injuries caused by animals. Injury 22(4):295–298. https://doi.org/10.1016/0020-1383(91)90009-4

Boateng JS, Matthews KH, Stevens HN, Eccleston GM (2008) Wound healing dressings and drug delivery systems: a review. J Pharm Sci 97(8):2892–2923. https://doi.org/10.1002/jps.21210

Boniakowski AE, Kimball AS, Jacobs BN, Kunkel SL, Gallagher KA (2017) Macrophage-mediated inflammation in Normal and diabetic wound healing. J Immunol 199(1):17–24. https://doi.org/10.4049/jimmunol.1700223

Bonifant H, Holloway S (2019) A review of the effects of ageing on skin integrity and wound healing. Br J Commun Nurs 24(Suppl 3):S28–S33. https://doi.org/10.12968/bjcn.2019.24.Sup3.S28

Bowler PG, Duerden BI, Armstrong DG (2001) Wound microbiology and associated approaches to wound management. Clin Microbiol Rev 14(2):244–269. https://doi.org/10.1128/CMR.14.2.244-269.2001

Braem SG, Rooijakkers SH, van Kessel KP, de Cock H, Wosten HA, van Strijp JA, Haas PJ (2015) Effective neutrophil phagocytosis of Aspergillus fumigatus is mediated by classical pathway complement activation. J Innate Immun 7(4):364–374. https://doi.org/10.1159/000369493

Bratton DL, Henson PM (2011) Neutrophil clearance: when the party is over, clean-up begins. Trends Immunol 32(8):350–357. https://doi.org/10.1016/j.it.2011.04.009

Broder JS (2011) Chapter 13 - imaging of the pelvis and hip. In: Diagnostic imaging for the emergency physician, pp 706–747

Broughton G 2nd, Janis JE, Attinger CE (2006) A brief history of wound care. Plast Reconstr Surg 117(7 Suppl):6S–11S. https://doi.org/10.1097/01.prs.0000225429.76355.dd

Brown C, Watson D (2011) Lip augmentation utilizing allogenic acellular dermal graft. Facial Plast Surg 27(6):550–554. https://doi.org/10.1055/s-0031-1298780

Brownell I, Guevara E, Bai CB, Loomis CA, Joyner AL (2011) Nerve-derived sonic hedgehog defines a niche for hair follicle stem cells capable of becoming epidermal stem cells. Cell Stem Cell 8(5):552–565. https://doi.org/10.1016/j.stem.2011.02.021

Caughey GH (2007) Mast cell tryptases and chymases in inflammation and host defense. Immunol Rev 217:141–154. https://doi.org/10.1111/j.1600-065X.2007.00509.x

Ceradini DJ, Kulkarni AR, Callaghan MJ, Tepper OM, Bastidas N, Kleinman ME, Capla JM, Galiano RD, Levine JP, Gurtner GC (2004) Progenitor cell trafficking is regulated by hypoxic gradients through HIF-1 induction of SDF-1. Nat Med 10(8):858–864. https://doi.org/10.1038/nm1075

Cerqueira MT, Pirraco RP, Marques AP (2016) Stem cells in skin wound healing: are we there yet? Adv Wound Care (New Rochelle) 5(4):164–175. https://doi.org/10.1089/wound.2014.0607

Chawla K, Lamba AK, Tandon S, Faraz F, Gaba V (2016) Effect of low-level laser therapy on wound healing after depigmentation procedure: a clinical study. J Indian Soc Periodontol 20 (2):184–188. https://doi.org/10.4103/0972-124X.176393

Chen L, Deng H, Cui H, Fang J, Zuo Z, Deng J, Li Y, Wang X, Zhao L (2018a) Inflammatory responses and inflammation-associated diseases in organs. Oncotarget 9(6):7204–7218. https://doi.org/10.18632/oncotarget.23208

Chen YE, Fischbach MA, Belkaid Y (2018b) Skin microbiota-host interactions. Nature 553 (7689):427–436. https://doi.org/10.1038/nature25177

Cheung CY, Anseth KS (2006) Synthesis of immunoisolation barriers that provide localized immunosuppression for encapsulated pancreatic islets. Bioconjug Chem 17(4):1036–1042. https://doi.org/10.1021/bc060023o

Choi JE, Di Nardo A (2018) Skin neurogenic inflammation. Semin Immunopathol 40(3):249–259. https://doi.org/10.1007/s00281-018-0675-z

Chrissouli S, Pratsinis H, Velissariou V, Anastasiou A, Kletsas D (2010) Human amniotic fluid stimulates the proliferation of human fetal and adult skin fibroblasts: the roles of bFGF and PDGF and of the ERK and Akt signaling pathways. Wound Repair Regen 18(6):643–654. https://doi.org/10.1111/j.1524-475X.2010.00626.x

Clark RA (2003) Fibrin is a many splendored thing. J Invest Dermatol 121(5):xxi–xxii. https://doi.org/10.1046/j.1523-1747.2003.12575.x

Clark RAF (2014) Wound repair: basic biology to tissue engineering. In: Principles of tissue engineering, vol 4. Elsevier, New York, pp 1595–1617

Clark RAF, Henson PM (1996) The molecular and cellular biology of wound repair, vol 2. Kluwer Academic Plenum Publishers, New York, pp 339–354

Cogle CR, Wainman DA, Jorgensen ML, Guthrie SM, Mames RN, Scott EW (2004) Adult human hematopoietic cells provide functional hemangioblast activity. Blood 103(1):133–135. https://doi.org/10.1182/blood-2003-06-2101

Conway K, Price P, Harding KG, Jiang WG (2006) The molecular and clinical impact of hepatocyte growth factor, its receptor, activators, and inhibitors in wound healing. Wound Repair Regen 14 (1):2–10. https://doi.org/10.1111/j.1743-6109.2005.00081.x

Cordeiro JV, Jacinto A (2013) The role of transcription-independent damage signals in the initiation of epithelial wound healing. Nat Rev Mol Cell Biol 14(4):249–262

Costin GE, Hearing VJ (2007) Human skin pigmentation: melanocytes modulate skin color in response to stress. FASEB J 21(4):976–994. https://doi.org/10.1096/fj.06-6649rev

Crisan M, Yap S, Casteilla L, Chen CW, Corselli M, Park TS, Andriolo G, Sun B, Zheng B, Zhang L, Norotte C, Teng PN, Traas J, Schugar R, Deasy BM, Badylak S, Buhring HJ, Giacobino JP, Lazzari L, Huard J, Peault B (2008) A perivascular origin for mesenchymal stem cells in multiple human organs. Cell Stem Cell 3(3):301–313. https://doi.org/10.1016/j.stem.2008.07.003

Das A, Sinha M, Datta S, Abas M, Chaffee S, Sen CK, Roy S (2015) Monocyte and macrophage plasticity in tissue repair and regeneration. Am J Pathol 185(10):2596–2606. https://doi.org/10.1016/j.ajpath.2015.06.001

Davies LC, Jenkins SJ, Allen JE, Taylor PR (2013a) Tissue-resident macrophages. Nat Immunol 14 (10):986–995. https://doi.org/10.1038/ni.2705

Davies LC, Rosas M, Jenkins SJ, Liao CT, Scurr MJ, Brombacher F, Fraser DJ, Allen JE, Jones SA, Taylor PR (2013b) Distinct bone marrow-derived and tissue-resident macrophage lineages proliferate at key stages during inflammation. Nat Commun 4:1886. https://doi.org/10.1038/ncomms2877

Devaney L, Rowell KS (2004) Improving surgical wound classification—why it matters. AORN J 80(2):208–209. https://doi.org/10.1016/s0001-2092(06)60559-0

Di Nardo A, Yamasaki K, Dorschner RA, Lai Y, Gallo RL (2008) Mast cell cathelicidin antimicrobial peptide prevents invasive group A Streptococcus infection of the skin. J Immunol 180 (11):7565–7573. https://doi.org/10.4049/jimmunol.180.11.7565

Diegelmann RF, Evans MC (2004) Wound healing: an overview of acute, fibrotic and delayed healing. Front Biosci 9:283–289. https://doi.org/10.2741/1184

DiPietro LA (2016) Angiogenesis and wound repair: when enough is enough. J Leukoc Biol 100 (5):979–984. https://doi.org/10.1189/jlb.4MR0316-102R

DiPietro LA, Polverini PJ, Rahbe SM, Kovacs EJ (1995) Modulation of JE/MCP-1 expression in dermal wound repair. Am J Pathol 146(4):868–875

Dixon KB, Davies SS, Kirabo A (2017) Dendritic cells and isolevuglandins in immunity, inflammation, and hypertension. Am J Physiol Heart Circ Physiol 312(3):H368–H374. https://doi.org/10.1152/ajpheart.00603.2016

Donati G, Rognoni E, Hiratsuka T, Liakath-Ali K, Hoste E, Kar G, Kayikci M, Russell R, Kretzschmar K, Mulder KW, Teichmann SA, Watt FM (2017) Wounding induces dedifferentiation of epidermal Gata6(+) cells and acquisition of stem cell properties. Nat Cell Biol 19 (6):603–613. https://doi.org/10.1038/ncb3532

Easley-Neal C, Foreman O, Sharma N, Zarrin AA, Weimer RM (2019) CSF1R ligands IL-34 and CSF1 are differentially required for microglia development and maintenance in White and Gray matter brain regions. Front Immunol 10:2199. https://doi.org/10.3389/fimmu.2019.02199

Egozi EI, Ferreira AM, Burns AL, Gamelli RL, Dipietro LA (2003) Mast cells modulate the inflammatory but not the proliferative response in healing wounds. Wound Repair Regen 11 (1):46–54. https://doi.org/10.1046/j.1524-475x.2003.11108.x

Eming SA, Martin P, Tomic-Canic M (2014) Wound repair and regeneration: mechanisms, signaling, and translation. Sci Transl Med 6(265):265sr266. https://doi.org/10.1126/scitranslmed.3009337

Evans BJ, Haskard DO, Sempowksi G, Landis RC (2013) Evolution of the macrophage CD163 phenotype and cytokine profiles in a human model of resolving inflammation. Int J Inflam 2013:780502. https://doi.org/10.1155/2013/780502

Falanga V (2005) Wound healing and its impairment in the diabetic foot. Lancet 366 (9498):1736–1743. https://doi.org/10.1016/S0140-6736(05)67700-8

Faurschou M, Borregaard N (2003) Neutrophil granules and secretory vesicles in inflammation. Microbes Infect 5(14):1317–1327. https://doi.org/10.1016/j.micinf.2003.09.008

Ferrero-Miliani L, Nielsen OH, Andersen PS, Girardin SE (2007) Chronic inflammation: importance of NOD2 and NALP3 in interleukin-1beta generation. Clin Exp Immunol 147 (2):227–235. https://doi.org/10.1111/j.1365-2249.2006.03261.x

Fife CE, Carter MJ (2012) Wound care outcomes and associated cost among patients treated in US outpatient wound centers: data from the US wound registry. Wounds 24(1):10–17

Foley TT, Ehrlich HP (2013) Through gap junction communications, co-cultured mast cells and fibroblasts generate fibroblast activities allied with hypertrophic scarring. Plast Reconstr Surg 131(5):1036–1044. https://doi.org/10.1097/PRS.0b013e3182865c3f

Fuchs E, Horsley V (2011) Ferreting out stem cells from their niches. Nat Cell Biol 13(5):513–518. https://doi.org/10.1038/ncb0511-513

Fujiwara H, Ferreira M, Donati G, Marciano DK, Linton JM, Sato Y, Hartner A, Sekiguchi K, Reichardt LF, Watt FM (2011) The basement membrane of hair follicle stem cells is a muscle cell niche. Cell 144(4):577–589. https://doi.org/10.1016/j.cell.2011.01.014

Galli SJ, Borregaard N, Wynn TA (2011) Phenotypic and functional plasticity of cells of innate immunity: macrophages, mast cells and neutrophils. Nat Immunol 12(11):1035–1044. https://doi.org/10.1038/ni.2109

Galliot B, Crescenzi M, Jacinto A, Tajbakhsh S (2017) Trends in tissue repair and regeneration. Development 144(3):357–364. https://doi.org/10.1242/dev.144279

Gallo RL (2017) Human skin is the largest epithelial surface for interaction with microbes. J Invest Dermatol 137(6):1213–1214. https://doi.org/10.1016/j.jid.2016.11.045

Gamelli RL, He LK (2003) Incisional wound healing. Model and analysis of wound breaking strength. Methods Mol Med 78:37–54. https://doi.org/10.1385/1-59259-332-1:037

Gaudry M, Bregerie O, Andrieu V, El Benna J, Pocidalo MA, Hakim J (1997) Intracellular pool of vascular endothelial growth factor in human neutrophils. Blood 90(10):4153–4161

Geer DJ, Andreadis ST (2003) A novel role of fibrin in epidermal healing: plasminogen-mediated migration and selective detachment of differentiated keratinocytes. J Invest Dermatol 121 (5):1210–1216. https://doi.org/10.1046/j.1523-1747.2003.12512.x

Gerhardt H, Golding M, Fruttiger M, Ruhrberg C, Lundkvist A, Abramsson A, Jeltsch M, Mitchell C, Alitalo K, Shima D, Betsholtz C (2003) VEGF guides angiogenic sprouting utilizing endothelial tip cell filopodia. J Cell Biol 161(6):1163–1177. https://doi.org/10.1083/jcb.200302047

Ghosh AK, Vaughan DE (2012) PAI-1 in tissue fibrosis. J Cell Physiol 227(2):493–507. https://doi.org/10.1002/jcp.22783

Giannouli CC, Kletsas D (2006) TGF-beta regulates differentially the proliferation of fetal and adult human skin fibroblasts via the activation of PKA and the autocrine action of FGF-2. Cell Signal 18(9):1417–1429. https://doi.org/10.1016/j.cellsig.2005.11.002

Gillitzer R, Goebeler M (2001) Chemokines in cutaneous wound healing. J Leukoc Biol 69 (4):513–521

Godo S, Shimokawa H (2017) Endothelial functions. Arterioscler Thromb Vasc Biol 37(9):e108–e114. https://doi.org/10.1161/ATVBAHA.117.309813

Golebiewska EM, Poole AW (2015) Platelet secretion: from haemostasis to wound healing and beyond. Blood Rev 29(3):153–162. https://doi.org/10.1016/j.blre.2014.10.003

Gonzalez AC, Costa TF, Andrade ZA, Medrado AR (2016) Wound healing - a literature review. An Bras Dermatol 91(5):614–620. https://doi.org/10.1590/abd1806-4841.20164741

Gordon S, Martinez-Pomares L (2017) Physiological roles of macrophages. Pflugers Arch 469 (3-4):365–374. https://doi.org/10.1007/s00424-017-1945-7

Gordon S, Taylor PR (2005) Monocyte and macrophage heterogeneity. Nat Rev Immunol 5 (12):953–964. https://doi.org/10.1038/nri1733

Gordon S, Pluddemann A, Martinez Estrada F (2014) Macrophage heterogeneity in tissues: phenotypic diversity and functions. Immunol Rev 262(1):36–55. https://doi.org/10.1111/imr.12223

Gorvetzian JW, Epler KE, Schrader S, Romero JM, Schrader R, Greenbaum A, McKee R (2018) Operating room staff and surgeon documentation curriculum improves wound classification accuracy. Heliyon 4(8):e00728. https://doi.org/10.1016/j.heliyon.2018.e00728

Gosain A, DiPietro LA (2004) Aging and wound healing. World J Surg 28(3):321–326. https://doi.org/10.1007/s00268-003-7397-6

Goto S (2008) Blood constitution: platelet aggregation, bleeding, and involvement of leukocytes. Rev Neurol Dis 5(Suppl 1):S22–S27

Gottrup F, Leaper D (2004) Wound healing: historical aspects. EWMA J 4(2)

Gravitz L (2018) Skin. Nature 563(7732):S83. https://doi.org/10.1038/d41586-018-07428-4

Greaves NS, Ashcroft KJ, Baguneid M, Bayat A (2013) Current understanding of molecular and cellular mechanisms in fibroplasia and angiogenesis during acute wound healing. J Dermatol Sci 72(3):206–217. https://doi.org/10.1016/j.jdermsci.2013.07.008

Greenhalgh DG (1998) The role of apoptosis in wound healing. Int J Biochem Cell Biol 30 (9):1019–1030. https://doi.org/10.1016/s1357-2725(98)00058-2

Greter M, Lelios I, Pelczar P, Hoeffel G, Price J, Leboeuf M, Kundig TM, Frei K, Ginhoux F, Merad M, Becher B (2012) Stroma-derived interleukin-34 controls the development and maintenance of langerhans cells and the maintenance of microglia. Immunity 37 (6):1050–1060. https://doi.org/10.1016/j.immuni.2012.11.001

Gronert K (2005) Lipoxins in the eye and their role in wound healing. Prostaglandins Leukot Essent Fatty Acids 73(3-4):221–229. https://doi.org/10.1016/j.plefa.2005.05.009

Guo S, Dipietro LA (2010) Factors affecting wound healing. J Dent Res 89(3):219–229. https://doi.org/10.1177/0022034509359125

Gurtner GC, Evans GR (2000) Advances in head and neck reconstruction. Plast Reconstr Surg 106 (3):672–682

Gurtner GC, Werner S, Barrandon Y, Longaker MT (2008) Wound repair and regeneration. Nature 453(7193):314–321. https://doi.org/10.1038/nature07039

Gurtner GC, Dauskardt RH, Wong VW, Bhatt KA, Wu K, Vial IN, Padois K, Korman JM, Longaker MT (2011) Improving cutaneous scar formation by controlling the mechanical environment: large animal and phase I studies. Ann Surg 254(2):217–225. https://doi.org/10.1097/SLA.0b013e318220b159

Hall CN, Reynell C, Gesslein B, Hamilton NB, Mishra A, Sutherland BA, O'Farrell FM, Buchan AM, Lauritzen M, Attwell D (2014) Capillary pericytes regulate cerebral blood flow in health and disease. Nature 508(7494):55–60. https://doi.org/10.1038/nature13165

Halliwell B (2006) Reactive species and antioxidants. Redox biology is a fundamental theme of aerobic life. Plant Physiol 141(2):312–322. https://doi.org/10.1104/pp.106.077073

Hamdan TA (2006) Missile injuries of the limbs: an Iraqi perspective. J Am Acad Orthop Surg 14 (10 Spec No.):S32–S36. https://doi.org/10.5435/00124635-200600001-00007

Han G, Ceilley R (2017) Chronic wound healing: a review of current management and treatments. Adv Ther 34(3):599–610. https://doi.org/10.1007/s12325-017-0478-y

Heath WR, Carbone FR (2013) The skin-resident and migratory immune system in steady state and memory: innate lymphocytes, dendritic cells and T cells. Nat Immunol 14(10):978–985. https://doi.org/10.1038/ni.2680

Hellstrom M, Phng LK, Hofmann JJ, Wallgard E, Coultas L, Lindblom P, Alva J, Nilsson AK, Karlsson L, Gaiano N, Yoon K, Rossant J, Iruela-Arispe ML, Kalen M, Gerhardt H, Betsholtz C (2007) Dll4 signalling through Notch1 regulates formation of tip cells during angiogenesis. Nature 445(7129):776–780. https://doi.org/10.1038/nature05571

Herman TF, Bordoni B (2020) Wound classification. In: StatPearls [Internet]. StatPearls Publishing, Treasure Island, FL

Hikosaka O, Wurtz RH (1989) The basal ganglia. Rev Oculomot Res 3:257–281

Hinz B, Gabbiani G (2003) Cell-matrix and cell-cell contacts of myofibroblasts: role in connective tissue remodeling. Thromb Haemost 90(6):993–1002. https://doi.org/10.1160/TH03-05-0328

Hoeffel G, Ginhoux F (2018) Fetal monocytes and the origins of tissue-resident macrophages. Cell Immunol 330:5–15. https://doi.org/10.1016/j.cellimm.2018.01.001

Hu Y, Liang D, Li X, Liu HH, Zhang X, Zheng M, Dill D, Shi X, Qiao Y, Yeomans D, Carvalho B, Angst MS, Clark JD, Peltz G (2010) The role of interleukin-1 in wound biology. Part II: in vivo and human translational studies. Anesth Analg 111(6):1534–1542. https://doi.org/10.1213/ANE.0b013e3181f691eb

Hughes EL, Becker F, Flower RJ, Buckingham JC, Gavins FNE (2017) Mast cells mediate early neutrophil recruitment and exhibit anti-inflammatory properties via the formyl peptide receptor 2/lipoxin A4 receptor. Br J Pharmacol 174(14):2393–2408. https://doi.org/10.1111/bph.13847

Hume DA (2008) Macrophages as APC and the dendritic cell myth. J Immunol 181(9):5829–5835. https://doi.org/10.4049/jimmunol.181.9.5829

Hunt TK, Hopf H, Hussain Z (2000) Physiology of wound healing. Adv Skin Wound Care 13 (2 Suppl):6–11

Ikpeme IA, Ngim NE, Ikpeme AA (2010) Diagnosis and treatment of pyogenic bone infections. Afr Health Sci 10(1):82–88

Ishida Y, Kondo T, Takayasu T, Iwakura Y, Mukaida N (2004) The essential involvement of cross-talk between IFN-gamma and TGF-beta in the skin wound-healing process. J Immunol 172 (3):1848–1855. https://doi.org/10.4049/jimmunol.172.3.1848

Ito M, Liu Y, Yang Z, Nguyen J, Liang F, Morris RJ, Cotsarelis G (2005) Stem cells in the hair follicle bulge contribute to wound repair but not to homeostasis of the epidermis. Nat Med 11 (12):1351–1354. https://doi.org/10.1038/nm1328

Iyer S, Balasubramanian D (2012) Management of radiation wounds. Indian J Plast Surg 45 (2):325–331. https://doi.org/10.4103/0970-0358.101311

Jacinto A, Martinez-Arias A, Martin P (2001) Mechanisms of epithelial fusion and repair. Nat Cell Biol 3(5):E117–E123. https://doi.org/10.1038/35074643

Jacobsona LK, Johnson MB, Dedhia RD, Bienia SN, Wong AK (2017) Impaired wound healing after radiation therapy: a systematic review of pathogenesis and treatment. JPRAS Open 13:92–105

Jameson J, Ugarte K, Chen N, Yachi P, Fuchs E, Boismenu R, Havran WL (2002) A role for skin gammadelta T cells in wound repair. Science 296(5568):747–749. https://doi.org/10.1126/science.1069639

Jameson JM, Cauvi G, Witherden DA, Havran WL (2004) A keratinocyte-responsive gamma delta TCR is necessary for dendritic epidermal T cell activation by damaged keratinocytes and maintenance in the epidermis. J Immunol 172(6):3573–3579. https://doi.org/10.4049/jimmunol.172.6.3573

John GM, McClain CD, Mooney DP (2014) Global surgery and anesthesia manual: providing care in resource-limited settings. CRC Press, Boca Raton, FL

Johnston A, Gudjonsson JE, Aphale A, Guzman AM, Stoll SW, Elder JT (2011) EGFR and IL-1 signaling synergistically promote keratinocyte antimicrobial defenses in a differentiation-dependent manner. J Invest Dermatol 131(2):329–337. https://doi.org/10.1038/jid.2010.313

Jones LM (2015) A short history of the development of wound care dressings. Br J Radiol 9(10)

Jones PH, Harper S, Watt FM (1995) Stem cell patterning and fate in human epidermis. Cell 80 (1):83–93. https://doi.org/10.1016/0092-8674(95)90453-0

Jorch SK, Kubes P (2017) An emerging role for neutrophil extracellular traps in noninfectious disease. Nat Med 23(3):279–287. https://doi.org/10.1038/nm.4294

Jun JI, Kim KH, Lau LF (2015) The matricellular protein CCN1 mediates neutrophil efferocytosis in cutaneous wound healing. Nat Commun 6:7386. https://doi.org/10.1038/ncomms8386

Kabashima K, Honda T, Ginhoux F, Egawa G (2019) The immunological anatomy of the skin. Nat Rev Immunol 19(1):19–30. https://doi.org/10.1038/s41577-018-0084-5

Kalesnikoff J, Galli SJ (2008) New developments in mast cell biology. Nat Immunol 9 (11):1215–1223. https://doi.org/10.1038/ni.f.216

Kankuri E, Cholujova D, Comajova M, Vaheri A, Bizik J (2005) Induction of hepatocyte growth factor/scatter factor by fibroblast clustering directly promotes tumor cell invasiveness. Cancer Res 65(21):9914–9922. https://doi.org/10.1158/0008-5472.CAN-05-1559

Kaur G, Singh N, Jaggi AS (2017) Mast cells in neuropathic pain: an increasing spectrum of their involvement in pathophysiology. Rev Neurosci 28(7):759–766. https://doi.org/10.1515/revneuro-2017-0007

Kischer CW, Bunce H 3rd, Shetlah MR (1978) Mast cell analyses in hypertrophic scars, hypertrophic scars treated with pressure and mature scars. J Invest Dermatol 70(6):355–357. https://doi.org/10.1111/1523-1747.ep12543553

Kohale BR, Agrawal AA, Raut CP (2018) Effect of low-level laser therapy on wound healing and patients' response after scalpel gingivectomy: a randomized clinical split-mouth study. J Indian Soc Periodontol 22(5):419–426. https://doi.org/10.4103/jisp.jisp_239_18

Komi DEA, Khomtchouk K, Santa Maria PL (2020) A review of the contribution of mast cells in wound healing: involved molecular and cellular mechanisms. Clin Rev Allergy Immunol 58 (3):298–312. https://doi.org/10.1007/s12016-019-08729-w

Korting HC, Schollmann C, White RJ (2011) Management of minor acute cutaneous wounds: importance of wound healing in a moist environment. J Eur Acad Dermatol Venereol 25 (2):130–137. https://doi.org/10.1111/j.1468-3083.2010.03775.x

Kosaraju R, Rennert RC, Maan ZN, Duscher D, Barrera J, Whittam AJ, Januszyk M, Rajadas J, Rodrigues M, Gurtner GC (2016) Adipose-derived stem cell-seeded hydrogels increase endogenous progenitor cell recruitment and neovascularization in wounds. Tissue Eng Part A 22 (3-4):295–305. https://doi.org/10.1089/ten.tea.2015.0277

Koster MI, Roop DR (2007) Mechanisms regulating epithelial stratification. Annu Rev Cell Dev Biol 23:93–113. https://doi.org/10.1146/annurev.cellbio.23.090506.123357

Krawczyk WS (1971) A pattern of epidermal cell migration during wound healing. J Cell Biol 49 (2):247–263. https://doi.org/10.1083/jcb.49.2.247

Kufareva I, Salanga CL, Handel TM (2015) Chemokine and chemokine receptor structure and interactions: implications for therapeutic strategies. Immunol Cell Biol 93(4):372–383. https://doi.org/10.1038/icb.2015.15

Kuhajda I, Zarogoulidis K, Kougioumtzi I, Huang H, Li Q, Dryllis G, Kioumis I, Pitsiou G, Machairiotis N, Katsikogiannis N, Papaiwannou A, Lampaki S, Zaric B, Branislav P, Dervelegas K, Porpodis K, Zarogoulidis P (2014) Penetrating trauma. J Thorac Dis 6(Suppl 4):S461–S465. https://doi.org/10.3978/j.issn.2072-1439.2014.08.51

Kujath P, Michelsen A (2008) Wounds – from physiology to wound dressing. Dtsch Arztebl Int 105 (13):239–248. https://doi.org/10.3238/arztebl.2008.0239

Kumar I, Staton CA, Cross SS, Reed MW, Brown NJ (2009) Angiogenesis, vascular endothelial growth factor and its receptors in human surgical wounds. Br J Surg 96(12):1484–1491. https://doi.org/10.1002/bjs.6778

Kuprash DV, Nedospasov SA (2016) Molecular and cellular mechanisms of inflammation. Biochemistry (Mosc) 81(11):1237–1239. https://doi.org/10.1134/S0006297916110018

Kwiecien K, Zegar A, Jung J, Brzoza P, Kwitniewski M, Godlewska U, Grygier B, Kwiecinska P, Morytko A, Cichy J (2019) Architecture of antimicrobial skin defense. Cytokine Growth Factor Rev 49:70–84. https://doi.org/10.1016/j.cytogfr.2019.08.001

Lammermann T, Afonso PV, Angermann BR, Wang JM, Kastenmuller W, Parent CA, Germain RN (2013) Neutrophil swarms require LTB4 and integrins at sites of cell death in vivo. Nature 498 (7454):371–375. https://doi.org/10.1038/nature12175

Lansdown AB (2002) Calcium: a potential central regulator in wound healing in the skin. Wound Repair Regen 10(5):271–285. https://doi.org/10.1046/j.1524-475x.2002.10502.x

Larouche J, Sheoran S, Maruyama K, Martino MM (2018) Immune regulation of skin wound healing: mechanisms and novel therapeutic targets. Adv Wound Care (New Rochelle) 7 (7):209–231. https://doi.org/10.1089/wound.2017.0761

Lau R, Paus R, Tiede S, Day P, Bayat A (2009) Exploring the role of stem cells in cutaneous wound healing. Exp Dermatol 18(11):921–933. https://doi.org/10.1111/j.1600-0625.2009.00942.x

Leavitt T, Hu MS, Marshall CD, Barnes LA, Lorenz HP, Longaker MT (2016) Scarless wound healing: finding the right cells and signals. Cell Tissue Res 365(3):483–493. https://doi.org/10.1007/s00441-016-2424-8

Levy V, Lindon C, Zheng Y, Harfe BD, Morgan BA (2007) Epidermal stem cells arise from the hair follicle after wounding. FASEB J 21(7):1358–1366. https://doi.org/10.1096/fj.06-6926com

Levy SM, Holzmann-Pazgal G, Lally KP, Davis K, Kao LS, Tsao K (2013) Quality check of a quality measure: surgical wound classification discrepancies impact risk-stratified surgical site infection rates in pediatric appendicitis. J Am Coll Surg 217(6):969–973. https://doi.org/10.1016/j.jamcollsurg.2013.07.398

Li J, Chen J, Kirsner R (2007) Pathophysiology of acute wound healing. Clin Dermatol 25(1):9–18. https://doi.org/10.1016/j.clindermatol.2006.09.007

Lim Y, Lee H, Woodby B, Valacchi G (2019) Ozonated oils and cutaneous wound healing. Curr Pharm Des 25(20):2264–2278. https://doi.org/10.2174/1381612825666190702100504

Liu L, Xu G, Dou H, Deng GM (2016) The features of skin inflammation induced by lupus serum. Clin Immunol 165:4–11. https://doi.org/10.1016/j.clim.2016.02.007

Lone RA, Wani MA, Hussain Z, Dar AM, Sharma ML, Bhat MA, Ahangar AG (2009) Missile cardiac injuries: review of 16 years' experience. Ulus Travma Acil Cerrahi Derg 15(4):353–356

Losquadro WD (2017) Anatomy of the skin and the pathogenesis of nonmelanoma skin cancer. Facial Plast Surg Clin North Am 25(3):283–289. https://doi.org/10.1016/j.fsc.2017.03.001

Lund LR, Green KA, Stoop AA, Ploug M, Almholt K, Lilla J, Nielsen BS, Christensen IJ, Craik CS, Werb Z, Dano K, Romer J (2006) Plasminogen activation independent of uPA and tPA maintains wound healing in gene-deficient mice. EMBO J 25(12):2686–2697. https://doi.org/10.1038/sj.emboj.7601173

Luo L, Tanaka R, Kanazawa S, Lu F, Hayashi A, Yokomizo T, Mizuno H (2017) A synthetic leukotriene B4 receptor type 2 agonist accelerates the cutaneous wound healing process in diabetic rats by indirect stimulation of fibroblasts and direct stimulation of keratinocytes. J Diabetes Complicat 31(1):13–20. https://doi.org/10.1016/j.jdiacomp.2016.09.002

Lynch MD, Watt FM (2018) Fibroblast heterogeneity: implications for human disease. J Clin Invest 128(1):26–35. https://doi.org/10.1172/JCI93555

Mackay AJ (1995) Bailey and Love's short practice of surgery. In: Mann CV, Russell RCG, Williams NS (eds) 278 × 216 mm, 22nd edn. Chapman and Hall, London, p 1041

Madhyastha HK, Radha KS, Nakajima Y, Omura S, Maruyama M (2008) uPA dependent and independent mechanisms of wound healing by C-phycocyanin. J Cell Mol Med 12 (6B):2691–2703. https://doi.org/10.1111/j.1582-4934.2008.00272.x

Mahdavian Delavary B, van der Veer WM, van Egmond M, Niessen FB, Beelen RH (2011) Macrophages in skin injury and repair. Immunobiology 216(7):753–762. https://doi.org/10.1016/j.imbio.2011.01.001

Malissen B, Tamoutounour S, Henri S (2014) The origins and functions of dendritic cells and macrophages in the skin. Nat Rev Immunol 14(6):417–428. https://doi.org/10.1038/nri3683

Maqsood MI (2018) Classification of wounds: know before research and clinical practice. J Gene Cell 4

Margenthaler JA, Longo WE, Virgo KS, Johnson FE, Oprian CA, Henderson WG, Daley J, Khuri SF (2003) Risk factors for adverse outcomes after the surgical treatment of appendicitis in adults. Ann Surg 238(1):59–66. https://doi.org/10.1097/01.SLA.0000074961.50020.f8

Martin P (1997) Wound healing: aiming for perfect skin regeneration. Science 276(5309):75–81. https://doi.org/10.1126/science.276.5309.75

Martin P, Nunan R (2015) Cellular and molecular mechanisms of repair in acute and chronic wound healing. Br J Dermatol 173(2):370–378. https://doi.org/10.1111/bjd.13954

Mascre G, Dekoninck S, Drogat B, Youssef KK, Brohee S, Sotiropoulou PA, Simons BD, Blanpain C (2012) Distinct contribution of stem and progenitor cells to epidermal maintenance. Nature 489(7415):257–262. https://doi.org/10.1038/nature11393

Mathieu D, Linke JC, Wattel F (2006) Non-healing wounds. In: Handbook on hyperbaric medicine. Springer, Dordrecht, pp 401–427

McCarty SM, Percival SL (2013) Proteases and delayed wound healing. Adv Wound Care (New Rochelle) 2(8):438–447. https://doi.org/10.1089/wound.2012.0370

McNeil PL, Steinhardt RA (2003) Plasma membrane disruption: repair, prevention, adaptation. Annu Rev Cell Dev Biol 19:697–731. https://doi.org/10.1146/annurev.cellbio.19.111301.140101

Medrado AR, Pugliese LS, Reis SR, Andrade ZA (2003) Influence of low level laser therapy on wound healing and its biological action upon myofibroblasts. Lasers Surg Med 32(3):239–244. https://doi.org/10.1002/lsm.10126

Medzhitov R (2010) Inflammation 2010: new adventures of an old flame. Cell 140(6):771–776. https://doi.org/10.1016/j.cell.2010.03.006

Mestas J, Hughes CC (2004) Of mice and not men: differences between mouse and human immunology. J Immunol 172(5):2731–2738. https://doi.org/10.4049/jimmunol.172.5.2731

Michopoulou A, Rousselle P (2015) How do epidermal matrix metalloproteinases support re-epithelialization during skin healing? Eur J Dermatol 25(Suppl 1):33–42. https://doi.org/10.1684/ejd.2015.2553

Mijnheer G, van Wijk F (2019) T-cell compartmentalization and functional adaptation in autoimmune inflammation: lessons from pediatric rheumatic diseases. Front Immunol 10:940. https://doi.org/10.3389/fimmu.2019.00940

Mioton LM, Jordan SW, Hanwright PJ, Bilimoria KY, Kim JY (2013) The relationship between preoperative wound classification and postoperative infection: a multi-institutional analysis of 15,289 patients. Arch Plast Surg 40(5):522–529. https://doi.org/10.5999/aps.2013.40.5.522

Monagle P, Massicotte P (2011) Developmental haemostasis: secondary haemostasis. Semin Fetal Neonatal Med 16(6):294–300. https://doi.org/10.1016/j.siny.2011.07.007

Montella E, Schiavone D, Apicella L, Di Silverio P, Gaudiosi M, Ambrosone E, Moscaritolo E, Triassi M (2014) Cost-benefit evaluation of a preventive intervention on the biological risk in health: the accidental puncture during the administration of insulin in the University Hospital "Federico II" of Naples. Ann Ig 26(3):272–278. https://doi.org/10.7416/ai.2014.1985

Morasso MI, Tomic-Canic M (2005) Epidermal stem cells: the cradle of epidermal determination, differentiation and wound healing. Biol Cell 97(3):173–183. https://doi.org/10.1042/BC20040098

Moreo K (2005) Understanding and overcoming the challenges of effective case management for patients with chronic wounds. Case Manag 16(2):62–63., 67. https://doi.org/10.1016/j.casemgr.2005.01.014

Morton LM, Phillips TJ (2016) Wound healing and treating wounds: differential diagnosis and evaluation of chronic wounds. J Am Acad Dermatol 74(4):589–605. https://doi.org/10.1016/j.jaad.2015.08.068

Nakano T, Sonoda T, Hayashi C, Yamatodani A, Kanayama Y, Yamamura T, Asai H, Yonezawa T, Kitamura Y, Galli SJ (1985) Fate of bone marrow-derived cultured mast cells after intracutaneous, intraperitoneal, and intravenous transfer into genetically mast cell-deficient W/Wv mice. Evidence that cultured mast cells can give rise to both connective tissue type and mucosal mast cells. J Exp Med 162(3):1025–1043. https://doi.org/10.1084/jem.162.3.1025

Nathan C, Ding A (2010) Nonresolving inflammation. Cell 140(6):871–882. https://doi.org/10.1016/j.cell.2010.02.029

Negut I, Grumezescu V, Grumezescu AM (2018) Treatment strategies for infected wounds. Molecules 23(9). https://doi.org/10.3390/molecules23092392

Nieuwenhuis E, Barnfield PC, Makino S, Hui CC (2007) Epidermal hyperplasia and expansion of the interfollicular stem cell compartment in mutant mice with a C-terminal truncation of Patched1. Dev Biol 308(2):547–560. https://doi.org/10.1016/j.ydbio.2007.06.016

Nishimura EK, Jordan SA, Oshima H, Yoshida H, Osawa M, Moriyama M, Jackson IJ, Barrandon Y, Miyachi Y, Nishikawa S (2002) Dominant role of the niche in melanocyte stem-cell fate determination. Nature 416(6883):854–860. https://doi.org/10.1038/416854a

Nording H, Langer HF (2018) Complement links platelets to innate immunity. Semin Immunol 37:43–52. https://doi.org/10.1016/j.smim.2018.01.003

Obagi Z, Damiani G, Grada A, Falanga V (2019) Principles of wound dressings: a review. Surg Technol Int 35:50–57

Odland G, Ross R (1968) Human wound repair. I Epidermal regeneration. J Cell Biol 39(1):135–151. https://doi.org/10.1083/jcb.39.1.135

Ohtani T, Mizuashi M, Ito Y, Aiba S (2007) Cadexomer as well as cadexomer iodine induces the production of proinflammatory cytokines and vascular endothelial growth factor by human macrophages. Exp Dermatol 16(4):318–323. https://doi.org/10.1111/j.1600-0625.2006.00532.x

Oishi Y, Manabe I (2018) Macrophages in inflammation, repair and regeneration. Int Immunol 30(11):511–528. https://doi.org/10.1093/intimm/dxy054

Okonkwo UA, DiPietro LA (2017) Diabetes and wound angiogenesis. Int J Mol Sci 18(7). https://doi.org/10.3390/ijms18071419

Oliveira THC, Marques PE, Proost P, Teixeira MMM (2018) Neutrophils: a cornerstone of liver ischemia and reperfusion injury. Lab Investig 98(1):51–62. https://doi.org/10.1038/labinvest.2017.90

Ono S, Kabashima K (2016) The role of dendritic cells and macrophages in the skin immunity. Nihon Rinsho Meneki Gakkai Kaishi 39(5):448–454. https://doi.org/10.2177/jsci.39.448

Onyekwelu I, Yakkanti R, Protzer L, Pinkston CM, Tucker C, Seligson D (2017) Surgical wound classification and surgical site infections in the orthopaedic patient. J Am Acad Orthop Surg Glob Res Rev 1(3):e022. https://doi.org/10.5435/JAAOSGlobal-D-17-00022

Oshima H, Rochat A, Kedzia C, Kobayashi K, Barrandon Y (2001) Morphogenesis and renewal of hair follicles from adult multipotent stem cells. Cell 104(2):233–245. https://doi.org/10.1016/s0092-8674(01)00208-2

Papapetropoulos A, Fulton D, Mahboubi K, Kalb RG, O'Connor DS, Li F, Altieri DC, Sessa WC (2000) Angiopoietin-1 inhibits endothelial cell apoptosis via the Akt/survivin pathway. J Biol Chem 275(13):9102–9105. https://doi.org/10.1074/jbc.275.13.9102

Park JE, Barbul A (2004) Understanding the role of immune regulation in wound healing. Am J Surg 187(5A):11S–16S. https://doi.org/10.1016/S0002-9610(03)00296-4

Pastar I, Stojadinovic O, Yin NC, Ramirez H, Nusbaum AG, Sawaya A, Patel SB, Khalid L, Isseroff RR, Tomic-Canic M (2014) Epithelialization in wound healing: a comprehensive review. Adv Wound Care (New Rochelle) 3(7):445–464. https://doi.org/10.1089/wound.2013.0473

Percival JN (2002) Classification of wounds and their management. Gen Surg 20(5)

Pergialiotis V, Vlachos D, Protopapas A, Pappa K, Vlachos G (2014) Risk factors for severe perineal lacerations during childbirth. Int J Gynaecol Obstet 125(1):6–14. https://doi.org/10.1016/j.ijgo.2013.09.034

Petreaca M, Green MM (2014) The dynamics of cell-ECM interactions, with implications for tissue engineering. In: Principles of tissue engineering, vol 4. Elsevier, New York, pp 161–187

Povoleri GAM, Nova-Lamperti E, Scotta C, Fanelli G, Chen YC, Becker PD, Boardman D, Costantini B, Romano M, Pavlidis P, McGregor R, Pantazi E, Chauss D, Sun HW, Shih HY, Cousins DJ, Cooper N, Powell N, Kemper C, Pirooznia M, Laurence A, Kordasti S, Kazemian M, Lombardi G, Afzali B (2018) Human retinoic acid-regulated CD161(+) regulatory T cells support wound repair in intestinal mucosa. Nat Immunol 19(12):1403–1414. https://doi.org/10.1038/s41590-018-0230-z

Powers JG, Higham C, Broussard K, Phillips TJ (2016) Wound healing and treating wounds: chronic wound care and management. J Am Acad Dermatol 74(4):607–625. https://doi.org/10.1016/j.jaad.2015.08.070

Purcell D (2016) Minor injuries E-book: a clinical guide. Churchill Livingstone, London

Putnam LR, Levy SM, Holzmann-Pazgal G, Lally KP, Lillian SK, Tsao K (2015) Surgical wound classification for pediatric appendicitis remains poorly documented despite targeted interventions. J Pediatr Surg 50(6):915–918. https://doi.org/10.1016/j.jpedsurg.2015.03.008

Qin Y (2016) Functional wound dressings. Medical textile materials. Woodhead Publishing, Cambridge, pp 89–107

Quirinia A (2000) Ischemic wound healing and possible treatments. Drugs Today (Barc) 36 (1):41–53. https://doi.org/10.1358/dot.2000.36.1.566626

Raja SK, Garcia MS, Isseroff RR (2007) Wound re-epithelialization: modulating keratinocyte migration in wound healing. Front Biosci 12:2849–2868. https://doi.org/10.2741/2277

Rajantie I, Ilmonen M, Alminaite A, Ozerdem U, Alitalo K, Salven P (2004) Adult bone marrow-derived cells recruited during angiogenesis comprise precursors for periendothelial vascular mural cells. Blood 104(7):2084–2086. https://doi.org/10.1182/blood-2004-01-0336

Raje N, Dinakar C (2015) Overview of immunodeficiency disorders. Immunol Allergy Clin N Am 35(4):599–623. https://doi.org/10.1016/j.iac.2015.07.001

Ramasastry SS (2005) Acute wounds. Clin Plast Surg 32(2):195–208. https://doi.org/10.1016/j.cps.2004.12.001

Rao KN, Brown MA (2008) Mast cells: multifaceted immune cells with diverse roles in health and disease. Ann N Y Acad Sci 1143:83–104. https://doi.org/10.1196/annals.1443.023

Raymond K, Kreft M, Song JY, Janssen H, Sonnenberg A (2007) Dual role of alpha6beta4 integrin in epidermal tumor growth: tumor-suppressive versus tumor-promoting function. Mol Biol Cell 18(11):4210–4221. https://doi.org/10.1091/mbc.e06-08-0720

Reeves EP, Lu H, Jacobs HL, Messina CG, Bolsover S, Gabella G, Potma EO, Warley A, Roes J, Segal AW (2002) Killing activity of neutrophils is mediated through activation of proteases by K+ flux. Nature 416(6878):291–297. https://doi.org/10.1038/416291a

Richmond JM, Harris JE (2014) Immunology and skin in health and disease. Cold Spring Harb Perspect Med 4(12):a015339. https://doi.org/10.1101/cshperspect.a015339

Rinnerthaler M, Bischof J, Streubel MK, Trost A, Richter K (2015) Oxidative stress in aging human skin. Biomol Ther 5(2):545–589. https://doi.org/10.3390/biom5020545

Roberts RE, Elumalai GL, Hallett MB (2018) Phagocytosis and motility in human neutrophils is competent but compromised by pharmacological inhibition of Ezrin phosphorylation. Curr Mol Pharmacol 11(4):305–315. https://doi.org/10.2174/1874467211666180516100613

Robson MC, Steed DL, Franz MG (2001) Wound healing: biologic features and approaches to maximize healing trajectories. Curr Probl Surg 38(2):72–140. https://doi.org/10.1067/msg.2001.111167

Rodrigues M, Kosaric N, Bonham CA, Gurtner GC (2019) Wound healing: a cellular perspective. Physiol Rev 99(1):665–706. https://doi.org/10.1152/physrev.00067.2017

Romani N, Holzmann S, Tripp CH, Koch F, Stoitzner P (2003) Langerhans cells – dendritic cells of the epidermis. APMIS 111(7-8):725–740. https://doi.org/10.1034/j.1600-0463.2003.11107805.x

Rousselle P, Braye F, Dayan G (2019) Re-epithelialization of adult skin wounds: cellular mechanisms and therapeutic strategies. Adv Drug Deliv Rev 146:344–365. https://doi.org/10.1016/j.addr.2018.06.019

Rumbaut RE, Thiagarajan P (2010) Platelet-vessel wall interactions in hemostasis and thrombosis. In: Integrated systems physiology: from molecule to function to disease. Morgan & Claypool Life Sciences, San Rafael, CA

Sadiq A, Shah A, Jeschke MG, Belo C, Qasim Hayat M, Murad S, Amini-Nik S (2018) The role of serotonin during skin healing in post-thermal injury. Int J Mol Sci 19(4). https://doi.org/10.3390/ijms19041034

Sahni A, Francis CW (2000) Vascular endothelial growth factor binds to fibrinogen and fibrin and stimulates endothelial cell proliferation. Blood 96(12):3772–3778

Said A, Weindl G (2015) Regulation of dendritic cell function in inflammation. J Immunol Res 2015:743169. https://doi.org/10.1155/2015/743169

Sandulache VC, Parekh A, Li-Korotky HS, Dohar JE, Hebda PA (2006) Prostaglandin E2 differentially modulates human fetal and adult dermal fibroblast migration and contraction: implication for wound healing. Wound Repair Regen 14(5):633–643. https://doi.org/10.1111/j.1743-6109.2006.00156.x

Santoro MM, Gaudino G (2005) Cellular and molecular facets of keratinocyte reepithelization during wound healing. Exp Cell Res 304(1):274–286. https://doi.org/10.1016/j.yexcr.2004.10.033

Sarabahi S, Tiwari VK (2012) Principles and practice of wound care. Jaypee Borthers Medical Publishers, New Delhi, India

Schaefer TJ, Tannan SC (2020) Thermal burns. In: StatPearls. Treasure Island, FL

Schafer BM, Maier K, Eickhoff U, Todd RF, Kramer MD (1994) Plasminogen activation in healing human wounds. Am J Pathol 144(6):1269–1280

Schultz G, Rotatori DS, Clark W (1991) EGF and TGF-alpha in wound healing and repair. J Cell Biochem 45(4):346–352. https://doi.org/10.1002/jcb.240450407

Sedmak DD, Orosz CG (1991) The role of vascular endothelial cells in transplantation. Arch Pathol Lab Med 115(3):260–265

Seifert AW, Maden M (2014) New insights into vertebrate skin regeneration. Int Rev Cell Mol Biol 310:129–169

Sen CK (2019) Human wounds and its burden: an updated compendium of estimates. Adv Wound Care (New Rochelle) 8(2):39–48. https://doi.org/10.1089/wound.2019.0946

Sgonc R, Gruber J (2013) Age-related aspects of cutaneous wound healing: a mini-review. Gerontology 59(2):159–164. https://doi.org/10.1159/000342344

Shah JB (2011) The history of wound care. J Am Col Certif Wound Spec 3(3):65–66. https://doi.org/10.1016/j.jcws.2012.04.002

Sharp LL, Jameson JM, Cauvi G, Havran WL (2005) Dendritic epidermal T cells regulate skin homeostasis through local production of insulin-like growth factor 1. Nat Immunol 6(1):73–79. https://doi.org/10.1038/ni1152

Siebenhaar F, Syska W, Weller K, Magerl M, Zuberbier T, Metz M, Maurer M (2007) Control of *Pseudomonas aeruginosa* skin infections in mice is mast cell-dependent. Am J Pathol 170 (6):1910–1916. https://doi.org/10.2353/ajpath.2007.060770

Simon LV, Lopez RA, King KC (2020) Blunt force trauma. In: StatPearls. Treasure Island, FL

Singer AJ, Clark RA (1999) Cutaneous wound healing. N Engl J Med 341(10):738–746. https://doi.org/10.1056/NEJM199909023411006

Singh S, Young A, McNaught CE (2017) The physiology of wound healing. Basic Sci 35(9)

Skapenko A, Leipe J, Lipsky PE, Schulze-Koops H (2005) The role of the T cell in autoimmune inflammation. Arthritis Res Ther 7(Suppl 2):S4–S14. https://doi.org/10.1186/ar1703

Slauch JM (2011) How does the oxidative burst of macrophages kill bacteria? Still an open question. Mol Microbiol 80(3):580–583. https://doi.org/10.1111/j.1365-2958.2011.07612.x

Smith SA, Travers RJ, Morrissey JH (2015) How it all starts: initiation of the clotting cascade. Crit Rev Biochem Mol Biol 50(4):326–336. https://doi.org/10.3109/10409238.2015.1050550

Snippert HJ, Haegebarth A, Kasper M, Jaks V, van Es JH, Barker N, van de Wetering M, van den Born M, Begthel H, Vries RG, Stange DE, Toftgard R, Clevers H (2010) Lgr6 marks stem cells in the hair follicle that generate all cell lineages of the skin. Science 327(5971):1385–1389. https://doi.org/10.1126/science.1184733

Solanas G, Benitah SA (2013) Regenerating the skin: a task for the heterogeneous stem cell pool and surrounding niche. Nat Rev Mol Cell Biol 14(11):737–748. https://doi.org/10.1038/nrm3675

Sonoda T, Kitamura Y, Haku Y, Hara H, Mori KJ (1983) Mast-cell precursors in various haematopoietic colonies of mice produced in vivo and in vitro. Br J Haematol 53(4):611–620. https://doi.org/10.1111/j.1365-2141.1983.tb07312.x

Sorg H, Tilkorn DJ, Hager S, Hauser J, Mirastschijski U (2017) Skin wound healing: an update on the current knowledge and concepts. Eur Surg Res 58(1-2):81–94. https://doi.org/10.1159/000454919

Spanholtz TA, Theodorou P, Amini P, Spilker G (2009) Severe burn injuries: acute and long-term treatment. Dtsch Arztebl Int 106(38):607–613. https://doi.org/10.3238/arztebl.2009.0607

Stanley ER, Chitu V (2014) CSF-1 receptor signaling in myeloid cells. Cold Spring Harb Perspect Biol 6(6). https://doi.org/10.1101/cshperspect.a021857

Staton CA, Valluru M, Hoh L, Reed MW, Brown NJ (2010) Angiopoietin-1, angiopoietin-2 and Tie-2 receptor expression in human dermal wound repair and scarring. Br J Dermatol 163 (5):920–927. https://doi.org/10.1111/j.1365-2133.2010.09940.x

Steenfos HH (1994) Growth factors and wound healing. Scand J Plast Reconstr Surg Hand Surg 28 (2):95–105. https://doi.org/10.3109/02844319409071186

Steinman RM, Cohn ZA (1973) Identification of a novel cell type in peripheral lymphoid organs of mice. I. Morphology, quantitation, tissue distribution. J Exp Med 137(5):1142–1162. https://doi.org/10.1084/jem.137.5.1142

Steinman RM, Gutchinov B, Witmer MD, Nussenzweig MC (1983) Dendritic cells are the principal stimulators of the primary mixed leukocyte reaction in mice. J Exp Med 157(2):613–627. https://doi.org/10.1084/jem.157.2.613

Stenn KSL, Depalma L (1988) The molecular and cellular biology of wound repair. Springer, New York, pp 321–335

Strid J, Hourihane J, Kimber I, Callard R, Strobel S (2004) Disruption of the stratum corneum allows potent epicutaneous immunization with protein antigens resulting in a dominant systemic Th2 response. Eur J Immunol 34(8):2100–2109. https://doi.org/10.1002/eji.200425196

Strid J, Tigelaar RE, Hayday AC (2009) Skin immune surveillance by T cells: a new order? Semin Immunol 21(3):110–120. https://doi.org/10.1016/j.smim.2009.03.002

Suchting S, Freitas C, le Noble F, Benedito R, Breant C, Duarte A, Eichmann A (2007) The notch ligand Delta-like 4 negatively regulates endothelial tip cell formation and vessel branching. Proc Natl Acad Sci U S A 104(9):3225–3230. https://doi.org/10.1073/pnas.0611177104

Sun T, Adra S, Smallwood R, Holcombe M, MacNeil S (2009) Exploring hypotheses of the actions of TGF-beta1 in epidermal wound healing using a 3D computational multiscale model of the human epidermis. PLoS One 4(12):e8515. https://doi.org/10.1371/journal.pone.0008515

Sun BK, Siprashvili Z, Khavari PA (2014) Advances in skin grafting and treatment of cutaneous wounds. Science 346(6212):941–945. https://doi.org/10.1126/science.1253836

Suri C, McClain J, Thurston G, McDonald DM, Zhou H, Oldmixon EH, Sato TN, Yancopoulos GD (1998) Increased vascularization in mice overexpressing angiopoietin-1. Science 282 (5388):468–471. https://doi.org/10.1126/science.282.5388.468

Suzuki-Inoue K, Tsukiji N, Shirai T, Osada M, Inoue O, Ozaki Y (2018) Platelet CLEC-2: roles beyond hemostasis. Semin Thromb Hemost 44(2):126–134. https://doi.org/10.1055/s-0037-1604090

Sveen L, Karlsen C, Ytteborg E (2020) Mechanical induced wounds in fish – a review on models and healing mechanisms. Rev Aquacult

Takahashi T, Kalka C, Masuda H, Chen D, Silver M, Kearney M, Magner M, Isner JM, Asahara T (1999) Ischemia- and cytokine-induced mobilization of bone marrow-derived endothelial progenitor cells for neovascularization. Nat Med 5(4):434–438. https://doi.org/10.1038/7434

Takeo M, Lee W, Ito M (2015) Wound healing and skin regeneration. Cold Spring Harb Perspect Med 5(1):a023267. https://doi.org/10.1101/cshperspect.a023267

Takeuchi O, Akira S (2010) Pattern recognition receptors and inflammation. Cell 140(6):805–820. https://doi.org/10.1016/j.cell.2010.01.022

Tamoutounour S, Guilliams M, Montanana Sanchis F, Liu H, Terhorst D, Malosse C, Pollet E, Ardouin L, Luche H, Sanchez C, Dalod M, Malissen B, Henri S (2013) Origins and functional specialization of macrophages and of conventional and monocyte-derived dendritic cells in mouse skin. Immunity 39(5):925–938. https://doi.org/10.1016/j.immuni.2013.10.004

Tan JH, Mohamad Y, Tan CLH, Kassim M, Warkentin TE (2018) Concurrence of symmetrical peripheral gangrene and venous limb gangrene following polytrauma: a case report. J Med Case Rep 12(1):131. https://doi.org/10.1186/s13256-018-1684-1

Taylor PR, Martinez-Pomares L, Stacey M, Lin HH, Brown GD, Gordon S (2005) Macrophage receptors and immune recognition. Annu Rev Immunol 23:901–944. https://doi.org/10.1146/annurev.immunol.23.021704.115816

Thieblemont N, Wright HL, Edwards SW, Witko-Sarsat V (2016) Human neutrophils in autoimmunity. Semin Immunol 28(2):159–173. https://doi.org/10.1016/j.smim.2016.03.004

Tomaiuolo M, Brass LF, Stalker TJ (2017) Regulation of platelet activation and coagulation and its role in vascular injury and arterial thrombosis. Interv Cardiol Clin 6(1):1–12. https://doi.org/10.1016/j.iccl.2016.08.001

Tonnesen MG, Feng X, Clark RA (2000) Angiogenesis in wound healing. J Investig Dermatol Symp Proc 5(1):40–46. https://doi.org/10.1046/j.1087-0024.2000.00014.x

Toriseva M, Kahari VM (2009) Proteinases in cutaneous wound healing. Cell Mol Life Sci 66 (2):203–224. https://doi.org/10.1007/s00018-008-8388-4

Trabucchi E, Radaelli E, Marazzi M, Foschi D, Musazzi M, Veronesi AM, Montorsi W (1988) The role of mast cells in wound healing. Int J Tissue React 10(6):367–372

Tran PO, Hinman LE, Unger GM, Sammak PJ (1999) A wound-induced [Ca2+]i increase and its transcriptional activation of immediate early genes is important in the regulation of motility. Exp Cell Res 246(2):319–326. https://doi.org/10.1006/excr.1998.4239

Tumbar T, Guasch G, Greco V, Blanpain C, Lowry WE, Rendl M, Fuchs E (2004) Defining the epithelial stem cell niche in skin. Science 303(5656):359–363. https://doi.org/10.1126/science.1092436

VanHoy TB, Metheny H, Patel BC (2020) Chemical burns. In: StatPearls. Treasure Island, FL

Vasudevan B (2014) Venous leg ulcers: pathophysiology and classification. Indian Dermatol Online J 5(3):366–370. https://doi.org/10.4103/2229-5178.137819

Velnar T, Bailey T, Smrkolj V (2009) The wound healing process: an overview of the cellular and molecular mechanisms. J Int Med Res 37(5):1528–1542. https://doi.org/10.1177/147323000903700531

Vidyarthi K, Gupta S (2003) Current trends in wound management. Indian J Pediatr 70(Suppl 1): S51–S56

Vojacek JF (2017) Should we replace the terms intrinsic and extrinsic coagulation pathways with tissue factor pathway? Clin Appl Thromb Hemost 23(8):922–927. https://doi.org/10.1177/1076029616673733

Vu LT, Nobuhara KK, Lee H, Farmer DL (2009) Conflicts in wound classification of neonatal operations. J Pediatr Surg 44(6):1206–1211. https://doi.org/10.1016/j.jpedsurg.2009.02.026

Wang Z, Lai Y, Bernard JJ, Macleod DT, Cogen AL, Moss B, Di Nardo A (2012) Skin mast cells protect mice against vaccinia virus by triggering mast cell receptor S1PR2 and releasing antimicrobial peptides. J Immunol 188(1):345–357. https://doi.org/10.4049/jimmunol.1101703

Wang PH, Huang BS, Horng HC, Yeh CC, Chen YJ (2018) Wound healing. J Chin Med Assoc 81 (2):94–101. https://doi.org/10.1016/j.jcma.2017.11.002

Watson N, Ding B, Zhu X, Frisina RD (2017) Chronic inflammation - inflammaging - in the ageing cochlea: a novel target for future presbycusis therapy. Ageing Res Rev 40:142–148. https://doi.org/10.1016/j.arr.2017.10.002

Watt FM (2002a) Role of integrins in regulating epidermal adhesion, growth and differentiation. EMBO J 21(15):3919–3926. https://doi.org/10.1093/emboj/cdf399

Watt FM (2002b) The stem cell compartment in human interfollicular epidermis. J Dermatol Sci 28 (3):173–180. https://doi.org/10.1016/s0923-1811(02)00003-8

Watt FM (2014) Mammalian skin cell biology: at the interface between laboratory and clinic. Science 346(6212):937–940. https://doi.org/10.1126/science.1253734

Werner S, Grose R (2003) Regulation of wound healing by growth factors and cytokines. Physiol Rev 83(3):835–870. https://doi.org/10.1152/physrev.2003.83.3.835

Werner S, Krieg T, Smola H (2007) Keratinocyte-fibroblast interactions in wound healing. J Invest Dermatol 127(5):998–1008. https://doi.org/10.1038/sj.jid.5700786

Wernig G, Chen SY, Cui L, Van Neste C, Tsai JM, Kambham N, Vogel H, Natkunam Y, Gilliland DG, Nolan G, Weissman IL (2017) Unifying mechanism for different fibrotic diseases. Proc Natl Acad Sci U S A 114(18):4757–4762. https://doi.org/10.1073/pnas.1621375114

Westby MJ, Norman G, Watson REB, Cullum NA, Dumville JC (2020) Protease activity as a prognostic factor for wound healing in complex wounds. Wound Repair Regen. https://doi.org/10.1111/wrr.12835

Wilgus TA, Roy S, McDaniel JC (2013) Neutrophils and wound repair: positive actions and negative reactions. Adv Wound Care (New Rochelle) 2(7):379–388. https://doi.org/10.1089/wound.2012.0383

Wilkins RG, Unverdorben M (2013) Wound cleaning and wound healing: a concise review. Adv Skin Wound Care 26(4):160–163. https://doi.org/10.1097/01.ASW.0000428861.26671.41

Wilson EB, Brooks DG (2013) Inflammation makes T cells sensitive. Immunity 38(1):5–7. https://doi.org/10.1016/j.immuni.2013.01.001

Witte MB, Barbul A (1997) General principles of wound healing. Surg Clin North Am 77 (3):509–528. https://doi.org/10.1016/s0039-6109(05)70566-1

Witte MB, Barbul A (2002) Role of nitric oxide in wound repair. Am J Surg 183(4):406–412. https://doi.org/10.1016/s0002-9610(02)00815-2

Wong R, Geyer S, Weninger W, Guimberteau JC, Wong JK (2016) The dynamic anatomy and patterning of skin. Exp Dermatol 25(2):92–98. https://doi.org/10.1111/exd.12832

Workalemahu G, Foerster M, Kroegel C, Braun RK (2003) Human gamma delta-T lymphocytes express and synthesize connective tissue growth factor: effect of IL-15 and TGF-beta 1 and comparison with alpha beta-T lymphocytes. J Immunol 170(1):153–157. https://doi.org/10.4049/jimmunol.170.1.153

Wulff BC, Wilgus TA (2013) Mast cell activity in the healing wound: more than meets the eye? Exp Dermatol 22(8):507–510. https://doi.org/10.1111/exd.12169

Xie J, Bian H, Qi S, Xu Y, Tang J, Li T, Liu X (2008) Effects of basic fibroblast growth factor on the expression of extracellular matrix and matrix metalloproteinase-1 in wound healing. Clin Exp Dermatol 33(2):176–182. https://doi.org/10.1111/j.1365-2230.2007.02573.x

Xu J, Zgheib C, Liechty KW (2015) miRNAs in bone marrow–derived mesenchymal stem cells. In: MicroRNA in regenerative medicine. Academic Press, Amsterdam, pp 111–136

Xu XR, Zhang D, Oswald BE, Carrim N, Wang X, Hou Y, Zhang Q, Lavalle C, McKeown T, Marshall AH, Ni H (2016) Platelets are versatile cells: new discoveries in hemostasis, thrombosis, immune responses, tumor metastasis and beyond. Crit Rev Clin Lab Sci 53(6):409–430. https://doi.org/10.1080/10408363.2016.1200008

Yang L, Scott PG, Dodd C, Medina A, Jiao H, Shankowsky HA, Ghahary A, Tredget EE (2005) Identification of fibrocytes in postburn hypertrophic scar. Wound Repair Regen 13(4):398–404. https://doi.org/10.1111/j.1067-1927.2005.130407.x

Yen YH, Pu CM, Liu CW, Chen YC, Liang CJ, Hsieh JH, Huang HF, Chen YL (2018) Curcumin accelerates cutaneous wound healing via multiple biological actions: the involvement of TNF-alpha, MMP-9, alpha-SMA, and collagen. Int Wound J 15(4):605–617. https://doi.org/10.1111/iwj.12904

Zhan L, Li J, Wei B (2018) Autophagy in endometriosis: friend or foe? Biochem Biophys Res Commun 495(1):60–63. https://doi.org/10.1016/j.bbrc.2017.10.145

Zhang Y, Gu W, Sun B (2014) TH1/TH2 cell differentiation and molecular signals. Adv Exp Med Biol 841:15–44. https://doi.org/10.1007/978-94-017-9487-9_2

Mechanisms of Collective Cell Migration in Wound Healing: Physiology and Disease

Chaithra Mayya, Sumit Kharbhanda, Ashadul Haque, and Dhiraj Bhatia

1 Introduction

Wounds or scars can arise due to physical injuries or as an outcome of a disease process and can have an accidental or intentional etiology in living organisms (Sen et al. 2009). The disruption of the integrity of various organs and tissues often leads to the formation of wounds. For example, skin being an outer barrier of the body is constantly affected by various external stress factors and has therefore developed a set of complex mechanisms to protect itself (Hunt 1988). Because of its accessibility and extraordinary capability to restore tissue integrity, the skin is considered to be the most preferred organ for exploring response mechanisms toward tissue damage and its repair (Lawrence 1998). Studies from skin injury and its repair mechanism have revealed the novel fundamental principles of tissue regeneration and regenerative biology (Skover 1991). Right after an injury, wound healing is triggered which involves an extensive communication between various cellular constituents and their extracellular matrix (Winter and Scales 1963). The identification of potential therapeutic targets for chronic and non-healing wound treatment is extremely challenging because the physiological complexity and the molecular processes of wound restoration are poorly understood (Strecker-McGraw et al. 2007). Future investigations must concentrate on developing an integrated research approach combining the

C. Mayya · S. Kharbhanda · A. Haque
Biological Engineering, Indian Institute of Technology Gandhinagar, Gandhinagar, Gujarat, India

D. Bhatia (✉)
Biological Engineering, Indian Institute of Technology Gandhinagar, Gandhinagar, Gujarat, India

Center for Biomedical Engineering, Indian Institute of Technology Gandhinagar, Gandhinagar, Gujarat, India
e-mail: dhiraj.bhatia@iitgn.ac.in

expertise from genetics, cellular, biochemical differences of normal, and chronic wounds, which may not only help in understanding the reason for delayed wound healing but at the same time help the community design better therapeutics for the same.

In this book chapter, we shed light on the fundamental steps of wound healing process and try to understand how different external factors affect the wound healing process. We will also focus on collective cell migration which plays a key role in migration of adjacent cells to the site of wound. We compare the mechanisms and factors affecting wound healing in normal cells compared to diseased ones and we conclude with an insight on the need for combined research efforts in wound healing for better understanding the process and design molecular tools and technologies for its repair.

2 Basic Steps in Wound Healing

A complex dynamic mechanism involving a series of highly coordinated physiological events is natural wound healing. A cascade of precisely regulated interactions among various cellular and immunological systems operate for healing a wound (Hunt 1988; Sen et al. 2009). Wound healing in healthy individuals consists of three phases: Coagulation and hemostasis, Inflammatory phase, Proliferative phase and wound Remodeling phase (Velnar 2009). The process of wound healing begins with hemostasis and the stimulation of several inflammatory cells. In the intermediate stage, proliferation and migration of cells, matrix deposition, and angiogenesis occur. The late stage of wound healing involves remodeling of extracellular matrix (ECM), forming scarred tissues. Specialized tissues such as liver and skeletal tissues, however, have distinctive regeneration and tissue repair forms and adopt various pathways (Lawrence 1998). Additionally, the supportive microenvironment at the wound surface plays a major role in maximizing the healing potential. However, by causing disruptions in tissue repair processes that are otherwise precisely regulated, many local and systemic factors may influence the healing process, resulting in non-healing or chronic wounds (Eming et al. 2014; Reinke and Sorg 2012; Velnar 2009) (Fig. 1).

2.1 Coagulation and Hemostasis

Blood coagulation, also known as clotting is a process where the blood in liquid form turns into semisolid blood clot. This is essential to avoid loss of blood. The coagulation cascade is essential for hemostasis to occur. Immediately after an injury, the process of coagulation and hemostasis begins. The main purpose of this is to protect the vascular system so that, after damage, the usual physiological functions of the vital organs remain intact (Broughton et al. 2006a). Another long-term goal is

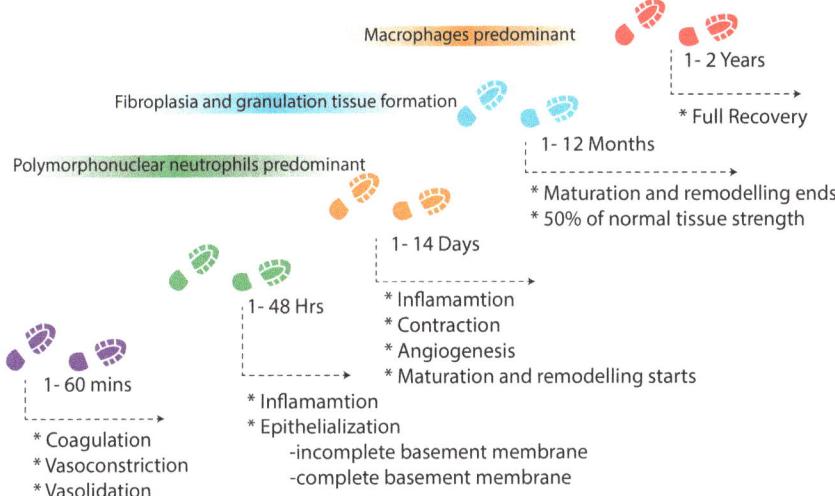

Fig. 1 Timeline for wound healing. The diagram shows main wound healing phases such as coagulation, inflammation, proliferation, and remodeling. The early stages of healing process would take place within 60 min up to 48 h, while the later stages of healing is a slow process requiring few days to months for the wound to repair completely producing scar tissue

providing a scaffold for the invading migratory cells that are needed in the later stages of wound healing process (Pool 1977). In order to minimize the loss of blood after an injury, the damaged ends of the blood vessels constrict rapidly. The reflex vasoconstriction can incidentally reduce the amount of bleeding or even may cause a complete cessation of blood leakage at the wound site. The blood clotting process is elicited by the initiation of coagulation cascade. The platelets then would adhere and aggregate at the wound site. Two major pathways operate in the coagulation cascade: the extrinsic pathway and the intrinsic pathway (Smith et al. 2015).

2.1.1   The Extrinsic Pathway

This pathway is also called as the Tissue Factor (TF) pathway, named after a cell surface integral membrane protein (Morrissey et al. 2012). TF is also called as Tissue thromboplastin or Coagulation factor III or CD142 (Morrissey et al. 1987). This single-pass membrane protein is in its glycosylated form which is 46 kDa in size. Also, it does not require proteolysis for its activity during coagulation. Tissue Factor is also implicated to play a role during thrombogenesis, inflammation, cancer, angiogenesis, cellular immune response, and embryogenesis as well (Butenas 2012). They are found abundantly in outermost layers surrounding the blood vessels, keratinocytes of skin, and more abundantly in certain organs such as kidney and brain. Their expression is comparatively lesser in synovial tissues and skeletal muscle tissues (Drake et al. 1989; Fleck et al. 1990; Smith et al. 2015). But, it is not expressed in vascular endothelial cells and circulating cells, but can be induced by certain inflammatory factors (Butenas 2012) and hypoxia (Yan et al. 1999). This protein is expressed only when there is mechanical or chemical injury to the vasculature exposing it to blood flow. This leads to the binding of Sub-endothelial TF to circulatory plasma factor VIIa, which is also a glycosylated protein, forming a complex. This complex formation requires calcium ions as well. This then triggers blood clotting by activating Zymogen factor IX and factor X into serine proteases, Factor IXa and Factor Xa. A surge of thrombin is induced, which also is a serine protease. A cross-linked insoluble fibrin clot is formed as an end product by multiple feedback reactions of thrombin, which involves efficient cleavage of fibrinogen and activation of Factor XIII via proteolysis (Butenas 2012; Smith et al. 2015). Tissue Factor Pathway Inhibitor (TFPI), a plasma serine protease inhibitor inhibits the TF-fVIIa complex through its direct binding to fXa (Piro and Broze 2005).

2.1.2 The Intrinsic Pathway

Also called the contact pathway, is activated without the involvement of TF. This pathway initiates when the plasma comes in contact with negatively charged biological or artificial surfaces (Maas and Renné 2018; Smith et al. 2015). The initiation of this pathway is through Factor XII (fXII) stimulation, also involving Kininogen (HK) and Plasma Kallikrein (PK) as its subsequent downstream activators. Unlike the extrinsic pathway, this does not require Calcium ions for the initiation of cascade. When blood encounters artificial or biological surfaces there is a change in the conformation of zymogen fXII converting it to an active form of fXII (fXIIa), which is a serine protease (Smith et al. 2015). A positive feedback loop occurs where fXIIa activates plasma prekallikrein to Kallikrein and fXI to fXIa, in turn hydrolyzing fXII by PK, this amplifying fXIIa. PK further cleaves high molecular weight Kininogen (HK) from Bradykinin from its precursor protein, further leading to the activation of thrombin and formation of fibrin clot (Naudin et al. 2017). A major inhibitor of fXIIa, fXIa, and PK is known as Serpin C1 esterase inhibitor (C1INH), plays a crucial role in regulating intrinsic pathway (Maas et al. 2011).

Thrombin is a potent platelet activator. At the site of injury after the initiation of extrinsic or intrinsic pathway, platelets aggregate and cause vasoconstriction which results in hypoxia, increase in glycolysis, and change in pH (Landén et al. 2016). After a certain time, vasodilation occurs allowing platelets, keratinocytes, endothelial cells, fibroblasts, and leukocytes to reach the clot. The above cells release various chemotactic factors like different growth factors and cytokines to activate inflammatory processes (Reinke and Sorg 2012). Various growth factors such as transforming growth factor-beta (TGF-β), platelet-derived growth factor (PDGF), insulin-like growth factor-1 (IGF-1), simple fibroblast growth factor (bFGF), and vascular endothelial growth factor (VEGF) are released from the platelet alpha granules. PDGF and TGF-β help in engaging neutrophils and monocytes into the wound site. Meanwhile, VEGF, TGF-α, and bFGF are required for angiogenesis through the activation of endothelial cells. PDGF also recruits fibroblasts to the

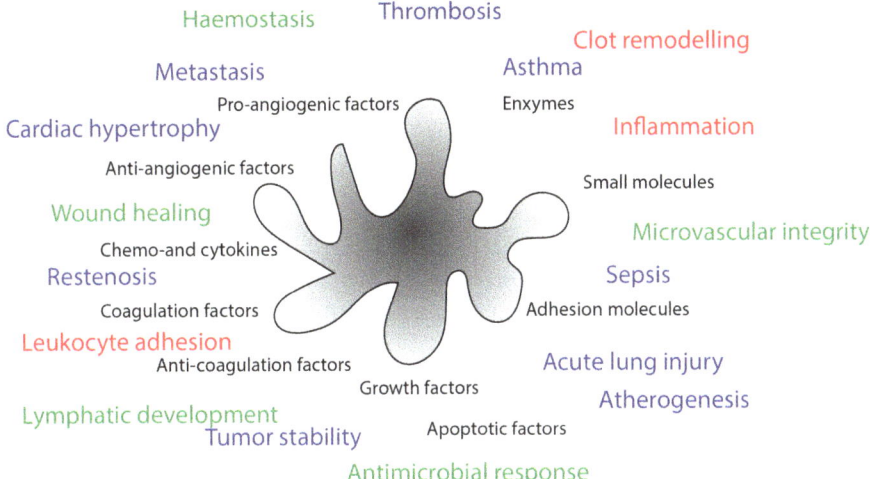

Fig. 2 Platelet activation. Platelets can recruit various other biomolecules such as cytokines, anticoagulation factors, antiangiogenic factors, and growth factors to the wounded area to accelerate wound healing. Adapted and modified from reference Golebiewska and Poole (2015)

injury site for collagen and extracellular matrix production, which enable cellular migration in the subsequent phases of the healing process (Pool 1977) (Fig. 2).

2.2 Inflammatory Phase of Wound Healing

The characteristics of inflammation are redness, swelling, heat, pain, and loss of tissue function. Inflammation is the second stage of the wound healing mechanism which starts within a day of injury and lasts for several days to weeks. In case of chronic non-healing wounds, the inflammatory phase may last significantly longer. The major goal of inflammation is establishing an immune barrier against the invading pathogens. The inflammatory phase is divided into two phases, early and late inflammatory phases (Hart 2002). These symptoms in turn reflect increased permeability of the vascular endothelium due to the influx of inflammatory cells and proangiogenic factors by chemokines and immune cells at the site of the wound. This step is essential in order to remove the cell debris and foreign bodies from the wound (Ridiandries et al. 2018; Takeuchi and Akira 2010). Rapid migration of neutrophils and monocytes, as well as macrophages, keratinocytes, mast, and dendritic cells to the site of injury is a characteristic of this phase. The innate immune system plays a fundamental role in causing acute immune response thereafter promoting the activation of acquired immune response. The inflammatory response can be of two types, Pattern Associated Molecular Patterns (PAMPs) and Damage Associated Molecular Patterns (DAMPs) (Landén et al. 2016).

Pattern Associated Molecular Patterns The type of response triggered by pathogens such as bacterial polysaccharides and polynucleotides, foreign to the host cells is called as PAMPs. They provide ligands for TLRs (Strbo et al. 2014).

The signals are recognized by C-type Lectin Receptors (CLRs), Toll-Like Receptors (TLRs), NOD-like Receptors (NLRs), and Retinoic Acid-inducible genes (RIG)-I-like Receptors (RLRs). This induces the activation of downstream pathways such as NFkB and MAPK pathways, leading to cytokines, chemokines, and antimicrobials gene expression (Landén et al. 2016; Takeuchi and Akira 2010).

Damage Associated Molecular Patterns This kind of response is stimulated when host stressed cells such as necrotic cells release certain proteins, DNA, and RNA. They can activate inflammasomes (Strbo et al. 2014).

2.2.1 Early Inflammatory Phase

The recruitment of neutrophils at the injury site is the first step in response to the chemokines in order to prevent infection by invasion of pathogens. This corresponds to the early inflammatory phase. Neutrophils attack and remove invading pathogens, foreign particles, and damaged tissues by phagocytosis. This phagocytic activity is crucial because wounds with microbial infection would not heal (Robson 1997). Infiltration of neutrophils into the site of wound takes place within 24–36 h. Various chemoattractants, such as formyl methionyl peptides and TGF-β secreted by platelets and invading bacteria, helps in recruiting neutrophils to the wound site (Robson et al. 2001). Due to biochemical alterations on the membrane surface of Neutrophils, they become adhesive in nature. This allows them to bind to the endothelial cells that cover the wound in the post-capillary venules (Hart 2002; Lawrence 1998). The neutrophils pass through the endothelial surface layer. Chemokines are secreted by the endothelial cells activating a stronger adhesion system, which are Integrin mediated (Flanagan 2000). Additionally, chemokines can induce the expression of Vascular Cell Adhesion Molecule (VCAM), Intracellular adhesion molecule 1 (ICAM1), P-Selectin, and E-Selectin on endothelial cells to promote the adhesion of neutrophils onto the endothelial cells, which leads to neutrophil extravasation from the blood vessels to the wound site (Vestweber 2015). Neutrophils are thus exposed to various chemokines and also release various cytokines like Inter-Leukin (IL) 1β, IL6, and Tumor Necrosis Factor (TNF)-α to intensify inflammatory response, further stimulating Vascular Endothelial Growth Factor (VEGF) and IL8 (Reinke and Sorg 2012).

2.2.2 Late Inflammatory Phase

Monocytes migrate toward the site of inflammation after 3 days of injury, differentiating them into macrophages. Like neutrophils, macrophages get into the site of injury by a myriad of chemoattractants, secreted by platelets and invading

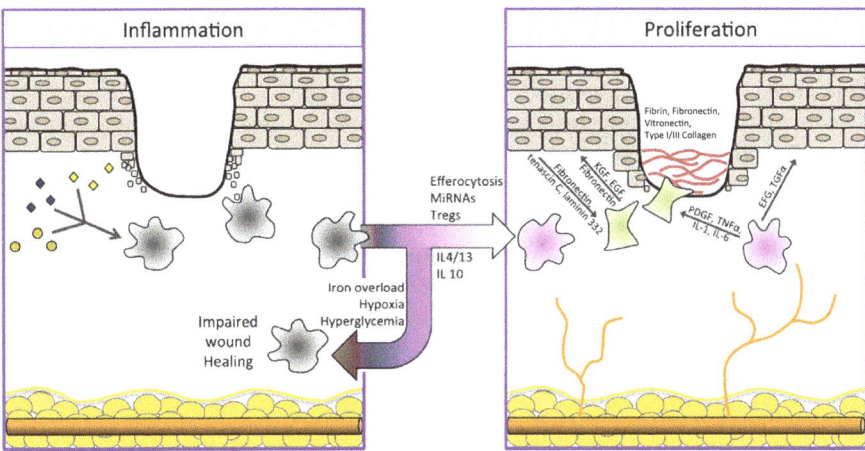

Fig. 3 Progression from inflammation to proliferation. This step marks the reduction of pro-inflammatory cytokines and rise in anti-inflammatory cytokines. The anti-inflammatory macrophages are reduced tissue inflammation, contributing toward angiogenesis and accelerated wound healing. Adapted and modified from Ellis et al. (2018)

pathogens. Compared to neutrophils, macrophages have a longer life span and can work at lower pH (Pierce et al. 1991; Ramasastry 2005). Macrophages act as key regulators and provide a pool of tissue growth factors, such as TGF-β, and other biochemical mediators like collagenase and fibroblast growth factor (FGF), which are essential for the activation of fibroblasts, keratinocytes, and endothelial cells (Broughton et al. 2006b; Diegelmann and Evans 2004; Hunt et al. 2000; Ramasastry 2005). Macrophages remove cell debris and bacteria through phagocytosis and Reactive Oxygen Species (ROS) production. They can even stimulate angiogenesis, fibroplasia and ECM production, and progression toward proliferative phase of wound healing. Macrophages can also stimulate adaptive immune cells such as lymphocytes to secrete chemokines and amplify inflammatory response as well as progress toward the proliferative phase (Landén et al. 2016). At the very end of the inflammatory phase, lymphocytes get into the wound site by the combined action of interleukin-1 (IL-1), immunoglobulin G (IgG), and complement immune system components (Broughton et al. 2006b; Hart 2002; Hunt et al. 2000). The IL-1 plays a crucial role in collagenase regulation during collagen remodeling, as well as generation and degeneration of ECM components (Hart 2002; Hunt 1988). This step is also known to induce an adaptive immune system, however, their role in wound healing process is yet to be explored (Fig. 3).

2.3   Proliferative Phase of Wound Healing

This phase begins with subsiding inflammation and promoting re-epithelialization, neovascularization, and formation of granulated tissue. Macrophages play a crucial role in subsiding inflammation and inducing proliferation. They play a vital role in secreting growth factors such as PDGF, FGF, TGF-α, TGF-β, VEGF, and many more. These growth factors promote Extracellular Matrix (ECM) formation (Reinke and Sorg 2012). Such growth factors also induce fibroblast migration towards the wound site. They proliferate rapidly as fibroblasts enter the wound site and begin to develop essential extracellular matrix proteins such as fibronectin, hyaluronic acid, type I and type III collagen, and proteoglycans, which will facilitate further wound healing processes (Ramasastry 2005; Robson et al. 2001; Witte and Barbul 1997).

2.3.1   Re-epithelialization

A new connective tissue is formed, controlled by regulatory cytokines such as IFN-γ and TGF-β. These cytokines stimulate fibroblasts to synthesize collagen, fibronectin, and other substances required for ECM formation, closure of gaps, and restoring mechanical strength of the wound. The epithelial and non-epithelial cells release variety of growth factors such as KGF, EGF, NGF, and IGF-1 (Werner and Grose 2003). This phase is also marked by the decrease in contact inhibition by altering cell adhesion molecules such as desmosomes, hemi-desmosomes, tight junction and focal adhesion proteins, activation of membrane-associated kinases, increased permeability to calcium ions, and cytoskeletal rearrangements in the cells. All these characteristics act in a synergistic fashion in order to facilitate cell migration. Dissolution of the fibrin clot is through collagenase and elastase (Jacinto et al. 2001). In case of skin injury, keratinocytes migrate toward chemotactic gradient established by cytokines such as IL1 which is present over the fibronectin matrix (Clark et al. 1982). The migration of cells ceases when the cells come in contact with each other. Small GTPases such as RhoGTPases modulate cytoskeletal structures, mainly actin, thereby creating a shift in migration and contact inhibition (Reinke and Sorg 2012).

2.3.2   Neovascularization

Neovascularization or angiogenesis is a complex phenomenon involving cellular, molecular, and humoral cues. Angiogenesis takes place via two steps: vessel sprouting and vessel anastomosis. Vessel sprouting requires the orchestration of anti-inflammatory cytokines and pro-repair macrophages. The pro-repair macrophages secrete VEGF, PDGF, bFGF, and serine protease thrombin. The endothelial cells in the existing blood vessels bind to these growth factors to initiate signaling cascades. This enables the endothelial cells to secrete various proteases, which dissolve basal lamina and proliferate to the wound site. This process is called vessel sprouting. Matrix metalloproteases further assist in dissolving the lamina to increase the propagation of the endothelial cells. The newly formed vessel sprouts surrounding the tissue form tubular canals which can interconnect with other vessels and differentiate into mature blood vessels through pericytes and smooth muscle cells. The initial blood flow would complete the angiogenesis process (Ellis et al. 2018).

2.3.3 Formation of Granulated Tissue

This is the last step of the proliferation stage. Fibroblasts, granulocytes, and macrophages are mainly involved during this phase. The congregated fibroblasts contract the wounded region by pulling the wounded edges together mediated by integrins and cytoskeletal machinery (Li et al. 2005). The fibronectin/fibrin-rich matrix is slowly replaced by loose bundles of fibroblasts, type III collagen, and new blood vessels (Ellis et al. 2018; Reinke and Sorg 2012). The major cells playing a role during this phase are fibroblasts. Fibroblasts secrete hyaluronic acid, collagen, glycosaminoglycans, proteoglycans, and fibronectin, which are the constituents of the ECM. Fibroblasts are also known to secrete PDGF-βb, TNF-α, and IL6, which would also induce re-epithelialization (Takehara 2000). Fibroblasts are influenced by TGF-β secreted by Pre-repair macrophages. Re-epithelization is initiated by EGF, KGF in case of keratinocytes and TGF-α as well (Schultz et al. 1991). Since a dense matrix is formed as a provisional wound matrix, this might lead to scar formation. This tissue is highly vascular, which appears red and can be traumatized easily (Reinke and Sorg 2012). The end of this phase is marked by the replacement of maturing fibroblasts by differentiating myofibroblasts (Desmouliere et al. 2014; Hinz 2007).

2.4 Remodeling Phase of Wound Healing

The final stage of wound healing is the remodeling phase that takes place from day 21 to a year or more, depending on the severity of the wound. During this process, fresh epithelium growth, and final scar tissue formation occur (Velnar 2009). The granulation of the tissue stops by apoptosis of fibroblast cells. With less metabolic activity, the wound is now avascular and non-cellular (Greenhalgh 1998). Smooth muscle actin (SMA) are secreted by myofibroblast cells are responsible for contractile force generation. They also release MMPs to degrade collagen, which was formed during granulation formation (Desmouliere et al. 2014; Hinz et al. 2012). The ECM goes through several changes during the final stage of wound healing, the notable one being the substitution of Type III collagen with stronger Type I collagen in small parallel bundles (Ellis et al. 2018; Reinke and Sorg 2012). The end result of this process would be a fully formed mature scar formed with high tensile strength, however, the original strength of the tissue is forever lost (Velnar 2009).

3 Single Cell Migration

Single cell migration is regulated by the molecular mechanism inducing single cell polarization. The small GTPase protein belonging to RHO family of proteins plays an important role in establishing cell polarity. Along with this small GTPase protein, RAC and CDC42 activate the cytoskeleton rearrangement from the leading edge, which includes active actin polymerization. The formation of filopodia and lamellipodia is due to actin polymerization. Further, this also promotes the interaction of integrin with ECM. In the back end of cell, RHO protein starts distinct signaling which results in actomyosin contraction (Pollard and Cooper 2009).

There are multiple factors involved in the single cell migration such as Intracellular signaling and extracellular mechanical cues. Cell migration requires actin polymerization at the leading edge assisted by integrins and focal adhesion molecules in order to bind to the 2D matrix. However, conditions are totally different in 3D matrix which resembles in vivo conditions. Actin polymerization starts in the front and myosin-driven contraction forces lead to the detachment of the cell at the rear position (Ridley 2003). Growth factors such as insulin, EGF, VEGF, TGF-alpha/beta, and ECM ligands also affect cell migration.

New adhesions form adhering to their new attachment sites at the leading edge of the cells. These adhesions form complexes that become more stable and constitute integrins, focal adhesion proteins which bind to the ECM in the matrix (Hood and Cheresh 2002). RhoA protein activates the myosin II motor which produces contractile forces from the leading edge of the migratory cell. This results in the net forward directional movement of the cell (Ridley 2003).

Extracellular mechanical cues affect the cell migration and are mainly studied in a physiological condition which 3D matrix. In 3D matrix, cell behaves differently with respect to the 2D matrix. Cells present in 3D matrix display a morphological distinct phenotype when compared to 2D. The main components of the 3D matrix are collagen and fibrin which make the matrix dense, affecting cell migration. Matrix metalloproteinases (MMPs) are the class of enzymes present in the plasma membrane, for example, Basigin/CD147. This family of enzymes cleaves the ECM matrix, which generates cell-scaled tracks helping in cell migration (Wolf and Friedl 2009). Without proteolytic activity, cells experience exceptional distortions and barely get through ECM pores in 3D lattices (Wyckoff et al. 2006). This results in retardation of migration of cell as to cross small pores, nucleus has to be deformed so that cell is able to migrate. As we know there are multiple types of cell in a multicellular organism, all distinct type of cells shows different migration patterns such as nerve cell is long and thin which can be modelled in 1D while in tumor cell with proteolytic activity can migrate easily in 3D matrix while if Proteolytic activity is lost, cells need to deform its nucleus for cell migration.

Stiffness of the matrix also affects the cell migration. One can make cells move in a unidirectional way just by modelling the stiffness of the matrix. The stiffness of 3D matrix decreases due to cross linking collagen by lysyl oxidase resulting in higher cell migration and metastasis in tumor cells (Levental et al. 2009).

4   Collective Cell Migration

Wound healing involves movement of a large number of cells in a synchronized way resulting in new tissue formation. It is a similar process seen during the developmental process of multicellular organisms. However, Collective cell migration is also seen in tumor formation and angiogenesis. Hence, an in-depth knowledge of the wound healing mechanism is necessary. In collective cell migration, cells move in one direction and show a similar pattern of movement which is not seen in single cell migration. Cells also respond to the environment they are in; this helps cells to form better coordination between 2 cells so that all cells follow similar movement which is required in wound healing.

If cells are closely bound/ attached to each other, they can show collective migration but if they are distinct apart, they are free to move in any direction which results in single cell migration. In the cohesive cell group, the outer cells lead the other cells present in the bunch in a particular direction. These cells sense the microenvironment which helps in the migration of whole cell cluster. The cells which are present outside in the cluster get a higher level of signals such as chemoattractants. These outermost cells also play a vital role in ECM remodeling which is required for collective cell migration. Cells present inside the cluster rely on the cell–cell interaction for the movement and to collectively polarize. Further, cells present inside the cluster can also influence the actions of the outer cells of cluster, which can affect the collective cell migration (Fig. 4).

5   Mechanism of Collective Cell Migration

In a cluster of cells, cells are interconnected by mechanical integrity, cell–cell cohesion, and intercellular signaling. Cell junctions play a vital role in collective cell migration.

5.1   Adherens Junctions

Coupling of the migratory cells is mediated by cadherins and immunoglobulin family proteins residing on the plasma membrane. In cancer cells, the suppression of E-cadherin combined with the overexpression of N-cadherin and neural cell adhesion molecules triggers the aggregate movement in which intercellular interactions are maintained. This approach is referred to as the fragmented transformation of epithelial–mesenchymal transition (Lee et al. 2006).

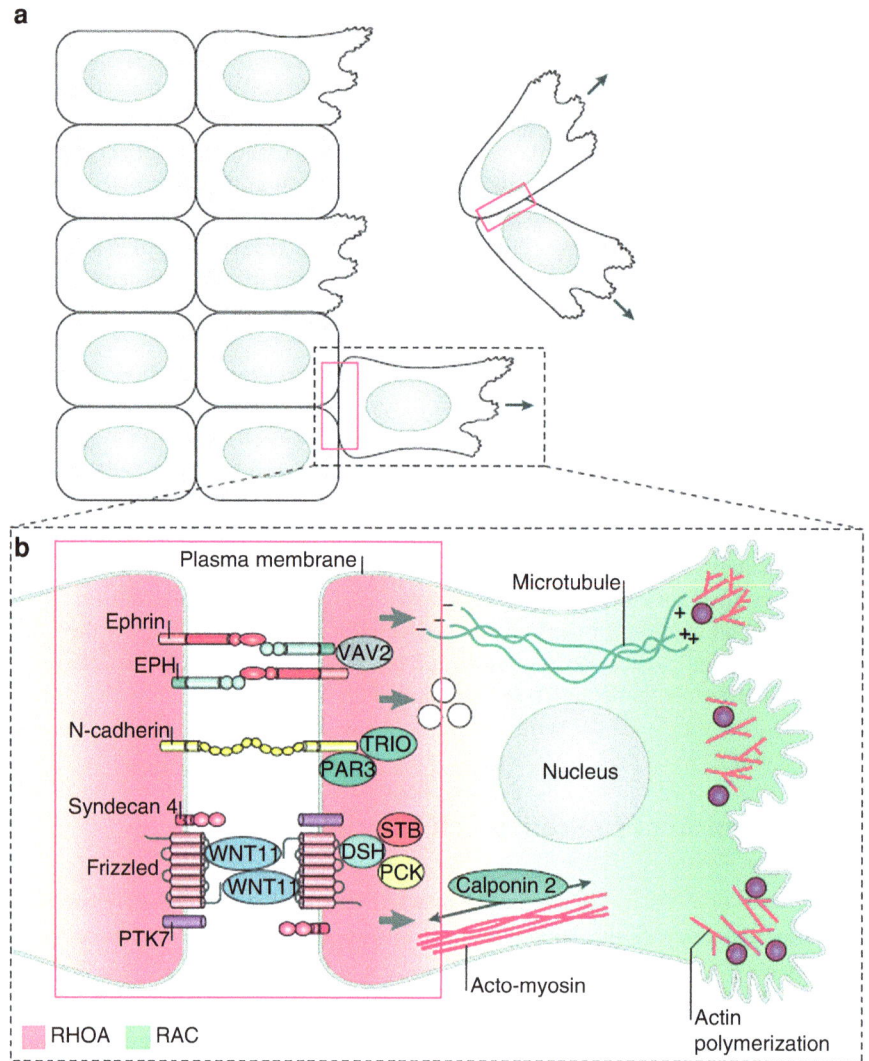

Fig. 4 Front and rear mechanics of cell migration

5.2 Desmosomes

During wound healing, migration of keratinocytes retains cell–cell junction (Shaw and Martin 2009). Furthermore, desmosomal proteins localized in membrane interact during migration of the aggregate in advanced epithelial tumor (Christiansen and Rajasekaran 2006).

5.3   Integrins

Integrins are the plasma membrane proteins, which help the cell in cell–ECM interaction. Recently, it was shown that alpha5beta1 integrin interacts with fibronectin which results in cell migration (Casey et al. 2001). If inactivates the beta1-integrin, cell–cell cohesion is lost which results in loss of collective migration and results in the migration of single cell (Hegerfeldt et al. 2002).

5.4   Growth Factors and Chemokines

Single cell polarization, migration, and persistence of migration are influenced by a variety of growth factors and chemokines as well. There are multiple growth factors involved in collective cell migration and wound healing such as TGF-alpha and beta, PDGF, VEGF, Serotonin, TNF-alpha, Leukotrienes, Interleukin-1, Lipoxins, and Interferon gamma. These growth factors are produced by various cells responsible for wound healing for paracrine and autocrine signaling of the cluster of cells (Steed 1997).

5.5   Cell–Cell and Cell–ECM Interactions

Intercellular adhesion and cell–ECM interaction are responsible for the morphological change in the tissue. Extracellular domain of Cadherin receptor is significant for intercellular adhesion (Shapiro and Weis 2009). Heterodimeric integrin receptor helps the cell to interact with the ECM. Integrin after interacting with ECM generates mechanical and chemical signals. ECM fibers also control the migration of cells and provide the directional cues. When the cluster of cells is migrating, cells present in the outermost receives the mechanical cues. Cadherins promote the signaling for the actin polymerization and also form adherent junctions, which maintain the integrity of the cluster of cells. Inactivation of Cadherins alters the collective cell dynamics (Shapiro and Weis 2009). If cadherins are inactivated and Integrins show the active binding to ECM, this will result in the breakdown of cluster and random movement of cells will be observed.

Cadherin plays a significant role in cell chemotaxis and interprets the chemical cues. Cadherin is required for the cell polarization and transmits the signal to the cells present inside the cluster. The cells present outside in the cluster are responsible for the direction and migration path. These cells adapt the matrix in 3D and produce MMPs, which cleaves the proteins present in ECM and make a path for the migration of cells.

Once the cells start migrating from the cluster, the cells present inside are polarized and push the cells from interior of the cluster for the movement.

Actomyosin contraction along with focal adhesion maturation at the leading edge of the cells enables the cells to contract from the rear edge. Further, the signal is transmitted to the cells present inside the cluster from outside and activation of small GTPase protein belonging to RHO family results in the activation of actomyosin. This allows cells to contract themselves at the rear edge.

Mechanosensing is necessary for the cells which are present inside a cluster. During cell migration, cells sense the forces exerted by the preceding cells. Cells sense the physical properties through their microenvironment via focal adhesions and adherens junctions. Mechanotransducers are the key protein, which help cells in mechanosensing. Mechanotransducers experience a conformational change after extending, uncovering a new protein connection domain and initiating biochemical flagging, which thus can adjust the strength of adhesion (Das et al. 2015). Under tension, α-catenin experiences a conformational change which brings about expanded junctional steadiness (Yonemura 2011). At the time a fraction of cells can participate in the pulling force and stop the cluster to migrate. There can be "tug-of-war" between the cells in the cluster during migration which can even break the focal adhesion and junction between the cells. This suggests that cells present outside the cluster are not solely responsible for the movement. The individual cells present inside the cluster sense gradient and chemotaxis and determine the polarity of the cluster. If the size of cluster is small, cells have the freedom to polarize in any direction and can show migration in different directions.

Cells change direction during migration upon collision with other cells where the cell–cell contacts induce inhibitory actions over migratory cells (top). This leads to change in cytoskeletal rearrangement ultimately leading to change in cell polarity and migration (bottom). Many membrane resident molecules interact at cell–cell contact sites triggering the activation of reaction cascades ultimately activating RHOA activation and RAC inhibition. Opposite phenotype happens on the leading edge of cells, i.e., RAC is activated, microtubules and actin machinery are organized preparing cells for protrusions leading to directed cell migration. Adapted and modified from Mayor and Etienne-Manneville (2019).

6   Factors in Wound Healing Process (Fig. 5)

6.1   Nutrition

It is well known that nutrition plays a significant role in wound healing. In 1747, James Lind, a Scottish Surgeon showed that wound repair is enhanced by taking citrus fruits. Angiogenesis and collagen deposition are reduced, which affects the fibroblast function in the malnutrition state of the organism. Vitamin A, Carbohydrates, and omega-3 fatty acids play a vital role during wound healing (Campos et al. 2008). Carbohydrates being the source of energy for the cell play an essential role in repairing the wound. Glucose is the major energy source required for the organism

Fig. 5 Multiple physiological and body-related factors affect the mechanisms of efficiency of wound healing. We have briefly summarized the effects of some of the key factors such as Nutrition, Hypoxia, Age, Diabetes, Smoking, Stress, etc. in the limit of the scope of this book chapter. Multiple other factors are as well affecting the process of wound healing and the readers are requested to refer to dedicated reviews for the elaborative description of same

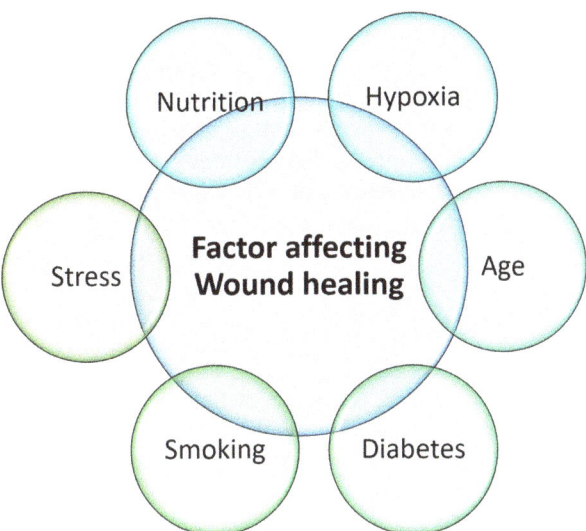

which produces ATP. ATP provides energy required for angiogenesis and formation of new tissue in wound healing.

Proteins are one of the major factors which affect the wound healing process. Fibroblast proliferation, collagen synthesis, wound remodeling can be altered by the deficiency in proteins. In addition, protein deficiency will also alter the activation of immune cells. There are multiple amino acids such as arginine, proline, and lysine, which play a vital role in collagen synthesis which provide the outcome as wound healing.

Vitamin A promotes wound healing process by an increased proliferation of fibroblasts and collagen synthesis. Vitamin C is also important for the wound healing process as its deficiency leads to improper immune response, which increases the risk of wound infection. Vitamin C deficiency may lead to decreased angiogenesis and increase in capillary fragility.

6.2 Hypoxia

As the local vascular supply is disrupted in wound, they become hypoxic to some extent. For re-epithelization, sufficient oxygen is required which induces wound healing. The elder patients going through surgery or having peripheral vascular disease have bad healing capacity whereas hyperbaric oxygen induces wound repair. However, oxygen is also essential for the deposition of collagen in ECM as it acts as a substrate for the hydroxylation of proline and lysine amino acids. It has also shown that proper supply of oxygen reduces the risk of wound infection (Kurz et al. 1996).

6.3   Smoking

Smoking reduces the fibroblast migration and proliferation and also affects the immune system function, downregulation and collagen deposition. In addition, smoking shows effects like chemotaxis, migration dysregulation and oxidative bactericidal mechanism which results in poor wound healing (Sørensen 2012).

6.4   Stress

Many diseases such as cancer, diabetes, cardiovascular disease are linked to stress. Delayed wound healing is shown in the student going under academic's stress such as examination. Stress leads to anxiety and depression. Due to the anxiety and depression, a person can get into poor sleep patterns, improper eating habits, less exercise, and sometimes consume alcohol. Further, in stressed state, there is a change in concentration of adrenal and pituitary hormones. These conditions play a negative role during wound repair.

6.5   Diabetes

Diabetes dysregulates the cellular function such as defect in chemotaxis of leukocyte, phagocytosis, impaired T-Cell immunity, and dysfunction in fibroblasts and epithelial cells which affect wound healing capacity. In addition, Neuropathy caused by diabetes can also affect the healing of the wound (Gary Sibbald and Woo 2008). VEGF is required for the wound healing process. However, during diabetic state, VEGF concentration is reduced which can impair wound healing process (Quattrini et al. 2008).

6.6   Age

Wound healing capacity is inversely proportional to the age, i.e., with an increase in age, wound healing capacity decreases, and vice versa. This is due to the altered inflammatory response and delayed T-cell activation and alteration in the production of chemokines. It has been shown that if an aged person performs regular exercise, wound healing can be fast due to the production of anti-inflammatory molecules due to exercise (Keylock et al. 2008).

7 Conclusions

Complete understanding of mechanisms involved in wound healing has direct implications in clinical relevance. The primary focus for epithelial cells is to migrate collectively and restore the barrier generated or broken during wounds. Therapeutic approaches can only be developed once we fully uncover the molecular mechanisms underlying different aspects of collective cell migration at the wound site. One special focus is on growth factors that stimulate or enhance the migration of cells. Possible perspective therapeutics might involve the incorporation of cell migration enhancing growth factors in bandages and applied at the site of wound that would promote cell migration of epithelial layers to heal the wounds. For example, applying prostaglantin E2 (PGE2) on mouse skin wound enhanced the cell migration leading to better wound healing efficiencies.

Multiple fundamental cellular processes in development, health, physiology, and disease involve largely collective cell migration. Despite tremendous progress in dissecting mechanisms of it our knowledge of collective cell migration still remains limited. The larger questions which will involve the integrated approaches would be how various signals—biological, chemical, physical, mechanical are integrated and regulate cellular movements. What makes or keeps the leader cells as leaders in terms of cytoskeletal machinery and membrane organization. What kind of forces operate at different layers of cells and how are they regulated during the process of wound healing. What makes cells behave differently in collective migration than individual migrations. The kind of questions can be answered by better coordination between experimental biologists and mathematical modelling of the same to dissect out the patterns and parameters involved in cell migration leading to wound healing. Collective cell migration has been recognized to play pivotal roles in processes like tissue regeneration, morphogenesis, and disease biology with focus on cancer biology. Understanding the molecular mechanisms and ground rules in wound healing will not only help us understand the process better, but we can apply those observations in designing better therapeutics for the same.

Acknowledgments We thank all the members of DB lab for critical inputs over this review. DB thanks IITGN for startup grant and SERB-DST for Ramanujan fellowship. CM, SK, AH thank MHRD for research fellowships.

Conflict of Interest The authors declare no conflict of interest.

References

Broughton G, Janis JE, Attinger CE (2006a) The basic science of wound healing. Plast Reconstr Surg 117:12S–34S

Broughton G, Janis JE, Attinger CE (2006b) Wound healing: an overview. Plast Reconstr Surg 117:1e-S–32e-S

Butenas S (2012) Tissue factor structure and function. Scientifica 2012:1–15

Campos AC, Groth AK, Branco AB (2008) Assessment and nutritional aspects of wound healing. Curr Opin Clin Nutr Metab Care 11:281–288

Casey RC, Burleson KM, Skubitz KM, Pambuccian SE, Oegema TR, Ruff LE, Skubitz APN (2001) β1-integrins regulate the formation and adhesion of ovarian carcinoma multicellular spheroids. Am J Pathol 159:2071–2080

Christiansen JJ, Rajasekaran AK (2006) Reassessing epithelial to mesenchymal transition as a prerequisite for carcinoma invasion and metastasis. Cancer Res 66:8319–8326

Clark RAF, Lanigan JM, DellaPelle P, Manseau E, Dvorak HF, Colvin RB (1982) Fibronectin and fibrin provide a provisional matrix for epidermal cell migration during wound reepithelialization. J Invest Dermatol 79:264–269

Das T, Safferling K, Rausch S, Grabe N, Boehm H, Spatz JP (2015) A molecular mechanotransduction pathway regulates collective migration of epithelial cells. Nat Cell Biol 17:276–287

Desmouliere A, Darby IA, Laverdet B, Bonté F (2014) Fibroblasts and myofibroblasts in wound healing. Clin Cosmet Investig Dermatol 7:301

Diegelmann RF, Evans MC (2004) Wound healing: an overview of acute, fibrotic and delayed healing. Front Biosci 9:283–289

Drake TA, Morrissey JH, Edgington TS (1989) Selective cellular expression of tissue factor in human tissues. Implications for disorders of hemostasis and thrombosis. Am J Pathol 134:1087–1097

Ellis S, Lin EJ, Tartar D (2018) Immunology of wound healing. Curr Dermatol Rep 7:350–358

Eming SA, Martin P, Tomic-Canic M (2014) Wound repair and regeneration: mechanisms, signaling, and translation. Sci Transl Med 6:265sr6

Flanagan M (2000) The physiology of wound healing. J Wound Care 9:299–300

Fleck RA, Rao LVM, Rapaport SI, Varki N (1990) Localization of human tissue factor antigen by immunostaining with monospecific, polyclonal anti-human tissue factor antibody. Thromb Res 59:421–437

Gary Sibbald R, Woo KY (2008) The biology of chronic foot ulcers in persons with diabetes. Diabetes Metab Res Rev 24:S25–S30

Golebiewska EM, Poole AW (2015) Platelet secretion: from haemostasis to wound healing and beyond. Blood Rev 29:153–162

Greenhalgh DG (1998) The role of apoptosis in wound healing. Int J Biochem Cell Biol 30:1019–1030

Hart J (2002) Inflammation 1: its role in the healing of acute wounds. J Wound Care 11:205–209

Hegerfeldt Y, Tusch M, Bröcker E-B, Friedl P (2002) Collective cell movement in primary melanoma explants: plasticity of cell-cell interaction, beta1-integrin function, and migration strategies. Cancer Res 62:2125–2130

Hinz B (2007) Formation and function of the myofibroblast during tissue repair. J Invest Dermatol 127:526–537

Hinz B, Phan SH, Thannickal VJ, Prunotto M, Desmoulière A, Varga J, De Wever O, Mareel M, Gabbiani G (2012) Recent developments in myofibroblast biology. Am J Pathol 180:1340–1355

Hood JD, Cheresh DA (2002) Role of integrins in cell invasion and migration. Nat Rev Cancer 2:91–100

Hunt TK (1988) The physiology of wound healing. Ann Emerg Med 17:1265–1273

Hunt TK, Hopf H, Hussain Z (2000) Physiology of wound healing. Adv Skin Wound Care 13:6–11

Jacinto A, Martinez-Arias A, Martin P (2001) Mechanisms of epithelial fusion and repair. Nat Cell Biol 3:E117–E123

Keylock KT, Vieira VJ, Wallig MA, DiPietro LA, Schrementi M, Woods JA (2008) Exercise accelerates cutaneous wound healing and decreases wound inflammation in aged mice. Am J Phys Regul Integr Comp Phys 294:R179–R184

Kurz A, Sessler DI, Lenhardt R (1996) Perioperative normothermia to reduce the incidence of surgical-wound infection and shorten hospitalization. N Engl J Med 334:1209–1216

Landén NX, Li D, Ståhle M (2016) Transition from inflammation to proliferation: a critical step during wound healing. Cell Mol Life Sci 73:3861–3885

Lawrence WT (1998) Physiology of the acute wound. Clin Plast Surg 25:321–340

Lee JM, Dedhar S, Kalluri R, Thompson EW (2006) The epithelial–mesenchymal transition: new insights in signaling, development, and disease. J Cell Biol 172:973–981

Levental KR, Yu H, Kass L, Lakins JN, Egeblad M, Erler JT, Fong SFT, Csiszar K, Giaccia A, Weninger W, Yamauchi M, Gasser DL, Weaver VM (2009) Matrix crosslinking forces tumor progression by enhancing integrin signaling. Cell 139:891–906

Li S, Huang NF, Hsu S (2005) Mechanotransduction in endothelial cell migration. J Cell Biochem 96:1110–1126

Maas C, Renné T (2018) Coagulation factor XII in thrombosis and inflammation. Blood 131:1903–1909

Maas C, Oschatz C, Renné T (2011) The plasma contact system 2.0. Semin Thromb Hemost 37:375–381

Mayor R, Etienne-Manneville S (2019) The front and rear of collective cell migration. Nat Rev Mol Cell Biol 17:97–109

Morrissey JH, Fakhrai H, Edgington TS (1987) Molecular cloning of the cDNA for tissue factor, the cellular receptor for the initiation of the coagulation protease cascade. Cell 50:129–135

Morrissey JH, Tajkhorshid E, Sligar SG, Rienstra CM (2012) Tissue factor/factor VIIa complex: role of the membrane surface. Thromb Res 129:S8–S10

Naudin C, Burillo E, Blankenberg S, Butler L, Renné T (2017) Factor XII contact activation. Semin Thromb Hemost 43:814–826

Pierce GF, Mustoe TA, Altrock BW, Deuel TF, Thomason A (1991) Role of platelet-derived growth factor in wound healing. J Cell Biochem 45:319–326

Piro O, Broze GJ (2005) Comparison of cell-surface TFPIalpha and beta. J Thromb Haemost 3:2677–2683

Pollard TD, Cooper JA (2009) Actin, a central player in cell shape and movement. Science 326:1208–1212

Pool JG (1977) Normal hemostatic mechanisms: a review. Am J Med Technol 43:776–780

Quattrini C, Jeziorska M, Boulton AJM, Malik RA (2008) Reduced vascular endothelial growth factor expression and intra-epidermal nerve fiber loss in human diabetic neuropathy. Diabetes Care 31:140–145

Ramasastry SS (2005) Acute wounds. Clin Plast Surg 32:195–208

Reinke JM, Sorg H (2012) Wound repair and regeneration. Eur Surg Res 49:35–43

Ridiandries A, Tan J, Bursill C (2018) The role of chemokines in wound healing. Int J Mol Sci 19:3217

Ridley AJ (2003) Cell migration: integrating signals from front to Back. Science 302:1704–1709

Robson MC (1997) Wound infection. Surg Clin North Am 77:637–650

Robson MC, Steed DL, Franz MG (2001) Wound healing: biologic features and approaches to maximize healing trajectories. Curr Probl Surg 38:A1–A140

Schultz G, Clark W, Rotatori DS (1991) EGF and TGF-α in wound healing and repair. J Cell Biochem 45:346–352

Sen CK, Gordillo GM, Roy S, Kirsner R, Lambert L, Hunt TK, Gottrup F, Gurtner GC, Longaker MT (2009) Human skin wounds: a major and snowballing threat to public health and the economy. Wound Repair Regen 17:763–771

Shapiro L, Weis WI (2009) Structure and biochemistry of cadherins and catenins. Cold Spring Harb Perspect Biol 1:a003053–a003053

Shaw TJ, Martin P (2009) Wound repair at a glance. J Cell Sci 122:3209–3213

Skover GR (1991) Cellular and biochemical dynamics of wound repair. Wound environment in collagen regeneration. Clin Podiatr Med Surg 8:723–756

Smith SA, Travers RJ, Morrissey JH (2015) How it all starts: initiation of the clotting cascade. Crit Rev Biochem Mol Biol 50:326–336

Sørensen LT (2012) Wound healing and infection in surgery: the pathophysiological impact of smoking, smoking cessation, and nicotine replacement therapy. Ann Surg 255:1069–1079

Steed DL (1997) The role of growth factors in wound healing. Surg Clin North Am 77:575–586

Strbo N, Yin N, Stojadinovic O (2014) Innate and adaptive immune responses in wound epithelialization. Adv Wound Care 3:492–501

Strecker-McGraw MK, Jones TR, Baer DG (2007) Soft tissue wounds and principles of healing. Emerg Med Clin North Am 25:1–22

Takehara K (2000) Growth regulation of skin fibroblasts. J Dermatol Sci 24:S70–S77

Takeuchi O, Akira S (2010) Pattern recognition receptors and inflammation. Cell 140:805–820

Velnar T (2009) The wound healing process: an overview of the cellular and molecular mechanisms. J Int Med Res 37:1528–1542

Vestweber D (2015) How leukocytes cross the vascular endothelium. Nat Rev Immunol 15:692–704

Werner S, Grose R (2003) Regulation of wound healing by growth factors and cytokines. Physiol Rev 83:835–870

Winter GD, Scales JT (1963) Effect of air drying and dressings on the surface of a wound. Nature 197:91–92

Witte MB, Barbul A (1997) General principles of wound healing. Surg Clin North Am 77:509–528

Wolf K, Friedl P (2009) Mapping proteolytic cancer cell-extracellular matrix interfaces. Clin Exp Metastasis 26:289–298

Wyckoff JB, Pinner SE, Gschmeissner S, Condeelis JS, Sahai E (2006) ROCK- and myosin-dependent matrix deformation enables protease-independent tumor-cell invasion in vivo. Curr Biol 16:1515–1523

Yan S-F, Mackman N, Kisiel W, Stern DM, Pinsky DJ (1999) Hypoxia/hypoxemia-induced activation of the procoagulant pathways and the pathogenesis of ischemia-associated thrombosis. Arterioscler Thromb Vasc Biol 19:2029–2035

Yonemura S (2011) A mechanism of mechanotransduction at the cell-cell interface: emergence of α-catenin as the center of a force-balancing mechanism for morphogenesis in multicellular organisms. BioEssays 33:732–736

Part II
Natural Products in the Management of Infected Wounds

Natural Products as Wound Healing Agents

Eman A. Khalil, Sara S. Abou-Zekry, Diana G. Sami, and Ahmed Abdellatif

1 Introduction

Chronic wounds are of increasing importance due to the economic burden of prolonged patient care and associated cost. Poor circulation and oxygenation due to venous and arterial insufficiency are major contributing factors to the incidence of chronic and non-healing wounds (Emanuelli et al. 2016; Jhamb et al. 2016; Guest et al. 2018). Diabetes and its complications are other precipitating factors for chronic wounds. Pressure ulcers or Bedsores are especially a problem for immune-compromised, elderly and bedridden patients leading to costly and prolonged care.

With a cost above USD10 billion annually (Raghav et al. 2018), there is a considerable need for effective wound care products worldwide, especially in developing countries where the healthcare systems are struggling to cope with other health problems. Therefore, wound management and prevention remains a challenge for healthcare (Sibbald et al. 2012). Inappropriate use of antibiotics causes antimicrobial resistance which often leads to poor wound healing, chronicity, and poor general condition. With the rise of bacterial resistance, natural products provide an excellent alternative to commercial antibiotics and other traditional wound care products.

E. A. Khalil · S. S. Abou-Zekry · D. G. Sami · A. Abdellatif (✉)
Department of Biology, School of Sciences and Engineering, American University in Cairo, New Cairo, Egypt
e-mail: ahmed.abdellatif@aucegypt.edu

© The Author(s), under exclusive license to Springer Nature Singapore Pte Ltd. 2021   77
P. Kumar, V. Kothari (eds.), *Wound Healing Research*,
https://doi.org/10.1007/978-981-16-2677-7_3

2 Wound Management

Preventive actions are essential to prevent the incidence of chronic wounds in diabetic and bedridden patients. The prevention of chronic wounds includes proper skincare and nutrition as well as awareness of healthcare workers to ensure proper patient mobility to prevent pressure ulcers (Langemo et al. 2015).

Debridement and cleaning of necrotic tissues from the wound to decrease infection are usually performed mechanically or biologically. Enzymes used in biological debridement may cause inflammation, and surgical techniques often remove both healthy and infected necrotic tissues, therefore, slowing the healing process (Falabella and Chen 2009). In some cases, larvae are used for maggot debridement, which removes only necrotic tissues. Skin grafts are sometimes used to cover resistant ulcers with healthy tissue (Han and Ceilley 2017). Tissue substitutes have gained interest recently due to their availability, biodegradability to promote healing (Han and Ceilley 2017; Maarof et al. 2016). The drawback of these scaffolds is their high cost (Han and Ceilley 2017).

Several growth factors are sometimes used to promote tissue regeneration and wound healing. Platelet-derived growth factor (PDGF), epidermal and fibroblast growth factors, show promising results (Barrientos et al. 2014). Hyperbaric oxygen, negative pressure, and Vacuum-Assisted Closure are non-invasive techniques used to enhance blood flow and oxygenation in the wound area to accelerate healing (Nain et al. 2011). These devices are noisy and expensive and are not recommended for patients on an anticoagulant.

The first line of treatment for wounds should be topical treatment with antibiotics. The chronic use and abuse of antibiotics, especially in low-income countries, leads to bacterial resistance and complicates wound care (Tsourdi et al. 2013). The topical application of drugs is useful in wound healing as it provides a higher concentration of the active ingredient at the wound site. Unfortunately, wound covering and multiple drug applications are required.

Different types of dressings are used for wound treatment; the choice of the kind of wound dressing hinge on the location and status of the wound. Vaseline gauze is cheap but may cause damage to the granulation tissue due to the need to frequently change the dressing, films are occlusive and retain moisture. Hydrocolloids cannot be used on infected wounds, and hydrogels are commonly used on dry wounds. Other synthetic polymers such as foams and hydro-fibers minimize trauma to the wound and have an absorptive capacity to remove exudate (Han and Ceilley 2017).

Antibiotic resistance and the cost associated with synthetic dressings can contribute to poor healing and chronic wounds. Many natural products are available at a reasonable cost. They are commonly used by folk medical practitioners in different cultures may provide an alternative to synthetic dressings and commonly used antibiotics. Below we discuss some examples of these natural products.

3 Natural Products for Wound Care

3.1 Phytotherapy

Natural medicinal plants are widely used topically for wound healing. Ginkgo, Green tea, *Aloe vera*, Ginseng, and olive oil, as well as other plants, are effective in wound healing (Pazyar et al. 2014). Phytochemicals help in treating inflammatory conditions and might aid in skin tissue regeneration. Plant bioactive components (phytochemicals) may improve tissue regeneration by removing oxidative stress and reducing inflammation by inhibition of Nuclear Factor Kappa B (NF-κB). More research is needed to illustrate the molecular targets of such active components and the development of effective wound care formulations (Shah and Amini-Nik 2017). We will discuss how some phytochemicals could help in wound healing.

Turmeric from the rhizome of *Curcuma longa*. Turmeric has antioxidant, anti-inflammatory, antidiabetic, antimicrobial, and wound healing potential (Gupta et al. 2012). The main component of turmeric is curcumin, which is useful in wound healing (Sami et al. 2020).

Curcumin inhibits arachidonic acid metabolism and downregulates lipoxygenase, Cyclooxygenase 2, and inducible nitric oxide synthase (iNOS) enzymes (Rao 2007). As a result, curcumin has anti-inflammatory and anti-oxidative properties; and prevents oxidative damage in human keratinocytes and fibroblasts. Proteases in curcumin might aid in stopping bleeding, and it could be a potential natural therapy to control severe pain associated with burns (Cheppudira et al. 2013). Further studies are needed to reveal the specific molecular targets of different Turmeric extract components.

Turmeric is used for many skin conditions, such as psoriasis. Wound dressing made from turmeric is effective as antibacterial, anti-inflammatory, and promotes rapid wound healing in diabetic and non-diabetic animals (Meizarini et al. 2018). It also accelerated the healing of alveolar osteitis after tooth extraction (Lone et al. 2018).

Oregano is an essential oil extracted from *Origanum vulgare* family *Lamiaceae* (Olmedo et al. 2014). Oregano contains mainly phenol isomers carvacrol and thymol, as well as their precursor monoterpenes γ-terpinene and *p*-cymene. Oregano shows antimicrobial, antifungal, anti-inflammatory, and antioxidant effects (Boateng and Catanzano 2015; Boateng and Diunase 2015).

Pomegranate juice and peel contain a considerable amount of bioactive polyphenols, such as tannins, flavanols, and gallic acid. The peel which represents half of the fruit weight has higher polyphenolic compounds with promising wound healing potential (Ismail et al. 2012; Mo et al. 2014).

Pomegranate peel extract gel is effective as a wound healing agent in experimental animals. Rats treated with 5.0% gel showed wound closure in 10 days, while those treated with lower concentration (2.5%) took longer. While control rats' wounds healed in 16–18 days (Chidambara Murthy et al. 2004). Pomegranate peel based-ointment improved wound closure and epithelialization (Hayouni et al. 2011). In diabetic rats, pomegranate gel promoted rapid healing, collagen regeneration, fibroblast infiltration, vascularization, and epithelialization (Yan et al. 2013).

Overall, the healing promoting potential of plant extracts was been verified in animal models of wound healing. The safety and efficacy of these plant extracts must be verified in human-controlled studies to be able to accept them as a standard of care in wound treatment. Such plants improved collagen deposition, epithelialization, and vascularization in diabetic and non-diabetic animal models. The list of plants is extensive (for a full review see Agyare et al. (2016)). Here are some examples; *Catharanthus roseus* (Nayak and Pinto Pereira 2006), *Calotropis gigantea* (Deshmukh et al. 2009), *Ageratum conyzoides* (Oladejo et al. 2003). *Chromolaena odorata* (Koca et al. 2009), *Kigelia Africana* (Agyare et al. 2013a), *Carica papaya* (Higashimori et al. 2005; Nayak et al. 2007; Gurung and Skalko-Basnet 2009). *Combretum mucronatum* (Kisseih et al. 2015), *Jatropha curcas* (Shetty et al. 2006), *Occimum sanctum* (Goel et al. 2010). Table 1 shows a summary of some other plants used in wound models.

3.2 Apitherapy (Honey Bee Products) for Wound Healing

3.2.1 Honey

Honey is used in folk medicine for burns and wounds due to its remarkable benefits. Honey contains many bioactive components such as vitamins, amino acids, and enzymes (Bagde et al. 2013). Honey differs in its chemical and biological properties depending on the environment, season, and type of plants that bees feed on. Honey is considered to be a better choice in the treatment of burns and pressure ulcers relative to the amniotic membrane, silver sulfadiazine, and nitrofurazone dressings (Günes and Eser 2007).

Honey-based wound dressings possess potent antimicrobial activities on both Gram-positive and negative bacteria (Fawole et al. 2012; Sarhan and Azzazy 2015; Abou Zekry et al. 2020). The antimicrobial effect of honey is due to the presence of methylglyoxal (MGO) (Adams et al. 2008; Mavric et al. 2008; Israili 2014). Honey also lowers the pH of the wound area, which provides skin fibroblasts with better growth conditions, and hinders bacterial growth (Lee et al. 2011; Oryan et al. 2016; Gethin et al. 2008). Honey also maintains its antibacterial activity in the presence of catalase, which is usually present in chronic wounds. Honey also has an anti-inflammatory role, and it promotes angiogenesis (Molan 2006; Tonks et al. 2007). Manuka honey also stimulates monocyte toll-like receptor 4, which leads to the activation of TNF-α, and interleukins 1-β, and 6. Such factors are critical for wound repair and tissue regeneration (Tonks et al. 2007).

Animal studies showed that wounds treated with honey have improved epithelialization, vascularization, and granulation tissue formation (Haryanto et al. 2012; Alizadeh et al. 2011). Folk medicine and clinical studies proved the effectiveness of honey over the years. Scientific studies also proved the benefits of honey as an analgesic, and in reducing swelling, and skin discoloration (Vijaya and Nishteswar 2012; Nikpour et al. 2014).

Table 1 Plant extracts used in wound healing models

Plant	Wound model	Outcome	
Abies cilicica, and *Cedrus libani*	Incision wounds in rats, treated with 1% ointment Excision wound in mice	Improved tensile strength at day 10, 33–40% compared to control 2% Enhanced wound closure	Tumen et al. (2011)
Blumea balsamifera	Excision wound in mice treated with 10 and 20% in olive oil daily	Enhanced wound healing Reduced the inflammatory cells and increased collagen formation	Pang et al. (2014)
Centella asiatica	0.2% topical solution of asiaticoside in Streptozotocin-diabetic rats 1 mg/kg oral in guinea pigs excision wounds	Significant increase in hydroxyproline, and tensile strength Enhanced collagen deposition and epithelialization	Agyare et al. (2013a)
Croton adamantinus	Excision wound in mice, treated with 1% for 3 days	Enhanced wound closure, and epithelialization. Improved collagen deposition	Ximenes et al. (2013)
Croton zehntneri	Excision wound in mice treated with 20% oil twice daily	Improved wound healing Enhanced fibroblasts and collagen deposition	Cavalcanti et al. (2012)
Cymbapogon nardus	Excision wound in diabetic mice, treated with essential oil	Accelerated wound healing Reduced inflammatory cytokines	Kandimalla et al. (2016)
Eucalyptus globulus	Excision wound in rats treated with 16.6% nano-emulsion	Improved wound healing	Sugumar et al. (2014)
Justicia flava	Excision wound treated with 7.5% extract	Improved angiogenesis, collagenation, and re-epithelialization	Agyare et al. (2013b)
Lannea welwitschii	Excision wound in rats treated with 7.5% extract	Improved healing and tensile strength Enhanced angiogenesis, collagen deposition, and wound re-epithelialization	Agyare et al. (2013b)
Lavandula angustifolia	Excision wound treated with 1% essential oil	Decreased wound area in the early phase of healing Increased transforming growth factor B, and myofibroblasts	Mori et al. (2016)
Lavandula spicata	Excision wound in rats, treated with lavender ointment 4%	Enhanced wound closure Well organized dermis and faster keratinization	Ben Djemaa et al. (2016)
Origanum vulgare and *Curcuma longa*	Excision wound in diabetic rats, treated with ointment and nanomaterial formulations	Enhanced wound healing, lower infection, improved collagen deposition by day 10 compared to day 21 in control	Sami et al. (2020)
Origanum vulgare and *Salvia triloba*	Incision wound in rats and excision wound in mice, treated once/day with	Improved tensile strength at day 10, 45% compared to 14% in control	Suntar et al. (2011)

(continued)

Table 1 (continued)

Plant	Wound model	Outcome	
	ointment, Origanum 12.5%, and Salvia 7.5%, with olive oil	Accelerated wound healing and re-epithelialization	
Pinus halepensis and *Pinus pinea*	Excision wound in mice, treated with 1% ointment Incision wound in rats treated with 1% ointment	40–54% better wound contraction 80–125% higher collagen content after 7 days of treatment Better tensile strength at 10 days	Suntar et al. (2012)
Plectranthus tenuiflorus	Excision wound in rats, treated with 10% solution	Complete re-epithelialization by day 14 and healing by day 18	Khorshid et al. (2010)
Pupalia lappacea	Excision wound treated with 1% to 20% extract ointment	Increased collagen deposition, re-epithelialization, granulation tissue formation, and angiogenesis	Udegbunam et al. (2014)
Rosmarinus officinalis	Excision wound in diabetic rats treated with oil for 3 days	Enhanced wound closure Dense well organized collagen Higher vascularization, and extracellular matrix deposition by day 15	Abu-Al-Basal (2010)

Modified from Perez-Recalde et al. (2018)

3.2.2 Bee Venom

Bee Venom is medically used for many conditions such as chronic pain, arthritis, tumors, and wound healing. It also has antimicrobial and anti-inflammatory properties (Abou Zekry et al. 2020; Lee et al. 2005, 2010).

The efficacy of Bee Venom on wound healing has been proven on deep wounds on diabetic and non-diabetic mice (Han et al. 2011; Hozzein et al. 2018) using traditional gauze dressing as well as in rats using nanomaterial dressing (Abou Zekry et al. 2020). Collagen deposition and wound closure rates were significantly improved by bee venom. Bee venom also reduced fibronectin, transforming growth factor-β1 (TGF-β1), and vascular endothelial growth factor (VEGF) and promoted healing by inhibiting fibrosis, and increasing epithelialization. Bee venom could be viewed as a potential treatment for enhancing wound healing, especially in diabetic wounds (Hozzein et al. 2018).

3.2.3 Propolis

Propolis or bee glue is the material used by bees for the construction of the beehive. It is composed mainly of resin and wax (80%), and essential oils (10%) (Gomez-Caravaca et al. 2006; Huang et al. 2014). It is commonly used in skincare products due to its anti-inflammatory, antioxidant, and antibacterial properties. It is useful in the treatment of acne vulgaris, and it improved wound healing in human diabetic ulcers and increased collagen content in wounds significantly (Henshaw et al. 2014; Olczyk et al. 2014).

3.2.4 Royal Jelly

Royal jelly is a jelly-like secretion of the worker bees that contains proteins (18%), carbohydrates (15%), and lipids (3–6%). Royal jelly contains many bioactive compounds (Sugiyama et al. 2012). It has proven its effectiveness in both wound-healing models (Kim et al. 2010), and in diabetic foot ulcers, possibly due to its effect on the wound blood vessels and collagen regeneration, as well as its antimicrobial activity (Siavash et al. 2015).

4 Marine Extracts for Wound Care

The ocean is rich in an extraordinarily diverse range of natural bioactive compounds. Marine organisms and their bioactive compounds are parts of thousands of promising pharmaceutical substitutes. Many studies confirmed that marine extracted macromolecules combined with bioactive molecules or synthetic polymers could promote skin healing and tissue regeneration (Chandika et al. 2015). Compounds such as chitin, fucoidan, collagen, carrageenan, alginate, chitosan, and other molecules have been successfully used to improve wound healing (Yeo et al. 2012).

4.1 Marine Collagen

Marine collagen is an abundant protein in the body of vertebrates; it also forms more than 50% of the skin and plays a crucial role in providing a scaffold for different cell types. It is involved in cell proliferation, differentiation, migration, and attachment (Anand et al. 2013; Kasoju et al. 2013). Collagen can be obtained from many sources, which, unfortunately, carries a risk of potential disease spread (Song et al. 2006).

Recently, collagen from marine organisms has grabbed the attention in tissue regeneration development due to its high potential in clinical applications plus its low risk in the transmission of diseases to human recipients (Hoyer et al. 2014). Salmon collagen showed significant wound healing properties in research studies (Shen et al. 2008). Collagen films implanted in rats improved the morphological appearance of the rat dermis and significantly enhanced new blood vessel formation. Other marine organisms (Fig. 1) are promising sources of marine collagen.

4.2 Marine Chitin

Marine chitin and Chitosan are abundant in nature. The exoskeleton of marine crustaceans (Fig. 2) is the major source of chitin. Byproducts of the food industry

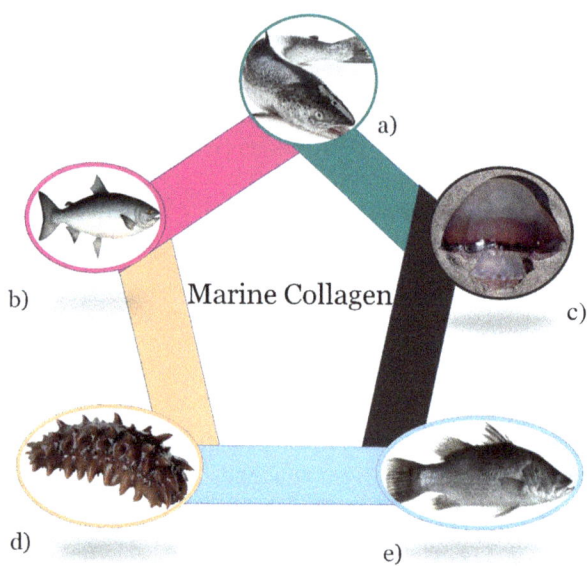

Fig. 1 Marine sources of collagen. (**a**) *Salmo salar* (Atlantic Salmon), image source https://www.alibaba.com/product-detail/Scottish-Salmon-salmo-salar_137741088.html (**b**) *Oncorhynchus keta* (Chum salmon), image source https://knowyour.fish/resources/species-profiles/chum-salmon (**c**) *Stomolophus meleagris* (jellyfish), image sources https://www.carolinanature.com/pix/jelly fish0394.jpg (**d**) *Stichopus Japonicus*, image source https://www.shutterstock.com/search/stichopus+japonicus (**e**) *Lates calcarifer*, image source http://www.marinespecies.org/aphia.php?p=taxdetails&id=278957

such as crab and shrimp shells are readily available and provide an endless source of chitin. Chitin and chitosan are effective in wound healing, and as antimicrobials, and they were used with other polymers for wound dressing and skin tissue engineering (Hu et al. 2007; Kurita 2006; Nwe et al. 2009). Hydrogel wound dressing made from crab shell chitin enhanced the healing environment, and blood vessel formation, and increased granulation tissue in deep wounds (Murakami et al. 2010).

4.3 Marine Fatty acids

Marine fatty acids, are primary precursors of several mediators and take part in the synthesis of membrane phospholipids. Therefore, they have essential functions in inflammation, chemotaxis, hemostasis, and stimulating epithelial cell proliferation (Ziboh et al. 2000; Calder 2006).

Marine extracted fatty acids (Fig. 3) from different sources, were examined for their wound healing potential. Fatty acids extracted from mollusks (*Mytilus galloprovincialis*) and gastropods (*Rapana venosa*) (Badiu et al. 2010), and sea cucumber (*Stichopus chloronotus*) (Fredalina et al. 1999a), showed

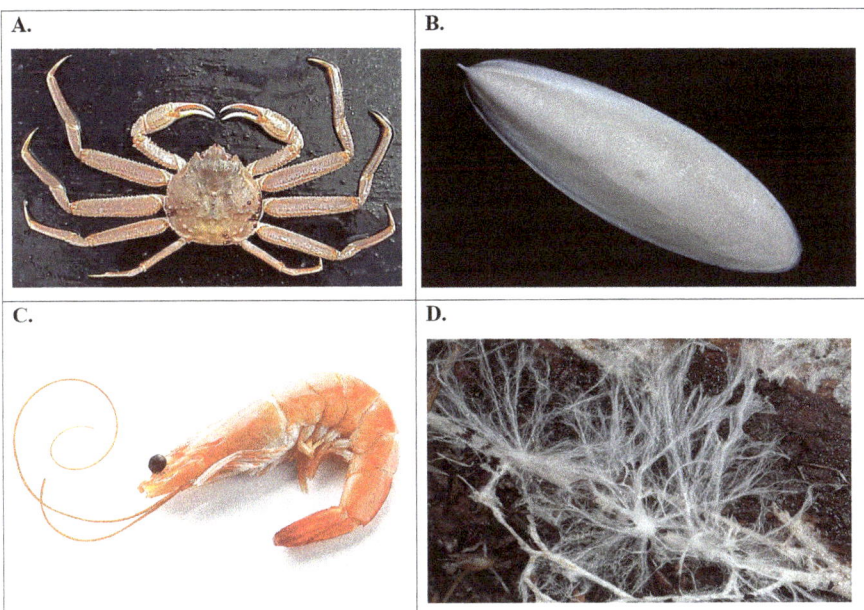

Fig. 2 Marine sources of chitin. (**a**) Snow Crab, image source https://chefs-resources.com/seafood/ shellfish/crab-species/snow-crab/ (**b**) Squid bone, image source https://en.wikipedia.org/wiki/ Cuttlebone#/media/File:Cuttlefish-Cuttlebone2.jpg (**c**) Shrimp Shells, image source https://energis. com.au/2015/12/shrimp-shells-to-solar-cells/#post/0 (**d**) Fungal Mycelium, image source https:// alexhyde.photoshelter.com/image/I0000Pm3ZvENDy.k

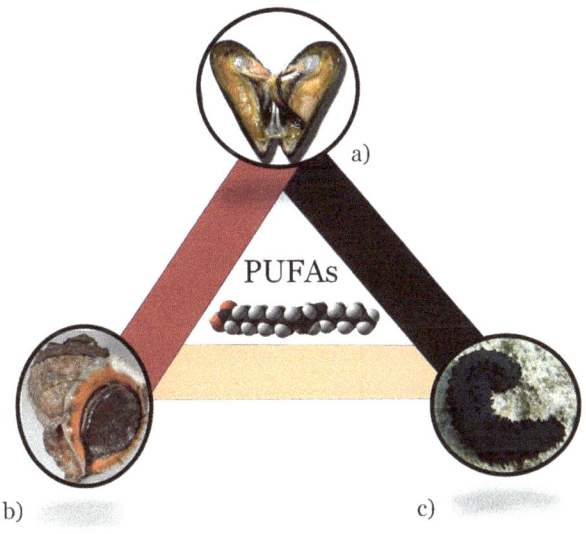

Fig. 3 Marine polyunsaturated fatty acids (PUFAs) extracted from gastropods, sea cucumber, and mollusks. (**a**) *Mytilus galloprovincialis* (**b**) *Rapana venosa*, image source https://www.flickr.com/ photos/117184384@N07/29527502712 (**c**) *Stichopus chloronotus* (Greenfish Sea Cucumber), image source https://www.flickr.com/photos/animaliaproject/8375094218

anti-inflammatory effects and reduced the healing time, as well as, increased vasculature and collagen formation in wound models.

Sea cucumbers are cylindrical echinoderms rich in fatty acids and are unique for their regenerative power. There are approximately 1500 species of these marine invertebrates, which are commonly used for traditional medicinal purposes in Southeast Asia. Several animal studies showed that wounds treated with topical application of sea cucumber extract showed improved healing in comparison to controls (Fredalina et al. 1999b; Masre et al. 2012).

4.4 Marine Alginate

Marine Alginates are polysaccharides extracted from brown seaweed, such as *Sargassum, Eclonia maxima, Ascophyllum nodosum, Macrocystis pyrifera*, and *Laminaria japonica* (d'Ayala et al. 2008; George and Abraham 2006). Alginate wound dressings are well accepted in wound management, due to their antibacterial properties, ability to maintain a moist physiological environment, facilitating wound healing by absorbing excess wound fluid. Additionally, alginates can be manufactured in different forms such as films, foams, gels, wafers, and nanofibers (d'Ayala et al. 2008; Gombotz and Wee 1998). They are also used in deep wounds as an alternative to grafting.

5 Fish Skin Grafting

Split Skin Grafts are one of the standards of care for deep wounds and full-thickness burns (Puri et al. 2016). When autologous grafts are a problem due to the large size of burns, doctors often rely on cadaveric and pigskin for temporary coverage. Unfortunately, cadaveric and pigskin grafts carry the risk of infection and immune response (Moss et al. 1994).

Acellular fish skin is an alternative source of xenograft, providing a safe and effective, skin substitute, (Yang et al. 2016; Baldursson et al. 2015). Acellular fish skin has also been used in chronic wounds (Yang et al. 2016). The skin of some species (e.g., Atlantic cod) is rich in omega-3 polyunsaturated fatty acids, and eicosapentaenoic and docosahexaenoic acids, which are effective anti-inflammatory and antibacterial agents (Magnusson et al. 2015). Fish skin is an ideal choice in treating wounds where cadaveric skin is not available due to its availability and the minimal processing required in the preparation of fish skin to maintain its three-dimensional structure (Alam and Jeffery 2019).

6   Modern Delivery Systems for Natural Products

Several new delivery systems were developed to improve the therapeutic potential of natural products, such as including hydrogels, nano-scaffolds, and nanoparticles (Johnson and Wang 2015; Sarabahi 2012). Hydrogels are water-based gels that are effective in rehydrating necrotic tissue allowing its removal without damage to neighboring healthy tissues (Sarabahi 2012). Hydrogels may be conjugated or loaded with other bioactive materials and other nano-formulations such as curcumin and oregano (Sami et al. 2020) as well as other material such as hyaluronic acid (HA) (Sharma et al. 2018), these hydrogels were effective in reducing infection and improving wound healing but were inferior to ointment preparations in animal studies (Sami et al. 2020).

7   Nanomaterial in Wound Care

Nanomaterials provide an excellent addition to wound care products as they enhance drug penetration, provide controlled release, increase their stability and protect them from being degraded; therefore, minimal amounts of drugs can be used to avoid toxicity (Goyal et al. 2016; Kalashnikova et al. 2015). Nanomaterials for wound care are superior to conventional dressings, due to their physical characteristics such as the large surface-to-volume ratio, and high porosity, which enhance exudate absorption and achieve better wound permeation (Miguel et al. 2018). Nanomaterials absorption through the skin is hypothesized to occur through inter and intracellular routes as well as via skin appendages (Fig. 4) (Palmer and DeLouise 2016). The only limitation of nanomaterials, either nanofibers or nanoparticles, is the cost and the difficulty of large-scale commercial production (Chen et al. 2018).

Chitosan is a widely used biopolymer for nanoparticle preparation due to its availability, biodegradability, and biocompatibility (Kamat et al. 2016). The antibacterial and antifungal nature of chitosan makes it a good wound healing agent (Katas et al. 2013). Chitosan nanoparticles are promising nanocarriers for drug delivery due to their safety as they are synthesized using non-toxic solvents (Agnihotri and Aminabhavi 2004). They also have high antimicrobial activity due to their spherical character and positive charge, which interacts with the negative charge of the bacterial cell membrane resulting in membrane disruption, and cell death (Divya et al. 2018; Jayakumar et al. 2010).

Nanofibers offer great advantages over current commercial dressings, as they have a structure similar to the extracellular matrix. Nanofibers may be loaded with growth factors and other bioactive materials for wound healing to promote angiogenesis, decrease infection, and accelerate wound healing rate (Rath et al. 2016). Polyvinyl alcohol is a common chemical used for the synthesis of nanofibers due to

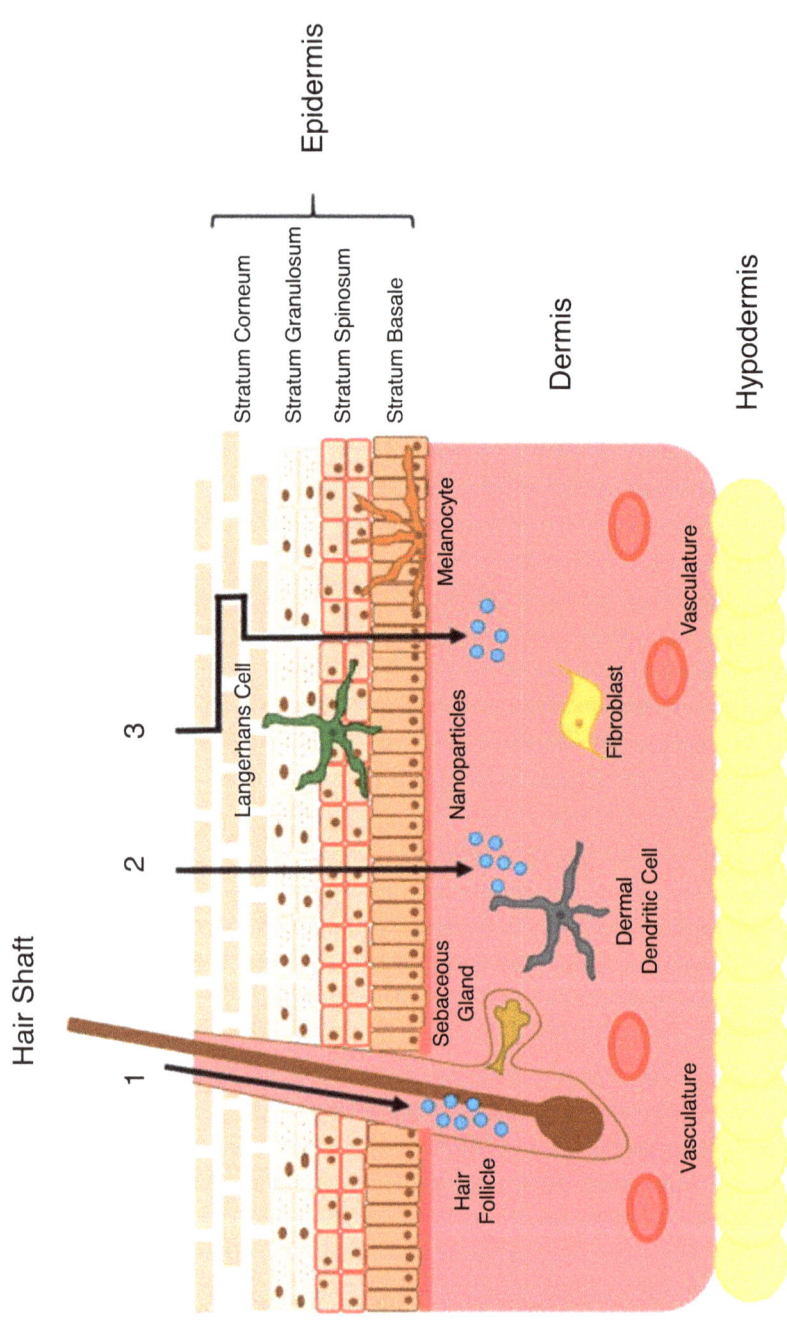

Fig. 4 Mechanisms of Nanoparticle penetration through the skin. Nanoparticles are absorbed based on the particle size, charge, morphology, and type of polymer used. Nanoparticles can penetrate the skin via the (1) *Appendageal route*, as hair follicles, sweat glands. (2) *The intracellular pathway* through corneocytes. (3) *Intercellular route* where particles pass between corneocytes. From Palmer and DeLouise (2016), open access Creative Common CC BY

its proven biocompatibility (Abou Zekry et al. 2020; Sarhan et al. 2016; Puppi et al. 2011). It is usually combined with many natural products to improve wound healing in human and animal studies (Sami et al. 2020; Abou Zekry et al. 2020; Sarhan et al. 2016).

8   Conclusions

Folk medicine remains a rich source for potential wound healing products. In times of antibiotic resistance and economic difficulties, many societies around the world go back to their roots and use traditional natural products for wound care. Evidence-based biomedical research proved many of these natural products as effective wound healing products either in their raw form or as modern formulations. New drug delivery tools increase the local availability of the drug and enhance wound healing. Nanomaterials are promising tools of drug delivery for wound healing, however, they remain limited due to the high cost and difficulty of industrial-scale production of such nanomaterial. More clinical trials are needed to ensure the safety and efficacy of these products for human use in mainstream wound care. The challenge remains to identify, isolate, and characterize the active ingredients from these natural products, and to identify their mechanisms of action.

References

Abou Zekry SS, Abdellatif A, Azzazy HME (2020) Fabrication of pomegranate/honey nanofibers for use as antibacterial wound dressings. Wound Med 28:100181

Abu-Al-Basal MA (2010) Healing potential of *Rosmarinus officinalis* L. on full-thickness excision cutaneous wounds in alloxan-induced-diabetic BALB/c mice. J Ethnopharmacol 131 (2):443–450

Adams CJ et al (2008) Isolation by HPLC and characterisation of the bioactive fraction of New Zealand Manuka (*Leptospermum scoparium*) honey. Carbohydr Res 343(4):651–659

Agnihotri SA, Aminabhavi TM (2004) Formulation and evaluation of novel tableted chitosan microparticles for the controlled release of clozapine. J Microencapsul 21(7):709–718

Agyare C et al (2013a) Antimicrobial, antioxidant, and wound healing properties of *Kigelia africana* (Lam.) Beneth. and *Strophanthus hispidus* DC. Adv Pharmacol Sci 2013:692613

Agyare C et al (2013b) Evaluation of antimicrobial and wound healing potential of *Justicia flava* and *Lannea welwitschii*. Evid Based Complement Alternat Med 2013:632927

Agyare C et al (2016) Review: African medicinal plants with wound healing properties. J Ethnopharmacol 177:85–100

Alam K, Jeffery SLA (2019) Acellular fish skin grafts for management of split thickness donor sites and partial thickness burns: a case series. Mil Med 184(Suppl 1):16–20

Alizadeh AM et al (2011) The effect of *Teucrium polium* honey on the wound healing and tensile strength in rat. Iran J Basic Med Sci 14(6):499–505

Anand S et al (2013) Biochemical and thermo-mechanical analysis of collagen from the skin of Asian Sea bass (*Lates calcarifer*) and Australasian Snapper (*Pagrus auratus*), an alternative for mammalian collagen. Eur Food Res Technol 236(5):873–882

Badiu DL et al (2010) Amino acids from *Mytilus galloprovincialis* (L.) and *Rapana venosa* molluscs accelerate skin wounds healing via enhancement of dermal and epidermal neoformation. Protein J 29(2):81–92

Bagde A et al (2013) Therapeutic and nutritional values of honey (Madhu). Int Res J Pharm 4 (3):19–22

Baldursson BT et al (2015) Healing rate and autoimmune safety of full-thickness wounds treated with fish skin acellular dermal matrix versus porcine small-intestine submucosa: a noninferiority study. Int J Low Extrem Wounds 14(1):37–43

Barrientos S et al (2014) Clinical application of growth factors and cytokines in wound healing. Wound Repair Regen 22(5):569–578

Ben Djemaa FG et al (2016) Antioxidant and wound healing activity of *Lavandula aspic* L. ointment. J Tissue Viability 25(4):193–200

Boateng J, Catanzano O (2015) Advanced therapeutic dressings for effective wound healing—a review. J Pharm Sci 104(11):3653–3680

Boateng J, Diunase KN (2015) Comparing the antibacterial and functional properties of Cameroonian and Manuka honeys for potential wound healing-have we come full cycle in dealing with antibiotic resistance? Molecules 20(9):16068–16084

Calder PC (2006) n−3 polyunsaturated fatty acids, inflammation, and inflammatory diseases. Am J Clin Nutr 83(6):1505S–1519S

Cavalcanti JM et al (2012) The essential oil of *Croton zehntneri* and trans-anethole improves cutaneous wound healing. J Ethnopharmacol 144(2):240–247

Chandika P, Ko S-C, Jung W-K (2015) Marine-derived biological macromolecule-based biomaterials for wound healing and skin tissue regeneration. Int J Biol Macromol 77:24–35

Chen H et al (2018) A novel wound dressing based on a Konjac glucomannan/silver nanoparticle composite sponge effectively kills bacteria and accelerates wound healing. Carbohydr Polym 183:70–80

Cheppudira B et al (2013) Curcumin: a novel therapeutic for burn pain and wound healing. Expert Opin Investig Drugs 22(10):1295–1303

Chidambara Murthy K et al (2004) Study on wound healing activity of *Punica granatum* peel. J Med Food 7(2):256–259

d'Ayala GG, Malinconico M, Laurienzo P (2008) Marine derived polysaccharides for biomedical applications: chemical modification approaches. Molecules 13(9):2069–2106

Deshmukh PT et al (2009) Wound healing activity of *Calotropis gigantea* root bark in rats. J Ethnopharmacol 125(1):178–181

Divya K, Smitha V, Jisha MS (2018) Antifungal, antioxidant and cytotoxic activities of chitosan nanoparticles and its use as an edible coating on vegetables. Int J Biol Macromol 114:572–577

Emanuelli T, Burgeiro A, Carvalho E (2016) Effects of insulin on the skin: possible healing benefits for diabetic foot ulcers. Arch Dermatol Res 308(10):677–694

Falabella CA, Chen W (2009) Cross-linked hyaluronic acid films to reduce intra-abdominal postsurgical adhesions in an experimental model. Dig Surg 26(6):476–481

Fawole OA, Makunga NP, Opara UL (2012) Antibacterial, antioxidant and tyrosinase-inhibition activities of pomegranate fruit peel methanolic extract. BMC Complement Altern Med 12 (1):200

Fredalina B et al (1999a) Fatty acid compositions in local sea cucumber. Gen Pharmacol Vasc S 33 (4):337–340

Fredalina BD et al (1999b) Fatty acid compositions in local sea cucumber, *Stichopus chloronotus*, for wound healing. Gen Pharmacol 33(4):337–340

George M, Abraham TE (2006) Polyionic hydrocolloids for the intestinal delivery of protein drugs: alginate and chitosan—a review. J Control Release 114(1):1–14

Gethin GT, Cowman S, Conroy RM (2008) The impact of Manuka honey dressings on the surface pH of chronic wounds. Int Wound J 5(2):185–194

Goel A et al (2010) Wound healing potential of *Ocimum sanctum* Linn. with induction of tumor necrosis factor-alpha. Indian J Exp Biol 48(4):402–406

Gombotz WR, Wee S (1998) Protein release from alginate matrices. Adv Drug Deliv Rev 31 (3):267–285

Gomez-Caravaca AM et al (2006) Advances in the analysis of phenolic compounds in products derived from bees. J Pharm Biomed Anal 41(4):1220–1234

Goyal R et al (2016) Nanoparticles and nanofibers for topical drug delivery. J Control Release 240:77–92

Guest JF, Singh H, Vowden P (2018) Potential cost-effectiveness of using a collagen-containing dressing in managing diabetic foot ulcers in the UK. J Wound Care 27(3):136–144

Günes ÜY, Eser I (2007) Effectiveness of a honey dressing for healing pressure ulcers. J Wound Ostomy Cont Nurs 34(2):184–190

Gupta SC et al (2012) Discovery of curcumin, a component of golden spice, and its miraculous biological activities. Clin Exp Pharmacol Physiol 39(3):283–299

Gurung S, Skalko-Basnet N (2009) Wound healing properties of *Carica papaya* latex: in vivo evaluation in mice burn model. J Ethnopharmacol 121(2):338–341

Han G, Ceilley R (2017) Chronic wound healing: a review of current management and treatments. Adv Ther 34(3):599–610

Han S et al (2011) Biological effects of treatment of an animal skin wound with honeybee (*Apis mellifera* L) venom. J Plast Reconstr Aesthet Surg 64(3):e67–e72

Haryanto H et al (2012) Effectiveness of Indonesian honey on the acceleration of cutaneous wound healing: an experimental study in mice. Wounds 24(4):110–119

Hayouni E et al (2011) Hydroalcoholic extract based-ointment from *Punica granatum* L. peels with enhanced in vivo healing potential on dermal wounds. Phytomedicine 18(11):976–984

Henshaw FR et al (2014) Topical application of the bee hive protectant propolis is well tolerated and improves human diabetic foot ulcer healing in a prospective feasibility study. J Diabetes Complicat 28(6):850–857

Higashimori H et al (2005) Peripheral axon caliber and conduction velocity are decreased after burn injury in mice. Muscle Nerve 31(5):610–620

Hoyer B et al (2014) Jellyfish collagen scaffolds for cartilage tissue engineering. Acta Biomater 10 (2):883–892

Hozzein WN et al (2018) Bee venom improves diabetic wound healing by protecting functional macrophages from apoptosis and enhancing Nrf2, Ang-1 and Tie-2 signaling. Mol Immunol 103:322–335

Hu X et al (2007) Solubility and property of chitin in NaOH/urea aqueous solution. Carbohydr Polym 70(4):451–458

Huang S et al (2014) Recent advances in the chemical composition of propolis. Molecules 19 (12):19610–19632

Ismail T, Sestili P, Akhtar S (2012) Pomegranate peel and fruit extracts: a review of potential anti-inflammatory and anti-infective effects. J Ethnopharmacol 143(2):397–405

Israili ZH (2014) Antimicrobial properties of honey. Am J Ther 21(4):304–323

Jayakumar R et al (2010) Novel chitin and chitosan nanofibers in biomedical applications. Biotechnol Adv 28(1):142–150

Jhamb S, Vangaveti VN, Malabu UH (2016) Genetic and molecular basis of diabetic foot ulcers: clinical review. J Tissue Viability 25(4):229–236

Johnson NR, Wang Y (2015) Drug delivery systems for wound healing. Curr Pharm Biotechnol 16 (7):621–629

Kalashnikova I, Das S, Seal S (2015) Nanomaterials for wound healing: scope and advancement. Nanomedicine (Lond) 10(16):2593–2612

Kamat V, Bodas D, Paknikar K (2016) Chitosan nanoparticles synthesis caught in action using microdroplet reactions. Sci Rep 6:22260

Kandimalla R et al (2016) Chemical composition and anti-candidiasis mediated wound healing property of *Cymbopogon nardus* essential oil on chronic diabetic wounds. Front Pharmacol 7:198

Kasoju N et al (2013) Exploiting the potential of collagen as a natural biomaterial in drug delivery. J Proteins Proteomics 1(1):9–14

Katas H, Raja MA, Lam KL (2013) Development of chitosan nanoparticles as a stable drug delivery system for protein/siRNA. Int J Biomater 2013:146320

Khorshid F, Shaker SA, Alsofyani T, Albar H (2010) *Plectranthus tenuiflorus* (Shara) promotes wound healing: in vitro and in vivo studies. Int J Bot 6:69–80

Kim J et al (2010) Royal jelly enhances migration of human dermal fibroblasts and alters the levels of cholesterol and sphinganine in an in vitro wound healing model. Nutr Res Pract 4(5):362–368

Kisseih E et al (2015) Phytochemical characterization and in vitro wound healing activity of leaf extracts from *Combretum mucronatum* Schum. & Thonn.: Oligomeric procyanidins as strong inductors of cellular differentiation. J Ethnopharmacol 174:628–636

Koca U et al (2009) In vivo anti-inflammatory and wound healing activities of *Centaurea iberica* Trev. ex Spreng. J Ethnopharmacol 126(3):551–556

Kurita K (2006) Chitin and chitosan: functional biopolymers from marine crustaceans. Mar Biotechnol 8(3):203–226

Langemo D, Spahn J, Snodgrass L (2015) Accuracy and reproducibility of the wound shape measuring and monitoring system. Adv Skin Wound Care 28(7):317–323

Lee JY et al (2005) Inhibitory effect of whole bee venom in adjuvant-induced arthritis. In Vivo 19 (4):801–805

Lee W-R et al (2010) Bee venom reduces atherosclerotic lesion formation via anti-inflammatory mechanism. Am J Chin Med 38(06):1077–1092

Lee DS, Sinno S, Khachemoune A (2011) Honey and wound healing. Am J Clin Dermatol 12 (3):181–190

Lone PA et al (2018) Role of turmeric in management of alveolar osteitis (dry socket): a randomised clinical study. J Oral Biol Craniofac Res 8(1):44–47

Maarof M et al (2016) Secretion of wound healing mediators by single and bi-layer skin substitutes. Cytotechnology 68(5):1873–1884

Magnusson S et al (2015) [Decellularized fish skin: characteristics that support tissue repair]. Laeknabladid 101(12): 567–573

Masre SF et al (2012) Quantitative analysis of sulphated glycosaminoglycans content of Malaysian sea cucumber *Stichopus hermanni* and *Stichopus vastus*. Nat Prod Res 26(7):684–689

Mavric E et al (2008) Identification and quantification of methylglyoxal as the dominant antibacterial constituent of Manuka (*Leptospermum scoparium*) honeys from New Zealand. Mol Nutr Food Res 52(4):483–489

Meizarini A et al (2018) Anti-inflammatory properties of a wound dressing combination of zinc oxide and turmeric extract. Vet World 11(1):25–29

Miguel SP et al (2018) Electrospun polymeric nanofibres as wound dressings: a review. Colloids Surf B: Biointerfaces 169:60–71

Mo J et al (2014) Wound healing activities of standardized pomegranate rind extract and its major antioxidant ellagic acid in rat dermal wounds. J Nat Med 68(2):377–386

Molan PC (2006) The evidence supporting the use of honey as a wound dressing. Int J Low Extrem Wounds 5(1):40–54

Mori HM et al (2016) Wound healing potential of lavender oil by acceleration of granulation and wound contraction through induction of TGF-beta in a rat model. BMC Complement Altern Med 16:144

Moss A Jr et al (1994) Analysis of the human anti-pig cellular immune response using the Hu-scid mouse-porcine skin graft model. Transplant Proc 26(3):1209

Murakami K et al (2010) Hydrogel blends of chitin/chitosan, fucoidan and alginate as healing-impaired wound dressings. Biomaterials 31(1):83–90

Nain PS et al (2011) Role of negative pressure wound therapy in healing of diabetic foot ulcers. J Surg Tech Case Rep 3(1):17–22

Nayak BS, Pinto Pereira LM (2006) *Catharanthus roseus* flower extract has wound-healing activity in Sprague Dawley rats. BMC Complement Altern Med 6:41

Nayak SB, Pinto Pereira L, Maharaj D (2007) Wound healing activity of *Carica papaya* L. in experimentally induced diabetic rats. Indian J Exp Biol 45(8):739–743

Nikpour M et al (2014) The effect of honey gel on abdominal wound healing in cesarean section: a triple blind randomized clinical trial. Oman Med J 29(4):255–259

Nwe N, Furuike T, Tamura H (2009) The mechanical and biological properties of chitosan scaffolds for tissue regeneration templates are significantly enhanced by chitosan from *Gongronella butleri*. Materials 2(2):374–398

Oladejo OW et al (2003) A comparative study of the wound healing properties of honey and *Ageratum conyzoides*. Afr J Med Med Sci 32(2):193–196

Olczyk P et al (2014) Propolis modulates fibronectin expression in the matrix of thermal injury. Biomed Res Int 2014:748101

Olmedo R, Nepote V, Grosso NR (2014) Antioxidant activity of fractions from oregano essential oils obtained by molecular distillation. Food Chem 156:212–219

Oryan A, Alemzadeh E, Moshiri A (2016) Biological properties and therapeutic activities of honey in wound healing: a narrative review and meta-analysis. J Tissue Viability 25(2):98–118

Palmer BC, DeLouise LA (2016) Nanoparticle-enabled transdermal drug delivery systems for enhanced dose control and tissue targeting. Molecules 21(12):1719

Pang Y et al (2014) Effect of volatile oil from *Blumea Balsamifera* (L.) DC. leaves on wound healing in mice. J Tradit Chin Med 34(6):716–724

Pazyar N et al (2014) Skin wound healing and phytomedicine: a review. Skin Pharmacol Physiol 27 (6):303–310

Perez-Recalde M, Ruiz Arias IE, Hermida EB (2018) Could essential oils enhance biopolymers performance for wound healing? A systematic review. Phytomedicine 38:57–65

Puppi D et al (2011) Poly (vinyl alcohol)-based electrospun meshes as potential candidate scaffolds in regenerative medicine. J Bioact Compat Polym 26(1):20–34

Puri V et al (2016) Comparative analysis of early excision and grafting vs delayed grafting in burn patients in a developing country. J Burn Care Res 37(5):278–282

Raghav A, Ahmad J, Alam K (2018) Preferential recognition of advanced glycation end products by serum antibodies and low-grade systemic inflammation in diabetes mellitus and its complications. Int J Biol Macromol 118(Pt B):1884–1891

Rao CV (2007) Regulation of COX and LOX by curcumin. Adv Exp Med Biol 595:213–226

Rath G et al (2016) Collagen nanofiber containing silver nanoparticles for improved wound-healing applications. J Drug Target 24(6):520–529

Sami DG, Abdellatif A, Azzazy HME (2020) Turmeric/oregano formulations for treatment of diabetic ulcer wounds. Drug Dev Ind Pharm 46:1613–1621

Sarabahi S (2012) Recent advances in topical wound care. Indian J Plast Surg 45(2):379–387

Sarhan WA, Azzazy HM (2015) High concentration honey chitosan electrospun nanofibers: biocompatibility and antibacterial effects. Carbohydr Polym 122:135–143

Sarhan WA, Azzazy HM, El-Sherbiny IM (2016) Honey/chitosan nanofiber wound dressing enriched with *Allium sativum* and *Cleome droserifolia*: enhanced antimicrobial and wound healing activity. ACS Appl Mater Interfaces 8(10):6379–6390

Shah A, Amini-Nik S (2017) The role of phytochemicals in the inflammatory phase of wound healing. Int J Mol Sci 18(5):1068

Sharma M et al (2018) Wound healing activity of curcumin conjugated to hyaluronic acid: in vitro and in vivo evaluation. Artif Cells Nanomed Biotechnol 46(5):1009–1017

Shen X et al (2008) Development of salmon milt DNA/salmon collagen composite for wound dressing. J Mater Sci Mater Med 19(12):3473–3479

Shetty BS et al (2006) Effect of *Centella asiatica* L (Umbelliferae) on normal and dexamethasone-suppressed wound healing in Wistar Albino rats. Int J Low Extrem Wounds 5(3):137–143

Siavash M et al (2015) The efficacy of topical royal jelly on healing of diabetic foot ulcers: a double-blind placebo-controlled clinical trial. Int Wound J 12(2):137–142

Sibbald RG et al (2012) A global perspective of Wound Care(c). Adv Skin Wound Care 25 (2):77–86

Song E et al (2006) Collagen scaffolds derived from a marine source and their biocompatibility. Biomaterials 27(15):2951–2961

Sugiyama T, Takahashi K, Mori H (2012) Royal jelly acid, 10-hydroxy-trans-2-decenoic acid, as a modulator of the innate immune responses. Endocr Metab Immune Disord Drug Targets 12 (4):368–376

Sugumar S et al (2014) Ultrasonic emulsification of eucalyptus oil nanoemulsion: antibacterial activity against *Staphylococcus aureus* and wound healing activity in Wistar rats. Ultrason Sonochem 21(3):1044–1049

Suntar I et al (2011) Investigating wound healing, tyrosinase inhibitory and antioxidant activities of the ethanol extracts of *Salvia cryptantha* and *Salvia cyanescens* using in vivo and in vitro experimental models. J Ethnopharmacol 135(1):71–77

Suntar I et al (2012) Appraisal on the wound healing and anti-inflammatory activities of the essential oils obtained from the cones and needles of Pinus species by in vivo and in vitro experimental models. J Ethnopharmacol 139(2):533–540

Tonks AJ et al (2007) A 5.8-kDa component of Manuka honey stimulates immune cells via TLR4. J Leukoc Biol 82(5):1147–1155

Tsourdi E et al (2013) Current aspects in the pathophysiology and treatment of chronic wounds in diabetes mellitus. Biomed Res Int 2013:385641

Tumen I et al (2011) Wound repair and anti-inflammatory potential of essential oils from cones of Pinaceae: preclinical experimental research in animal models. J Ethnopharmacol 137 (3):1215–1220

Udegbunam SO et al (2014) Wound healing and antibacterial properties of methanolic extract of *Pupalia lappacea* Juss in rats. BMC Complement Altern Med 14:157

Vijaya KK, Nishteswar K (2012) Wound healing activity of honey: a pilot study. Ayu 33 (3):374–377

Ximenes RM et al (2013) Antinociceptive and wound healing activities of *Croton adamantinus* Mull. Arg. essential oil. J Nat Med 67(4):758–764

Yan H et al (2013) Effect of pomegranate peel polyphenol gel on cutaneous wound healing in alloxan-induced diabetic rats. Chin Med J 126(9):1700–1706

Yang CK, Polanco TO, Lantis JC 2nd (2016) A prospective, postmarket, compassionate clinical evaluation of a novel acellular fish-skin graft which contains omega-3 fatty acids for the closure of hard-to-heal lower extremity chronic ulcers. Wounds 28(4):112–118

Yeo M, Jung W-K, Kim G (2012) Fabrication, characterisation and biological activity of phlorotannin-conjugated PCL/β-TCP composite scaffolds for bone tissue regeneration. J Mater Chem 22(8):3568–3577

Ziboh VA, Miller CC, Cho Y (2000) Metabolism of polyunsaturated fatty acids by skin epidermal enzymes: generation of antiinflammatory and antiproliferative metabolites. Am J Clin Nutr 71 (1):361s–366s

Wound Healing Agents from Natural Sources

Subramani Parasuraman and Pandurangan Perumal

1 Introduction

Wounds are injuries including cuts, scratches, scrapes, and punctured skin that break the skin or other tissues of the body. Wound healing is the body's response to injury in an effort to restore normal structure and function through regeneration and repair. The wound healing process starts in the moment of injury and comprises various steps including hemostasis, inflammation, proliferation, re-epithelialization, and angiogenesis. Wound healing can be accomplished by either a primary union or a secondary union.

Wound healing is a complex process. The failure in wound healing process leads to the progression of nonhealing chronic wounds. The wound healing process is affected by many factors including age and sex hormones, nutrition, oxygenation, stress, diabetes, obesity, infection, medications, alcoholism, and smoking and it leads to the development of chronic wounds.

The wound can be managed by hyperbaric oxygen therapy, debridement (removal of dead or inflamed tissue), dressing (films, gauze, hydrogel dressings, hydrocolloid dressings, dressings containing silver or alginates, and foam dressings), medications, ultrasound and electromagnetic therapy, negative pressure wound therapy, skin grafts, and patient education (Wound Care Treatment Methods; Cologne 2006).

Anti-inflammatory agents, analgesics, antimicrobials agents, and hemostatic agents are the commonly used medications for the management of wounds. Natural

S. Parasuraman (✉)
Department of Pharmacology, Faculty of Pharmacy, AIMST University, Bedong, Kedah, Malaysia
e-mail: parasuraman@aimst.edu.my

P. Perumal
Department of Pharmaceutical chemistry, Faculty of Pharmacy, AIMST University, Bedong, Kedah, Malaysia

substances (e.g., silver, honey, and plant extracts) are also used for the management of the wound. *Centella asiatica* and *Aloe vera* are commonly known for their wound healing activity (Kumar 2017). But, in folk medicine, various plant extracts are used for the management of the wound which is not well documented. In this chapter, the importance of natural sources for the discovery of drug candidates for the management of wounds is discussed.

2 Drug

A World Health Organization (WHO) Scientific Group defined a drug as "any substance or product that is used or intended to be used to modify or explore physiological systems or pathological states for the benefit of the recipient" (World Health Organization 1972).

3 Sources of Drugs

The drugs were discovered by trial and error method, until the end of the nineteenth century. Basically, drugs are obtained from plants, animals, marine organisms, microbiological, minerals, synthetic and semisynthetic sources, and recombinant deoxyribonucleic acid (r-DNA) technologies.

Plant sources of drugs are obtained from plant parts and plant phytoconstituents. Plant sources are considered the oldest sources. Medicinal plants are generally used as substances to treat or prevent diseases. The early documented records of herbal medicines dating back to 5000 years in India and China indicates the importance of medicinal plants in the healthcare system (Parasuraman 2018). In ancient times, most drugs were derived from plants. Examples of plant parts that can be used are leaves, flowers, fruits, seeds, roots, barks, and stems. For example, digitoxin and digoxin are obtained from the leaves of *Digitalis purpurea* (Plantaginaceae) which are cardiac glycosides. Vincristine and vinblastine are the anticancer drugs obtained from the flowers of *Vinca rosea* (Apocynaceae). Strychnine is obtained from the seeds of *Strychnos nux-vomica* (Loganiaceae) which acts as a central nervous system stimulant. Reserpine is obtained from the roots of *Rauwolfia serpentina* (Apocynaceae) which is a hypotensive agent. Atropine is obtained from the bark of *Atropa belladonna* (Solanaceae) which is an anticholinergic (Rates 2001).

In the aspect of animal sources, hormones play a very vital role in the treatment of diseases or deficiencies by replacement therapy. Insulin is the most commonly used hormone to treat diabetes mellitus to maintain the blood sugar level. Insulin is obtained from the pancreas of bovine or porcine (Aronoff et al. 2004). The cod liver oil contains high levels of omega-3 fatty acids, vitamin A and D which are obtained from the liver of codfish.

Mineral sources are obtained from some minerals including acids, bases, and salts such as potassium chloride. For example, iron therapy is used to treat anemia, kaolin and activated charcoal are used as antidiarrheal agent.

Synthetic drugs are chemically synthesized in the laboratory. Some of the drugs are obtained earlier from natural sources and nowadays synthesized in the laboratory. The main advantages of chemically synthesized drugs are that the quality can be controlled, the process is easier and cheaper, more safe and potent, and it is suitable for large-scale production. Examples of synthetic sources are meperidine (analgesic), diphenoxylate (antidiarrheal), and co-trimoxazole (antimicrobial). Semisynthetic drugs are normally obtained from natural drugs and their chemical structures are altered to improve their pharmacological properties. Examples of semisynthetic drugs are diacetylmorphine, ethinyl estradiol, ampicillin, and atropine bromide (Rates 2001).

DNA recombinant technology involved restriction endonucleases enzymes that cleavage of DNA at specific sequences, and the gene of interest is coupled to rapidly replicating DNA. The new genetic combination has been inserted into bacterial cultures and allows the production of genetic material. For example, human insulin can be produced by using recombinant DNA technology by the modification of porcine insulin. Due to the variation in the amino acid chain of porcine and human insulin, the amino acids alanine of porcine insulin at the 30th position of the B chain is changed with threonine to convert it into human insulin (Khan and Firoz 2015).

4 Current Perspectives on Drug Discovery from Natural Sources

The increase in the global population has overburdened the existing resources for the discovery of new drugs. The rise of new diseases and increasing resistance of microbes to the currently available drugs is also increasing at an alarming rate. Factors such as food habits, sedentary lifestyles, and changing environments have been attributed to change in the normal disease patterns. New pathways of tolerance are evolving and spreading globally, threatening our ability to manage widespread infectious diseases, resulting in prolonged sickness, disability, and death (Venkateskumar et al. 2019). Hence, the discovery of new drugs or searching the new leads from natural sources is essential.

Globally, natural products are playing an important role in preventing and treating human disease. Natural product drugs are obtained from various sources including terrestrial microorganisms, marine organisms, terrestrial vertebrates, invertebrates, and terrestrial plants. Between 1981 and 2002, about 28% of the new chemical entities launched on the market are natural products or natural product-derived drugs. From 2000 to 2005, over 20 new drugs introduced onto the market were from natural sources (Chin et al. 2006). Nearly 75,000 higher plant species have been estimated to exist on Earth, and only 10% have been used in traditional medicine. Around 1–5% is tested for its biological activity and is considered to have a therapeutic value (Koparde et al. 2019).

According to the World Health Organization (WHO), about 80% of the world's population is relying on plant-based systems of drugs for their primary healthcare needs, and plants contribute 80% of the raw materials in the traditional medical system. The traditional medicine system accounts for 40% of all healthcare provided in China, and the usage of traditional medicine serves the demands of developed countries (Parasuraman 2018). Globally, the market for and acceptance of herbal products has grown from day to day. The adverse effect of allopathic drugs contributes to a rise in the usage of herbal medications that led to the rapid growth of the herbal medicines industry.

There are many newly emerged diseases in these few decades such as acquired immune deficiency syndrome (AIDS), Ebola, and coronavirus disease 2019 and still do not have any cure for them. Tuberculosis is reemerged diseases that are resistant to chemotherapeutic agents that are available nowadays. However, synthetic chemists have failed to solve these health challenges. Thus, drugs from natural sources are another alternative way to solve these health challenges, as they are less costly, widely accepted, and less toxic when compared to allopathic medicine. The search for active constituents present in the plant began in the nineteenth century. Later many compounds were isolated from the plant including morphine, codeine, atropine, and digoxin. Many anticancer drugs are obtained from natural sources which include doxorubicin, etoposide, irinotecan, and vinblastine (Moshi 2005).

The multidisciplinary study approach has provided an impetus to find specific pharmacophores as the site of action other than the extension of therapeutic armament and herbal therapy. Different approaches to herbal medicines contribute to further drug development efforts in both single molecules of drug and polyherbal formulation (Parasuraman and Perumal 2020).

5 Natural Sources as Wound Healing Agents

5.1 Plant Sources of Wound Healing Agents

The use of plant and plant-based substances has a long history in the treatment of various diseases. Many plants are used for the treatment of wounds in folk medicine and many plants are experimentally tested for their wound healing activity. In Table 1, the plants which are screened for their wound healing activity are listed. The clinical efficacy of *Ageratina pichinchensis* (Asteraceae), *Angelica sinensis* (Apiaceae), *Alchemilla vulgaris* (Rosaceae), *Aloe vera* (Liliaceae), *Calendula officinalis* (Asteraceae), *Lavandula stoechas* (Lamiaceae), *Mimosa tenuiflora* (Leguminosae), *Origanum vulgare* (Labiatae), *Radix astragali* (Leguminosae), *Rehmanniae radix* (Scrophulariaceae), and *Salvia miltiorrhiza* (Lamiaceae) are reported elsewhere (Lordani et al. 2018). The plants' secondary metabolites such as alkaloids, glucosinolates terpenoids, and phenolics possess wound healing activity.

(+)-*Epi*-α-bisabolol: It is a sesquiterpene alcohol present in the leaves of *Lippia dulcis* (Verbenaceae), *Peperomia galioides* (Piperaceae), and *Matricaria*

Table 1 List of the plants that are screened for their wound-healing activity

S. No.	Plant name	Family	Plant part studied	Experimental model	Reference(s)
1	*Achillea kellalensis*	Acanthaceae	Ethanolic extract of flowers	Excision wound model	Pirbalouti et al. (2010)
2	*Adhatoda vasica* or *Justicia adhatoda*	Acanthaceae	Methanolic extract of leaves	Excision wound model	Barth et al. (2015)
3	*Andrographis paniculata*	Acanthaceae	10% Aqueous extracts of leaves	Excision wound model	Al-Bayaty et al. (2012)
4	*Blepharis maderaspatensis*	Acanthaceae	Ethanolic extract of leaves	Excision and incision wound models	Rajasekaran et al. (2012)
5	*Crossandra infundibuliformis*	Acanthaceae	Ethanolic extract of flowers	Excision wound model	Gundamaraju and Verma (2012)
6	*Dyschoriste littoralis*	Acanthaceae	5% Ethanolic extract	Excision wound model	Subha et al. (2017)
7	*Hemigraphis colorata* (*Hemigraphis alternata*)	Acanthaceae	Water extract of leaves	In vitro	Edwin and Nair (2016)
8	*Justicia flava*	Acanthaceae	Methanolic extract of leaves	Excision wound model	Agyare et al. (2013a)
9	*Strobilanthes crispus*	Acanthaceae	Ethanolic extract of leaves	Excision wound model	Al-Henhena et al. (2011)
10	*Acanthus polystachyus*	Acanthaceae	Methanolic extracts of leaves	Excision wound model	Demilew et al. (2018)
11	*Acorus calamus*	Acoraceae	Ethanolic extract of rhizomes	Excision wound model	Ponrasu et al. (2014)
12	*Sambucus ebulus*	Adoxaceae	Methanolic extract of leaves	Linear incision and circular excision wound models	Süntar et al. (2010)
13	*Achyranthes aspera*	Amaranthaceae	Methanolic extract of leaves	Excision and incision wound models	Fikru et al. (2012)
14	*Alternanthera sessilis*	Amaranthaceae	Chloroform extract of leaves	Excision, incision and dead space wound models	Jalalpure et al. (2008)
15	*Celosia argentea*	Amaranthaceae	Alcoholic extract of leaves	Burn wound model	Priya et al. (2004)
16	*Chenopodium botrys*	Amaranthaceae	6% Essential oil of *C. botrys*	Excision wound model	Sayyedrostami et al. (2018)
17	*Pupalia lappacea*	Amaranthaceae	Methanolic extract of leaves	Excision wounds	Udegbunam et al. (2014)
18	*Alternanthera brasiliana*	Amaranthaceae	Methanolic extract of leaves	Excision wounds	Barua et al. (2013)

(continued)

Table 1 (continued)

S. No.	Plant name	Family	Plant part studied	Experimental model	Reference(s)
19	*Allium cepa* (Onion)	Amaryllidaceae	Aqueous extract of bulbs	Excision, incision, and dead space wound models	Shenoy et al. (2009a)
20	*Allium sativum*	Amaryllidaceae	Aqueous extract of leaves	Excision wound model	Tahvilian et al. (2019)
21	*Allium sativum* (Garlic)	Amaryllidaceae	Hydroethanolic extract of plant	Excision wound model	Farahpour et al. (2017)
22	*Crinum zeylanicum*	Amaryllidaceae	Methanolic extract of bulbs	Excision wound model	Yahaya et al. (2012)
23	*Buchanania lanzan*	Anacardiaceae	Alcoholic extract of whole plant	Excision, incision, and dead space wound models	Chitra et al. (2009)
24	*Lannea coromandelica*	Anacardiaceae	Ethanol and acetone extracts of barks	Excision wound model	Sathish et al. (2010)
25	*Lannea welwitschii*	Anacardiaceae	Methanolic leaf extracts	Excision wound model	Agyare et al. (2013a)
26	*Pistacia atlantica*	Anacardiaceae	Methanolic extract whole plant	Excision wound model	Tohidi et al. (2011)
27	*Annona muricata*	Annonaceae	Alcoholic extract of stem bark	Open wound method	Paarakh et al. (2009)
28	*Annona squamosa*	Annonaceae	Petroleum ether extracts of whole plant	Excision wound model	Shenoy et al. (2009b)
29	*Anethum graveolens*	Apiaceae	Essential oil of the plant	Wound infection model	Manzuoerh et al. (2019)
30	*Centella asiatica*	Apiaceae	Aqueous extract of the whole plant	Punch wound model	Shukla et al. (1999)
31	*Falcaria vulgaris*	Apiaceae	Aqueous extract of the whole plant	Excision wound model	Goorani et al. (2019)
32	*Allamanda cathartica*	Apocynaceae	Aqueous extract of leaves	Excision and incision wound models	Nayak et al. (2006)
33	*Calotropis gigantea*	Apocynaceae	Ethanolic extract of root bark	Excision, incision and dead space wound models	Deshmukh et al. (2009)
34	*Carissa spinarum*	Apocynaceae	Methanolic extract of root	Burn wound model	Sanwal and Chaudhary (2011)

35	*Catharanthus roseus*	Apocynaceae	Ethanolic extract of leaves	Excision, incision and dead space wound models	Nayak et al. (2007a)
36	*Hemidesmus indicus*	Apocynaceae	Methanolic extract of root	Excision wound model	Ganesan et al. (2012)
37	*Saba florida*	Apocynaceae	Methanolic extract of leaves	Excision and incision models	James and Victoria (2010)
38	*Thevetia peruviana*	Apocynaceae	Fruit rind water extract	Excision, incision, and dead space wound models	Rahman et al. (2017)
39	*Typhonium trilobatum*	Araceae	Methanolic and ethyl acetate extract	Excision and incision wound models	Roy et al. (2012)
40	*Panax ginseng*	Araliaceae	Aqueous extract of leaves	Excision wound model	Kim et al. (2013)
41	*Elaeis guineensis*	Arecaceae	Crude extract of the oil palm leaf	Excision wound model	Sasidharan et al. (2010)
42	*Aristolochia bracteolata*	Aristolochiaceae	Ethanolic extract of the shade-dried leaves	Excision, incision and dead space wound models	Shirwaikar et al. (2003a)
43	*Aristolochia saccata*	Aristolochiaceae	Methanolic extract of leaves	In vitro	Bolla et al. (2019)
44	*Asparagus racemosus*	Asparagaceae	Water extract of roots	Excision and incision wound models	Kumar et al. (2011a)
45	*Polygonatum odoratum*	Asparagaceae	Ethanolic extract of leaves	Excision wound model	Mughrabi et al. (2014)
46	*Aloe vera*	Asphodelaceae	Aqueous extract of leaves	In vitro (mouse dermal fibroblast cells)	Ghayempour et al. (2016)
47	*Aloe littoralis*	Asphodelaceae	Mucilaginous gel	Linear incisional and thermal burn wound models	Hajhashemi et al. (2012)
48	*Aloe megalacantha*	Asphodelaceae	Leaf latex	Excision and incision wound models	Gebremeskel et al. (2018)
49	*Achillea biebersteinii*	Asterace	n-Hexane extract of whole plant	Excision and incision wound models	Akkol et al. (2009)
50	*Achillea millefolium*	Asteraceae	Aqueous extract of leaves	Excision, incision, and dead space wound models	Nirmala and Karthiyayini (2011)
51	*Ageratum conyzoides*	Asteraceae	Roots	Excision wound model	Jain et al. (2009)
52	*Bellis perennis*	Asteraceae	Ethanolic extract of flowers	Circular excision wound model	Karakaş et al. (2012)
53	*Bidens pilosa*	Asteraceae	Ethanolic extract of leaves	Excision wound model	Hassan et al. (2011)
54	*Blumea balsamifera*	Asteraceae	Volatile oil of leaves	Mouse wound model	Pang et al. (2014)

(continued)

Table 1 (continued)

S. No.	Plant name	Family	Plant part studied	Experimental model	Reference(s)
55	*Calendula officinalis*	Asteraceae	Ethanolic extract of flowers	Excision wound model	Preethi and Kuttan (2009)
56	*Carthamus tinctorius*	Asteraceae	Methanolic extract of leaves	Excision, incision and dead space wound models	Paramesha et al. (2015)
57	*Centaurea iberica*	Asteraceae	Aqueous methanol extract of aerial parts	Excision and incision wound models	Koca et al. (2009)
58	*Centaurea sadleriana*	Asteraceae	Methanolic extract of aerial parts	Burn wound model	Csupor et al. (2010)
59	*Centratherum anthelminticum*	Asteraceae	Ethanolic extract of whole plant	Excision and incision wound models	Sahoo et al. (2012)
60	*Cichorium intybus*	Asteraceae	Methanolic extract of roots	Linear incision and circular excision wound models	Süntar et al. (2012)
61	*Elephantopus scaber*	Asteraceae	Ethanolic extract of leaves	Excision, incision and dead space wound models	Singh et al. (2005)
62	*Eupatorium odoratum* (*Chromolaena odorata*)	Asteraceae	Aqueous extract of leaves	In vitro	Phan et al. (1996)
63	*Gynura procumbens*	Asteraceae	Ethanolic extract of leaves	Excision wound model	Zahra et al. (2011)
64	*Helichrysum graveolens*	Asteraceae	Flowers	Excision and incision wound models	Süntar et al. (2013)
65	*Matricaria chamomilla* (*Chamomilla recutita*, *Matricaria recutita*)	Asteraceae	Aqueous extract of the whole plant	Excision, incision and dead space wound models	Nayak et al. (2007c)
66	*Mikania scandens*	Asteraceae	Aqueous and methanolic extracts of leaves	Punch wounds in rabbits on either side of the depilated dorsum	Pattanayak et al. (2015)
67	*Neurolaena lobata*	Asteraceae	Ethanolic extract of leaves	Excision wound model	Nayak et al. (2014)
68	*Roman chamomile* (*Chamaemelum nobile*)	Asteraceae	Ethanolic extract of flowers	Wound infection model	Kazemian et al. (2018)
69	*Siegesbeckia pubescens*	Asteraceae	Methanolic extract of aerial parts	Excision and incision wound models	Wang et al. (2011a)

70	*Sphaeranthus amaranthoides*	Asteraceae	Whole plant	Excision wound model	Geethalakshmi et al. (2013)
71	*Sphaeranthus indicus*	Asteraceae	Alcoholic extract of flower	Excision wound model	Chopda et al. (2010)
72	*Stevia rebaudiana*	Asteraceae	Aqueous crude extract	Excision and incision wound models	Das (2013)
73	*Tagetes erecta*	Asteraceae	Hydroalcoholic extract of leaves	Excision, incision and dead space wound models	Chatterjee et al. (2011)
74	*Tridax procumbens*	Asteraceae	Extract of leaves	Dead space wound model	Udupa et al. (1995)
75	*vernonia arborea*	Asteraceae	Aqueous and methanol extracts of leaves	Excision, incision and dead space wound models	Manjunatha et al. (2005)
76	*Vernonia scorpioides*	Asteraceae	Ethanolic extract of leaves	Excision wound model	Dalazen et al. (2005)
77	*Wedelia chinensis*	Asteraceae	Ethanolic extract of whole plant	Excision, incision and dead space wound models	Verma et al. (2008)
78	*Wedelia trilobata*	Asteraceae	Ethanolic extract of leaves	In vitro wound healing assays	Balekar et al. (2012)
79	*Ageratina pichinchensis*	Asteraceae	Aqueous extract of aerial parts	Diabetic foot ulcer rat model	Romero-Cerecero et al. (2014)
80	*Plagiochasma appendiculatum*	Aytoniaceae	Extract of thalli	Excision and incision wound models	Singh et al. (2006)
81	*Anredera cordifolia*	Basellaceae	Ethanolic extract of leaves	Excision wound model	Miladiyah and Prabowo (2012)
82	*Berberis lyceum*	Berberidaceae	Aqueous and methanolic extracts of roots	Excision, incision and dead space wound models	Asif et al. (2007)
83	*Arrabidaea chica Verlot*	Bignoniaceae	Methanolic extract of leaves	Excision wound model	Jorge et al. (2008)
84	*Kigelia Africana*	Bignoniaceae	Methanolic extract of leaves	Excision wound model	Agyare et al. (2013b)
85	*Spathodea campanulata*	Bignoniaceae	Methanolic extract of barks	Burn wound model	Sy et al. (2005)
86	*Stereospermum colais*	Bignoniaceae	Chloroform and ethanolic extracts of leaves	Excision wound model	Bharathi et al. (2010)
87	*Tecoma stans*	Bignoniaceae	Ethanolic extract of flowers	Excision, incision and burn wound models	Kameshwaran et al. (2014)

(continued)

Table 1 (continued)

S. No.	Plant name	Family	Plant part studied	Experimental model	Reference(s)
88	*Tecomaria capensis* (*Tecoma capensis*)	Bignoniaceae	Methanolic extract of leaves	Excision and incision wound models	Saini et al. (2012)
89	*Blechnum orientale*	Blechnaceae	Methanolic extract of leaves	Excision wound model	Lai et al. (2011)
90	*Alkanna tinctoria*	Boraginaceae	Aqueous extract of whole plant	Burn wound model	Ogurtan et al. (2002)
91	*Arnebia densiflora*	Boraginaceae	*n*-Hexane extract of roots	Incision wound model	Kosger et al. (2009)
92	*Arnebia euchroma*	Boraginaceae	Diethyl ether extract of roots	Burn wound model	Pirbalouti et al. (2009)
93	*Carmona retusa*	Boraginaceae	Alcoholic extract of whole plant	Excision wound model	Mageswari et al. (2012)
94	*Cordia dichotoma*	Boraginaceae	Solvent fractions of ethanolic extract of fruits	Excision, incision and dead space wound models	Kuppast and Nayak (2006)
95	*Heliotropium indicum*	Boraginaceae	Methanolic and aqueous extracts leaves	Excision and incision wound models	Dash and Murthy (2011b)
96	*Onosma hispidum*	Boraginaceae	Methanolic extract of roots	Excision and incision wound models	Kumar and Gupta (2010)
97	*Coronopus didymus*	Brassicaceae	Ethanolic and aqueous extracts of the whole plant	Excision wound model	Prabhakar et al. (2002)
98	*Lepidium meyenii*	Brassicaceae	Hydroalcoholic extract of roots	Excision wound model	Bramara et al. (2017)
99	*Commiphora guidottii*	Burseraceae	Essential oil and oleo-gum-resin	Incision wound model	Gebrehiwot et al. (2015)
100	*Commiphora myrrha*	Burseraceae	Aqueous extracts of the whole plant	Excision and dead space wound models	Elzayat et al. (2018)
101	*Hylocereus undatus*	Cactaceae	Aqueous extracts of leaves, rind, fruit pulp and flowers	Excision and incision wound models	Perez et al. (2005)
102	*Opuntia ficus-indica*	Cactaceae	Methanolic extract of stems	Incision wound model	Park and Chun (2001)

103	Pereskia aculeata	Cactaceae	Methanolic extract of leaves	Excision wound model	Pinto et al. (2016)
104	Calophyllum inophyllum	Calophyllaceae	Methanolic extract of bark	Excision, incision and dead space wound models	Jawaid et al. (2016)
105	Mesua ferrea	Calophyllaceae	Ethanolic extract of aerial parts	Excision and incision wound models	Choudhary (2012)
106	Capparis ovata	Capparaceae	Methanolic extract of fruit	Excision wound model	Okur et al. (2018)
107	Capparis spinosa	Capparaceae	Ethanolic extract of roots	Excision wound model	Amiri et al. (2015)
108	Lonicera japonica	Caprifoliaceae	Ethanolic extract of flowering aerial parts	Excision wound model	Chen et al. (2012)
109	Carica candamarcensis (Vasconcellea pubescens)	Caricaceae	Proteolytic fraction	Third-degree burn wound model	Gomes et al. (2010)
110	Carica papaya	Caricaceae	Aqueous extract of leaves	Excision wound model	Mahmood et al. (2005)
111	Gymnosporia emarginata	Celastraceae	Methanolic extract of the plant	Excision wound model	Hemamalini et al. (2011)
112	Ceratophyllum demersum	Ceratophyllaceae	Methanolic extract of whole plant	Excision wound model	Taranhalli et al. (2011)
113	Cleome rutidosperma	Cleomaceae	Chloroform, aqueous and methanolic extracts of roots	Excision and incision wound models	Mondal and Suresh (2012)
114	Anogeissus acuminata (Terminalia phillyreifolia)	Combretaceae	Methanolic extract of the plant	Excision wound model	Hemamalini et al. (2011)
115	Anogeissus latifolia	Combretaceae	Ethanolic extract of barks	Excision and incision wound models	Govindarajan et al. (2004)
116	Anogeissus leiocarpus	Combretaceae	Methanolic extract of leaves	Excision wound model	Barku et al. (2013)
117	Combretum mucronatum	Combretaceae	Leaf extracts	In vitro	Kisseih et al. (2015)
118	Terminalia arjuna	Combretaceae	Ethanolic extract of bark	Excision and incision wound models	Chaudhari and Mengi (2006)
119	Terminalia avicennioides	Combretaceae	Ethanolic extract of roots	Excision and incision wound models	Mann et al. (2011)
120	Terminalia bellirica	Combretaceae	Ethanolic extract of fruits	Excision and incision wound models	Choudhary (2008)
121	Terminalia catappa	Combretaceae	Ethanolic extract of leaves	Incision wound model	Nugroho et al. (2019)

(continued)

Table 1 (continued)

S. No.	Plant name	Family	Plant part studied	Experimental model	Reference(s)
122	*Terminalia chebula*	Combretaceae	50% Aqueous alcoholic extract of fruits	Excision and dead space wound models	Singh and Sharma (2009)
123	*Terminalia coriacea*	Combretaceae	Methanolic extract of stem barks	Excision wound model	Khan et al. (2012)
124	*Tragopogon graminifolius*	Compositae	Ethanolic extract of aerial parts	Burn wound model	Heidari et al. (2018)
125	*Argyreia nervosa*	Convolvulaceae	Ethanolic extract of leaves	Excision wound model	Singhal et al. (2011)
126	*Ipomoea batatas* (Sweet potato)	Convolvulaceae	Methanolic extract of peel of tubers and Peel Bandage of tubers	Excision wound model	Panda et al. (2011)
127	*Alangium salvifolium*	Cornaceae	Ethanolic extract of leaves	Excision, incision and dead space wound model	Karigar et al. (2010)
128	*Bryophyllum pinnatum*	Crassulaceae	Petroleum ether, alcohol and water extracts of leaves	Excision, resutured incision and dead space wound models	Khan et al. (2004)
129	*Kalanchoe pinnata* (*Bryophyllum pinnatum*)	Crassulaceae	Ethanolic extract of leaves	Excision wound model	Nayak et al. (2010b)
130	*Rhodiola imbricata*	Crassulaceae	Ethanolic extrac of rhizomes	Excision wound model	Gupta et al. (2007)
131	*Citrullus colocynthis*	Cucurbitaceae	Methanolic extract of fruit pulp	Excision wound model	Gupta et al. (2018)
132	*Coccinia indica*	Cucurbitaceae	Ethanolic and aqueous extracts of fruits	Excision and incision wound model	Bambal et al. (2011)
133	*Cucurbita moschata*	Cucurbitaceae	70% ethanolic extract of fruit peel	Second degree burn wound model	Bahramsoltani et al. (2017)
134	*Cucurbita pepo*	Cucurbitaceae	Oil from pumpkin seeds	Excision wound model	Bardaa et al. (2016)
135	*Luffa cylindrica*	Cucurbitaceae	Chloroform extract of whole plant	Excision wound model	Abirami et al. (2011)

136	*Momordica charantia*	Cucurbitaceae	Fruit powder	Excision, incision and dead space wound models	Prasad et al. (2006)
137	*Trichosanthes dioica*	Cucurbitaceae	Methanolic extract of fruits	Excision and incision wound models	Shivhare et al. (2010)
138	*Cyperus rotundus*	Cyperaceae	Alcoholic extract of tuber parts	Excision, incision and dead space wound models	Puratchikody et al. (2006)
139	*Dioscorea bulbifera*	Dioscoreaceae	Tubers	Excision wound model	Panduraju et al. (2010)
140	*Shorea robusta*	Dipterocarpaceae	Aqueous extracts of the whole plant	Excision, incision and dead space wound models	Mukherjee et al. (2013)
141	*Hippophae rhamnoides*	Elaeagnaceae	Sea buckthorn flavone	Cutaneous wound	Gupta et al. (2006)
142	*Bergia ammannioides*	Elatinaceae	Ethanolic extract of the aerial parts of the plant	Excision wound model	Ezzat et al. (2016)
143	*Ephedra alata*	Ephedraceae	Aqueous extracts of the whole plant	Deep wound and full-thickness skin burn	Kittana et al. (2017)
144	*Equisetum arvense*	Equisetaceae	Ointment of *Equisetum arvense*	Diabetic wounds	Ozay et al. (2013)
145	*Acalypha indica*	Euphorbiaceae	Ethanolic extract of leaves	Excision and incision wound models	Laut et al. (2019)
146	*Baliospermum Montanum*	Euphorbiaceae	Ethanolic extract of roots	Excision wound model	Kumar et al. (2011b)
147	*Croton adamantinus*	Euphorbiaceae	Essential of the plant	Excision and dead space wound models	Ximenes et al. (2013)
148	*Croton bonplandianum*	Euphorbiaceae	Alcoholic and aqueous extracts of leaves	Excision wound model	Divya et al. (2011)
149	*Croton macrostachyus*	Euphorbiaceae	Methanolic extract of leaves	Excision and incision wound models	Mechesso et al. (2016)
150	*Mallotus philippinensis*	Euphorbiaceae	Ethanolic extract of fruit	Excision and incision wound models	Gangwar et al. (2015)
151	*Euphorbia caducifolia*	Euphorbiaceae	Latex of the plant	Excision and incision wound models	Goyal et al. (2012)
152	*Euphorbia hirta*	Euphorbiaceae	Ethanolic extract of whole plant	Burn wound model	Jaiprakash et al. (2006)
153	*Euphorbia neriifolia*	Euphorbiaceae	Ethanolic extract of leaves	Excision and dead space wound models	Bigoniya and Rana (2007)
154	*Euphorbia nivulia*	Euphorbiaceae	Latex of the plant	Excision wound model	Badgujar et al. (2009)

(continued)

Table 1 (continued)

S. No.	Plant name	Family	Plant part studied	Experimental model	Reference(s)
155	*Jatropha curcas*	Euphorbiaceae	Methanolic extract of leaves	Excision wound model	Esimone et al. (2008)
156	*Jatropha tanjorensis*	Euphorbiaceae	Methanolic extract of the whole plant	Excision and incision wound models	Gowdu Viswanathan et al. (2018)
157	*Pedilanthus tithymaloides (Euphorbia tithymaloides)*	Euphorbiaceae	Methanolic extract of the whole plant	Excision, incision and dead space wound models	Ghosh et al. (2012)
158	*Tragia involucrata*	Euphorbiaceae	Methanolic extract of roots	*Staphylococcus aureus*-induced excision wound	Perumal Samy et al. (2006)
159	*Acalypha langiana*	Euphorbiaceae	Aqueous extract of leaves	Excision and incision wound models	Perez Gutierrez and Vargas (2006)
160	*Abrus cantoniensis*	Fabaceae	Fraction of ethanolic extract	Excision and incision wound models	Zeng et al. (2016)
161	*Acacia leucophloea*	Fabaceae	Ethanolic extract of barks	Excision and incision wound models	Suriyamoorthy et al. (2012)
162	*Anadenanthera colubrina*	Fabaceae	Alcoholic extract of barks	Excision model	Pessoa et al. (2012)
163	*Caesalpinia mimosoides (Hultholia mimosoides)*	Fabaceae	Aqueous and ethanolic extracts of tender shoots (along with leaves)	Excision and incision wound models	Bhat et al. (2016)
164	*Cassia fistula*	Fabaceae	Alcoholic extract of leaves	Infected wound model	Kumar et al. (2006)
165	*Cassia occidentalis (Senna occidentalis)*	Fabaceae	Methanolic extract of leaves	Excision, incision and dead space wound models	Sheeba et al. (2009)
166	*Copaifera langsdorffii*	Fabaceae	Oleo-resin	Excision and incision wound models	Paiva et al. (2002)
167	*Crotalaria verrucosa*	Fabaceae	Aqueous extract of leaves	Excision, incision and dead space wound models	Kumari et al. (2010)
168	*Desmodium triquetrum*	Fabaceae	Ethanolic extract of leaves	Excision, incision and dead space wound models	Shirwaikar et al. (2003b)
169	*Indigofera enneaphylla (Indigofera linnaei)*	Fabaceae	Ethanol extract of whole plant	Excision and incision wound models	Sivagamy et al. (2012)

170	*Mimosa pudica*	Fabaceae	Aqueous extract of roots, methanolic extract of leaves	Excision and incision wound models	Kokane et al. (2009) and Venkateswarlu et al. (2011)
171	*Pongamia pinnata*	Fabaceae	Ethanolic extract of barks	Excision and incision wound models	Bhandirge et al. (2015)
172	*Prosopis cineraria*	Fabaceae	Ethanolic extract of leaves	Excision wound model	Gupta et al. (2015)
173	*Pterocarpus santalinus*	Fabaceae	Wood	Burn wound model	Biswas et al. (2004)
174	*Radix astragali*	Fabaceae	Water extracts	Experimental open wound	Han et al. (2009)
175	*Sesbania grandiflora*	Fabaceae	Methanolic extract of barks	Excision wound model	Karthikeyan et al. (2011)
176	*Tephrosia purpurea*	Fabaceae	Ethanolic extract of aerial parts	Excision, incision and dead space wound models	Lodhi et al. (2006)
177	*Trigonella foenum-graecum*	Fabaceae	Water extract of seeds	Excision, incision and dead space wound models	Taranalli and Kuppast (1996)
178	*Quercus brantii*	Fagaceae	Hydro-alcoholic extract of crude plant power	Deep skin ulcer model	Hemmati et al. (2015)
179	*Quercus coccifera*	Fagaceae	Water extract of stem	In vitro	Anlas et al. (2019)
180	*Quercus infectoria*	Fagaceae	Ethanolic extract of galls	Excision, incision and dead space wound models	Umachigi et al. (2008)
181	*Gentiana lutea*	Gentianaceae	Alcohol and petrol ether extracts of rhizomes	Excision, resutured incision and dead space wound models	Mathew et al. (2004)
182	*Hypericum patulum*	Hypericaceae	Methanolic extract of leaves	Excision and incision wound models	Mukherjee et al. (2000)
183	*Hypericum perforatum*	Hypericaceae	Alcoholic extract of aerial parts	In vitro	Öztürk et al. (2007)
184	*Lavandula aspic*	Lamiaceae	Essential oil	Excision wound model	Djemaa et al. (2016)
185	*Clerodendrum infortunatum*	Lamiaceae	Petroleum ether, chloroform, and ethanol extracts of roots	Excision and incision wound models	Kuluvar et al. (2009)
186	*Gmelina arborea*	Lamiaceae	Methanolic extract of aerial parts	Excision wound model	Prakashbabu et al. (2017)
187	*Hoslundia opposita*	Lamiaceae	Methanolic extract	Excision and incision wound models	Annan and Dickson (2008)

(continued)

Table 1 (continued)

S. No.	Plant name	Family	Plant part studied	Experimental model	Reference(s)
188	*Hyptis suaveolens* (*Mesosphaerum suaveolens*)	Lamiaceae	Ethanolic extract of leaves	Excision, incision and dead space wound models	Shirwaikar et al. (2003c)
189	*Leucas hirta*	Lamiaceae	Aqueous and methanolic leaf extracts	Excision, incision and dead space wound models	Manjunatha et al. (2006)
190	*Leucas lanata*	Lamiaceae	Ethanolic extract of whole plant	Excision wound model	Dixit et al. (2015)
191	*Marrubium vulgare*	Lamiaceae	80/20 Methanol/water extract of leaves	In vitro	Amri et al. (2017)
192	*Ocimum basilicum*	Lamiaceae	Aqueous extract	Cutaneous wound model	Zangeneh et al. (2019)
193	*Ocimum kilimandscharicum*	Lamiaceae	Aqueous extract of leaves	Excision, incision and dead space wound models	Paschapur et al. (2009)
194	*Ocimum sanctum*	Lamiaceae	Alcoholic and aqueous extract of leaves	Excision, incision and dead space wound models	Shetty et al. (2008)
195	*Ocimum suave*	Lamiaceae	Ethanol extract of leaves	Excision wound model	Hassan et al. (2011)
196	*Origanum vulgare*	Lamiaceae	Aqueous extract of leaves engineered titaniumdioxide nanoparticles	Excision wound model	Sankar et al. (2014)
197	*Rosmarinus officinalis* (Rosemary)	Lamiaceae	Aqueous extract and essential oil of the aerial parts	Excision wound model	Abu-Al-Basal (2010)
198	*Salvia officinalis*	Lamiaceae	Hydroethanolic extract of leaves	Excision and linear incision wound models	Karimzadeh and Farahpour (2017)
199	*Stachys hissarica*	Lamiaceae	Phytoecdysteroid compounds	Linear incision wound model	Ramazanov et al. (2017)
200	*Stachys lavandulifolia*	Lamiaceae	Aqueous extract of flowers	Excision wound model	Pirbalouti and Koohpyeh (2011)
201	*Teucrium polium*	Lamiaceae	*Teucrium polium* honey	Excision and incision wound models	Alizadeh et al. (2011)

202	*Vitex altissima*	Lamiaceae	Ethanolic extract of leaves	Excision, incision and dead space wound models	Manjunatha et al. (2007a)
203	*Vitex negundo*	Lamiaceae	Methanolic extract of leaves	Excision, incision and dead space wound models	Roosewelt et al. (2011)
204	*Vitex trifolia*	Lamiaceae	Ethanolic extract of leaves	Excision, incision and dead space wound models	Manjunatha et al. (2007a)
205	*Cinnamomum zeylanicum*	Lauraceae	Aqueous extracts	Excision wound model	Farahpour et al. (2012)
206	*Litsea glutinosa*	Lauraceae	Methanolic extract of whole plant	Excision and incision wound models	Devi and Meera (2010)
207	*Napoleona vogelii*	Lecythidaceae	Methanolic extract of leaves	Incision model	Adiele et al. (2014)
208	*Albizzia lebbeck*	Leguminosae	Ethanolic extract of roots	Excision and incision wound models	Joshi et al. (2013)
209	*Poincianella pluviosa*	Leguminosae	50% Ethanolic Extract of barks	Excision wound model	Bueno et al. (2016)
210	*Acacia catechu*	Leguminoseae	Ethanolic extract of nuts	Burn wound model	Bharat et al. (2014)
211	*Linum usitatissimum*	Linaceae	Flaxseed seed oil	Excision wound model	Farahpour et al. (2011)
212	*Dendrophthoe falcata*	Loranthaceae	Ethanolic extract of aerial parts	Excision and incision wound models	Pattanayak and Sunita (2008)
213	*Struhanthus vulgaris*	Loranthaceae	Ethanolic extract leaves and branches	In vitro scratch wound assay	Vittorazzi et al. (2016)
214	*Lycopodium serratum* (*Huperzia serrata*)	Lycopodiaceae	Ethanolic extract of leaves	Excision and incision wound models	Manjunatha et al. (2007b)
215	*Ammannia baccifera*	Lythraceae	Ethanolic extract of leaves	Excision and resutured incision wound models	Rajasekaran et al. (2012)
216	*Lawsonia inermis* (*Lawsonia alba*)	Lythraceae	Ethanolic extract of leaves	Excision and incision wound models	Nayak et al. (2007b)
217	*Punica granatum*	Lythraceae	Methanolic extract of dried *P. granatum* peels	Excision wound model	Chidambara Murthy et al. (2004)
218	*Michelia champaca* (*Magnolia champaca*)	Magnoliaceae	Ethanolic extract of flowers	Burn wound model	Shanbhag et al. (2011)

(continued)

Table 1 (continued)

S. No.	Plant name	Family	Plant part studied	Experimental model	Reference(s)
219	*Flabellaria paniculata*	Malpighiaceae	Chloroform fractions of the methanolic extract of leaves	Excision wound model	Olugbuyiro et al. (2010)
220	*Abutilon Indicum*	Malvaceae	Ethanolic extract of whole plant	Excision and incision wound models	Roshan et al. (2008)
221	*Althaea officinalis*	Malvaceae	Aqueous extract of whole plant	Excision wound model	Rezaei et al. (2015)
222	*Gossypium arboreum*	Malvaceae	Methanolic extract of leaves	Excision, incision and dead space wound models	Velmurugan et al. (2012)
223	*Hibiscus rosa-sinensis*	Malvaceae	Ethanolic extract flowers	Excision, incision and dead space wound models	Bhaskar and Nithya (2012)
224	*Hibiscus sabdariffa*	Malvaceae	Methanolic extract of dry calyces	Excision wound model	Builders et al. (2013)
225	*Malva sylvestris*	Malvaceae	Aqueous extract of flowers	Cutaneous cut wound model	Afshar et al. (2015)
226	*Pterospermum acerifolium*	Malvaceae	Ethanolic extract of flowers	Excision wound model	Senapati et al. (2011)
227	*Sida cordifolia*	Malvaceae	Ethanolic extract whole plant	Excision, incision and burn wound models	Pawar et al. (2013)
228	*Sida Corymbosa*	Malvaceae	Aqueous extract of leaves	Excision wound model	John-Africa et al. (2014)
229	*Melastoma malabathricum*	Melastomataceae	Methanolic extract of leaves	Excision and incision wound models	Anbu et al. (2008)
230	*Azadirachta indica* (Neem)	Meliaceae	Water extract of stem bark	Excision and incision wound models	Maan et al. (2017)
231	*Carapa guianensis*	Meliaceae	Ethanolic extract of barks	Excision, incision and dead space wound models	Nayak et al. (2010a)
232	*Coscinium fenestratum*	Menispermaceae	Ethanolic extract of stem	Excision and incision wound healing models	Thangathirupathi and Santharam (2011)
233	*Tinospora cordifolia*	Menispermaceae	Methanolic extract of leaves	Excision and incision wound healing models	Barua et al. (2010)
234	*Acacia caesia*	Mimosoideae	Ethanolic extract of bark	Excision and incision wound healing models	Suriyamoorthy et al. (2014)

235	Artocarpus heterophyllus (Jackfruit)	Moraceae	Ethanolic extract of leaves	Excision, incision and dead space wound models	Patil et al. (2005)
236	Ficus benghalensis	Moraceae	Ethanolic and aqueous extracts of bark	Excision and incision wound models	Garg and Paliwal (2011)
237	Ficus deltoidea	Moraceae	Aqueous extract of whole plant	Excision wound model	Abdulla et al. (2010)
238	Ficus hispida	Moraceae	Methanolic extract of leaves	Excision wound model	Singh et al. (2014)
239	Ficus racemosa	Moraceae	Ethanolic extract of roots	Excision and incision wound healing models	Murti and Kumar (2012)
240	Ficus religiosa	Moraceae	Hydroalcoholic extract of leaves	Excision and incision wound healing models	Roy et al. (2009)
241	Moringa oleifera	Moringaceae	Aqueous leaves extract	Excision, resutured incision and dead space wound models	Rathi et al. (2006)
242	Musa acuminata	Musaceae	Hydroalcoholic extract of fruit's peel	Excision wound model	Tamri et al. (2016)
243	Musa paradisiaca	Musaceae	Methanolic and hexanoic extracts of fruit's peel	Longitudinal incision model	Padilla-Camberos et al. (2016)
244	Musa sapientum	Musaceae	Aqueous extract of fruit's peel	Excision, incision and dead space wound models	Agarwal et al. (2009)
245	Myristica andamanica	Myristicaceae	Methanolic extract of leaves	Excision wound model	Arunachalam and Subhashini (2011)
246	Eucalyptus (genus)	Myrtaceae	Essential oil	Excision wound model	Alam et al. (2018)
247	Eugenia jambolana (Syzygium cumini)	Myrtaceae	Ethanolic extract of barks	Incision wound model	Palanimuthu et al. (2011)
248	Plinia peruviana	Myrtaceae	Hydroalcoholic extract of fruit peels	In vitro (Scratch assay)	Pitz et al. (2016)
249	Boerhavia diffusa	Nyctaginaceae	Methanolic extract of leaves	Excision wound model	Juneja et al. (2020)
250	Pisonia grandis	Nyctaginaceae	Methanolic extract of leaves	Excision and incision wound models	Prabu et al. (2008)

(continued)

Table 1 (continued)

S. No.	Plant name	Family	Plant part studied	Experimental model	Reference(s)
251	*Jasminum auriculatum*	Oleaceae	Petroleum ether, chloroform, ethanolic and aqueous extract of leaves	Excision and incision wound models	Arun et al. (2016)
252	*Jasminum grandiflorum*	Oleaceae	Methanolic extract of leaves	Cutaneous wound healing model	Chaturvedi et al. (2013)
253	*Jasminum sambac*	Oleaceae	Aqueous and ethanolic extract	Excision wound model	Sabharwal et al. (2012)
254	*Nyctanthes arbor-tristis*	Oleaceae	Methanolic extract of leaves	Excision and incision wound models	Bharti et al. (2011)
255	*Ophioglossum vulgatum*	Ophioglossaceae	Galactoglycerolipids	Scratch wound assay	Clericuzio et al. (2014)
256	*Prosthechea michuacana*	Orchidaceae	Aqueous suspension of extracts of aerial parts and bulbs	Excision and incision wound models	Gutierrez and Solis (2009)
257	*Vanda roxburghii (Vanda tessellata)*	Orchidaceae	Leaves (paste prepared with water)	Excision wound model	Nayak et al. (2005b)
258	*Argemone Mexicana*	Papaveraceae	Chloroform, methanol, and aqueous extracts of leaves	Dead space wound model	Dash and Murthy (2011a)
259	*Sesamum indicum*	Pedaliaceae	Seed oil (Sesame oil)	Excision, incision, burn and dead space wound models	Kiran and Asad (2008)
260	*Emblica officinalis*	Phyllanthaceae	Aqueous and ethanolic extracts of barks	Excision and linear incision wound models	Talekar et al. (2012)
261	*Piper betle*	Piperaceae	Organic solvent (ethanol and methanol) extracts of leaves	In vitro (Scratch wound healing assays using Fibroblast cells) and in vivo (burn and excision wound) models	Lien et al. (2015)
262	*Piper cubeba*	Piperaceae	Essential oil	Excision wound model	Shakeel et al. (2019)
263	*Piper nigrum*	Piperaceae	Methanolic and ethanolic extracts of black berries	In vitro	Wong and Ling (2014)
264	*Plagiochila beddomei*	Plagiochilaceae	Methanolic and aqueous extracts of fresh thallus	Excision and incision wound models	Manoj and Murugan (2012)

265	*Bacopa monniera*	Plantaginaceae	Ethanolic extract of whole plant	Excision, incision and dead space wound models	Murthy et al. (2013)
266	*Plantago australis*	Plantaginaceae	Hydroethanolic extract of leaves	In vitro and in vivo (Excision wound model)	de Moura Sperotto et al. (2018)
267	*Plantago major*	Plantaginaceae	Ethanol- and water-based extracts of leaves	Ex vivo	Zubair et al. (2016)
268	*Ribwort plantain (Plantago lanceolata)*	Plantaginaceae	Cold aqueous extract of flowers	Excision wound model	Ismayilnajadteymurabadi et al. (2012)
269	*Plumbago zeylanica*	Plumbaginaceae	Methanolic extract of root	Excision wound model	Kodati et al. (2011)
270	*Cynodon dactylon*	Poaceae	Aqueous extract of whole plant	Punch wound model	Biswas et al. (2017)
271	*Polygonum cuspidatum*	Polygonaceae	Ethanolic extract	Excision wound model	Wu et al. (2012b)
272	*Rheum officinale*	Polygonaceae	Ethanolic extract	Excision wound model	Yang et al. (2017)
273	*Rumex abyssinicus*	Polygonaceae	Methanolic extract of rhizomes	Excision and incision wound models	Mulisa et al. (2015)
274	*Portulaca oleracea*	Portulacaceae	Crude aerial parts	Excision wound	Rashed et al. (2003)
275	*Embelia ribes*	Primulaceae	Ethanolic extract of leaves	Excision, incision and dead space wound models	Swamy et al. (2007)
276	*Adiantum capillus-veneris*	Pteridaceae	Methanolic extract of aerial parts of *Adiantum capillus-veneris*	In vitro (human dermal fibroblast cell line)	Nilforoushzadeh et al. (2014)
277	*Radix paeoniae*	Ranunculaceae	Aqueous extract of roots	Excision, incision and dead space wound models	Malviya and Jain (2009)
278	*Ziziphus jujube*	Rhamnaceae	Ethanol (70%) extract of leaves	Burn wound model	Hovanet et al. (2016)
279	*Ziziphus Mauritiana*	Rhamnaceae	Aqueous and Ethanolic extracts of leaves	Excision wound model	Rajan et al. (2013)
280	*zyziphus oenolpia*	Rhamnaceae	Ethanolic and aqueous extracts of roots	Excision, incision and dead space wound models	Majumder (2012)

(continued)

Table 1 (continued)

S. No.	Plant name	Family	Plant part studied	Experimental model	Reference(s)
281	*Agrimonia eupatoria*	Rosaceae	Aqueous extract of areal parts of plant	Excision and incision wound healing models	Ghaima (2013)
282	*Alchemilla mollis* (*Alchemilla xanthochlora*)	Rosaceae	Aqueous methanolic of extracts of aerial parts and roots	Circular excision and linear incision wound models	Öz et al. (2016)
283	*Potentilla fulgens*	Rosaceae	Ethanolic extract roots	Excision and incision wound models	Kundu et al. (2016)
284	*Rubus fairholmianus*	Rosaceae	Acetone extract of roots	Excision, infected and burn wound models	George et al. (2014)
285	*Rubus sanctus*	Rosaceae	Aerial parts	Excision and linear incision wound models	Süntar et al. (2011)
286	*Sanguisorba officinalis*	Rosaceae	Polysaccharide	Burn wound healing	Zhang et al. (2018)
287	*Alchemilla persica*	Rosaceae	Aqueous methanolic of extracts of aerial parts and roots	Linear incision and circular excision wound models	Öz et al. (2016)
288	*Alchemilla vulgaris*	Rosaceae	Methanolic of extracts of aerial parts and roots	Excision and incision wound models	Choi et al. (2018)
289	*Anthocephalus cadamba*	Rubiaceae	Hydroalcoholic extract of whole plant	Excision and incision wound models	Umachigi et al. (2007)
290	*Canthium coromandelicum*	Rubiaceae	Ethanolic extract of leaves	Excision wound model	Mohan et al. (2014)
291	*Galium odoratum*	Rubiaceae	Methanolic and aqueous extract of aerial parts	Burn wound model	Kahkeshani et al. (2013)
292	*Ixora coccinea*	Rubiaceae	Alcoholic extract of flowers	Dead space wound model	Nayak et al. (1999)
293	*Morinda citrifolia*	Rubiaceae	Ethanolic extract of leaves	Excision, incision and burn wound models	Nayak et al. (2009b)
294	*Pentas lanceolata*	Rubiaceae	Ethanolic extract of flowers	Excision wound model	Nayak et al. (2005a)
295	*Rubia cordifolia*	Rubiaceae	Ethanolic extract of roots	Excision wound model	Karodi et al. (2009)

296	*Aegle marmelos*	Rutaceae	Methanolic extract of whole plant	Excision and incision wound models	Arunachalam et al. (2012)
297	*Clausena excavata*	Rutaceae	Methanolic extract of leaves	Excision wound model	Albaayit et al. (2015)
298	*Glycosmis arborea*	Rutaceae	Ethanolic extract of leaves	Excision, incision and dead space wound models	Silambujanaki et al. (2011)
299	*Limonia acidissima*	Rutaceae	Methanolic extract of fruit pulp	Excision, incision and dead space wound models	Ilango and Chitra (2010)
300	*Zanthoxylum bungeanum*	Rutaceae	Seed oil	Burn wound model	Li et al. (2017)
301	*Zanthoxylum leprieurii*	Rutaceae	Methanol-water extract of stem bark	Excision wound model	Agyare et al. (2014)
302	*Paullinia pinnata*	Sapindaceae	Methanolic extract of roots	Excision and incision wound models	Annan et al. (2010)
303	*Madhuca longifolia*	Sapotaceae	Ethanolic extract of leaves	Excision and incision wound models	Sharma et al. (2010)
304	*Mimusops elengi*	Sapotaceae	Methanolic extract of leaves	Excision and incision wound models	Aleti et al. (2015)
305	*Scrophularia striata*	Scrophulariaceae	Methanolic extract of aerial parts	Excision wound model	Ghashghaii et al. (2017)
306	*Cestrum nocturnum*	Solanaceae	Ethanolic extract of leaves	Excision and incision wound models	Nagar et al. (2016)
307	*Lycium depressum*	Solanaceae	Methanolic extract of leaves	Excision and incision wound models	Naji et al. (2017)
308	*Solanum xanthocarpum*	Solanaceae	Methanolic extract of fruits	Excision and incision wound models	Kumar et al. (2010)
309	*Camellia sinensis* (Tea plant)	Theaceae	Ethanolic extract Green tea	Incision wound model	Asadi et al. (2013)
310	*Phaleria macrocarpa*	Thymelaeaceae	Ethanolic extract of fruits	Excision wound model	Abood et al. (2015)
311	*Holoptelea integrifolia*	Ulmaceae	Methanolic extract of leaves and stems	Excision and incision wound models	Reddy et al. (2008)
312	*Cecropia peltata*	Urticaceae	Ethanolic extract of leaves	Excision wound model	Nayak (2006)
313	*Urtica urens*	Urticaceae	Aqueous and ethanolic extracts	Infectious wound model	Taqa et al. (2014)
314	*Lantana camara*	Verbenaceae	Aqueous extract of leaves	Excision wound model	Nayak et al. (2009a)
315	*Leea Asiatica*	Vitaceae	Methanolic extract of whole plant	Excision wound model	Nair et al. (2014)

(continued)

Table 1 (continued)

S. No.	Plant name	Family	Plant part studied	Experimental model	Reference(s)
316	*Vitis Vinifera*	Vitaceae	Whole grape	Excision wound model	Nayak et al. (2010c)
317	*Curcuma longa* (Turmeric)	Zingiberaceae	Ethanolic extract of rhizomes	Excision wound model	Purohit et al. (2013)
318	*Curcuma mangga*	Zingiberaceae	Ethanolic extract of rhizomes	In vitro (Cells proliferation by MTT assay)	Srirod and Tewtrakul. (2019)
319	*Kaempferia galangal*	Zingiberaceae	Ethanolic extract of rhizomes	Excision, incision and burn wound models	Shanbhag et al. (2006)
320	*Kaempferia marginata*	Zingiberaceae	Ethanolic extract of rhizomes	In vitro (human dermal fibroblasts)	Muthachan and Tewtrakul (2019)
321	*Zingiber officinale* (Ginger)	Zingiberaceae	Aqueous extract of roots	Excision and incision wound models	Pourali and Yahyaei (2019)
322	*Balanites aegyptiaca*	Zygophyllaceae	Methanolic extract of stem bark	Incision wound	Annan and Dickson (2008)
323	*Fagonia schweinfurthii*	Zygophyllaceae	Alcoholic extract	Excision wound model	Alqasoumi et al. (2011)
324	*Tribulus terrestris*	Zygophyllaceae	Aqueous extract of leaves	Excision, incision and burn wound models	Wesley et al. (2009)

chamomilla (Asteraceae). Villegas et al. (2001) isolated the (+)-*epi*-α-bisabolol from *Peperomia galioides* and reported its wound healing activity. α-bisabolol is known for its antioxidant, anti-inflammatory, anti-infective, and antiplasmodial activities. And it also inhibits melanogenesis by lowering intracellular cyclic Adenosine monophosphate (cAMP) levels in dermal hyperpigmentation (Kamatou and Viljoen 2010).

4-Hydroxybenzaldehyde (4-HBA): 4-HBA is phytoconstituent of *Gastrodia elata* (Orchidaceae) and it accelerates acute wound healing by promoting keratinocyte cell migration (Ha et al. 2000; Kang et al. 2017).

Alkannins/Shikonins: Alkannins/shikonins are chiral-pairs of naturally occurring isohexenylnaphthazarins and found in the roots of *Alkanna tinctoria* (Boraginaceae), *Onosma echioides* (Boraginaceae), *Lithospermum erythrorhizon* (Boraginaceae), and *Echium italicum* (Boraginaceae). Alkannins/shikonins are present in more than 150 species that belong to genera of Alkanna, Arnebia, Echium, Lithospermum, and Onosma of the Boraginaceae family (Papageorgiou et al. 2008; Sagratini et al. 2008). Karayannopoulou et al. (2011) and Nikita et al. (2015) reported wound healing activities of alkannins/shikonins in animal models. Alkannins/shikonins are known for their wound healing activity (Papageorgiou et al. 2008).

Arnebin-1: It is a naphthoquinone derivative from the plant root of *Arnebia nobilis* (Boraginaceae). *Arnebia nobilis* root extract is commonly used to treat wounds in India (Thangapazham et al. 2016). It has a pro-angiogenic effect and promotes the wound healing process by increasing Endothelial nitric oxide synthase (eNOS), Vascular endothelial growth factor (VEGF), and Hypoxia-inducible factor 1-alpha (HIF-1α) expression levels in diabetic rats (Zeng and Zhu 2014; Zeng et al. 2015).

Asiatic Acid: It is a phytoconstituent of edible plant *Centella asiatica* belongs to the Apiaceae family. *Centella asiatica* has been identified as a medicinal herb that finds traditional applications in the treatment of leprosy and wound. It was mentioned as a wound-healing plant in the French pharmacopeia in 1884 and has been used in ancient traditional Chinese culture for the past 2000 years, and also has a history connected with the Ayurvedic medicine that dates back to about 3000 years ago (Nabi et al. 2019). Asiatic acid is a triterpenic acid and showed numerous pharmacological activities including anticancer, anti-inflammatory, antioxidant, cardioprotective, gastroprotective, hepatoprotective, and neuroprotective properties. It also showed beneficial effects in the management of Alzheimer's, cerebral ischemia, dementia, hyperglycemia, liver fibrosis, metabolic syndrome, Parkinson's diseases, obesity, and wound (Nagoor Meeran et al. 2018). In in vitro experiments, Asiatic acid showed concentration-dependent enhancement of collagen type I and type III synthesis in fibroblasts cells and the action is mediated through the Transforming growth factor-beta (TGF-β)/Smad pathway (Wu et al. 2012a). Asiatic acid derivatives namely "ethoxymethyl 2-oxo-3,23-isopropylidene-asiatate" also showed the strongest and the fastest wound healing activity in in silico experiments (Jeong 2006).

Asiaticoside: It is a triterpenoid saponins of *Centella asiatica* (Apiaceae) (Sh Ahmed et al. 2019). Asiaticoside exhibited wound healing activity in rabbits by increasing the formation of granulation tissue and collagenase triggered re-epithelialization (Ozdemir et al. 2016). Asiaticoside also induced type-I collagen synthesis in human dermal fibroblast cells (Mukherjee et al. 2015).

Baicalin: It is a flavone glycoside and is one of the main phytoconstituents responsible for the pharmacological actions of *Scutellaria baicalensis* (Lamiaceae), a Chinese herb used for the treatment of psoriasis. This compound exhibited antiviral activity, anti-inflammatory activity, and photoprotective effects. In psoriatic patients, baicalin blocks the pathological changes of keratinocytes (Bonesi et al. 2018). The nanohydrogel preparation of baicalin exhibited wound healing activity and this effect may be due to its antioxidant and anti-inflammatory properties. Baicalin has the ability to inhibit nitric oxide and Tumor necrosis factor-α (TNF-α) which is playing a key role in the inflammatory pathway (Manconi et al. 2018).

Calophyllolide: It is neoflavone isolated from the seeds of *Calophyllum inophyllum* (Calophyllaceae). It exhibited wound healing activity in the incision wound model in mice. Calophyllolide (6 mg/animal; topical application) accelerates wound closer rate by around 80% and 97% at day 7 and day 14 post-treatment whereas povidone-iodine (100 mg/animal; topical application) promoted closure of the wound area by around 71% and 93% at day 7 and day 14 post-treatment, respectively (Nguyen et al. 2017).

Chlorogenic Acid: It is an ester of caffeic acid with quinic acid (phenolic acids), found in apples, carrots, pears, sweet potatoes, tomatoes, coffee, and tea. A high concentration of chlorogenic acid is found in coffee (Gagliardini et al. 2017). In both in vitro and in vivo experiments, chlorogenic acid exhibited significant wound healing activity (Moghadam et al. 2017; Chen et al. 2013).

Curcumin: Curcumin, a polyphenol from the rhizome of *Curcuma longa* (turmeric plant), belongs to the Zingiberaceae family. The medicinal values of turmeric have been known for thousands of years. In Asian countries, turmeric is traditionally used as an antioxidant, anti-inflammatory, antimutagenic, antimicrobial, and anticancer agent (Hewlings and Kalman 2017). Curcumin and its preparations are well reported for its wound healing activity. The wound-healing effect of curcumin is attributed to its anti-inflammatory, anti-infectious, and antioxidant properties (Akbik et al. 2014). Mohanty et al. (2012) reported the wound healing properties of curcumin-loaded oleic acid-based polymeric bandages in *Sprague Dawley* male rats.

Deoxyelephantopin: It is a sesquiterpene lactone present in *Elephantopus scaber* and *Elephantopus carolinianus* (Asteraceae). Deoxyelephantopin exhibited anticancer activity against breast, cervical, colon, liver, lung, and nasopharyngeal cancer (Mehmood et al. 2017). It also showed hepatoprotective activity against lipopolysaccharide/d-galactosamine (LPS/D-GalN)-induced fulminant hepatitis in rodents (Huang et al. 2013). Singh et al. (2005) reported the wound healing activity of deoxyelephantopin. It showed a significant wound healing activity in excision

(0.2% gell; topical application), incision (0.2% gel; topical application), and dead space wound (4 mg/kg, b.w.; systemic application/oral) models in rodents.

Ellagic Acid: It is a phenol compound present in several fruits including pomegranates and vegetables. Primarizky et al. (2017) reported the wound healing activity of ellagic acid in incision wound model in rats. In another study, ellagic acid not effectively increases the collagen levels in dermal wound rats and also not inhibited neutrophil infiltration (Mo et al. 2014). Mo et al. studied the wound healing effect of ellagic acid using linear incision, excision, and burn wound models and found that ellagic acid has wound healing activity, but does not increase the collagen levels. Ellagic acid is also not associated with any deformations or exuberant granulation that might have occurred in the healing process (Mo et al. 2014).

Emodin: It is a trihydroxyanthraquinone and phytoconstituent present in *Reynoutria japonica* (Polygonaceae), *Ventilago leiocarpa* (Rhamnaceae), *Cassia nigricans* (Caesalpinaceae), *Cassia obtusifolia* (Fabaceae), *Embelia ribes* (Myrsinaceae), *Rheum officinale* (Polygonaceae), and *Aspergillus ochraceus* (Trichocomaceae). Emodin is known to have anti-allergic, antibacterial, anticancer, antidiabetic, anti-inflammatory, anti-osteoporotic, antiviral, hepatoprotective, immunosuppressive, and neuroprotective activities (Dong et al. 2016). Tang et al. (2007) reported wound healing activity of emodin (at the dose levels of 100, 200, and 400 µg/ml; topical application) using excisional wound model in rats.

Entagenic Acid: It is a phytoconstituents from the seed kernel of *Entada pursaetha* or *Entada rheedii* (Leguminosae). This plant is indigenous to Africa, Asia, Australia, and Madagascar. Entagenic acid exhibited wound healing activity in rats. Entagenic acid also showed inhibitory activity against Glycogen synthase kinase 3 beta (GSK3B) protein (target for wound healing activity through Wnt signaling pathway) with minimum binding (-8.79 kJ mol^{-1}) and docking (-8.95 kJ mol^{-1}) energy in in silico studies (Vidya et al. 2012).

Ferulic Acid: Ferulic acid is a natural antioxidant and an abundant phenolic phytochemical found in vegetables and fruits, such as tomatoes, rice bran, and sweet corn. Ghaisas et al. studied the wound healing activity of ferulic acid in diabetic rats using the excision wound model and the results indicate that ferulic acid can promote wound healing and inhibiting oxidative stress in diabetic rats (Ghaisas et al. 2014). Isoferulic acid is an isomer of ferulic acid that exhibited anti-inflammatory action (Gadallah et al. 2020).

Gallic Acid: It is a phenolic compound present in number of land plants. Gallic acid is a strong antioxidant and anti-inflammatory agent. In in vitro experiments, gallic acid exhibited wound healing activity by accelerating the cell migration of keratinocytes and fibroblasts (Yang et al. 2016). Preparation of gallic acid also exhibited the wound healing activity and substantially increases TGF-β expression. In wound healing process, inflammation is a crucial phase followed by the proliferation and maturation phase which is inhibited by gallic acid by blocking the nuclear factor κB (NF-κB) pathway (Karatas and Gevrek 2019).

Honey: Honey is a sweet and viscous food substance made by honey bees. Bees collect honey from the sugary secretions of plants (floral nectar) or from secretions of other insects (honeydew). Vijaya and Nishteswar (2012) reported the clinical effect (wound healing) of honey. It promotes the natural healing process and also possesses anti-inflammatory action.

Juglone: Juglone is a toxic isomer of lawsone present in almost all of the parts of the walnut tree (Chudhary et al. 2020). Juglone ameliorates wound healing by increasing the activation and/or expression of Cell division control protein 42 homolog (Cdc42), Ras-related C3 botulinum toxin substrate 1 (Rac1), and Alpha-p21-activated kinases (α-PAK) (Wahedi et al. 2016).

Kaempferol: It is a flavonoid and found in many edible vegetables including apples, blackberries, broccoli, brussels sprouts, cucumbers, grapes, green beans, green tea, lettuce, onions, peaches, potatoes, raspberries, spinach, squash, and tomatoes (Calderón-Montaño et al. 2011; Liu 2013). Kaempferol is known to have anticancer, anti-inflammatory, antioxidant, cardioprotective, hepatoprotective, and neuroprotective activities (Ren et al. 2019). Kaempferol exhibited wound healing activity in both diabetic and nondiabetic animals using excisional and nondiabetic incisional wound models in rats at a concentration of 1% w/w ointment (Özay et al. 2019). Ambiga et al. isolated kaempferol from flowers of *Ipomoea carnea* (Convolvulaceae) and studied its wound healing activity using incision and excision wound models in rats. In both models, kaempferol exhibited significant wound healing activity by the increased weight of the granulation tissue and hydroxyproline content in the incision wound model and by the reduction in the wound area, a faster rate of epithelialization, an increased dry weight of the tissue, and increased hydroxyproline content in excision wound model (Ambiga et al. 2007).

Lawsone: It is also known as hennotannic acid. Lawsone is a naphthoquinone and is one of the phytoconstituents of Hina/*Lawsonia inermis* (Lythraceae). Hennotannic acid is a red-orange dye and known to have antibacterial, antifungal, antiparasitic, antitumor, and antiviral activities (López López et al. 2014). Mandawgade and Patil reported wound healing activity of lawsone in rodents using excision and incision model at the dose levels of 50 ± 3 mg/kg (per oral) and 0.1 w/w (topical/ointment). In both models, lawsone exhibited significant wound healing activity (Mandawgade and Patil 2003). Lawsone preparations also showed a significant wound healing activity in rodents (Bascha et al. 2016).

Luteolin (3′,4′,5,7-Tetrahydroxy Flavone): It is a naturally occurring flavonoid present in flowering plants including *Martynia annua* (Martyniaccae) and *Reseda luteola* (Resedaceae) (Lodhi et al. 2016; Cristea et al. 2003). In Chinese traditional medicine, plants rich in luteolin have been used for the treatment of cancer, hypertension, and inflammatory disorders (Lin et al. 2008). Lodhi and Singhai (2013) reported the wound healing activity of luteolin and the effect is due to its capacity to enhance tissue antioxidant levels.

Madecassic Acid: It is a natural pentacyclic triterpenoid and first isolated from *Centella asiatica* belonging to the Apiaceae family. Various extracts of *Centella asiatica* are well reported for their wound healing activity (Somboonwong et al. 2012). Asiatic acid (aglycone), madecassic acid (aglycone), asiaticoside (glycosides), and madecassoside (glycosides) are triterpenoid compounds present in *Centella asiatica* and these compounds are ingredients with pharmacological activities (Wu et al. 2012a). Asiatic acid and madecassic acid is also stimulating collagen I secretion in dermal fibroblast cells (Bonte et al. 1994).

Mannose-6-Phosphate: It is the major sugar in *Aloe vera* (Asphodelaceae). In rabbits, mannose-6-phosphate effectively prevented flexor tendon adhesion formation after anastomosis by inhibiting transforming growth factor β1 (Xia et al. 2012). A randomized, interventional clinical trial (Phase II) was carried out in 2008 to investigate the efficacy and safety of mannose-6-phosphate in accelerating the healing of split-thickness skin graft donor sites (Bush 2008). This compound is also used for the treatment of fibrotic disorders (US6093388A) (Ferguson and BTG International Ltd 2000).

Oleanolic Acid: Oleanolic acid is one of the common pentacyclic triterpenoid compounds commonly found in apple, pomegranate, lemon, mandarin, bilberries, pears, grapes, persimmon, jujube, and olives (Žiberna et al. 2017). Moura-Letts et al. reported the wound healing activity of oleanolic acid, isolated from *Anredera diffusa* (Family: Basellaceae) (Moura-Letts et al. 2006). Oleanolic acid also exhibited antidiabetic, anticancer, and antioxidant activates (Wang et al. 2011a, b; Žiberna et al. 2017).

Picroliv: It is an active hepatoprotective principle isolated from *Picrorhiza kurroa* (Plantaginaceae) (Girish and Pradhan 2012). Picroliv improved re-epithelialization, neovascularization and migration in endothelial, dermal myofibroblasts, and fibroblasts cells into the wound bed and upregulates the expression of VEGF and α-Smooth muscle actin (α-SMA) (Singh et al. 2007).

Quercetin: It is a bioflavonoid present in more than 20 plants including *Allium cepa* (Amaryllidaceae), *Allium fistulosum* (Amaryllidaceae), *Asparagus officinalis* (Asparagaceae), *Apium graveolens* (Apiaceae), *Brassica oleracea* (Brassicaceae), *Calamus scipionum* (Calamoideae), *Camellia sinensis* (Theaceae), *Capparis spinosa* (Capparaceae), *Centella asiatica* (Apiaceae), *Coriandrum sativum* (Apiaceae), *Hypericum hircinum* (Clusiaceae), *Hypericum perforatum* (Hypericaceae), *Lactuca sativa* (Asteraceae), *Malus domestica* (Rosaceae), *Morus alba* (Moraceae), *Moringa oleifera* (Moringaceaei), *Nasturtium officinale* (Brassicaceae), *Solanum lycopersicum* (Solanaceae), *Prunus avium* (Rosaceae), *Prunus domestica* (Rosaceae), and *Vaccinium oxycoccus* (Ericaceae). Quercetin is known for its antiatherosclerotic, antihypercholesterolemic, antihypertensive, anti-inflammatory, and antiobesity activities (Anand David et al. 2016). Quercetin and its preparations exhibited wound healing activity in preclinical studies (Gomathi et al. 2003; Jangde et al. 2018).

Taspine: It is an alkaloid present in *Croton lechleri* (Euphorbiaceae) and *Magnolia x soulangiana* (Magnoliaceae). Porras-Reyes et al. reported the wound healing activity of taspine (at dose levels of 10–250 µg/rat with body weight of 280–320 g) using a linear incision model in rodents. Taspine at 250 µg showed significant increases in mean breaking strength and wound tensile strength (Porras-Reyes et al. 1993).

Ursolic Acid: Ursolic acid is a lipophilic pentacyclic triterpenoid, found in rosemary, lavender, marjoram, thyme, organum, apple fruit peel, and berries. It exhibits anti-inflammatory, antioxidant, anti-carcinogenic, antiobesity, antidiabetic, cardioprotective, neuroprotective, hepatoprotective, anti-skeletal muscle atrophy, and thermogenic effects (Seo et al. 2018; Parasuraman et al. 2020). Pravez and Patel (2014) reported the wound healing activity of stearoyl glucoside of ursolic acid using excision and incision wound models. Naika et al., reported wound healing activity of ursolic acid isolated from *Clematis gouriana* using in silico (against glycogen synthase kinase 3-β) and in vivo (using excision and dead space wound models) studies (Naika et al. 2016).

5.2 Wound Healing Potential of Marine Organisms

In recent years, the study of marine organisms for their bioactive potential has expanded, and now one-third of the available medicines are either natural products or have been produced based on nature-borne lead materials, and almost 60% of drugs approved/and pre-New Drug Application (NDA) candidates (not including biologicals) for the pharmacotherapy of cancer are of natural origin (Porras-Reyes et al. 1993). Approximately 15,000 pharmacologically active substances have been extracted from marine organisms, many of which are structurally unique and absent in terrestrial organisms. The ocean provides an opportunity to explore newer lead molecules as it has more than 13,000 molecules and these are secondary metabolites. The majority of bioactive molecules have been isolated from benthic animals, such as ascidians, bryozoans, cnidarians, echinoderms, mollusks, polychaetes, and sponges. The reports say that only 5% of our oceans have been studied and current research trends focus on the intensive search of this vast and diversified marine ecosystem for the exploration of potent bioactive compounds. In the past few years, researchers and pharmaceutical firms have been making efforts to discover new pharmacological agents from marine sources (Venkateskumar et al. 2019).

The red pigment isolate of *Vibrio* sps, bioactive compounds from *Micrococcus* sp. OUS9, low molecular weight fucoidan of *Undaria pinnatifida* (Alariaceae), organic and aqueous extracts of *Ceratothoa oestroides* (Cymothoidae), extract of *Fucus vesiculosus* (Fucaceae), methanolic extract of *Padina gymnospora* (Dictyotaceae), methanol-water layer and n-hexane layer fractions of Starfish, compounds of *Astragalus species* (cycloastragenol, astragaloside IV, cyclocephaloside I and cyclocanthoside E) and Sea-star coelomic fluid (SCF) extract of Sea-star

Astropecten indicus (Astropectinidae) were exhibited wound healing activity in preclinical studies (Baveja et al. 2018; Sevimli-Gür et al. 2011; Shiva Krishna et al. 2019; Kumari et al. 2020; Park et al. 2017; Sofrona et al. 2020; Baliano et al. 2016; Craciun et al. 2015; Dai et al. 2016). The marine sources are the most unexplored area for the search of lead for drug discovery. Hence, exploration of marine sources may give the new lead for drug discovery.

5.3 Antioxidant Therapies for Wound Healing

The wound is an injury and several factors such as accidental traumas or surgery contribute to wound generation. Wound healing is the natural process that results in the restoration of the structural and functional integrity of damaged tissues. Several biochemical and cellular pathways involving in the wound healing process and deficiencies in this pathway may result in the failure of the wound healing process. Additionally, oxygen- and nitrogen-centered reactive species are known to play vital roles in regulating healing. In wound sites, reactive species may be in high concentrations which can cause harmful effects on cells and tissues and promote oxidative stress (Barku 2019). This Reactive oxygen species (ROS) play an essential role in the mechanism of the normal wound-healing response (Dunnill et al. 2017).

Physiologic antioxidant defenses include the ROS-detoxifying enzymes superoxide dismutase, catalase, glutathione peroxidases, and peroxiredoxins; endogenous and exogenous low-molecular-weight antioxidants such as vitamin C, vitamin E, glutathione, and phenolic are nonenzymatic defenders against ROS (Fitzmaurice et al. 2011). Green leafy vegetables and fruits are rich sources of low-molecular-weight antioxidants, which can enhance and accelerate the wound healing process by inhibiting free radicals and neutralizing reactive oxygen species.

Herbal wound products are remarkably growing and it induces the healing by modulating inflammatory mediators such as cytokines, eicosanoids, nitric oxide, and other growth factors. In the traditional medicine system, combinations of herbs such as polyherbal/herbomineral formulations are used to have better pharmacotherapeutic activity (Aslam et al. 2016). The plants have complex enzymatic (superoxide dismutase, catalase, glutathione peroxidase, and glutathione reductase) and nonenzymatic (ascorbic acid, glutathione, proline, carotenoids, phenolic acids, flavonoids, and tannins) antioxidant defense systems to prevent the toxic effects of free radicals (Barku 2019). The antioxidant properties of the herbs may enhance and accelerate the wound healing process. MEDIHONEY® Gel contains 100% active Leptospermum honey in a hydrocolloid suspension and indicated for Wound and Burn dressing. Honey is known for its antioxidant, antibacterial, and wound healing activates and is approved by the United States Food and Drug Administration (FDA) for wound healing applications (Fitzmaurice et al. 2011).

6 Future Prospective

Skin is the protective barrier and wound on the skin may lead to serious morbidity and mortality. The wound-healing process is affected by numerous factors including humidity and oxygenation, infection, stress, increasing prevalence of acute and chronic wounds, and patient-related factors such as age, lifestyle, and health status. Wound healing is an important physiological process; when it fails, the quality of a patent's life becomes worst, especially in patients with chronic metabolic disorders. Currently, one-third of the adult population is living with diabetes and 6.5 million cases of chronic skin ulcers annually (Dreifke et al. 2015). In 2016, the global Wound Healing market was valued at approximately US$22.01 billion and is projected to register a cumulative annual growth rate of over 3.7% by 2022 (Polerà et al. 2019). The cost of the management for chronic and nonhealing wounds is increasing which may affect the patient's quality of life. Hence, the search for new lead or drugs from natural sources is essential to reduce the cost of the treatment. In the traditional medicine system, numerous herbs are reported to have wound-healing properties and those plants are used in folk medicine to treat wounds. The plant and marine sources are extensively studied for their wound-healing activity (preclinical studies) and further clinical studies are required to explore the possible effect on humans.

Nature always has been a valuable source of drugs and about 80% of the world's population is dependent on plant-based systems of medicine for their primary healthcare needs. During the past 30 years, up to 50% of the approved drugs by the regulatory agencies are from natural products (plant products or their derivatives). In the area of cancer, around the 1940s to date, 85 out of 175 small molecules are being either from natural products or their derivatives (Veeresham 2012). The utility of natural products as sources of new drugs/lead is still alive and well. Hence, exploring the natural sources or their derivatives for new drug discovery may give numerous drug molecules for the management of wounds, as well as other diseases.

References

Abdulla MA, Ahmed KA, Abu-Luhoom FM, Muhanid M (2010) Role of *Ficus deltoidea* extract in the enhancement of wound healing in experimental rats. Biomed Res 21(3):241–245

Abirami MS, Indhumathy R, Devi GS, Kumar DS (2011) Evaluation of the wound healing and anti-inflammatory activity of whole plant of *Luffa cylindrica* (Linn). in Rats. Pharmacologyonline 3:281–285

Abood WN, Al-Henhena NA, Najim Abood A, Al-Obaidi MM, Ismail S, Abdulla MA et al (2015) Wound-healing potential of the fruit extract of *Phaleria macrocarpa*. Bosn J Basic Med Sci 15 (2):25–30. https://doi.org/10.17305/bjbms.2015.39

Abu-Al-Basal MA (2010) Healing potential of *Rosmarinus officinalis* L. on full-thickness excision cutaneous wounds in alloxan-induced-diabetic BALB/c mice. J Ethnopharmacol 131 (2):443–450. https://doi.org/10.1016/j.jep.2010.07.007

Adiele LC, Adiele RC, Enye JC (2014) Wound healing effect of methanolic leaf extract of *Napoleona vogelii* (Family: Lecythidaceae) in rats. Asian Pac J Trop Med 7(8):620–624

Afshar M, Ravarian B, Zardast M, Moallem SA, Fard MH, Valavi M (2015) Evaluation of cutaneous wound healing activity of *Malva sylvestris* aqueous extract in BALB/c mice. Iran J Basic Med Sci 18(6):616–622

Agarwal PK, Singh A, Gaurav K, Goel S, Khanna HD, Goel RK (2009) Evaluation of wound healing activity of extracts of plantain banana (*Musa sapientum* var. paradisiaca) in rats. Indian J Exp Biol 47:32–40

Agyare C, Bempah SB, Boakye YD, Ayande PG, Adarkwa-Yiadom M, Mensah KB (2013a) Evaluation of antimicrobial and wound healing potential of *Justicia flava* and *Lannea welwitschii*. Evid Based Complement Alternat Med 2013. https://doi.org/10.1155/2013/632927

Agyare C, Dwobeng AS, Agyepong N, Boakye YD, Mensah KB, Ayande PG et al (2013b) Antimicrobial, antioxidant, and wound healing properties of *Kigelia africana* (Lam.) Beneth. and *Strophanthus hispidus* DC. Adv Pharmacol Sci 2013:692613. https://doi.org/10.1155/2013/692613

Agyare C, Kisseih E, Kyere IY, Ossei PP (2014) Medicinal plants used in wound care: assessment of wound healing and antimicrobial properties of *Zanthoxylum leprieurii*. Issues Biol Sci Pharm Res 2(8):81–89. https://doi.org/10.15739/ibspr.002

Akbik D, Ghadiri M, Chrzanowski W, Rohanizadeh R (2014) Curcumin as a wound healing agent. Life Sci 116(1):1–7. https://doi.org/10.1016/j.lfs.2014.08.016

Akkol EK, Koca U, Pesin I, Yilmazer D (2009) Evaluation of the wound healing potential of *Achillea biebersteinii* Afan. (Asteraceae) by *in vivo* excision and incision models. Evid Based Complement Alternat Med. https://doi.org/10.1093/ecam/nep039

Alam P, Shakeel F, Anwer MK, Foudah AI, Alqarni MH (2018) Wound healing study of eucalyptus essential oil containing nanoemulsion in rat model. J Oleo Sci 67:ess18005. https://doi.org/10.5650/jos.ess18005

Albaayit SF, Abba Y, Rasedee A, Abdullah N (2015) Effect of *Clausena excavata* Burm. f. (Rutaceae) leaf extract on wound healing and antioxidant activity in rats. Drug Des Devel Ther 9:3507–3518. https://doi.org/10.2147/DDDT.S84770

Al-Bayaty FH, Abdulla MA, Hassan MI, Ali HM (2012) Effect of *Andrographis paniculata* leaf extract on wound healing in rats. Nat Prod Res 26(5):423–429. https://doi.org/10.1080/14786419.2010.496114

Aleti S, Swapna Reddy M, Sneha JA, Suvarchala NV (2015) Wound healing activity of *Mimusops elengi* leaves. Iran J Pharmacol Ther 13(1):13–18

Al-Henhena N, Mahmood AA, Al-Magrami A, Syuhada AN, Zahra AA, Summaya MD et al (2011) Histological study of wound healing potential by ethanol leaf extract of *Strobilanthes crispus* in rats. J Med Plant Res 5(16):3660–3666

Alizadeh AM, Sohanaki H, Khaniki M, Mohaghgheghi MA, Ghmami G, Mosavi M (2011) The effect of *Teucrium polium* honey on the wound healing and tensile strength in rat. Iran J Basic Med Sci 14(6):499–505

Alqasoumi SI, Yusufoglu HS, Alam A (2011) Anti-inflammatory and wound healing activity of *Fagonia schweinfurthii* alcoholic extract herbal gel on albino rats. Afr J Pharm Pharmacol 5(17):1996–2001

Ambiga S, Narayanan R, Gowri D, Sukumar D, Madhavan S (2007) Evaluation of wound healing activity of flavonoids from *Ipomoea carnea* Jacq. Anc Sci Life 26(3):45–51

Amiri IA, Moslemi HR, Tehrani-Sharif M, Kafshdouzan K (2015) Wound healing potential of *Capparis spinosa* against cutaneous wounds infected by *Escherichia coli* in a rat model. Herba Polonica 61(2):63–72. https://doi.org/10.1515/hepo-2015-0016

Amri B, Martino E, Vitulo F, Corana F, Kaâb LB, Rui M et al (2017) *Marrubium vulgare* L. leave extract: phytochemical composition, antioxidant and wound healing properties. Molecules 22(11):1851

Anand David AV, Arulmoli R, Parasuraman S (2016) Overviews of biological importance of quercetin: a bioactive flavonoid. Pharmacogn Rev 10(20):84–89. https://doi.org/10.4103/0973-7847.194044

Anbu J, Jisha P, Varatharajan R, Muthappan M (2008) Antibacterial and wound healing activities of *Melastoma malabathricum* linn. Afr J Infect Dis 2(2):68–73

Anlas C, Bakirel T, Ustun-Alkan F, Celik B, Baran MY, Ustuner O et al (2019) *In vitro* evaluation of the therapeutic potential of Anatolian kermes oak (*Quercus coccifera* L.) as an alternative wound healing agent. Ind Crop Prod 137:24–32

Annan K, Dickson R (2008) Evaluation of wound healing actions of *Hoslundia opposita* Vahl, *Anthocleista nobilis* G. Don. and *Balanites aegyptiaca* L. J Sci Technol (Ghana) 28(2):26–35

Annan K, Govindarajan R, Kisseih E (2010) Wound healing and cytoprotective actions of *Paullinia pinnata* L. Pharmacogn J 2(10):345–350

Aronoff SL, Berkowitz K, Shreiner B, Want L (2004) Glucose metabolism and regulation: beyond insulin and glucagon. Diabetes Spectr 17(3):183–190

Arun M, Satish S, Anima P (2016) Evaluation of wound healing, antioxidant and antimicrobial efficacy of *Jasminum auriculatum* Vahl. leaves. Avicenna J Phytomed 6(3):295–304

Arunachalam KD, Subhashini S (2011) Preliminary phytochemical investigation and wound healing activity of *Myristica andamanica* leaves in Swiss albino mice. J Med Plant Res 5(7):1095–1106

Arunachalam KD, Subhashini S, Annamalai SK (2012) Wound healing and antigenotoxic activities of *Aegle marmelos* with relation to its antioxidant properties. J Pharm Res 5(3):1492–1502

Asadi SY, Parsaei P, Karimi M, Ezzati S, Zamiri A, Mohammadizadeh F et al (2013) Effect of green tea (*Camellia sinensis*) extract on healing process of surgical wounds in rat. Int J Surg 11(4):332–337. https://doi.org/10.1016/j.ijsu.2013.02.014

Asif A, Kakub G, Mehmood S, Khunum R, Gulfraz M (2007) Wound healing activity of root extracts of *Berberis lyceum* royle in rats. Phytother Res 21(6):589–591. https://doi.org/10.1002/ptr.2110

Aslam MS, Ahmad MS, Mamat AS, Ahmad MZ, Salam F (2016) Antioxidant and wound healing activity of polyherbal fractions of *Clinacanthus nutans* and *Elephantopus scaber*. Evid Based Complement Alternat Med 2016:4685246. https://doi.org/10.1155/2016/4685246

Badgujar SB, Mahajan RT, Chopda MZ (2009) Wound healing activity of latex of *Euphorbia nivulia* Buch.-Ham. in mice. RJPPD 1(2):90–92

Bahramsoltani R, Farzaei MH, Abdolghaffari AH, Rahimi R, Samadi N, Heidari M et al (2017) Evaluation of phytochemicals, antioxidant and burn wound healing activities of *Cucurbita moschata* Duchesne fruit peel. Iran J Basic Med Sci 20(7):798–805. https://doi.org/10.22038/IJBMS.2017.9015

Balekar N, Katkam NG, Nakpheng T, Jehtae K, Srichana T (2012) Evaluation of the wound healing potential of *Wedelia trilobata* (L.) leaves. J Ethnopharmacol 141(3):817–824. https://doi.org/10.1016/j.jep.2012.03.019

Baliano AP, Pimentel EF, Buzin AR, Vieira TZ, Romão W, Tose LV et al (2016) Brown seaweed *Padina gymnospora* is a prominent natural wound-care product. Rev Bras Farm 26(6):714–719. https://doi.org/10.1016/j.bjp.2016.07.003

Bambal VC, Wyawahare NS, Turaskar AO, Deshmukh TA (2011) Evaluation of wound healing activity of herbal gel containing the fruit extract of *Coccinia indica* wight and arn. (cucurbitaceae). Int J Pharm Pharm Sci 3(4):319–324

Bardaa S, Ben Halima N, Aloui F, Ben Mansour R, Jabeur H, Bouaziz M, Sahnoun Z (2016) Oil from pumpkin (*Cucurbita pepo* L.) seeds: evaluation of its functional properties on wound healing in rats. Lipids Health Dis 15:73. https://doi.org/10.1186/s12944-016-0237-0

Barku VY (2019) Wound healing: contributions from plant secondary metabolite antioxidants. In: Wound healing-current perspectives. IntechOpen. https://doi.org/10.5772/intechopen.81208

Barku VY, Boye A, Ayaba S (2013) Phytochemical screening and assessment of wound healing activity of the leaves of *Anogeissus leiocarpus*. Eur J Exp Biol 3(4):18–25

Barth A, Hovhannisyan A, Jamalyan K, Narimanyan M (2015) Antitussive effect of a fixed combination of *Justicia adhatoda*, *Echinacea purpurea* and *Eleutherococcus senticosus* extracts in patients with acute upper respiratory tract infection: a comparative, randomized, double-blind, placebo-controlled study. Phytomedicine 22(13):1195–1200. https://doi.org/10.1016/j.phymed.2015.10.001

Barua CC, Talukdar A, Barua AG, Chakraborty A, Sarma RK, Bora RS (2010) Evaluation of the wound healing activity of methanolic extract of *Azadirachta Indica* (Neem) and *Tinospora cordifolia* (Guduchi) in rats. Pharmacologyonline 1:70–77

Barua CC, Begum SA, Pathak DC, Bora RS (2013) Wound healing activity of *Alternanthera brasiliana* Kuntze and its anti-oxidant profiles in experimentally induced diabetic rats. J Appl Pharm Sci 3(10):161–166

Bascha J, Murthy BR, Likhitha PR, Ganesh Y, Bai BG, Rani RJ et al (2016) *In vitro* and *in vivo* assessment of lawsone microsphere loaded chitosan scaffolds. Int J Phytopharm 6(4):74–84. https://doi.org/10.7439/ijpp.v6i4.3529

Baveja M, Sarkar A, Chakrabarty D (2018) Hemotoxic and wound healing potential of coelomic fluid of sea-star *Astropecten indicus*. J Basic Appl Zool 79(1):27. https://doi.org/10.1186/s41936-018-0038-2

Bhandirge SK, Tripathi AS, Bhandirge RK, Chinchmalatpure TP, Desai HG, Chandewar AV (2015) Evaluation of wound healing activity of ethanolic extract of *Pongamia pinnata* bark. Drug Res 65(6):296–299

Bharat M, Verma DK, Shanbhag V, Rajput RS, Nayak D, Amuthan A (2014) Ethanolic extract of oral *Areca catechu* promotes burn wound healing in rats. Int J Pharm Sci Rev Res 25(2):145–148

Bharathi RV, Veni BK, Suseela L, Thirumal M (2010) Antioxidant and wound healing studies on different extracts of *Stereospermum colais* leaf. Int J Res Pharm sci 1(4):435–439

Bharti M, Saxena RC, Baghel OS, Saxena R, Apte KG (2011) Wound healing activity of leaf of *Nyctanthes arbor-trisitis* (linn.). Int J Pharm Sci Res 2(10):2694–2698

Bhaskar A, Nithya V (2012) Evaluation of the wound-healing activity of *Hibiscus rosa sinensis* L (Malvaceae) in Wistar albino rats. Indian J Pharmacol 44(6):694–698. https://doi.org/10.4103/0253-7613.103252

Bhat PB, Hegde S, Upadhya V, Hegde GR, Habbu PV, Mulgund GS (2016) Evaluation of wound healing property of *Caesalpinia mimosoides* Lam. J Ethnopharmacol 193:712–724

Bigoniya P, Rana AC (2007) Wound healing activity of *Euphorbia neriifolia* leaf ethanolic extract in rats. J Nat Remedies 7(1):94–101

Biswas TK, Maity LN, Mukherjee B (2004) Wound healing potential of *Pterocarpus santalinus* Linn: a pharmacological evaluation. Int J Low Extrem Wounds 3(3):143–150

Biswas TK, Pandit S, Chakrabarti S, Banerjee S, Poyra N, Seal T (2017) Evaluation of *Cynodon dactylon* for wound healing activity. J Ethnopharmacol 197:128–137. https://doi.org/10.1016/j.jep.2016.07.065

Bolla SR, Al-Subaie AM, Al-Jindan RY, Balakrishna JP, Ravi PK, Veeraraghavan VP et al (2019) *In vitro* wound healing potency of methanolic leaf extract of *Aristolochia saccata* is possibly mediated by its stimulatory effect on collagen-1 expression. Heliyon 5(5):e01648. https://doi.org/10.1016/j.heliyon.2019.e01648

Bonesi M, Loizzo MR, Menichini F, Tundis R (2018) Flavonoids in treating psoriasis. In: Immunity and inflammation in health and disease. Academic Press, London, pp 281–294

Bonte F, Dumas M, Chaudagne C, Meybeck A (1994) Influence of asiatic acid, madecassic acid, and asiaticoside on human collagen I synthesis. Planta Med 60(2):133–135. https://doi.org/10.1055/s-2006-959434

Bramara BV, Vasavi HS, Sudeep HV, Prasad S (2017) Hydroalcoholic extract from *Lepidium meyenii* (Black Maca) root exerts wound healing activity in Streptozotocin-induced diabetic rats. Wound Med 19:75–81. https://doi.org/10.1016/j.wndm.2017.10.003

Bueno FG, Moreira EA, Morais GR, Pacheco IA, Baesso ML, Leite-Mello EV et al (2016) Enhanced cutaneous wound healing *in vivo* by standardized crude extract of *Poincianella pluviosa*. PLoS One 11(3):e0149223. https://doi.org/10.1371/journal.pone.0149223

Builders PF, Kabele-Toge B, Builders M, Chindo BA, Anwunobi PA, Isimi YC (2013) Wound healing potential of formulated extract from *Hibiscus sabdariffa* calyx. Indian J Pharm Sci 75 (1):45–52. https://doi.org/10.4103/0250-474X.113549

Bush JA (2008) Double blind, placebo controlled trial to investigate the efficacy and safety of two concentrations of juvidex (mannose-6-phosphate) in accelerating the healing of split thickness skin graft donor sites using different dosing regimes. ClinicalTrials.gov Identifier: NCT00664352. Available at https://www.clinicaltrials.gov/ct2/show/record/NCT00664352. Accessed 26 Jul 2020

Calderón-Montaño JM, Burgos-Morón E, Pérez-Guerrero C, López-Lázaro M (2011) A review on the dietary flavonoid kaempferol. Mini Rev Med Chem 11(4):298–344. https://doi.org/10.2174/138955711795305335

Chatterjee S, Prakash T, Kotrsha D, Rao NR, Goli D (2011) Comparative efficacy of *Tagetes erecta* and *Centella asiatica* extracts on wound healing in albino rats. Chin Med 2(4):138–142. https://doi.org/10.4236/cm.2011.24023

Chaturvedi AP, Kumar M, Tripathi YB (2013) Efficacy of *Jasminum grandiflorum* L. leaf extract on dermal wound healing in rats. Int Wound J 10(6):675–682. https://doi.org/10.1111/j.1742-481X.2012.01043.x

Chaudhari M, Mengi S (2006) Evaluation of phytoconstituents of *Terminalia arjuna* for wound healing activity in rats. Phytother Res 20(9):799–805. https://doi.org/10.1002/ptr.1857

Chen WC, Liou SS, Tzeng TF, Lee SL, Liu IM (2012) Wound repair and anti-inflammatory potential of *Lonicera japonica* in excision wound-induced rats. BMC Complement Altern Med 12:226. https://doi.org/10.1186/1472-6882-12-226

Chen WC, Liou SS, Tzeng TF, Lee SL, Liu IM (2013) Effect of topical application of chlorogenic acid on excision wound healing in rats. Planta Med 79(8):616–621. https://doi.org/10.1055/s-0032-1328364

Chidambara Murthy KN, Reddy VK, Veigas JM, Murthy UD (2004) Study on wound healing activity of *Punica granatum* peel. J Med Food 7(2):256–259. https://doi.org/10.1089/1096620041224111

Chin YW, Balunas MJ, Chai HB, Kinghorn AD (2006) Drug discovery from natural sources. AAPS J 8(2):E239–E253. https://doi.org/10.1007/BF02854894

Chitra V, Dharani PP, Pavan KK, Alla NR (2009) Wound healing activity of alcoholic extract of *Buchanania lanzan* in Albino rats. Int J ChemTech Res 1(4):1026–1031

Choi J, Park YG, Yun MS, Seol JW (2018) Effect of herbal mixture composed of *Alchemilla vulgaris* and Mimosa on wound healing process. Biomed Pharmacother 106:326–332. https://doi.org/10.1016/j.biopha.2018.06.141

Chopda MZ, Patole SS, Mahajan RT (2010) Wound healing activity of *Sphaeranthus indicus* (Linn) in albino rats. In: Kulkarni GK, Pandey BN, Joshi BD (eds) Bioresources for rural livelihood. Narendra Publishing House I, Delhi, pp 239–244

Choudhary GP (2008) Wound healing activity of the ethanol extract of *Terminalia bellirica* Roxb. fruits. Nat Prod Radiance 7(1):19–21

Choudhary GP (2012) Wound healing activity of the ethanolic extract of *Mesua ferrea* Linn. Int J Adv Pharm Biol Chem 1(3):369–371

Chudhary Z, Khera RA, Hanif MA, Ayub MA, Hamrouni L (2020) Walnut. In: Medicinal plants of South Asia. Elsevier, London, pp 671–684

Clericuzio M, Burlando B, Gandini G, Tinello S, Ranzato E, Martinotti S, Cornara L (2014) Keratinocyte wound healing activity of galactoglycerolipids from the fern *Ophioglossum vulgatum* L. J Nat Med 68(1):31–37. https://doi.org/10.1007/s11418-013-0759-y

Cologne (2006) InformedHealth.org [Internet]. In Institute for Quality and Efficiency in Health Care (IQWiG). Available in https://www.ncbi.nlm.nih.gov/books/NBK326436/; Last assessed on 25 Aug 2020

Craciun L, Chavan MM, BASF Corp (2015) Marine plants extract for wound healing. United States patent application US 14/700,326

Cristea D, Bareau I, Vilarem G (2003) Identification and quantitative HPLC analysis of the main flavonoids present in weld (*Reseda luteola* L.). Dyes Pigments 57(3):267–272. https://doi.org/10.1016/S0143-7208(03)00007-X

Csupor D, Blazsó G, Balogh A, Hohmann J (2010) The traditional Hungarian medicinal plant *Centaurea sadleriana* Janka accelerates wound healing in rats. J Ethnopharmacol 127 (1):193–195. https://doi.org/10.1016/j.jep.2009.09.049

Dai Y, Prithiviraj N, Gan J, Zhang XA, Yan J (2016) Tissue extract fractions from starfish undergoing regeneration promote wound healing and lower jaw blastema regeneration of Zebrafish. Sci Rep 6:38693. https://doi.org/10.1038/srep38693

Dalazen P, Molon A, Biavatti MW, Kreuger MR (2005) Effects of the topical application of the extract of *Vernonia scorpioides* on excisional wounds in mice. Rev Bras Farm 15(2):82–87. https://doi.org/10.1590/S0102-695X2005000200002

Das K (2013) Wound healing potential of aqueous crude extract of *Stevia rebaudiana* in mice. Rev Bras Farm 23(2):351–357. https://doi.org/10.1590/S0102-695X2013005000011

Dash GK, Murthy PN (2011a) Evaluation of *Argemone mexicana* Linn. leaves for wound healing activity. J Nat Prod Plant Resour 1(1):46–56

Dash GK, Murthy PN (2011b) Studies on wound healing activity of *Heliotropium indicum* Linn. leaves on rats. ISRN Pharmacol 2011:847980. https://doi.org/10.5402/2011/847980

de Moura Sperotto ND, Steffens L, Veríssimo RM, Henn JG, Péres VF, Vianna P et al (2018) Wound healing and anti-inflammatory activities induced by a *Plantago australis* hydroethanolic extract standardized in verbascoside. J Ethnopharmacol 225:178–188. https://doi.org/10.1016/j.jep.2018.07.012

Demilew W, Adinew GM, Asrade S (2018) Evaluation of the wound healing activity of the crude extract of leaves of *Acanthus polystachyus* Delile (Acanthaceae). Evid Based Complement Alternat Med 2018. https://doi.org/10.1155/2018/2047896

Deshmukh PT, Fernandes J, Atul A, Toppo E (2009) Wound healing activity of *Calotropis gigantea* root bark in rats. J Ethnopharmacol 125(1):178–181. https://doi.org/10.1016/j.jep.2009.06.007

Devi P, Meera R (2010) Study of antioxdant, antiinflammatory and woundhealing activity of extracts of *Litsea glutinosa*. J Pharm Sci Res 2(3):155–163

Divya S, Naveen Krishna K, Ramachandran S, Dhanaraju MD (2011) Wound healing and in vitro antioxidant activities of *Croton bonplandianum* leaf extract in rats. Global J Pharmacol 5 (3):159–163

Dixit V, Verma P, Agnihotri P, Paliwal AK, Rao CV, Husain T (2015) Antimicrobial, antioxidant and wound healing properties of *Leucas lanata* Wall. ex Benth. J Phytopharmacol 4(1):9–16

Djemaa FG, Bellassoued K, Zouari S, El Feki A, Ammar E (2016) Antioxidant and wound healing activity of *Lavandula aspic* L. ointment. J Tissue Viability 25(4):193–200

Dong X, Fu J, Yin X, Cao S, Li X, Lin L, Huyiligeqi, Ni J (2016) Emodin: a review of its pharmacology, toxicity and pharmacokinetics. Phytother Res 30(8):1207–1218. https://doi.org/10.1002/ptr.5631

Dreifke MB, Jayasuriya AA, Jayasuriya AC (2015) Current wound healing procedures and potential care. Mater Sci Eng C Mater Biol Appl 48:651–662. https://doi.org/10.1016/j.msec.2014.12.068

Dunnill C, Patton T, Brennan J, Barrett J, Dryden M, Cooke J et al (2017) Reactive oxygen species (ROS) and wound healing: the functional role of ROS and emerging ROS-modulating technologies for augmentation of the healing process. Int Wound J 14(1):89–96. https://doi.org/10.1111/iwj.12557

Edwin BT, Nair PD (2016) *In vitro* evaluation of wound healing property of *Hemigraphis alternata* (Burm. F) t. Anders using fibroblast and endothelial cells. Biosci Biotechnol Res Asia 8 (1):185–193

Elzayat EM, Auda SH, Alanazi FK, Al-Agamy MH (2018) Evaluation of wound healing activity of henna, pomegranate and myrrh herbal ointment blend. Saudi Pharm J 26(5):733–738. https://doi.org/10.1016/j.jsps.2018.02.016

Esimone CO, Nworu CS, Jackson CL (2008) Cutaneous wound healing activity of a herbal ointment containing the leaf extract of *Jatropha curcas* L. (Euphorbiaceae). Int J Appl Res Nat Prod 1(4):1–4

Ezzat SM, Choucry MA, Kandil ZA (2016) Antibacterial, antioxidant, and topical anti-inflammatory activities of *Bergia ammannioides*: a wound-healing plant. Pharm Biol 54 (2):215–224. https://doi.org/10.3109/13880209.2015.1028079

Farahpour MR, Taghikhani H, Habibi M (2011) Wound healing activity of flaxseed *Linum usitatissimum* L. in rats. Afr J Pharm Pharmacol 5(21):2386–2389. https://doi.org/10.5897/AJPP11.258

Farahpour MR, Amniattalab A, Hajizadeh H (2012) Evaluation of the wound healing activity of *Cinnamomum zeylanicum* extract on experimentally induced wounds in rats. Afr J Biotechnol 11(84):15068–15071

Farahpour MR, Hesaraki S, Faraji D, Zeinalpour R, Aghaei M (2017) Hydroethanolic *Allium sativum* extract accelerates excision wound healing: evidence for roles of mast-cell infiltration and intracytoplasmic carbohydrate ratio. Braz J Pharm Sci 53(1):e15079. https://doi.org/10.1590/s2175-97902017000115079

Ferguson MW, BTG International Ltd (2000) Mannose-6-phosphate composition and its use in treating fibrotic disorders. United States patent US 6,093,388

Fikru A, Makonnen E, Eguale T, Debella A, Abie MG (2012) Evaluation of *in vivo* wound healing activity of methanol extract of *Achyranthes aspera* L. J Ethnopharmacol 143(2):469–474. https://doi.org/10.1016/j.jep.2012.06.049

Fitzmaurice SD, Sivamani RK, Isseroff RR (2011) Antioxidant therapies for wound healing: a clinical guide to currently commercially available products. Skin Pharmacol Physiol 24 (3):113–126. https://doi.org/10.1159/000322643

Gadallah AS, Mujeeb-Ur-Rehman, Atta-Ur-Rahman, Yousuf S, Atia-Tul-Wahab, Jabeen A et al (2020) Anti-Inflammatory principles from *Tamarix aphylla* L.: a bioassay-guided fractionation study. Molecules 25(13):E2994. https://doi.org/10.3390/molecules25132994

Gagliardini E, Benigni A, Perico N (2017) Pharmacological induction of kidney regeneration. In: Kidney transplantation, bioengineering and regeneration. Academic Press, London, pp 1025–1037

Ganesan S, Parasuraman S, Maheswaran SU, Gnanasekar N (2012) Wound healing activity of *Hemidesmus indicus* formulation. J Pharmacol Pharmacother 3(1):66–67. https://doi.org/10.4103/0976-500X.92516

Gangwar M, Gautam MK, Ghildiyal S, Nath G, Goel RK (2015) *Mallotus philippinensis* Muell. Arg fruit glandular hairs extract promotes wound healing on different wound model in rats. BMC Complement Altern Med 15:123. https://doi.org/10.1186/s12906-015-0647-y

Garg VK, Paliwal SK (2011) Wound-healing activity of ethanolic and aqueous extracts of *Ficus benghalensis*. J Adv Pharm Technol Res 2(2):110–114. https://doi.org/10.4103/2231-4040.82957

Gebrehiwot M, Asres K, Bisrat D, Mazumder A, Lindemann P, Bucar F (2015) Evaluation of the wound healing property of *Commiphora guidottii* Chiov ex Guid. BMC Complement Altern Med 15:282. https://doi.org/10.1186/s12906-015-0813-2

Gebremeskel L, Bhoumik D, Sibhat GG, Tuem KB (2018) *In vivo* wound healing and anti-inflammatory activities of leaf latex of *Aloe megalacantha* baker (Xanthorrhoeaceae). Evid Based Complement Alternat Med 2018:5037912. https://doi.org/10.1155/2018/5037912

Geethalakshmi R, Sakravarthi C, Kritika T, Arul Kirubakaran M, Sarada DV (2013) Evaluation of antioxidant and wound healing potentials of *Sphaeranthus amaranthoides* Burm.f. Biomed Res Int 2013:607109. https://doi.org/10.1155/2013/607109

George BP, Parimelazhagan T, Chandran R (2014) Anti-inflammatory and wound healing properties of *Rubus fairholmianus* Gard. root-An *in vivo* study. Ind Crop Prod 54:216–225. https://doi.org/10.1016/j.indcrop.2014.01.037

Ghaima KK (2013) Antibacterial and wound healing activity of some *Agrimonia eupatoria* extracts. Baghdad Sci J 10(1):152–160

Ghaisas MM, Kshirsagar SB, Sahane RS (2014) Evaluation of wound healing activity of ferulic acid in diabetic rats. Int Wound J 11(5):523–532. https://doi.org/10.1111/j.1742-481X.2012.01119.x

Ghashghaii A, Hashemnia M, Nikousefat Z, Zangeneh MM, Zangeneh A (2017) Wound healing potential of methanolic extract of *Scrophularia striata* in rats. Pharm Sci 23(4):256–263. https://doi.org/10.15171/PS.2017.38

Ghayempour S, Montazer M, Mahmoudi RM (2016) Encapsulation of *Aloe Vera* extract into natural tragacanth Gum as a novel green wound healing product. Int J Biol Macromol 93 (Pt A):344–349. https://doi.org/10.1016/j.ijbiomac.2016.08.076

Ghosh S, Samanta A, Mandal NB, Bannerjee S, Chattopadhyay D (2012) Evaluation of the wound healing activity of methanol extract of *Pedilanthus tithymaloides* (L.) Poit leaf and its isolated active constituents in topical formulation. J Ethnopharmacol 142(3):714–722. https://doi.org/10.1016/j.jep.2012.05.048

Girish C, Pradhan SC (2012) Hepatoprotective activities of picroliv, curcumin, and ellagic acid compared to silymarin on carbon-tetrachloride-induced liver toxicity in mice. J Pharmacol Pharmacother 3(2):149–155. https://doi.org/10.4103/0976-500X.95515

Gomathi K, Gopinath D, Rafiuddin Ahmed M, Jayakumar R (2003) Quercetin incorporated collagen matrices for dermal wound healing processes in rat. Biomaterials 24(16):2767–2772. https://doi.org/10.1016/s0142-9612(03)00059-0

Gomes FS, Spínola Cde V, Ribeiro HA, Lopes MT, Cassali GD, Salas CE (2010) Wound-healing activity of a proteolytic fraction from *Carica candamarcensis* on experimentally induced burn. Burns 36(2):277–283. https://doi.org/10.1016/j.burns.2009.04.007

Goorani S, Zangeneh MM, Koohi MK, Seydi N, Zangeneh A, Souri N et al (2019) Assessment of antioxidant and cutaneous wound healing effects of *Falcaria vulgaris* aqueous extract in Wistar male rats. Comp Clin Pathol 28(2):435–445. https://doi.org/10.1007/s00580-018-2866-3

Govindarajan R, Vijayakumar M, Venkateshwara Rao CH, Shirwaikar A, Mehrotra S, Pushpangadan P (2004) Healing potential of *Anogeissus latifolia* for dermal wounds in rats. Acta Pharma 54(4):331–338

Gowdu Viswanathan MB, Ananthi JD, Raja NL, Venkateshan N (2018) Wound healing activity of *Jatropha tanjorensis* leaves. Pharm Res 3(4):24–30

Goyal M, Nagori BP, Sasmal D (2012) Wound healing activity of latex of *Euphorbia caducifolia*. J Ethnopharmacol 144(3):786–790. https://doi.org/10.1016/j.jep.2012.10.006

Gundamaraju R, Verma TM (2012) Evaluation of wound healing activity of *Crossandra infundibuliformis* flower extract on albino rats. Int J Pharm Sci Res 3(11):4545

Gupta A, Kumar R, Pal K, Singh V, Banerjee PK, Sawhney RC (2006) Influence of sea buckthorn (*Hippophae rhamnoides* L.) flavone on dermal wound healing in rats. Mol Cell Biochem 290 (1-2):193–198. https://doi.org/10.1007/s11010-006-9187-6

Gupta A, Kumar R, Upadhyay NK, Pal K, Kumar R, Sawhney RC (2007) Effects of *Rhodiola imbricata* on dermal wound healing. Planta Med 73(8):774. https://doi.org/10.1055/s-2007-981546

Gupta A, Verma S, Gupta AK, Jangra M, Pratap R (2015) Evaluation of *Prosopis cineraria* (Linn.) Druce leaves for wound healing activity in rats. Ann Pharm Res 3:70–74

Gupta SC, Tripathi T, Paswan SK, Agarwal AG, Rao CV, Sidhu OP (2018) Phytochemical investigation, antioxidant and wound healing activities of *Citrullus colocynthis* (bitter apple). Asian Pac J Trop Biomed 8:418–424. https://doi.org/10.4103/2221-1691.239430

Gutierrez RM, Solis RV (2009) Anti-inflammatory and wound healing potential of *Prosthechea michuacana* in rats. Pharmacogn Mag 5(19):219

Ha JH, Lee DU, Lee JT, Kim JS, Yong CS, Kim JA et al (2000) 4-Hydroxybenzaldehyde from *Gastrodia elata* B1. is active in the antioxidation and GABAergic neuromodulation of the rat brain. J Ethnopharmacol 73(1-2):329–333

Hajhashemi V, Ghannadi A, Heidari AH (2012) Anti-inflammatory and wound healing activities of *Aloe littoralis* in rats. Res Pharm Sci 7(2):73–78

Han DO, Lee HJ, Hahm DH (2009) Wound-healing activity of *Astragali Radix* in rats. Methods Find Exp Clin Pharmacol 31(2):95–100

Hassan KA, Deogratius O, Nyafuono JF, Francis O, Engeu OP (2011) Wound healing potential of the ethanolic extracts of *Bidens pilosa* and *Ocimum suave*. Afr J Pharm Pharmacol 5(2):132–136

Heidari M, Bahramsoltani R, Abdolghaffari AH, Rahimi R, Esfandyari M, Baeeri M et al (2018) Efficacy of topical application of standardized extract of *Tragopogon graminifolius* in the healing process of experimental burn wounds. J Tradit Complement Med 9(1):54–59. https://doi.org/10.1016/j.jtcme.2018.02.002

Hemamalini K, Ramu A, Mallu G, Srividya VV, Sravani V, Deepak P et al (2011) Evaluation of wound healing activity of different crude extracts of *Anogeissus acuminata* and *Gymnosporia emerginata*. Rasayan J Chem 4:466–471

Hemmati AA, Houshmand G, Nemati M, Bahadoram M, Dorestan N, Rashidi-Nooshabadi MR et al (2015) Wound healing effects of persian Oak (*Quercus brantii*) ointment in rats. Jundishapur J Nat Pharm Prod 10(4):e25508

Hewlings SJ, Kalman DS (2017) Curcumin: a review of its' effects on human health. Foods 6 (10):92. https://doi.org/10.1016/S0378-8741(00)00313-5

Hovanet MV, Oprea E, Ancuceanu RV, Dutu LE, Budura EA, Şeremet O et al (2016) Wound healing properties of *Ziziphus jujuba* Mill. leaves. Rom Biotechnol Lett 21(5):11842–11849

Huang CC, Lin KJ, Cheng YW, Hsu CA, Yang SS, Shyur LF (2013) Hepatoprotective effect and mechanistic insights of deoxyelephantopin, a phyto-sesquiterpene lactone, against fulminant hepatitis. J Nutr Biochem 24(3):516–530. https://doi.org/10.1016/j.jnutbio.2012.01.013

Ilango K, Chitra V (2010) Wound healing and anti-oxidant activities of the fruit pulp of *Limonia acidissima* Linn (Rutaceae) in rats. Trop J Pharm Res 9(3):223–230

Ismayilnajadteymurabadi H, Farahpour MR, Amniattalab A (2012) Histological evaluation of *Plantago lanceolata* L. extract in accelerating wound healing. J Med Plant Res 6 (34):4844–4847. https://doi.org/10.5897/JMPR11.516

Jain S, Jain N, Tiwari A, Balekar N, Jain DK (2009) Simple evaluation of wound healing activity of polyherbal formulation of roots of *Ageratum conyzoides* Linn. Asian J Res Chem 2(2):135–138

Jaiprakash B, Chandramohan, Reddy DN (2006) Burn wound healing activity of *Euphorbia hirta*. Anc Sci Life 25(3-4):16–18

Jalalpure SS, Agrawal N, Patil MB, Chimkode R, Tripathi A (2008) Antimicrobial and wound healing activities of leaves of *Alternanthera sessilis* Linn. IJGP 2(3):141–144

James O, Victoria IA (2010) Excision and incision wound healing potential of *Saba florida* (Benth) leaf extract in *Rattus novergicus*. Int J Pharm Biomed Res 1(4):101–107

Jangde R, Srivastava S, Singh MR, Singh D (2018) *In vitro* and *in vivo* characterization of quercetin loaded multiphase hydrogel for wound healing application. Int J Biol Macromol 115:1211–1217. https://doi.org/10.1016/j.ijbiomac.2018.05.010

Jawaid T, Kamal M, Mallik T (2016) Wound healing potential of methanolic extract of *Calophyllum inophyllum* Linn. bark. Bull. Env Pharmacol Life Sci 5:27–32

Jeong BS (2006) Structure-activity relationship study of asiatic acid derivatives for new wound healing agent. Arch Pharm Res 29(7):556–562. https://doi.org/10.1007/BF02969264

John-Africa LB, Yahaya TA, Isimi CY (2014) Anti-ulcer and wound healing activities of *Sida corymbosa* in rats. Afr J Tradit Complement Altern Med 11(1):87–92

Jorge MP, Madjarof C, Gois Ruiz AL, Fernandes AT, Ferreira Rodrigues RA, de Oliveira Sousa IM et al (2008) Evaluation of wound healing properties of *Arrabidaea chica Verlot* extract. J Ethnopharmacol 118(3):361–366. https://doi.org/10.1016/j.jep.2008.04.024

Joshi A, Sengar N, Prasad SK, Goel RK, Singh A, Hemalatha S (2013) Wound-healing potential of the root extract of *Albizzia lebbeck*. Planta Med 79(09):737–743. https://doi.org/10.1055/s-0032-1328539

Juneja K, Mishra R, Chauhan S, Gupta S, Roy P, Sircar D (2020) Metabolite profiling and wound-healing activity of *Boerhavia diffusa* leaf extracts using *in vitro* and *in vivo* models. J Tradit Complement Med 10(1):52–59. https://doi.org/10.1016/j.jtcme.2019.02.002

Kahkeshani N, Farahanikia B, Mahdaviani P, Abdolghaffari A, Hassanzadeh G, Abdollahi M et al (2013) Antioxidant and burn healing potential of *Galium odoratum* extracts. Res Pharm Sci 8 (3):197–203

Kamatou GP, Viljoen AM (2010) A review of the application and pharmacological properties of α-bisabolol and α-bisabolol-rich oils. J Am Oil Chem Soc 87(1):1–7. https://doi.org/10.1007/s11746-009-1483-3

Kameshwaran S, Senthilkumar R, Thenmozhi S, Dhanalakshmi M (2014) Wound healing potential of ethanolic extract of *Tecoma stans* flowers in rats. Pharmacologia 5(6):215–221. https://doi.org/10.5567/pharmacologia.2014.215.221

Kang CW, Han YE, Kim J, Oh JH, Cho YH, Lee EJ (2017) 4-Hydroxybenzaldehyde accelerates acute wound healing through activation of focal adhesion signalling in keratinocytes. Sci Rep 7 (1):14192. https://doi.org/10.1038/s41598-017-14368-y

Karakaş FP, Karakaş A, Boran Ç, Türker AU, Yalçin FN, Bilensoy E (2012) The evaluation of topical administration of *Bellis perennis* fraction on circular excision wound healing in Wistar albino rats. Pharm Biol 50(8):1031–1037. https://doi.org/10.3109/13880209.2012.656200

Karatas O, Gevrek F (2019) Gallic acid liposome and powder gels improved wound healing in wistar rats. Ann Med Res 26(12):2720–2727. https://doi.org/10.5455/annalsmedres.2019.05.301

Karayannopoulou M, Tsioli V, Loukopoulos P, Anagnostou TL, Giannakas N, Savvas I et al (2011) Evaluation of the effectiveness of an ointment based on Alkannins/Shikonins on second intention wound healing in the dog. Can J Vet Res 75(1):42–48

Karigar AA, Shariff WR, Sikarwar MS (2010) Wound healing property of alcoholic extract of leaves of *Alangium salvifolium*. J Pharm Res 3(2):267–269

Karimzadeh S, Farahpour MR (2017) Topical application of *Salvia officinalis* hydroethanolic leaf extract improves wound healing process. Indian J Exp Biol 55:98–106

Karodi R, Jadhav M, Rub R, Bafna A (2009) Evaluation of the wound healing activity of a crude extract of *Rubia cordifolia* L. (Indian madder) in mice. Int J Appl Res Nat Prod 2(2):12–18

Karthikeyan P, Suresh V, Suresh A (2011) Wound healing activity of *Sesbania grandiflora* (L.) poir. Bark. Int J of Pharm Res Dev 3(2):87–93

Kazemian H, Ghafourian S, Sadeghifard N, Houshmandfar R, Badakhsh B, Taji A et al (2018) *In vivo* antibacterial and wound healing activities of roman chamomile (*Chamaemelum nobile*). Infect Disord Drug Targets 18(1):41–45. https://doi.org/10.2174/1871526516666161230123133

Khan A, Firoz N (2015) Drug discovery and development, a new hope. Int J Adv Res Comput Sci Softw Eng 5(1):337–345

Khan M, Patil PA, Shobha JC (2004) Influence of *Bryophyllum pinnatum* (Lim.) leaf extract on wound healing in albino rats. J Nat Remedies 4(1):41–46

Khan MS, Mat Jais AM, Zakaria ZA, Mohtarrudin N, Ranjbar M, Khan M et al (2012) Wound healing potential of leathery murdah, *Terminalia coriacea* (Roxb.) Wight and Arn. Phytopharmacology 3:158–168

Kim WK, Song SY, Oh WK, Kaewsuwan S, Tran TL, Kim WS et al (2013) Wound-healing effect of ginsenoside Rd from leaves of *Panax ginseng via* cyclic AMP-dependent protein kinase pathway. Eur J Pharmacol 702(1-3):285–293. https://doi.org/10.1016/j.ejphar.2013.01.048

Kiran K, Asad M (2008) Wound healing activity of *Sesamum indicum* L seed and oil in rats. Indian J Exp Biol 46:777–782

Kisseih E, Lechtenberg M, Petereit F, Sendker J, Zacharski D, Brandt S et al (2015) Phytochemical characterization and *in vitro* wound healing activity of leaf extracts from *Combretum mucronatum* Schum. & Thonn.: *Oligomeric procyanidins* as strong inductors of cellular differentiation. J Ethnopharmacol 174:628–636. https://doi.org/10.1016/j.jep.2015.06.008

Kittana N, Abu-Rass H, Sabra R, Manasra L, Hanany H, Jaradat N et al (2017) Topical aqueous extract of *Ephedra alata* can improve wound healing in an animal model. Chin J Traumatol 20 (2):108–113. https://doi.org/10.1016/j.cjtee.2016.10.004

Koca U, Süntar IP, Keles H, Yesilada E, Akkol EK (2009) *In vivo* anti-inflammatory and wound healing activities of *Centaurea iberica* Trev. ex Spreng. J Ethnopharmacol 126(3):551–556. https://doi.org/10.1016/j.jep.2009.08.017

Kodati DR, Burra S, Kumar GP (2011) Evaluation of wound healing activity of methanolic root extract of *Plumbago zeylanica* L. in wistar albino rats. Asian J Plant Sci Res 1(2):26–34

Kokane DD, More RY, Kale MB, Nehete MN, Mehendale PC, Gadgoli CH (2009) Evaluation of wound healing activity of root of *Mimosa pudica*. J Ethnopharmacol 124(2):311–315

Koparde AA, Doijad RC, Magdum CS (2019) Natural products in drug discovery. In: Pharmacognosy-medicinal plants. IntechOpen

Kosger HH, Ozturk M, Sokmen A, Bulut E, Ay S (2009) Wound healing effects of *Arnebia densiflora* root extracts on rat palatal mucosa. Eur J Dent 3(2):96–99

Kuluvar G, Mahmood R, Ahamed BM, Babu PS, Krishna V (2009) Wound-healing activity of *Clerodendrum infortunatum* L. root extracts. Int J Biomed Pharm Sci 3(1):21–25

Kumar SA (2017) Wound healing: current understanding and future prospect. Int J Drug Discov 8 (1):240–246

Kumar N, Gupta AK (2010) Wound-healing activity of *Onosma hispidum* (Ratanjot) in normal and diabetic rats. Int J Geogr Inf Syst 15(4):342–351. https://doi.org/10.1080/10496470903507924

Kumar MS, Sripriya R, Raghavan HV, Sehgal PK (2006) Wound healing potential of *Cassia fistula* on infected albino rat model. J Surg Res 131(2):283–289

Kumar N, Prakash D, Kumar P (2010) Wound healing activity of *Solanum xanthocarpum* Schrad. & Wendl. fruits. Indian J Nat Prod Resour 1(4):470–475

Kumar S, Rajput R, Patil V, Udupa AL, Gupta S, Rathnakar UP et al (2011a) Wound healing profile of *Asparagus racemosus* (Liliaceae) wild. Curr Pharm Res 1(2):111–114

Kumar H, Jain SK, Singh N, Dixit V, Singh P (2011b) Wound healing activity of the plant of *Baliospermum montanum* willd. Int J Pharm Sci Res 2(4):1073–1076

Kumari M, Eesha BR, Amberkar M, Kumar N (2010) Wound healing activity of aqueous extract of *Crotalaria verrucosa* in Wistar albino rats. Asian Pac J Trop Med 3(10):783–787

Kumari KS, Shivakrishna P, Ganduri VR (2020) Wound healing of the bioactive compounds from micrococcus sp. ous9 isolated from marine waters. Saudi J Biol Sci 27(9):2398–2402. https://doi.org/10.1016/j.sjbs.2020.05.007

Kundu A, Ghosh A, Singh NK, Singh GK, Seth A, Maurya SK et al (2016) Wound healing activity of the ethanol root extract and polyphenolic rich fraction from *Potentilla fulgens*. Pharm Biol 54 (11):2383–2393. https://doi.org/10.3109/13880209.2016.1157192

Kuppast IJ, Nayak PV (2006) Wound healing activity of *Cordia dichotoma* Forst. F fruits. Nat Prod Radiance 5(2):99–102

Lai HY, Lim YY, Kim KH (2011) Potential dermal wound healing agent in *Blechnum orientale* Linn. BMC Complement Altern Med 11:62. https://doi.org/10.1186/1472-6882-11-62

Laut M, Ndaong NA, Utami T (2019) Cutaneous wound healing activity of herbal ointment containing the leaf extract of *Acalypha indica* L. on mice (*Mus musculus*). Int J Phys Conf Ser 1146:012025

Li XQ, Kang R, Huo JC, Xie YH, Wang SW, Cao W (2017) Wound-healing activity of *Zanthoxylum bungeanum* maxim seed oil on experimentally burned rats. Pharmacogn Mag 13 (51):363–371. https://doi.org/10.4103/pm.pm_211_16

Lien LT, Tho NT, Ha DM, Hang PL, Nghia PT, Thang ND (2015) Influence of phytochemicals in piper *Betle linn* leaf extract on wound healing. Burns Trauma 3:s41038-015-0023-7. https://doi.org/10.1186/s41038-015-0023-7

Lin Y, Shi R, Wang X, Shen HM (2008) Luteolin, a flavonoid with potential for cancer prevention and therapy. Curr Cancer Drug Targets 8(7):634–646. https://doi.org/10.2174/156800908786241050

Liu RH (2013) Health-promoting components of fruits and vegetables in the diet. Adv Nutr 4 (3):384S–392S. https://doi.org/10.3945/an.112.003517

Lodhi S, Singhai AK (2013) Wound healing effect of flavonoid rich fraction and luteolin isolated from *Martynia annua* Linn. on streptozotocin induced diabetic rats. Asian Pac J Trop Med 6 (4):253–259. https://doi.org/10.1016/S1995-7645(13)60053-X

Lodhi S, Pawar RS, Jain AP, Singhai AK (2006) Wound healing potential of *Tephrosia purpurea* (Linn.) Pers. in rats. J Ethnopharmacol 108(2):204–210

Lodhi S, Jain A, Jain AP, Pawar RS, Singhai AK (2016) Effects of flavonoids from *Martynia annua* and *Tephrosia purpurea* on cutaneous wound healing. Avicenna J Phytomed 6(5):578–591

López López LI, Nery Flores SD, Silva Belmares SY, Sáenz GA (2014) Naphthoquinones: biological properties and synthesis of lawsone and derivatives-a structured review. Vitae 21 (3):248–258

Lordani TVA, de Lara CE, FBP F, de Souza Terron Monich M, Mesquita da Silva C, Felicetti Lordani CR et al (2018) Therapeutic effects of medicinal plants on cutaneous wound healing in humans: a systematic review. Mediators Inflamm 2018:7354250. https://doi.org/10.1155/2018/7354250

Maan P, Yadav KS, Yadav NP (2017) Wound healing activity of *Azadirachta indica* A. Juss stem bark in mice. Pharmacogn Mag 13(Suppl 2):S316–S320. https://doi.org/10.4103/0973-1296.210163

Mageswari S, Karpagam S, Reddy GA (2012) Evaluation of wound healing activity of the plant *Carmona retusa* (Vahl.) Masam., in mice. Int J Intell Inf Technol 4:1–4

Mahmood AA, Sidik K, Salmah I (2005) Wound healing activity of *Carica papaya* L. aqueous leaf extract in rats. Int J Mol Med Adv Sci 1(4):398–401

Majumder P (2012) Evaluation of wound healing potential of crude extracts of *Zyziphus oenolpia* l. Mill (Indian jujuba) in wistar rats. Int J Phytomed 4(3):419–428

Malviya N, Jain S (2009) Wound healing activity of aqueous extract of *Radix paeoniae* root. Acta Pol Pharm 66(5):543–547

Manconi M, Manca ML, Caddeo C, Cencetti C, di Meo C, Zoratto N, Nacher A, Fadda AM, Matricardi P (2018) Preparation of gellan-cholesterol nanohydrogels embedding baicalin and evaluation of their wound healing activity. Eur J Pharm Biopharm 127:244–249. https://doi.org/10.1016/j.ejpb.2018.02.015

Mandawgade SD, Patil KS (2003) Wound healing potential of some active principles of *Lawsonia alba* Lam. leaves. Indian J Pharm Sci 65(4):390–394

Manjunatha BK, Vidya SM, Rashmi KV, Mankani KL, Shilpa HJ, Singh SJ (2005) Evaluation of wound-healing potency of *Vernonia arborea* Hk. Indian J Pharmacol 37:223–226. https://doi.org/10.4103/0253-7613.16567

Manjunatha BK, Vidya SM, Krishna V, Mankani KL (2006) Wound healing activity of *Leucas hirta*. Indian J Pharm Sci 68(3):380–384

Manjunatha BK, Vidya SM, Krishna V, Mankani KL, Singh SD, Manohara YN (2007a) Comparative evaluation of wound healing potency of *Vitex trifolia* L. and *Vitex altissima* L. Phytother Res 21(5):457–461. https://doi.org/10.1002/ptr.2094

Manjunatha BK, Krishna V, Vidya SM, Mankani KL, Manohara YN (2007b) Wound healing activity of *Lycopodium serratum*. Indian J Pharm Sci 69(2):283. https://doi.org/10.4103/0250-474X.33159

Mann A, Ajiboso OS, Ajeigbe S, Gbate M, Isaiah S (2011) Evaluation of the wound healing activity of ethanol extract of *Terminalia avicennioides* root bark on two wound models in rats. Int J Med Arom Plants 1(2):95–100

Manoj GS, Murugan K (2012) Wound healing activity of methanolic and aqueous extracts of *Plagiochila beddomei* Steph. thallus in rat model. Indian J Exp Biol 50:551–558

Manzuoerh R, Farahpour MR, Oryan A, Sonboli A (2019) Effectiveness of topical administration of *Anethum graveolens* essential oil on MRSA-infected wounds. Biomed Pharmacother 109:1650–1658. https://doi.org/10.1016/j.biopha.2018.10.117

Mathew A, Taranalli AD, Torgal SS (2004) Evaluation of anti-inflammatory and wound healing activity of *Gentiana lutea* rhizome extracts in animals. Pharm Biol 42(1):8–12. https://doi.org/10.1080/13880200390502883

Mechesso AF, Tadese A, Tesfaye R, Tamiru W, Eguale T (2016) Experimental evaluation of wound healing activity of *Croton macrostachyus* in rat. Afr J Pharm Pharmacol 10 (39):832–838. https://doi.org/10.5897/AJPP2015.4454

Mehmood T, Maryam A, Ghramh HA, Khan M, Ma T (2017) Deoxyelephantopin and isodeoxyelephantopin as potential anticancer agents with effects on multiple signaling pathways. Molecules 22(6):1013. https://doi.org/10.3390/molecules22061013

Miladiyah I, Prabowo BR (2012) Ethanolic extract of *Anredera cordifolia* (Ten.) Steenis leaves improved wound healing in guinea pigs. Univ Med 31(1):4–11

Mo J, Panichayupakaranant P, Kaewnopparat N, Nitiruangjaras A, Reanmongkol W (2014) Wound healing activities of standardized pomegranate rind extract and its major antioxidant ellagic acid in rat dermal wounds. J Nat Med 68(2):377–386. https://doi.org/10.1007/s11418-013-0813-9

Moghadam SE, Ebrahimi SN, Salehi P, Moridi Farimani M, Hamburger M, Jabbarzadeh E (2017) Wound healing potential of chlorogenic acid and myricetin-3-O-β-rhamnoside isolated from *Parrotia persica*. Molecules 22(9):1501. https://doi.org/10.3390/molecules22091501

Mohan SC, Sasikala K, Anand T, Vengaiah PC, Krishnaraj S (2014) Antimicrobial and wound healing potential of *Canthium coromandelicum* leaf extract-a preliminary study. Res J Phytochem 8(2):35–41. https://doi.org/10.3923/rjphyto.2014.35.41

Mohanty C, Das M, Sahoo SK (2012) Sustained wound healing activity of curcumin loaded oleic acid based polymeric bandage in a rat model. Mol Pharm 9(10):2801–2811. https://doi.org/10.1021/mp300075u

Mondal S, Suresh P (2012) Wound healing activity of *Cleome rutidosperma* DC. roots. Int Curr Pharm J 1(6):151–154

Moshi MJ (2005) Current and future prospects of integrating traditional and alternative medicine in the management of diseases in Tanzania. Tanzan Health Res Bull 7(3):159–167. https://doi.org/10.4314/thrb.v7i3.14254

Moura-Letts G, Villegas LF, Marçalo A, Vaisberg AJ, Hammond GB (2006) *In vivo* wound-healing activity of oleanolic acid derived from the acid hydrolysis of *Anredera diffusa*. J Nat Prod 69 (6):978–979. https://doi.org/10.1021/np0601152

Mughrabi FF, Hashim H, Mahmood AA, Suzy SM, Salmah I, Zahra AA et al (2014) Acceleration of wound healing activity by *Polygonatum odoratum* leaf extract in rats. J Med Plant Res 8 (13):523–528. https://doi.org/10.5897/JMPR10.451

Mukherjee PK, Verpoorte R, Suresh B (2000) Evaluation of *in-vivo* wound healing activity of *Hypericum patulum* (Family: hypericaceae) leaf extract on different wound model in rats. J Ethnopharmacol 70(3):315–321. https://doi.org/10.1016/s0378-8741(99)00172-5

Mukherjee H, Ojha D, Bharitkar YP, Ghosh S, Mondal S, Kaity S et al (2013) Evaluation of the wound healing activity of *Shorea robusta*, an Indian ethnomedicine, and its isolated constituent (s) in topical formulation. J Ethnopharmacol 149(1):335–343. https://doi.org/10.1016/j.jep.2013.06.045

Mukherjee PK, Bahadur S, Chaudhary SK, Harwansh RK, Nema NK (2015) Validation of medicinal herbs for skin aging. In: Evidence-based validation of herbal medicine. Elsevier, pp 119–147

Mulisa E, Asres K, Engidawork E (2015) Evaluation of wound healing and anti-inflammatory activity of the rhizomes of *Rumex abyssinicus* J. (Polygonaceae) in mice. BMC Complement Altern Med 15(1):341. https://doi.org/10.1186/s12906-015-0878-y

Murthy S, Gautam MK, Goel S, Purohit V, Sharma H, Goel RK (2013) Evaluation of *in vivo* wound healing activity of *Bacopa monniera* on different wound model in rats. Biomed Res Int 2013. https://doi.org/10.1155/2013/972028

Murti K, Kumar U (2012) Enhancement of wound healing with roots of *Ficus racemosa* L. in albino rats. Asian Pac J Trop Biomed 2(4):276–280. https://doi.org/10.1016/S2221-1691(12)60022-7

Muthachan T, Tewtrakul S (2019) Anti-inflammatory and wound healing effects of gel containing *Kaempferia marginata* extract. J Ethnopharmacol 240:111964. https://doi.org/10.1016/j.jep. 2019.111964

Nabi B, Rehman S, Baboota S, Ali J (2019) Natural antileprotic agents: a boon for the management of leprosy. In: Discovery and development of therapeutics from natural products against neglected tropical diseases. Elsevier, Amsterdam, pp 351–372

Nagar HK, Srivastava AK, Srivastava R, Kurmi ML, Chandel HS, Ranawat MS (2016) Pharmacological investigation of the wound healing activity of *Cestrum nocturnum* (L.) ointment in Wistar albino rats. J Pharm 2016:9249040. https://doi.org/10.1155/2016/9249040

Nagoor Meeran MF, Goyal SN, Suchal K, Sharma C, Patil CR, Ojha SK (2018) Pharmacological properties, molecular mechanisms, and pharmaceutical development of asiatic acid: a pentacyclic triterpenoid of therapeutic promise. Front Pharmacol 9:892. https://doi.org/10. 3389/fphar.2018.00892

Naika HR, Bhavana S, Da Silva JA, Lingaraju K, Mohan VC, Krishna V (2016) *In silico* and *in vivo* wound healing studies of ursolic acid isolated from *Clematis gouriana* against GSK-3 beta. Nus Biosci 8(2):232–244. https://doi.org/10.13057/nusbiosci/n080216

Nair S, Nair M, Nair D, Juliet S, Padinchareveetil S, Samraj S et al (2014) Wound healing, anti-inflammatory activity and toxicological studies of *Leea Asiatica* (L.) Ridsdale. Int J Biol Pharm Res 5(9):745–749

Naji S, Zarei L, Pourjabali M, Mohammadi R (2017) The extract of *Lycium depressum* stocks enhances wound healing in streptozotocin-induced diabetic rats. Int J Low Extrem Wounds 16 (2):85–93

Nayak BS (2006) *Cecropia peltata* L (Cecropiaceae) has wound-healing potential: a preclinical study in a Sprague Dawley rat model. Int J Low Extrem Wounds 5(1):20–26. https://doi.org/10. 1177/1534734606286472

Nayak BS, Udupa AL, Udupa SL (1999) Effect of *Ixora coccinea* flowers on dead space wound healing in rats. Fitoterapia 70(3):233–236. https://doi.org/10.1016/S0367-326X(99)00025-8

Nayak BS, Vinutha B, Geetha B, Sudha B (2005a) Experimental evaluation of *Pentas lanceolata* flowers for wound healing activity in rats. Fitoterapia 76(7-8):671–675. https://doi.org/10.1016/ j.fitote.2005.08.007

Nayak BS, Suresh R, Rao AV, Pillai GK, Davis EM, Ramkissoon V et al (2005b) Evaluation of wound healing activity of *Vanda roxburghii* R. Br (Orchidacea): a preclinical study in a rat model. Int J Low Extrem Wounds 4(4):200–204. https://doi.org/10.1177/1534734605282994

Nayak S, Nalabothu P, Sandiford S, Bhogadi V, Adogwa A (2006) Evaluation of wound healing activity of *Allamanda cathartica* L and *Laurus nobilis* L extracts on rats. BMC Complement Altern Med 6:12. https://doi.org/10.1186/1472-6882-6-12

Nayak BS, Anderson M, Pereira LP (2007a) Evaluation of wound-healing potential of *Catharanthus roseus* leaf extract in rats. Fitoterapia 78(7–8):540–544. https://doi.org/10.1016/ j.fitote.2007.06.008

Nayak BS, Isitor G, Davis EM, Pillai GK (2007b) The evidence based wound healing activity of *Lawsonia inermis* Linn. Phytother Res 21(9):827–831. https://doi.org/10.1002/ptr.2181

Nayak BS, Raju SS, Rao AV (2007c) Wound healing activity of *Matricaria recutita* L. extract. J Wound Care 16(7):298–302. https://doi.org/10.12968/jowc.2007.16.7.27061

Nayak BS, Sandiford S, Maxwell A (2009a) Evaluation of the wound-healing activity of ethanolic extract of *Morinda citrifolia* L leaf. Evid Based Complement Alternat Med 6(3):351–356

Nayak BS, Raju SS, Eversley M, Ramsubhag A (2009b) Evaluation of wound healing activity of *Lantana camara* L.-a preclinical study. Phytother Res 23(2):241–245. https://doi.org/10.1002/ ptr.2599

Nayak BS, Kanhai J, Milne DM, Swanston WH, Mayers S, Eversley M et al (2010a) Investigation of the wound healing activity of *Carapa guianensis* L.(Meliaceae) bark extract in rats using excision, incision, and dead space wound models. J Med Food 13(5):1141–1146. https://doi.org/10.1089/jmf.2009.0214

Nayak BS, Marshall JR, Isitor G (2010b) Wound healing potential of ethanolic extract of *Kalanchoe pinnata* Lam. Leaf-a preliminary study. Indian J Exp Biol 48:572–576

Nayak BS, Ramdath DD, Marshall JR, Isitor GN, Eversley M, Xue S et al (2010c) Wound-healing activity of the skin of the common grape (*Vitis Vinifera*) variant, Cabernet Sauvignon. Phytother Res 24(8):1151–1157. https://doi.org/10.1002/ptr.2999

Nayak BS, Ramlogan S, Chalapathi Rao A, Maharaj S (2014) *Neurolaena lobata* L. promotes wound healing in Sprague Dawley rats. Int J Appl Basic Med Res 4(2):106–110. https://doi.org/10.4103/2229-516X.136791

Nguyen VL, Truong CT, Nguyen BCQ, Vo TV, Dao TT, Nguyen VD et al (2017) Anti-inflammatory and wound healing activities of calophyllolide isolated from *Calophyllum inophyllum* Linn. PLoS One 12(10):e0185674. https://doi.org/10.1371/journal.pone.0185674

Nikita G, Vivek P, Chhaya G (2015) Wound-healing activity of an oligomer of alkannin/shikonin, isolated from root bark of *Onosma echioides*. Nat Prod Res 29(16):1584–1588. https://doi.org/10.1080/14786419.2014.986126

Nilforoushzadeh MA, Javanmard SH, Ghanadian M, Asghari G, Jaffary F, Yakhdani AF et al (2014) The effects of *Adiantum capillus-veneris* on wound healing: an experimental *in vitro* evaluation. Int J Prev Med 5(10):1261–1268

Nirmala S, Karthiyayini T (2011) Wound healing activity on the leaves of *Achillea millefolium* L. by excision, incision, and dead space model on adult Wistar albino rats. Int Res J Pharm 2 (3):240–245

Nugroho RA, Utami D, Aryani R, Nur FM, Sari YP, Manurung H (2019) *In vivo* wound healing activity of ethanolic extract of *Terminalia catappa* L. leaves in mice (*Mus musculus*). J Phys 1277(1):012031

Ogurtan Z, Hatipoglu F, Ceylan C (2002) The effect of *Alkanna tinctoria* Tausch on burn wound healing in rabbits. Dtsch Tierarztl Wochenschr 109(11):481–485

Okur ME, Ayla Ş, Çiçek Polat D, Günal MY, Yoltaş A, Biçeroğlu Ö (2018) Novel insight into wound healing properties of methanol extract of *Capparis ovata* Desf. var. *palaestina Zohary* fruits. J Pharm Pharmacol 70(10):1401–1413. https://doi.org/10.1111/jphp.12977

Olugbuyiro JA, Abo KA, Leigh OO (2010) Wound healing effect of *Flabellaria paniculata* leaf extracts. J Ethnopharmacol 127(3):786–788. https://doi.org/10.1016/j.jep.2009.10.008

Öz BE, Ilhan M, Ozbilgin S, Akkol EK, Acikara ÖB, Saltan G et al (2016) Effects of *Alchemilla mollis* and *Alchemilla persica* on the wound healing process. Bangladesh J Pharmacol 11 (3):577–584. https://doi.org/10.3329/bjp.v11i3.26024

Ozay Y, Kasim Cayci M, Guzel-Ozay S, Cimbiz A, Gurlek-Olgun E, Sabri OM (2013) Effects of *Equisetum arvense* ointment on diabetic wound healing in rats. Wounds 25(9):234–241

Özay Y, Güzel S, Yumrutaş Ö, Pehlivanoğlu B, Erdoğdu İH, Yildirim Z, Türk BA, Darcan S (2019) Wound healing effect of kaempferol in diabetic and nondiabetic rats. J Surg Res 233:284–296. https://doi.org/10.1016/j.jss.2018.08.009

Ozdemir O, Ozkan K, Hatipoglu F, Uyaroglu A, Arican M (2016) Effect of asiaticoside, collagenase, and alpha-chymotrypsin on wound healing in rabbits. Wounds 28(8):279–286

Öztürk N, Korkmaz S, Öztürk Y (2007) Wound-healing activity of St. John's Wort (*Hypericum perforatum* L.) on chicken embryonic fibroblasts. J Ethnopharmacol 111(1):33–39

Paarakh PM, Chansouria JP, Khosa RL (2009) Wound healing activity of *Annona muricata* extract. J Pharm Res 2(3):404–406

Padilla-Camberos E, Flores-Fernández JM, Canales-Aguirre AA, Barragán-Álvarez CP, Gutiérrez-Mercado Y, Lugo-Cervantes E (2016) Wound healing and antioxidant capacity of *Musa paradisiaca* Linn. peel extracts. J Pharm Pharmacogn Res 4(5):165–173

Paiva LA, de Alencar Cunha KM, Santos FA, Gramosa NV, Silveira ER, Rao VS (2002) Investigation on the wound healing activity of oleo-resin from *Copaifera langsdorffi* in rats. Phytother Res 16(8):737–739

Palanimuthu P, Nandagopal S, Jalaludeen MD, Subramonian K, Ganthi AS, Sankaram S (2011) Wound healing activities of *Eugenia jambolana* lam. Bark extracts in albino rats. Int J App Biol Pharm Technol 2(1):112–116

Panda V, Sonkamble M, Patil S (2011) Wound healing activity of *Ipomoea batatas* tubers (sweet potato). Funct Food Health Dis 1(10):403–415

Panduraju T, Bitra VR, Vemula SK, Reddy PR (2010) Wound healing activity of *Dioscorea bulbifera* Linn. J Pharm Res 3(12):3138–3139

Pang Y, Wang D, Hu X, Wang H, Fu W, Fan Z et al (2014) Effect of volatile oil from *Blumea Balsamifera* (L.) DC. leaves on wound healing in mice. J Tradit Chin Med 34(6):716–724. https://doi.org/10.1016/s0254-6272(15)30087-x

Papageorgiou VP, Assimopoulou AN, Ballis AC (2008) Alkannins and shikonins: a new class of wound healing agents. Curr Med Chem 15(30):3248–3267. https://doi.org/10.2174/092986708786848532

Paramesha M, Ramesh CK, Krishna V, Swamy HM, Rao SA, Hoskerri J (2015) Effect of dehydroabietylamine in angiogenesis and GSK3-β inhibition during wound healing activity in rats. Med Chem Res 24(1):295–303. https://doi.org/10.1007/s00044-014-1110-1

Parasuraman S (2018) Herbal drug discovery: challenges and perspectives. Curr Pharm Person Med 16(1):63–68. https://doi.org/10.2174/1875692116666180419153313

Parasuraman S, Perumal P (2020) Siddha, an indigenous medical system of peninsular India. In: Herbal medicine in India. Springer, Singapore, pp 9–21. https://doi.org/10.1007/978-981-13-7248-3_2

Parasuraman S, Zhen KM, Wen LE, Hean CK, Balamurugan S, Christapher PV, Banik U (2020) Effect of ursolic acid on olanzapine induced weight gain in Sprague Dawley rats. Indian J Exp Biol 58(11):760–769

Park EH, Chun MJ (2001) Wound healing activity of *Opuntia ficus-indica*. Fitoterapia 72 (2):165–167. https://doi.org/10.1016/S0367-326X(00)00265-3

Park JH, Choi SH, Park SJ, Lee YJ, Park JH, Song PH et al (2017) Promoting wound healing using low molecular weight fucoidan in a full-thickness dermal excision rat model. Mar Drugs 15 (4):112. https://doi.org/10.3390/md15040112

Paschapur MS, Patil MB, Kumar R, Patil SR (2009) Evaluation of aqueous extract of leaves of *Ocimum kilimandscharicum* on wound healing activity in albino wistar rats. Int J PharmTech Res 1:544–550

Patil KS, Jadhav AG, Joshi VS (2005) Wound healing activity of leaves of *Artocarpus heterophyllus*. Indian J Pharm Sci 67(5):629–632

Pattanayak SP, Sunita P (2008) Wound healing, anti-microbial and antioxidant potential of *Dendrophthoe falcata* (Lf) Ettingsh. J Ethnopharmacol 120(2):241–247. https://doi.org/10.1016/j.jep.2008.08.019

Pattanayak S, Das P, Mandal TK, Debnath PK, Bandyopadhyay SK (2015) A study on comparative antimicrobial and wound healing efficacy of solvent extracts and succulent leaf extract of *Mikania scandens* (L.) Willd. Am J Phytomed Clinic Therapeut 3(4):346–362

Pawar RS, Chaurasiya PK, Rajak H, Singour PK, Toppo FA, Jain A (2013) Wound healing activity of *Sida cordifolia* Linn. in rats. Indian J Pharmacol 45(5):474–478. https://doi.org/10.4103/0253-7613.117759

Perez Gutierrez RM, Vargas SR (2006) Evaluation of the wound healing properties of *Acalypha langiana* in diabetic rats. Fitoterapia 77(4):286–289. https://doi.org/10.1016/j.fitote.2006.03.011

Perez GRM, Vargas SR, Ortiz HYD (2005) Wound healing properties of *Hylocereus undatus* on diabetic rats. Phytother Res 19(8):665–668. https://doi.org/10.1002/ptr.1724

Perumal Samy R, Gopalakrishnakone P, Sarumathi M, Ignacimuthu S (2006) Wound healing potential of *Tragia involucrata* extract in rats. Fitoterapia 77(4):300–302. https://doi.org/10.1016/j.fitote.2006.04.001

Pessoa WS, Estevão LR, Simões RS, Barros ME, Mendonça FD, Baratella-Evêncio L et al (2012) Effects of angico extract (*Anadenanthera colubrina* var. cebil) in cutaneous wound healing in rats. Acta Cir Bras 27(10):655–670

Phan TT, Hughes MA, Cherry GW, Le TT, Pham HM (1996) An aqueous extract of the leaves of *Chromolaena odorata* (formerly *Eupatorium odoratum*) (Eupolin) inhibits hydrated collagen lattice contraction by normal human dermal fibroblasts. J Altern Complement Med 2 (3):335–343. https://doi.org/10.1089/acm.1996.2.335

Pinto NC, Cassini-Vieira P, Souza-Fagundes EM, Barcelos LS, Castañon MC, Scio E (2016) *Pereskia aculeata* Miller leaves accelerate excisional wound healing in mice. J Ethnopharmacol 194:131–136. https://doi.org/10.1016/j.jep.2016.09.005

Pirbalouti AG, Koohpyeh A (2011) Wound healing activity of extracts of *Malva sylvestris* and *Stachys lavandulifolia*. Int J Biol 3(1):174–179

Pirbalouti AG, Yousefi M, Nazari H, Karimi I, Koohpayeh A (2009) Evaluation of burn healing properties of *Arnebia euchroma* and *Malva sylvestris*. Electron J Biol 5(3):62–66

Pirbalouti AG, Koohpayeh A, Karimi I (2010) The wound healing activity of flower extracts of *Punica granatum* and *Achillea kellalensis* in Wistar rats. Acta Pol Pharm 67(1):107–110

Pitz HD, Pereira A, Blasius MB, Voytena AP, Affonso RC, Fanan S et al (2016) *In vitro* evaluation of the antioxidant activity and wound healing properties of Jaboticaba (*Plinia peruviana*) fruit peel hydroalcoholic extract. Oxidative Med Cell Longev 2016. https://doi.org/10.1155/2016/3403586

Polerà N, Badolato M, Perri F, Carullo G, Aiello F (2019) Quercetin and its natural sources in wound healing management. Curr Med Chem 26(31):5825–5848. https://doi.org/10.2174/0929867325666180713150626

Ponrasu T, Madhukumar KN, Ganeshkumar M, Iyappan K, Sangeethapriya V, Gayathri VS et al (2014) Efficacy of *Acorus calamus* on collagen maturation on full thickness cutaneous wounds in rats. Pharmacogn Mag 10(Suppl 2):S299–S305. https://doi.org/10.4103/0973-1296.133283

Porras-Reyes BH, Lewis WH, Roman J, Simchowitz L, Mustoe TA (1993) Enhancement of wound healing by the alkaloid taspine defining mechanism of action. Proc Soc Exp Biol Med 203 (1):18–25. https://doi.org/10.3181/00379727-203-43567

Pourali P, Yahyaei B (2019) The healing property of a bioactive wound dressing prepared by the combination of bacterial cellulose (BC) and *Zingiber officinale* root aqueous extract in rats. 3 Biotech 9(2):59. https://doi.org/10.1007/s13205-019-1588-9

Prabhakar KR, Srinivasan KK, Rao PG (2002) Chemical investigation, anti-inflammatory and wound healing properties of *Coronopus didymus*. Pharm Biol 40(7):490–493. https://doi.org/10.1076/phbi.40.7.490.14684

Prabu D, Nappinnai M, Ponnudurai K, Prabhu K (2008) Evaluation of wound-healing potential of *Pisonia grandis* R. Br: a preclinical study in Wistar rats. Int J Low Extrem Wounds 7(1):21–27. https://doi.org/10.1177/1534734607314051

Prakashbabu BC, Vijay D, George S, Kodiyil S, Nair SN, Gopalan AK et al (2017) Wound healing and anti-inflammatory activity of methanolic extract of *Gmelina arborea* and *Hemigraphis colorata* in rats. Int J Curr Microbiol App Sci 6(8):3116–3122

Prasad V, Jain V, Girish D, Dorle AK (2006) Wound-healing property of *Momordica charantia* L. fruit powder. J Herb Pharmacother 6(3-4):105–115. https://doi.org/10.1080/J157v06n03_05

Pravez M, Patel AK (2014) Wound healing activity of ursolic acid stearoyl glucoside (uasg) isolated from *Lantana camara* l. Int J Pharm Sci Res 5(10):4439–4444

Preethi KC, Kuttan R (2009) Wound healing activity of flower extract of *Calendula officinalis*. J Basic Clin Physiol Pharmacol 20(1):73–79

Primarizky H, Yuniarti WM, Lukiswanto BS (2017) Ellagic acid activity in healing process of incision wound on male albino rats (*Rattus norvegicus*). KnE Life Sci 3:224–233. https://doi.org/10.18502/kls.v3i6.1131

Priya KS, Arumugam G, Rathinam B, Wells A, Babu M (2004) *Celosia argentea* Linn. leaf extract improves wound healing in a rat burn wound model. Wound Repair Regen 12(6):618–625

Puratchikody A, Devi CN, Nagalakshmi G (2006) Wound healing activity of *Cyperus rotundus* linn. Indian J Pharm Sci 68(1):97–101. https://doi.org/10.4103/0250-474X.22976

Purohit SK, Solanki R, Mathur V, Mathur M (2013) Evaluation of wound healing activity of ethanolic extract of *Curcuma longa* rhizomes in male albino rats. Asian J Pharm Biol Res 3 (2):79–81

Rahman N, Rahman H, Haris M, Mahmood R (2017) Wound healing potentials of *Thevetia peruviana*: Antioxidants and inflammatory markers criteria. J Tradit Complement Med 7 (4):519–525. https://doi.org/10.1016/j.jtcme.2017.01.005

Rajan DS, Rajkumar M, Kumarappan CT, Kumar KS (2013) Wound healing activity of an herbal ointment containing the leaf extract of *Ziziphus Mauritiana* Lam. Afr J Pharm Pharmacol 7 (4):98–103. https://doi.org/10.5897/AJPP12.795

Rajasekaran A, Sivakumar V, Darlinquine S (2012) Evaluation of wound healing activity of *Ammannia baccifera* and *Blepharis maderaspatensis* leaf extracts on rats. Rev Bras Farmacogn 22(2):418–427. https://doi.org/10.1590/S0102-695X2011005000207

Ramazanov NS, Bobayev ID, Yusupova UY, Aliyeva NK, Egamova FR, Yuldasheva NK et al (2017) Phytoecdysteroids-containing extract from *Stachys hissarica* plant and its wound-healing activity. Nat Prod Res 31(5):593–597. https://doi.org/10.1080/14786419.2016.1205058

Rashed AN, Afifi FU, Disi AM (2003) Simple evaluation of the wound healing activity of a crude extract of *Portulaca oleracea* L.(growing in Jordan) in *Mus musculus* JVI-1. J Ethnopharmacol 88(2-3):131–136. https://doi.org/10.1016/S0378-8741(03)00194-6

Rates SM (2001) Plants as source of drugs. Toxicon 39(5):603–613. https://doi.org/10.1016/s0041-0101(00)00154-9

Rathi BS, Bodhankar SL, Baheti AM (2006) Evaluation of aqueous leaves extract of *Moringa oleifera* Linn for wound healing in albino rats. Indian J Exp Biol 44:898–901

Reddy BS, Reddy RK, Naidu VG, Madhusudhana K, Agwane SB, Ramakrishna S et al (2008) Evaluation of antimicrobial, antioxidant and wound-healing potentials *of Holoptelea integrifolia*. J Ethnopharmacol 115(2):249–256. https://doi.org/10.1016/j.jep.2007.09.031

Ren J, Lu Y, Qian Y, Chen B, Wu T, Ji G (2019) Recent progress regarding kaempferol for the treatment of various diseases. Exp Ther Med 18(4):2759–2776. https://doi.org/10.3892/etm.2019.7886

Rezaei M, Dadgar Z, Noori-Zadeh A, Mesbah-Namin SA, Pakzad I, Davodian E (2015) Evaluation of the antibacterial activity of the *Althaea officinalis* L. leaf extract and its wound healing potency in the rat model of excision wound creation. Avicenna J Phytomed 5(2):105–112

Romero-Cerecero O, Zamilpa A, Díaz-García ER, Tortoriello J (2014) Pharmacological effect of *Ageratina pichinchensis* on wound healing in diabetic rats and genotoxicity evaluation. J Ethnopharmacol 156:222–227. https://doi.org/10.1016/j.jep.2014.09.002

Roosewelt C, Vincent S, Sujith K, Darwin CR (2011) Wound healing activity of methanolic extract of Vitex negundo leaves in albino Wistar rats. J Pharm Res 4(8):2553–2555

Roshan S, Ali S, Khan A, Tazneem B, Purohit MG (2008) Wound Healing activity of *Abutilon Indicum*. Pharmacogn Magaz 4(15):s85–s88

Roy K, Shivakumar H, Sarkar S (2009) Wound healing potential of leaf extracts of *Ficus religiosa* on Wistar albino strain rats. Int J Pharm Tech Res 1:506–508

Roy SK, Mishra PK, Nandy S, Datta R, Chakraborty B (2012) Potential wound healing activity of the different extract of *Typhonium trilobatum* in albino rats. Asian Pac J Trop Biomed 2(3): S1477–S1486. https://doi.org/10.1016/S2221-1691(12)60441-9

Sabharwal S, Aggarwal S, Vats M, Sardana S (2012) Preliminary phytochemical investigation and wound healing activity of *Jasminum sambac* (Linn) Ait.(Oleaceae) leaves. Int J Pharmacogn Phytochem Res 2(3):146–150

Sagratini G, Cristalli G, Giardinà D, Gioventù G, Maggi F, Ricciutelli M et al (2008) Alkannin/ shikonin mixture from roots of *Onosma echioides* (L.) L.: extraction method study and quantification. J Sep Sci 31(6-7):945–952. https://doi.org/10.1002/jssc.200700408

Sahoo HB, Sagar R, Patel VK (2012) Wound healing activity of *Centratherum anthelminticum* Linn. Mol Clin Pharmacol 3:1–7

Saini NK, Singhal M, Srivastava B (2012) Evaluation of wound healing activity of *Tecomaria capensis* leaves. Chin J Nat Med 10(2):138–141. https://doi.org/10.3724/SP.J.1009.2012.00138

Sankar R, Dhivya R, Shivashangari KS, Ravikumar V (2014) Wound healing activity of *Origanum vulgare* engineered titanium dioxide nanoparticles in Wistar Albino rats. J Mater Sci Mater Med 25(7):1701–1708

Sanwal R, Chaudhary AK (2011) Wound healing and antimicrobial potential of *Carissa spinarum* Linn. in albino mice. J Ethnopharmacol 135(3):792–796. https://doi.org/10.1016/j.jep.2011.04.025

Sasidharan S, Nilawatyi R, Xavier R, Latha LY, Amala R (2010) Wound healing potential of *Elaeis guineensis* Jacq leaves in an infected albino rat model. Molecules 15(5):3186–3199. https://doi.org/10.3390/molecules15053186

Sathish R, Ahmed MH, Natarajan K, Lalitha KG (2010) Evaluation of wound healing and antimicrobial activity of *Lannea coromandelica* (Houtt) Merr. J Pharm Res 3(6):1225–1228

Sayyedrostami T, Pournaghi P, Vosta-Kalaee SE, Zangeneh MM (2018) Evaluation of the wound healing activity of *Chenopodium botrys* leaves essential oil in rats (a short-term study). J Essent Oil Bear Plants 21(1):164–174

Senapati AK, Giri RK, Panda DS, Satyanarayan S (2011) Wound healing potential of *Pterospermum acerifolium* wild. with induction of tumor necrosis factor-α. J Basic Clin Pharm 2(4):203–206

Seo DY, Lee SR, Heo JW, No MH, Rhee BD, Ko KS et al (2018) Ursolic acid in health and disease. Korean J Physiol Pharmacol 22(3):235–248. https://doi.org/10.4196/kjpp.2018.22.3.235

Sevimli-Gür C, Onbaşılar I, Atilla P, Genç R, Cakar N, Deliloğlu-Gürhan I et al (2011) *In vitro* growth stimulatory and *in vivo* wound healing studies on cycloartane-type saponins of *Astragalus genus*. J Ethnopharmacol 134(3):844–850. https://doi.org/10.1016/j.jep.2011.01.030

Sh Ahmed A, Taher M, Mandal UK, Jaffri JM, Susanti D, Mahmood S et al (2019) Pharmacological properties of Centella asiatica hydrogel in accelerating wound healing in rabbits. BMC Complement Altern Med 19(1):213. https://doi.org/10.1186/s12906-019-2625-2

Shakeel F, Alam P, Anwer MK, Alanazi SA, Alsarra IA, Alqarni MH (2019) Wound healing evaluation of self-nanoemulsifying drug delivery system containing *Piper cubeba* essential oil. 3 Biotech 9(3):1–9. https://doi.org/10.1007/s13205-019-1630-y

Shanbhag TV, Sharma C, Adiga S, Bairy LK, Shenoy S, Shenoy G (2006) Wound healing activity of alcoholic extract of *Kaempferia galanga* in Wistar rats. Indian J Physiol Pharmacol 50(4):384–390

Shanbhag T, Kodidela S, Shenoy S, Amuthan A, Kurra S (2011) Effect of *Michelia champaca* Linn flowers on burn wound healing in wistar rats. Int J Pharm Sci Rev Res 7(2):112–115

Sharma S, Sharma MC, Kohli DV (2010) Wound healing activity and formulation of etherbenzene-95% ethanol extract of herbal drug *Madhuca longifolia* leaves in albino rats. J Optoelectron Biomed Mater 1(1):13–15

Sheeba M, Emmanuel S, Revathi K, Ignacimuthu S (2009) Wound healing activity of *Cassia occidentalis* L. in albino Wistar rats. Int J Integr Biol 8(1):1–6

Shenoy C, Patil MB, Kumar R, Patil S (2009a) Preliminary phytochemical investigation and wound healing activity of *Allium cepa* Linn (Liliaceae). Int J Pharm Pharm Sci 2(2):167–175

Shenoy C, Patil MB, Kumar R (2009b) Antibacterial and wound healing activity of the leaves of *Annona squamosa* Linn.(Annonaceae). Res J Pharmacogn Phytochem 1(1):44–50

Shetty S, Udupa S, Udupa L (2008) Evaluation of antioxidant and wound healing effects of alcoholic and aqueous extract of *Ocimum sanctum* Linn in rats. Evid Based Complement Alternat Med 5(1):95–101

Shirwaikar A, Somashekar AP, Udupa AL, Udupa SL, Somashekar S (2003a) Wound healing studies of *Aristolochia bracteolata* Lam. with supportive action of antioxidant enzymes. Phytomedicine 10(6-7):558–562. https://doi.org/10.1078/094471103322331548

Shirwaikar A, Jahagirdar S, Udupa AL (2003b) Wound healing activity of *Desmodium triquetrum* leaves. Indian J Pharm Sci 65(5):461–464

Shirwaikar A, Shenoy R, Udupa AL, Udupa SL, Shetty S (2003c) Wound healing property of ethanolic extract of leaves of *Hyptis suaveolens* with supportive role of antioxidant enzymes. Indian J Exp Biol 41(3):238–241

Shiva Krishna P, Sudha S, Reddy KA, Al-Dhabaan FA, Meher, Prakasham RS et al (2019) Studies on wound healing potential of red pigment isolated from marine *Bacterium Vibrio* sp. Saudi J Biol Sci 26(4):723–729. https://doi.org/10.1016/j.sjbs.2017.11.035

Shivhare Y, Singour PK, Patil UK, Pawar RS (2010) Wound healing potential of methanolic extract of *Trichosanthes dioica* Roxb (fruits) in rats. J Ethnopharmacol 127(3):614–619. https://doi.org/10.1016/j.jep.2009.12.015

Shukla A, Rasik AM, Jain GK, Shankar R, Kulshrestha DK, Dhawan BN (1999) *In vitro* and *in vivo* wound healing activity of asiaticoside isolated from *Centella asiatica*. J Ethnopharmacol 65 (1):1–11. https://doi.org/10.1016/s0378-8741(98)00141-x

Silambujanaki P, Chandra CB, Kumar KA, Chitra V (2011) Wound healing activity of *Glycosmis arborea* leaf extract in rats. J Ethnopharmacol 134(1):198–201. https://doi.org/10.1016/j.jep.2010.11.046

Singh MP, Sharma CS (2009) Wound healing activity of *Terminalia chebula* in experimentally induced diabetic rats. Int J Pharmtech Res 1(4):1267–1270

Singh SD, Krishna V, Mankani KL, Manjunatha BK, Vidya SM, Manohara YN (2005) Wound healing activity of the leaf extracts and deoxyelephantopin isolated from *Elephantopus scaber* Linn. Indian J Pharm 37(4):238–242. https://doi.org/10.4103/0253-7613.16570

Singh M, Govindarajan R, Nath V, Rawat AK, Mehrotra S (2006) Antimicrobial, wound healing and antioxidant activity of *Plagiochasma appendiculatum* Lehm. et Lind. J Ethnopharmacol 107(1):67–72. https://doi.org/10.1016/j.jep.2006.02.007

Singh AK, Sharma A, Warren J, Madhavan S, Steele K, RajeshKumar NV et al (2007) Picroliv accelerates epithelialization and angiogenesis in rat wounds. Planta Med 73(3):251–256. https://doi.org/10.1055/s-2007-967119

Singh R, Thakur P, Semwal A, Kakar S (2014) Wound healing activity of leaf methanolic extract of *Ficus hispida* Linn. Afr J Pharm Pharmacol 8(1):21–23

Singhal A, Gupta H, Bhati V (2011) Wound healing activity of *Argyreia nervosa* leaves extract. Int J Appl Basic Med Res 1(1):36–39. https://doi.org/10.4103/2229-516X.81978

Sivagamy M, Jeganathan MR, Manavalan R, Senthamarai R (2012) Wound healing activity of *Indigofera enneaphylla* Linn. Int J Adv Pharm Chem Biol 1(2):211–214

Sofrona E, Tziveleka LA, Harizani M, Koroli P, Sfiniadakis I, Roussis V et al (2020) *In vivo* evaluation of the wound healing activity of extracts and bioactive constituents of the marine isopod *Ceratothoa oestroides*. Mar Drugs 18(4):219. https://doi.org/10.3390/md18040219

Somboonwong J, Kankaisre M, Tantisira B, Tantisira MH (2012) Wound healing activities of different extracts of *Centella asiatica* in incision and burn wound models: an experimental animal study. BMC Complement Altern Med 12:103. https://doi.org/10.1186/1472-6882-12-103

Srirod S, Tewtrakul S (2019) Anti-inflammatory and wound healing effects of cream containing *Curcuma mangga* extract. J Ethnopharmacol 238:111828. https://doi.org/10.1016/j.jep.2019.111828

Subha S, Ganthi SA, Hemalatha T (2017) Study of in vivo wound healing activity of *Dyschoriste littoralis* extract in rats models. RJPPD 9(4):181–185

Süntar IP, Akkol EK, Yalçin FN, Koca U, Keleş H, Yesilada E (2010) Wound healing potential of *Sambucus ebulus* L. leaves and isolation of an active component, quercetin 3-O-glucoside. J Ethnopharmacol 129(1):106–114. https://doi.org/10.1016/j.jep.2010.01.051

Süntar I, Koca U, Keleş H, Akkol EK (2011) Wound healing activity of *Rubus sanctus* Schreber (Rosaceae): preclinical study in animal models. Evid Based Complement Alternat Med 2011. https://doi.org/10.1093/ecam/nep137

Süntar I, Küpeli Akkol E, Keles H, Yesilada E, Sarker SD, Baykal T (2012) Comparative evaluation of traditional prescriptions from *Cichorium intybus* L. for wound healing: stepwise isolation of an active component by *in vivo* bioassay and its mode of activity. J Ethnopharmacol 143 (1):299–309. https://doi.org/10.1016/j.jep.2012.06.036

Süntar I, Küpeli Akkol E, Keles H, Yesilada E, Sarker SD (2013) Exploration of the wound healing potential of *Helichrysum graveolens* (Bieb.) Sweet: isolation of apigenin as an active component. J Ethnopharmacol 149(1):103–110. https://doi.org/10.1016/j.jep.2013.06.006

Suriyamoorthy S, Subramaniam K, Wahab F, Karthikeyan G (2012) Evaluation of wound healing activity of *Acacia leucophloea* bark in rats. Rev Bras Farmacogn 22(6):1338–1343. https://doi.org/10.1590/S0102-695X2012005000121

Suriyamoorthy S, Subramaniam K, Durai SJ, Wahaab F, Chitraselvi RP (2014) Evaluation of wound healing activity of *Acacia caesia* in rats. Wound Med 7:1–7. https://doi.org/10.1016/j.wndm.2015.03.001

Swamy HK, Krishna V, Shankarmurthy K, Rahiman BA, Mankani KL, Mahadevan KM et al (2007) Wound healing activity of embelin isolated from the ethanol extract of leaves of *Embelia ribes* Burm. J Ethnopharmacol 109(3):529–534. https://doi.org/10.1016/j.jep.2006.09.003

Sy GY, Nongonierma RB, Ngewou PW, Mengata DE, Dieye AM, Cisse A et al (2005) Activite cicatrisante de l'extrait methanolique des ecorces de *Spathodea campanulata* Beauv (Bagnoniaceae) sur un modele de brulures experimentales chez le rat [Healing activity of methanolic extract of the barks of *Spathodea campanulata* Beauv (Bignoniaceae) in rat experimental burn model]. Dakar Med 50(2):77–81. French

Tahvilian R, Zangeneh MM, Falahi H, Sadrjavadi K, Jalalvand AR, Zangeneh A (2019) Green synthesis and chemical characterization of copper nanoparticles using *Allium saralicum* leaves and assessment of their cytotoxicity, antioxidant, antimicrobial, and cutaneous wound healing properties. Appl Organomet Chem 33(12):e5234. https://doi.org/10.1002/aoc.5234

Talekar Y, Das B, Paul T, Talekar D, Apte K, Parab P (2012) Wound healing activity of aqueous and ethanolic extract of bark of *Emblica officinalis* in Wistar rats. Inventi Impact Planta Activa 4:1–5

Tamri P, Hemmati A, Amirahmadi A, Zafari J, Mohammadian B, Dehghani M et al (2016) Evaluation of wound healing activity of hydroalcoholic extract of banana (*Musa acuminata*) fruit's peel in rabbit. Pharmacologyonline 3:203–208

Tang T, Yin L, Yang J, Shan G (2007) Emodin, an anthraquinone derivative from *Rheum officinale* Baill, enhances cutaneous wound healing in rats. Eur J Pharmacol 567(3):177–185. https://doi.org/10.1016/j.ejphar.2007.02.033

Taqa GA, Mustafa EA, Al-Haliem SM (2014) Evaluation of anti-bacterial and efficacy of plant extract (*Urtica urens*) on skin wound healing in rabbit. Int J Enhanc Res Sci Technol Eng 3 (1):64–70

Taranalli AD, Kuppast IJ (1996) Study of wound healing activity of seeds of *Trigonella foenum graecum* in rats. Indian J Pharm Sci 58(3):117–119

Taranhalli AD, Kadam AM, Karale SS, Warke YB (2011) Evaluation of antidiarrhoeal and wound healing potentials of *Ceratophyllum demersum* Linn. whole plant in rats. Lat Am J Pharm 30 (2):297–303

Thangapazham RL, Sharad S, Maheshwari RK (2016) Phytochemicals in wound healing. Adv Wound Care (New Rochelle) 5(5):230–241. https://doi.org/10.1089/wound.2013.0505

Thangathirupathi A, Santharam B (2011) Evaluation of wound healing activity of *Coscinium fenestratum* (Gaertn.) Colebr in Albino rats. RJPPD 3(2):81–87

Tohidi M, Khayami M, Nejati V, Meftahizade H (2011) Evaluation of antibacterial activity and wound healing of *Pistacia atlantica* and *Pistacia khinjuk*. J Med Plant Res 5(17):4310–4314. https://doi.org/10.5897/JMPR.9000613

Udegbunam SO, Udegbunam RI, Muogbo CC, Anyanwu MU, Nwaehujor CO (2014) Wound healing and antibacterial properties of methanolic extract of *Pupalia lappacea* Juss in rats. BMC Complement Altern Med 14(1):157

Udupa AL, Kulkarni DR, Udupa SL (1995) Effect of *Tridax procumbens* extracts on wound healing. Int J Pharmacogn 33(1):37–40. https://doi.org/10.3109/13880209509088145

Umachigi SP, Kumar GS, Jayaveera KN, Dhanapal R (2007) Antimicrobial, wound healing and antioxidant activities of *Anthocephalus cadamba*. Afr J Tradit Complement Altern Med 4 (4):481–487

Umachigi SP, Jayaveera KN, Kumar CA, Kumar GS, Kumar DK (2008) Studies on wound healing properties of *Quercus infectoria*. Trop J Pharm Res 7(1):913–919

Veeresham C (2012) Natural products derived from plants as a source of drugs. J Adv Pharm Technol Res 3(4):200–201. https://doi.org/10.4103/2231-4040.104709

Velmurugan C, Venkatesh S, Sandhya K, Lakshmi SB, Vardhan RR, Sravanthi B (2012) Wound healing activity of methanolic extract of leaves of *Gossypium herbaceum*. Centr Eur J Exp Biol 1(1):7–10

Venkateskumar K, Parasuraman S, Chuen LY, Ravichandran V, Balamurgan S (2019) Exploring antimicrobials from the flora and fauna of marine: opportunities and limitations. Curr Drug Discov Technol 17(4):507–514. https://doi.org/10.2174/1570163816666190819141344

Venkateswarlu G, Vijayahhaskar K, Pavankumar G, Kirankumar P, Harishbabu K, Malothu R (2011) Wound healing activity of *Mimosa pudica* in albino wistar rats. J Chem Pharm Res 3 (5):56–60

Verma N, Khosa RL, Garg VK (2008) Wound healing activity of *Wedelia chinensis* leaves. Pharmacologyonline 2(3):139–145

Vidya SM, Krishna V, Manjunatha BK, Bharath BR, Rajesh KP, Manjunatha H et al (2012) Wound healing phytoconstituents from seed kernel of *Entada pursaetha* DC. and their molecular docking studies with glycogen synthase kinase 3-β. Med Chem Res 21(10):3195–3203. https://doi.org/10.1007/s00044-011-9860-5

Vijaya KK, Nishteswar K (2012) Wound healing activity of honey: a pilot study. Ayu 33 (3):374–377. https://doi.org/10.4103/0974-8520.108827

Villegas LF, Marçalo A, Martin J, Fernández ID, Maldonado H, Vaisberg AJ et al (2001) (+)-epi-Alpha-bisabolol [correction of bisbolol] is the wound-healing principle of *Peperomia galioides*: investigation of the in vivo wound-healing activity of related terpenoids. J Nat Prod 64 (10):1357–1359. https://doi.org/10.1021/np0102859

Vittorazzi C, Endringer DC, Andrade TU, Scherer R, Fronza M (2016) Antioxidant, antimicrobial and wound healing properties of *Struthanthus vulgaris*. Pharm Biol 54(2):331–337. https://doi.org/10.3109/13880209.2015.1040515

Wahedi HM, Park YU, Moon EY, Kim SY (2016) Juglone ameliorates skin wound healing by promoting skin cell migration through Rac1/Cdc42/PAK pathway. Wound Repair Regen 24 (5):786–794. https://doi.org/10.1111/wrr.12452

Wang JP, Ruan JL, Cai YL, Luo Q, Xu HX, Wu YX (2011a) *In vitro* and *in vivo* evaluation of the wound healing properties of *Siegesbeckia pubescens*. J Ethnopharmacol 134(3):1033–1038. https://doi.org/10.1016/j.jep.2011.02.010

Wang X, Li YL, Wu H, Liu JZ, Hu JX, Liao N et al (2011b) Antidiabetic effect of oleanolic acid: a promising use of a traditional pharmacological agent. Phytother Res 25(7):1031–1040. https://doi.org/10.1002/ptr.3385

Wesley JJ, Christina AJ, Chidambaranathan N, Ravikumar K (2009) Wound healing activity of the leaves of *Tribulus terrestris* (Linn) aqueous extract in rats. J Pharm Res 2(5):841–843

Wong CM, Ling JJ (2014) *In vitro* study of wound healing potential in black pepper (*Piper nigrum* L.). UK J Pharm Biosci 2(4):05–09. https://doi.org/10.20510/ukjpb/2/i4/91104

World Health Organization (1972) International drug monitoring: the role of national centres, report of a WHO meeting [held in Geneva from 20 to 25 September 1971]. World Health Organization

Wound Care Treatment Methods. Available in https://www.midmichigan.org/conditions-treatments/woundtreatmentcenter/treatmentmethods/. Last assessed on 25 Aug 2020

Wu F, Bian D, Xia Y, Gong Z, Tan Q, Chen J, Dai Y (2012a) Identification of major active ingredients responsible for burn wound healing of *Centella asiatica* herbs. Evid Based Complement Alternat Med 2012:848093. https://doi.org/10.1155/2012/848093

Wu XB, Luo XQ, Gu SY, Xu JH (2012b) The effects of *Polygonum cuspidatum* extract on wound healing in rats. J Ethnopharmacol 141(3):934–937. https://doi.org/10.1016/j.jep.2012.03.040

Xia C, Zuo J, Wang C, Wang Y (2012) Tendon healing *in vivo*: effect of mannose-6-phosphate on flexor tendon adhesion formation. Orthopedics 35(7):e1056–e1060. https://doi.org/10.3928/01477447-20120621-21

Ximenes RM, de Morais NL, Cassundé NM, Jorge RJ, dos Santos SM, Magalhães LP et al (2013) Antinociceptive and wound healing activities of *Croton adamantinus* Müll. Arg. essential oil. J Nat Med 67(4):758–764. https://doi.org/10.1007/s11418-012-0740-1

Yahaya TA, Adeola SO, Jaiyeoba GL, Christian MC, Adamu MA, Christianah IA et al (2012) Wound healing activity of *Crinum zeylanicum* L. (Amaryllidaceae). Phytopharmacology 3:319–325

Yang DJ, Moh SH, Son DH, You S, Kinyua AW, Ko CM et al (2016) Gallic acid promotes wound healing in normal and hyperglucidic conditions. Molecules 21(7):899. https://doi.org/10.3390/molecules21070899

Yang WT, Ke CY, Wu WT, Harn HJ, Tseng YH, Lee RP (2017) Effects of *Angelica dahurica* and rheum officinale extracts on excisional wound healing in rats. Evid Based Complement Alternat Med 2017:1583031. https://doi.org/10.1155/2017/1583031

Zahra AA, Kadir FA, Mahmood AA, Suzy SM, Sabri SZ, Latif II et al (2011) Acute toxicity study and wound healing potential of *Gynura procumbens* leaf extract in rats. J Med Plant Res 5 (12):2551–2558

Zangeneh MM, Zangeneh A, Seydi N, Moradi R (2019) Evaluation of cutaneous wound healing activity of *Ocimum basilicum* aqueous extract ointment in rats. Comp Clin Pathol 28 (5):1447–1454

Zeng Z, Zhu BH (2014) Arnebin-1 promotes the angiogenesis of human umbilical vein endothelial cells and accelerates the wound healing process in diabetic rats. J Ethnopharmacol 154 (3):653–662. https://doi.org/10.1016/j.jep.2014.04.038

Zeng Z, Huang WD, Gao Q, Su ML, Yang YF, Liu ZC et al (2015) Arnebin-1 promotes angiogenesis by inducing eNOS, VEGF and HIF-1α expression through the PI3K-dependent pathway. Int J Mol Med 36(3):685–697. https://doi.org/10.3892/ijmm.2015.2292

Zeng Q, Xie H, Song H, Nie F, Wang J, Chen D, Wang F (2016) *In vivo* wound healing activity of *Abrus cantoniensis* extract. Evid Based Complement Alternat Med 2016. https://doi.org/10.1155/2016/6568528

Zhang H, Chen J, Cen Y (2018) Burn wound healing potential of a polysaccharide from *Sanguisorba officinalis* L. in mice. Int J Biol Macromol 112:862–867. https://doi.org/10.1016/j.ijbiomac.2018.01.214

Žiberna L, Šamec D, Mocan A, Nabavi SF, Bishayee A, Farooqi AA et al (2017) Oleanolic Acid alters multiple cell signaling pathways: implication in cancer prevention and therapy. Int J Mol Sci 18(3):643. https://doi.org/10.3390/ijms18030643

Zubair M, Nybom H, Lindholm C, Brandner JM, Rumpunen K (2016) Promotion of wound healing by *Plantago major* L. leaf extracts–*ex-vivo* experiments confirm experiences from traditional medicine. Nat Prod Res 30(5):622–624. https://doi.org/10.1080/14786419.2015.1034714

Wound Healing: Understanding Honey as an Agent

Victor B. Oti

1 Introduction

Honey, is a supersaturated solution that is sweetened and viscous in nature, which is derived from the gathering and modification of nectar by honey bee (*Apis mellifera*) (Vallianou et al. 2014; Okpala 2019). Honey has been depicted to be a God gift to humanity and globally, it is one of the ancient items used as food. Honey has fructose (40%), glucose (40%), water (20%), vitamins, enzymes, organic acids, and minerals; its pH is 3.6 and a specific weight of 1.4 (Manyi-Loh et al. 2011). It is non-toxic, non-irritant, and can be available easily and not too costly, and has been playing role in healing wounds over thousands of years (Molan 2006; Majtan 2011; Aziz and Abdul Rasool Hassan 2017). There are many clinical evidences that reported the effectiveness of honey in this usage and application in the past few years ago but the efficacy behind honey was available in just recent times (Aziz and Abdul Rasool Hassan 2017; Medhi et al. 2008; Ede et al. 2017). The physical and biological properties of honey are understood in healing of wounds, having numerous bioactive compounds which have potential in enhancing wound healing processes (Frydman et al. 2020; MEDIHONEY®HCS 2013; Gambo et al. 2018). The therapeutic property of honey in healing of wounds is as an outcome of the synergistic role of chemical debridement of devitalized and dead tissues from ulcerations by engulfing of oedema and catalase which is usually done by the hygroscopic characteristics of it, the enhancement of epithelization and granulation from the edges of wounds, the microbial honey features, its nutritional features and the hydrogen peroxide production (Mandal and Mandal 2011; Clark and Adcock 2018). Treatment of wounds with honey is simple and does not cost much, it possesses bactericidal content, due to its high viscosity, it creates an obstacle physically, forming an ecosystem that is slightly wet which shows to be significant and enhances healing of wounds (Frydman et al.

V. B. Oti (✉)
Department of Microbiology, Nasarawa State University, Keffi, Nigeria

© The Author(s), under exclusive license to Springer Nature Singapore Pte Ltd. 2021
P. Kumar, V. Kothari (eds.), *Wound Healing Research*,
https://doi.org/10.1007/978-981-16-2677-7_5

2020; Jull et al. 2015; Samarghandian et al. 2017). Full healing has shown in an observational group study of sick people in 2 weeks, a positive result was shown in most situations, and in a clinical trial that is controlled, individuals showed betterment but the differs in recovery time when compared to the control groups (Jull et al. 2008a). Honey's physical features only will impact the wound healing domain positively and its processes, in a specific manner due to the acidic character of honey, its low pH and studies have agreed that wound surface acidification accelerates recovery by facilitating the oxygen that is giving out from haemoglobin (Clark and Adcock 2018). Furthermore, pH is less favourable for the activity of protease that helps reduce the destruction of the matrix that is used in repairing tissues. The increased honey osmolarity that is hinged on its enhanced content of glucose plays a significant function in wound healing processes, reasonable findings depict sugar pastes as favourable in wound healing (Okpala 2019; Maddock 2012). If blood movement under injury is enough to restore liquid that was cellularly unavailable, after which the role of the osmolarity of glucose on the outside literally develops an efflux of lymph in which such outflow is of high importance to the healing process (Hixon et al. 2019). Wounds are damage to the physiological compartment in the cells of the skin and an insult to its functionality in protecting and its connection to where organs and tissues lie underneath. Wounds that are acute are often caused by damage that is external, for example mechanical damage, Ultraviolet radiation, thermal radiation among others while chronic wounds are result from internal damages in the form of a circulatory compromise (Molan and Rhodes 2015; Sarheed et al. 2016). They include diabetic ulcer and leg ulcer. When circulation is not sufficient, tissue necrosis may result due to inadequate tissue nutrients and thus potentiates proinflammatory cytokines. Wound healing processes exhibit the overlapping of the haemostasis, inflammatory, growth or proliferation, and remodelling phases in specified durations (Sarheed et al. 2016). The process begins with the haemostasis which is achieved in few minutes, then the bacteria and debris, which get eliminated during the inflammation phase from the wound. After the inflammation phase, invasion of blood vessels and creation of new epithelium and connective tissues, as well as contraction of the wound, this is the proliferation phase that takes place. The remodelling phase is the stage where collagens are aligned on tension lines, and the remaining tissues are eliminated through apoptosis (Hixon et al. 2019). Militating factors in the healing process include old age, poor circulation, and infection among others. Microbial load control of a wound is one of the most significant aspects to look into when you want an optimal environment for healing to occur (Ruttermann et al. 2013). A bacterial level that is higher than 105 bacteria in 1 g in tissues of wound was reported to impact negatively in the process of wound healing, especially chronic and surgical wounds (Ede et al. 2017). Honey-treated wounds heal within a mean of 13 days faster than the silver sulfadiazine treatment in a study (Aziz and Abdul Rasool Hassan 2017). Significant findings from pre-clinical, animal model studies, and clinical researches have reported honey's ability in healing wounds (Aziz and Abdul Rasool Hassan 2017; Medhi et al. 2008; Gambo et al. 2018; Jull et al. 2008a; Hussain 2018; Hosseini et al.

2020; Smaropoulos and Cremers 2020). This chapter will look at honey as an agent for wound healing.

2 Brief Explanation on Wound

Wounds are disarrangement or disruption of the skin cells integrity and its operations and a nuisance to its functionality in the protection and connection of organs and tissues that are underlined (Sarheed et al. 2016; Boateng et al. 2008). Wounds can be primarily as a result of accidental tear, cut, scratch, extreme temperature, pressure, chemicals, and electrical current, infection and/or disease (which might include: ulcers, diabetes, cancers among others) (Gottrup et al. 2005; Song and Salcido 2011). Wounds can be either superficial, partial thickness, or complete thickness in nature (Sarheed et al. 2016; Alavi et al. 2016).

2.1 Wound Types

2.1.1 Acute Wounds

Wounds that are acute can heal at a short time void of complications and has a skin integrity loss feature that suddenly happens (Sarheed et al. 2016). The acute wounds may be either traumatic or surgical in nature. Fibroblasts, platelets, keratinocytes among others facilitate healing and restored the integrity of tissues (Boateng et al. 2008; Gottrup et al. 2005; Janis et al. 2010).

2.1.2 Chronic Wounds

Prolonged wounds heal in a longer time and has predictors associated with weak epidermal and dermal tissue integrity (Sarheed et al. 2016; Gottrup et al. 2005; Song and Salcido 2011). Predisposing factors either change the balance between the immune system of the patients and wound bioburden or damage the cycle of the healing process (Janis et al. 2010; Moffatt 2005). With respect to duration, if the wound did not show any signs of healing or did not heal in 2 weeks, it is referred to as a wound that is chronic (Gottrup et al. 2005). Factors could hamper with perfusion of the vessel leading to wounds that are chronic (e.g. vascular ulcers), related to metabolic changes, e.g. diabetes which can cause ulcer of the foot due to diabetes (Alavi et al. 2016). Resistant microbial presence, prolonged healing and inflammation phase, and tissue with friable granulation among others are features of chronic wounds (Sarheed et al. 2016; Janis et al. 2010).

2.2 Aspect of Wound Healing Phases

The wound healing process is grouped into four (4) major interactive phases as discussed below (Fig. 1) (Bowler et al. 2011).

2.2.1 Phase of Haemostasis

The haemostasis phase is marked by microvascular wounds and the discharge of blood features at the site of the wound (Sarheed et al. 2016; Moffatt 2005). There is a touch between platelets and the damaged blood vessels and binds to the wall. Such adherence triggers the platelets to discharge the growth factors, cytokines, and various negotiators of inflammation, which brings about the collections of platelets and activating the inherent and external coagulation mechanisms shape a clot of fibre

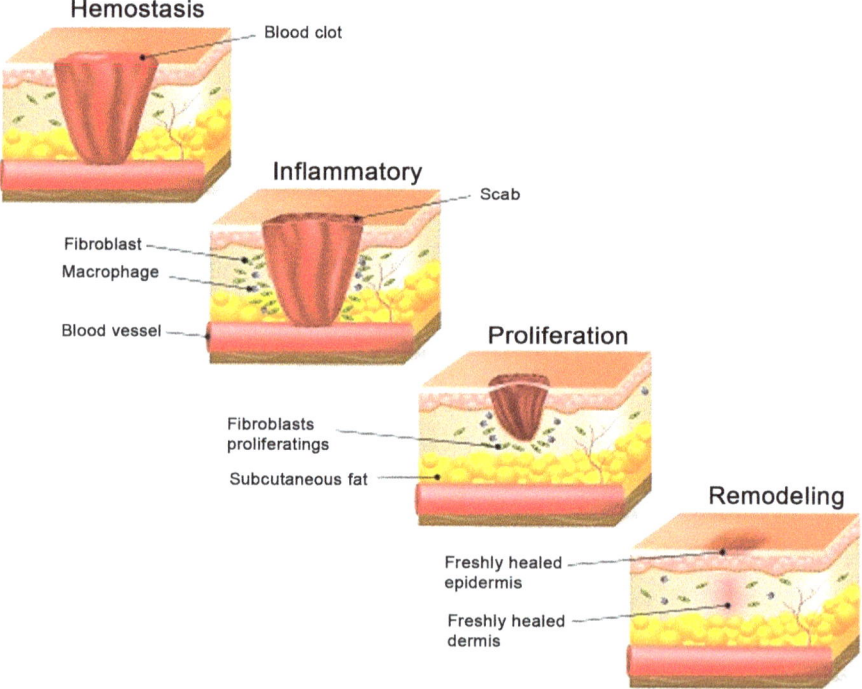

Fig. 1 Wound healing phases (Sarheed et al. 2016; Ruttermann et al. 2013; Bowler et al. 2011). Wound healing phases often begin with haemostasis where blood components are released at the site of the wound. The inflammation phase commences together with haemostasis but in some cases few minutes after injury and can last for about 3 days. In the proliferative phase, the inflammatory cells that remain generate cytokines to begin development of new blood vessels. The remodelling phase, rearrangement of collagen along strain lines, and tissues that are beyond limits are eliminated through cell death

that prevents post destruction of blood. The healing cascade is triggered by the growth factors released by the platelets (Reinke and Sorg 2012).

2.2.2   Phase of Inflammation

This phase begins together as haemostasis commences but in some cases a few minutes after injury up to 1 day and can stay up to around 3 days (Eming et al. 2007). Grouped platelets contain vasoactive amines like histamine and prostaglandins, other amines that is in granules extracted by the cellular mast which led to higher microvascular water content and dilation of blood vessels in reaction to injury, resulting in liquid hydrolysis into the intracellular room. This gives space for the movement of exudates that is rich in proteins and monocytes through and across the wound tissue which causes oedema (Bowler et al. 2011). These are typical signs of the inflammatory processes; within about 24 h sufferers begin to talk about distress at the site of injury (Sarheed et al. 2016; Eming et al. 2007).

2.2.3   Phase of Proliferation

The phase of proliferation starts just after the end of the inflammation. The surviving proinflammatory cytokines generate signalling molecules to introduce angiogenesis, that is necessary to preserve a substantial blood flow inside the bed of the wound (Sarheed et al. 2016). Red blood cells that are newly generated often cause the development of the epithelial cells (made up of extracellular matrix and collagen) and make available the nutrients that are needed (Wilkins et al. 2012).

2.2.4   Phase of Remodelling

In the phase of remodelling, the rearrangement of collagen along strain lines, and tissues that are beyond limits are eliminated through cell death (Sarheed et al. 2016; Ruttermann et al. 2013). This phase often happens when the wound is closed superficially.

2.3   Factors that Affect Wound Healing

Factors that affect wound healing are discussed below.

2.3.1   Oxygen Presence

The presence of oxygen is essential for tissue repair and infection resistance, and is used by adenosine triphosphate for cellular energy production (Gottrup 2004). It performs on various stages of recovering process by stimulating the development of new blood vessels, segmentation of keratinocytes, relocation, re-epithelialization, multiplication of fibroblasts and formulation of collagen, and encourages deceleration of wounds (Sarheed et al. 2016). Oxygen and transient hypoxia when injury occurs is important in stimulating the recovery process by triggering cytokine production and growth variables from fibroblasts, macrophages, and keratinocytes. Prolonged injuries that have oxygen tissue strain of not more than 20 mmHg is usually hypoxic relative to natural rates of 30–50 mmHg (Rodriguez et al. 2008). Through impaired vascular flow, predictors of prolonged injury like diabetes or advanced age could trigger weak oxygen presence (Guo and Dipietro 2010). Integrative revascularization treatments have been used in diabetic foot ulcers to reverse the hypoxic conditions. Thus, it was also found that such methods can cause harsh effects to diabetes patients (Sarheed et al. 2016).

2.3.2   Bioburden from Wound

Intact skin acts on the surface of the skin itself to regulate the microbial community. The subcutaneous tissue is uncovered as dignity is compromised due to injury, creating a suitable space for microbial colonization and growth (Gardner and Frantz 2008). But this does not necessarily lead to infection due to the bridge between the immune system and the wound's bioburden (Sarheed et al. 2016).

2.3.3   Wound Infection

Microflora of the skin is available in approximately 105 microbial cultures, and has no challenges clinically (Moffatt 2005). However, when tempered, microbes grow and invade tissues, initiating a chain of microbial mechanisms that trigger a response that is inflammatory in nature which can result in a deferred healing process and destruction of tissues (Sarheed et al. 2016; Bowler et al. 2011; Eming et al. 2007). Infection occurs when it damages host tissues. Among other negative effects of wound infection is inflammation that is delayed as a result of long-term rise in cytokines of inflammatory cells. This makes the injury proceed to a prolonged phase and healing is halted within the 3 months stipulated time (Bowler et al. 2011). The long-term swelling is linked to elevated matrix metalloproteinases stages that have the ability to degrade extracellular matrix, a major feature of the wound healing's proliferation stage which happens because elevated stages of proteases come at the risk of reduced levels of protease inhibitors that happens in nature. From a microbiological point of view, infection of wounds is the presence of microbes that multiplies at the site of the injury, which suppresses the immune system of the host. It hinders injury to heal by the discharge of microbial poisons and shows infection which is symptomatic and sign that is active (Reinke and Sorg 2012).

2.3.4 Wounds Associated Bacteria

Open wounds that are infected can accommodate a myriad of bacteria, including *Escherichia coli, Pseudomonas aeruginosa*, and methicillin-resistant *Staphylococcus aureus* (MRSA) (Weese et al. 2007). Overall, wounds that are infected are polymicrobial in nature and most often are infected with pathogenic bacteria which are in the domain, mucosal indigenous microorganisms, and adjacent skin microflora (Moffatt 2005; Weese and van Duijkeren 2010). Microbes of the bacterial origin are the major organism that leads to infections of wounds apart from other microbes that use skin as their habitat, but in certain mixed infections other microbes such as fungi have been found. Gram-positive bacteria like; *S. aureus* and *E. coli* predominate in the first phase of formation of chronic wound (Sarheed et al. 2016). In the last step, *Pseudomonas* species are generally seen and can penetrate layers that are deep in wound, leading to serious injury of the tissues. Others might include aerobic bacteria, for example *Staphylococcus* (such as methicillin-resistant *S. pseudintermedius*) and *Streptococcus* species (*S. pyogenes*), and anaerobic bacteria which are seen in about 50% of injuries that are prolonged (Weese et al. 2007; Weese and van Duijkeren 2010).

2.3.5 Prolonged Wounds and Formation of Biofilm

A biofilm is an anchoring community of microorganisms characterized by cells embedded in a secreted extracellular polymeric substances (EPS) matrix that adhere in an irreversible manner to a substratum or interface or to each other, and it shows a changed phenotype with regards to the transcription of genes (Garrett et al. 2008). First, a film that is conditioned is formed and consists of starch and proteins stick on the area that is solid. It prepares the surface to take in the first cellular biofilm of the incipient. Second, bacteria approach and begin to adhere to the area by Van der Waals forces and the likes as well as the electrostatic current on the surface of the bacteria which is negative (Donlan and Costerton 2002). The attached bacteria are wrapped in a polymer matrix called EPS. The organism adsorption causes quorum sensing that is involved in the gene expression and its regulation with regards to changes in the density of the cellular numbers (Bowler et al. 2011). This results to biofilm change phenotypically, producing response of harmful factors to signals from other organisms in the biofilm. Such factors that have a coverage that is generated from EPS lead to an enhanced antimicrobial resistance. Recommendations that EPS may spontaneously react with antimicrobial agent has been suggested which halts their approach to bacteria and exerting antibacterial activity (Sarheed et al. 2016). Biofilms also protect bacteria from host defences by coating the sugar coating, while they emit substances in the film that impairs phagocyte entry. This knowledge is paramount to intervention modalities for chronic wounds. In management of clinical injury, cleaning the wound quickly is imperative and the elimination of foreign bodies and necrotic tissue from the area that surrounds the injury to increase the potential for healing of wounds, which is also called debridement (Wolcott et al. 2008).

3   Aspect of Honey

Honey is a sweet, viscous, supersaturated solution derived by the gathering and modification of sap by honeybees (Bogdanov et al. 2008). Bees are known to produce honey, which is a winter food source. Bees can soar about 55,000 miles approximately to get sap from about two million floras and create around honey of one pound. Bees of the worker's class throw up honey and partially digest it beforehand packing them in a honeybunch (Ajibola et al. 2012). In the honeycomb, they spread the honey with their extensions to properly disperse the sap and prevent fermentation of the solution. Honeybees inject the solution's antibacterial function during pollen collection and ripening. The difference in honey depends on the origin of the plant, and hundreds of variations of them have been documented (Wang and Li 2011). Since honey is highly variable and the substance structure is hinged on the bud from where it is manufactured as many products of plant origin, its antibacterial activity might differ depending on the type of honey (Khan et al. 2007). To research on honey's therapeutic properties and its mechanism of deed, it is important to confirm the arrangement of the ingredient. The diet reported in honey is low when relates to the actual intake per day, but its importance is hinged on its varied functional properties (Hussain 2018). Honey is mainly composed of 95% dehydrated bulk carbohydrates. Furthermore, it has substances other than those mention above like amino acids, proteins, aromatic compounds, organic acids, amino acids, vitamins, minerals, and polyphenols (Mandal and Mandal 2011). It is composed of a number of different properties, which gives honey its ideal ability to enhance injury recovery. Honey has a low water content, which provides a moist healing environment (Mandal and Mandal 2011; Mandal et al. 2011). The extraordinary thickness of it acts as a shielding obstacle against cross-contamination and subsequent infection. It possesses a 3.6–3.7 pH range in which almost all bacteria and most bacteria cannot thrive grow in such toxic domain (Samarghandian et al. 2017). Furthermore, it encourages the creation of lymphocytes that help the reactions of the body's immune system. One key antibacterial constituent in this liquid is H_2O_2. Although, despite low levels of hydrogen peroxide, some honey effectively control bacteria (Mandal et al. 2011). Rapid healing was noted, especially burns, and the long-standing ulcers healed due to honey's ability to accelerate the process of recovery (Song and Salcido 2011). This liquid does not possess any bad reactions in system unlike most antimicrobial agents do, but it has an operative antibacterial property against hardy strains, e.g. MRSA (Samarghandian et al. 2017; Ajibola et al. 2012). Honey also has excellent cleansing properties and is a unique way to reconstruct damaged tissue with minimal scarring. Honey possesses antioxidant and anti-inflammatory properties in addition to the aforementioned (Song and Salcido 2011; Janis et al. 2010). In recent years, the concept of its usage as a remedy has become brighter and more popular (Gambo et al. 2018; Samarghandian et al. 2017; Hussain 2018). Some studies have shown that ancient treatments once have considerable controlled proof to back the miraculous function of honey in the treatment of injuries and other diseases (Hussain 2018; Khan et al. 2007; Mandal et al. 2011).

3.1 Types of Honey

There are many different types of honey, including Manuka honey (New Zealand), Medihoney, Jerry Bush Honey (Australia), Kanuka Honey, Grass Honey, Beech Honey, Blueberry Honey, Sage Honey, Clover Honey, Buttercup Honey, Dew Honey among others (Fig. 2). It is known to inhibit more than 80 types of bacteria and other organisms (MEDIHONEY®HCS 2013; Bogdanov et al. 2008; Ajibola et al. 2012). Medihoney trademark is one of the initial medically certified honey as a professional injury management medical product in the America, Australia, and Europe (Manyi-Loh et al. 2011; MEDIHONEY®HCS 2013). As reported by Maddock, honey may interfere with the reaction of fibronectin and *Streptococcus pyogenes* on destroyed cell surface (Maddock 2012). Thus, the agent against bacteria which is inherent in honey is primarily due to osmotic pressure and the enzymatic

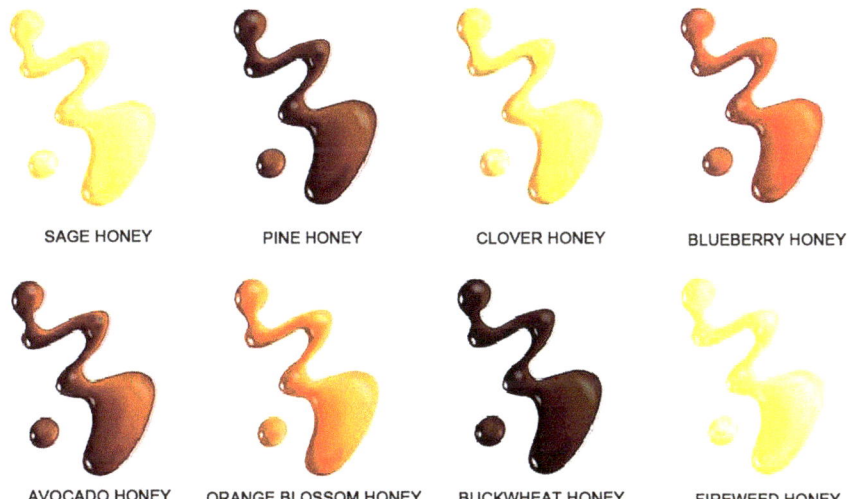

Fig. 2 Different types of honey (MEDIHONEY®HCS 2013; Clark and Adcock 2018; Bogdanov et al. 2008; Ajibola et al. 2012). Honey has different types, which is mostly based on their country of origin and plant species

production of H_2O_2 (MEDIHONEY®HCS 2013). The production of reduced levels of H_2O_2 has been shown to improve tissue oxygenation required for tissue re-development. Components other than H_2O_2 in honey that increase the activity of antibacterials include phenolics, lysosomes, flavonoids, and other components that are not discovered (Bernstein 2013). Increased osmolality (17% water, 83% sugar) and reduced pH (3.6–3.7) enhances the function of antibacterial by preventing bacterial multiplication and promoting cure. Although, there is a dire concern in surface antibacterial agents usage in managing wound infections (Hussain 2018; Molan 2011; Kateel et al. 2016).

3.2   The Role of Honey in the Wound Healing Process

Honey's role in the wound healing process cannot be overemphasized (Fig. 3). They are grouped under the following subheadings.

3.2.1   Cross-Contamination Prevention

The sticky nature of honey not only provides a wet injury domain and creates space for the re-growth of skin cells throughout the injury, but also gives a protective boundary that protects the patient by stopping cross-contamination (Manyi-Loh et al. 2011; Tan et al. 2009). Infections of the wound and subsequent colonization by bacteria are caused by microorganisms derived from the inner skin of the patient, flora of the respiratory and gastrointestinal tracts via interaction with outside environmental surfaces, air, water which are contaminated, and dirty palms of hospital practitioners can occur (Wilkins et al. 2012; Tan et al. 2009).

3.2.2   Stimulation of Tissue Growth

Tissue regrowth is imperative in process of managing injury. Honey causes the development of fresh capillaries, the fibroblasts generation which is used in place of connective tissue in the skin deeper layers, and synthesize collagen fibres that create energy for repair (Okpala 2019; Rozaini et al. 2004). Furthermore, it triggers regrowth of cells of the epithelium that create novel skin covers on the recovered wounds, impedes the development of keloid and scarring, and eliminates the want for skin grafts in large wounds (Maddock 2012; Kateel et al. 2016).

3.2.3   Self-degrading Debridement

Studies have reported that the process of wound healing, which creates the domain of such wet injury provided by honey, enhances self-degrading debridement processes

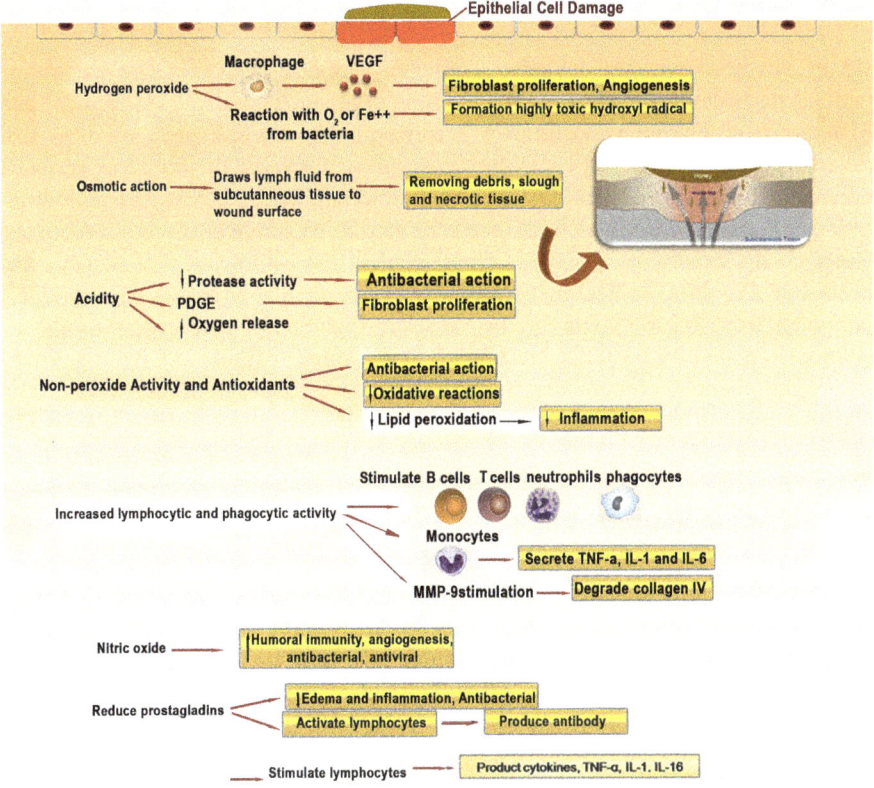

Fig. 3 The mechanisms of honey in wound healing processes (Vallianou et al. 2014; Samarghandian et al. 2017; Molan and Rhodes 2015). Firstly, due to its antibacterial activity, it produces an environment that is free from infection which enhances healing of wounds. Furthermore, honey possesses a great acidity level that is bactericidal in nature. Secondly, honey that promotes wound healing is its role in stimulating host cells to achieve healing. Honey triggers the generation of cytokines by monocytes. Interestingly, honey enhances the growth of T and B lymphocytes that help generate antibodies and enhance foreign cell phagocytosis

(Molan 2002, 2006; Wilkins et al. 2012). Lymph is often drawn from deeper tissues due to the increased osmotic pressure as well as usual washing of bed of the injury (Simon et al. 2009). In fact, the protease contained in the lymph contributes to debridement role. Odours occur in injury which is dominated by anaerobic organisms like *Clostridium* species and Bacteroides, and in rod-like bacilli like; *Proteus* and *Pseudomonas* due to the metabolic properties of proteins. As a result, it produces a bad odour chemicals like sulphur and ammonia substances in the final product (Vallianou et al. 2014; Manyi-Loh et al. 2011). It is interesting to know that honey gives bacteria a substituted sugar source and when processed, produces lactic acid.

3.2.4 Bioburden

Honey has been reported to have significant antibacterial action against myriad pathogenic bacteria of wounds and biofilms (Vallianou et al. 2014; Mandal and Mandal 2011; Garrett et al. 2008). Surprisingly, honey has been used to heal refractory wounds and has been shown to infect a myriad of multidrug-resistant organisms, like; vancomycin-resistant enterococci (VRE), MRSA, and multidrug-resistant *P. aeruginosa*, which has been reported to be active in in vitro researches. Research by Rendell et al. (2001) demonstrated that acidification of wounds has been shown to promote healing because of the low pH that promotes the sum of O_2 unburden from haemoglobin in the vessels. Apparently, acidification stops the ammonia created by bacterial breakdown, in addition to lysozyme, hydrogen peroxide, and phenolic compounds, from harming body tissues (Williams et al. 2009; Oyeleke et al. 2010).

3.2.5 Anti-inflammatory Effect

Honey has anti-inflammatory properties and has been shown in clinical and in vitro studies of human burns (Manyi-Loh et al. 2011; Hixon et al. 2019; Adebolu 2005). Prospective penalties of successfully dealing with swelling comprise quick drop of ache, oedema, and exudate. In addition, scars of the hypertrophics are reduced by evading long-term swelling that can lead to fibrosis (Manyi-Loh et al. 2011). Decreasing swelling then may reduce exudate production and dressing exchanges frequently, reducing dressing costs, control interval, and needless injury to the bed of the injury and the sufferer (Molan and Rhodes 2015).

3.3 Aspects of Honey Properties.

The aspects of the healing properties of honey with their useful effects are summarized in Table 1.

3.3.1 Antibacterial Action

Honey has exhibited diverse antibacterial activities on infected wounds (Ede et al. 2017). Various studies have found minimum inhibitory concentrations (MICs) of honey in wide classes of pathogenic bacteria that contaminate injuries (Okpala 2019; Ede et al. 2017; Gambo et al. 2018). The stage is often below 10%; honey's concentration is generally much lower than that existing beneath the injury. Reduced susceptibility is often recorded on pathogenic fungi wounds with honey, with an MIC of honey ranging from 10 to 50% (Gambo et al. 2018). Honey has shown to

Table 1 Overview of the healing properties of honey with their useful effects

Property	Attributed factors	Role	References
Antimicrobial	Acidity, increased osmolarity, H_2O_2, and non-peroxide components	Inhibitory and/or microbicidal, pain reduction, immunostimulatory	Okpala (2019), Manyi-Loh et al. (2011), Ede et al. (2017), and Gambo et al. (2018)
Antioxidant	Phenolic acids, flavonoids	Prevent formation of free radicals	Molan (2011) and Baltrusaityte et al. (2007)
Anti-inflammation	White blood cell (leucocytes)	Decreases inflammation, reduces scarring in wounds	Molan and Rhodes (2015), Hussain (2018), and Molan (2011)
Immunological	Macrophages, leucocytes	Cytokine synthesis, provides substrate for glycolysis	Manyi-Loh et al. (2011) and Tonks et al. (2007)
Malodour	Volatile acids	Reduces malodour in wounds	MEDIHONEY®HCS (2013) and Gethin (2011)

have a wide spectrum of antibacterial action and it halts both Gram-positive and Gram-negative bacteria from proliferating, and also aerobic and anaerobic organisms (Vallianou et al. 2014). Of specific choice to the professionals of injury management is their efficacy to bacteria that shows wide antimicrobial resistance like VRE, coagulase-negative Staphylococci, *Pseudomonas*, MRSA, *Stenotrophomonas maltophilia*, and *Acinetobacter baumannii* (Song and Salcido 2011). Medical professionals has shown interest in the outcomes of a prolonged concept "resistance training" experiment with four bacteria that contaminated injury in which there was definite perpetual reduction in honey's vulnerability and mutants of the isolates that shows resistant in honey were absent. The researchers conclude that as far as high levels of this agent honey are clinically kept, the danger of the organisms that gains resistance to this agent is minimized. Similar researches have reported that honey fight organisms in a biofilm environment, and wound healing with antibiotics and silver has proven ineffective (Garrett et al. 2008).

3.3.2 Anti-inflammatory Effect

Honey has long been used as an anti-inflammatory substance (Hussain 2018). Much evidence that honey has actions against swelling is exhibited from a myriad of studies. In a clinical point of view, many reports that this natural agent reduces exudate and swelling, minimizes damaging and also possess sedative activity when it is used to swollen burns and injuries (Khan et al. 2007; Nagane et al. 2004). Straight indication of anti-inflammatory actions in the hospital setting possesses biological and chemical components in the arrangement of reduced stages of lipid peroxide and malondialdehyde in biopsy samples in histologically burned honey clinical trials. A reduction in the amount of inflammatory cells presence has been observed (Molan 2002). There are scientific observations that this agent possesses a primary anti-swelling property and not an advanced action due to the activity against bacteria in the agent that eliminates the organisms that cause swelling is because of different of the anti-swelling effects reported in studies that involve burns and injury as the study population. In prototypes of animal origin, these aseptically generated wounds were found to have little or no bacteria (Mandal and Mandal 2011; Hixon et al. 2019). Honey's anti-swelling activity also reduced the severity of mucositis in head and neck radiotherapy, reduced the symptoms of dyspepsia, and in various clinical trials that reduced the number of bleeding sites in the gingiva in testing its use. It has been reported to treat gingivitis (Molan 2011; Subrahmanyam et al. 2001). It has also been shown to relieve various ophthalmological inflammatory conditions, relieve pain in unhealed lower extremity sores, and also in kids who were operated of tonsils. Findings retrieved from studies that involve animal origin also show anti-swelling effect: Triggered colitis chemically in mice is reduced and before administration of the natural agent to mice is associated with subsequent gastric inflammation due to ethanol administration prevented the occurrence (Molan 2011). Five hundred (500) microlitres of 50% injection of honey into the mice's paws in 60 min before lipopolysaccharide jab resulted in reduced inflammation, less reaction to discomfort, as well as reduced levels of prostaglandin E2 and nitric oxide (Kassim et al. 2010).

In a burn clinical trial in which silver sulfadiazine was related to this natural agent, honey reported a reduced stages of inflammatory markers, reduced malondialdehyde stages, and small numbers of swelling cells are shown in biopsy specimens (Molan and Rhodes 2015). Several findings of anti-swelling action in scientific burns and injuries with little or no bacteria present indicate that there are little or no bacteria due to the aseptically created wounds secondary effects of honey's antibacterial activity in eliminating inflammatory bacteria (Ede et al. 2017; Molan and Rhodes 2015; Bilsel et al. 2002; Prakash et al. 2008). The actions of anti-swelling by honey were also reported in hospital-based studies to minimize the extent of neck and head mucositis radiation therapy and to irritation of the gingiva therapy. The anti-swelling property of honey is demonstrated in hospital-based studies to cure dyspepsia and has been shown to be excellent in alleviating different swelling situations in ophthalmology. This agent is also shown to reduce ache that comes with the operation of the tonsils in kids as well as for uncured lower limb ulcers (Molan and Rhodes 2015). In Laboratory investigations with mice, this agent showed a key role in minimizing peritoneal bonds after caecum operation, which usefulness as a therapeutic agent for colitis that is triggered chemically and hinders irritation of the gastrointestinal tract (GIT), which is triggered by doses of alcohol. Injecting 50% honey inside mice paws before inoculating injecting lipopolysugars lead to reduced inflammation, decreased feeling of discomfort, and reduced levels of markers of inflammations (Kassim et al. 2010). Research that involves carrageenan-triggered oedema in mice paw, pre-treatment with honey resulted in a minimized dose-dependent oedema and clampdown of nitric oxide synthase induced enzyme, IL-6, COX-2, and tumour necrosis factor-alpha (TNF-α) gene expression. Anti-inflammatory activity of honey is due to the current herbal phenolic compounds. Although, there was no relationship between the levels of actions of anti-swelling seen in the phenolic compounds that are observed and in honey (Kassim et al. 2010). The phenolic component may be involved because it is a group of compounds that exhibits inflammatory cytokine TNF-α synthesis stoppage, but in recent times, it is an added main honey's non-phenolic constituent. Although, a unique constituent of anti-swelling was found, it is rather a protein that functions differently (Kassim et al. 2010). This honeybee-derived protein, aparbumin-1, has been found as an element in honey that halts macrophage phagocytosis, which is the initial stage in a series of swelling responses to necrotic soft tissue as well as microorganisms. Studies confirmed methylglyoxal, which is a compound seen only in large amounts in honey, acts in response with aparbumin-1 to saccharify. Glycated aparbumin-1 is a considerable tougher stopper of phagocytosis than the unchanged aparbumin-1 seen in honey. Solution of about 0.5% of honey resulted in the stoppage of phagocytosis of about 67% (Hussein et al. 2012).

3.3.3 Antioxidant Activity

Since ROS acts as a messenger and amplifies the feedback of the inflammatory response, honey's antioxidant activity may also contribute to its anti-inflammatory effect, a process that is blocked by phenolic antioxidants (Molan 2011). The use of antioxidants on wounds was shown to minimize swelling, and the key role of honey in making burn healing better was reported to be due to its antioxidant function (Molan 2011; Subrahmanyam et al. 2003). Phenolics and methyl syringate are abundant in honey and have been proven as intoxicating superoxide scavengers, which might be probably used to eliminate the key ROS messengers that amplify swelling (Inoue et al. 2005).

3.3.4 Immune Stimulating Effect

The elimination of honey's antibacterial infection can be further enhanced by its immunostimulatory effect (Molan and Rhodes 2015), and thus, the anti-inflammatory effect of honey always weakens it to some extent. If in any way immune reaction triggers back the cure of injuries by encouraging control contamination, it certainly contributes to curing via motivation and development of restored tissue (Molan and Rhodes 2015; Molan 2011). Frequent hospital-based findings show that a fast cure is obtained if injuries are healed with honey, and related reports were found in different studies of animal models (Molan 2011; Majtan et al. 2010; Majtan 2014). These findings may be considered the result of honey-suppressing infections. Thus, in laboratory observations with skin burn ulcer on pigs and mice under operations situations, the injuries were found devoid of bacterial cells and the application of honey further enhanced healing rate (Majtan et al. 2010; Majtan 2014).

In vitro studies have demonstrated that honey's immunostimulatory activity on white blood cells results in cytokines synthesis that leads to cell trigger and development (Molan and Rhodes 2015). At a 1% concentration of honey, it was reported to trigger monocyte discharge of TNF-α. However, the triggering of swelling is usually reflected as detrimental, the anti-swelling effects of honey modulate this swelling response (Molan 2006; Molan and Rhodes 2015). Thus, when 1% honey was applied to swollen macrophages, which are generated by activation of monocytes by opsonized zymosan and lipopolysaccharide, TNF-α release was not increased and honey formed in respiratory bursts. It suppressed the formation of reactive oxygen intermediates (Tonks et al. 2003). In a related development, at 1% concentration of honey, TNF-α from monocytes, a cytokine known to play a role in tissues in vivo, interleukin-1 beta (IL-1β), and found to stimulate IL-6 release repair (Majtan et al. 2010). Honey has also shown to trigger angiogenesis in vitro in the mice aortic ring test, with honey concentrations up to approximately 0.2% (Rossiter et al. 2010).

3.3.5 Debriding Action

Debridement with honey was reported in some hospital-based studies that involve burns (Molan 2011; Esmon 2004). One such study reported that honey has the ability to alter eschar development, but in situations where silver sulfadiazine was used, eschar was formed (Thomson 2011). Several researches that reported honey to be potent in debridement of injuries has been found in repository of hospital-based findings on the efficacy of honey as an agent of debridement (Molan 2009, 2011). Studies that are carried out in hospital settings also report honey as an excellent unusual choice for operational debridement for genital necrotizing fasciitis management (Molan 2011). Clinical trials comparing honey and hydrogels for wound debridement showed no statistically significant differences, but found that honey gave better debridement (Gethin and Cowman 2009). Outcomes when related to those online from different researches, and in the loss of wounds of the venous leg, honey was retarded than larval treatment, but greater to some cadexomer iodine, hydrogels, enzymes, and hydrocolloids. It was concluded that in experiments with adjacent rabbits on experimental wounds, honey-soaked gauze treated wounds were kept clean and saline-soaked gauze treated wounds resulted in concentrated dark scabs (Molan 2011). A report backed the insinuations that honey leads to wound debridement (Molan 2011). It has been postulated that honey enhances plasmin action, an enzyme that particularly degrades fibrin but is incapable of degrading collagen matrix required for tissue amendment. Treating cultures of inflamed macrophages can increase plasmin activity because honey skyrocketed plasmin action in the liquid and honey halt macrophage synthesis of plasminogen activator inhibitor (PAI) (Molan 2011; Esmon 2004). Plasminogen activator inhibitors usually block the transformation of plasminogen, the enzymatically sedentary originator of plasmin, to energize plasmin. Since swelling enhances PAI synthesis, it is expected that honey will reduce PAI production, as honey's anti-inflammatory action has been established (Esmon 2004).

3.3.6 Malodour Activity

Malodour of wounds are a result of combined products of anaerobic organisms, especially bacteria, e.g. Bacteroides, facultative anaerobes (*P. aeruginosa*), and also volatile fatty acids synthesized from dead tissue (Gethin and Cowman 2009). Honey possesses myriad bactericidal, damaging, and halting roles on bacterial cells (Molan 2006) and also makes use of its sugar to minimize volatile acids which possess the ability to minimize malodour of wounds (Pieper 2009), which is a sure clinical gain for patients (Gethin and Cowman 2009; Gethin 2011). A study by Gethin and Cowman (Gethin and Cowman 2005) found a case series of eight participants that involves patients aged 22–82 years old and an array of chronic and acute wounds which were around for about 18 months. All the injury except ulcer of mixed causal agents and rheumatoid was healed (94 and 54%, respectively) after a month of using honey as a therapy. Malodour of the wound occurs in three participants and pre-therapy with honey removed the odour fully.

3.3.7 Anti-pain Activity

Pain reduction is a significant part of quick wound healing (Wilkins et al. 2012). Reports by observed information show that treatment of wet injury with honey is able to increase the healing rate and decrease pain that is linked with injuries when related to injuries that are not wet (Hussain 2018). Different clinical results that were reported with respect to honey usage have healing properties, anti-pain activity (Okpala 2019; Pieper 2009), and scar minimizing are also reported (Dunford et al. 2000; Molan 2001). Antibacterial activity of honey negates wound infection and thereby retards recurring inflammation that is caused by infection. The double role gives a strong anti-pain effect and also enhances healing. Worthy to note is that honey also gives a domain that is wet for the tissues of the wound and creates a physical barrier between tissue and bandages. By guiding the strict use of dressings to bed of wounds, bandage changes lead to less pain and tissue destruction (Molan 2002). By meticulously evaluating the injury, the initial aim of healing can be investigated and precise honey application can be suggested.

3.4 Mechanism of Action of Honey

Honey's effectiveness largely relates to its therapeutic characteristics like antioxidant, anti-swelling, and antimicrobial actions among others (Samarghandian et al. 2017; Hixon et al. 2019). Furthermore, it strengthens the body's defence system and triggers the development of cells. Most of the features that make honey a great treatment agent have been described briefly above. Scientists have discussed the ingredients in honey that help enhance the antimicrobial activity (Gambo et al.

2018). The central ideas are low acidity, osmotic action, and H_2O_2 action. This natural agent is known for its composition of glucose and partly water, and these sugars and water molecules interact strongly to keep other microorganisms largely free of water molecules. Unrestricted H_2O is measured as water activity (aw), which is as low as 0.562–0.62 for this agent. Bacteria are suppressed in such a toxic domain. Although, the outcomes of researches that relates the efficacy of nature with the effects of adulterated honey, which has the same concentration of sugar and water, show that honey has additional elements that can inhibit bacteria (Vallianou et al. 2014). The low acidity of honey is due to the existence of gluconolactone that is synthesized due to the catalytic activity of maturing sap (Molan 2006). Most times, honey's reduced pH prevents a myriad of organisms from breeding at an optimal pH range. Thus, in the experimental environment, the culture liquid has a deactivating activity on the natural agent and prevents its stoppage. Hydrogen peroxide is synthesized by glucose oxidase (located in the hypopharynx of the honey bee) and, when diluted, its action is enhanced 2500–50,000 times. The level of this compound that is found in honey can be evaluated by the quantity of catalase and glucose oxidase (Mandal and Mandal 2011). At increased concentrations, this compound could lead to the destruction of proteins and cells in tissues by producing oxygen radicals. It is interesting to know that some dim honey positively suppressed organisms, notwithstanding the addition of catalase, which justified that extra non-peroxide features were effective (Bernstein 2013; Mittal et al. 2012). Honey displays non-peroxide antibacterial action and its major factors which destroy microbes have not yet been determined. However, it is believed that this ingredient comes from the unique floral source from which honey originated. Honey is labelled as "non-peroxide" when it exhibits antibacterial activity, even though it is exposed to catalase, which destroys hydrogen peroxide (Bittmann et al. 2010). Experiments by Snow and Manley-Harris (Snow and Harris 2004) relate the effects of a tenfold surplus of catalase with the usual quantity of catalase that is used to terminate H_2O_2. There was an insignificant association among H_2O_2 and non-peroxide, demonstrating that the non-peroxide action of the antibacterial agent was not because of the remaining H_2O_2 that honey possesses (Snow and Harris 2004). Although, the therapeutic mechanism of honey is not still known, nevertheless it is thought to be derived from a wide array of honey phytochemicals. Discovery was done recently which shows that H_2O_2 action only does not result to DNA strand scission, slightly it is the link attraction among H_2O_2 and the phenolic constituents existing in honey (Bernstein 2013). Further experiments have shown that elimination of H_2O_2 by catalase prevents bacterial DNA breakdown, but polyphenols pull out from honey are mediated by the Fenton reaction in the company of Cu (II) and H_2O_2. The plasmid DNA was fragmented to small quantity, honey polyphenols showed pro-oxidant action that damages DNA (Bernstein 2013). Therefore, oxidative stress induced by phenol/hydrogen peroxide describes the tool of this agent's antibacterial action and DNA destroying effects (Jaganathan and Mandal 2009). Inflammation is a key stage in treating injuries, and this natural agent has been shown to trigger intracellular monocytes to discharge the cellular messengers' cytokines TNF-a, 1, and IL-6 that mediate immune responses. Furthermore, honey encourages the synthesis of T, B, and neutrophils that help immunity. In chronic wounds, sustained swelling could stop restoration and leads to destruction (Mittal et al. 2012). In extreme swelling, white blood cells discharge prostaglandins, causing ache. Other biochemical messengers encourage inflammation, restricting blood movement via capillaries and starving the damaged tissue for the nutrients and O_2 it needs. Species of reactive O_2 are generated and can wear away body matters. Unnecessary fibroblast action results in scarring and fibrosis. Because honey has an enhanced anti-swelling effect, applying it on a wound reduces the effects of white blood cells, reducing ache, scarring, and injury exudate (Mandal and Mandal 2011). Anti-swelling effect of honey is related to its phenolic composition, the findings of experiments conducted on rat paw swelling show that extracts of honey methanol and ethyl acetate reduce inflammatory markers and signs. Through suppression of swelling, reduction of ache, and minimization of mediators of swelling examined (PGE2 and NO) (Clark and Adcock 2018). The presence of polyphenols and flavonoids in honey enhances its antioxidant action, it is known as honey's capacity to hunt free radicals. These free radicals tend to be detrimental to human system due to the fact that they create havoc to the body's proteins and DNA and lead to the destruction of cells. Honey's antioxidant action is also pointed towards its capacity to halt the early development of free radicals rather than its capacity to hunt free radicals. Swelling is the body's normal reaction to contamination and wound and produces superoxide. Superoxide is changed to H_2O_2, producing highly reactive peroxide radicals (Brudzynski and Lannigan 2012). Peroxide radicals are produced due to the Fenton reaction and are catalyzed by metal ions (Fe^{3+} and Cu^{2+}). The presence of polyphenols and flavonoids in honey appropriate these metal ions in composite with biological particles, and this process has made honey a very strong antioxidant. Polyphenols hinder severe prolonged diseases like carcinoma, diabetes, and cardiovascular disease which result due from oxidative trauma (Brudzynski et al. 2012). Specifically, polyphenols in honey have been shown to inhibit oxidative degradation reactions. Inoue et al. (2005) evaluated the antioxidant activity of different honey using a radical of 1, 1-diphenyl-2-picrylhydroazyl (DPPH) and (methyl (CH_3), hydroxyl (OH), and superoxide anion (O_2-)) hunting system. Honey of the buckwheat type showed the maximum hunting action against radicals of hydroxyl and DPPH, and Manuka honey hunts especially for superoxide anion radicals because of its increased methyl silicate amount. Apparently, the honey polyphenols are strongly associated with the antibacterial, antioxidant, and anti-swelling activity of honey (Inoue et al. 2005).

4 Honey and Its Role on Wounds

In the human system, honey is able to minimize oxidative reactions, considering the important antioxidant components characterized by its ability to feed on free radicals (Molan 2011). Furthermore, flavonoids and other polyphenols in honey are capable to impound ions of metals in the form of complexes in order to halt the free radicals formation (Baltrusaityte et al. 2007). Hence, honey would exert its antioxidant properties to stop free radicals formation triggered by ions of metals, for example Fe^{3+} and Cu^{2+}. Honey can as well trigger body immune system to fight infection. To be more specific, it encourages the multiplication of T and β lymphocytes in cells as well phage-producing H_2O_2, which is a major component of their bacteria destroying activity (Molan and Rhodes 2015; Molan 2011). Also, the mechanisms of action of honey include antimicrobial, antioxidant, and peroxide generator, which enhance its ability of inducing pro-inflammatory cytokine production, and retards biofilm formation, halt progression of bacterial cell cycle, together with its capacity to retard wound pH and modulate pain perception (Samarghandian et al. 2017; Hussain 2018; Bowler et al. 2011; Wolcott et al. 2008). Regardless of wounds severity, honey serves promising to occupy a convincing position especially in the processes of treating injury. Ingle et al. (2006) carried out a study that is randomly controlled on wound healing with honey. Honey among ancient wound treatment substances was shown to create moist healing environs, particularly harmless to (human) body (internal/external) tissues yet stimulating both healing processes and epithelialization. Increasingly gaining acceptance as mainline wound care, there still remains some concerns about honey's potency, stability, and contamination from natural sources, all of which have brought about some increases in products' licensing, regulation, and standardization (Ingle et al. 2006). How honey gets applied to a wound, and how it gets supplemented in wound dressing, is believed to influence its healing efficacy (Molan 2006, 2011). This is because wound exudates can either activate or dilute its healing properties. Indeed, advances continue to increase about honey's wound healing properties, how it is used, its product availability, and development around the globe. Yaghoobi et al. (2008) reviewed the evidence of using honey scientifically in treating injuries as an agent of anti-swelling, antioxidant, antibacterial, and against viruses. In its traditional perspective/sense, honey was reported to remain very promising in treating burn, wounds that are infected, and those that resist cure as well as boils, ulcers, pilonidal sinus, as well as diabetic foot ulcers (DFUs) (Yaghoobi et al. 2008). Honey was also shown to be efficacious in treating venous ulcers and malignant wounds as it would improve wound hygiene, especially when it is used in coating the dressing (Bernstein 2013). Oryan et al. (Oryan and Zaker 1998) researched both meta-analysis and narrative studies of biological features and therapeutics role of honey in curing injuries. Their study revealed that good quantities of lipids, amino acids, carbohydrates, proteins, minerals, and vitamins in honey help in wound healing, as it provides minimum trauma during redressing (Oryan and Zaker 1998). Also, medicinal values and wound-healing properties can differ across honey types and can improve the results of curing injuries by minimizing unnecessary scar development. Honey not only helps in both limiting and preventing bacterial infection, but also, reduces the bioburden of (emerging/existing) wound(s) (Vallianou et al. 2014). This function is derived from its biochemical properties associated with peroxide generation via intrinsic glucose oxidase activity. Honey possesses both hydroscopic and mechano-physical properties, by hampering biofilm development to limit the degree/extent of wound oedema (Hussain 2018; Vandamme et al. 2013). Besides, honey's wound healing potential prevails via mechanisms of antimicrobial action, immunologic modulation, and physiological mediation (Molan and Rhodes 2015). Primarily, an injury must have occurred before the processes of wound hypoxis can then start to take place, which subsequently gets followed by bacterial colonization specifically around the wound itself. After honey is applied, both clinical and (proposed) mechanisms can take place, through inflammatory, proliferations, and remodelling phases, after which the wound healing it would subsequently target to become accomplished (Clark and Adcock 2018; Maddock 2012). Besides, the clinical uses of honey would cut across acute, chronic, and mixed acute/chronic wounds. Majtan (2011) found several honey models and their biological roles, which are able to induce wound-healing capacities. The immunomodulatory action of honey on immune/cutaneous cells takes place during the process of wound healing (Majtan 2014). Upon its application to any given wound, honey basically begins to either inspire or hinder the discharge of some factors as MMP-9, cytokines, and Reactive Oxygen Species (ROS), from safe and cutaneous cells and this is very much dependent on the condition of wound (Majtan 2014). Moreover, honey can equally induce emitting of proinflammatory cytokines and MMP-9, which is agreed to take place during the inflammatory and proliferative wound-healing phases. By controlling inflammation of wounds, honey has shown great promise to bring an end to any of such prolonged inflammation of wounds, in the view to reduce the raised levels of proinflammatory cytokines, MMP-9, as well as ROS (Molan and Rhodes 2015; Majtan 2014). Honey's wound healing activity/potential involves a great deal of processes, from sanitizing the surrounding of wound location and indeed, the capacity to remain very efficacious in both healing and repair processes, so as to eventually reduce the scar formation. More promisingly, honey depicts great potential to help heal wounds that may seem medically complex to solve (MEDIHONEY®HCS 2013; Hixon et al. 2019).

5 Empirical Studies of Wound Healing using Honey

Evidence of using honey to heal wounds has been reported by several researchers (Okpala 2019; Clark and Adcock 2018; Molan 2011). Positive honey reports of wound healing were found in randomized controlled trials (RCTs) and clinical trials (Molan 2006, 2011; Malik et al. 2010). The profits of using honey to treat wounds have also been shown in several researches with multiple wounds, allowing honey to be compared with other treatments (Motallebnejad et al. 2008; Rashad et al. 2009). Researches that are clinically carried out include burns that are either partially thick or surface, surgical injuries that are infected, prolonged ulceration of the leg purulent myositis pus, and donor location from split-thickness skin grafts, a type of necrotizing fasciitis. Fournier's gangrene, and a catheter performed at the exit site of the central vein (Malik et al. 2010; Baghel et al. 2009; Sami et al. 2011). Evidence that honey is not harmful is found in the lack of hostile properties shown in quite a number of studies (Molan 2011; Sami et al. 2011). Patients are often reported to have severe pain in honey, when the wound is swollen and also because of the acidity content in honey (Molan 2011). Nerve endings of the nociceptor that notice sourness are sensitized by swelling. This is a scientific evaluation that honey susceptibility reduces in times when the amount of honey required to enable honey's anti-inflammatory action is retained in the wound bed explain inflammation (Vandamme et al. 2013). The justification for the honey's act to enhance destruction of wound infections and to promote recovery is obviously justified nevertheless poorly acknowledged due to the lack of advertising (Bernstein 2013; Molan 2011). What little is known about honey by wound care professionals is another bioactivity that is also associated with encouraging the recovery of injury: anti-swelling role, antioxidant role, stimulating self-degrading debridement, encouraging cell development for the healing of tissues and osmosis (Okpala 2019; Hussain 2018).

The study compared the activities of silver sulfadiazine and honey in the burns therapy (Aziz and Abdul Rasool Hassan 2017; Jull et al. 2015; Vandamme et al. 2013). They found that the thorough cure period for surface thick injuries had a usual healing time reduction of 5 days, indicating that honey was highly preferred. The percentage of wounds that were treated and the amount of injuries contaminated that undergone sterilization were as well considerably higher for honey related to less harmful silver sulfadiazine. Research by Jull et al. (2015) reported a poor value indication of a difference between orthodox cure and honey therapy of mild severe injuries. The average variance in the period of treatment between these two methods was statistically insignificant. Two studies on the time-to-heal outcome of venous leg ulcers treated with honey impregnation and normal care or hydrogel therapy were uncertain if honey enhances wound cure (Jull et al. 2015). Examination of every study showed harsh outcomes (whether treatment is in-line with therapy or without therapy) revealed considerably further trials stated among a set of honey, such as worsening discomfort and ulcers. There was uncertainty if this therapeutic agent could minimize the contamination rate of leg ulcers. Research by Kateel et al. (2016) posited that surface application of honey was greater in action than superior to orthodox therapeutic methods in the treatment of DFUs in three of the five studies, which are randomly controlled, but two of the five RCTs cured of two treatments and there was no difference. No adverse events were also reported. Jull et al. (2015) reported that there was no statistically significant association between honey cure and saline gauze after 4 months or the healing of undesirable wound pads after a month. There was inadequate proof to comment on the activity of honey on the infection of diabetic foot ulcers. One study likewise observed honey's activity on DFUs in RCTs of four persons. The report showed that the association is insignificant statistically in cure rates of regulated sets and honey (Tian et al. 2014). Vandamme et al. (2013) study did not find enough backup for anti-inflammatory effects, deodorant effects, debridement effects, or injury discomfort and found an insignificant good antibacterial activity that points honey as favourable. Studies have shown that it is uncertain if honey improved recovery in a prolonged injury of varied populace when compared to regular povidone-iodine and film dressing (Jull et al. 2015). The honey-treated wounds healed an average of 13 days faster than the silver sulfadiazine-treated wounds. Harsh outcomes like bulimia, irritation, contractures, and hypertrophic scarring were found in 4% of honey groups when related to the 28% of silver population (Jull et al. 2008b). The research determined the antibacterial action of honey from the source and adulterated ones on the development and biofilm establishment of *Streptococcus mutans*. The honey from the source showed reduced growth and larger zones of inhibition than the adulterated wells (Nassar et al. 2011). Various researches have reported that honey is active in limiting oral pathogens. *S. mutans* is the major bacterium that causes tooth decay, including other bacteria that produce microbial communities on the tooth surface, called tooth biofilms. Living organisms synthesize lactic acid that decalcifies teeth (Nassar et al. 2011). At 100% concentration, honey shows excellent outcomes counter to *P. aeruginosa*, *S. aureus*, and other organisms that is found in contaminated core canals. Recent researches by Gambo et al. (2018) and Ede et al. (2017) have found the antibacterial activity of honey showed antibacterial activity against a test bacterium in which *S. aureus* was more active than *P. aeruginosa* (Gambo et al. 2018), while, *P. aeruginosa* has the highest MIC (50 mg/ml) (Ede et al. 2017). They both conclude that honey activity is successful in inhibiting pathogenic bacterial isolates and should be considered as a therapeutic agent for wound infections (Ede et al. 2017; Gambo et al. 2018). Recent studies reported medical-grade honey (MDH) as having pro-healing and antimicrobial characteristics, especially for paediatric severe wounds (Smaropoulos and Cremers 2020). Hosseini et al. (2020) reported that honey accelerates wound healing and has a promoting effect on wounds more than any other natural product. In a 2020 study by Frydman et al. (2020), Manuka Honey Microneedles (MHM) have been proven to possess marvellous bactericidal effect especially against MRSA at concentrations below 10%, vacuum-treated honey was the most bactericidal, showing as high as bacterial concentrations reported to kill 8×10^7 CFU/ml. Treatment of wounds assays showed that at a concentration of 0.1%, cooked honey reported an incomplete closure of the wound, whereas vacuum-treated honey tended to have closure of wound that is faster (Frydman et al. 2020).

6 Conclusion

This chapter briefly described the use of honey as a potential wound healing agent. Since ancient times, honey has been used as an important nutritional medicine. It has antibacterial, anti-inflammatory, immune-stimulating, antioxidant, and malodorous properties that may help combat infectious agents and wound healing. With the advent of antibiotics, pharmaceutical companies have focused on the development of potentially harmful and exorbitant antibiotics. In the new era, antimicrobial resistance has resulted in researchers to look at the efficacy of old therapies. They are currently exploring honey's mechanism of action, an appreciable number of studies have succeeded in determining the biochemical reasons for their activity. Honey's effectiveness as a wound healing tool has been clearly shown, thus, the once abolished solution has recently developed general interest as an established therapeutic solution.

7 Future Prospects for Using Honey in Wound Healing

The future outlook of honey and its use in wound healing revolves around the following:

1. Clinical trials should focus on discovering new antibacterial properties of honey. This provides a variety of treatment options for different infectious agents.
2. An approach that is systematic is imperative in the assessment and identification of injury infections which will help clinicians treat patients with wound infections.
3. Honey researchers are now investigating into honey's mechanism of action, and most have recorded substantial breakthroughs in the validation of the biochemical explanations towards this regard.
4. Researchers have started to demonstrate the abundant composition of phenols which is in honey that halts malignant cell growth and gives antitumour function (Jaganathan and Mandal 2009).

References

Adebolu TT (2005) Effect of natural honey on local isolates of diarrhea- causing bacteria in South Western Nigeria. Afr J Biotechnol 4:1172–1174

Ajibola A, Chamunorwa JP, Erlwanger KH (2012) Nutraceutical values of natural honey and its contribution to human health and wealth. Nutr Metab (Lond) 9:61

Alavi A, Sibbald RG, Phillips TJ et al (2016) What's new: management of venous leg ulcers: approach to venous leg ulcers. J Am Acad Dermatol 174(4):627–640

Aziz Z, Abdul Rasool Hassan B (2017) The effects of honey compared to silver sulfadiazine for the treatment of burns: a systematic review of randomized controlled trials. Burns 43(1):50–57

Baghel PS, Shukla S, Mathur RK, Randa R (2009) A comparative study to evaluate the effect of honey dressing and silver sulfadiazine dressing on wound healing in burn patients. Indian J Plast Surg 42(2):176–181

Baltrusaityte V, Rimantas Venskutonis P, Čeksteryte V (2007) Antibacterial activity of honey and beebread of different origin against *S. aureus* and *S. epidermidis*. Food Technol Biotechnol 45 (2):201–208

Bernstein RC (2013) The scientific evidence validating the use of honey as a medicinal agent. Sci J Lander Coll Arts Sci 6(2):55–70

Bilsel Y, Bugra D, Yamaner S, Bulut T, Cevikbas U, Turkoglu U (2002) Could honey have a place in colitis therapy? Effects of honey, prednisolone, and disulfiram on inflammation, nitric oxide, and free radical formation. Dig Surg 19:306–312

Bittmann S, Luchter E, Thiel M, Kameda G, Hanano R, Langler A (2010) Does honey have a role in paediatric wound management? Br J Nurs 19(15):S19–S24

Boateng JS, Matthews KH, Stevens HN, Eccleston GM (2008) Wound healing dressings and drug delivery systems: a review. J Pharm Sci 97(8):2892–2923

Bogdanov S, Jurendic T, Sieber R, Gallmann P (2008) Honey for nutrition and health: a review. J Am Coll Nutr 27:677–689

Bowler PG, Duerden BI, Armstrong DG (2011) Wound microbiology and associated approaches to wound management. Clin Microbiol Rev 14(2):244–269

Brudzynski K, Lannigan R (2012) Mechanism of honey bacteriostatic action against MRSA and VRE involves hydroxyl radicals generated from honey's hydrogen peroxide. Front Microbiol 3 (36):1–8

Brudzynski K, Abubaker K, Miotto D (2012) Unraveling a mechanism of honey antibacterial action: polyphenol/H2O2-induced oxidative effect on bacterial cell growth and on DNA degradation. Food Chem 133:329–336

Clark M, Adcock L (2018) Honey for wound management: a review of clinical effectiveness and guidelines. CADTH (CADTH rapid response report: summary with critical appraisal), Ottawa

Donlan RM, Costerton JW (2002) Biofilms: survival mechanisms of clinically relevant microorganisms. Clin Microbiol Rev 15(2):167–193

Dunford C, Cooper R, White RJ, Molan P (2000) The use of honey as a dressing for infected skin lesion. Nurs Times 96(14):7–9

Ede FR, Sheyin Z, Essien UC, Bigwan EI, Okwchukwu OE (2017) In vitro antibacterial activity of honey on some bacteria isolated from wound. World J Pharm Pharm Sci 6(3):77–84

Eming SA, Krieg T, Davidson JM (2007) Inflammation in wound repair: molecular and cellular mechanisms. J Invest Dermatol 127(3):514–525

Esmon CT (2004) Crosstalk between inflammation and thrombosis. Maturitas 47(4):305–314

Frydman GH, Olaleye D, Annamalai D, Layne K, Yang I, Kaafarani HMA, Fox JG (2020) Manuka honey microneedles for enhanced wound healing and the prevention and/or treatment of Methicillin-resistant *Staphylococcus aureus* (MRSA) surgical site infection. Sci Rep 10:13229

Gambo SE, Ali M, Disso SU, Abubakar NS (2018) Antibacterial activity of honey against *Staphylococcus aureus* and *Pseudomonas aeruginosa* isolated from infected wound. Arch Pharm Pharmacol Res 1(2):APPR.MS.ID.000506

Gardner SE, Frantz RA (2008) Wound bioburden and infection-related complications in diabetic foot ulcers. Biol Res Nurs 10(1):44–53

Garrett TR, Bhakoo M, Zhang Z (2008) Bacterial adhesion and biofilms on surfaces. Prog Natl Sci 18(9):1049–1056

Gethin G (2011) Management of malodour in palliative wound care. Br J Commun Nurs 16(5): S28–S36. (supplement)

Gethin G, Cowman S (2005) Case series of use of Manuka honey in leg ulceration. Int Wound J 2:10–15

Gethin G, Cowman S (2009) Manuka honey vs. hydrogelVa prospective, open label, multicentre, randomised controlled trial to compare desloughing efficacy and healing outcomes in venous ulcers. J Clin Nurs 18:466–474

Gottrup F (2004) Oxygen in wound healing and infection. World J Surg 28(3):312–315

Gottrup F, Melling A, Hollander DA (2005) An overview of surgical site infections: aetiology, incidence and risk factors. EWMA J 5(2):11–15

Guo S, Dipietro LA (2010) Factors affecting wound healing. J Dent Res 89(3):219–229

Hixon KR, Klein RC, Eberlin CT et al (2019) A critical review and perspective of honey in tissue engineering and clinical wound healing. Adv Wound Care 8(8):403–415

Hosseini SM, Fekrazad R, Malekzadeh H, Farzadinia P, Hajiani M (2020) Evaluation and comparison of the effect of honey, milk and combination of honey–milk on experimental induced second-degree burns of Rabit. J Family Med Prim Care 9:915–920

Hussain MB (2018) Role of honey in topical and systemic bacterial infections. J Altern Complement Med 24(1):15–24

Hussein SZ, Mohd Yusoff K, Makpol S, Mohd Yusof YA (2012) Gelam honey inhibits the production of Proinflammatory, mediators NO, PGE(2), TNF-Î±, and IL-6 in carrageenan-induced acute paw edema in rats. Evid Based Complement Alternat Med 2012:109636

Ingle R, Levin J, Polinder K (2006) Wound healing with honey—a randomised controlled trial. S Afr Med J 96(9):831–835

Inoue K, Murayama S, Seshimo F, Takeba K, Yoshimura Y, Nakazawa H (2005) Identification of phenolic compound in manuka honey as specific superoxide anion radical scavenger using electron spin resonance (ESR) and liquid chromatography with colormetric array detection. J Sci Food Agric 85(5):872–878

Jaganathan SK, Mandal M (2009) Antiproliferative effects of honey and of its polyphenols: a review. J Biomed Biotechnol **2009**:830616

Janis JE, Kwon RK, Lalonde DH (2010) A practical guide to wound healing. Plast Reconstr Surg 125(6):230e–244e

Jull AB, Rodgers A, Walker N (2008a) Honey as a topical treatment for wounds. Cochrane Database Syst Rev 4:CD005083

Jull A, Walker N, Parag V et al (2008b) Randomized clinical trial of honey-impregnated dressings for venous leg ulcers. Br J Surg 95:175–182

Jull AB, Cullum N, Dumville JC, Westby MJ, Deshpande S, Walker N (2015) Honey as a topical treatment for wounds. Cochrane Database Syst Rev (3):Cd005083 https://doi.org/10.1002/14651858.CD005083.pub4

Kassim M, Achoui M, Mansor M, Yusoff KM (2010) The inhibitory effects of Gelam honey and its extracts on nitric oxide and prostaglandin E (2) in inflammatory tissues. Fitoterapia 81:1196–1201

Kateel R, Adhikari P, Augustine AJ, Ullal S (2016) Topical honey for the treatment of diabetic foot ulcer: a systematic review. Complement Ther Clin Pract 24:130–133

Khan FR, Abadin Z, Rauf N (2007) Honey: nutritional and medicinal value. Int J Clin Pract 61:1705–1707

Maddock SE (2012) Honey could be effective at treating and preventing wound infection. Society for General Microbiology. www.sciencedaily.com/releases/01/1013105919.htm

Majtan J (2011) Methylglyoxal-a potential risk factor of manuka honey in healing of diabetic ulcers. Evid Based Complement Alternat Med 2011:295494

Majtan J (2014) Honey: an immunomodulator in wound healing. Wound Rep Regen 22(2):187–192

Majtan J, Kumar P, Majtan T, Walls AF, Klaudiny J (2010) Effect of honey and its major royal jelly protein 1 on cytokine and MMP-9 mRNA transcripts in human keratinocytes. Exp Dermatol 19:e73–e79

Malik KI, Malik MAN, Aslam A (2010) Honey compared with silver sulphadiazine in the treatment of superficial partial-thickness burns. Int Wound J 7(5):413–417

Mandal MD, Mandal S (2011) Honey: its medicinal property and antibacterial activity. Asian Pac J Trop Biomed 1:154–160

Mandal S, Deb Mandal M, Pal NK, Saha K (2011) Antibacterial activity of honey against clinical isolates of *Escherichia coli, Pseudomonas aeruginosa* and *Salmonella enterica serovar typhi*. Asian Pac J Trop Med 1:154–160

Manyi-Loh C, Anna M, Clarke A, Ndip R (2011) An overview of honey: therapeutic properties and contribution in nutrition and human health. Afr J Micro Res 5(8):844–852

Medhi B, Puri A, Upadhyay S, Kaman L (2008) Topical application of honey in the treatment of wound healing: a metaanalysis. JK Sci 10(4):166–169

MEDIHONEY®HCS (2013) A next generation honey dressing: London: Wounds UK 9(4). Supplement. Available to download from: www.wounds-uk.com

Mittal L, Kakkar P, Verma A, Dixit KK, Mehrotra MM (2012) Antimicrobial activity of honey against various endodontic microorganisms: an in vitro study. J Int Dent Med Res 5(1):9–13

Moffatt C (2005) Identifying criteria for wound infection. EWMA position document: 1–5 [Internet] http://ewma.org/fileadmin/user_upload/EWMA/pdf/Position_Documents/2005__Wound_Infection_/English_pos_doc_final.pdf. Accessed 17 Mar 2016

Molan PC (2001) Why honey is effective as a medicine 2. The scientific explanation of its effects. Bee World 82:22–40

Molan PC (2002) Re-introducing honey in the management of wounds and ulcers-theory and practice. Ostomy. Wound Manag 48:28–40

Molan PC (2006) The evidence supporting the use of honey as a wound dressing. Int J Low Extrem Wounds 5(1):40–54

Molan PC (2009) Debridement of wounds with honey. J Wound Technol 5:12–17

Molan PC (2011) The evidence and the rationale for the use of honey as a wound dressing. Wound Pract Res 19(4):204–221

Molan P, Rhodes T (2015) Honey: a biologic wound dressing. Wounds 27(6):141–151

Motallebnejad M, Akram S, Moghadamnia A, Moulana Z, Omidi S (2008) The effect of topical application of pure honey on radiation-induced mucositis: a randomized clinical trial. J Contemp Dent Pract 9(3):040–047

Nagane NS, Ganu JV, Bhagwat VR, Subramanium M (2004) Efficacy of topical honey therapy against silver sulphadiazine treatment in burns: a biochemical study. Indian J Clin Biochem 19 (2):173–176

Nassar HM, Li M, Gregory RL (2011) Effect of honey on *Streptococcus mutans* growth and biofilm formation. Appl Environ Microbiol 78(2):536–540

Okpala COR (2019) Honey for healthy living and wound healing for all? A concise human function discourse. EC Pharmacol Toxicol 7(10):1057–1066

Oryan A, Zaker SR (1998) Effects of topical application of honey on cutaneous wound healing in rabbits. J Vet Med A 45:181–188

Oyeleke SB, Dauda BEN, Jimoh T, Musa SO (2010) Nutritional analysis and antibacterial effect of honey on bacterial wound pathogens. J Appl Sci Res 6(11):1561–1565

Pieper B (2009) Honey-based dressings and wound care: an option for care in the United States. J Wound Ostomy Continence Nurs 36(1):60–66

Prakash A, Medhi B, Avti PK et al (2008) Effect of different doses of Manuka honey in experimentally induced inflammatory bowel disease in rats. Phytother Res 22:1511–1519

Rashad UM, Al-Gezawy SM, El-Gezawy E, Azzaz AN (2009) Honey as topical prophylaxis against radiochemotherapy-induced mucositis in head and neck cancer. J Laryngol Otol 123 (2):223–228

Reinke JM, Sorg H (2012) Wound repair and regeneration. Eur Surg Res 49(1):35–43

Rendel M, Mayer C, Weninger W, Tshachler E (2001) Topically applied lactic acid increases spontaneous secretion of vascular endothelial growth factor by human constructed epidermis. Br J Dermatol 145:3–9

Rodriguez PG, Felix FN, Woodley DT, Shim EK (2008) The role of oxygen in wound healing: a review of the literature. Dermatol Surg 34(9):1159–1169

Rossiter K, Cooper AJ, Voegeli D, Lwaleed BA (2010) Honey promotes angiogenic activity in the rat aortic ring assay. J Wound Care 19(10):440–446

Rozaini MZ, Zuki ABZ, Noordin M, Norimah Y, Nazrul-Hakim A (2004) The effects of different types of honey on tensile strength evaluation of burn wound tissue healing. Int J Appl Res Vet Med 2(4):290296

Ruttermann M, Maier-Hasselmann A, Nink-Grebe B, Burckhardt M (2013) Local treatment of chronic wounds: in patients with peripheral vascular disease, chronic venous insufficiency, and diabetes. Dtsch Arztebl Int 110(3):25–31

Samarghandian S, Farkhondeh T, Samini F (2017) Honey and health: a review of recent clinical research. Pharm Res 9(2):121–127

Sami AN, Mehmood N, Qureshi MA, Zeeshan HK, Malik IA, Khan MI (2011) A comparative study to evaluate the effect of honey dressing and silver sulfadiazine dressing on wound healing in burn patients. Ann Pak Inst Med Sci 7(1):22–25

Sarheed O, Ahmed A, Shouqair D, Boateng J (2016) Antimicrobial dressings for improving wound healing, chapter 17. In: Wound healing—new insights into ancient challenge, pp 374–378

Simon A, Traynor K, Santos K, Blaser G, Bode U, Molan P (2009) Medical honey for wound care-still the "latest resort"? Evid Based Complement Alternat Med 6:165–173

Smaropoulos E, Cremers NAJ (2020) Treating severe wounds in paediatrics with medical grade honey: a case series. Clin Case Rep 8(3):469–476

Snow MJ, Harris MM (2004) On the nature of non-peroxide antibacterial activity in New Zealand manuka honey. Food Chem 84:145–147

Song JJ, Salcido R (2011) Use of honey in wound care: an update. Adv Skin Wound Care 24(1):40–44. quiz 45-46

Subrahmanyam M, Sahapure AG, Nagane NS, Bhagwat VR, Ganu JV (2001) Effects of topical application of honey on burn wound healing. Ann Burns Fire Disasters 14(3):143–145

Subrahmanyam M, Sahapure AG, Nagane NS, Bhagwat VR, Ganu JV (2003) Free radical control—the main mechanism of the action of honey in burns. Ann Burns Fire Disasters 16 (3):135–138

Tan HT, Rahman RA, Gan SH et al (2009) The antibacterial properties of Malaysian tualang honey against wound and enteric microorganisms in comparison to manuka honey. BMC Complement Altern Med 9:34. https://doi.org/10.1186/1472-6882-9-34

Thomson CH (2011) Biofilms: do they affect wound healing? Int Wound J 8(1):63–67

Tian X, Yi L-J, Ma L, Zhang L, Song G-M, Wang Y (2014) Effects of honey dressing for the treatment of DFUs: a systematic review. Int J Nurs Sci 1(2):224–231

Tonks AJ, Cooper RA, Jones KP, Blair S, Parton J, Tonks A (2003) Honey stimulates inflammatory cytokine production from monocytes. Cytokine 21:242–247

Tonks A, Dudley E, Porter N et al (2007) A 5.8kDa component of manuka honey stimulates immune cells via TLR4. J Leukoc Biol 82:1147–1155

Vallianou NG, Gounari P, Skourtis A, Panagos J, Kazazis C (2014) Honey and its anti-inflammatory, anti-bacterial and anti-oxidant properties. Gen Med 2:132. https://doi.org/10.4172/2327-5146.1000132

Vandamme L, Heyneman A, Hoeksema H, Verbelen J, Monstrey S (2013) Honey in modern wound care: a systematic review. Burns 39(8):1514–1525

Wang J, Li QX (2011) Chemical composition, characterization, and differentiation of honey botanical and geographical origins. Adv Food Nutr Res 62:89–137

Weese JS, van Duijkeren E (2010) Methicillin-resistant *Staphylococcus aureus* and *Staphylococcus pseudintermedius* in veterinary medicine. Vet Microbiol 140:418–429

Weese JS, Faires M, Rousseau J, Bersenas A, Mathews K (2007) Cluster of methicillin-resistant *Staphylococcus aureus* colonization in a small animal intensive care unit. J Am Vet Med Assoc 231:1361–1364

Wilkins RG, Minnich KE, Unverdorben M (2012) Wound cleaning and wound healing—a concise review. Desert foot 9th annual high risk diabetic foot conference, USA. Braun

Williams ET, Jeffrey J, Barminas JT, Toma I (2009) Studies on the effects of the honey of two floral types (*Ziziphus* spp. and *Acelia* spp.) on organism associated with burn wound infections. Afr J Pure Appl Chem 3:98–101

Wolcott RD, Rhoads DD, Dowd SE (2008) Biofilms and chronic wound inflammation. J Wound Care 17(8):333–341

Yaghoobi N, Al-Waili N, Ghayour-Mobarhan M et al (2008) Natural honey and cardiovascular risk factors; effects on blood glucose, cholesterol, triacylglycerol, CRP, and body weight compared with sucrose. Sci World J 8:463–469

Role of Medicinal Plants in Wound Healing: An Ethnopharmacological Approach

Foram Patel, A. Doshi Ankita, and Darshee Baxi

1 Introduction

Wound is defined as the anatomical rupture of the tissue that can be due to chemical, microbial, thermal, physical, or immunological changes (Ikobi et al. 2012). Restoration of the structural and functional integrity of the wound is a dynamic process involving a cascade of various phases like inflammation process, tissue formation, and tissue remodeling. The wound healing capacity may vary based on the wound environment and the individual's health status (Kerstein 1997). As per the retrospective survey carried out by Sen in 2019 ∼8.2 million people have wounds with or without infections that results in \$28.1 billion to \$96.8 billion Medicare cost. Treatment cost of surgical wounds, diabetic foot ulcers, and other chronic wounds with physiological complications like aging, metabolic disorders, obesity imposes huge challenges at economic, clinical, and social front. Economically, the annual usage of wound care products is expected to reach \$15–22 billion by 2024. Looking to the increasing number of cases and economic burden, "WOUND" is incorporated as one of the categories by The National Institutes of Health's (NIH) Research Portfolio Online Reporting Tool (RePORT) (Sen 2019).

There are various agents available for restoring and healing the wound. To name a few, antibiotics and antiseptics, desloughing agents, chemical debridement (hydrogen peroxide, Eusol), collagenase ointment, wound healing promoters, etc. (Raina et al. 2008). However, owing to the higher treatment expenses, side effects, and drug resistance, public attention is moved towards utilization of the ethnobotanical knowledge as primary healthcare to address their healthcare needs and concerns

F. Patel · A. Doshi Ankita (✉) · D. Baxi
Division of Biomedical and Life Sciences, School of Science, Navrachana University, Vadodara, Gujarat, India
e-mail: ankitad@nuv.ac.in

© The Author(s), under exclusive license to Springer Nature Singapore Pte Ltd. 2021
P. Kumar, V. Kothari (eds.), *Wound Healing Research*,
https://doi.org/10.1007/978-981-16-2677-7_6

(Robinson and Zhang 2011). As per the World Health Organization (WHO), approximately 70–80% of the world's populations depend on nonconventional medicines that are mainly of herbal origin for healthcare (Stephen et al. 2014). The most ancient and traditional medicinal system of India, "THE VEDIC MEDICINE," attributes to a holistic approach to health and personalized medicine (Semwal et al. 2015).

This chapter aims to discuss various traditional plants used for the treatment and management of different types of wounds.

2   Types of Wound and Pathophysiology

Owing to the increasing complexities of wound repair, choosing appropriate diagnosis and to know about the related risk, classifying various types of the wound is very imperative (Devaney and Rowell 2004; Irfan-Maqsood 2018). Wounds that have been caused by burn usually classified based on its cause, colour, location, etiology, depth of injury, degree of contamination, wound healing duration, etc. (Baranoski and Ayello 2008; Chard 2008; Gomez and Cancio 2007; Tiwari 2012). Cutaneous wounds are formed due to some external factors like accidental injuries and internal factors like prolonged diseased conditions, e.g., diabetes. Wounds are basically classified as chronic wounds and acute wounds, depending on their healing period.

Chronic Wound Due to aging, obesity, diabetes, and cardiovascular diseases, rate of the chronic wounds is expected to rise in the future (Powers et al. 2016). This type of wound does not follow timely and orderly healing process (Mustoe 2004; Moreo 2005). Such wounds are categorized into (1) venous/vascular ulcers: It is very common type of wound and more prevalent in legs (Nelson et al. 2006; Jones and Nelson 2007). The major events consist of activation of polymorphonuclear cells (PMN), endothelial cells, platelet aggregation, and internal edema (Brem et al. 2004). (2) Diabetic ulcers: Majorly the small wound injuries in the diabetic patient remain unnoticed for longer period of time owing to compromised neurological actions. Additionally, partially functional immune system will not allow the speedy recovery of the injury that leads to diabetic ulcers (Mustoe 2004; Moreo 2005). (3) Pressure ulcers: Paralytic patients are most prone to such conditions due to the immobility of the body. The ratio of the blood pressure, the blood capillaries, and blood flow disturbs markedly. (4) Ischemic wounds: In such wounds, the blood supply gets restricted to the injured part that results in less or no supply of oxygen and glucose, and hence the metabolism and repair process get either slow or halted (Xue et al. 2009). Apart from all these conditions, microbial colonization at wound cause inflammation in the chronic condition wherein the proteinase-induced tissue degradation caused due to anaerobic condition lead to the severity of the chronic wound (Darenfed et al. 1999).

Acute Wound It is defined as the type of wound caused by an external injury like trauma or surgery. Such wounds heal in stipulated time and order as it follows an accurate balance of generation and degradation of cells and extracellular matrix (Bowler 2002). Types of acute injuries are briefly described as follows; superficial wounds that may occur due to puncturing of the skin by nails or knives. Scrapes that occurs due to rubbing of skin with some rough surfaces. Avulsions or Contusions are produced by pulling any body part or forceful strike. Crush or cut occurs when person has been injured with some sharp instrument like a knife or when some heavy object result in crushing of the tissue. Childbirth is an example of the wound known as lacerations wherein tremendous force is required in order to tear the tissues. Velocity wounds are caused by gunshots or ballistic trauma where high-speed objects enter the body, injuring the tissue integrity (Muhammad et al. 2016).

3 Mechanism of Wound Healing

Compromised wound healing leads to the poor quality of life that results in less active lifestyle (Hopman et al. 2009; Edwards et al. 2013). Chronic wounds are characterized by a longer wound healing period with prolonged or halted proliferation phase. The USA claims approximately 25 billion dollars on wound management annually (Sen et al. 2011). Wound healing is a crucial physiological collaborative process that depends on the critical interplay between cellular and biochemical processes (Shaw and Rognoni 2020). The normal wound healing process is basically divided into four steps, i.e., hemostasis, inflammation, proliferation and repair, and lastly remodeling (Reinke and Sorg 2012; Morton and Phillips 2016) (Fig. 1). Any disturbances in these cascadual events lead to the formation of chronic ulcers and/or excessive scarring (Landén et al. 2016).The major steps of the wound healing process are as follow;

Hemostasis This phase of the repair begins within minutes to hours of injury (Queen et al. 2004). The very first response to the injury is the constriction of the blood vessels, but the spasm relaxes soon. In order to seal the blood vessel, the army of the platelet reaches to the wound and starts secreting vasoconstrictors, i.e., serotonin along with initialization of the blood clotting events (Reinke and Sorg 2012). In ATP dependent fashion, platelet adheres to Type I collagen for forming the hemostatic plug and also activate proteins like fibronectin, vitronectin, and thrombospondins which collectively aid for migration of keratinocyte, fibroblasts, and other immune cells (Balaji et al. 2015). Eventually, platelet degranulation stimulates the release of several immune-modulators like IL-1α, IL-1β, IL-6, IF-γ and tumor necrosis factor (TNF)-α to activate the complement system that in turn releases histamine, which causes capillary dilation. This event allows the switching of the hemostasis phase to the inflammation phase by accelerating the migration of immune cells to the wound site (Ellis et al. 2018).

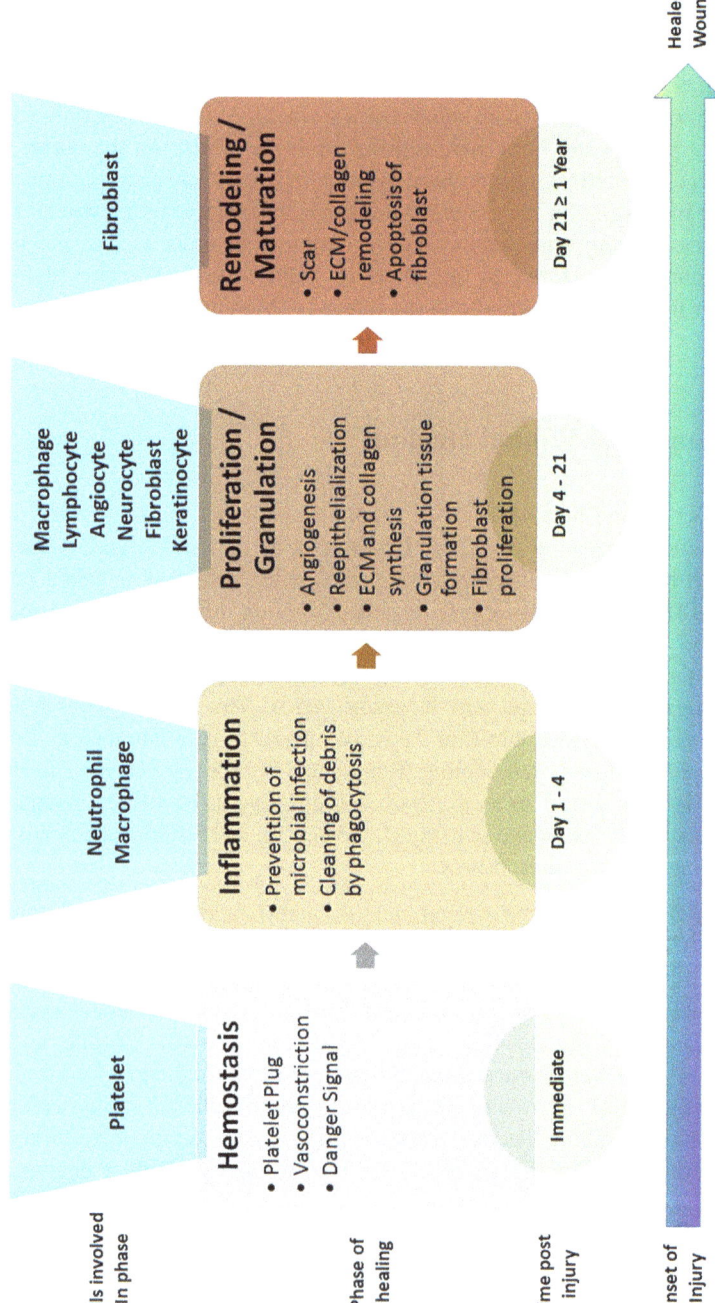

Fig. 1 Major events of the wound healing process

Inflammation The inflammatory stage of the wound healing represents initially by erythema, swelling, and warmth. This phase generally lasts for 3–4 days. However, the events greatly overlay with phase one. The events of this phase are very tightly regulated by the release of various factors by timely cellular communication. Briefly, the inflammatory phase consists of the following events. This phase facilitates infiltration of neutrophils and monocyte to the injury site, wherein they repair tissue and engulf foreign organisms (Bianchi and Manfredi 2009). The broken fibrin gets digested by fibrinogen. The key regulators like TNF-α, IL-1β, IL-6, CXCL8, CXCL2, and monocyte chemo-attractant protein-1 (MCP-1) level increases at greater extent in order to stimulate angiogenesis, boosting keratinocyte and fibroblast levels and also aid in adhesion of keratinocytes to dermal layer (Theilgaard-Mönch et al. 2004). Macrophage secretes many forms of the matrix metalloproteases (MMPs) that helps to remove dead tissues (Field and Kerstein 1994).

Proliferation and Repair Onset of the proliferation phase will occur on day 4 of the injury, and it usually last till 21 days. This phase is characterized by angiogenesis, collagen deposition, tissue granulation, and re-epithelializing of the wound surface. Briefly, Fibroblast is the main player of tissue granulation along with type III collagen. The re-establishment of the vascular channels, i.e., angiogenesis is carried out by pericyte cells by regenerating the outer lining of the endothelial cells. The epithelialization is initiated by the proliferation and migration of the keratinocytes from the wound edge to the center by migration of the epithelial stem cells. Keratinocytes, macrophage, platelets are known to secrete epidermal growth factor (EGF), keratinocyte growth factor (KGF), and transforming growth factor-α (TGF-α) that may help promoting re-epithelialization (Iorio et al. 2015).

Wound Remodeling Remodeling usually commences around 2 weeks post-injury and continues till approximately 1 year, depending upon the individual's health status. This phase is marked by scar formation (Landén et al. 2016) and the slowing of angiogenesis. During this process, the protein collagen type III is replaced by the collagen type I, which is stronger in nature (Wynn 2004). Lastly, the contraction of the wound is promoted by the myofibroblasts (Yannas et al. 2017) that eventually differentiate into fat cells (Plikus et al. 2017).

4 Traditional Approach for Enhancing Wound Healing and Its Ethnopharmacological Validation

Day by day, the process of drug discovery is becoming extremely costly, riskier, and inefficient. Failure of developed drugs and serious innovation deficit is the major concerns of big pharmaceutical companies. As a result of which, there has been a remarkable shift to multi-targeted drugs based on the traditional Medicare knowledge. Reports state the usage of specific plant parts for the treatment of wounds. In this chapter, we have presented various medicinal plants and polyherbal formulations that display wound healing properties. Table 1 summarizes the list of the plants, their scientific and common name, part of the plant used for the healing purpose, status of their pharmacological validation, and inform about the In Vivo and In Vitro studies that have been carried out.

All the plants listed in Tables 1 and 2 display the wound healing activity for different types of wounds when used alone or as polyherbal decoction respectively. Elevated wound contraction, increased tensile strength, proliferation of fibroblast, angiogenesis, differentiation of keratinocytes, re-epithelization, and remodeling are the key features of the wound healing process that have been observed during the treatment with plant extracts. The biomarkers of the wound healing process, i.e., hydroxyproline, hexuronic acid and hexosamines are found to be increased upon treatment using listed medicinal plants. Chronic wounds are characterized by the prolonged period of wound recovery. Microbial infections at wound site increase complications in the wound healing in such conditions. Tables 1 and 2 mention about the plants with antimicrobial and antifungal activities that might be added advantage provided by these plants for treatment of the microbial infection at the wound site. Reactive oxygen species are the toxic products that cause oxidative stress during the inflammatory phase of the wound healing process. Plant secondary metabolites such as alkaloids, phenols, triterpene, etc. are known to have free radical scavenging abilities, and hence such plants are considered as beneficial for wound care. As per our observations, listed members in Tables 1 and 2 have been demonstrated powerful antioxidant properties that makes them utilized for wound treatment. The plant extract of Acanthaceae members reported significantly reduced wound size in rat excision wound model by enhanced collagenation, angiogenesis, and re-epithelialization (Agyare et al. 2013a). *A. aspera* and *A. sessilis* leaves are known to use in wound care in the Ethiopian region. The results of the scientific studies proved the remarkable increase in neovascularization, epithelization, and in numbers of fibrocytes in order to organize epidermal layer. Ointment of methanol leaf extract, ethanol, and chloroform extract cream of *P. lappacea* is a well-known African folklore medicine for the treatment of chronic wounds, skin infections, and boils (Agyare et al. 2009). The methanolic root extract of the *Buchanania lanzan* Spreng displayed increase in the tensile strength of the incision and excision wound model as compared to the methanolic extract of the fruit (Chitra et al. 2009; Pattnaik et al. 2013). Some of the studies have reported the isolation of potent bioactive compounds from plant extracts for wound healing. For instance, *Centella asiatica* (L.) is known for the treatment of burns and post-operative hypertrophic scars. Isolates from this plant, i.e., asiaticoside and triterpene (asiatic acid, madecassic acid, and madecassoside) are the prime components responsible for wound healing (Somboonwong et al. 2012). Protocatechuic acid from the n-butanol fraction of *Trianthema portulacastrum* L. (Yadav et al. 2017), calophyllolide from *Calophyllum inophyllum* (Nguyen et al. 2017), Flavonoids [hyperoside, isoquercitrin, rutin and (−)-epicatechin] and naphthoquinones (hypericins) from *Hypericum perforatum* (Suntar et al. 2010a), leutolin from *Martynia annua* (Lodhi and Singhai 2013), emblin compound from *Embelia ribes* Burm. (Swamy et al. 2007), quinone compound, embelin from *Embelia ribes* Burm. (Swamy et al. 2007)

Table 1 Ethnobotanical information on plants used in the treatment of various types of wound

Plant family	Scientific name	Common name	Plant part analyzed for wound healing property	Type of extract	Dose	Ethno-pharmacological validation using specific Study model	Type of wound studied	Other properties	References
Adoxaceae	Sambucus ebulus L.	Danewort Daneweed Dwarf elderberry	Leaves	Methanol	–	Wistar albino rats	Excision Incision	AM, AO, AI	Suntar et al. (2010b)
Acanthaceae	Justicia flava (Forssk.) Vahl	Afema	Leaves	Methanol	7.5% w/w	Sprague Dawley rats (both sex)	Excision	AM, AF, AO	Agyare et al. (2013a, 2009)
	Hemigraphis colorata	Red flame ivy, metal leaf	Leaves	Aqueous suspension	1 g/kg body weight	Swiss albino mice (male)	Excision		Subramonium et al. (2001)
	Acanthus polystachyus	Acanthus	Leaves	Methanol	20 g ointment	Swiss albino mice	Excision Incision		Demilew et al. (2018)
Acoraceae	Acorus calamus L.	Sweet flag, calamus	Leaves	Ethanol	40% w/w	Wistar albino rats	Incision Excision	AO	Jain et al. (2010)
Amaranthaceae	Achyranthes aspera L.	Chaff flower	Leaves	Methanol	2.5% and 5% ointment	Albino rats (either of sex)	Circular excision Linear incision		Fikru et al. (2012)
	Alternanthera brasiliana Kuntze	Brazilian joyweed, metal weed	Leaves	Methanol	–	Sprague Dawley rats	Excision dermal cutaneous		Barua et al. (2012)
	Pupalia lappacea	Forest Burr	Leaves	Methanol	20% w/w	Rats	Excision Incision Dead space		Udegbunam et al. (2014)

(continued)

Table 1 (continued)

Plant family	Scientific name	Common name	Plant part analyzed for wound healing property	Type of extract	Dose	Ethno-pharmacological validation using specific Study model	Type of wound studied	Other properties	References
Anacardiaceae	*Spondias mombin*	Hog plum	Fruits	Decoction	1 mg/kg body weight	Wistar albino rats (male)	External	AM, AO, AF	Villegas et al. (1997)
	Semecarpus anacardium L.	Marking nut	Stem bark	Methanol	100 mg/kg body weight	Albino rats	Excision Incision Dead space		Lingaraju et al. (2012)
	Buchanania lanzan Spreng.	Calumpong nut, Chironji	Root	Methanol	10% w/w	Wistar albino rats	Circular excision Linear incision		Pattnaik et al. (2013)
	Lannea welwitschii (Hiern) Engl.	Kumbi	Leaves	Methanol	7.5% w/w	Wistar albino rats	Excision		Agyare et al. (2013a, 2009)
	Anacardium occidentale L.	Cashew	Ripe/unripe fruit	Juice	–	Sprague Dawley rats (both sex)	Excision		da Silveira vasconcelos et al. (2015)
Apiaceae	*Centella asiatica*	Indian pennywort	Aerial parts	Methanol	–	Swiss mice	Incision burn		Somboonwong et al. (2012)
	Catharanthus roseus/Vinca rosea	Periwinkle	Leaves	Ethanol	100 mg/kg body weight	Sprague Dawley rats (male)	Excision Incision Circular	–	Nayak et al. (2007)

Family	Plant name	Common name	Part used	Extract	Dose	Animal model	Wound model		Reference
Apocynaceae	*Allamanda cathartica*	Golden trumpet, trumpetvine, yellow allamanda	Leaves	Aqueous	—	Sprague Dawley rats (both sex)	Excision Incision	AM, AI	Nayak et al. (2006)
	Calotropis gigantea R.Br	Crown flower	Root bark	Ethanol	400 mg/kg body weight	Wistar Albino rats (either sex)	Excision Incision Dead space		Deshmukh et al. (2009)
	Himatanthus sucuuba	Sucuba	Woody cortex	Decoction	1 mg/kg body weight	Albino rats	External wounds		Villegas et al. (1997)
	Strophanthus hispidus DC	Brown strophanthus	Leaves roots	Methanol	7.5% w/w	Sprague Dawley rats (female)	Excision		Agyare et al. (2013b)
	Carissa spinarium L.	Conkerberry, bush plum	Roots	Methanol	2.5% w/w	Swiss albino mice (either of sex) Wistar rats (either of sex)	Burn		Sanwal and Chaudhary (2011)
Araceae	*Elaeis guineensis*	African oil palm	Leaves	Methanol	5 g	Sprague Dawley albino rats	Full thickness	AM, AO, AI	Sasidharan et al. (2012)
	Ageratina pichinchensis Kunth	Ageratina	Aerial part	Aqueous Hexane-ethyl acetate	—	Sprague Dawley rats (male)	Excision		Romero-Cerecero et al. (2014)
	Ageratum conyzoides L.	Goat weed, chickweed, white weed	Leaves	Methanol	—	Albino rats	Circular incision		Chah et al. (2006)
	Mutisia acuminata	—	Leave Stem	Decoction	1 mg/kg body weight	Albino rats	External wounds		Villegas et al. (1997)

(continued)

Table 1 (continued)

Plant family	Scientific name	Common name	Plant part analyzed for wound healing property	Type of extract	Dose	Ethno-pharmacological validation using specific Study model	Type of wound studied	Other properties	References
	Podospermum canum	–	Aerial parts	Methanol	–		Linear incision Circular excision		Bahadır-Acıkara et al. (2018)
	Vernonia arborea	–	Leaves	Aqueous Methanol	–	Swiss Wistar rats (either of sex)	Excision Incision Dead space		Manjunatha et al. (2005)
	Tridax procumbens	Coat buttons	Leaves	Ethanol	–	Wistar rats (male)	Burn		Kakade et al. (2017)
	Wedelia trilobata L.	Trailing daisy	Leaves	Hexane Ethyl acetate Chloroform Methanol	2.5–0.08 μg/ml	In vitro L929 fibroblast cells	Scratch		Balekar et al. (2012)
Asparagaceae	Asparagus racemosus	Shatavari	Roots	Aqueous	400 mg/kg body weight	Albino rats	Excision Incision	AM, AO, AI	Kumar et al. (2011)
Aytoniaceae	Plagiochasma appendiculatum	Patharshali	Thallus or thalli	Ethanol	250 mg/kg body weight	MPAE Sprague Dawley rats	Excision Incision	AM, AO, AI	Singh et al. (2006)

Aizoaceae	Trianthema portulacastrum L.	Desert horse purslane, giant pigweed	Leaves	n-butanol	10% w/w	Wistar Albino rats (male)	Full thickness excision Full thickness incision	AO, AI	Yadav et al. (2017)
Basellaceae	Anredera diffusa	Madeira vines	Leaves	Decoction	1 mg/kg body weight	Albino rats	External wounds	–	Villegas et al. (1997)
Berberidaceae	Berberis lyceum royle	Indian Lycium Daruhaldi (local name)	Root	Aqueous Methanol (more effective)	–	In vivo	Excision Incision Dead space	–	Asif et al. (2007)
Bignoniaceae	Kigelia africana Beneth.	Sausage tree	Stem Bark Leaves	Methanol	–	Sprague Dawley rats (female)	Excision	AM, AO	Agyare et al. (2013b)
	Pyrostegia venusta (Ker Gawl) Miers.	Flamevine	Flowers	Methanol	100 mg/kg body weight	Wistar rats	Full thickness excision Linear incision		Roy et al. (2012)
Burseraceae	Commiphora guidottii Chiov. ex.Guid.	Scented myrrh	Whole plant	Methanol Oleo-gum-resin	5% w/w	Swiss albino mice (either of sex)	Excision Incision	AM, AO, AI	Gebrehiwot et al. (2015)
Calophyllaceae	Calophyllum inophyllum	Beauty leaf, Alexandrian laurel	Seeds	Ethanol	6 mg/animal	Human keratinocyte cell line (HacaT) RAW264.7	Incision	AO, AI	Nguyen et al. (2017)

(continued)

Table 1 (continued)

Plant family	Scientific name	Common name	Plant part analyzed for wound healing property	Type of extract	Dose	Ethno-pharmacological validation using specific Study model	Type of wound studied	Other properties	References
Caprifoliaceae	Lonicera japonica Thunb.	Japanese honeysuckle	Flowering aerial part	Ethanol	10% w/w	Wistar rats (male)	Excision	–	Villegas et al. (1997)
			Seeds of ripenend fruits	Ethanol	–	Albino rats	Full thickness circular excision	–	Chen et al. (2012)
Caricaceae	Carica papaya L.	Papaya	Latex	Gel	2.5% w/w	Sprague Dawley rats (male)	Burn	–	Gurung and Škalko-Basnet (2009)
			Green & ripe epicarp	Powder	–	Swiss albino mice (male)	Excision Incision	–	Anuar et al. (2008); Nayak et. al. (2012)
	Carica candamarcensis	Mountain papaya	Latex from immature fruit	Protein fraction	0.1%	ICR strain mice	Burn	–	Gomes et al. (2010)
Cecropiaceae	Cecropia peltata	Trumpet tree	Leaves	Aqueous Ethanol	–	Male and female mice	Excision	–	Nayak (2006)
Cleomaceae	Cleome rutidosperma	Fringed spider flower Purple cleome	Roots	Methanol	10% w/w	Sprague Dawley rats	Excision Incision	–	Mondal and Suresh (2012)
Compositae	Aspilia africana (Pers.) C.D.Adams	Wild sunflower	Leaves	Methanol	200 mg & 400 mg	Wistar albino rats (either of sex)	Excision	AM	Okoli et al. (2007)
	Anogeissus latifolia	Axle wood	Bark	Ethanol	33.3% w/w	Swiss albino mice and rats Guinea pigs	Excision Incision	–	Govindarajan et al. (2004)

Combretaceae	*Terminalia arjuna*	Arjuna	Bark	Hydro-alcohol	–	Sprague Dawley rats (male)	Excision Incision	AM, AO, AI	Chaudhari and Mengi (2006)
	Terminalia bellerica Roxb.	Belleric Myrobalan	Fruit	Ethanol	4 g	In vivo	Excision Incision	AM, AO, AI	Choudhary (2008)
	Terminalia chebula	Black or chebulic Myrobalan	Leaves	Alcohol	200 µl	Wistar albino rats	Excision Incision		Suguna et al. (2002)
Crassulaceae	*Rhodiola imbricata*	Golden root	Rhizomes	Ethanol	1% w/v	Sprague Dawley rats (male)	Full thickness cutaneous wound	–	Gupta et al. (2007)
Cucurbitaceae	*Momordica charantia*	Bitter gourd	Fruit	Ethanol	50 mg	Sprague Dawley rats (male)	Full thickness excision	–	Teoh et al. (2009)
	Trichosanthes diocia Roxb.	Pointed guard	Fruit	Methanol	5 g	Albino rats (either of sex)	Full thickness excision Incision	–	Shivhare et al. (2010)
Ericaceae	*Vaccinium macrocarpon*	Cranberry	Fruit	Oil	100 mg/kg body weight	Rats	Excision	–	Shivananda Nayak et al. (2011)
Elaeagnaceae	*Hippophae rhamnoides* L.	Seabuckthorn	Leaves	Aqueous	1%	Sprague Dawley rats (male)	Full thickness cutaneous excision	–	Gupta et al. (2005)
Elatinaceae	*Bergia ammannioides*	Ammannia Waterwort	Aerial parts	Ethanol Hexane fraction Ethyl acetate Butanol	5% w/w 10% w/w	Sprague Dawley rats (male)	Full thickness circular excision	AM, AO, AI	Ezzat et al. (2016)

(continued)

Table 1 (continued)

Plant family	Scientific name	Common name	Plant part analyzed for wound healing property	Type of extract	Dose	Ethno-pharmacological validation using specific Study model	Type of wound studied	Other properties	References
Euphorbiaceae	*Pedilanthus tithymaloides* L. Poit.	Devil's backbone	Leaves	Petroleum ether Chloroform Methanol (most potent)	5 g	Sprague Dawley rats (male)	Excision Incision Dead space	AO, AI	Ghosh et al. (2012)
	Jatropha curcas L.	Physic nut, barbados nut, poison nut, bubble bush, purging nut	Bark	Juice	–	Albino rats (male)	Excision Incision Dead space		Shetty et al. (2006a, b)
	Euphorbia characias subsp.wulfenii	Mediterranean spurge, Albian spurge	Aerial parts	Methanol	–	Sprague Dawley rats (male) Swiss albino mice	Linear incision Circular excision		Özbilgin et al. (2018)
	Euphorbia hirta	Asthma plant	Whole plant	Ethanol	–	Wistar albino rats (male)	Burn		Jaiprakash and Chandramohan (2006)
	Bridelia ferruginea	Bredila	Leaves	Aqueous Ethanol	5 µg/ml	Human dermal fibroblasts (F55)	Scratch		Adetutu et al. (2011)

Family	Plant	Common name	Part used	Solvent	Dose	Animal model	Wound model		Reference
	Mallotus philippinensis Muell.Arg	Kamala tree	Fruits	Ethanol	200 mg/kg body weight	Charles-Foster albino rats	Full thickness circular excision / Full thickness linear incision		Gangwar et al. (2015)
	Acalypha indica	Indian copperleaf	Leaves	Ethanol	40 mg/kg body weight	Wistar rats (male)	Excision / Incision		Ganeshkumar et al. (2012)
	Acalypha langiana	Arlomo	Leaves	Aqueous	–	Wistar rats	Excision / Incision		Gutierrez (2006)
	Pterocarpus santalinus L.	Red sandalwood	Leaves Stem	Ethanol	–	Wistar albino rats	Excision / Incision		Nagori and Solanki (2011)
Fabaceae	Prosopis africana Taubert	African mesquite	Stem bark	Methanol	–	Swiss albino rats and mice	Excision / Dead space	AM, AO	Ezike et al. (2010)
	Trigonella foenum graecum	Fenugreek	Seeds	Aqueous	500 mg/kg body weight	Albino rats (male)	Excision / Incision / Dead space		Taranalli and Kuppast (1996)
	Albizzia lebbeck L. Benth	Flea tree	Roots	Ethanol	–	Charles-Foster albino rats	Excision / Incision		Joshi et al. (2013)
	Pongamia pinnata	Pongam tree, Indian beech, Pongame oil tree, Karanj	Leaves	Ethanol	100 mg/kg body weight	Wistar rats	Excision / Incision	AM, AO	Dwivedi et al. (2017)
			Stem Bark	Ethanol	10% w/w	Wistar albino strain (both the sex)	Excision / Incision		Bhandirge et al. (2015)
	Caesalpinia mimosoides lam.	Thorn mimosa	Tender shoots	Aqueous Ethanol	5% w/w	Wistar albino rats	Circular excision / Linear incision		Bhat et al. (2016)

(continued)

Table 1 (continued)

Plant family	Scientific name	Common name	Plant part analyzed for wound healing property	Type of extract	Dose	Ethno-pharmacological validation using specific Study model	Type of wound studied	Other properties	References
Fagaceae	*Castanea mollissima* Bl.	Chest nut	Shell	Ethanol	3% w/w 5% w/w	Wistar rats (male)	Circular excision Linear incision	AI	Luo et al. (2018)
	Hypericum perforatum L.	St. John's wort	Flowering aerial parts	Ethanol	–	Sprague Dawley rats (male) Swiss albino mice	Linear Incision Circular excision		Süntar et al. (2010a)
Hypericaceae	*Hypericum hookerianum*	Hooker's St. John's wort	Leaves Stem	Methanol	10% w/w	Wistar albino rats	Excision Incision	–	Mukherjee and Suresh (2000)
	Hypericum scabrum L.	–	Flowering aerial parts	Ethanol	–	Sprague Dawley rats (male) Swiss albino mice	Linear incision Circular excision	–	Vafi et al. (2016)
Iridaceae	*Eleutherine bulbosa*	Dayak onion	Woody cortex	Decoction	1 mg/kg body weight	Albino rats	External wounds	–	Villegas et al. (1997)
	Eleutherine indica L.	–	Bulb	Methanol	2.5% w/w	Swiss albino mice Wistar rats (male)	Cutaneous	–	Upadhyay et al. (2013)

Lamiaceae	Ocimum sanctum	Holy basil, ram tulsi	Leaves	Ethanol	–	Albino rats	Excision Incision Dead space	–	Udupa et al. (2006)
	Ocimum gratissimum Briq	Wild basil Shyam tulsi	Leaves	Methanol	20 mg/ml	Albino rats	Circular incision	–	Chah et al. (2006)
	Hyptis suaveolens	Pignut Vilayati-Tulsi	Leaves	Ethanol	–	Wistar rats	Excision Incision Dead space	–	Shirwaikar et al. (2003)
	Clerodendrum serratum	Blue glory Bettle killer	Roots Leaves	Ethanol	–	Albino rats	Excision	–	Nagori and Solanki (2011)
	Rosmarinus officnalis L.	Rosemary	Aerial parts	Oil	–	BALB/c mice (male)	Circular excision	–	Abu-al-basal (2010)
Lateolabracidae	Lateolabra maculatus	Sea bass	Edible part	Aqueous	4.5 g/kg body weight	RAW 264.7 fibroblast cells L929 fibroblast cells	Scratch	–	Chen et al. (2019)
Lauraceae	Persea americana	Avocado	Extract	Paste	300 mg/kg dose	Sprague Dawley rats (both sex) Wistar rats	Full thickness excision Dead space	–	Nayak et al. (2008)
	Laurus nobilis L.	Sweet Bay	Leaves	Aqueous	–	Sprague Dawley rats	Excision Incision	–	Ghosh and Gaba (2013)
	Cinnamomum zeylanicum	Cinnamon	Bark	Ethanol	–	Charles-Foster albino rats (either of sex)	Excision Incision Dead space	–	Kamath et al. (2003)

(continued)

Table 1 (continued)

Plant family	Scientific name	Common name	Plant part analyzed for wound healing property	Type of extract	Dose	Ethno-pharmacological validation using specific Study model	Type of wound studied	Other properties	References
Leeaceae	Leea macrophylla	Hastikarnapalasa	Root tubers	Ethanol	–	Albino rats	Excision Incision	AO	Joshi et al. (2016)
Lecythidaceae	Napoleonaea imperialis P. Beauv	Napoleon's hat	Leaves	Methanol	20 mg/ml	Albino Wistar rats	Circular incision	–	Chah et al. (2006)
Leguminosae	Prosopis cineria L. Druce	Golden tree, Ghaf	Leaves	Ethanol	–	Goto-Kakizaki diabetic rats	Excision	–	Gupta et al. (2015)
Liliaceae	Aloe barbadensis	Aloe vera	Mucilages	Gel	40 mg/kg	RAW 264.7 fibroblast cells L929 fibroblast cells	Cutaneous	–	Atiba et al. (2010)
	Aloe ferox Milller	Bitter aloe	Mucilages	Gel		Wistar rats (male) New Zealand white rabbits	Cutaneous	–	Atiba et al. (2010)
	Aloe arborescens Miller	Krantz aloe Candelabra aloe	Leaves	Juice	–	New Zealand white rabbits	Linear incision	–	Jia et al. (2008)
	Allium ascalonicum L.	Shallot (Thai onions)	Bulb	Ethanol	10%, 20% w/w	Sprague Dawley rats (male)	Excision	–	Saenthaweesuk et al. (2015)

Family	Plant	Common name	Plant part	Extract	Dose	Model	Wound model		Reference
Loasaceae	*Mentzelia cordifolia*	Blazing stars, stick-leafs, moonflowers	Woody cortex	Decoction	1 mg/kg body weight	Albino rats	External wounds	–	Villegas et al. (1997)
Loganiaceae	*Anthocleista djalonensis* A. chev	–	Leaves	Methanol	20 mg/ml	Albino rats	Circular incision	–	Chah et al. (2006)
Loranthaceae	*Dendrophthoe falcata*	Honeysuckle mistletoe Neem mistletoe	Aerial parts	Ethanol	10% w/w	Wistar Albino rats (either of sex)	Full thickness excision Incision	–	Pattanayak and Sunita (2008)
Lycopodiaceae	*Lycopodium serratum*	Club moss	Leaves	Aqueous Ethanol	–	Wistar rats	Excision Incision Dead space	–	Manjunatha et al. (2007)
Lygodiaceae	*Lygodium flexuosum*	Japanese climbing fern, vine-like fern	Leaves	Ethanol	400 mg/kg body weight	Wistar Albino rats	Excision Incision Dead space	AM, AO	Chandra et al. (2015)
	Gossypium arboretum L.	Tree cotton	Whole plant	Aqueous	–	Human dermal fibroblast	Sratch		Annan and Houghton (2008)
Lythraceae	*Woodfordia fruticosa* Kurz.	Fire flame bush	Flowers	Ethanol	500 mg/kg body weight	Wistar Albino rats (male)	Excision Incision Dead space	AO	Verma et al. (2013)
	Lawsonia inermis L.	Heena	Leaves	Ethanol	–	Sprague Dawley rats (male)	Excision Incision Dead space		Nayak et al. (2007)
Malvaceae	*Hibiscus rosa Sinensis* L.	Chinese hibiscus, China rose, Hawaiian hibiscus,	Flowers	Ethanol	–	Wistar Albino rats	Excision Incision Dead space	AO, AI	Bhaskar and Nithya (2012)

(continued)

Table 1 (continued)

Plant family	Scientific name	Common name	Plant part analyzed for wound healing property	Type of extract	Dose	Ethno-pharmacological validation using specific Study model	Type of wound studied	Other properties	References
	Hibiscus micranthus L.	Shoeblack plant, rose mallow, tiny hibiscus flower	Leaves	Methanol	5 g and 10 g	Wistar Albino rats	Excision		Begashaw et al. (2017) and Jayapal et al. (2014)
	Malva sylvestris	Common mallow	Flowers	Diethyl ether	200 mg/kg body weight	Wistar rats (male)	Full thickness excision	AO, AI	Pirbalouti et al. (2012)
	Sida cordifolia L.	Country mallow	Whole plant	Ethanol	10% w/w	Wistar Albino rats (either of sex)	Full thickness excision Full thickness incision		Pawar et al. (2013)
Martyniaceae	Martynia annua	Tiger's claw	Leaves	Methanol	–	Wistar Albino rats	Excision	AM	Lodhi and Singhai (2013)
Meliaceae	Azadirachta indica	Neem	Stem bark	Aqueous	–	Swiss Albino mice (male)	Excision Incision	AM, AF, AO, AI	Mann et al. (2017)
	Carapa guianensis	Andiroba, crabwood	Stem bark	Ethanol	200 mg/kg body weight	Sprague Dawley rats	Full thickness excision Incision Dead space		Nayak et al. (2010)

Family	Plant	Common name	Part	Solvent	Dose	Animal model	Wound type	Activity	Reference
Mimosaceae	Mimosa pudica	Touch-me-not	Roots	Aqueous Methanol	2% w/w	Wistar Albino rats (either of sex)	Excision Incision	–	Kokane et al. (2009)
Moraceae	Ficus amplissima	Indian bat fig	Leaves	Acetone	100 mg/kg body weight	Wistar Albino rats (male)	Excision Incision	AM, AO, AI	Arunachalam and Parimelazhagan (2013)
	Ficus asperifolia	Sandpaper tree	Whole plant	Aqueous	50 µg/ml	Human dermal fibroblast	Scratch		Annan and Houghton (2008)
	Ficus benghalensis L.	Indian banayan fig	Roots	Aqueous Ethanol	–	Wistar Albino rats (either of sex)	Incision Full thickness circular excision		Murti et al. (2011)
	Ficus racemosa L.	Cluster fig	Roots	Aqueous Ethanol	–	Swiss Albino mice (male)	Excision Incision	–	Murti and Kumar (2012)
Moringaceae	Moringa oleifera	Drum-stick	Leaves	Aqueous	300 mg/kg body weight	Swiss Albino rats (either of sex)	Excision Resutured incision Dead space	AM, AI	Rathi et al. (2006)
Myrsinaceae	Embelia ribes Burm.	False black pepper	Leaves	Ethanol	–	Charles-Foster Albino rats (male)	Excision Incision Dead space	AM, AO, AI	Swamy et al. (2007)
Musaceae	Musa sapientum var. paradisiaaca	Plantain Banana	Fruit	Aqueous	100 mg/kg	Albino rats	Excision Incision Dead space	–	Agarwal et al. (2009)

(continued)

Table 1 (continued)

Plant family	Scientific name	Common name	Plant part analyzed for wound healing property	Type of extract	Dose	Ethno-pharmacological validation using specific Study model	Type of wound studied	Other properties	References
Myrtaceae	*Psidium guajava*	Guava	Leaves	Methanol	20 mg/ml	Wistar Albino rats (male)	Circular incision	AM	Chah et al. (2006)
Oleaceae	*Jasminum grandiflorum* L.	Jasmine	Flowers	Ethanol	250 mg/kg body weight	Wistar rats (male)	Full thickness Excision Dead space	AM, AI	Nayak and Mohan (2007)
	Schrebera swietenioides	Weaver beam tree	Bark	Aqueous Ethanol	300 mg/kg body weight	Wistar Albino rats (either of sex)	Excision Incision Dead space		Rasal et al. (2009)
Orchidaceae	*Vanda roxburghii* R.Br	Vanda orchid	Leaves	Paste	–	Sprague Dawley rats (female)	Full thickness excision	–	Nayak et al. (2005)
Paeonaceae	*Radix paeoniae*	Peony root	Roots	Aqueous	–	Wistar rats	Excision Incision Dead space	–	Malviya and Jain (2009)
Papaveraceae	*Argemone mexicana*	Mexican poppy	Stem Leaves Aerial parts	Methanol	–	Wistar Albino rats	Excision Incision	–	Dev et al. (2019)
Papilionaceae	*Butea monosperma*	Flame of the forest, bastard teak	Bark	Alcohol	200 µl	Albino Wistar rats (male)	Cutaneous excision wound	–	Sumitra et al. (2005)

Pedaliaceae	*Sesamum indicum* L.	Sesame	Seeds	Oil	250 mg/kg body weight	Albino Wistar rats	Excision Incision Dead space Burn	–	Kiran and Asad (2008)
Piperaceae	*Piper cubeba*	Tailed pepper	Whole plant	Essential oil	0.82 mg/kg	Albino Wistar rats (female)	Excision	–	Shakeel et al. (2019)
Polygonaceae	*Muehlenbeckia tamnifolia*	Maiden hair	Woody cortex	Decoction	1 mg/kg body weight	Albino rats	External wounds	AI	Villegas et al. (1997)
	Rumex abyssinicus	Spinach rhubarb	Rhizomes	Methanol	500 mg/kg & 750 mg/kg body weight	Swiss Albino mice (either of sex)	Excision Incision	–	Mulisa et al. (2015)
Plantaginaceae	*Plantago major* subsp.major L.	Broadleaf plantain, white man's	Aerial parts	Methanol	–	In vitro	Excision	–	Genc et al. (2020)
Plumbaginaceae	*Plumbago zeylanica*	Ceylon leadwort	Stem Leaves Aerial parts	Methanol	–	Wistar Albino rats	Excision Incision	–	Kotadi et al. (2011)
Punicaceae	*Punica granatum* L.	Pomegranate	Flowers	Diethyl ether	–	Wistar rats (male)	Full thickness excision	AM, AO, AI	Pirbalouti et al. (2012)
Rosaceae	*Hagenia abyssinica* (Bruce)	African redwood	Flower	Methanol	–	Swiss Albino mice	Excision Incision	AM, AO, AI	Belachew et al. (2020)
	Potentilla fulgens Wall.ex Hook	Himalayan cinquefoil	Root	Ethanol Ethyl acetate	200–400 mg/kg 75 mg/kg	Charles-Foster Albino rats (either of sex)	Excision Incision		Kundu et al. (2016)

(continued)

Table 1 (continued)

Plant family	Scientific name	Common name	Plant part analyzed for wound healing property	Type of extract	Dose	Ethno-pharmacological validation using specific Study model	Type of wound studied	Other properties	References
	Rubus sanctus Schreber	Holy bamble	Aerial parts	*n*-hexane Chloroform Ethyl acetate Methanol	–	Sprague Dawley rats (male)	Excision Linear incision		Süntar et al. (2011)
Rubiceae	*Morinda citrifolia* L.	Indian mulberry	Leaves	Ethanol	–	Sprague Dawley rats (male)	Excision Dead space	AM, AO, AI	Nayak et al. (2009)
	Anthocephalus cadamba	Wild cinchona, Kadam tree	Whole plant	Hydro alcohol	–	Wistar rats (either of sex)	Excision Incision		Umachigi et al. (2007)
	Ixora coccinea	Ixora, jungle gernium	Roots	Ethanol	1.5 g	Wistar rats (either of sex)	Excision Incision		Selvaraj et al. (2011) and Latha et al. (2012)
	Rubia cordifolia L.	Indian maddar	Roots	Ethanol	1% gel	Wistar Albino rats (male)	Excision		Karodi et al. (2009)
Scrophulariaceae	*Bacopa monniera*	Water hyssop	Whole plant	Ethanol	25 mg/kg body weight	Swiss Albino mice (male)	Excision Incision Dead space	AM	Murthy et al. (2013)
Solanaceae	*Atropa belladonna*	Belladonna Deadly nightshade	Leaves	Aqueous	1% w/w	Charles-Foster Albino rats (either of sex)	Full thickness incision	AM	Gal et al. (2009)

Family	Species	Common name	Part used	Solvent	Dose	Animal model	Wound type	Assay	Reference
	Brugmansia suaveolens	Snowy angel's trumpet	Leaves, aerial parts, flowers	Ethanol	—	Sprague Dawley rats (male)	Linear		Schmidt et al. (2009)
	Datura stramonium	Thorn apple	Stem Leaves Aerial parts	Methanol	—	Swiss 3T3 Albino mouse fibroblasts	Excision Incision		Dev et al. (2019)
	Lycium depressum	Desert-thorn	Leaves	Methanol	4 g	Wistar Albino rats	Excision Incision		Naji et al. (2017)
Sapotaceae	*Mimusops elengi* L.	Spanish cherry, medlar, bullet wood	Stem Bark	Methanol	5% w/w	Wistar Diabetic rats (male)	Excision Incision Dead space	AM	Gupta and Jain (2011)
Tectariaceae	*Tectaria cicutaria*	Button fern	Rhizomes	Aqueous	200 mg/kg body weight	Sprague Dawley rats Swiss Albino mice	Excision Incision	AM, AI	Choudhari et al. (2013)
Thymelaeaceae	*Daphne oleoides* subsp. *kurdica*	—	Aerial parts	Methanol	—	Sprague-Dawley rats (male) Swiss Albino rats	Linear incision Circular excision	AM, AI, AO	Suntar et al. (2012)
	Stellera chamaejasme L.	Himalayan stellar	Whole plant	Ethanol	3% w/w	Sprague Dawley rats	Cutaneous		Kim et al. (2017)
	Phaleria macrocarpa	God's crown	Fruits	Ethanol	—	Sprague Dawley rats	Full thickness excision		Abood et al. (2015)
Tiliacea	*Grewia tillifolia*	Dhaman	Stem bark	Methanol	—	Wistar rats	Excision Incision Dead space	AM, AO, AI	Ahamed et al. (2009)

(continued)

Table 1 (continued)

Plant family	Scientific name	Common name	Plant part analyzed for wound healing property	Type of extract	Dose	Ethno-pharmacological validation using specific Study model	Type of wound studied	Other properties	References
Typhaceae	*Typha elephantina* (Roxb.)	Cattail, elephant grass, bulrush	Infloresence	Methanol	3% w/w 5% w/w gel	Wistar Albino rats (either of sex)	Excision	AM, AI	Panda and Thakur (2014)
Umbelliferae	*Centella asiatica* L.	Indian penny wort Asia penny wort	Leaves	Ethanol	–	Albino Wistar rats	Excision Incision Dead space	AM	Shetty et al. (2006a, b)
Urticaceae	*Holoptelea integrifolia* (Roxb.)	Indian elm	Leaves Stem	Methanol	50 mg/wound area	Wistar rats (male)	Full thickness excision Incision	AM	Reddy et al. (2008)
Vitaceae	*Vitis vinifera*	Grapes	Fruits	Hydro alcoholic	–	Iranian rabbits (either of sex)	Full thickness excision	AM, AO, AI	Hemmati et al. (2011)
				Oil	–	Sprague Dawley rats (male)	Excision		Shivananda Nayak et al. (2011)
Zingiberaceae	*Curcuma longa*	Haridra, Turmeric	Rhizome	Ethanol	10% w/w	Albino rats	Excision Incision	AM, AF, AO, AI	Pawar et al. (2015)
	Kaempferia galanga	Aromatic ginger	Rhizome	Alcohol	–	Wistar rats	Excision Incision Dead space		Shanbhag et al. (2006)

Boesenbergia rotunda L. Mansf.	Chinese ginger Finger roots	Rhizomes	Ethanol	–	Sprague Dawley rats (male)	Excision Linear incision	Jitvaropas et al. (2012)
Boesenbergia longiflora (wall.)	Boesenbergia	Rhizomes	Ethanol	–	Swiss Albino mice (male and female) Wistar rats (male)	Excision Incision	Sudsai et al. (2016)

AM Antimicrobial, *AF* Antifungal, *AO* Antioxidant, *AI* Anti-Inflammatory

Table 2 Ethanobotanical information on Polyherbal Formulation used in the treatment of wounds

Ingredients	Common name	Plant part analysed for wound healing property	Type of extract	Dose	Ethnopharmacological validation using specific study model	Type of wound studied	Other properties	References
Azadirachta indica	Neem tree	Leaves	Ethanol		Wistar rats (male)	Excision	–	Kakade et al. (2017)
Tridax procumbens	Coatbuttons							
Honey								
Plumbago zeylanica L.	Ceyleon leadwort	Aerial parts	Methanol (4:4:2)	5% w/w	Wistar Albino rats	Excision Incision	AM, AI	Dev et al. (2019)
Datura stramonium L.	Thorn apple							
Argemone mexicana L.	Mexican prickly poppy							
Hippophae rhamnoides L.	Sea buckthorn	Leaves	Ethanol (1:7:1)		Diabetic Sprague Dawley rats (male)	Full thickness cutaneous wound	AI	Gupta et al. (2008)
A.barbadensis Miller	Aloe Vera	Rhizomes						
Curcuma longa L.	Turmeric							
Centella asiatica	Indian pennywort	Leaves	Ethanol	2% w/w	Wistar Albino rats	Excision	AM, AO, AI	Patel et al. (2011)
Curcuma longa	Turmeric	Rhizomes						
Terminalia arjuna	Arjun tree	Bark						

				Wistar rat (male)	Second-degree burn wound	AM, AO, AI	Soni et al. (2010)
Azadirachta indica	Neem tree	—	—				
Berberis aristata	Indianberberry						
Curcuma longa	Turmeric						
Glycyrrhiza glabra	Licorice						
Jasminum officinal	Common jasmine						
Pongamia Pinnata	Indian beech						
Rubia cordifolia	Indian madder						
Terminalia chebula	Chebulic myrobalan						
Trichosanthes dioica	Pointed guard						
Symplocos racemosa	Lodh tree						
Ichnocarpus frutescens	Black creeper						
Capsicum abbreviata	Capsicum						
Nymphaea lotus	Waterlily						

(continued)

Table 2 (continued)

Ingredients	Common name	Plant part analysed for wound healing property	Type of extract	Dose	Ethnopharmacological validation using specific study model	Type of wound studied	Other properties	References
Malva sylvestris	Common mallow	Flower	–		Wistar rat (male)	Burn	AI, AM, Carminative	Pirbaolouti et al. (2012)
Punica granatum	Pomegranate	Flower						
Amygdalus communis	Common almond	Leaves						
Arnebia euchroma	Pink arnebia	Root						
Scrophularia deserti	Desert figwort	Stem						
Cassia auriculata	Avaram	Whole plant	Aqueous	500 mg/kg	Wistar Albino rats	Excision Incision Dead space	AO	Majumder and Paridhavi (2019)
Mangifera indica	Mango							
Ficus banghalensis	Indian banayan							
Cinnamomum tamala	Indian bay leaf							
Curcuma longa	Turmeric	–	Ethanol	250 mg/kg body weight 500 mg/kg body weight	Wistar Albino rats	Excision Incision Dead space Burn	AO	Nasir et al. (2016)
Eclipta alba	False daisy							
Tridax procumbens	Coatbuttons							

Scientific name	Common name	Part used	Solvent/extract	Dose	Animal model	Wound type	Activity	Reference
Clinacanthus nutans	Snake grass	Leaves	Ethanol		Swiss Albino rats	Excision Incision Burn		Aslam et al. (2016)
Elephantopus scaber	Elephant foot							
Vitex negundo L.	Negundo Chastetree	Leaves	Aqueous		Wistar rats	Excision	AO	Talekar et al. (2017)
Emblica officinalis Gaertn	Indian gooseberry	Bark						
Tridax procumbens L.	Coatbuttons	Whole plant						
Citrus aurantium	Sweet orange	Peel	Ethanol	500 mg/ kg body weight	Rats	Excision	AI	Soujanya et al. (2020)
Curcuma longa	Turmeric	Rhizomes						
Emblica officinalis	Amla	Fruits						
Zingiber officinalis	Ginger	Rhizomes						
Aloe barbadensis	Aloe Vera	Leaves						
Ricinus communis	Castor oil	Oil						
Tectona grandis	Teak	Leaves	Methanol		Sprague Dawley rats	Excision Incision	AI	Kavitha et al. (2013)
Ficus religiosa	Sacred Fig							
Caesalpinia pulcherrima	Peacock flower							

(AM Antimicrobial, AF Antifungal, AO Antioxidant, AI Anti Inflammatory) and calceorioside B, homoplantaginin (hispidulin-7-O-glucoside), and plantamajoside from the aerial parts of *Plantago major subsp. major* L. (Genc et al. 2020) have also demonstrated their active role in the wound healing process. The ethanolic extract of the *Woodfordia fruticosa* flower found immune-modulatory effects during the primary phase of wound healing. Levels of pro-inflammatory cytokines TNF-α, IL-6, and anti-inflammatory cytokine IL-10 levels were found to be up-regulated (Verma et al. 2013). Amongst polyherbal preparations, ampucare was found to be most effective.

5 Conclusion

This chapter displays a vast variety of medicinal plants and polyherbal formulations showing wound healing properties that are tested and validated using In Vivo and In Vitro models. In spite of the rich knowledge regarding the traditional Medicare system, there is a lack of clinical evidences for most of these formulations. Other major challenge includes variation in the scientific proofs from the observations of the tribal communities. This may be because, the researchers screen the pharmaco-logical compound using organic solvent systems and sometimes water, whereas traditionally it is being used as paste or juice of the fresh plant material and sometimes they also being mixed with some other plant species so as to impart synergistic effects. Our observations also confirm that the polyherbal formulations are more effective for wound care even at low doses. Most of the available studies display the extent to which specific plants component or polyherbal formulation is effective in the wound healing process. It, therefore, behooves the strong need of the hour to find out novel bioactive compound(s) of the available formulations by characterizing the plant extracts in order to study its molecular targets, which can be beneficial to all for wound care.

References

Abood WN, Al-Henhena NA, Abood AN, Al-Obaidi MMJ, Ismail S, Abdulla MA, Al Batran R (2015) Wound-healing potential of the fruit extract of *Phaleria macrocarpa*. Bosn J Basic Med Sci 15(2):25

Abu-Al-Basal MA (2010) Healing potential of *Rosmarinus officinalis* L. on full-thickness excision cutaneous wounds in alloxan-induced-diabetic BALB/c mice. J Ethnopharmacol 131 (2):443–450

Adetutu A, Morgan WA, Corcoran O (2011) Antibacterial, antioxidant and fibroblast growth stimulation activity of crude extracts of *Bridelia ferruginea* leaf, a wound-healing plant of Nigeria. J Ethnopharmacol 133(1):116–119

Agarwal PK, Singh A, Gaurav K, Goel S, Khanna HD, Goel RK (2009) Evaluation of wound healing activity of extracts of plantain banana (*Musa sapientum var. paradisiaca*) in rats. Indian J Exp Biol 47(01):32–40

Agyare C, Asase A, Lechtenberg M, Niehues M, Deters A, Hensel A (2009) An ethnopharmacological survey and in vitro confirmation of ethnopharmacological use of medicinal plants used for wound healing in Bosomtwi-Atwima-Kwanwoma area, Ghana. J Ethnopharmacol 125(3):393–403

Agyare C, Bempah SB, Boakye YD, Ayande PG, Adarkwa-Yiadom M, Mensah KB (2013a) Evaluation of antimicrobial and wound healing potential of Justicia flava and Lannea welwitschii. Evid Based Complement Alter Med. Article ID 632927. https://doi.org/10.1155/2013/632927

Agyare C, Dwobeng AS, Agyepong N, Boakye YD, Mensah KB, Ayande PG, Adarkwa-Yiadom M (2013b) Antimicrobial, antioxidant, and wound healing properties of Kigelia africana (Lam.) Beneth. and Strophanthus hispidus DC. Adv Pharmacol Sci 2013:692613

Ahamed BMK, Krishna V, Malleshappa KH (2009) In vivo wound healing activity of the methanolic extract and its isolated constituent, gulonic acid γ-lactone, obtained from Grewia tiliaefolia. Planta Med 75(05):478–482

Annan K, Houghton PJ (2008) Antibacterial, antioxidant and fibroblast growth stimulation of aqueous extracts of Ficus asperifolia Miq. and Gossypium arboreum L., wound-healing plants of Ghana. J Ethnopharmacol 119(1):141–144

Anuar NS, Zahari SS, Taib IA, Rahman MT (2008) Effect of green and ripe Carica papaya epicarp extracts on wound healing and during pregnancy. Food Chem Toxicol 46(7):2384–2389

Arunachalam K, Parimelazhagan T (2013) Anti-inflammatory, wound healing and in-vivo antioxidant properties of the leaves of Ficus amplissima Smith. J Ethnopharmacol 145(1):139–145

Asif A, Kakub G, Mehmood S, Khunum R, Gulfraz M (2007) Wound healing activity of root extracts of Berberis lyceum Royle in rats. Phytother Res 21(6):589–591

Aslam MS, Ahmad MS, Mamat AS, Ahmad MZ, Salam F (2016) Antioxidant and wound healing activity of polyherbal fractions of Clinacanthus nutans and Elephantopus scaber. Evid Based Complement Altern Med. 2016:4685246

Atiba A, Ueno H, Uzuka Y (2010) The effect of aloe vera oral administration on cutaneous wound healing in type 2 diabetic rats. J Vet Med Sci 73(5):583

Bahadır-Acıkara Ö, Özbilgin S, Saltan-İşcan G, Dall'Acqua S, Rjašková V, Özgökçe F, Šmejkal K (2018) Phytochemical analysis of Podospermum and Scorzonera n-hexane extracts and the HPLC quantitation of triterpenes. Molecules 23(7):1813

Balaji S, Watson CL, Ranjan R, King A, Bollyky PL, Keswani SG (2015) Chemokine involvement in fetal and adult wound healing. Adv Wound Care 4(11):660–672

Balekar N, Nakpheng T, Katkam NG, Srichana T (2012) Wound healing activity of ent-kaura-9 (11), 16-dien-19-oic acid isolated from Wedelia trilobata (L.) leaves. Phytomedicine 19(13):1178–1184

Baranoski S, Ayello EA (2008) Wound care essentials: practice principles. Lippincott Williams & Wilkins, Philadelphia

Barua CC, Begum SA, Sarma DK, Pathak DC, Borah RS (2012) Healing efficacy of methanol extract of leaves of Alternanthera brasiliana Kuntze in aged wound model. J Basic Clin Pharm 3(4):341

Begashaw B, Mishra B, Tsegaw A, Shewamene Z (2017) Methanol leaves extract Hibiscus micranthus Linn exhibited antibacterial and wound healing activities. BMC Complement Altern Med 17(1):337

Belachew TF, Asrade S, Geta M, Fentahun E (2020) In vivo evaluation of wound healing and anti-inflammatory activity of 80% methanol crude flower extract of Hagenia abyssinica (Bruce) JF Gmel in mice. Evid Based Complement Altern Med 2020:9645792

Bhandirge SK, Tripathi AS, Bhandirge RK, Chinchmalatpure TP, Desai HG, Chandewar AV (2015) Evaluation of wound healing activity of ethanolic extract of Pongamia pinnata bark. Drug Res 65(6):296–299

Bhaskar A, Nithya V (2012) Evaluation of the wound-healing activity of Hibiscus rosa sinensis L. (Malvaceae) in Wistar albino rats. Indian J Pharmacol 44(6):694

Bhat PB, Hegde S, Upadhya V, Hegde GR, Habbu PV, Mulgund GS (2016) Evaluation of wound healing property of *Caesalpinia mimosoides* Lam. J Ethnopharmacol 193:712–724

Bianchi ME, Manfredi AA (2009) Dangers in and out. Science 323(5922):1683–1684

Bowler PG (2002) Wound pathophysiology, infection and therapeutic options. Ann Med 34 (6):419–427

Brem H, Sheehan P, Boulton AJ (2004) Protocol for treatment of diabetic foot ulcers. Am J Surg 187(5):S1–S10

Chah KF, Eze CA, Emuelosi CE, Esimone CO (2006) Antibacterial and wound healing properties of methanolic extracts of some Nigerian medicinal plants. J Ethnopharmacol 104(1–2):164–167

Chandra P, Yadav E, Mani M, Ghosh AK, Sachan N (2015) Protective effect of *Lygodium flexuosum* (family: Lygodiaceae) against excision, incision and dead space wounds models in experimental rats. Toxicol Ind Health 31(3):274–280

Chard R (2008) Wound classifications. AORN J 88(1):108–110

Chaudhari M, Mengi S (2006) Evaluation of phytoconstituents of *Terminalia arjuna* for wound healing activity in rats. Phytother Res 20(9):799–805

Chen WC, Liou SS, Tzeng TF, Lee SL, Liu IM (2012) Wound repair and anti-inflammatory potential of *Lonicera japonica* in excision wound-induced rats. BMC Complement Altern Med 12(1):226

Chen J, Jayachandran M, Xu B, Yu Z (2019) Sea bass (*Lateolabrax maculatus*) accelerates wound healing: a transition from inflammation to proliferation. J Ethnopharmacol 236:263–276

Chitra V, Dharani PP, Pavan KK, Alla NR (2009) Wound healing activity of alcoholic extract of *Buchanania lanzan* in Albino rats. Int J ChemTech Res 1(4):1026–1031

Choudhari AS, Raina P, Deshpande MM, Wali AG, Zanwar A, Bodhankar SL, Kaul- Ghanekar, R. (2013) Evaluating the anti-inflammatory potential of *Tectaria cicutaria* L. rhizome extract in vitro as well as in vivo. J Ethnopharmacol 150(1):215–222

Choudhary GP (2008) Wound healing activity of the ethanol extract of *Terminalia bellirica* Roxb. fruits. Indian J Nat Prod Res 7(1):19–21

da Silveira Vasconcelos M, Gomes-Rochette NF, de Oliveira MLM, Nunes-Pinheiro DCS, Tomé AR, Maia de Sousa FY et al (2015) Anti-inflammatory and wound healing potential of cashew apple juice (*Anacardium occidentale* L.) in mice. Exp Biol Med 240(12):1648–1655

Darenfed H, Grenier D, Mayrand D (1999) Acquisition of plasmin activity by *Fusobacterium nucleatum* subsp. nucleatum and potential contribution to tissue destruction during periodontitis. Infect Immun 67(12):6439–6444

Demilew W, Adinew GM, Asrade S (2018) Evaluation of the wound healing activity of the crude extract of leaves of *Acanthus polystachyus* Delile (Acanthaceae). Evid Based Complement Altern Med 2018:1–9

Deshmukh PT, Fernandes J, Atul A, Toppo E (2009) Wound healing activity of *Calotropis gigantea* root bark in rats. J Ethnopharmacol 125(1):178–181

Dev SK, Choudhury PK, Srivastava R, Sharma M (2019) Antimicrobial, anti-inflammatory and wound healing activity of polyherbal formulation. Biomed Pharmacother 111:555–567

Devaney L, Rowell KS (2004) Improving surgical wound classification—why it matters. AORN J 80(2):208–223

Dwivedi D, Dwivedi M, Malviya S, Singh V (2017) Evaluation of wound healing, anti-microbial and antioxidant potential of Pongamia pinnata in wistar rats. J Tradit Complement Med 7 (1):79–85

Edwards H, Finlayson K, Courtney M, Graves N, Gibb M, Parker C (2013) Health service pathways for patients with chronic leg ulcers: identifying effective pathways for facilitation of evidence based wound care. BMC Health Serv Res 13(1):86

Ellis S, Lin EJ, Tartar D (2018) Immunology of wound healing. Curr Dermatol Rep 7(4):350–358

Ezike AC, Akah PA, Okoli CO, Udegbunam S, Okwume N, Okeke C, Iloani O (2010) Medicinal plants used in wound care: a study of *Prosopis africana* (Fabaceae) stem bark. Indian J Pharm Sci 72(3):334

Ezzat SM, Choucry MA, Kandil ZA (2016) Antibacterial, antioxidant, and topical anti-inflammatory activities of *Bergia ammannioides*: a wound-healing plant. Pharm Biol 54 (2):215–224

Field CK, Kerstein MD (1994) Overview of wound healing in a moist environment. Am J Surg 167 (1):S2–S6

Fikru A, Makonnen E, Eguale T, Debella A, Mekonnen GA (2012) Evaluation of in vivo wound healing activity of methanol extract of *Achyranthes aspera* L. J Ethnopharmacol 143 (2):469–474

Gál P, Toporcer T, Grendel T, Vidová Z, Smetana K Jr, Dvořánková B et al (2009) Effect of *Atropa belladonna* L. on skin wound healing: biomechanical and histological study in rats and in vitro study in keratinocytes, 3T3 fibroblasts, and human umbilical vein endothelial cells. Wound Repair Regen 17(3):378–386

Ganeshkumar M, Ponrasu T, Krithika R, Iyappan K, Gayathri VS, Suguna L (2012) Topical application of *Acalypha indica* accelerates rat cutaneous wound healing by up-regulating the expression of type I and III collagen. J Ethnopharmacol 142(1):14–22

Gangwar M, Gautam MK, Ghildiyal S, Nath G, Goel RK (2015) *Mallotus philippinensis* Muell.Arg fruit glandular hairs extract promotes wound healing on different wound model in rats. BMC Complement Altern Med 15(1):123

Gebrehiwot M, Asres K, Bisrat D, Mazumder A, Lindemann P, Bucar F (2015) Evaluation of the wound healing property of *Commiphora guidottii* Chiov. ex. Guid. BMC Complement Altern Med 15(1):282

Genc Y, Dereli FTG, Saracoglu I, Akkol EK (2020) The inhibitory effects of isolated constituents from *Plantago major subsp. major* L. on collagenase, elastase and hyaluronidase enzymes: potential wound healer. Saudi Pharm J 28(1):101–106

Ghosh PK, Gaba A (2013) Phyto-extracts in wound healing. J Pharm Pharm Sci 16(5):760–820

Ghosh S, Samanta A, Mandal NB, Bannerjee S, Chattopadhyay D (2012) Evaluation of the wound healing activity of methanol extract of *Pedilanthus tithymaloides* (L.) Poit leaf and its isolated active constituents in topical formulation. J Ethnopharmacol 142(3):714–722

Gomes FS, Spínola CDV, Ribeiro HA, Lopes MT, Cassali GD, Salas CE (2010) Wound- healing activity of a proteolytic fraction from *Carica candamarcensis* on experimentally induced burn. Burns 36(2):277–283

Gomez R, Cancio LC (2007) Management of burn wounds in the emergency department. Emerg Med Clin North Am 25(1):135–146

Govindarajan R, Vijayakumar M, Chadana Venkateshwar R, Shirwaikar A, Mehrotra S, Pushpangadan P (2004) Healing potential of *Anogeissus latifolia* for dermal wounds in rats. Acta Pharma 54(4):331–338

Gupta N, Jain UK (2011) Investigation of wound healing activity of methanolic extract of stem bark of *Mimusops elengi* Linn. Afr J Tradit Complement Altern Med 8(2):98–103

Gupta A, Kumar R, Pal K, Banerjee PK, Sawhney RC (2005) A preclinical study of the effects of seabuckthorn (*Hippophae rhamnoides* L.) leaf extract on cutaneous wound healing in albino rats. Int J Low Extrem Wounds 4(2):88–92

Gupta A, Kumar R, Upadhyay NK, Pal K, Kumar R, Sawhney RC (2007) Effects of *Rhodiola imbricata* on dermal wound healing. Planta Med 73(8):774

Gupta A, Upadhyay NK, Sawhney RC, Kumar R (2008) A polyherbal formulation accelerates normal and impaired diabetic wound healing. Wound Repair Regen 16(6):784–790

Gupta A, Verma S, Gupta AK, Jangra M, Pratap R (2015) Evaluation of *Prosopis cineraria* (Linn.) Druce leaves for wound healing activity in rats. Ann Pharm Res 3:70–74

Gurung S, Škalko-Basnet N (2009) Wound healing properties of *Carica papaya* latex: in vivo evaluation in mice burn model. J Ethnopharmacol 121(2):338–341

Gutierrez RP (2006) Evaluation of the wound healing properties of *Acalypha langiana* in diabetic rats. Fitoterapia 77(4):286–289

Hemmati AA, Aghel N, Rashidi I, Gholampur Aghdami A (2011) Topical grape *(Vitis vinifera)* seed extract promotes repair of full thickness wound in rabbit. Int Wound J 8(5):514–520

Hopman WM, Harrison MB, Coo H, Friedberg E, Buchanan M, Van DenKerkhof EG (2009) Associations between chronic disease, age and physical and mental health status. Chronic Dis Can 29(3):108–116

Ikobi E, Igwilo CI, Azubuike CP, Awodele O (2012) Antibacterial and wound healing properties of methanolic extract of dried fresh *Gossypium barbadense* leaves. Asain J Biomed Pharm Sci 21:32–39

Iorio V, Troughton LD, Hamill KJ (2015) Laminins: roles and utility in wound repair. Adv Wound Care 4(4):250–263

Irfan-Maqsood M (2018) Classification of wounds: know before research and clinical practice. J Genes Cells 4(1):1–4

Jain N, Jain R, Jain A, Jain DK, Chandel HS (2010) Evaluation of wound-healing activity of *Acorus calamus* Linn. Nat Prod Res 24(6):534–541

Jaiprakash B, Chandramohan D (2006) Burn wound healing activity of *Euphorbia hirta*. Anc Sci Life 25(3–4):16

Jayapal J, Tangavelou AC, Panneerselvam A (2014) Studies on the plant diversity of Muniandavar sacred groves of Thiruvaiyaru, Thanjavur, Tamil Nadu, India. Hygeia 6(1):48–62

Jia Y, Zhao G, Jia J (2008) Preliminary evaluation: the effects of *Aloe ferox* Miller and *Aloe arborescens* Miller on wound healing. J Ethnopharmacol 120(2):181–189

Jitvaropas R, Saenthaweesuk S, Somparn N, Thuppia A, Sireeratawong S, Phoolcharoen W (2012) Antioxidant, antimicrobial and wound healing activities of *Boesenbergia rotunda*. Nat Prod Commun 7(7):901–912

Jones JE, Nelson EA (2007) Skin grafting for venous leg ulcers. Cochrane Database Syst Rev (2). https://doi.org/10.1002/14651858.CD001737.pub3

Joshi A, Sengar N, Prasad SK, Goel RK, Singh A, Hemalatha S (2013) Wound-healing potential of the root extract of *Albizzia lebbeck*. Planta Med 79(09):737–743

Joshi A, Joshi VK, Pandey D, Hemalatha S (2016) Systematic investigation of ethanolic extract from *Leea macrophylla*: implications in wound healing. J Ethnopharmacol 191:95–106

Kakade AS, Pagore RR, Biyani KR (2017) Evaluation of wound healing activity of polyherbal gel formulation. World J Pharm Res 6:501–509

Kamath JV, Rana AC, Roy Chowdhury A (2003) Pro-healing effect of *Cinnamomum zeylanicum* bark. Phytother Res 17(8):970–972

Karodi R, Jadhav M, Rub R, Bafna A (2009) Evaluation of the wound healing activity of a crude extract of *Rubia cordifolia* L (Indian madder) in mice. Int J Appl Res Nat Prod 2(2):12–18

Kavitha AN, Deepthi V, Nayeem N (2013) Design, formulation and evaluation of a polyherbal ointment for its wound healing activity. Pharmacophore 4(5):175–180

Kerstein MD (1997) The scientific basis of healing. Adv Skin Wound Care 10(3):30–36

Kim M, Lee HJ, Randy A, Yun JH, Oh SR, Nho CW (2017) *Stellera chamaejasme* and its constituents induce cutaneous wound healing and anti-inflammatory activities. Sci Rep 7(1):1–12

Kiran K, Asad M (2008) Wound healing activity of *Sesamum indicum* L seed and oil in rats. Indian J Exp Biol 46(11):777–782

Kodati DR, Burra S, Kumar GP (2011) Evaluation of wound healing activity of methanolic root extract of *Plumbago zeylanica* L. in wistar albino rats. Asian J Plant Sci Res 1(2):26–34

Kokane DD, More RY, Kale MB, Nehete MN, Mehendale PC, Gadgoli CH (2009) Evaluation of wound healing activity of root of *Mimosa pudica*. J Ethnopharmacol 124(2):311–315

Kumar S, Rajput R, Patil V, Udupa AL, Gupta S, Rathnakar UP et al (2011) Wound healing profile of *Asparagus racemosus* (Liliaceae) wild. Curr Pharm Res 1(2):111–114

Kundu A, Ghosh A, Singh NK, Singh GK, Seth A, Maurya SK et al (2016) Wound healing activity of the ethanol root extract and polyphenolic rich fraction from *Potentilla fulgens*. Pharm Biol 54(11):2383–2393

Landén NX, Li D, Ståhle M (2016) Transition from inflammation to proliferation: a critical step during wound healing. Cell Mol Life Sci 73(20):3861–3885

Latha LY, Darah I, Jain K, Sasidharan S (2012) Pharmacological screening of methanolic extract of Ixora species. Asian Pac J Trop Biomed 2(2):149–151

Lingaraju GM, Krishna V, Joy Hoskeri H, Pradeepa K, Venkatesh, Babu PS (2012) Wound healing promoting activity of stem bark extract of *Semecarpus anacardium* using rats. Nat Prod Res 26 (24):2344–2347

Lodhi S, Singhai AK (2013) Wound healing effect of flavonoid rich fraction and luteolin isolated from *Martynia annua* Linn. on streptozotocin induced diabetic rats. Asian Pac J Trop Med 6 (4):253–259

Luo P, Li X, Ye Y, Shu X, Gong J, Wang J (2018) *Castanea mollissima* shell prevents an over expression of inflammatory response and accelerates the dermal wound healing. J Ethnopharmacol 220:9–15

Maan P, Yadav KS, Yadav NP (2017) Wound healing activity of *Azadirachta indica* A. juss stem bark in mice. Pharmacogn Mag 13(Suppl 2):S316

Majumder P, Paridhavi M (2019) A novel poly-herbal formulation hastens diabetic wound healing with potent antioxidant potential: a comprehensive pharmacological investigation. Pharm J 11 (2):324–331

Malviya N, Jain S (2009) Wound healing activity of aqueous extract of *Radix paeoniae* root. Acta Pol Pharm 66(5):543–547

Manjunatha BK, Vidya SM, Rashmi KV, Mankani KL, Shilpa HJ, Singh SJ (2005) Evaluation of wound-healing potency of *Vernonia arborea* Hk. Indian J Pharmacol 37(4):223

Manjunatha BK, Krishna V, Vidya SM, Mankani KL, Manohara YN (2007) Wound healing activity of *Lycopodium serratum*. Indian J Pharm Sci 69(2):283

Mondal S, Suresh P (2012) Wound healing activity of *Cleome rutidosperma* DC. roots. Int Curr Pharm J 1(6):151–154

Moreo K (2005) Understanding and overcoming the challenges of effective case management for patients with chronic wounds. Case Manager 16(2):62–67

Morton LM, Phillips TJ (2016) Wound healing and treating wounds: differential diagnosis and evaluation of chronic wounds. J Am Acad Dermatol 74(4):589–605

Muhammad AA, Arulselvan P, Cheah PS, Abas F, Fakurazi S (2016) Evaluation of wound healing properties of bioactive aqueous fraction from *Moringa oleifera* Lam on experimentally induced diabetic animal model. Drug Des Devel Ther 10:1715

Mukherjee PK, Suresh B (2000) The evaluation of wound-healing potential of *Hypericum hookerianum* leaf and stem extracts. J Altern Complement Med 6(1):61–69

Mulisa E, Asres K, Engidawork E (2015) Evaluation of wound healing and anti-inflammatory activity of the rhizomes of *Rumex abyssinicus* J.(Polygonaceae) in mice. BMC Complement Altern Med 15(1):341

Murthy S, Gautam MK, Goel S, Purohit V, Sharma H, Goel RK (2013) Evaluation of in vivo wound healing activity of *Bacopa monniera* on different wound model in rats. Biomed Res Int 2013:972028

Murti K, Kumar U (2012) Enhancement of wound healing with roots of *Ficus racemosa* L. in albino rats. Asian Pac J Trop Biomed 2(4):276–280

Murti K, Kumar U, Panchal M (2011) Healing promoting potentials of roots of *Ficus benghalensis* L. in albino rats. Asian Pac J Trop Med 4(11):921–924

Mustoe T (2004) Understanding chronic wounds: a unifying hypothesis on their pathogenesis and implications for therapy. Am J Surg 187(5):S65–S70

Nagori BP, Solanki R (2011) Role of medicinal plants in wound healing. Res J Med Plant 5 (4):392–405

Naji S, Zarei L, Pourjabali M, Mohammadi R (2017) The extract of *lycium depressum* stocks enhances wound healing in streptozotocin-induced diabetic rats. Int J Low Extrem Wounds 16 (2):85–93

Nasir MM, Mahammed NL, Roshan S, Ahmed MW (2016) Wound healing activity of poly herbal formulation in albino rats using excision wound model, incision wound model, dead space wound model and burn wound model. Int J Res Dev Pharm Life Sci 5:2080–2087

Nayak BS (2006) *Cecropia peltata* L (Cecropiaceae) has wound-healing potential: a preclinical study in a Sprague Dawley rat model. Int J Low Extrem Wounds 5(1):20–26

Nayak BS, Mohan K (2007) Short communication influence of ethanolic extract of *jasminum Grandflorum* linn flower on wound healing activity in rats. Indian J Physiol Pharmacol 51 (2):189–194

Nayak BS, Suresh R, Rao AVC, Pillai GK, Davis EM, Ramkissoon V, McRae A (2005) Evaluation of wound healing activity of *Vanda roxburghii* R. Br (Orchidacea): a preclinical study in a rat model. Int J Low Extrem Wounds 4(4):200–204

Nayak S, Nalabothu P, Sandiford S, Bhogadi V, Adogwa A (2006) Evaluation of wound healing activity of *Allamanda cathartica* L. and *Laurus nobilis* L. extracts on rats. BMC Complement Altern Med 6(1):12

Nayak BS, Anderson M, Pereira LP (2007) Evaluation of wound-healing potential of *Catharanthus roseus* leaf extract in rats. Fitoterapia 78(7–8):540–544

Nayak BS, Raju SS, Chalapathi Rao AV (2008) Wound healing activity of *Persea americana* (avocado) fruit: a preclinical study on rats. J Wound Care 17(3):123–125

Nayak BS, Sandiford S, Maxwell A (2009) Evaluation of the wound-healing activity of ethanolic extract of *Morinda citrifolia* L. leaf. Evid Based Complement Altern Med 6:351–356

Nayak BS, Kanhai J, Milne DM, Swanston WH, Mayers S, Eversley M, Rao AC (2010) Investigation of the wound healing activity of *Carapa guianensis* L.(Meliaceae) bark extract in rats using excision, incision, and dead space wound models. J Med Food 13(5):1141–1146

Nayak BS, Ramdeen R, Adogwa A, Ramsubhag A, Marshall JR (2012) Wound-healing potential of an ethanol extract of *Carica papaya* (Caricaceae) seeds. Int Wound J 9(6):650–655

Nelson EA, Cullum N, Jones J (2006) Venous leg ulcers. Clin Evid 15:2607–2626

Nguyen VL, Truong CT, Nguyen BCQ, Vo TNV, Dao TT, Nguyen VD, Bui CB (2017) Anti-inflammatory and wound healing activities of calophyllolide isolated from *Calophyllum inophyllum* Linn. PLoS One 12(10):0185674

Okoli CO, Akah PA, Okoli AS (2007) Potentials of leaves of *Aspilia africana* (Compositae) in wound care: an experimental evaluation. BMC Complement Altern Med 7(1):1–7

Özbilgin S, Acıkara ÖB, Akkol EK, Süntar I, Keleş H, İşcan GS (2018) In vivo wound-healing activity of *Euphorbia characias subsp. wulfenii*: isolation and quantification of quercetin glycosides as bioactive compounds. J Ethnopharmacol 224:400–408

Panda V, Thakur T (2014) Wound healing activity of the inflorescence of *Typha elephantina* (Cattail). Int J Low Extrem Wounds 13(1):50–57

Patel NA, Patel M, Patel RP (2011) Formulation and evaluation of polyherbal gel for wound healing. Int Res J Pharm 1(1):15–20

Pattanayak SP, Sunita P (2008) Wound healing, anti-microbial and antioxidant potential of *Dendrophthoe falcata (Lf)* Ettingsh. J Ethnopharmacol 120(2):241–247

Pattnaik A, Sarkar R, Sharma A, Yadav KK, Kumar A, Roy P et al (2013) Pharmacological studies on *Buchanania lanzan* Spreng. A focus on wound healing with particular reference to anti-biofilm properties. Asian Pac J Trop Biomed 3(12):967–974

Pawar RS, Chaurasiya PK, Rajak H, Singour PK, Toppo FA, Jain A (2013) Wound healing activity of *Sida cordifolia* Linn. in rats. Indian J Pharm 45(5):474

Pawar RS, Toppo FA, Mandloi AS, Shaikh S (2015) Exploring the role of curcumin containing ethanolic extract obtained from *Curcuma longa* (rhizomes) against retardation of wound healing process by aspirin. Indian J Pharm 47(2):160

Pirbalouti AG, Azizi S, Koohpayeh A (2012) Healing potential of Iranian traditional medicinal plants on burn wounds in alloxan-induced diabetic rats. Rev Bras 22(2):397–403

Plikus MV, Guerrero-Juarez CF, Ito M, Li YR, Dedhia PH, Zheng Y et al (2017) Regeneration of fat cells from myofibroblasts during wound healing. Science 355(6326):748–752

Powers JG, Higham C, Broussard K, Phillips TJ (2016) Wound healing and treating wounds: chronic wound care and management. J Am Acad Dermatol 74(4):607–625

Queen D, Orsted H, Sanada H, Sussman G (2004) A dressing history. Int Wound J 1(1):59–77

Raina R, Parwez S, Verma PK, Pankaj NK (2008) Medicinal plants and their role in wound healing. Online Vet J 3(1):21

Rasal AS, Nayak PG, Baburao K, Shenoy RR, Mallikarjuna Rao C (2009) Evaluation of the healing potential of *Schrebera swietenioides* in the dexamethasone-suppressed wound healing in rodents. Int J Low Extrem Wounds 8(3):147–152

Rathi BS, Bodhankar SL, Baheti AM (2006) Evaluation of aqueous leaves extract of *Moringa oleifera* Linn for wound healing in albino rats. Indian J Exp Biol 44(11):898–901

Reddy BS, Reddy RKK, Naidu VGM, Madhusudhana K, Agwane SB, Ramakrishna S, Diwan PV (2008) Evaluation of antimicrobial, antioxidant and wound-healing potentials of *Holoptelea integrifolia*. J Ethnopharmacol 115(2):249–256

Reinke JM, Sorg H (2012) Wound repair and regeneration. Eur Surg Res 49(1):35–43

Robinson MM, Zhang X (2011) The world medicines situation 2011. WHO, Geneva

Romero-Cerecero O, Zamilpa A, Díaz-García ER, Tortoriello J (2014) Pharmacological effect of *Ageratina pichinchensis* on wound healing in diabetic rats and genotoxicity evaluation. J Ethnopharmacol 156:222–227

Roy P, Amdekar S, Kumar A, Singh R, Sharma P, Singh V (2012) In vivo antioxidative property, antimicrobial and wound healing activity of flower extracts *of Pyrostegia venusta* (Ker Gawl) Miers. J Ethnopharmacol 140(1):186–192

Saenthaweesuk S, Jitvaropas R, Somparn N, Thuppia A (2015) An investigation of antimicrobial and wound healing potential of *Allium ascalonicum* Linn. J Med Assoc Thai 98:S22–S27

Sanwal R, Chaudhary AK (2011) Wound healing and antimicrobial potential of *Carissa spinarum* Linn. in albino mice. J Ethnopharmacol 135(3):792–796

Sasidharan S, Logeswaran S, Latha LY (2012) Wound healing activity of *Elaeis guineensis* leaf extract ointment. Int J Mol Sci 13(1):336–347

Schmidt C, Fronza M, Goettert M, Geller F, Luik S, Flores EMM et al (2009) Biological studies on Brazilian plants used in wound healing. J Ethnopharmacol 122(3):523–532

Selvaraj N, Lakshmanan B, Mazumder PM, Karuppasamy M, Jena SS, Pattnaik AK (2011) Evaluation of wound healing and antimicrobial potentials of *Ixora coccinea* root extract. Asian Pac J Trop Med 4(12):959–963

Semwal DK, Mishra SP, Chauhan A, Semwal RB (2015) Adverse health effects of tobacco and role of Ayurveda in their reduction. J Med Sci 15(3):139

Sen CK (2019) Human wounds and its burden: an updated compendium of estimates. Adv Wound Care 8(2):39–48

Sen S, Chakraborty R, De B (2011) Challenges and opportunities in the advancement of herbal medicine: India's position and role in a global context. J Herbal Med 1(3–4):67–75

Shakeel F, Alam P, Anwer MK, Alanazi SA, Alsarra IA, Alqarni MH (2019) Wound healing evaluation of self-nanoemulsifying drug delivery system containing *Piper cubeba* essential oil. 3Biotech 9(3):1–9

Shanbhag TV, Sharma C, Adiga S, Bairy LK, Shenoy S, Shenoy G (2006) Wound healing activity of alcoholic extract of *Kaempferia galanga* in Wistar rats. Indian J Physiol Pharmacol 50 (4):384–390

Shaw TJ, Rognoni E (2020) Dissecting fibroblast heterogeneity in health and fibrotic disease. Curr Rheumatol Rep 22(8):1–10

Shetty BS, Udupa SL, Udupa AL, Somayaji SN (2006a) Effect of *Centella asiatica* L (Umbelliferae) on normal and dexamethasone-suppressed wound healing in Wistar Albino rats. Int J Low Extrem Wounds 5(3):137–143

Shetty S, Udupa SL, Udupa AL, Vollala VR (2006b) Wound healing activities of bark extract of *Jatropha curcas* Linn in albino rats. Saudi Med J 27(10):1473–1476

Shirwaikar A, Shenoy R, Udupa AL, Udupa SL, Shetty S (2003) Wound healing property of ethanolic extract of leaves of *Hyptis suaveolens* with supportive role of antioxidant enzymes. Indian J Exp Biol 41(3):238–241

Shivananda Nayak B, Dan Ramdath D, Marshall JR, Isitor G, Xue S, Shi J (2011) Wound-healing properties of the oils of *Vitis vinifera* and *Vaccinium macrocarpon*. Phytother Res 25 (8):1201–1208

Shivhare Y, Singour PK, Patil UK, Pawar RS (2010) Wound healing potential of methanolic extract of *Trichosanthes dioica* Roxb (fruits) in rats. J Ethnopharmacol 127(3):614–619

Singh M, Govindarajan R, Nath V, Rawat AKS, Mehrotra S (2006) Antimicrobial, wound healing and antioxidant activity of *Plagiochasma appendiculatum* Lehm. et Lind. J Ethnopharmacol 107(1):67–72

Somboonwong J, Kankaisre M, Tantisira B, Tantisira MH (2012) Wound healing activities of different extracts of *Centella asiatica* in incision and burn wound models: an experimental animal study. BMC Complement Altern Med 12(1):1–7

Soni A, Dwivedi VK, Chaudhary M, Shrivastava SM, Naithani V (2010) Efficacy of ampucare: a novel herbal formulation for burn wound healing versus other burn medicines. Asian J Biol Sci 3(1):18–27

Soujanya K, Reddy KS, Kumaraswamy D, Reddy GV, Girija P, Sirisha K (2020) Evaluation of wound healing and antiinflammatory activities of new poly-herbal formulations. Indian J Pharm Sci 82(1):174–179

Stephen-Haynes J, Gibson E, Greenwood M (2014) Chitosan: a natural solution for wound healing. J Community Nur 28(1):48–53

Subramoniam A, Evans DA, Rajasekharan S, Nair GS (2001) Effect of *Hemigraphis colorata* (Blume) HG Hallier leaf on wound healing and inflammation in mice. Indian J Pharm 33 (4):283–285

Sudsai T, Wattanapiromsakul C, Tewtrakul S (2016) Wound healing property of isolated compounds from *Boesenbergia kingii* rhizomes. J Ethnopharmacol 184:42–48

Suguna L, Singh S, Sivakumar P, Sampath P, Chandrakasan G (2002) Influence of *Terminalia chebula* on dermal wound healing in rats. Phytother Res 16(3):227–231

Sumitra M, Manikandan P, Suguna L (2005) Efficacy of *Butea monosperma* on dermal wound healing in rats. Int J Biochem Cell Biol 37(3):566–573

Süntar IP, Akkol EK, Yalçın FN, Koca U, Keleş H, Yesilada E (2010a) Wound healing potential of *Sambucus ebulus* L. leaves and isolation of an active component, quercetin 3-O-glucoside. J Ethnopharmacol 129(1):106–114

Süntar IP, Akkol EK, Yılmazer D, Baykal T, Kırmızıbekmez H, Alper M, Yeşilada E (2010b) Investigations on the in vivo wound healing potential of *Hypericum perforatum* L. J Ethnopharmacol 127(2):468–477

Süntar I, Koca U, Keleş H, Akkol EK (2011) Wound healing activity of *Rubus sanctus* Schreber (Rosaceae): preclinical study in animal models. Evid Based Complement Altern Med 2011:816156

Süntar I, Akkol EK, Keles H, Yesilada E, Sarker SD, Arroo R, Baykal T (2012) Efficacy of *Daphne oleoides subsp. kurdica* used for wound healing: identification of active compounds through bioassay guided isolation technique. J Ethnopharmacol 141(3):1058–1070

Swamy HK, Krishna V, Shankarmurthy K, Rahiman BA, Mankani KL, Mahadevan KM et al (2007) Wound healing activity of embelin isolated from the ethanol extract of leaves of Embelia ribes Burm. J Ethnopharmacol 109(3):529–534

Talekar YP, Apte KG, Paygude SV, Tondare PR, Parab PB (2017) Studies on wound healing potential of polyherbal formulation using in vitro and in vivo assays. J Ayurveda Integr Med 8 (2):73–81

Taranalli AD, Kuppast IJ (1996) Study of wound healing activity of seeds of *Trigonella foenum graecum* in rats. Indian J Pharm Sci 58(3):117

Teoh SL, Latiff AA, Das S (2009) The effect of topical extract of *Momordica charantia* (bitter gourd) on wound healing in nondiabetic rats and in rats with diabetes induced by streptozotocin. Clin Exp Dermatol 34(7):815–822

Theilgaard-Mönch K, Knudsen S, Follin P, Borregaard N (2004) The transcriptional activation program of human neutrophils in skin lesions supports their important role in wound healing. J Immunol 172(12):7684–7693

Tiwari VK (2012) Burn wound: how it differs from other wounds? Indian J Plast Surg 45(2):364

Udegbunam SO, Udegbunam RI, Muogbo CC, Anyanwu MU, Nwaehujor CO (2014) Wound healing and antibacterial properties of methanolic extract of *Pupalia lappacea* Juss in rats. BMC Complement Altern Med 14(1):157

Udupa SL, Shetty S, Udupa AL, Somayaji SN (2006) Effect of *Ocimum sanctum* Linn. on normal and dexamethasone suppressed wound healing. Indian J Exp Biol 44(1):49–54

Umachigi SP, Kumar GS, Jayaveera KN, Dhanapal R (2007) Antimicrobial, wound healing and antioxidant activities of *Anthocephalus cadamba*. Afr J Tradit Complement Altern Med 4 (4):481–487

Upadhyay A, Chattopadhyay P, Goyary D, Mazumder PM, Veer V (2013) *Eleutherine indica* L. accelerates in vivo cutaneous wound healing by stimulating Smad-mediated collagen production. J Ethnopharmacol 146(2):490–494

Vafi F, Bahramsoltani R, Abdollahi M, Manayi A, Hossein Abdolghaffari A, Samadi N, Amin G, Hassanzadeh G, Jamalifar H, Baeeri M, Heidari M, Khanavi M (2016) Burn wound healing activity of lythrum salicaria l. and hypericum scabrum l. wounds. WNDS20160929-2. Epub ahead of print. PMID:27701123

Verma N, Amresh G, Sahu PK, Mishra N, Rao CV, Singh AP (2013) Wound healing potential of flowers extracts of *Woodfordia fruticosa* Kurz. Indian J Biochem Biophys 50(4):296–304

Villegas LF, Fernández ID, Maldonado H, Torres R, Zavaleta A, Vaisberg AJ, Hammond GB (1997) Evaluation of the wound-healing activity of selected traditional medicinal plants from Peru. J Ethnopharmacol 55(3):193–200

Wynn TA (2004) Fibrotic disease and the TH 1/TH 2 paradigm. Nat Rev Immunol 4(8):583–594

Xue C, Friedman A, Sen CK (2009) A mathematical model of ischemic cutaneous wounds. Proc Natl Acad Sci 106(39):16782–16787

Yadav E, Singh D, Yadav P, Verma A (2017) Attenuation of dermal wounds via downregulating oxidative stress and inflammatory markers by protocatechuic acid rich n-butanol fraction of *Trianthema portulacastrum* Linn. in Wistar Albino rats. Biomed Pharmacother 96:86–97

Yannas IV, Tzeranis DS, So PT (2017) Regeneration of injured skin and peripheral nerves requires control of wound contraction, not scar formation. Wound Repair Regen 25(2):177–191

Mainstreaming Traditional Practices for Wound Management

Bharat Patel, Vijay Kothari, and Niyati Acharya

1 Introduction

A significant public health issue remains to be the successful management of wounds (Järbrink et al. 2016), and delay to cure, or elongation of the wound healing cycle contributes to the additional economic and social burden on health care facilities, staff, and patients. A wound can be defined as a disruption in the continuity of the epithelial lining of the skin or mucosa. Injury, due to surgery or accident, results in destruction of tissue, disruption of blood vessels and extravasations of blood constituents, and hypoxia. Wound healing is a very complex and dynamic process and has three phases (inflammatory phase, proliferative phase, and maturation phase) which works by replacing devitalized and damaged cellular structures and layers of tissue. The damaged tissue is repaired by many underlying biochemical events integrated into an organized cascade of processes. Many developing countries have witnessed the prevalence and tremendous rise of different types of wounds such as chronic and acute wounds, pressure ulcers, venous stasis ulcers, and diabetic ulcers. Developing countries have exponentially increased average lifespan, which has contributed to the development of many degenerative diseases (Frykberg and Banks 2015).

Curing of wounds has been a matter of interest for many, starting from the days of *Sushruta* in Ancient India. *Sushruta Samhita* provides two different sections concerned with the healing of such wounds and mentions more than 100 plants both individually and in combination for the treatment of wounds. Sushruta not only listed techniques and medications for getting a clean wound supported by healing,

B. Patel · N. Acharya (✉)
Institute of Pharmacy, Nirma University, Ahmedabad, Gujarat, India
e-mail: niyati.acharya@nirmauni.ac.in

V. Kothari
Institute of Science, Nirma University, Ahmedabad, Gujarat, India

but also the medicines for managing scars associated with wounds (Kumar et al. 1994; Udupa et al. 1970). Traditional medicine (also known as indigenous or folk medicine or alternative medicine) knowledge database has been established over generation to generation in different cultures before the context of modern medicine has been used to manage various diseases. Traditional medical practices in human cultures before the introduction of modern science have been ancient and culturally-based medical practices. In accordance with the social and cultural heritage of particular nations, traditional medicine practices differ widely. The challenges of sustaining health and managing diseases are faced by the establishment of a medical system in every human society.

Around 80% of the population in Asian countries depends on traditional medicine for their basic health care-related needs. The uses of herbal, Ayurveda, Siddha, Unani, Acupuncture, and some other non-medical expertise and procedures are known as traditional or indigenous traditional medicines (World Health Organization 2002). The World Health Organization (WHO) defines natural remedies as "public health, strategies, values and skills that incorporate medicines based on plants, animals and minerals, spiritual medications, home remedies and exercises, used individually or jointly to cure, evaluate as well as prevent disease or establish quality of life-being" (World Health Organization 2019). Many traditional systems of medicine (TSM) like Ayurveda and Siddha from India, Traditional Chinese herbal medicine (TCM) from China, Traditional African Medicine (TAM) from Africa and Native American medicine are the classical examples of successfully adopted TSM practices. For certain traditional medicine procedures worldwide, the usage of plant species in the treatment of acute and chronic wounds is very popular. Depending on it, several species have been tested for their wound healing action throughout the tropical and subtropical parts of the world. But several medicinal plants and promising bioactives still need to be tested in terms of modern, more powerful, and cost-effective wound healing product development. The use of medicinal plants, animal products, methods of preparation, type of care and attention given to patients vary among different cultures in the traditional approach to wound care; however, severity and cause of the wound are some significant considerations. Generally, traditional medicines were overlooked by the modern medical systems after allopathy medicine was introduced to the world. Ayurveda, the Indian traditional system of medicine, is the most ancient system but still facing evidence-based issues with respect to global acceptance. Nevertheless, recent research and developments have arisen a positive wave in the perceptions of traditional medicines with respect to conventional medicine. In terms of wounds, modern medicine is slightly different from traditional practices. Traditional methodologies concentrate on natural resources such as water, plants, animals, and minerals, and the majority of the world's population continues to value and practice them widely. In modern medicine, the most frequently used antiseptic drugs for preliminary care of wounds contain povidone-iodine/or its combination, chlorhexidine, hydrogen peroxide-based preparation, silver nitrate formulations, silver sulfadiazine, and topical antibiotics. The routine clinical settings are typically used for the burns, cuts, and scratches of acute skin infections. Nevertheless, a continuous or overuse of topical antibiotics can trigger problems related to antimicrobial resistance amongst pathogens. Thus, consideration should be given to find an alternative methodology that improves the healing process, whether it generally progresses or is inhibited by different agents such as corticosteroids or anti-inflammatory agents. Natural bioactive research is always focused on ethno-botanical knowledge, and several drugs used today have been developed from such traditional medicinal plants. Traditional medicinal plant preparations have been successful in wound management because of their multifactorial advantages, including disinfection, debridement, and the provision of an appropriate environment for natural healing processes (Ghadi et al. 2016).

Evidence-based scientific claims on natural alternatives help to promote herbal cures as a complementary or alternative treatment. Traditional medication has begun to be accepted by people to fulfill their health needs through natural ways. The Indian government has also expressed their commitment and ability to facilitate the correct usage of traditional medicines. The Indian Medical Research Council has been releasing many publications and has established several scientific advisory panels working on conventional medicines and also promoting alternative therapies for the better management of various diseases (Tandon and Yadav 2017). Involvement of the alternative approach in mainstream therapy can only be possible with the wider acceptance of such practices with scientific evidence. Issues regarding standardization and commercialization, clinical evidence, and regulatory framework for such treatments must be resolved in order to ensure the safety and efficacy of such products globally.

2 Traditional Medicines and Wound Therapy

Majority of traditional systems of medicine advocate the use of plant-, animal-, and mineral-based products. They offer a wide range of medicinal plants to treat skin conditions, including cutting, injuries, and burns. Wound healing is a dynamic and orderly followed series of overlapping, interacting physiological processes with the four most important phases like coagulation, inflammation, proliferation/migration/re-epithelialization/granulation, and remodeling/maturation (Reinke and Sorg 2012; Xue and Jackson 2015). Many medicinal plant-based preparations have been reported to be used for the management of wounds because of their incredible ability to influence the process of wound healing at all different phases. Extensive research in the area of wound healing and management through plant-derived medicinal products has been carried out, and most of the studies involve screening of plant extracts and fractions for wound healing activity followed by bioactivity guided isolation of active principles. Many plant drugs from Ayurveda and TCM have well established their use in wound healing due to the presence of phenolics, flavonoids, essential oils, and polysaccharides. Some plant species like *Aloe vera*, *Centella asiatica*, *Panax ginseng*, *Curcuma longa*, *Shorea robusta*, *Carthamus tinctorius*, *Argyreia speciosa*, *Angelica sinensis*, *Azadirachta indica*, *Hypericum perforatum* have been reported with promising healing potential, and many of them are present in some popular products in several global markets. The combination of antioxidant, antibacterial, anti-inflammatory, and analgesic properties of many of the herbal products used for wounds. Products may be taken by mouth or applied directly to the wounds.

Some of the plants which were evaluated pharmacologically for their wound healing activity with possible underlying mechanisms have been listed in Table 1. Some available marketed formulations were also complied with for their activity in wound healing (Table 2).

2.1 Advantages of Drug Delivery System for Traditional Medicines

Many of the bioactives having wound healing potential like curcumin, asiaticosides, bacosides, and wide variety of flavonoids belong to the class of compounds having poor solubility and permeability, which ultimately results in poor bioavailability. Poor bioavailability of phytochemicals might also be due to their high intrinsic activity, poor absorption, rapid metabolism, and clearance from the body. With the advent of novel drug delivery systems, target-based delivery of bioactive have differing benefits over traditional wound formulations. Drug delivery system technology is effective in supplying wound healing medication at a fixed rate and delivery of actives at the target site, reducing the toxic effects with better drug bioavailability. Incorporation of bioactive into the suitably delivery system often significantly improves solubility, stability, reduced toxicity, enhanced biologic function, improved tissue macrophage transportation, stable absorption, and lessen physiochemical degradation (Reddy et al. 2012; Martins et al. 2015). The advanced nano formulations have smaller particles and a large volume-to-surface ratio that raises the probability of biological contact and absorption at wound location (World Health Organization 2019). These are suitable for repeated topical distribution of medications, provoking cell to cell reactions, cell proliferation, vascularization, cell signaling, and biomolecules needed for successful wound healing (Mihai et al. 2019). Micelles, hydrogel, nanoparticles, nanofibrer, and nanoemulsions are widely explored approaches for the target delivery of different bioactives like aloemannans, curcumin, epigallocatechin gallate, and dihydroquercetin (Marziyeh et al. 2018).

2.2 Quality Issues and Analytical Challenges for the Traditional Medicines

The detection and verification of bioactives and the safety of the subsequent herbal medicines are one of the main concerns in modern therapeutics. The analysis and standardization of traditional medicine are subjective but phytochemical screening

Table 1 Important traditional medicinal plants with wound healing actions

Botanical name	Vernacular name of the plant	Details	Wound healing activity	References
Achyranthes aspera Linn.	Aghedo	Aqueous extract of leaves at 800 mg/kg in Swiss albino mice	Anti-inflammatory activity, antioxidant activity	Bhosale et al. (2012)
Acacia catechu Wild	Kher (Baval)	Heartwood extract	Antioxidant activity, anti-inflammatory action, tissue protectant, and analgesic activities	Stohs and Bagchi (2015)
Abies webbiana Lindl	Tallish patra	Leaves extract at 400 mg/kg p.o. in Sprague Dawley rat	Anti-inflammatory activity	Nayak et al. (2004)
Azadirachta indica	Neem	5% ointment of leaves extract on rat	Anti-inflammatory, antibacterial, antifungal, and antiviral properties	Barua et al. (2010)
Tinospora cordifolia	Galo/Giloy	5% ointment of leaves extract on rat	Anti-inflammatory, antibacterial properties	Barua et al. (2010)
Asparagus racemosus	Shatavari	Whole plant extract at 500 mg/kg	Anti-inflammatory activity and prevention of ulceration	Vema et al. (2017)
Berberis aristata	Daru Haridra	Hydro-alcoholic extract of bark at 200 mg/kg in Wistar albino rats	Anti-granuloma activity and anti-inflammatory	Kumar et al. (2016)
Caesalpinia bonducella	Kaanchakaa	Hydroethanolic extract of seeds in Wistar rats	Anti-inflammatory	Ra et al. (2019)
Terminalia browniii	Saaj	Methanolic extracts of bark at 150 mg/kg in wistar rats	Anti-inflammatory	Mbiri et al. (2016)
Curcuma longa	Haldar	20% of ethanolic plant extract in albino rats	Inhibited the growth of wound associated pathogens and increased the rate of wound healing	Dons and Soosairaj (2018)
Cynodon dactylon	Drow	Aqueous extract of the plant (15% w/w) on Wister rats	Wound healing property through its antioxidative activity	Biswas et al. (2017)
Emblica officinalis	Amla	Bark extract on Wistar rats	Fast contraction of wound and antioxidant	Talekar et al. (2017)
Vitex negundo	Nirgundi/Nagod	Leaves extract on Wistar rats	Fast contraction of the wound and antioxidant activity	Talekar et al. (2017)

(continued)

Table 1 (continued)

Botanical name	Vernacular name of the plant	Details	Wound healing activity	References
Euphorbia characias subsp. wulfenii	Thor	1% ointment of hexane, ethyl acetate, and methanol extract of aerial parts of on Swiss albino mice	Antioxidant, anti-inflammatory, and wound healing activity	Özbilgin et al. (2018)
Apis cerana indica	Madh	Medical grade honey on rats	Reduced inflammation and cleared infection	Maruhashi (2020)
Terminalia chebula	Harde/ Haritaki	10% of fruit extracts in Wistar rats	Anti-inflammatory activity	Nasiri et al. (2015)
Terminalia bellirica	Baheda	Extract of leaves and fruits at 300 mg/kg in Wistar rat	Potent anti-inflammatory activity and promotes ulcer protection	Akter et al. (2019)
Rubia cordifolia	Manjistha	Methanolic extract of roots and fruits in Wistar rat	Anti-inflammatory, antibacterial, and antioxidant effects	Meena and Chaudhary (2015)
Glycyrrhiza glabra	Jethi Madh	3% aqueous extract of *G. glabra* in Sprague Dawley rats	Promoting accelerated wound healing activity	Zangeneh et al. (2019)
Luffa echinata	Kukadvel	Hydro-alcoholic extract of fruit in in vitro *activity*	Antioxidant and anti-inflammatory activity	Bhatt et al. (2019)
Operculina turpethum	Nasotar	Hydro-alcoholic leaf extract in in vitro activity	Antioxidant and anti-inflammatory activity	
Sphaeranthus indicus	Gorakhmundi	Hydro-alcoholic fruit extract in in vitro activity	Antioxidant and anti-inflammatory activity	
Cressa cretica	Rudanti	Hydro-alcoholic leaf extract in in vitro activity	Antioxidant and anti-inflammatory activity	
Corchorus depressus	Bahufali	Hydro-alcoholic root extract in in vitro activity	Antioxidant and anti-inflammatory activity	Bhatt et al. (2019)
Cassia absus	Chimed	Hydro-alcoholic seed extract in in vitro activity	Antioxidant and anti-inflammatory activity	
Acalypha indica	Vanchhi-kanto	Leaves and flowers extract on Sprague Dawley rats	Antioxidant activity also used for bedsores and wounds treatment	Yeng et al. (2019)
Aloe Vera	Kuvarpathu/ Dhritkumari	Pulp of leaves of Wistar albino rats	Antiseptic, antiulcer, antibacterial, antioxidant, anti-inflammatory, and wound healing properties	Purohit et al. (2012)

(continued)

Table 1 (continued)

Botanical name	Vernacular name of the plant	Details	Wound healing activity	References
Bryophyllum pinnatum	Ghaimari	Petroleum ether, alcohol (400 mg/kg orally), and water extract (2% suspension topically) of leaves of the plant on Wistar albino rats	Anti-inflammatory and wound healing property	Khan et al. (2004)
Calotropis gingantea	Aakdo	Methanolic extracts of the root on Wistar albino rats	Analgesic and anti-inflammatory activity	Maiti et al. (2017)
Centella asiatica	Brahmi	In vitro scratch assay on human dermal fibroblast (HDF) and human dermal keratinocyte (HaCaT) In vivo activity of asiaticoside rich fraction of leaves extract in rabbit (40%, 10%, and 2.5%, w/w topically)	Antioxidative activity, wound healing activity, and ability to induce collagen synthesis	Azis et al. (2017)
Elephantopus scaber Linn.	Galjibhi	Ethanolic extract whole plant of in vitro	Antioxidant and anti-inflammatory activity	Qi et al. (2020a)
Euphorbia nerrifolia Linn.	Bhungara Thor	Hydro-alcoholic extract of aerial parts of on in vitro activity	Analgesic effect and anti-inflammatory	Qi et al. (2020b)
Jatropha curcas	Ratanjot	10 and 15% cream from latex on mice	Angiogenesis activity	Balqis et al. (2018)
Ocimum sanctum	Tulsi	Aqueous extract of leaves on rabbits	Anti-inflammatory, immune modulatory and free radical scavenging activity	Gupta et al. (2016)
Ficus benghalensis	Banyan	Aqueous extract of leaves on Sprague Dawley rats	Antioxidant, wound healing	Chowdhary et al. (2014)
Ficus religiosa	Pipal	Aqueous extract of leaves on Sprague Dawley rats	Antioxidant, wound healing	

Table 2 Preclinical reports on wound healing activity of some herbal formulations

Formulation	Ingredients	Animal species under study	Wound healing actions	References
Polyherbal cream	5% aqueous extracts of *Chenopodium album, Coccinia indica, Momordica dioica, Praecitrullus fistulosus,* and *Trichosanthes dioica*	Wistar rats	Increased fibroblasts cells collagenation and angiogenesis	Shivhare and Jain (2020)
Gel and butter	5% ethanolic extract of Juca (*Libidibia ferrea*)	Dogs	Dermal healing through wound fibroplasia also exhibited antimicrobial activity	Américo et al. (2020)
Nano dispersion ointment	1% grape seed extract (proanthocyanidins)	Wistar rats	Inhibits the inflammatory response also improves the cell adhesion and proliferation	Rajakumari et al. (2020)
Chitosan-based gel formulation	Vitexin	Sprague Dawley rats	Increased cell proliferation and provide re-epithelization	Bektas et al. (2020)
Castor oil-based ointments	2, 4, and 6% w/w of extracts of *Zingiber officinale, Curcuma longa, Aloe barbadensis, Citrus aurantium,* and *Emblica officinalis*	Wistar rats	Anti-inflammatory and wound healing effects	Soujanya et al. (2020)
Jatyadi Ghrita (Ayurvedic dosage forms)	Jati—*Jasminum sambac*; Nimbapatra—*Azadirachta indica*; Patolapatra—*Trichosanthes dioica*; Katuka—*Picrorrhiza kurroa*; Darvi—*Berberis aristata*, Nisha—*Curcuma longa*; Sariva—*Hemidesmus indicus*; Manjishta—*Rubia cordifolia*; Abhaya—*Terminalia chebula*; Madhuka—*Glycyrrhiza glabra*; Naktahva—*Pongamia pinnata*; Siktaka-Honey bee wax; Tuttha—purified blue	Wistar rats	The period of epithelization of the burn wound significantly decreased, improvement in the percentage of wound contraction	Dhande Priti et al. (2012)

(continued)

Table 2 (continued)

Formulation	Ingredients	Animal species under study	Wound healing actions	References
	vitriol; Sarpi-Ghee			
Jatyadi Taila (Ayurvedic dosage forms)	Jati—*Myristica fragrans*; Nimba—*Azadirachta indica*; Patolapatra—*Stereospermum suaveolens*; Naktamala—*Pongamia pinnata;* Sikta-Honey bee wax; Madhuka—*Glycyrrhiza glabra*; Kushta—*Saussurea lappa*; Haridra—*Curcuma longa*; Daruharidra—*Berberis aristata*; Manjishta-*Rubia cordifolia*; Katurohini—*Picrorhiza kurroa*; Padmaka—*Prunus puddum*; Lodhra—*Symplocos racemosa*; Abhaya*Terminalia chebula*; Nilotpala—*Nymphaea stellata*; Tutthaka—copper sulfate; Sariva—*Hemidesmus indicus*; Naktamala beeja—seeds of *Pongamia pinnata*; Taila and water	Wistar rats	The period of epithelization of the burn wound significantly decreased, improvement in the percentage of wound contraction	Dhande Priti et al. (2012)
Cannabidiol with 2% tween 80 in saline	5 and 10 mg/kg given i.p	Wistar rat	Anti-inflammatory effect in the early phase of wound healing process in oral wounds	Klein et al. (2018)
Electrospun curcumin/gelatin-blended nanofibrous mats	1 g of gelatin, 0.1 g of curcumin per 10 ml of trifluoroethanol	Sprague Dawley rat	Persistent inflammatory response inhibition and decreased monocyte chemoattractant protein-1 expression by fibroblasts, mobilization of wound site fibroblasts by activating the Wnt signaling pathway	Dai et al. (2017)

(continued)

Table 2 (continued)

Formulation	Ingredients	Animal species under study	Wound healing actions	References
Polyherbal 2% and 5% gel	Methanolic extract of aerial parts with 4:4:2 proportion of each of *Datura stramonium* Linn., *Plumbago zeylanica* Linn., *Argemone Mexicana* Linn. respectively	Wistar albino rats	Antimicrobial activity, promoted the wound healing process and accelerated remodeling of damaged tissue	Dev et al. (2019)

methods followed by marker specific identification and characterization is the most common approach to ensure the quality of raw material and finished products.

Following substantial developments in modern scientific production, the quality assurance techniques applicable to herbal goods also have a large void. For research and development, the focus was put on utilizing single, fast, and therefore economical (i.e., TLC, HPTLC) methods for primary qualitative analysis or by using colloquial techniques (i.e., HPLC, UV, NMR) to allow the quantitative analysis of target or indicator molecules as well. UV spectroscopic research is being used to identify the marker molecules from the traditional herbs qualitatively and quantitatively. Infrared spectroscopy, NMR, and Mass Spectroscopy are being used to elucidate the structure of biologically active components. For standardization of herbal formulations, the usage of spectrophotometric techniques and marker molecules has its own merits and demerits. Certain chromatographic techniques like HPLC, HPTLC, and GC have broad applicability in the characterization of bioactives but are not cost-effective and lacking in real time application for polyherbal formulation standardization. (Laguerre et al. 2007; Mukherjee et al. 2011; Butt et al. 2018).

Innovations in genetic analysis have led to many advances in DNA marker techniques as an important tool for standardization and material validation for wound healing herbs (Ganie et al. 2015). The usage of DNA barcoding allows the recognition of individual bioactive using small standard DNA areas, defined as a barcode system of DNA. DNA barcoding is commonly used by the science community and industry for molecular recognition to address a broad variety of problems of taxonomy, molecular phylogenetic, population genetics, and biogeography. Work on DNA barcoding has made major progress in current years in the field of traditional medicines, as examined by Qiu et al. (2017). With the help of LC-MS and next-generation DNA sequencing techniques are used to detect the presence of adulterants and contaminants in place of the authentic herbal drugs in market formulations. cDNA microarray strategies have been proposed to understand the ability of bio actives like asiaticoside to induce functional gene expression in human dermal fibroblasts in in vitro study.

2.3 Regulatory Concerns

Traditional medicine is a great source for the new drug discovery with the principles of ethnopharmacology as a successful strategy to be followed. Although a plethora of natural products have been tested for their wound healing properties in vitro and in vivo, very few have reached the level of clinical trials and even rarer have been launched successfully in the market as approved medicines. Assimopoulou and Karapanagioti (2016) has reviewed many traditional plants from different cultures and reported that around only 6% of plants have been systematically investigated pharmacologically for their wound healing potential. In the perspective of this, more intensified efforts and emerging development are required to exploit the full potential of nature to design novel wound management dosage forms (Assimopoulou and Karapanagioti (2016). Further drug development is a very time-consuming process, and many natural therapeutics have not sufficiently complied with the global standards for wider acceptance. Country-specific regulations and lack of harmonized regulatory guidance for marketing and sale of safe and efficacious herbal products are some of the major features hampering the mainstreaming of such therapeutics.

2.4 Safety Aspects

Traditional medicines are usually deemed safe, focusing upon their long-term usage in different traditions. Scientific progress has helped to develop numerous wound healing products to help cure wounds and associated problems. But following administration, instances of significant adverse effects on the skin through allergic reaction, rashes, and other skin-related problems have been found with some natural products. The toxicity was linked in many instances to pollutants and to use of adulterant and non-authenticated sources of the drug. Many of the TWM (traditional wound medicine) being used as natural remedies, though, may also be highly poisonous in some instances. Ultimately, if not correctly tested, wound medicine can pose a chance of harmful consequences on the skin and surrounding mucosa. Therefore, the determination of the safety aspects and clinical toxicity studies are some of the major concerns (Wilson 2005). For an instance, honey is considered safe to be given as a wound healing agent, rarely leading to allergic reactions or adverse effects. However, some clinical studies reported that the use of honey may result in itching, as well as the interaction between honey and the wound site may be painful because of its acidic nature. However, with the growing number of clinical studies on the safety and therapeutic efficacy of herbal products, many more herbs can be implemented with full range of safety information being used in wound management.

3 Bridging the Ayurveda–Allopathy Gap for Wound Management

Modern medicine practitioners are normally informed of the framework of western medicine, their activity and sign of usage, drawbacks, and their role in medical care. Though traditional medicine practices mention quite a large number of herbal/herbo-mineral formulations for use in wound management, their widespread use in the modern world will be possible only after these formulations are characterized in terms of their chemical constituents and dosage, their wound healing potential is validated through carefully designed experiments, and the molecular mechanisms associated with their biological activity are elucidated. As a demonstration of how the modern scientific approach can be applied for studying traditional polyherbal formulations, we are presenting two examples here: *Panchvalkal* and Herboheal.

Panchvalkal is a polyherbal Ayurvedic formulation prescribed for diverse indi-cations, such as burns, bedsores, non-healing ulcers, etc. (Joshi and Vaidya 2013). Though it was since long assumed that its wound healing potential in part may be dependent on its antimicrobial properties, it was pending since long to elucidate the molecular mechanisms associated with its possible activity against wound-infec-tive bacteria. The first report describing the quorum-modulatory effect of this ancient formulation on different pathogenic bacteria viz. *Serratia marcescens*, *Chromobacterium violaceum*, and *Staphylococcus aureus* appeared in 2018 (Patel et al. 2018); wherein *Panchvalkal* at 250–750 microgram/ml, besides altering the production of the quorum sensing-regulated pigments in these bacteria, also exerted in vitro effect on antibiotic susceptibility, hemolytic potential, and catalase activity of the pathogens. In vivo assay did confirm the protective effect of this *panchvalkal* formulation on the nematode host *Caenorhabditis elegans* challenged with the pathogenic bacteria. Repeated exposure of *S. aureus* to *panchvalkal* was not found to induce resistance in this bacterium. Besides being capable of interfering with bacterial intercellular communication, this formulation also was reported to possess moderate prebiotic property. More details on its clinical use are available in Palep et al. (2016) and the monograph on this formulation edited by Joshi and Vaidya (2019). Recently a group led by one of the authors of the current chapter identified molecular targets of this formulation in the notorious pathogenic bacterium *Pseu-domonas aeruginosa* (Patel et al. (2019a)). This formulation in vitro was found to be capable of affecting QS-regulated traits (biofilm, pyocyanin, pyoverdine) of *Pseu-domonas aeruginosa*. It raised the susceptibility of the test bacterium to cephalexin and tetracycline antibiotics. Even repeated exposure of the bacterium to PF did not cause reversal of the effect of *Panchvalkal* formulation (PF) on QS-regulated traits of *P. aeruginosa*. In vivo efficacy of PF was demonstrated in the model worm host *Caenorhabditis elegans*, wherein PF-treated bacteria could kill lesser worms than their PF-unexposed counterparts. The whole transcriptome study indicated ~14% of the *P. aeruginosa* genome getting expressed differently under the influence of PF. Major mechanisms through which *Panchvalkal* seemed to display its anti-virulence effect are the generation of nitrosative and oxidative stress, and disturbing metal (molybdenum and iron) homeostasis, besides interfering with QS machinery.

Another polyherbal wound-care formulation, whose example we are presenting here is "Herboheal" whose formulation traces back to the folklore use of its crude form by certain tribal populations. Its antipathogenic efficacy against three multidrug-resistant strains of gram-negative bacterial pathogens often associated with wound infections was investigated by Patel et al. (2019a). Herboheal at $\geq 0.1\%$ v/v inhibited in vitro production of QS-regulated pigments in *C. violaceum*, *S. marcescens*, and *P. aeruginosa* by 19–55%. It seemed to disturb bacterial QS by acting as a signal-response inhibitor. This formulation suppressed the hemolytic activity of all three bacteria by $\sim$18–69%, while eliciting their catalase activity by $\sim$8–21%. Herboheal inhibited *P. aeruginosa* biofilm formation up to 40%, reduced surface hydrophobicity of *P. aeruginosa* cells by $\sim$9%, and increased (25%) their susceptibility to lysis in presence of human serum. Exposure of these test pathogens to Herboheal ($\geq 0.025\%$ v/v) could effectively reduce their virulence towards the nematode host *C. elegans*.

Repeated subculturing of *P. aeruginosa* on the Herboheal-supplemented growth medium was not found to induce resistance to Herboheal in this notorious pathogen, and this polyherbal preparation also exerted a post-extract effect (https://doi.org/10. 32388/359873) on *P. aeruginosa*, wherein virulence of the Herboheal-unexposed daughter cultures (of the Herboheal-exposed parent culture) was also attenuated. Molecular mechanisms associated with anti-infective property of Herboheal against another important pathogen, *Staphylococcus aureus*, were investigated by Patel et al. (2019b). This formulation had an inhibitory effect on bacterial growth and QS-regulated pigment (staphyloxanthin) production at $\geq 0.025\%$v/v. It not only inhibited *S. aureus* biofilm formation, but eradicated pre-formed biofilm too effectively. This formulation raised bacterial susceptibility to human serum heavily while compromising its hemolytic potential. Herboheal-treated bacteria expressed significantly reduced virulence towards the nematode worm *C. elegans*. Even repeated exposure of *S. aureus* to this polyherbal formulation was not observed to give rise to resistant phenotype. Whole transcriptome analysis indicated genes associated with virulence, quorum sensing, hemolysis, enzyme activity, transport, basic cellular processes, and transcriptional regulators as the major targets of Herboheal in *S. aureus*.

Above mentioned studies validate the traditional use of two different polyherbal formulations in wound care by demonstrating their efficacy against pathogenic bacteria commonly present in infected wounds. Information on the molecular targets of both above polyherbal formulations identified in two different pathogens can be useful in the development of novel anti-infectives against wound-infective bacteria. Such investigations on traditional formulations from a modern science perspective are a good demonstration of the therapeutic utility of the "polyherbalism" concept, inherent to many traditional medicinal practices of Asian origin. It also exemplifies the utility of modern "omics" tools for validation of the traditional medicine.

4   Latest Technologies for Wound Management

With the advent of the modern nano material concept, wound treatment materials are currently being marketed in the form of nano scaffolds, gels, films, and woven fabrics. The discovery of electrospinning is an advanced and effective method that produces fine fibers using electrical fields that can be reduced to nanometers that produce fiber pads for wound care. Because of the scaffolding design, these nano fibrous pads help in cell proliferation and can facilitate absorption of nutrients, exchange of gases, and excretion of waste (Zhang et al. 2007). Publications on electrospun matrices combined with allopathic drugs with various fiber phenotypes are known in different fields of study (Wilson 2005). Electrospinning matrices introduced for traditional medicines have generated considerable attention in the medical profession. This is considered to be made of complex carbohydrates, hemi-celluloses, anthraquinones, and antioxidants that can be spread on wounds for topical use (Marziyeh et al. 2018). It is also well known that curcumin spontaneously facilitates the production of collagen during the wound healing process. Chitosan nanoparticles loaded with curcumin impregnated in collagen-alginate scaffolds have been used successfully for diabetic wound healing with better tissue regeneration (Karri et al. 2016).

Chronic skin wound healing is a dynamic physical phenomenon that is controlled by various types of cells, growth factors, chemokines, and cytokines. Traditionally, wound healing mechanisms are categorized into four synchronized stages of inflammation, swelling, multiplication, and renovating how every wound needs to go through to recover naturally (Williams et al. 1950). Active components such as growth factors (GFs) and wound healing peptides have drawn many concerns based on their mechanisms capable of facilitating one or more stages of healing. Numerous formulations loaded with specific GFs were formulated for localized delivery, i.e., epidermal growth factor (EGF), vascular endothelial growth factor (VEGF), nerve growth factor (NGF), platelet-derived growth factor (PDGF-BB), etc. Many hydrogel-based formulas have been reported for their tissue engineering potential in wound therapy (Malafaya et al. 2007; Hasnain et al. 2010).

These days, dressings on most wounds are still unable to solve problems of resistance to bacterial infection, protein absorption, and elevated incidence of exudates. The key to satisfactorily solving the problem may be the use of inorganic substances like clay minerals, metal cations, zeolites, etc. Such products have demonstrated biocompatibility and the ability to promote cell adhesion, proliferation, cell differentiation, and absorption. Inorganic ingredients offer improved physicochemical and biological properties of wound dressings and have proved to be promising and easily accessible products in the treatment of chronic types of wounds (García-Villén et al. 2020). Hydrogel with natural inorganic clay material and spring water has been reported to accelerate wound healing with better biocompatibility in comparison to powder samples at the same concentration (Saghazadeh et al. 2018).

Researchers also have contributed significant attention to lipid nanoparticles (LNPs) because of their high flexibility, increased adhesion to skin and film formation, allowing hydration and skin integrity to be maintained, as well as more successful absorption through the skin barriers. Owing to the natural adhesive properties and ability to promote the healing process, some natural and biocompatible polysaccharides, including alginate, are commonly used as a wound dressing (Lee and Mooney 2012).

5 Preclinical and Clinical Research

The growing interest for Traditional Wound Healing Medicine (TWHM) by the consumer and the immense potential of natural products has resulted in the combined and holistic approach towards treatment aspects. TWHM therapy is based on largely traditional use and clinical experience (Dorai 2012). Comprehensive accounts on usefulness and observations from generation to generation provide some evidence of the efficacy of wound medicine. Nevertheless, there is a requirement for a clinical study to get further proof of its protection and efficacy. Investigating any function of traditional medicine in maintaining health, establishing a rational approach to policymaking in traditional medicine and, finally, focusing on how to harmonize classical medicine with modern medicine is the need of time. Several plant species are being medically tested by various preclinical pharmacological methods and later on patients in clinical trials.

Preclinical reports on plant drug extracts rich in phytochemicals, and bioactives like curcumin, asiaticoside, quercetin, arnebin-1, etc. are promising to offer cost-effective therapeutics for chronic wound healing and skin regeneration (Thangapazham et al. 2016). Unraveling possible modes of action with exact molecular mechanisms and targets of natural products can offer new insights into the therapeutics benefits of such bioactives in wound healing in the skin and other tissues.

Following the principles of ethnopharmacology, systematic screening of plant constituents may lead to a successful strategy for drug development. Assimopoulou and Karapanagioti (2016) has reported that the majority of approved medicines for wound healing are based on references of the traditional medicine systems, supporting this strategy as one of the most promising one to guide the drug development process. However, only 6% of plants have been systematically investigated pharmacologically for their wound healing potential, and more intensified efforts and emerging advancements are needed to exploit the potentials of nature for the development of novel medicines (Assimopoulou and Karapanagioti 2016).

The total number of natural products that successfully underwent clinical trials in patients with wounds, ulcers, and burns is relatively very small compared with preclinical trials reported. This could be due to the clinical trial failures attributed to poor bioavailability, difficulties in administration and efficacy, instability, and safety-related factors.

Majority of trials reported for the formulations have used topical application of such bioactives for external wounds, burns, and ulcers. Further, combining such bioactives with any biomaterials can help solve bioavailability-related constraints and can deliver the actives to the target sites more effectively. With the help of strong preclinical and clinical background only, substantial therapeutic claims for the traditional medicines can be proposed. Many traditional wound care formulations have reached the world market by testing the ethno-medicinal arguments of various ancient traditions.

Many plants from Ayurveda and from TCM have been evaluated and reported to have wound healing effects owing to their antimicrobial, antioxidant, anti-inflammatory, and angiogenesis inducing effects. The active ingredients, bioactivities, clinical uses, formulations, methods of preparation, and clinical value of 36 medical plant species have been described by Shedoeva et al. (2019). So many species are famous wound healing products used by many cultures and ethnic groups around the globe, such as *Centella asiatica*, *Curcuma longa*, and *Paeonia suffruticosa*. Out of those listed plants, many of them, for e.g., *Arctium lappa*, *Angelica sinensis*, *Paeonia Suffruticosa*, *Centella asiatica*, *Lithospermum erythrorhizon* have been reportedly available and used in the form of ointment successfully for wound care. *Aloe vera* and *Sophora flavescens* gels and essential oils from sources like *Blumea Balsamifera*, *Boswellia sacra*, *Calendula Officinalis*, *Camellia Sinensis*, *Carthamus tinctorius*, *Cinnamonum cassia*, *Commiphora myrrha*, *Lonicera japonica* have been extensively utilized for the preparation of spray and oils for wound management (Shedoeva et al. 2019).

We have enlisted some of the most studied traditional plants with the wound healing effects reported along with the preclinical details in Table 1. Preclinical evaluation of the various wound healing formulations with their mechanism of action is listed in Table 2. While clinical evaluation of different natural products and formulations have been listed along with clinical outcome in Table 3.

6 Limitations

Heavy metal intoxication and the presence of some endotoxins and aflatoxins in raw material are two major issues associated with traditional medicines and or Herbal Medicinal Products (HMPs).

Long-term use of traditional medicine has been found to contain heavy metal toxins, which can cause serious problems. The use of Ayurvedic herbal medicine products has been linked with lead, mercury, and arsenic poisoning. Estimated values on daily ingestion of such heavy metals have to be established compared with global regulatory standards.

Out of more than 6000 Ayurvedic medicines, it is estimated that 35–40% of the Ayurvedic medicines intentionally contain at least one metal. These metal-containing medicines are prepared through specific detoxification processes called the *Shodhana* process, involving multiple heating/cooling cycles and the addition of specific herbs and liquid medium.

Table 3 Clinical studies and clinical trials of natural products for Wound healing properties

Study type	Sample Size	Plant/formulations details	Mode of administration	Clinical outcome	References
Randomized clinical trial	90	*Salvia miltiorrhiza*	Intravenously every 12 h upto 3 days after mastectomy for breast carcinoma	Reduced wound complications with the reduction in skin flap ischemia and necrosis after mastectomy	Chen et al. (2010)
Randomized double-blind placebo-controlled study	16	Two traditional herb of Chinese herbs (NF3), comprised of *Astragali Radix* and *Radix Rehmanniae* extracts in ratio of 2:1	Powder granules were formulated and given two sachets daily (5 g/sachet)	Improved healing and sensation of wounds accompanied by concerted modifications in gene expression after 6 months treatment in diabetic foot ulcer	Ko et al. (2014)
Prospective non-randomized controlled study	57	*Calendula officinalis*	Plenusdermax, a bioactive extract of *Calendula officinalis* was converted into spray and applied twice a day for 30 weeks	Fourfold increase in percentage healing velocity per week in venous leg ulcers with complete epithelialization and wound contraction	Buzzi et al. (2016)
Single-arm, interventional trial	30	Herbal gel with Jasmine- *Jasminum grandiflorum* Licorice—*Glycyrrhiza glabra* Punarnava—*Boerhaavia diffusa* Triphala (*Terminalia chebula*, *Terminalia bellerica*, *Emblica officinalis*)	Topical application at the surgical site for 6 weeks post-surgery	Promote healing after gingivectomy	Koduganti et al. (2019)
Clinical studies	93	3% procyanidin (*Mimosa tenuiflora* and *Artemisia vulgaris*) extracts were prepared in glycerol and honey	The solution was applied to the wound surface area to form a thin film	Vital role in deep wound healing and also healing rate was faster	Shrivastava (2011)

(continued)

Table 3 (continued)

Study type	Sample Size	Plant/formulations details	Mode of administration	Clinical outcome	References
Randomized clinical trial	20	MEBO is USA patented pure herbal formula, containing *Phellodendron amurense, scutellaria baicalensis, Coptis chinensis, Pheretima aspergillum*, beeswax, and sesame oil	Moist exposed burn ointment (MEBO) at skin grafted sites	Improving the healing speed and re-epithelization process of skin grafted sites	Mabrouk et al. (2012), NCT02737943 (2019)
Randomized pilot study	28	Organic herbs: *Curcuma longa* root, *Hemidesmus indicus* root, *Rubia cordifolia* root, *Azadirachta indica* leaf, *Centella asiatica* leaf, *Tinospora cordifolia* stem, *Phyllanthus amarus* herb, *Phyllanthus emblica* fruit, and *Glycyrrhiza glabra* root	500 mg of turmeric and tablets of turmeric containing polyherbal combination given twice daily for 4 weeks	Decrease facial redness	Vaughn et al. (2019)
Randomized phase-III clinical trial	200	Made up of 5 components (*Radix Scutellariae, Rhizoma Coptidis, Cortex Phellodendri, Pheritima, Pericarpium Papaveris, Oleum Sesami*)	Moist exposed burn ointment (MEBO) once a day for 14 days	Evaluation of the analgesic effect, increased healing rate, and wound healing quality	CTRI/2010/091/000189 (2010)
Randomized clinical trial Phase3/4	90	Neutral Oral cream with active zinc chloride (0.1%) and antioxidants	Topical application, twice daily for 14 days	Wound healing time, healing rate, and wound healing quality	CTRI/2010/091/001242 (2010)
Phase I/II interventional trail	30	Vranaropana karma of shikari plant (*Cordia macleodii*)	Local application of *Ghrita* of *Cordia macleodii* leaves for 21 days local application	Wound healing and antimicrobial activities	CTRI/2011/12/002311 (2010)

Phase IV trial	12	*Centella Asiatica*	*C. asiatica* sheet (fibro heal) applied to gauze dressing for 21 days	Efficacy in the split-thickness in skin graft sites and partial thickness. Decrease healing time and scar formation	CTRI/2014/07/004732 (2013)
CT	100	Poly(3-hydroxybutyrate-co-3-hydroxy valerate) (PHBV) nano fibrous scaffolds of biodegradable natural polymer	The PHBV nano fibrous mat of 0.2 mm thick applied to the wounds	For burns, post-traumatic and diabetic wounds	CTRI/2013/11/004141 (2013)
Interventional phase-III	59	Absorbable gelatin sponge dressing	2 cm broad absorbable gelatin sponge over the incision site	For use to reduce pain and bleeding during dressing removal after neurosurgical procedures	CTRI/2014/05/004643 (2014)
Interventional post marketing surveillance	100	*Vranari Guggulu* and Vrana lepa	*Vranari Guggulu* 500 mg twice daily orally and *Karanjadi Vrana lepa* as external application for 45 days	Relief in wound area, pain and improve skin re-epitheliazation in diabetic foot ulcer	CTRI/2015/02/005587 (2015)
Interventional phase 2/phase 3	100	*Three Ayurvedic formulations Saubhagya sunthi pak, Satavari kalpa, and Balant kadha*	Twice a day with tab ziprax 200; *Saubhagya sunthi pak* 10 gm at Rasayankaal, i.e., 6 am, *Satavari Kalpa* 10 gm with milk, *Balanat Kadha* 10 gm with water	Reduced post-cesarean pain and boosting immunity	CTRI/2016/09/007314 (2016)

Many herbal extracts/products are found contamin wated with endotoxins (Yang et al. 2002) depending upon the method of preparation, storage, and experimental conditions. Endotoxins are lipopolysaccharides (LPS) correlated with gram-negative bacteria's cell membrane, which have distinct biological consequences.

Yang et al. (2002) reported that aqueous extract from *Chromolaena odorata* and polyphenolic extract from *Cudrania cochinchinensis* has mitogenic effects due to the presence of such LPS. The herbal extracts also would need to be screened for endotoxin and often should be examined for the impact of an LPS from *E. Coli* in wound healing effects and skin cell development effects in in vitro and in vivo models. Such LPS of *E. coli* origin on both normal human skin fibroblasts and keratinocytes in culture have shown stimulatory effects.

7 Future Aspects Towards Wound Care and Management

The growing usage of traditional drugs and treatments needs more clinically relevant support for the concepts underlying interventions and treatment efficacy. Recent wound care developments in the theoretical and biological sciences, coupled with advancements in genomics and proteomics, may play a major role in the validity of such therapies. Current rehabilitation involves healing medication and its delivery to the location of concern and fulfilling that aim in clinical care. Gene and stem cell therapies are evolving as a modern and successful path for improving wound healing pathology. Gene encoding for growth factors or cytokines has demonstrated the strongest ability to promote the healing of wounds. New approaches to functional genomics may help in the understanding of the cellular processes associated with tissue morphogenesis and enable molecular signals and routes to be classified (Eming et al. 2014). Genomics and proteomics tools will be used to profile the individual wounds to customize the treatment for patients. Gene therapy or cell-based therapy that delivers angiogenesis growth factors and recombinant protein medications can bring benefits over other tissue repair therapies. New growth factor requirements, such as angiogenic gene sutures, autologous stem cell transplantation, genetically altered tissue-engineered structures, and growth factor impregnated dressings or sprays, are currently under study as promising future trends in wound healing therapeutics (Yamakawa and Hayashida 2019).

The new drug delivery vehicles, created by nanotechnology, raise the innovative and stimulating potential of controlled and continuous delivery of drugs through the impenetrable skin barrier is a splendid clinical application of wound treatment options. Small size is a core aspect, but in order to obtain effectiveness as a topical delivery method, nanoparticles must also possess some intrinsic properties. Consequently, such ingredients must be capable of adapting as a portion of their design and targeted to applicable pathological variations.

The reactive oxidative species play a major role in significant cell damage at high amounts and may result in cellular proliferation, which also hinders the recovery process by disrupting cell components, DNA, proteins, and lipids (André-Lévigne et al. 2017). Therefore, if herbal extract has antioxidant ability and an increased antimicrobial function, it may be a successful therapeutic agent to speed up the cycle of wound healing. Oxidative trauma frequently takes a significant part in the cure of compromised wounds. This oxidative stress can trigger harm to developing tissue (Amini-Nik et al. 2018; Nethi et al. 2019).

Microcirculation is typically impaired in the case of burning wounds, due to a reduction or interruption of blood supply, which in effect induces ischemia and eventually reperfusion, which therefore creates oxidative stress. In theory, therapy that might improve blood supply or scavenge the free radicals could mitigate the harmful effects of oxidative stress. Therefore, by regulating oxidative stress, antioxidants rich herbal extracts may be predicted to encourage rapid epithelization. Botanicals with good antioxidant or free radical scavenging operation may play an important role in tissue regeneration also (Hasnain et al. 2010).

In specific in vivo tests, both pharmaceutically and biologically, the herbal extracts facilitate faster wound healing than control and non-medicated classes. Numerous herbal extracts encourage the healing process that consists of active agents such as triterpenes, alkaloids, flavonoids, tannins, saponins, anthraquinones, and other biomolecules. In experimental animals, a number of secondary plant isolates like alkaloids, terpenoids, phenolics have been found as active constituents responsible for implementing tissue repair (Thakur et al. 2011), as a result, many medicinal plants and bioactives were cited and known to have effects on wound healing and antioxidant properties.

The capacity to precisely interpret the microbiota has been transformed with the introduction of DNA barcoding techniques. Yet, knowledge of how the cutaneous microbiome progressively shifts during typical stages of wound healing is limited. The wound microbiota comprises living microbial communities ("microbiota") which communicate in the damaged tissues to each other and to their environments. The wound microbiota's function in compromised healing and development to problems associated with infection is a field of ongoing research that has gained from advancements in next-generation sequencing technologies. Finding alternative ways to modulate the microbiome in the hope of improving wound healing is of utmost importance, with increasing concerns about antibiotic over-prescription and the development of pharmacological resistance (Burmeister et al. 2018).

8 Conclusion

In this chapter, we have highlighted wound healing potential of traditional medicines with respect to the conventional modern medicines available for wound treatment. Factors affecting the development of promising wound healing agents have also been discussed so as to offer such bioactives as one of the mainstream therapeutic options for wound management. Different approaches for the target delivery of such potential classes of compounds can provide wider acceptance and better clinical outcomes. Quality issues and analytical techniques, safety and regulatory concerns along with the clinical reports on traditional medicines and formulations have been discussed to suggest the better future implementation of such therapies.

We present all observations in the assumption that we have had a lot to consider from existing approaches, some of which might certainly offer different ingredients and strategies for the clinical complexity of today. Combining traditional and modern expertise will produce better wound healing treatments with less adverse effects.

References

Akter S, Begum T, Begum R, Tamanna S, Tonny M, Yasmin S, Shifa F, Afroze F (2019) Phytochemical analysis and investigation of anti-inflammatory and anti-ulcer activity of *Terminalia bellirica* leaves extract. Int J Pharmacogn 6:54–65. https://doi.org/10.13040/IJPSR.0975-8232.IJP.6(2).54-65

Américo ÁVLDS, Nunes KM, de Assis FFV, Dias SR, Passos CTS, Morini AC, de Araújo JA, Castro KCF, da Silva SKR, Barata LES, Minervino AHH (2020) Efficacy of phytopharmaceuticals from the Amazonian plant *Libidibia ferrea* for wound healing in dogs. Front Vet Sci 7:244. https://doi.org/10.3389/fvets.2020.00244

Amini-Nik S, Yousuf Y, Jeschke MG (2018) Scar management in burn injuries using drug delivery and molecular signaling: current treatments and future directions. Adv Drug Deliv Rev 123:135–154

André-Lévigne D, Modarressi A, Pepper MS, Pittet-Cuénod B (2017) Reactive oxygen species and NOX enzymes are emerging as key players in cutaneous wound repair. Int J Mol Sci 18:2149

Assimopoulou AN, Karapanagioti EG (2016) Naturally occurring wound healing agents: an evidence-based review. Curr Med Chem 23:3285–3321. https://doi.org/10.2174/0929867323666160517120338

Azis HA, Taher M, Ahmed AS, Sulaiman WMAW, Susanti D, Chowdhury SR, Zakaria ZA (2017) *In vitro* and *In vivo* wound healing studies of methanolic fraction of *Centella asiatica* extract. S Afr J Bot 108:163–174. https://doi.org/10.1016/j.sajb.2016.10.022

Balqis U, Darmawi, Iskandar CD, Salim MN (2018) Angiogenesis activity of *Jatropha curcas* L. latex in cream formulation on wound healing in mice. Vet World 11:939–943. https://doi.org/10.14202/vetworld.2018.939-943

Barua CC, Talukdar A, Barua AG, Chakraborty A, Sarma RK, Bora RS (2010) Evaluation of the wound healing activity of methanolic extract of *Azadirachta Indica* (Neem) and *Tinospora cordifolia* (Guduchi) in rats. Pharmacologyonline 1:70–77

Bektas N, Şenel B, Yenilmez E, Özatik O, Arslan R (2020) Evaluation of wound healing effect of chitosan-based gel formulation containing vitexin. Saudi Pharm J 28:87–94. https://doi.org/10.1016/j.jsps.2019.11.008

Bhatt PR, Pandya KB, Patel UD, Modi CM, Patel HB, Javia BB (2019) Antidiabetic, antioxidant and anti-inflammatory activity of medicinal plants collected from the nearby area of Junagadh, Gujarat. Ann Phytomed 8(2):75–84. https://doi.org/10.21276/ap.2019.8.2.8

Bhosale U, Pophale P, Somani R, Yegnanarayan R (2012) Effect of aqueous extracts of *Achyranthes aspera* Linn. on experimental animal models for inflammation. Anc Sci Life 31:202. https://doi.org/10.4103/0257-7941.107362

Biswas TK, Pandit S, Chakrabarti S, Banerjee S, Poyra N, Seal T (2017) Evaluation of *Cynodon dactylon* for wound healing activity. J Ethnopharmacol 197:128–137. https://doi.org/10.1016/j.jep.2016.07.065

Burmeister DM, Taylor RJ, Gómez I, Matthew KM, Dubick MA, Christy RJ, Nicholson SE (2018) The cutaneous microbiome and wounds: new molecular targets to promote wound healing. Int J Mol Sci 19(9):2699. https://doi.org/10.3390/ijms19092699

Butt J, Ishtiaq S, Ijaz B, Mir ZA, Arshad S, Awais S (2018) Authentication of polyherbal formulations using PCR technique. Ann Phytomed 7(1):131–139. https://doi.org/10.21276/ap.2018.7.1.16

Buzzi M, De Freitas F, De Barros WM (2016) Therapeutic effectiveness of a *Calendula officinalis* extract in venous leg ulcer healing. J Wound Care 25(12):732–739. https://doi.org/10.12968/jowc.2016.25.12.732-739

Chen J, Lv Q, Yu M, Zhang X, Gou J (2010) Randomized clinical trial of Chinese herbal medications to reduce wound complications after mastectomy for breast carcinoma. Br J Surg 97:1798–1804. https://doi.org/10.1002/bjs.7227

Chowdhary N, Mohanjit K, Singh A, Kumar B (2014) Wound healing activity of aqueous extracts of *Ficus religiosa and Ficus benghalensis* leaves in rats. Indian J Res Pharm Biotechnol 2:2071–2081

CTRI/2010/091/000189 (2010) Safety and efficacy of topically applied moist exposed burn ointment (MEBO) compared to silver sulphadiazine (SSD) phase-iii clinical study protocol. http://www.ctri.nic.in/Clinicaltrials/pmaindet2.php?trialid=1402

CTRI/2010/091/001242 (2010) To evaluate and compare the efficacy and safety of neutral oral cream to heal mucosal wound against placebo on healthy subjects over a 14 day period of home use. http://www.ctri.nic.in/Clinicaltrials/pmaindet2.php?trialid=20391–3

CTRI/2011/12/002311 (2010) Vranaropana karma of shikari plant which is a folklore medicine of Orissa. A pharmacoclinical study of wound healing effect of *Cordia macleodii* leaf. Ghrit

CTRI/2013/11/004141 (2013) Electrospun nanofibrous scaffold for skin tissue engineering. http://14.139.13.47:8080/jspui/bitstream/10603/74068/9/09_abstract.pdf. Accessed 23 Aug 2020

CTRI/2014/05/004643 (2014) Is absorbable gelatin sponge a better post operative dressing than traditional dressings after surgery on the head? pp 1–5. http://www.who.int/trialsearch/Trial2.aspx?TrialID=CTRI/2014/05/004643

CTRI/2014/07/004732 (2013) Multicentric, open-label, safety and efficacy study of Fibroheal® (Centella Asiatica herbal wound healing sheet) in split- thickness skin graft donor site treatment and partial thickness wounds caused by Trauma/burns

CTRI/2015/02/005587 (2015) Effect of herbal medicine Vranari Guggulu and Vrana lepa for Diabetic Foot wound. pp 1–7. http://www.who.int/trialsearch/Trial2.aspx?TrialID=CTRI/2015/02/005587

CTRI/2016/09/007314 (2016) Collaboration approach of ayurveda and allopathic after post cesarean. pp 1–3. http://www.who.int/trialsearch/Trial2.aspx?TrialID=CTRI/2016/09/007314

Dai X, Liu J, Zheng H, Wichmann J, Hopfner U, Sudhop S, Prein C, Shen Y, Machens HG, Schilling AF (2017) Nano-formulated curcumin accelerates acute wound healing through Dkk-1-mediated fibroblast mobilization and MCP-1-mediated anti-inflammation. NPG Asia Mater 9:368. https://doi.org/10.1038/am.2017.31

Dev SK, Choudhury PK, Srivastava R, Sharma M (2019) Antimicrobial, anti-inflammatory and wound healing activity of polyherbal formulation. Biomed Pharmacother 111:555–567. https://doi.org/10.1016/j.biopha.2018.12.075

Dhande Priti P, Simpy R, Kureshee Nargis I, Sanghavi Dhara R, Pandit Vijaya A (2012) Burn wound healing potential of *Jatyadi* formulations in rats. Res J Pharm Biol Chem Sci 3:747–754

Dons T, Soosairaj S (2018) Evaluation of wound healing effect of herbal lotion in albino rats and its antibacterial activities. Clin Phytosci 4:6. https://doi.org/10.1186/s40816-018-0065-z

Dorai AA (2012) Wound care with traditional, complementary and alternative medicine. Indian J Plast Surg 45:418–424

Eming SA, Martin P, Tomic-Canic M (2014) Wound repair and regeneration: mechanisms, signaling, and translation. Sci Transl Med 6:265sr6

Frykberg RG, Banks J (2015) Challenges in the treatment of chronic wounds. Adv Wound Care 4:560–582. https://doi.org/10.1089/wound.2015.0635

Ganie SH, Upadhyay P, Das S, Prasad Sharma M (2015) Authentication of medicinal plants by DNA markers. Plant Gene 4:83–99

García-Villén F, Souza IMS, de Melo BR, Borrego-Sánchez A, Sánchez-Espejo R, Ojeda-Riascos S, Iborra CV (2020) Natural inorganic ingredients in wound healing. Curr Pharm Des 26:621–641. https://doi.org/10.2174/1381612826666200113162114

Ghadi R, Jain A, Khan W, Domb AJ (2016) Microparticulate polymers and hydrogels for wound healing. In: Wound healing biomaterials. Elsevier, Amsterdam, pp 203–225

Gupta V, Pathak S, Jain M (2016) Evaluation of burn wound healing property of *Ocimum sanctum* by monitoring of period of re-epithelization in rabbits. Int J Basic Clin Pharmacol 5:146–148. https://doi.org/10.18203/2319-2003.ijbcp20160117

Hasnain MS, Nayak AK, Ahmad F, Singh RK (2010) Emerging trends of natural-based polymeric systems for drug delivery in tissue engineering applications. Sci J UBU 1:1–13

Järbrink K, Ni G, Sönnergren H, Schmidtchen A, Pang C, Bajpai R, Car J (2016) Prevalence and incidence of chronic wounds and related complications: a protocol for a systematic review. Syst Rev 5:152. https://doi.org/10.1186/s13643-016-0329-y

Joshi J, Vaidya R (eds) (2013) Panchvalkal: a monograph. CCRAS, Ministry of Health & Family Welfare, Government of India, Delhi

Joshi J, Vaidya R (2019) Monograph: Panchavalkal (modified) for the treatment of leucorrhoea. Kasturba Health Society, Mumbai

Karri VVSR, Kuppusamy G, Talluri SV, Mannemala SS, Kollipara R, Wadhwani AD, Mulukutla S, Raju KRS, Malayandi R (2016) Curcumin loaded chitosan nanoparticles impregnated into collagen-alginate scaffolds for diabetic wound healing. Int J Biol Macromol 93:1519–1529. https://doi.org/10.1016/j.ijbiomac.2016.05.038

Khan M, Patil PA, Shobha JC (2004) Influence of *Bryophyllum pinnatum* (Lam.) leaf extract on wound healing in albino rats. J Nat Remedies 4:41–46. https://doi.org/10.18311/jnr/2004/380

Klein M, de Quadros De Bortolli J, Guimarães FS, Salum FG, Cherubini K, de Figueiredo MAZ (2018) Effects of cannabidiol, a *Cannabis sativa* constituent, on oral wound healing process in rats: clinical and histological evaluation. Phytother Res 32:2275–2281. https://doi.org/10.1002/ptr.6165

Ko CH, Yi S, Ozaki R, Cochrane H, Chung H, Lau W, Koon CM, Hoi SW, Lo W, Cheng KF, Lau CB, Chan WY, Leung PC, Chan JC (2014) Healing effect of a two-herb recipe on foot ulcers in Chinese patients with diabetes: a randomized double-blind placebo-controlled study. J Diabetes 6:323–334. https://doi.org/10.1111/1753-0407.12117

Koduganti RR, Reddy SP, Reddy PV, Prasanna JS, Gireddy H, Dasari R, Ambati M, Chandra GB (2019) Efficacy of low-level laser therapy, hyaluronic acid gel, and herbal gel as adjunctive tools in Gingivectomy wound healing: a randomized comparative clinical and histological study. Cureus 11(12):e6438. https://doi.org/10.7759/cureus.6438

Kumar S, Akhila A, Naqvi AA, Farooqi AH, Singh AK, Singh D, Uniyal GC, Srivastava GN, Gupta MM, Bindra RL, Aasan SA (1994) Medicinal plants in skin care. Central Institute of Medicinal and Aromatic Plants, Lucknow

Kumar R, Gupta YK, Singh S (2016) Anti-inflammatory and anti-granuloma activity of *Berberis aristata* DC in experimental models of inflammation. Indian J Pharmacol 48:155–161. https://doi.org/10.4103/0253-7613.178831

Laguerre M, Lecomte J, Villeneuve P (2007) Evaluation of the ability of antioxidants to counteract lipid oxidation: existing methods, new trends and challenges. Prog Lipid Res 46:244–282

Lee KY, Mooney DJ (2012) Alginate: properties and biomedical applications. Prog Polym Sci 37:106–126

Mabrouk A, Boughdadi NS, Helal HA, Zaki BM, Maher A (2012) Moist occlusive dressing (Aquacel ® ag) versus moist open dressing (MEBO ®) in the management of partial-thickness facial burns: a comparative study in Ain Shams University. Burns 38:396–403. https://doi.org/10.1016/j.burns.2011.09.022

Maiti PP, Ghosh N, Kundu A, Panda S, De B, Mandal SC (2017) Evaluation of anti-inflammatory and antinociceptive activity of methanol extract of *Calotropis gigantea* root. Int J Green Pharm 11:198–205. https://doi.org/10.22377/IJGP.V11I03.1126

Malafaya PB, Silva GA, Reis RL (2007) Natural-origin polymers as carriers and scaffolds for biomolecules and cell delivery in tissue engineering applications. Adv Drug Deliv Rev 59:207–233

Martins JT, Ramos ÓL, Pinheiro AC, Bourbon AI, Silva HD, Rivera MC, Cerqueira MA, Pastrana L, Malcata FX, González-Fernández Á, Vicente AA (2015) Edible bio-based nanostructures: delivery, absorption and potential toxicity. Food Eng Rev 7:491–513

Maruhashi E (2020) Honey in wound healing. In: Therapeutic dressings and wound healing applications. Wiley, New York, pp 235–254

Marziyeh H, Tewari D, Eduardo SS, Seyed MN, Mohammad HF, Mohammad A (2018) Natural product-based nanomedicines for wound healing purposes: therapeutic targets and drug delivery systems. Int J Nanomedicine 13:5023–5043

Mbiri JW, Kasili S, Patrick K, Mbinda W, Piero NM (2016) Anti-inflammatory properties of methanolic bark extracts of *Terminalia brownii* in Wistar albino rats. Int J Curr Pharm Rev Res 8:100–104

Meena V, Chaudhary AK (2015) Manjistha (*Rubia cordifolia*)—a helping herb in cure of acne. J Ayurveda Holist Med 3:11–17

Mihai MM, Dima MB, Dima B, Holban AM (2019) Nanomaterials for wound healing and infection control. Materials (Basel) 12(13):2176. https://doi.org/10.3390/ma12132176

Mukherjee PK, Ponnusankar S, Venkatesh P, Gantait A, Pal BC (2011) Marker profiling: an approach for quality evaluation of Indian medicinal plants. Ther Innov Regul Sci 45:1–14. https://doi.org/10.1177/009286151104500101

Nasiri E, Hosseinimehr SJ, Azadbakht M, Akbari J, Enayati-Fard R, Azizi S (2015) The effect of *Terminalia chebula* extract vs. silver sulfadiazine on burn wounds in rats. J Complement Integr Med 12:127–135. https://doi.org/10.1515/jcim-2014-0068

Nayak SS, Ghosh AK, Debnath B, Vishnoi SP, Jha T (2004) Synergistic effect of methanol extract of *Abies webbiana* leaves on sleeping time induced by standard sedatives in mice and anti-inflammatory activity of extracts in rats. J Ethnopharmacol 93:397–402. https://doi.org/10.1016/j.jep.2004.04.014

NCT02737943 (2019) Effect of MEBO dressing versus standard care on managing donor and recipient sites of split-thickness skin graft (EMG-SCZ-SGS)

Nethi SK, Das S, Patra CR, Mukherjee S (2019) Recent advances in inorganic nanomaterials for wound-healing applications. Biomater Sci 7:2652–2674

Özbilgin S, Acıkara ÖB, Akkol EK, Süntar I, Keleş H, İşcan GS (2018) *In vivo* wound-healing activity of *Euphorbia characias* subsp. wulfenii: isolation and quantification of quercetin glycosides as bioactive compounds. J Ethnopharmacol 224:400–408. https://doi.org/10.1016/j.jep.2018.06.015

Palep H, Kothari V, Patil S (2016) Quorum sensing inhibition: a new antimicrobial mechanism of Panchvalkal, an ayurvedic formulation. Bombay Hosp J 58(2):198–204

Patel P, Joshi C, Palep H, Kothari V (2018) Anti-infective potential of a quorum modulatory polyherbal extract (*Panchvalkal*) against certain pathogenic bacteria. J Ayurveda Integr Med 11 (3):336–343. https://doi.org/10.1101/172056

Patel P, Joshi C, Kothari V (2019a) Antipathogenic potential of a polyherbal wound-care formulation (herboheal) against certain wound-infective gram-negative bacteria. Adv Pharmacol Sci 2019:1739868. https://doi.org/10.1155/2019/1739868

Patel P, Joshi C, Kothari V (2019b) Anti-pathogenic efficacy and molecular targets of a polyherbal wound-care formulation (Herboheal) against *Staphylococcus aureus*. Infect Disord Drug Targets 19(2):193–206. https://doi.org/10.2174/1871526518666181022112552

Purohit SK, Solanki R, Soni MK, Mathur V (2012) Experimental evaluation of Indian Aloe (*Aloe Vera*) leaves pulp as topical medicament on wound healing. Int J Pharmacol Res. https://doi.org/10.7439/ijpr.v2i3.702

Qi R, Li X, Zhang X, Huang Y, Fei Q, Han Y, Cai R, Gao Y, Qi Y (2020a) Ethanol extract of *Elephantopus scaber* Linn. Attenuates inflammatory response via the inhibition of NF-κB signaling by dampening p65-DNA binding activity in lipopolysaccharide-activated macrophages. J Ethnopharmacol 250:112499. https://doi.org/10.1016/j.jep.2019.112499

Qi WY, Gao XM, Ma ZY, Xia CL, Xu HM (2020b) Antiangiogenic activity of terpenoids from *Euphorbia neriifolia* Linn. Bioorg Chem 96:103536. https://doi.org/10.1016/j.bioorg.2019.103536

Qiu D, Cook CE, Yue Q, Hu J, Wei X, Chen J, Liu D, Wu K, Adamowicz S (2017) Species-level identification of the blowfly *Chrysomya megacephala* and other Diptera in China by DNA barcoding. Genome 60:158–168. https://doi.org/10.1139/gen-2015-0174

Ra J, Somkuwar AP, Bhoye SK, Sarode KG, Limsay RP (2019) *In vivo* anti-inflammatory activity and GC-MS analysis of hydroethanolic extract of *Caesalpinia bonducella* seeds. J Pharmacogn Phytochem 8:929–934

Rajakumari R, Volova T, Oluwafemi OS, Rajeshkumar S, Thomas S, Kalarikkal N (2020) Nano formulated proanthocyanidins as an effective wound healing component. Mater Sci Eng C 106:110056. https://doi.org/10.1016/j.msec.2019.110056

Reddy LH, Arias JL, Nicolas J, Couvreur P (2012) Magnetic nanoparticles: design and characterization, toxicity and biocompatibility, pharmaceutical and biomedical applications. Chem Rev 112:5818–5878

Reinke JM, Sorg H (2012) Wound repair and regeneration. Eur Surg Res 49:35–43

Saghazadeh S, Rinoldi C, Schot M, Kashaf SS, Sharifi F, Jalilian E, Nuutila K, Giatsidis G, Mostafalu P, Derakhshandeh H, Yue K, Swieszkowski W, Memic A, Tamayol A, Khademhosseini A (2018) Drug delivery systems and materials for wound healing applications. Adv Drug Deliv Rev 127:138–166

Shedoeva A, Leavesley D, Upton Z, Fan C (2019) Wound healing and the use of medicinal plants. Evid Based Complement Alternat Med:1–30. https://doi.org/10.1155/2019/2684108

Shivhare Y, Jain AP (2020) Potential emphasis of formulated herbal cream on wound healing. Asian J Pharm Res Dev 8:73–77. https://doi.org/10.22270/ajprd.v8i1.639

Shrivastava R (2011) Clinical evidence to demonstrate that simultaneous growth of epithelial and fibroblast cells is essential for deep wound healing. Diabetes Res Clin Pract 92:92–99. https://doi.org/10.1016/j.diabres.2010.12.021

Soujanya K, Srinivas Reddy K, Kumaraswamy D, Vishwanath Reddy G, Girija P, Sirisha K (2020) Evaluation of wound healing and antiinflammatory activities of new poly-herbal formulations. Indian J Pharm Sci 82:174–179. https://doi.org/10.36468/pharmaceutical-sciences.636

Stohs SJ, Bagchi D (2015) Antioxidant, anti-inflammatory, and chemoprotective properties of *Acacia catechu* heartwood extracts. Phytother Res 29:818–824

Talekar YP, Apte KG, Paygude SV, Tondare PR, Parab PB (2017) Studies on wound healing potential of polyherbal formulation using *in vitro* and *in vivo* assays. J Ayurveda Integr Med 8:73–81. https://doi.org/10.1016/j.jaim.2016.11.007

Tandon N, Yadav SS (2017) Contributions of Indian Council of Medical Research (ICMR) in the area of medicinal plants/traditional medicine. J Ethnopharmacol 197:39–45

Thakur R, Jain N, Pathak R, Sandhu SS (2011) Practices in wound healing studies of plants. Evid Based Complement Altern Med 2011:17

Thangapazham RL, Sharad S, Maheshwari RK (2016) Phytochemicals in wound healing. Adv Wound Care 5:230–241. https://doi.org/10.1089/wound.2013.0505

Udupa K, Chaturvedi G, Tripathi S (1970) Advances in research in Indian medicine, Published by Banaras Hindu University, Varanasi.

Vaughn AR, Pourang A, Clark AK, Burney W, Sivamani RK (2019) Dietary supplementation with turmeric poly herbal formulation decreases facial redness: a randomized double-blind controlled pilot study. J Integr Med 17:20–23. https://doi.org/10.1016/j.joim.2018.11.004

Vema BK, Mukerjee A, Verma A, Bhushan S (2017) Preclinical screening of antiulcer activity of *Asparagus racemosus* extract on phenylbutazone induced ulceration in experimental animals. J Med Plant Res 5(2):348–352

Williams RW, Mason LB, Bradshaw HH (1950) Factors affecting wound healing. Surg Forum:410–417. https://doi.org/10.1177/0022034509359125

Wilson V (2005) Assessment and management of fungating wounds: a review. Br J Community Nurs 10(3):28–34

World Health Organization (2002) Traditional medicine: growing needs and potential. No. WHO/EDM/2002.4, vol. 2. Geneva: World Health Organization, p. 6

World Health Organization (2019) WHO global report on traditional and complementary medicine

Xue M, Jackson CJ (2015) Extracellular matrix reorganization during wound healing and its impact on abnormal scarring. Adv Wound Care 4:119–136. https://doi.org/10.1089/wound.2013.0485

Yang H, Kaneko M, He C, Hughes MA, Cherry GW (2002) Effect of a lipopolysaccharide from E. coli on the proliferation of fibroblasts and keratinocytes in vitro. Phytother Res 16 (1):43–47

Yamakawa S, Hayashida K (2019) Advances in surgical applications of growth factors for wound healing. Burns Trauma 7:10. https://doi.org/10.1186/s41038-019-0148-1

Yeng NK, Shaari R, Nordin ML, Sabri J (2019) Investigation of wound healing effect of *Acalypha indica* extract in Sprague Dawley rats. Biomed Pharmacol J 12:1857–1865. https://doi.org/10.13005/bpj/1816

Zangeneh A, Pooyanmehr M, Zangeneh MM, Moradi R, Rasad R, Kazemi N (2019) Therapeutic effects of *Glycyrrhiza glabra* aqueous extract ointment on cutaneous wound healing in Sprague Dawley male rats. Comp Clin Pathol 28:1507–1514. https://doi.org/10.1007/s00580-019-03007-9

Zhang YZ, Su B, Venugopal J, Ramakrishna S, Lim CT (2007) Biomimetic and bioactive nanofibrous scaffolds from electrospun composite nanofibers. Int J Nanomedicine 2:623–638

Traditional Probiotics, Next-Generation Probiotics and Engineered Live Biotherapeutic Products in Chronic Wound Healing

Shilpa Deshpande Kaistha and Neelima Deshpande

1 Introduction

Wounds are injuries that break the skin and other body tissues (Bryant and Nix 2012). These may be caused by cuts, abrasions, chemical, radiation or temperature (hot or cold) burns, gunshots, surgery or an underlying metabolic condition such as diabetes mellitus, atherosclerosis, arteriosclerosis or renal damage (Martin and Nunan 2015). Based on their ability to heal, wounds are broadly classified as acute wounds that typically follow the natural progression of healing irrespective of the severity of injury typically within 4 weeks of injury onset. On the contrary, 'wounds that fail to go through the normal healing phases in an orderly and time bound manner, usually taking more than three months to heal despite adequate diagnosis and intervention measures are classified as chronic wounds' (Frykberg and Banks 2015). Chronic wounds may occur superficially in the skin epidermis and dermis or may be present in deep tissues. All acute wounds may progress into chronic wounds based on infection type, inadequate blood supply, pressure on wound or incorrect hygiene and wound care. Microbial infections are the largest cause of progression of acute wounds into chronic wounds and strategies to prevent pathogenic infections that hinder the wounds healing process are important to address (Drago et al. 2019).

Usual wound treatment strategies include tissue debridement, compression, hyperbaric oxygen therapy, restoration of arterial inflow, removal of pressure, management of underlying systemic conditions and the use of antibiotic regimens (Han and Ceilley 2017). Despite the application of such integrated management, chronic wounds often appear to be arrested in the inflammatory stage without further

S. D. Kaistha (✉)
Department of Microbiology, Institute of Biosciences and Biotechnology, CSJM University, Kanpur, UP, India

N. Deshpande
BOUMS Health Clinic, Pune, MH, India

© The Author(s), under exclusive license to Springer Nature Singapore Pte Ltd. 2021   247
P. Kumar, V. Kothari (eds.), *Wound Healing Research*,
https://doi.org/10.1007/978-981-16-2677-7_8

progression to the curative proliferation and remodelling phases. Typically, an aggressive proinflammatory milieu and the presence of pathogenic microorganisms perpetuate the fight mode of the system without allowing a moderation of events in favour of healing (Frykberg and Banks 2015). Polymicrobial biofilms hosting a myriad of pathogens are capable of orchestrating a cellular and biochemical storm disrupting wound healing events (Scalise et al. 2015; Drago et al. 2019). *Staphylococcus aureus*, *Pseudomonas aeruginosa* and *Enterococcus faecalis*, are some of the main biofilm-forming organisms associated with wound infections. Biofilms harbouring multiple drug-resistant (MDR) pathogens further decrease the effectiveness of antimicrobial therapy (Scalise et al. 2015). The complex interplay of biofilm-forming pathogens and chronic inflammatory milieu in recurrent chronic wounds challenges the current wound treatment efficacy and alternative methods are increasingly being explored (Scalise et al. 2015). The increasing global antibiotic resistance threat further demands that we approach the problem by looking at out-of-the-box solutions.

A non-conventional strategy for chronic wound care would be to address one of the root causes of poor wound healing that involves bringing to balance the natural host–microbe interaction (Kadam et al. 2019). Live biotherapeutics (LBP) consist of biological products containing living organisms that can help in prevention, treatment and cure of disease or conditions (Charbonneau et al. 2020). LBP encompass within them different types of biotics and their products. Amongst these are (a) traditional probiotics are 'micro-organisms originally derived from fermented foods and faecal matter, scientifically proven to confer health benefits on host' (b) Next generation probiotics are defined as 'live commensal microorganisms identified on the basis of comparative microbiome analyses that, when administered in adequate amounts, confer a health benefit on the host', (c) Genetically engineered LBPs include probiotics genetically engineered to be safe and those that can act as biosensors, delivery agents etc. in addition to their general health benefits (Martín and Langella 2019).

Beneficial microorganisms inhabiting the human body popularly known as human microbiota play a crucial function in our health (Drago et al. 2019). Cutaneous dysbiosis is hence considered one of the primary causes of disruptive pathogenesis and delayed wound healing in chronic wounds (Sanford and Gallo 2013). Recent metagenomics studies comparing microbiome of healthy and non-healthy subjects have provided evidence that the use of beneficial microbiota or replenishment of unhealthy tissues with beneficial microbiota can help in reversing disease onset and severity (Sanford and Gallo 2013). The 'gut-brain-skin axis' model gives credence to the interrelationship between intestinal microbiome, emotional well-being, systemic as well local inflammations (Arck et al. 2010). The skin encompasses complex neuroendocrine and associated lymphoid tissues as well as phylogenetically diverse microbial communities defining its unique physiology and wellness (Slominski 2005). Gut microbiota influences the cutaneous tissue by enhancing absorption of nutrients with systemic immunomodulatory and hormonal effects (Cani 2018). Hence oral probiotics have distal wellness effects on cutaneous tissues while topical applications have also shown promising results (Yu et al. 2020).

In this chapter, the non-vaccine preventive and therapeutic application of micro-organisms and their products in the form of Live biotherapeutic products (LBP), which includes traditional probiotics, next generation probiotics and engineered live biotherapeutics administered orally or directly to the infected site for the treatment of cutaneous chronic wounds will be discussed. Some relevant terminologies related to LBP are defined in Box 1.

Box 1 List of Terminologies

Live Biotherapeutic Product (LBP): 'A biological product that contains live organisms and is applicable to the prevention, treatment or cure of a disease or condition of human beings; and is not a vaccine, virus or delivery vector' (Martín and Langella 2019).

Probiotic: 'Live microorganisms that, when administered in adequate amounts, confer a health benefit to the host' (Martín and Langella 2019).

Prebiotic: 'A non-digestible compound that, through its metabolization by microorganisms in the gut, modulates composition and/or activity of the gut microbiota, thus conferring a beneficial physiological effect on the host' (Martín and Langella 2019).

Synbiotic/Conbiotic: 'Dietary supplements that are a combination of probiotics and prebiotics that benefit the host by improving the survival and implantation of live microbial dietary supplements in the gastrointestinal tract, by selectively stimulating the growth and/or by activating the metabolism of one or a limited number of health-promoting bacteria' (Martín and Langella 2019).

Parabiotic: 'Non-living cells which when administered in sufficient amounts confer benefits to consumers' (Martín and Langella 2019).

Postbiotics: 'Intracellular soluble products or metabolites secreted by live cells which confer physiological health benefit and replace administration of live or non-viable bacteria' (Martín and Langella 2019).

Genobiotic: 'Probiotics that has as main mechanism of action that brings to a clinical improvement the change in gene expression' (Gorreja 2019).

Metagenomics: 'Collective genome of microorganisms from an environmental sample obtained directly by isolating the genome and circumventing culture techniques prior to using sequencing methodologies' (Escobar-Zepeda et al. 2015).

Metabolomics: 'Comprehensive analysis of all the metabolites present in a sample' (Escobar-Zepeda et al. 2015).

Microbiome: 'Complete collection of micro-organisms with their genome present in an environment' (Berg et al. 2020).

Microbiota: 'Set of micro-organisms inhabiting an environment or host' (Berg et al. 2020).

(continued)

Box 1 (continued)

Next Generation Sequencing: 'High throughput, massively parallel genome sequencing technologies that low cost and very fast compared to traditional DNA sequencing methods' (Escobar-Zepeda et al. 2015).

Next Generation Probiotics (*NGP*): 'Live microorganisms identified on the basis of comparative microbiota analyses that, when administered in adequate amounts, confer a health benefit on the host' (Martín and Langella 2019).

Engineered Live Biotherapeutic Product: 'Live Biotherapeutic organisms genetically engineered or modified for the prevention, treatment or cure of a disease or condition of human beings' (Ozdemir et al. 2018).

2 Chronic Wounds and Factors Impeding Their Healing

Chronic wounds are the cause of significant morbidity, mortality and high economic costs for patients (Olsson et al. 2019). Lack of proper wound healing in an orderly and timely fashion is characterized by chronic inflammation with increased levels of reactive oxygen species (ROS), protein degrading enzymes and cytokines leading to cellular senescence and microbial colonization (Morton and Phillips 2016; Zhao et al. 2016; Rahim et al. 2017). Multiple systemic and local factors can contribute to non-healing chronic wounds, i.e. age, vascular supply, nutritional factors, immune function and metabolic disorders (Balsa and Culp 2015). These types of wounds last for approximately 12 months and may recur in 60–70% of the patients resulting in high rates of morbidity (Frykberg and Banks 2015). Based on differential diagnosis, chronic wounds may be categorized as arterial/ischemic ulcers, diabetic ulcers, pressure/decubitus ulcers, venous ulcers and post-operative wound complications in patients with co-morbidities (Frykberg and Banks 2015).

Arterial/Ischemic Ulcers These occur due to inadequate oxygen and nutritionally rich blood supply (perfusion) to the affected body part resulting in tissue necrosis and ulceration. The common etiological factors include diabetes mellitus and its complications including vasculitis, peripheral neuropathy, peripheral artery disease, arteriosclerosis, atherosclerosis, joint immobility, hypertension, trauma injury or injury due to improper footwear, renal failure etc. These are usually observed on lower extremities such as lower legs, ankles, heels or toes (Morton and Phillips 2016). Arterial ulcers are typically characterized by well-defined rounded margins with a punched-out crater. Outer skin or nails have pale, taut, dry appearance with hair loss, grey or yellow fibrotic base, low temperature, poor pulse and underlying deep wounds. Wounds are typically painful and when untreated can result in complications such as tissue necrosis and require amputation in severe conditions (Morton and Phillips 2016).

Diabetic Ulcers These are major complications of uncontrolled diabetes mellitus reported in about 15% of patients being treated for ulceration and gangrene in the lower extremities (Frykberg and Banks 2015). Frequently underlying aetiologies include neuropathy, trauma and peripheral artery disease. Diabetic ulcers are graded based on severity as grade 0 (intact skin), grade 1 (superficial ulcer), grade 2 (deep ulcer to tendon, bone, or joint), grade 3 (deep ulcer with abscess or osteomyelitis), grade 4 (forefoot gangrene) and grade 5 (whole foot gangrene). In almost 85% of such infections, amputation is the final outcome (Zhao et al. 2016).

Pressure/Decubitus Ulcers When undue pressure is exerted on body parts such as joints, back or heel for extended periods of time due to impaired movement, it causes restriction of blood flow to the tissue parts resulting in sores (Morton and Phillips 2016). This is typically observed in hospital patients undergoing post-surgical operative complications, rehabilitation centres and home care bed-rest prescribed populations.

Venous Ulcers Chronic venous diseases such as varicose veins and chronic venous insufficiency often result from improper functioning of the venous valves. This causes blood to pool in veins resulting in inflammation, pain and subsequent ulceration. These are observed in lower extremities such as lower leg or ankles and usually present as large, superficial wounds with irregular margins. The associated oedema and atrophy causes loss of skin pigmentation. Venous leg ulcers (VLU) is a type of recurrent chronic wound estimated to affect quality of life style, work productivity and high health care costs (Morton and Phillips 2016).

Post-operative Wound Complications in Patients with Co-morbidities Certain patients show an increased risk of post-operative wound infection complication and dehiscence due to co-morbidities such as poorly controlled diabetes, hypothyroidism, nutritional deficiencies, peripheral vascular diseases, immunosuppressive conditions, cancer and tobacco smoking (Delmore et al. 2017). Nutrition deficiencies cause slow metabolic processes leading to poor wound healing post-surgery. Reduced blood circulation leads to ischemia and compromised oxygen delivery creates hypoxic conditions conducive to chronic non-healing surgical wounds. Dysregulated and suppressed healing inflammatory conditions in such patients also lead to delayed repair and increased incidence of wound infections and dehiscence following surgery.

2.1 Pathophysiology of Chronic Wound Healing

Normal wound healing consists of four phases which include homeostasis, inflammation, proliferation and remodelling (Fig. 1) (Morton and Phillips 2016). Homeostasis sets in an hour after injury and involves blood clotting and vasoconstriction. Platelets and epithelial cells release chemokines, cytokines and growth factors which help recruit leucocytes to areas of injury for the healing process (Ridiandries et al. 2018). Inflammation phase may occur from 5 to 7 days and is the interplay of the immune cells and any ensuing infectious organisms. Phagocytic cells such as

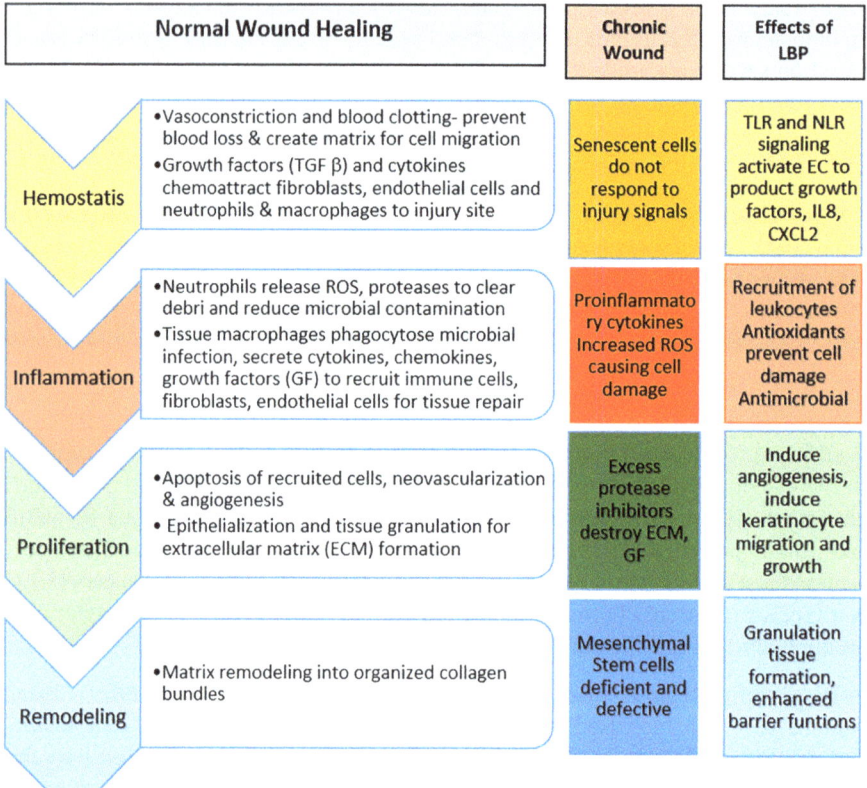

Fig. 1 Normal wound healing phases, chronic wound events and reversal effects by Live Biotherapeutic application. (**a**) Normal wound healing: Depiction of pathophysiological phases and events in normal wound healing process which include homeostasis, inflammation, proliferation and remodelling. (**b**) Chronic wounds: Corresponding events occurring in chronic wounds at each of the phases leading to the development of non-healing wound. (**c**) Effect of LBP administration in reversing the events in each phase leading to wound healing

macrophages and neutrophils are recruited to pathogen-associated molecular patterns (PAMP) of wound microbiota to prevent bacterial growth and phagocytose any cellular debris. Wound macrophages also secrete growth factors to stimulate vascularization and formation of a provisional extracellular matrix (ECM). Complement and plasma proteins including antibodies enhance phagocytosis and antibody-mediated cytotoxicity. Proliferation phase may last from 2 to 20 days based on severity and is characterized by tissue granulation, angiogenesis and epithelialization. Angiogenesis in response to increased cytokines and growth factors such as epidermal growth factor (EGF), vascular endothelial growth factor (VEGF) and tumour growth factor-β (TGF-β) and fibroplasia helps in the synthesis of granulation tissue. In the healing phase, modulators such as interleukin 4 (IL4), interleukin 10 (IL10), glucocorticoids, prostaglandins induce reparative M2 macrophage phenotype. The M2 macrophages secrete growth factors that induce cellular prolif-eration, anti-inflammatory cytokines and phagocytose neutrophils in the wounds. In the remodelling phase, provisional ECM matrix is replaced by proteoglycan and collagen molecules (Bryant and Nix 2012; Martin and Nunan 2015).

In chronic wounds, this normal wound healing process remains disrupted and does not usually proceed beyond the inflammation or proliferation phase despite intervening wound management strategies (Baczako et al. 2019). Several factors contribute to the delayed wound process that stretches beyond the normal 3-month period observed for acute or healing wounds (Morton and Phillips 2016; Han and Ceilley 2017). Co-morbidity factors such as diabetes, obesity, vascular insufficiency, recalcitrant microbial infections, nutritional deficiency, ageing and factors such as pressure and oedema in the wound environment significantly hamper the process (Frykberg and Banks 2015). Increased pro-inflammatory cytokines, toxic ROS and proteases lead to tissue degradation impairing wound healing. Inflammatory milieu represented by an abundance of neutrophils is considered a biological marker for chronic wound inflammation (Zhao et al. 2016). Neutrophils are one of the major causes of an overproduction of pro-inflammatory cytokines such as interleukin-1 β (IL1β) and Tumour Necrosis Factor α (TNFα), ROS and proteases that cause direct damage to the extracellular matrix and cellular membranes leading to cell senes-cence. Levels of various matrix metalloproteinase (MMP) such as collagenase, gelatinase A, gelatinase B, stromelysin, neutrophil elastase and serine proteases are markedly elevated in pressure ulcers and chronic venous leg ulcer wounds in comparison to acute wounds (Lazaro et al. 2016). Wound fluid from chronic venous leg ulcers is found to be rich in inflammatory cytokines TNFα and IL1β produced by activated macrophages and neutrophils that increase MMP production with decrease in Tissue Inhibitor of Metallo Proteinases (Morton and Phillips 2016). Growth factors are signalling molecules that regulate cellular migration, differentiation as well as proliferation. Deregulation of various growth factors such as platelet-derived growth factor (PDGF), VEGF, EGF, basic fibroblast growth factor (bFGF) and granulocyte-macrophage colony-stimulating factors (GMCSF) has been reported in chronic wounds, events not observed in acute wounds. Lack of this anti-inflammatory-inducing environment results in poor resolution of the inflammation driving events and the future healing events of cellular proliferation remain in limbo (Baczako et al. 2019). In addition, wound environments such as hypoxia, wound desiccation, wound pressure and temperature provide multiple signals for sustaining the state of chronic inflammation and delaying the wound healing process (Han and Ceilley 2017). Diminished angiogenesis is a marked feature causing decreased cell migration of endothelium progenitor cells, keratinocytes and fibroblasts essential for tissue regeneration (Zhao et al. 2016). An important pathogenic factor is the presence of biofilm-forming wound pathogens that are refractory to the immune response and antibiotic regimens and perpetuates the non-healing wound saga (Rahim et al. 2017).

3   Chronic Wound Microbiology

Most microorganisms in clinical settings are found to exist in community living mode popularly referred to as microbial biofilms. These are typically found attached to substrata (biotic or abiotic, solid–liquid or air—liquid interface) wherein the community flourishes covered by a self-secreted exopolymeric matrix protecting its residents from the immune system or antimicrobial products (Donlan and Costerton 2002). In the human body, normal microbiota may consist of true pathogens, opportunistic pathogens or commensals, all of which are reported to exist as natural biofilms (Sanford and Gallo 2013). Infectious diseases including chronic wounds are largely a tug-of-war between the immune response and the virulence of the infectious agent (Scalise et al. 2015). Dysbiosis of natural microbiota plays a significant role in augmenting pathogenic biofilm infections that prolong the inflammatory stage of wound healing by activating neutrophils and macrophages and creating an inflammatory cytokine storm contributing to poor healing in chronic wounds (Metcalf and Bowler 2013; Drago et al. 2019).

Wounds can be contaminated with both normal skin microbiota and invading pathogens, such as the multiple drug-resistant biofilm-forming ESKAPE organisms: '*Enterococcus caecum, Staphylococcus aureus, Klebsiella pneumoniae, Acinetobacter baumannii, Pseudomonas aeruginosa and Enterobacter* spp' amongst other bacteria and fungi. Pathogenic biofilms in wounds have been found to be polymicrobial (widely influenced by location and environmental factors), highly resistant to immune factors and antimicrobial treatments which are typically the cause of persistent and recurrent infections (Wu et al. 2019). As a microbial biofilm matures it can adapt to the changes in its surrounding by quorum sensing signalling, allowing it to thrive in the pervading settings. Biofilms are reported to be present in only 6% of acute wounds but over 90% of chronic wounds (Drago et al. 2019). Scanning electron microscopy of 50 chronic wound specimens were found to contain biofilms in 6% of acute wounds and 60% biofilms in chronic wounds. Molecular sequence analyses with denaturing gradient gel electrophoresis showed polymicrobial communities and strict anaerobic bacteria not detected by culture methods (James et al. 2008).

The major evidence that biofilms physically impair wound healing comes from murine, porcine and rabbit ear wound models wherein the presence of pathogenic biofilms significantly delayed healing and required antibiotic or debridement as treatment regimens (Metcalf and Bowler 2013; Pereira et al. 2017). In the mouse chronic wound model, *Ps. aeruginosa* biofilm delayed wound healing without affecting the general health of the mice (Metcalf and Bowler 2013). Clinical evidence in several case studies also highlights that biofilms exist in wounds. A Global Wound Biofilm Expert Panel has provided a clinical focus on identification, management and care of biofilms in chronic wounds for improved patient care (Schultz et al. 2017).

The role of wound microbes and their compositional diversity in non-healing chronic wounds is presented by the following studies. In a recent report, a Bayesian statistical model was used to model patient-to-patient variability in the identification of microbiome changes in wounds versus healthy skin. No effect of tissue debridement was found in wound microbiome of 20 outpatients with chronic wounds while *Enterobacter*, a facultative anaerobic bacteria was significantly associated with reduced healing (Verbanic et al. 2020). Next generation DNA sequencing profiled Diabetic Foot Ulcers (DFU) microbiome of 39 patients who had not received any oral or topical antimicrobial in 14 days before the study. While short-duration DFU showed only the presence of *Streptococcus agalactiae* in two cases and *S. aureus* in three cases, respectively, longer-duration DFU ($\geq$6 weeks) were highly polymicrobial ranging from 19 to 125 different bacteria (average, 63) (Malone et al. 2017).

In a retrospective multicentre surveillance study with 792 diabetic foot patients from which 1803 causative organisms were isolated, 48.5% of the patients had polymicrobial infections. The pathogen profile included 22.2% *S. aureus*, 7.7% methicillin-resistant *S. aureus* (MRSA), 21.8% *Enterococcus* spp., 9.4% *Ps. aeruginosa*, 8.9% *Proteus mirabilis*, 7.9% *E. coli*, 7.5% *Klebsiella* spp., 6.6% Coagulase negative Staphylococci (CoNS), 5.9% anaerobics and 2.3% fungi. Of these 15.8% were moderate and 34% high biofilm formers, respectively (Al-Joufi et al. 2020).

Two thousand nine hundred sixty-three patients (chronic diabetic foot ulcers (DFU) $N = 901$; Venous leg ulcers (VLU) $N = 916$; decubitus ulcers $N = 767$; non-healing surgical wounds $N = 370$) were analyzed for the chronic wound microbiota by 16S rDNA pyrosequencing. A high prevalence of *Staphylococcus* (63%) and *Pseudomonas aeruginosa* (25%) was found in all wounds in addition to anaerobic and commensal bacteria (Wolcott et al. 2016). Anaerobes found in chronic wound microbiome include *Prevotella*, *Peptoniphilus*, *Peptostreptococcus*, *Anaerococcus* and *Fingoldia* (Choi et al. 2019). A more comprehensive study of 40 different chronic DFU using bacterial tag FLX amplicon pyrosequencing (bTEFAP) technology found that *Corynebacterium* sp. was the most prevalent amongst other obligate anaerobes such as *Bacteroides*, *Peptinophilus*, *Fingoldia*, *Anaerococcus* and *Peptostreptococcus* spp. (Dowd et al. 2008a). Other major genera included frequently cultured organisms such as *Streptococcus*, *Serratia*, *Staphylococcus* and *Enterococcus* spp. The authors also introduced the concept of functional equivalent pathogroups (FEP) as 'consortia of genotypically distinct bacteria that symbiotically produce a pathogenic community' (Dowd et al. 2008a). A survey of bacterial biofilm diversity in chronic wounds (DLU, VU and PU) using denaturing gradient gel electrophoresis (DGFE), pyrosequencing and full ribosome shotgun sequencing, reports that major populations included *Staphylococcus*, *Serratia*, *Stenotrophomonas*, *Pseudomonas*, *Peptoniphilus*, *Enterobacter* and *Finegoldia* spp. (Dowd et al. 2008b). In another study by the same group, a comparison of microbiome on contralateral intact skin versus diabetic ulcer reveals higher diversity on intact skin. Wounds showed higher incidence of anaerobic bacteria (*Peptoniphilus*, *Finegoldia*, *Anaerococcus*) and opportunistic pathogens such *Corynebacterium*, *Staphylococcus* and *E. coli*. Significant differences in diversity were found between intact skin and wound samples with distinct similarities between different wound samples (Gontcharova 2010).

A longitudinal study of 88 type II diabetics with chronic foot ulcers microbiome over 10 weeks using 16S rDNA PCR and sequencing revealed a significantly less biodiversity than control skin. The most abundant genera on diabetics included *Staphylococcus* followed by *Corynebacterium*, *Acinetobacter* and unclassified Enterobacteriaceae (Gardiner et al. 2017).

Polymicrobial nature of 40 separate venous leg ulcer infections were evaluated using bTEFAP titanium and metagenomics approaches (Wolcott et al. 2009). Majority of the uncharacterized bacteria belonged to *Bacteroides*, *Staphylococcus* and *Corynebacterium* species. *Streptococcus*, *Finegoldia*, *Peptoniphilu*s, *Proteus* and *Pseudomonas* were also found in the wounds. However, individual wounds were found to have distinct site-specific biofilm-based microbial footprints. Interestingly, in this study other microorganisms were mapped to Apicomplexa (related to *Plasmodium yoelii*) and fungi such as *Candida albicans*, *C. glabrata* and *Aspergillus* were also genetically mapped but not verified (Wolcott et al. 2009). Sequence reads closely related to dsDNA viruses such as human herpesvirus, human adenovirus, *Staphylococcus* phage, Bacteriophage B3, *Corynebacterium* phage and closely related retro transcribing virus amongst others were also identified using gene identification (Wolcott et al. 2009).

In a comprehensive study comparing microbiota of skin versus four types of chronic wounds, *S. aureus*, *Proteus*, *Helcococcus*, *Enterobacter* and *Pseudomonas* genera were found in pre-debridement wound samples, while *Micrococcus*, *Paracoccus* and *Kocuria* in control skin samples. A comparison of organisms based on their oxygen requirement revealed that the abundance of facultative anaerobes was $20.8 \pm 29.7\%$ (*Enterobacter*) in unhealed wounds versus $5.32 \pm 7.21\%$ (*Corynebacterium*) in healed wounds (Verbanic et al. 2020). *Fusobacterium* and *Actinobacillus* were found to be strongly associated in the initial phases of healing proving the resilience and stability of skin microbiota in an equine wound healing model (Kamus et al. 2018).

The role of mycobiota (fungi) or fungal-bacterial biofilms also plays a significant part in chronic non-healing wounds. A cross-sectional study of mixed aetiology without standardized treatment reported that 23% of chronic wounds contained fungi using molecular biology methods (Dowd et al. 2011). Nuclear ribosomal internal transcribed spacer (ITS1) sequencing was used to longitudinally profile 100 non-healing diabetic foot ulcers wherein 80% of the wounds contained fungi, while culture-based techniques only identified fungi in 5% of colonized wounds. *Cladosporium* spp. was the most abundant species followed by *Candida* spp. (*albicans* > *parasilopsis* > *tropicalis* > *glabrata* > *smithsonii*) (Kalan et al. 2016).

These studies suggest that dominant pathogenic organisms colonize most chronic wounds, which differ significantly in composition and diversity from commensal bacteria.

3.1 Commensals in the Cutaneous Microbiome

Based on culture studies and 16S rDNA metagenomic sequencing, majority of cutaneous microbiota can be grouped into four phyla, Firmicutes (24%), Bacteroidetes (7%), Proteobacteria (17%) and Actinobacteria (52%) and includes *Staphylococcus* spp., *Micrococcus* spp., *Acinetobacte*r spp., *Corynebacterium* spp. and *Propionibacteria* spp. (Martin et al. 2010). The sub-epidermal compartments contain higher numbers of Proteobacteria such as *Burkholderia* spp. and Pseudomonads; Actinobacteria and lower Firmicutes (Byrd et al. 2018). Moist areas of the skin (navel, axilla, groin, sole of feet, inner knee and inner elbow) are abundant in members of phyla Firmicutes (*Staphylococcus* spp.) and Actinobacteria (*Corynebacterium* spp.). Sebaceous sites (back area, forehead, nasal sides, behind ears) which form the anaerobic and lipid-rich environments are rich in *Actinobacteria* while dry areas (forearms, hands, legs etc.) of the skin have varying biodiversity of the four phyla (Sanford and Gallo 2013). Microbial inter-variability is often found between individuals while intra-variability is found between various body sites and niches. Even so, the dominant genera remain stable and include *Staphylococcus*, *Propionibacterium* and *Corynebacterium* followed by *Streptococcus* and *Pseudomonas* in interpersonal variability (The Human Microbiome Project Consortium 2012). Besides bacteria, fungi and viruses are also present as commensals. High throughput sequencing for pan mycobiome analysis shows that *Malassezia* genus is the most predominant represented by *M. restricta* followed by *Aspergillus*, *Candida* and *Cryptococcus* (Kalan et al. 2016). Metagenomic sequencing has identified Human Papilloma Virus and Polyomavirus as common species of cutaneous virome followed by Circoviruses and bacteriophages of the abundantly found bacterial genera (Hannigan et al. 2015). Analyses of human skin virome determined by purification of virus-like particles (VLPs), deep sequencing, 'clustered regularly interspaced short palindromic repeat' (CRISPR) identification and network analyses techniques shows the persistence of temperate cutaneous phages, which may contribute to novel gene transfer amongst the microbiome (Hannigan et al. 2015).

Commensal or mutualistic microbiota are known to confer health benefits by production of antimicrobial products, maintaining pH, competition for space and nutrition, immunomodulation and strengthening the epithelial barrier functions (The Human Microbiome Project Consortium 2012; Martín and Langella 2019). Surface level proliferation of non-pathogenic wound microbiota which do not cause invasion of underlying tissue are typically tolerated by the host and recent studies show that they may actually help in the wound healing and repair processes (Sanford and Gallo 2013).

4   Non-conventional Treatment Strategies of Chronic Wounds: Live Biotherapeutics

Typically, advanced wound care strategies are applied if failure of even 50% wound area reduction is observed after 4 weeks of conventional treatment (Frykberg and Banks 2015). A healing approach that brings balance to the system by replenishing the useful and removing the harmful elements can be the key to treating chronic wounds. A direct application of our understanding of healthy human microbiome has led to the exploration and development of a novel class of non-conventional treatment strategies known as live biotherapeutics.

Live biotherapeutics (LBP) includes both well-defined live naturally occurring and genetically engineered organisms for the prevention and treatment of a disease or condition (Martín and Langella 2019; Rouanet et al. 2020). The medicinal use of microorganisms as pharmaceutical products is being referred to as live biotherapeutics by the US Food and Drug Administration (US FDA) and the European Pharmacopeia (Ph.Eur). The FDA defined these as 'a biological product that (1) contains live organisms, such as bacteria (2) is applicable to the prevention, treatment or cure of a disease or condition of human beings; and (3) is not a vaccine' and by PhEur as 'medicinal products containing live micro-organisms (bacteria or yeasts) for human use' (Dreher-Lesnick et al. 2017; Rouanet et al. 2020). The desirable criteria in a LBP include safety for all age groups, non-toxicity, antibiotic and antifungal sensitivity, genetic stability, high survivability in acidic environments, biofilm forming so as to create an effective barrier against pathogens, high shelf life, ease of high biomass production and environmentally safe with low-risk biocontainment (Rouanet et al. 2020).

4.1   LBP Benefits and Their Modes of Action

LBP are considered as administration of safe organisms that can confer health benefits and curative effects to humans (Martín and Langella 2019). Several mechanisms proposed for the positive effect of live organisms are discussed below and depicted in Fig. 2.

4.1.1   Adhesion, Aggregation Ability and Colonization

The ability to form a protective lining over host surfaces is an important criterion for potential probiotics that results in transient colonization of epithelial lining and provides competition to pathogens (Pereira and Bártolo 2016). Adhesion, aggregation and biofilm formation ensures low rates of LBP clearance and an opportunity for the organisms to multiply and produce metabolites that promote immunomodulation and stimulate host metabolic pathways (Monteagudo-Mera et al. 2019). Adhesions

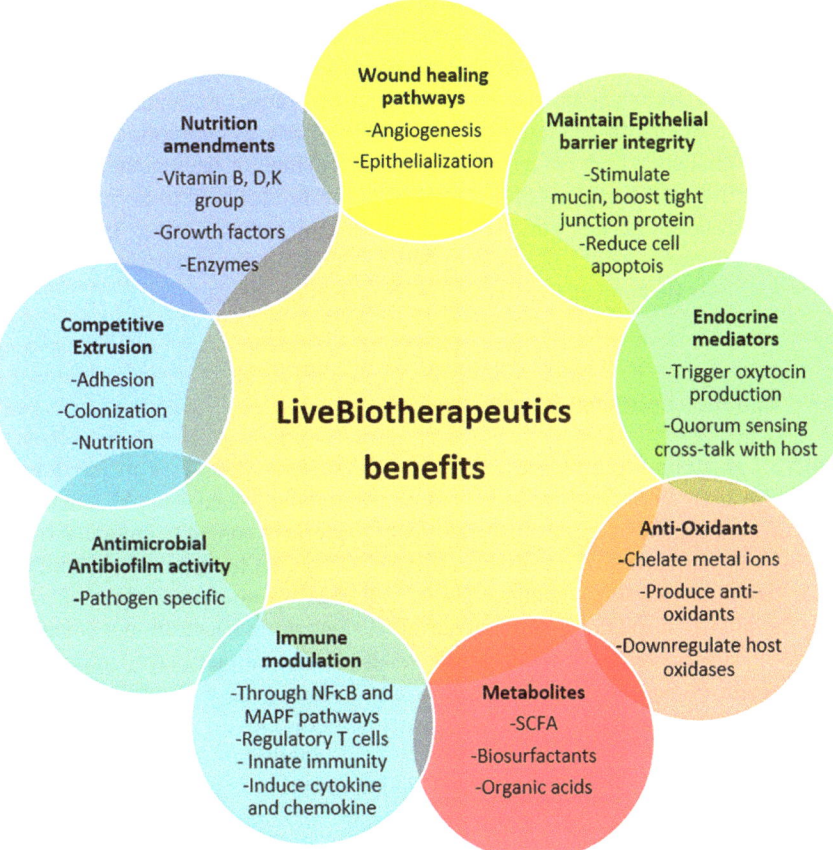

Fig. 2 Live biotherapeutics and their beneficial properties

depend on cell surface hydrophobicity as well as the presence of adhesion promoting surface components.

Cellular aggregation refers to the property of cells to aggregate and precipitate. *Lactobacillus* spp. secrete cell aggregation factors that mediate self-aggregation as well as act as pheromone-like factors that induce surface protein adhesins (Miljkovic et al. 2015). Moreover, recent studies also show that probiotics (*Lactobacillus rhamnosus*, *L. plantarum*, *L. reuteri* and *L. fermentum*) themselves are potent biofilm formers and secrete quorum sensing signalling autoinducer peptides (AIP) and quenching molecules that influence host cells as well as other biofilm-forming pathogens (Spangler et al. 2019). In vitro studies with different epithelial cell lines assess the adherence ability of probiotics and pathogen antagonism (Lopes et al. 2017; Rosignoli et al. 2018).

4.1.2 Competitive Exclusion

Beneficial probiotics have the ability to exert vigorous competition to incoming pathogens for ecological niche and nutritional resources. Reduction in pH and regulating the oxidation–reduction potential in the body by probiotic consumption alters environments which create colonization resistance for pathogenic organisms. The metabolites and local conditions created by probiotics help to normalize commensal microbiota in the gut epithelium (Abatenh et al. 2018). In a human keratinocyte culture model, *L. reuteri* ATCC 55730 reduced keratinocyte death due to *S. aureus* by competitive exclusion (Prince et al. 2012).

4.1.3 Enhance Epithelial Barrier Integrity

Probiotics are found to enhance epithelial cell barriers which is a major defence mechanism used to prevent invasiveness of pathogenic organisms (Abatenh et al. 2018). The mechanism pertains to the ability of bacteria to adhere to epithelial tissue, form biofilms and produce metabolites that help maintain homeostasis in epithelial barrier function. Probiotics enhance goblet cell expression and secretion of mucin along the intestinal tract and mucosal surfaces, preventing bacterial movement and provide a protective layer (Liu et al. 2020). The surface layer proteins of *L. plantarum* interact with TLR domains and induce gene expression of F-actin distribution increasing trans epithelial resistance (TER) levels and enhancing tight junction stability (Liu et al. 2011). They are also reported to increase extracellular signal-regulated kinases (ERK) phosphorylation resulting in reduced host cell apoptosis (Prado Acosta et al. 2016). Short-chain fatty acids (SCFA) particularly butyrate, which is the preferential energy source for colonic epithelial cells are produced by probiotics which also boost tight junction protein expression, secretion of mucin by goblet cells and reduce enterocyte senescence (Yan et al. 2007).

4.1.4 Antimicrobial Properties

The commensal microbiota contributes to host health by inhibiting pathogenic growth and creating colonization resistance by the production of antimicrobial products (Fijan 2016). Antimicrobial substances produced by probiotics include organic acids (formic acid, acetic acid, phenyl lactic acid, lactic acid, benzoic acid), acetaldehyde, acetoin, diacetyl and short-chain fatty acids (SCFA) as well as carbon dioxide, hydrogen peroxide and bacteriocins (Fijan 2016; Abatenh et al. 2018). Lantibiotics, a class I bacteriocins containing lanthionine/β methyllanthionine residues such as gallidermin, epidermin, hominicin are abundantly produced by *S. gallinarum*, *S. epidermidis* and *S. hominis*, respectively (Götz et al. 2014). Coagulase negative *Staphylococcus* species are reported to produce bacteriocins, heat-stable antimicrobial peptides as well quorum sensing signalling molecules for interspecies communication (O'Sullivan et al. 2019).

Topical application of human skin microbiota protects against *S. aureus* by the production of strain and species-specific antimicrobial peptides that are synergistic with defensins such as LL37 in pigskin and mice model (Nakatsuji et al. 2017). In a recent survey, 21 different bacteriocins produced in seven human body sites characterized by colony mass spectrophotometry, were found to inhibit opportunistic Gram-positive pathogens such as *Cutibacterium acne*, *S. epidermidis* and MRSA. 16S rDNA sequencing revealed *Bacillus* spp. (*licheniformis*, *endophyticus* and *safensis*) and Coagulase Negative Staphylococcus spp. (*capitis*, *hominis*, *epidermidis*, *simulans*, *warneri*) as the antimicrobial producing microbiota (O'Sullivan et al. 2019).

Staphylococcus epidermis, commensal bacteria-laden polyethylene glycol dimethacrylate (PEG-DMA) was used to decolonize Methicillin-Resistant *S. aureus* USA300 by the production of SCFA such as acetic, propionic acid and butyric in a skin wound mice model (Kao et al. 2017). Transcriptional analysis of commensal *S. epidermis* inhibiting *S. aureus* biofilm revealed that the non-proteinaceous small molecule products modulated several *S. aureus* gene expression including biofilm formation (Glatthardt et al. 2020). Phenol-soluble modulins (PSM) production by *S. epidermidis* can cause targeting cytolysis of *Streptococcus pyogenes* and *S. aureus* in conjunction with host immune antimicrobial peptides (AMP) (Cogen et al. 2010).

Lactobacillus plantarum culture supernatant was found to contain pro-healing and anti-pathogenic compounds such as antimicrobials (5-methyl-hydantoine, benzoic acid, mevalonolactone etc.); biosurfactants (distearin, di-palmitin, and 1,5-monolinolein); anesthetics (barbituric acid derivatives) and quorum sensing autoinducer (AI-2) precursors (4,5-dihydroxy-2,3-pentanedione and 2-methyl-2,3,3,4-tetrahydroxytetrahydrofurane) (Ramos et al. 2015).

4.1.5 Biosurfactants

Amphipathic molecules produced by microbes as secondary metabolites that can influence surface/interfacial tensions of fluids are biosurfactants. Probiotic biosurfactants possess antimicrobial, antiadhesive and antibiofilm properties (Hajfarajollah et al. 2018). *L. plantarum* and *Pediococcus acidilactici* produce biosurfactants that inhibit adhesion and biofilms in *S. aureus* CMSS 26003 by affecting the expression of quorum sensing and biofilm genes such as *cidA*, *sarA*, *icaA*, *dltB*, *sortase* A, agrA (Yan et al. 2019). Cell-derived biosurfactants *Lactobacillus jensenii* and *L. rhamnosus* demonstrated antimicrobial and antibiofilm activities against multiple drug-resistant *A. baumannii*, *S. aureus* and *E. coli* (Sambanthamoorthy et al. 2014).

4.1.6 Antimetabolites

Bacterial microbiome produces metabolites such as folates, indoles, serotonin, gamma aminobutyric acids, SCFA (acetate, propionate, butyrate) and trimethylamine N oxides which can bind to host membrane and nuclear receptors causing physiological changes that can be beneficial or be the cause of inflammation and disease (Abatenh et al. 2018; Liu et al. 2020). *Lactococcus lactis* and *L. reuteri* bacterial metabolites induce anti-inflammatory T regulatory cell differentiation by increasing transcriptional upregulation of Fox P3 and IL10 gene expression. The metabolites also suppressed CD11c and MHC Class II expression on dendritic cells (Fu et al. 2018). *Lactobacillus plantarum* DC400 co-cultured with *L. sanfranciscensis* DPPMA174 produced plantaricin (PlnA), a pheromone that shows antimicrobial activity, induces proliferation of human keratinocyte NCTC 2544 cells and increases keratinocyte growth factor 7 (FGF7), VEFG-A, transforming growth factor-β1 (TGF-β1), and interleukin-8 (IL-8) genes (Pinto et al. 2011).

4.1.7 Antioxidant Properties

Antioxidant properties of probiotics regulate oxidative damage to lipids, proteins and nucleic acid caused by excessive and unmanaged ROS and hydroxyl radicals generated in diseased conditions (Wang et al. 2017). The antioxidant properties of different probiotic strains are attributed to various mechanisms, which include (1) metal ion chelating ability and preventing ions from catalyzing oxidation reactions; (2) Production of antioxidant metabolites such as glutathione, butyrate and folate; (3) Downregulation of host ROS generating oxidases (NADPH, COX-2, Cytochrome P450); (4) Upregulating host antioxidant metabolites by overcoming vitamin B group deficiencies by probiotic supplementation and (5) Regulating host signalling pathways (Nrf2-Keap1-ARE transcription elements, NFκB, MAPK, PKC pathways) for promoting genes encoding antioxidant and detoxifying enzymes (Wang et al. 2017).

4.1.8 Endocrine Effectors

Microbe–host interaction influencing changes in hormonal levels is now being actively studied as microbial endocrinology (Neuman et al. 2015). Quorum sensing molecules such as autoinducers crosstalk with host hormone signalling pathways. Host hormones also affect bacterial growth and gene expression which in turn influences host behaviour. *L. reuteri* mediated vagus nerve mediated pathway controlled upregulation of hormone oxytocin resulted in enhanced wound healing. The bacteria triggered oxytocin also activated CD4+ FoxP3+ CD25+ T regulatory cells providing transplantable wound healing in naïve Rag 2-deficient mice (Poutahidis et al. 2013).

4.1.9 Angiogenic Activity

Angiogenesis is crucial for proper wound healing and probiotics such as *Bacillus polyfermenticus* (B.P) used for treatment of intestinal disorders cause an increase in cellular migration, permeability and tube formation in human intestinal microvascular endothelial cells. B.P treatment of mice colitis model showed increased IL8 production with reduced rectal bleeding and disease severity. The mechanism of angiogenesis induction was dissected to be NFκB/Interleukin 8/CXCR-2 dependant manner (Im et al. 2009).

4.1.10 Immunomodulatory

Recent studies show that probiotics have an effect on innate, humoral and cellular immunity (Clarke et al. 2010; Naik et al. 2012; Thaiss et al. 2016; Cheng et al. 2019; Delgado et al. 2020). Pathogen recognition receptors such as Toll-Like Receptors (TLR) and NOD-like receptors (NLR) are triggered by probiotic-associated molecular patterns (PAMP) such as flagella, pili, surface layer proteins (SLP), capsular polysaccharides, lipoteichoic acids and lipopolysaccharide (Thaiss et al. 2016). These, in turn, activate MyD88, MAPK, NF-kB and other signalling pathways that regulate growth factors, cytokine and chemokine-induced inflammation and enhance epithelial functions (Liu et al. 2020). *Lactobacilli* in various studies are shown to stimulate intra-epithelial T cells, NK cells, dendritic cells, macrophages and neutrophils (Salas-Jara et al. 2016; Abatenh et al. 2018).

Functional genomic studies demonstrate that probiotics play an immunomodulatory role by influencing inflammation related gene expression and are referred to as Genobiotics (Gorreja 2019). Probiotics play a major role in the T cell differentiation into Treg (Fog P3+)/Th1/Th2 type cells and the antibody isotypes being produced. *Lactobacillus reuteri* can indirectly boost wound healing via upregulation of oxytocin via T regulatory cells (Poutahidis et al. 2013).

Commensal bacteria such as *Roseburia intestinalis*, *Bacteroides fragilis*, *Akkermansia muciniphila* can cause an increase in anti-inflammatory IL10 and IL22 cytokine production (Gurung et al. 2020). The production of SCFA by skin commensals suppresses skin inflammation by upregulating Treg-specific transcription factors FoxP3 and IL10 (Schwarz et al. 2017).

4.1.11 Nutritional Amendments

Probiotics such as *Lactobacilli* and *Bifidobacterium* are natural producers of water-soluble vitamin B group, vitamin K and digestive enzymes such as esterases, lipases,

Co-enzymes A, Co-enzyme Q, NAD and NADP (Gu and Li 2016). Oral *L. reuteri* NCIMB 30242 was used in a placebo-controlled, double blind, randomised, parallel arm, multicentre study which showed increased serum 25-hydroxyvitamin D compared to placebo group (Di Marzio et al. 2008). Probiotics enhance the absorption of nutrients, vitamins and minerals and induce amino acids and organic acid production (Quigley 2019). *Bacteroides ovatus* is a probiotic species capable of digesting xyloglucans using enzymes not found in humans (Larsbrink et al. 2014). Probiotics hence play a major role in fulfilling nutritional deficiencies leading to a robust immune system capable of defending against pathogens and accelerating the wound healing process.

The above-mentioned LBP benefits are highly genus and strain specific. While many can be used for systemic health and disease amelioration, tailored LBP prescription seems to be necessitated based on a detailed analysis of the microbiome dysbiosis, metabolic profiling and requirements of a patient-specific disease condition (Lin et al. 2019). In Fig. 1, the role of LBP in chronic wound healing depicts how oral or topical application can reverse the destructive turn of events that are characteristic of chronic wounds and re-establish the normal wound healing process.

4.2 Live Biotherapeutic Products Types and Their Application for Chronic Wounds

Live biotherapeutic Products (LBP) is a terminology for a pharmaceutical product containing live organisms that can be used for the prevention/treatment/cure of diseases (Martín and Langella 2019). Topical LBP administration can affect the cutaneous wound environment while oral systemic administration can also help in wound healing. Gut microbiota influences skin immunity by enhancing a balanced systemic immune response (Salem et al. 2018). With disturbed gut immunity, intestinal pathogens and metabolites enter the bloodstream and gather in the cutaneous tissue causing disease. The 'gut-brain-skin' hypothesis suggests that gut microbiome modulation using probiotics can influence distal effects such as neuronal skin inflammation and even hair follicle development (Arck et al. 2010).

The following kinds of LBP will hence be considered in their role in chronic wound healing: traditional probiotics, next-generation probiotics and engineered live biotherapeutics.

4.2.1 Traditional Probiotics

'Probiotic' means 'for life' in the Greek language. The traditional definition of probiotics endorsed by World Health Organization (WHO), Food and Agriculture Organization and United Nations (FAO) and The International Associate for Probiotic and Prebiotics (ISAPP) is as follows: 'Probiotics are live organisms that, when administered in adequate amounts, confer a health effect on the host' (Hill et al. 2014). Around 1900, Russian scientist Elie Metchnikoff, showed that live bacteria (*Lactobacillus bulgaricus*) in fermented foods (yogurt and milk) improved the gastrointestinal tract functions. In 1965, Lilly and Stillwell introduced the term 'Probiotics' (Gasbarrini et al. 2016).

Traditional Probiotics (TP) include specific bacterial strains of Lactic Acid Bacteria (LAB) such as *Lactobacillus* (*rhamnosus, lactis, acidophilus, plantarum, casei, delbrueckii* subsp. *bulgaricus, gasseri, fermentum, reuteri, johnsonii, paracasei and salivarius*); *Bifidobacterium* (*animalis* subsp. *lactis, longumadolescentis, animalis, bifidum, breve*); *Propionibacterium acidilactici, Leuconostic mesenteroides, Enterococcus faecium, Streptococcus thermophiles* etc. *Saccharomyces boulardii*, a yeast probiotic has been compared with bacterial probiotics and found to be highly effective in treatment and prevention of diarrhoea and intestinal illness associated with long-term antibiotic usage, *Clostridium difficile* infections, and traveller's diarrhoea (Chamberlain and Lau 2016). Topical probiotics for skin also include *Nitrosomonas eutropha*, an ammonia oxidizing soil bacteria with antibacterial, anti-inflammatory properties used for removal of skin wrinkles (Notay et al. 2020). The lysate from *Vitreoscilla filiformis*, a non-pathogenic Gram negative bacteria isolated from thermal spa waters, is being used for beneficial skin effects with atopic dermatitis and seborrheic dermatitis symptoms (Gueniche et al. 2008).

The major origin of traditional probiotics is fermented non-digestible carbohydrates, dairy-based foods and intestinal origin microorganisms. GRAS (generally recognized as safe) status has been conferred on traditional probiotics due to their ubiquitous appearance, fermentative metabolism and contribution to host enteric health. A combination of prebiotics (foods supporting probiotics) and probiotics are referred to as conbiotic or synbiotic and now considered a type of functional food (Martín and Langella 2019; Charbonneau et al. 2020). Traditional probiotics are known to produce antimicrobial peptides, antioxidant compounds, regulate mucosal IgA production, maintain intestinal epithelial cell integrity and even modulate bile acid production and secretion (Abatenh et al. 2018). Therefore, such probiotics which provide a distinct health benefit may be included in the category of live biotherapeutics.

The other major category for probiotics are the normal commensal human body microbiota. Gut microbiome directly or indirectly affects metabolic functions, immune system and protection against pathogens (Cani 2018).

Recent research suggests that the microbiome is a regulator of the 'gut-brain- skin axis' influencing the three-way communication between the central, enteric and cutaneous system involving the nervous, immune and endocrine signalling (Arck et al. 2010; Salem et al. 2018). A hypothetical model of how the 'gut-brain-skin axis' can be a useful model to understand the role of LBP administrations in mitigating disease conditions as well as enhance wound healing is depicted in Fig. 3. For the treatment of chronic wound diseases, LBP may be administered orally (probiotic as dietary supplement), or topically at the site of the wound. A plethora of LBP benefits may be exerted at the intestinal levels which are communicated to the brain. A

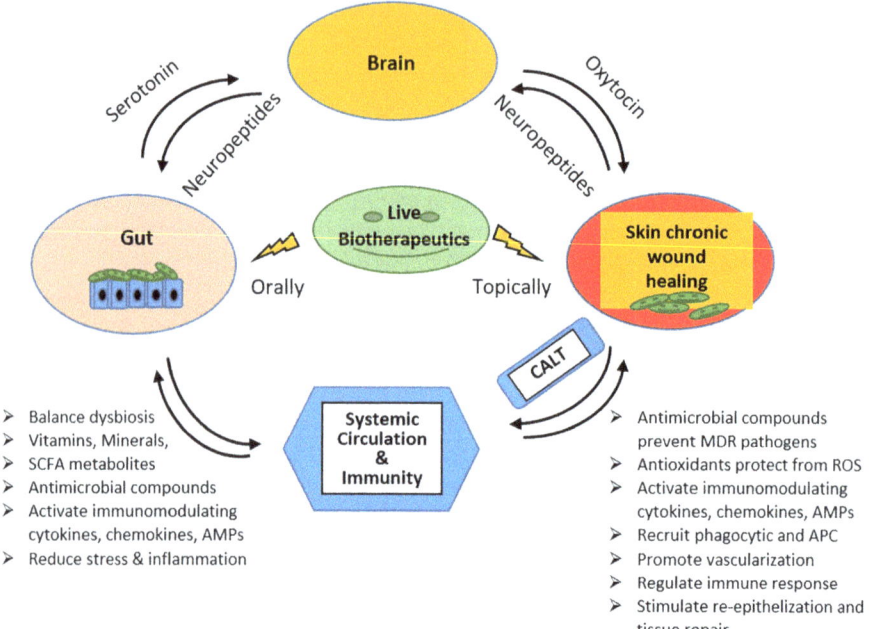

Fig. 3 Live bio-therapeutics and the 'Gut-brain-skin axis' for chronic wound healing and repair. Oral application of LBP leads to several benefits to gut enterocytes. These are shared to brain via neuropeptides (e.g. serotonin) and systemically to enhance repair in affected body tissues such as chronic wounds. A healthy brain provides signals of tissue regeneration through oxytocin. Topical application of LBP leads to increased wound healing through beneficial properties exerted at the site of wound as well as through the cutaneous associated lymphoid tissues (CALT)

release of stress-relieving neuropeptides can modulate the systemic inflammatory response including sites of chronic wounds and induce reparative processes. Oral probiotics can systemically downregulate inflammatory responses and compress wound repair healing processes by rapid collagen deposition as well as promoting skin allostasis (Salem et al. 2018). Topical application of specific probiotics leads to altering wound microbiome, enhanced re-epithelization and immune response modulation (Erdman and Poutahidis 2014).

Several in vitro, mice and human experiments suggest that probiotic supplementation has many beneficial effects on skin health as well as in healing of cutaneous wounds (Table 1). Treatment with probiotics *L. brevis* producing high exopolysaccharide showed accelerated progress from granulation formation to re-epithelialization in cutaneous wound site in Wistar rat model (Heydari Nasrabadi et al. 2011). LAB also produces lactic acid and other antimicrobial products that prevent the growth of pathogenic wound organisms (Liu et al. 2011; Ramos et al. 2012; Rosignoli et al. 2018; Yan et al. 2019; Sürmeli et al. 2019; Ong et al. 2020). A large number of in vitro, experimental and in vivo studies show efficient antagonistic activity of probiotics against wound pathogens *S. aureus*, *Ps. aeruginosa*,

Table 1 Effect of live biotherapeutics in chronic wound models

Wound infection study	Model	Probiotic genus	Pathogens	Study method	Potential benefits	References
Ischemic wound	New Zealand White rabbit	*Lactobacillus fermentum* on a wound healing patch (adhesive gas permeable membrane)	*S. aureus*	Local patch with lyophilized microbeads (10^6 cfu^a)	Nitro oxide gas production Increase blood flow Increased collagen deposition Reduced pathogen	Jones et al. (2012)
Diabetic wound	Diabetic rat model	Probiatop®, (1×10^9 CFU/g) *Lactobacillus paracasei* LPC-37®, *Bifidobacterium lactis* HN0019®, *Lactobacillus rhamnosus* HN001® and *Lactobacillus acidophilus* NCFM® or maltodextrin as control	Not done	46 rats divided into 4 groups (C3, P3, C10, P10) according to the treatment (C = control or P = probiotic, both orally administered) third or tenth postoperative days or euthanasia day. Probiotic or maltodextrin for 5 days prior to the skin excisional wound generation continued to euthanasia day	Perioperative supplementation of probiotics promotes weight loss, accelerated skin healing, increased collagen deposition and reduced inflammation in diabetic rats	Campos et al. (2020)
Hind-limb ischemia	Hyper Glycemic mice	*Lactobacillus reuteri* transformed with a plasmid encoding the chemokine CXCL12	Not done	Wounds treated daily with 10 µL control *L. reuteri*, or CXCL12-expressing *L. reuteri* or *L. lactis* or saline solution	Accelerated wound closure at 2 d in severely hyperglycemic mice, and at 1 d post wound induction in moderate hyperglycemic mice but not afterwards	Vågesjö et al. (2018)

(continued)

Table 1 (continued)

Wound infection study	Model	Probiotic genus	Pathogens	Study method	Potential benefits	References
					Normalized blood flow in wound	
Chronic ulcerative lesion	In vitro	SYNBIO formulation *L. rhamosus*: *L. paracasei* (1:1)	Isolates from chronic ulcerative lesion (16S Rdna) *Ps. aeruginosa, S. aureus, Corynebacterium, A. baumanii, P. mirabilis, E. faecalis, Candida parapsilosis*	Adhesion assay on HaCaT and fibroblast cell line co-aggregation assay Agar diffusion with wound dressing components	17–19% adhesion 50% inhibition of *E. faecalis* 100% inhibition of *Ps. aeruginosa* and *S. aureus* at 72 h	Coman et al. (2020)
Diabetic foot ulcer (Wagner grade 3)	Randomized, double-blind, placebo-controlled trial 60 human subjects (aged 40–85 years) with grade 3 DFU	Oral probiotic capsules contained *Lactobacillus acidophilus, Lactobacillus casei, Lactobacillus fermentum and Bifidobacterium bifidum*	Not done	Probiotic capsules daily for 12 weeks	Reduced ulcer length Reduced insulin, fasting glucose, inflammatory markers and lipid profile when compared with placebo	Mohseni et al. (2018)
Chronic venous leg ulcer	Prospective human study In 14 diabetic and 20 non-diabetic patients	*Lacobacillus plantarum* ATCC 10241 (10^5 cfu)	*S. aureus, Ps. aeruginosa. S. epidermidis, E. cloacae, K. pneumoniae, E. faecalis*	Topical application daily, 10 days No antibiotics	Decrease pathogen load in 10 days ($P < 0.001$) compared to day 1 treatment. At 30 days, debridement, granulation formation and healing in	Peral et al. (2010)

Chronic limb ischemic wound	Human case study 83 old diabetic woman	10^{11} cfu lyophilized probiotics bacteria of L. plantarum NCIBMB 43029 (20% w/w), L. acidophilus NCIBMB 43030 (20% w/w and Streptococcus thermophilus NCIMB 30438 (40% w/w)	MDR K. pneumonia, E. faecalis and P. mirabilis infections	Topically three times a week for 24 days with no systemic or topical antibiotic treatment	Negative for E. faecalis at 12 days and K. pneumoniae and P. mirabilis after 21 days. Wound stabilized. Reduced inflammatory metabolites using metabolomics profiling	50% non-diabetics and 43% diabetics	Venosi et al. (2019)
Post-operative wound in colorectal cancer surgery	Randomized, double-blind, placebo-controlled trial for 84 test and 80 control humans	L. acidophilus LA5, L. plantarum, B. lactis BB-12, Saccharomyces boulardii (5.5 × 10^9 cfu)	A. baumannii, Ps. aeruginosa, MRSA	Oral given 1 day before operation and 14 days post-surgery	7.1% wound infection in treated versus 20% in control. 2.4% post-operative pneumonia in treated versus 11.3% in control. Lower requirement of mechanical ventilation in treated group		Kotzampassi et al. (2015)

[a]cfu: colony forming units

A. baumanii and *E. coli* (Wieërs et al. 2019). In an in vitro study, antimicrobial efficacy of probiotic combination SYNBIO (1:1 *L. rhamnosus* IMC501 and *L. paracasei* IMC 502IN) was demonstrated against chronic ulcerative lesion isolated bacteria. Co-aggregation and adherence of SYNBIO to human keratinocytes (HaCaT cells) and human fibroblast (NHF) was also found which could create a protective shield preventing pathogen biofilm formation (Coman et al. 2020). In an in vivo human plasma biofilm model, wound pathogens *S. aureus*, *Ps. aeruginosa*, *S. epidermidis*, *E. faecium* and *C. albicans* were challenged to probiotics *B. lactis*, *L. plantarum* and commensal *S. cerevisiae*. *L. plantarum* was found to inhibit *Ps. aeruginosa* biofilm completely with differential effect on other pathogens.

Using in vitro human epidermis model, topical treatment with heat-treated *L. johnsonii* NCC 533 (HT La1) reduced radiolabelled *S. aureus* adhesion by 74% and induced the expression of antimicrobial peptide hBD-2 (Rosignoli et al. 2018). Further, in an open-label multicentre study, topical lotion application containing HT La1 twice daily for 3 weeks to 31 atopic dermatitis patients with *S. aureus* load was correlated with local clinical improvement as well as controlled *S. aureus* colonization (Blanchet-Réthoré et al. 2017).

Novel LAB probiotic wound healing patch containing LAB in an adhesive gas permeable membrane producing nitric oxide gas was tested on ischaemic wounds in a New Zealand white rabbit model over 21 days and showed statistically significant wound healing (Jones et al. 2012).

Effect of probiotics on re-epithelialization has also been recorded. Primary human keratinocyte monolayers were scratched and treated with lysates of *L. reuteri*, *L. rhamnosus* GG, *L. fermentum* or *L. plantarum*. Increased rates of re-epithelialization were recorded with *L. rhamnosus* GG lysate. Keratinocyte proliferation was increased upon treatment with *L. reuteri*. *L. rhamnosus* GG treated scratches revealed elevated expression of wound healing chemokines such as CXCL2 and its receptor CXCR2 using microarray analysis (Mohammedsaeed et al. 2015). Treatment of keratinocytes infected with *S. aureus* with 10^8cfu/ml of live *L. rhamnosus* GG/spent culture fluid was found to increase viability of infected keratinocytes by 65% and 57%, respectively. The mechanisms involved were competitive extrusion and growth inhibition by live bacteria and inhibition of adhesion to keratinocytes (Mohammedsaeed et al. 2014).

Cutaneous probiotics such as *S. thermophilus* increase amounts of ceramides and phosphorylcholine through sphingomyelinase production in keratinocytes to improve skin barrier functions and promote wound healing (Di Marzio et al. 2008). A meta-regression analysis of six animal studies with random effect model estimated that bacteria probiotic therapies (70% kefir gel, *L. fermentum*, *L. plantarum*, *L. reuteri*, *L. brevis*) are effective in treatment of cutaneous wounds (Blanchet-Réthoré et al. 2017).

In diabetes rat model, perioperative supplementation of Probiatop® (four probiotic strains consisting of *L. paracaei* LPC-37, *L. rhamnosus* HN001, *B. lactis* HN019 and *L. acidophilus* NCFM (dose 10^9 cfu) or maltodextrin as control provided 5 days prior to induction of second-intention wound resulted in better healing possibly due to increased vascularization, collagen type 1 deposition and reduced inflammatory response (Campos et al. 2020).

In an oral probiotic supplementation study with 60 humans (aged 40–85 years) suffering from grade 3 DFU, increased wound healing with a decrease in metabolic parameters such as fasting glucose levels, lipid profiles and inflammatory markers were recorded. Thirty individuals received probiotic capsules (*L. casei*, *L. acidophilus L. fermentum* and *B. bifidum*) while 30 were treated with placebo in addition to standard wound care treatment (Mohseni et al. 2018).

A case study of diabetic woman, age 83 years, suffering from chronic limb ischemic wound infected with MDR *K. pneumonia*, *E. faecalis* and *P. mirabilis* infections was reported. The wound was topically treated with a mixture of 100 billion cfu lyophilized probiotics bacteria of *L. plantarum* NCIBMB 43029 (20%), *L. acidophilus* NCIBMB 43030 (20%) and *Streptococcus thermophilus* NCIMB 30438 (40%) three times a week for 24 days with no systemic or topical antibiotic treatment. The wound condition stabilized in 2 weeks following topical probiotic application with tests negative for *E. faecalis* (day 12 post application) and *P. mirabilis* and *K. pneumoniae* (day 21 days post application), respectively. Metabolomic profiling by Nuclear Magnetic Resonance (NMR) analysis of the infected wound showed a change in at least 30% of the metabolites including decreased pyrimidine (indicative of infection resolution), polyamine putrescine (involved in protection from oxidative stress, phagolysosomes and biofilm formation,), lysine (reduced bacterial proteolytic activity) and 2,3-butenediol (*K. pneumonia* metabolite) levels (Venosi et al. 2019).

L. plantarum was topically applied on venous ulcer lesions of 14 diabetics and 20 non-diabetic patients for 30 days. Tissue debridement, granulation formation and complete healing were observed in 43% diabetics and 50% non-diabetics with decreased PMN, enhanced IL8 levels in ulcer bed cells, apoptosis/necrotic cell death and induced wound healing (Peral et al. 2010).

Pre-operative probiotic or synbiotic administrations can contribute to reduction in post-operative surgical wound dehiscence in patients with co-morbidities such as diabetes, hypertension, ageing, cancer surgery and pre-existing immunocompromised conditions. A meta-analysis investigating randomized controlled trials testing probiotic efficacy in controlling infection related complications in colorectal surgery concluded that multi-strain probiotics have reduced surgical site and non-surgical site infections. This was largely attributed to immunomodulation and pathogen inhibition by probiotic administration (Liu et al. 2017). In a randomized controlled double-blind study, efficacy of oral probiotic formulation containing *L. acidophilus* LA5, *B. lactis* BB-12, *L. plantarum*, *Saccharomyces boulardii* (5.5×10^9 cfu) given 1 day prior operation and continued for 15 days' post-operation was tested in colorectal surgery patients. Treated patients showed statistically significant lower wound infections, post-operative pneumonia and requirement of mechanical ventilators (Kotzampassi et al. 2015).

Nutritional benefits, reduced pathogenic load and further immunomodulation appear to be the major contribution of probiotic administration for wound healing and repair. Although several in vitro and animal studies indicate positive effects of probiotics on wound healing, there are far fewer human studies targeting chronic wounds.

4.2.2   Next Generation Probiotics (NGP)

Beneficial micro-organisms useful in prevention, cure or treatment of disease, identified using comparative genomics and next generation sequencing technology, are referred to as Next Generation Probiotics (NGP) (Martín and Langella 2019). Metagenomics approaches, next generation sequencing and computational technologies are used for screening and characterization of microbial isolates from specific sites and associated diseases, bypassing the traditional microbial culture and cloning methods. Using metagenomic approaches, nucleic acid material directly extracted from body tissues is amplified using polymerase chain reaction (PCR) with primers targeting conserved bacterial regions. The DNA sequence of the amplicons is then determined using high throughput Next Generation Sequencing (NGS) technology. Comparative genomics and computational softwares are used for phylogenetic analysis and identification of microbiome at particular body sites or tissues (Hodkinson and Grice 2015).

Increased awareness about the human microbiome using NGS technology and the association of disease with dysbiosis is used to identify beneficial commensal microbiota that can be used to restore tissue homeostasis. Identification of microbiota in wounds or disease conditions irrespective of the polymicrobial nature and heterogeneity in terms of spatial and nutritional requirements can help determine novel biomarkers of diseased tissue. Measurements of microbial population dynamics, inflammatory markers and wound healing progression can be conducted in consort with administration of probiotics or microbiome transplantation (Hodkinson and Grice 2015). Identification of previously unknown species can also be performed using advances in culturomics using high throughput culture conditions and mass spectrometry (MALDI-TOF MS) (Bilen et al. 2018). New information regarding the molecular burdens of pathogen and commensals at wound sites now offers a precise and targeted preventive and curative approach for disease amelioration (Chang et al. 2019).

Strategies for isolation of NGP require comprehensive information and analyses of microbiota composition, their metabolome and host responses from big data study groups involving healthy, disease and experimental sets. *Akkermansia muciniphila, Bacteroides fragilis, Faecalibacterium prausnitzii, Christensenella minuta, and Pasabacteriodes goldsteinii* are some of the NGP under investigation (Chang et al. 2019). NGP bacterial species such as *Bifidobacteirum longum, Eubacterium limosum, Enterococcus hirae, Enterococcus faecium, Collinsella aerofaceins* and *Burkholderia cepacia* show promising effects as anticancer immunotherapeutics (Cani and Van Hul 2015). *Akkermansia muciniphila* is a potential key NGP candidate which modulates endocannabinoid system which is a regulatory component in targeting obesity, type 2 diabetes and inflammation (Cani et al. 2014).

There are some encouraging reports of beneficial commensals identified using the next generation sequencing technologies in wound healing and care. Comparative metagenomics of healthy and diseased skin microbiome show distinct differences in the biodiversity and microbial loads and several such studies on chronic wounds are discussed previously in the section on wound microbiology. There are however no NGP related human studies with chronic wound infections. Promising data have been obtained with microbiome reconstitution studies in Atopic Dermatitis, a recurrent chronic skin inflammatory disease associated with *S. aureus* pathogenesis. Commensal Coagulase Negative Staphylococci (CoNS) isolated from AD patients were found to be deficient producers of antimicrobial activity against pathogenic *S. aureus* in comparison with CoNS isolated from skin of healthy volunteers. An antimicrobial peptide producing CoNS, *S. hominis* A9 (1×10^5 cfu) was used for autologous microbiome transplantation onto AD patients. Once a day application reduced *S. aureus* colonization while twice a day application for a week completely inhibited *S. aureus* growth in AD patients (Nakatsuji et al. 2017).

Treatment with *Roseomonas mucosa*, a commensal isolated from healthy volunteers improved AD outcome in mice and cell culture model while *R. mucosa* from AD patients worsened symptoms. In an open-label phase I/II safety and activity trial with 10 adults and 5 paediatric patients ('the Beginning Assessment of Cutaneous Treatment Efficacy for *Roseomonas* in Atopic Dermatitis trial; BACTERiAD I/II'), *R. mucosa* treatment resulted in a significant reduction in disease severity, *S. aureus* count and topical steroid requirement. No treatment complications or adverse effects were observed (Myles et al. 2018). Successful microbiome transplant studies in AD provide hope that such studies can be conducted for chronic wound healing using the appropriate microbiome diagnostics.

Most studies with NGP are in various stages of in vitro or in vivo animal studies with few targeted towards chronic wound healing. However, many NGP candidates are involved in regulating metabolic syndromes and disorders including inflammation and hence can be beneficial in wound healing and repair according to the 'gut-skin-brain axis' hypothesis (Arck et al. 2010). Our recent understanding of chronic wound microbiome can be translated into the oral or local application of identified NGP commensals for wound healing and repair. Skin dysbiosis hence can be effectively treated by administering topically or systemically distinctly identified beneficial probiotics tailored to a particular chronic wound (Tsiouris et al. 2017).

4.2.3 Engineered LBP

Current probiotics are designed for general health and wellness and do not particularly target treatment of a specific condition. Moreover, these typically form transient microbiota and are not retained for long durations (Abatenh et al. 2018). Even the identification and reconstitution of generalized microbiota does not provide guaranteed treatment of diseases. The use of genetically engineered live organisms for disease diagnosis and treatment has been envisaged by the combination of synthetic probiotic biology, system engineering and molecular metabolomics

(Ozdemir et al. 2018). Development of engineered probiotics producing specific substances relevant and customized to a certain condition either constitutively or upon induction promise to be next-gen pharmaceutical products (Sessions et al. 2017). Live engineered probiotics that also detect presence of biomarker chemicals indicative of an altered metabolism or condition will be useful as biosensors (Charbonneau et al. 2020). While most studies have focussed on engineering of *E. coli* and *L. lactis*, recent studies are focused on developing genetically engineered beneficial commensal microbiota as they are better at colonization and may increase success in treatment strategies (Praveschotinunt et al. 2019). Bioengineered probiotics offer great promise as delivery agents for antibiotics, bacteriocins, immune stimulants, growth factors as well as antagonists for harmless metabolites at the site of disease (Wang et al. 2017; Charbonneau et al. 2020). Other applications include diagnosis and detection of disease by introducing pathogen-related quorum sensing systems into probiotics (Zhou et al. 2020).

Engineered *Lactobacillus casei* BL23 producing superoxide dismutase (SOD) or catalase (CAT) treatment of mice with Crohn's disease had lower intestinal inflammation, increased gut enzymatic activity and faster recovery (LeBlanc et al. 2011). An auxotrophic strain of *S. epidermis* NRRL B-4268 containing three deletions of alanine biosynthetic genes (two Alanine racemases alr1, alr2 and d-alanine aminotransferase dat) was designed as a live LBP, the growth of which can be controlled using d-alanine in the absence of antibiotics or genes coding for antibiotic resistance. *S. epidermis* colonized on cultures human skin model in vitro and increased the expression of human β defensins when supplemented with d-alanine (Dodds et al. 2020).

Introducing biomolecules in chronic wound environments deficient in factors responsible for vascularization, re-epithelialization and immunomodulation serve in enhancing wound healing. A key role of chemokines in regulating wound healing by angiogenesis and leukocyte recruitment for the synthesis of growth factors and cytokines has been delineated (Ridiandries et al. 2018). Preclinical studies in mice administered with CCL2, CCL21, CXCL12 and CXCR4 antagonist were shown to improve wound healing. Recombinant CXCL12 delivering transformed *Lactobacillus reuteri* were administered topically to wounds in mice (Vågesjö et al. 2018). An increased proliferation of dermal cells and macrophages with TGF-β production was observed. *Lactobacillus* caused decreased local pH inhibiting peptidase CD26 activity with increase in local CXCL12. Treatment of hind limb ischemia model in mice with the CXCL-12 *Lactobacilli* accelerated wound closure. However, in severe hyperglycemic mice, rCXCL12 *Lactobacillus* was able to normalize blood flow in the wound but help in closure only at day 1 after induction and but not thereafter. Using an in vitro wound re-epithelialization model, CXCL12 producing *L. reuteri* treatment caused increased keratinocytes and macrophages proliferation. Treatment was found to be safe and only locally functional as neither bacteria nor the chemokine was found systemically (Vågesjö et al. 2018). The technology was used to develop ILP100, a novel investigational new product using a novel drug delivery technology by Ilya Pharma which is approved for first-in-human/Phase I trial (https://www.ilyapharma.se/).

Aurealis Therapeutics developed a genetically engineered *Lactococcus lactis* probiotic (AUP-16) producing three products human basic fibroblast growth factors (FGF2, bFGF), IL4 and macrophage colony stimulating factor (CSF1) for treatment of non-healing wounds and regenerative diseases, which has clinical trial application approval for DFU patient trial by German Health Authority Paul-Ehrlich-Institute (https://aurealistherapeutics.com/aurealis-therapeutics-receives-clinical-trial-applica tion-approval-for-aup-16-diabetic-foot-ulcer-patient-trial/).

Such studies provide the basis for the development of engineered biologics promoting chronic wound healing.

5 Regulatory and Intellectual Property Issues

According to Roots Analysis, a business research and consulting group, the LBP market is projected to exponentially expand at the rate of 38% annually with the USA market alone being worth 2.1 billion by the year 2030 (https://www. rootsanalysis.com/microbiome-contract-manufacturing/). Several pharmaceutical companies are interested in investing in the initiative and therefore regulatory and intellectual property issues need formulation. Stringent safety regulatory protocols are mandated due to the living nature and multifactorial mode of action of LBP (El Hage et al. 2017). It is imperative that LBP are tested for safety, efficacy, quality of manufacturing and consistency in well-controlled human clinical trials (Rouanet et al. 2020).

LBP are considered as pharmaceutical products in comparison to probiotics, which are essential dietary supplements, though many are now being considered under LBP. Traditional probiotics intended for use as dietary supplements are regulated by the USA Food and Drug Administration (FDA) Center for Food Safety and Applied Nutrition and do not require FDA approval before being sold as food products or supplements (Venugopalan et al. 2010). The safety and authenticity of the product are monitored by the Dietary Supplement Health and Education Act (DSHEA) of 1994 that also ensures Current Good Manufacturing Practices (GMP) requirements in probiotic production. The law allows manufacturers of probiotics to make health claims for the products which must be substantiated by experts in the field. They also can declare a structure/function claim that states that the product does not affect normal functioning of the human body. While in Europe, disease-specific health claims cannot be made with dietary supplements. Such claims (supported with scientific evidence) can be made with an FDA mandated disclaimer statement (http://internationalprobiotics.org/regulation-probiotics-usa/). A category of medicinal foods in the USA permits probiotic categories to be administered under physicians' supervision and intended for a specific disease condition (Lewis et al. 2019).

Unlike other Microbiotic Medicinal Products (MMP), LBP are clearly defined and identified by both the US FDA and European Pharmacopoeia. According to the European Union, LBP are 'medicinal products containing live micro-organisms (bacteria or yeasts) for human use' and exclude faecal microbiota transplants and gene therapy) as per Directive 2001/83/EC ('Council Directive (EC) 2001/83 of 6 November 2001 on the Community Code Relating to Medicinal Products for Human Use'. 2001). In the USA, the Center for Biologics Evaluation and Research (CBER) under the Food and Drug Administration (FDA) is yet to approve LBP as a medicinal product (Dreher-Lesnick et al. 2017). Any substance to be used as a drug must be approved by the FDA or an Investigational New Drug (IND) application must be sought. In 2016, regulatory considerations were put forth for conducting human clinical trials with LBP (U.S. Department of Health and Human Services 2016).

An engineered LBP as a therapeutic application must be with a well-characterized microorganism in the form of investigational new drug application (IND) (Charbonneau et al. 2020). Each engineered LBP strain is unique regarding properties related to by-products, colonization, distribution, persistence, clearance, pharmacokinetics and toxicity. The mode of application may differ being oral, topical or systemic depending on the type of disease or condition as in the case of chronic wounds. Additionally, issues regarding the stability of exogenously inserted genetic material as well horizontal transfer of antibiotic/virulence genes must be addressed. Biocontainment of engineered LBP within the target site as well as prevention of engineered LBP from contaminating the environment are some of the issues that regulatory bodies are currently focussing on (Lee et al. 2018). No engineered LBPs have yet been approved for human use, although several natural LBP are now in clinical development stage and proof of concept are awaited for the development of future protocols.

6   Challenges and Future Outlook

The foremost challenge for the development of LBP as mainstream therapeutics for chronic wound healing is that most of the current probiotic studies are with heterogeneous strains and often uncontrolled. This has created a general public perception linking probiotics to health providing fortified foods and nutritional supplements. There is, however, little clinical research-based evidence to correlate the two effects. A review of literature brings forth a gap between translation from in vitro and in vivo laboratory studies to clinical application which requires large-scale human clinical trials. Today, people have access to probiotics over the counter. A limited understanding of the beneficial role of the different formulations, strains, dosage requirements for a particular chronic wound healing condition remains. Even professional opinions of the health benefits about probiotics remain invalidated. Several health organizations are already pointing out that the indiscriminate use of probiotics as health supplements may lead to the junking of a potentially useful strategy due to lack of proper understanding of the host–LBP interaction. Change in this perception requires major investments from regulatory bodies, government health departments and physicians.

LBPs are currently perceived as adjuncts to conventional approaches to disease treatment. Major advantages of engineered LBP are that they can be used as delivery agents for drugs and antibodies. It is foreseen that LBP will be conjoined with alternative treatments in different bioformulations (herbal, antibiotics, cancer therapy, drug molecules, nanoparticles etc.) delivered in diverse pharmaceutical formats such as lozenges, capsules and wound dressings. In most cases, antibiotic-resistant probiotics are being used in order to sustain their colonization. This, however, remains a risk particularly in the use of probiotic strains in immunocompromised patients and must be addressed.

Even though several new species have been identified through metagenomic approaches, the majority of beneficial live organisms are difficult to culture in the current paradigm. Production of LBP as a product involves several more steps beyond screening and characterization. Current Intellectual Property laws for biologicals prevent the patenting of live and naturally occurring microorganisms but allow patenting of genetically engineered organisms. All LBP would have to go through a series of pharmaceutical clinical trials. The final phases of pharmaceutical product development are randomized, placebo-controlled, double-blind, clinical trials with well-designed study models that can provide evidence for the clinical benefits of prescribing LBP. Delivery systems and encapsulation need to be standardized and the risk associated with genetic mutations must be considered. Toxicity and environmental safety issues regarding its release and subsequent consequences must be worked out. The other challenges pertaining to commercialization of the products lies with developing adequate and expert manpower as well as infrastructure for LBP mass production, and formulating industry standards with regards to reproducibility issues and safe manufacturing practices.

The potential application of LBP in chronic wound healing can be revolutionary. A targeted approach of using next generation molecular techniques and systems biology to understand the deficient microbiome and metabolic profiles in chronic wound patients followed by tailored reconstitution with a live biotherapeutics product has far-reaching potential.

Conflict of Interest   There is no conflict of interest.

References

Abatenh E, Gizaw B, Tsegay Z et al (2018) Health benefits of probiotics. J Bacteriol Infect Dis 2:8–27

Al-Joufi FA, Aljarallah KM, Hagras SA et al (2020) Microbial spectrum, antibiotic susceptibility profile, and biofilm formation of diabetic foot infections (2014–18): a retrospective multicenter analysis. 3. Biotech 10:325. https://doi.org/10.1007/s13205-020-02318-x

Arck P, Handjiski B, Hagen E et al (2010) Is there a 'gut-brain-skin axis'? Exp Dermatol 19:401–405. https://doi.org/10.1111/j.1600-0625.2009.01060.x

Baczako A, Fischer T, Konstantinow A, Volz T (2019) Chronic wounds. MMW Fortschr Med 161:48–56. https://doi.org/10.1007/s15006-019-0006-x

Balsa IM, Culp WTN (2015) Wound care. Vet Clin North Am Small Anim Pract 45:1049–1065. https://doi.org/10.1016/j.cvsm.2015.04.009

Berg G, Rybakova D, Fischer D et al (2020) Microbiome definition re-visited: old concepts and new challenges. Microbiome 8:103. https://doi.org/10.1186/s40168-020-00875-0

Bilen M, Dufour J-C, Lagier J-C et al (2018) The contribution of culturomics to the repertoire of isolated human bacterial and archaeal species. Microbiome 6:94. https://doi.org/10.1186/s40168-018-0485-5

Blanchet-Réthoré S, Bourdès V, Mercenier A et al (2017) Effect of a lotion containing the heat-treated probiotic strain Lactobacillus johnsonii NCC 533 on *Staphylococcus aureus* colonization in atopic dermatitis. Clin Cosmet Investig Dermatol 10:249–257. https://doi.org/10.2147/CCID.S135529

Bryant RA, Nix DP (2012) Acute and chronic wounds : current management concepts. Elsevier/Mosby, St. Louis

Byrd AL, Belkaid Y, Segre JA (2018) The human skin microbiome. Nat Rev Microbiol 16:143–155. https://doi.org/10.1038/nrmicro.2017.157

Campos LF, Tagliari E, Casagrande TAC et al (2020) Effects of probiotic supplementation on skin wound healing in diabetic rats. Arq Bras Cir Dig 33:e1498. https://doi.org/10.1590/0102-672020190001e1498

Cani PD (2018) Human gut microbiome: hopes, threats and promises. Gut 67:1716–1725. https://doi.org/10.1136/gutjnl-2018-316723

Cani PD, Van Hul M (2015) Novel opportunities for next-generation probiotics targeting metabolic syndrome. Curr Opin Biotechnol 32:21–27. https://doi.org/10.1016/j.copbio.2014.10.006

Cani PD, Geurts L, Matamoros S et al (2014) Glucose metabolism: focus on gut microbiota, the endocannabinoid system and beyond. Diabetes Metab 40:246–257. https://doi.org/10.1016/j.diabet.2014.02.004

Chamberlain R, Lau C (2016) Probiotics are effective at preventing *Clostridium difficile*-associated diarrhea: a systematic review and meta-analysis. Int J Gen Med 9:27. https://doi.org/10.2147/IJGM.S98280

Chang C-J, Lin T-L, Tsai Y-L et al (2019) Next generation probiotics in disease amelioration. J Food Drug Anal 27:615–622. https://doi.org/10.1016/J.JFDA.2018.12.011

Charbonneau MR, Isabella VM, Li N, Kurtz CB (2020) Developing a new class of engineered live bacterial therapeutics to treat human diseases. Nat Commun 11:1738. https://doi.org/10.1038/s41467-020-15508-1

Cheng H-Y, Ning M-X, Chen D-K, Ma W-T (2019) Interactions between the gut microbiota and the host innate immune response against pathogens. Front Immunol 10:607. https://doi.org/10.3389/fimmu.2019.00607

Choi Y, Banerjee A, McNish S et al (2019) Co-occurrence of anaerobes in human chronic wounds. Microb Ecol 77:808–820. https://doi.org/10.1007/s00248-018-1231-z

Clarke TB, Davis KM, Lysenko ES et al (2010) Recognition of peptidoglycan from the microbiota by Nod1 enhances systemic innate immunity. Nat Med 16:228–231. https://doi.org/10.1038/nm.2087

Cogen AL, Yamasaki K, Sanchez KM et al (2010) Selective antimicrobial action is provided by phenol-soluble Modulins derived from *Staphylococcus epidermidis*, a normal resident of the skin. J Invest Dermatol 130:192–200. https://doi.org/10.1038/JID.2009.243

Coman MM, Mazzotti L, Silvi S et al (2020) Antimicrobial activity of SYNBIO ® probiotic formulation in pathogens isolated from chronic ulcerative lesions: *in vitro* studies. J Appl Microbiol 128:584–597. https://doi.org/10.1111/jam.14482

Delgado S, Sánchez B, Margolles A et al (2020) Molecules produced by probiotics and intestinal microorganisms with immunomodulatory activity. Nutrients 12:391. https://doi.org/10.3390/NU12020391

Delmore B, Cohens J, O'Neill D et al (2017) Reducing postsurgical wound complications: a critical review. Adv Ski Wound Care 30:272–285

Di Marzio L, Cinque B, Cupelli F et al (2008) Increase of skin-ceramide levels in aged subjects following a short-term topical application of bacterial sphingomyelinase from *Streptococcus thermophilus*. Int J Immunopathol Pharmacol 21:137–143. https://doi.org/10.1177/039463200802100115

Dodds D, Bose J, Deng M et al (2020) Controlling the growth of the skin commensal *Staphylococcus Epidermidis* using d-alanine auotrophy. mSphere 5:e00360-20. https://doi.org/10.1128/MSPHERE.00360-20

Donlan RM, Costerton JW (2002) Biofilms: survival mechanisms of clinically relevant microorganisms. Clin Microbiol Rev 15:167–193. https://doi.org/10.1128/cmr.15.2.167-193.2002

Dowd S, Wolcott R, Sun Y et al (2008a) Polymicrobial nature of chronic diabetic foot ulcer biofilm infections determined using bacterial tag encoded FLX amplicon pyrosequencing (bTEFAP). PLoS One 3:e3326. https://doi.org/10.1371/journal.pone.0003326

Dowd SE, Sun Y, Secor PR et al (2008b) Survey of bacterial diversity in chronic wounds using pyrosequencing, DGGE, and full ribosome shotgun sequencing. BMC Microbiol 8:43. https://doi.org/10.1186/1471-2180-8-43

Dowd SE, Delton Hanson J, Rees E et al (2011) Survey of fungi and yeast in polymicrobial infections in chronic wounds. J Wound Care 20:40–47. https://doi.org/10.12968/jowc.2011.20.1.40

Drago F, Gariazzo L, Cioni M et al (2019) The microbiome and its relevance in complex wounds. Eur J Dermatol 29(1):6–13. https://doi.org/10.1684/EJD.2018.3486

Dreher-Lesnick SM, Stibitz S, Carlson PE (2017) U.S. regulatory considerations for development of live biotherapeutic products as drugs. Microbiol Spectr 5. https://doi.org/10.1128/microbiolspec.BAD-0017-2017

El Hage R, Hernandez-Sanabria E, Van de Wiele T (2017) Emerging trends in "smart probiotics": functional consideration for the development of novel health and industrial applications. Front Microbiol 8:1889. https://doi.org/10.3389/fmicb.2017.01889

Erdman SE, Poutahidis T (2014) Probiotic "glow of health": it's more than skin deep. Benef Microbes 5:109–119. https://doi.org/10.3920/BM2013.0042

Escobar-Zepeda A, Vera-Ponce de León A, Sanchez-Flores A (2015) The road to metagenomics: from microbiology to DNA sequencing technologies and bioinformatics. Front Genet 6:348. https://doi.org/10.3389/fgene.2015.00348

Fijan S (2016) Antimicrobial effect of probiotics against common pathogens. In: Probiotics and prebiotics in human nutrition and health. InTech, Rijeka

Frykberg RG, Banks J (2015) Challenges in the treatment of chronic wounds. Adv Wound Care 4:560–582. https://doi.org/10.1089/wound.2015.0635

Fu RH, Wu DC, Yang W et al (2018) Probiotic metabolites promote anti-inflammatory functions of immune cells. FASEB J 31:1048.1–1048.1. https://doi.org/10.1096/FASEBJ.31.1_SUPPLEMENT.1048.1

Gardiner M, Vicaretti M, Sparks J et al (2017) A longitudinal study of the diabetic skin and wound microbiome. PeerJ 5:e3543. https://doi.org/10.7717/peerj.3543

Gasbarrini G, Bonvicini F, Gramenzi A (2016) Probiotics history. J Clin Gastroenterol 2015:S116–S119. https://doi.org/10.1097/MCG.0000000000000697

Glatthardt T, Campos J, Chamon R et al (2020) Small molecules produced by commensal *Staphylococcus Epidermidis* disrupt formation of biofilms by *Staphylococcus Aureus*. Appl Environ Microbiol 86:e02539. https://doi.org/10.1128/AEM.02539-19

Gontcharova V (2010) A comparison of bacterial composition in diabetic ulcers and contralateral intact skin. Open Microbiol J 4:8–19. https://doi.org/10.2174/1874285801004010008

Gorreja F (2019) Gene expression changes as predictors of the immune-modulatory effects of probiotics: towards a better understanding of strain-disease specific interactions. NFS J 14–15:1–5. https://doi.org/10.1016/J.NFS.2019.02.001

Götz F, Perconti S, Popella P et al (2014) Epidermin and gallidermin: *Staphylococcal lantibiotics*. Int J Med Microbiol 304:63–71. https://doi.org/10.1016/J.IJMM.2013.08.012

Gu Q, Li P (2016) Biosynthesis of vitamins by probiotic bacteria. In: Probiotics and prebiotics in human nutrition and health. InTech, Rijeka

Gueniche A, Knaudt B, Schuck E et al (2008) Effects of nonpathogenic gram-negative bacterium Vitreoscilla filiformis lysate on atopic dermatitis: a prospective, randomized, double-blind, placebo-controlled clinical study. Br J Dermatol 159:1357–1363. https://doi.org/10.1111/j. 1365-2133.2008.08836.x

Gurung M, Li Z, You H et al (2020) Role of gut microbiota in type 2 diabetes pathophysiology. EBioMedicine 51:102590. https://doi.org/10.1016/j.ebiom.2019.11.051

Hajfarajollah H, Eslami P, Mokhtarani B, Akbari Noghabi K (2018) Biosurfactants from probiotic bacteria: a review. Biotechnol Appl Biochem 65:768–783. https://doi.org/10.1002/bab.1686

Han G, Ceilley R (2017) Chronic wound healing: a review of current management and treatments. Adv Ther 34:599–610. https://doi.org/10.1007/s12325-017-0478-y

Hannigan GD, Meisel JS, Tyldsley AS et al (2015) The human skin double-stranded DNA virome: topographical and temporal diversity, genetic enrichment, and dynamic associations with the host microbiome. MBio 6:e01578–e01515. https://doi.org/10.1128/mBio.01578-15

Heydari Nasrabadi HM, Ebrahimi T et al (2011) Study of cutaneous wound healing in rats treated with *Lactobacillus plantarum* on days 1, 3, 7, 14 and 21. Afr J Pharm Pharmacol 5:2395–2401. https://doi.org/10.5897/AJPP11.568

Hill C, Guarner F, Reid G et al (2014) The International Scientific Association for Probiotics and Prebiotics consensus statement on the scope and appropriate use of the term probiotic. Nat Rev Gastroenterol Hepatol 11:506–514. https://doi.org/10.1038/nrgastro.2014.66

Hodkinson BP, Grice EA (2015) Next-generation sequencing: a review of technologies and tools for wound microbiome research. Adv Wound Care 4:50–58. https://doi.org/10.1089/wound. 2014.0542

Im E, Choi YJ, Kim CH et al (2009) The angiogenic effect of probiotic Bacillus polyfermenticus on human intestinal microvascular endothelial cells is mediated by IL-8. Am J Physiol Gastrointest Liver Physiol 297:G999. https://doi.org/10.1152/AJPGI.00204.2009

James GA, Swogger E, Wolcott R et al (2008) Biofilms in chronic wounds. Wound Repair Regen 16:37–44. https://doi.org/10.1111/j.1524-475X.2007.00321.x

Jones M, Ganopolsky JG, Labbé A et al (2012) Novel nitric oxide producing probiotic wound healing patch: preparation and in vivo analysis in a New Zealand white rabbit model of ischaemic and infected wounds. Int Wound J 9:330–343. https://doi.org/10.1111/j.1742-481X.2011.00889.x

Kadam S, Shai S, Shahane A, Kaushik KS (2019) Recent advances in non-conventional antimicrobial approaches for chronic wound biofilms: have we found the 'chink in the armor'? Biomedicine 7:35. https://doi.org/10.3390/biomedicines7020035

Kalan L, Loesche M, Hodkinson BP et al (2016) Redefining the chronic-wound microbiome: fungal communities are prevalent, dynamic, and associated with delayed healing. MBio 7:e01058–e01016. https://doi.org/10.1128/mBio.01058-16

Kamus LJ, Theoret C, Costa MC (2018) Use of next generation sequencing to investigate the microbiota of experimentally induced wounds and the effect of bandaging in horses. PLoS One 13:e0206989. https://doi.org/10.1371/journal.pone.0206989

Kao M-S, Huang S, Chang W-L et al (2017) Microbiome precision editing: using PEG as a selective fermentation initiator against methicillin-resistant *Staphylococcus aureus*. Biotechnol J 12:1600399. https://doi.org/10.1002/biot.201600399

Kotzampassi K, Stavrou G, Damoraki G et al (2015) Randomized, double-blind, placebo-controlled study of the efficacy of four probiotics to modify the risk for postoperative complications in colorectal surgery. Crit Care 19:P390. https://doi.org/10.1186/cc14470

Larsbrink J, Rogers TE, Hemsworth GR et al (2014) A discrete genetic locus confers xyloglucan metabolism in select human gut Bacteroidetes. Nature 506:498–502. https://doi.org/10.1038/nature12907

Lazaro JL, Izzo V, Meaume S et al (2016) Elevated levels of matrix metalloproteinases and chronic wound healing: an updated review of clinical evidence. J Wound Care 25:277–287. https://doi.org/10.12968/jowc.2016.25.5.277

LeBlanc JG, del Carmen S, Miyoshi A et al (2011) Use of superoxide dismutase and catalase producing lactic acid bacteria in TNBS induced Crohn's disease in mice. J Biotechnol 151:287–293. https://doi.org/10.1016/J.JBIOTEC.2010.11.008

Lee JW, Chan CTY, Slomovic S, Collins JJ (2018) Next-generation biocontainment systems for engineered organisms. Nat Chem Biol 14:530–537. https://doi.org/10.1038/s41589-018-0056-x

Lewis CA, Jackson MC, Bailey JR (2019) Understanding medical foods under FDA regulations. In: Nutraceutical and functional food regulations in the United States around world. Academic Press, London, pp 203–213. https://doi.org/10.1016/B978-0-12-816467-9.00015-0

Lin T-L, Shu C-C, Lai W-F et al (2019) Investiture of next generation probiotics on amelioration of diseases—strains do matter. Med Microecol 1–2:100002. https://doi.org/10.1016/J.MEDMIC.2019.100002

Liu Z, Shen T, Zhang P et al (2011) *Lactobacillus plantarum* surface layer adhesive protein protects intestinal epithelial cells against tight junction injury induced by enteropathogenic *Escherichia coli*. Mol Biol Rep 38:3471–3480. https://doi.org/10.1007/s11033-010-0457-8

Liu PC, Yan YK, Ma YJ et al (2017) Probiotics reduce postoperative infections in patients undergoing colorectal surgery: a systematic review and meta-analysis. Gastroenterol Res Pract 2017:1–9. https://doi.org/10.1155/2017/6029075

Liu Q, Yu Z, Tian F et al (2020) Surface components and metabolites of probiotics for regulation of intestinal epithelial barrier. Microb Cell Factories 19:23. https://doi.org/10.1186/s12934-020-1289-4

Lopes EG, Moreira DA, Gullón P et al (2017) Topical application of probiotics in skin: adhesion, antimicrobial and antibiofilm *in vitro* assays. J Appl Microbiol 122:450–461. https://doi.org/10.1111/jam.13349

Malone M, Johani K, Jensen SO et al (2017) Next generation DNA sequencing of tissues from infected diabetic foot ulcers. EBioMedicine 21:142–149. https://doi.org/10.1016/j.ebiom.2017.06.026

Martín R, Langella P (2019) Emerging health concepts in the probiotics field: streamlining the definitions. Front Microbiol 10:1047. https://doi.org/10.3389/fmicb.2019.01047

Martin P, Nunan R (2015) Cellular and molecular mechanisms of repair in acute and chronic wound healing. Br J Dermatol 173:370–378. https://doi.org/10.1111/bjd.13954

Martin JM, Zenilman JM, Lazarus GS (2010) Molecular microbiology: new dimensions for cutaneous biology and wound healing. J Invest Dermatol 130:38–48. https://doi.org/10.1038/jid.2009.221

Metcalf D, Bowler P (2013) Biofilm delays wound healing: a review of the evidence. Burn Trauma 1:5. https://doi.org/10.4103/2321-3868.113329

Miljkovic M, Strahinic I, Tolinacki M et al (2015) AggLb is the largest cell-aggregation factor from *Lactobacillus paracasei Subsp. paracasei* BGNJ1-64, functions in collagen adhesion, and pathogen exclusion in vitro. PLoS One 10:e0126387. https://doi.org/10.1371/journal.pone.0126387

Mohammedsaeed W, McBain AJ, Cruickshank SM, O'Neill CA (2014) *Lactobacillus rhamnosus* GG inhibits the toxic effects of *Staphylococcus aureus* on epidermal keratinocytes. Appl Environ Microbiol 80:5773–5781. https://doi.org/10.1128/AEM.00861-14

Mohammedsaeed W, Cruickshank S, McBain AJ, O'Neill CA (2015) *Lactobacillus rhamnosus* GG lysate increases re-epithelialization of keratinocyte scratch assays by promoting migration. Sci Rep 5:16147. https://doi.org/10.1038/srep16147

Mohseni S, Bayani M, Bahmani F et al (2018) The beneficial effects of probiotic administration on wound healing and metabolic status in patients with diabetic foot ulcer: a randomized, double-blind, placebo-controlled trial. Diabetes Metab Res Rev 34. https://doi.org/10.1002/dmrr.2970

Monteagudo-Mera A, Rastall RA, Gibson GR et al (2019) Adhesion mechanisms mediated by probiotics and prebiotics and their potential impact on human health. Appl Microbiol Biotechnol 103:6463–6472. https://doi.org/10.1007/s00253-019-09978-7

Morton LM, Phillips TJ (2016) Wound healing and treating wounds: differential diagnosis and evaluation of chronic wounds. J Am Acad Dermatol 74:589–605. https://doi.org/10.1016/J.JAAD.2015.08.068

Myles IA, Earland NJ, Anderson ED et al (2018) First-in-human topical microbiome transplantation with Roseomonas mucosa for atopic dermatitis. JCI Insight 3:120608. https://doi.org/10.1172/jci.insight.120608

Naik S, Bouladoux N, Wilhelm C et al (2012) Compartmentalized control of skin immunity by resident commensals. Science 337:1115–1119. https://doi.org/10.1126/science.1225152

Nakatsuji T, Chen TH, Narala S et al (2017) Antimicrobials from human skin commensal bacteria protect against Staphylococcus aureus and are deficient in atopic dermatitis. Sci Transl Med 9: eaah4680. https://doi.org/10.1126/scitranslmed.aah4680

Neuman H, Debelius JW, Knight R, Koren O (2015) Microbial endocrinology: the interplay between the microbiota and the endocrine system. FEMS Microbiol Rev 39:509–521. https://doi.org/10.1093/femsre/fuu010

Notay M, Saric-Bosanac S, Vaughn AR et al (2020) The use of topical Nitrosomonas eutropha for cosmetic improvement of facial wrinkles. J Cosmet Dermatol 19:689–693. https://doi.org/10.1111/jocd.13060

O'Sullivan JN, Rea MC, O'Connor PM et al (2019) Human skin microbiota is a rich source of bacteriocin-producing staphylococci that kill human pathogens. FEMS Microbiol Ecol 95. https://doi.org/10.1093/femsec/fiy241

Olsson M, Järbrink K, Divakar U et al (2019) The humanistic and economic burden of chronic wounds: a systematic review. Wound Repair Regen 27:114–125. https://doi.org/10.1111/wrr.12683

Ong JS, Taylor TD, Yong CC et al (2020) Lactobacillus plantarum USM8613 aids in wound healing and suppresses Staphylococcus aureus infection at wound sites. Probiotics Antimicrob Proteins 12:125–137. https://doi.org/10.1007/s12602-018-9505-9

Ozdemir T, Fedorec AJH, Danino T, Barnes CP (2018) Synthetic biology and engineered live biotherapeutics: toward increasing system complexity. Cell Syst 7:5–16. https://doi.org/10.1016/J.CELS.2018.06.008

Peral MC, Rachid MM, Gobbato NM et al (2010) Interleukin-8 production by polymorphonuclear leukocytes from patients with chronic infected leg ulcers treated with Lactobacillus plantarum. Clin Microbiol Infect 16:281–286. https://doi.org/10.1111/j.1469-0691.2009.02793.x

Pereira RF, Bártolo PJ (2016) Traditional therapies for skin wound healing. Adv Wound Care 5:208–229. https://doi.org/10.1089/wound.2013.0506

Pereira SG, Moura J, Carvalho E, Empadinhas N (2017) Microbiota of chronic diabetic wounds: ecology, impact, and potential for innovative treatment strategies. Front Microbiol 8:1791. https://doi.org/10.3389/fmicb.2017.01791

Pinto D, Marzani B, Minervini F et al (2011) Plantaricin A synthesized by Lactobacillus plantarum induces in vitro proliferation and migration of human keratinocytes and increases the expression of TGF-β1, FGF7, VEGF-A and IL-8 genes. Peptides 32:1815–1824. https://doi.org/10.1016/J.PEPTIDES.2011.07.004

Poutahidis T, Kearney SM, Levkovich T et al (2013) Microbial symbionts accelerate wound healing via the neuropeptide hormone oxytocin. PLoS One 8:e78898. https://doi.org/10.1371/journal.pone.0078898

Prado Acosta M, Ruzal SM, Cordo SM (2016) S-layer proteins from Lactobacillus sp. inhibit bacterial infection by blockage of DC-SIGN cell receptor. Int J Biol Macromol 92:998–1005. https://doi.org/10.1016/j.ijbiomac.2016.07.096

Praveschotinunt P, Duraj-Thatte AM, Gelfat I et al (2019) Engineered E. coli Nissle 1917 for the delivery of matrix-tethered therapeutic domains to the gut. Nat Commun 10:5580. https://doi.org/10.1038/s41467-019-13336-6

Prince T, McBain AJ, O'Neill CA (2012) Lactobacillus reuteri protects epidermal keratinocytes from Staphylococcus aureus-induced cell death by competitive exclusion. Appl Environ Microbiol 78:5119–5126. https://doi.org/10.1128/AEM.00595-12

Quigley EMM (2019) Prebiotics and probiotics in digestive health. Clin Gastroenterol Hepatol 17:333–344. https://doi.org/10.1016/J.CGH.2018.09.028

Rahim K, Saleha S, Zhu X et al (2017) Bacterial contribution in chronicity of wounds. Microb Ecol 73:710–721. https://doi.org/10.1007/s00248-016-0867-9

Ramos AN, Sesto Cabral ME, Noseda D et al (2012) Antipathogenic properties of *Lactobacillus plantarum* on *Pseudomonas aeruginosa*: the potential use of its supernatants in the treatment of infected chronic wounds. Wound Repair Regen 20(4):552–562. https://doi.org/10.1111/j.1524-475X.2012.00798.x

Ramos AN, Sesto Cabral ME, Arena ME et al (2015) Compounds from *Lactobacillus plantarum* culture supernatants with potential pro-healing and anti-pathogenic properties in skin chronic wounds. Pharm Biol 53:350–358. https://doi.org/10.3109/13880209.2014.920037

Ridiandries A, Tan JTM, Bursill CA (2018) The role of chemokines in wound healing. Int J Mol Sci 19:3217. https://doi.org/10.3390/ijms19103217

Rosignoli C, Thibaut de Ménonville S, Orfila D et al (2018) A topical treatment containing heat-treated Lactobacillus johnsonii NCC 533 reduces *Staphylococcus aureus* adhesion and induces antimicrobial peptide expression in an in vitro reconstructed human epidermis model. Exp Dermatol 27:358–365. https://doi.org/10.1111/exd.13504

Rouanet A, Bolca S, Bru A et al (2020) Live biotherapeutic products, a road map for safety assessment. Front Med 7:237. https://doi.org/10.3389/fmed.2020.00237

Salas-Jara MJ, Ilabaca A, Vega M, García A (2016) Biofilm forming Lactobacillus: new challenges for the development of probiotics. Microorganisms 4:1–14. https://doi.org/10.3390/MICROORGANISMS4030035

Salem I, Ramser A, Isham N, Ghannoum MA (2018) The gut microbiome as a major regulator of the gut-skin Axis. Front Microbiol 9:1459. https://doi.org/10.3389/fmicb.2018.01459

Sambanthamoorthy K, Feng X, Patel R et al (2014) Antimicrobial and antibiofilm potential of biosurfactants isolated from lactobacilli against multi-drug-resistant pathogens. BMC Microbiol 14:197. https://doi.org/10.1186/1471-2180-14-197

Sanford JA, Gallo RL (2013) Functions of the skin microbiota in health and disease. Semin Immunol 25:370–377. https://doi.org/10.1016/J.SMIM.2013.09.005

Scalise A, Bianchi A, Tartaglione C et al (2015) Microenvironment and microbiology of skin wounds: the role of bacterial biofilms and related factors. Semin Vasc Surg 28:151–159. https://doi.org/10.1053/j.semvascsurg.2016.01.003

Schultz G, Bjarnsholt T, James GA et al (2017) Consensus guidelines for the identification and treatment of biofilms in chronic nonhealing wounds. Wound Repair Regen 25:744–757. https://doi.org/10.1111/wrr.12590

Schwarz A, Bruhs A, Schwarz T (2017) The short-chain fatty acid sodium butyrate functions as a regulator of the skin immune system. J Invest Dermatol 137:855–864. https://doi.org/10.1016/J.JID.2016.11.014

Sessions JW, Armstrong DG, Hope S, Jensen BD (2017) A review of genetic engineering bio-technologies for enhanced chronic wound healing. Exp Dermatol 26:179–185. https://doi.org/10.1111/exd.13185

Slominski A (2005) Neuroendocrine system of the skin. Dermatology 211:199–208. https://doi.org/10.1159/000087012

Spangler JR, Dean SN, Leary DH, Walper SA (2019) Response of *Lactobacillus plantarum* WCFS1 to the gram-negative pathogen-associated quorum sensing molecule N-3-oxododecanoyl homoserine lactone. Front Microbiol 10:715. https://doi.org/10.3389/fmicb.2019.00715

Sürmeli M, Maçin S, Akyön Y, Kayikçioğlu AU (2019) The protective effect of *Lactobacillus plantarum* against meticillin-resistant *Staphylococcus aureus* infections: an experimental animal model. J Wound Care 28:s29–s34. https://doi.org/10.12968/jowc.2019.28.Sup3b.S29

Thaiss CA, Zmora N, Levy M, Elinav E (2016) The microbiome and innate immunity. Nature 535:65–74. https://doi.org/10.1038/nature18847

The Human Microbiome Project Consortium (2012) Structure, function and diversity of the healthy human microbiome. Nature 486:207–214. https://doi.org/10.1038/nature11234

Tsiouris CG, Kelesi M, Vasilopoulos G et al (2017) The efficacy of probiotics as pharmacological treatment of cutaneous wounds: Meta-analysis of animal studies. Eur J Pharm Sci 104:230–239. https://doi.org/10.1016/J.EJPS.2017.04.002

U.S. Department of Health and Human Services F and DA (2016) Early clinical trials with live biotherapeutic products: chemistry. Manufacturing and Control Information- Guidance for Industry

Vågesjö E, Öhnstedt E, Mortier A et al (2018) Accelerated wound healing in mice by on-site production and delivery of CXCL12 by transformed lactic acid bacteria. Proc Natl Acad Sci U S A 115:1895–1900. https://doi.org/10.1073/pnas.1716580115

Venosi S, Ceccarelli G, de Angelis M et al (2019) Infected chronic ischemic wound topically treated with a multi-strain probiotic formulation: a novel tailored treatment strategy. J Transl Med 17:364. https://doi.org/10.1186/s12967-019-2111-0

Venugopalan V, Shriner KA, Wong-Beringer A (2010) Regulatory oversight and safety of probiotic use. Emerg Infect Dis 16:1661–1665. https://doi.org/10.3201/eid1611.100574

Verbanic S, Shen Y, Lee J et al (2020) Microbial predictors of healing and short-term effect of debridement on the microbiome of chronic wounds. Biofilms and Microbiomes 6:21. https://doi.org/10.1038/s41522-020-0130-5

Wang Y, Wu Y, Wang Y et al (2017) Antioxidant properties of probiotic bacteria. Nutrients 9:521. https://doi.org/10.3390/NU9050521

Wieërs G, Belkhir L, Enaud R et al (2019) How probiotics affect the microbiota. Front Cell Infect Microbiol 9:454. https://doi.org/10.3389/fcimb.2019.00454

Wolcott RD, Gontcharova V, Sun Y, Dowd SE (2009) Evaluation of the bacterial diversity among and within individual venous leg ulcers using bacterial tag-encoded FLX and titanium amplicon pyrosequencing and metagenomic approaches. BMC Microbiol 9:226. https://doi.org/10.1186/1471-2180-9-226

Wolcott RD, Hanson JD, Rees EJ et al (2016) Analysis of the chronic wound microbiota of 2,963 patients by 16S rDNA pyrosequencing. Wound Repair Regen 24:163–174. https://doi.org/10.1111/wrr.12370

Wu Y-K, Cheng N-C, Cheng C-M (2019) Biofilms in chronic wounds: pathogenesis and diagnosis. Trends Biotechnol 37:505–517. https://doi.org/10.1016/j.tibtech.2018.10.011

Yan F, Cao H, Cover TL et al (2007) Soluble proteins produced by probiotic bacteria regulate intestinal epithelial cell survival and growth. Gastroenterology 132:562–575. https://doi.org/10.1053/j.gastro.2006.11.022

Yan X, Gu S, Cui X et al (2019) Antimicrobial, anti-adhesive and anti-biofilm potential of biosurfactants isolated from Pediococcus acidilactici and *Lactobacillus plantarum* against *Staphylococcus aureus* CMCC26003. Microb Pathog 127:12–20. https://doi.org/10.1016/j.micpath.2018.11.039

Yu Y, Dunaway S, Champer J et al (2020) Changing our microbiome: probiotics in dermatology. Br J Dermatol 182:39–46. https://doi.org/10.1111/bjd.18088

Zhao R, Liang H, Clarke E et al (2016) Inflammation in chronic wounds. Int J Mol Sci 17:1–14. https://doi.org/10.3390/ijms17122085

Zhou Z, Chen X, Sheng H et al (2020) Engineering probiotics as living diagnostics and therapeutics for improving human health. Microb Cell Factories 19:56. https://doi.org/10.1186/s12934-020-01318-z

Role of Probiotics in Wound Healing

Amandeep Singh, Arpna Devi, and Uttam Kumar Mandal

1 Introduction

A wound can be described as an injury or damages in the body part, particularly in which rupture is formed in the skin or tissue (Velnar et al. 2009). This may occur due to various reasons; among them, the most common are thermal, physical damage, or medical and physiological conditions. According to a retrospective analysis in 2018, Medicare beneficiaries identified that globally around 8.2 million people had wounds with or without infections (Sen 2019). As per the market analysis, the global wound care market has witnessed a steady increase. In 2014, the estimated global market was around $2.8 billion, and it is expected to reach up to $15 billion by 2022 and $22 billion by 2024 (Sen 2019). This sharp rise signifies the importance as well as the gravity of the problem, which needs special attention to address the problem of the wound and wound-related complications.

Various types of wound are classified as lacerated, contusion, abrasion, ulcer, incised wound, and burn wound. Laceration wound may be defined as a type of wound which is produced by shredding of soft body tissue. It is craggy and rugged in nature. This type of wound is often infected with bacteria from the object that causes a cut. Contusion is the injured and destructed blood vessels beneath the skin resulting from a rush in the skin. Abrasion, in line with the meaning of the word, is a type of wound which is produced by rubbing the skin against a rough surface.

A. Singh
Department of Pharmaceutical Sciences and Technology, Maharaja Ranjit Singh Punjab Technical University (MRSPTU), Bathinda, Punjab, India

Department of Pharmaceutics, I.S.F. College of Pharmacy, Moga, Punjab, India

A. Devi · U. K. Mandal (✉)
Department of Pharmaceutical Sciences and Technology, Maharaja Ranjit Singh Punjab Technical University (MRSPTU), Bathinda, Punjab, India
e-mail: uttam@mrsptu.ac.in

The ulcer is an open wound that is mainly produced on the skin in case of impaired blood circulation, pressure, and injury. If this type of wound is not treated in time, it can cause severe medical complexity and ultimately may take an extended time for healing. Incised wound results from cutting off the skin with a sharp object or incising that causes laceration. This type of wound is usually long and deep. Burn wound is another unique type of wound that is initiated by hot liquids, electricity, fire, extreme heat of the Sun, and chemicals. Burn wounds are classified as the first-degree, the second-degree, and the third-degree burn, where the degree of burn is evaluated on the bases of size and depth of the burn.

2 Wound Healing Procedure

Wound healing is a natural repairing process of damaged tissue. Usually, wound healing is a well-ordered and distinguished biological procedure that takes care on its own. In the wound healing process, there is the involvement of some mediators and several cellular components such as growth factor or cytokine, microvascular cells, fibroblasts, keratinocytes, immune surveillance cells, blood cells like WBC and platelets, extracellular matrix, parenchymal cells, etc. (Falanga 2005; Singer and Dagum 2008; Bunman et al. 2017). The whole process of wound healing is classified under four phases, namely hemostasis, inflammatory, proliferative, and maturation. Elsewhere they are alternatively described as coagulation/ inflammation, granulation and tissue formation, and matrix remodeling or scar formation phase.

2.1 Coagulation/Inflammatory Phase

Instantly afterward the injury, the platelets bind with the impaired blood vessels, a hemostatic reaction starts which increases the blood-clotting cascade, avoids extreme bleeding, and offers tentative defense to the injured area (Fig. 1a). Simultaneously, blood platelets deliver various cytokines, apoptosis-inducing agents, and growth factors (Weyrich and Zimmerman 2004). The main elements of platelets release transforming growth factors are platelet-derived growth factor (PDGF), TGF-A1 and TGF-2 and some proactive cells like macrophages, neutrophils, and leukocytes (Delavary et al. 2011). Phagocytic and leukocytes cells deliver reactive species and proteases that have antibacterial properties and protect the wound from unknown pathogens and harmful bacteria. The phase of inflammatory is followed by the apoptosis of proactive cells; it starts progressively in some days after the formation of the wound. The agreeable mechanism of action of inflammation is not much understandable. Many reported studies recommend that cytokines, such as interleukin 1, bioactive lipids, and TGF-A1 such as lipoxins, resolvins, and cyclopentenone prostaglandin are the major component in this process (Gilroy

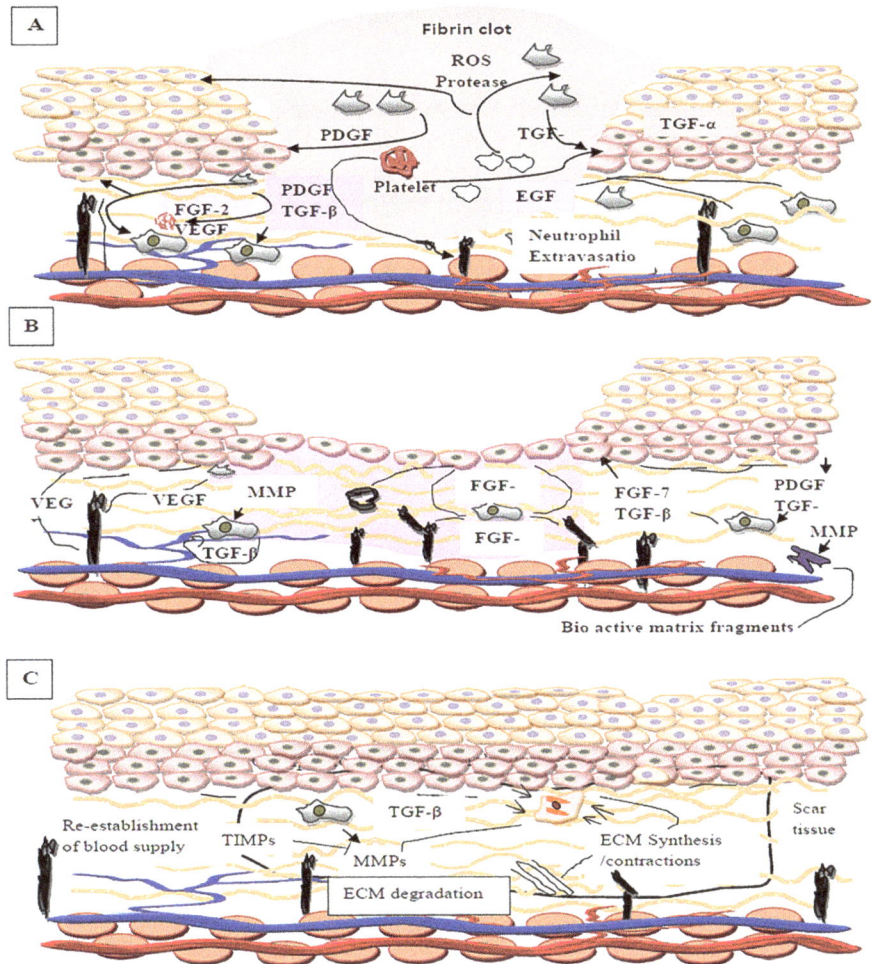

Fig. 1 Representation of various stages of wound healing. (**a**) Coagulation/inflammatory phase, (**b**) Proliferative phase: granulation and tissue formation, (**c**) Matrix remodeling and scar formation

et al. 2004; Eming et al. 2007). The main purpose of all these elements in the inflammatory process is still under investigation.

2.2 *Proliferative Phase: Granulation and Tissue Formation*

After the completion of the proactive phase, the new phase, i.e., the proliferative phase (Fig. 1a, b) starts. Growth factors formed by residual migration of epidermal, inflammatory cells and dermal cells act in juxtacrine, autocrine, and paracrine ways to produce or preserve its cellular proliferation and cellular migration. All actions are important in the development of granulation tissue and epithelialization (Falanga 2005).

The dermal and epidermal cells start to shift and propagate towards the wound site. The appropriate blood supply is required for the distribution of nutrients, gas, and metabolite exchange for the effective wound healing procedure. Initiation of an angiogenic response is equally important. The procedure of wound healing initiates instantly after the injury at the low level of oxygen, resulting in disturbance of blood vessels. This causes the creation of pro-angiogenic factors. The infected site is enriched by fibroblast PDGF, vascular endothelial growth factor (VEGF), and growth factor 2 (FGF-2) delivered by platelets and afterward by resident cells (Humar et al. 2002). These all are essential receptors for injury-induced angiogenic initiation. In the reaction of process, endothelial cells degrade basement covering that drifts to wound bed, proliferates and forms cell to cell associates and thus ultimately forming new blood vessels (Folkman and Klagsbrun 1987). Currently, it has been confirmed that endothelial progenitor cells (EPCs) also take part in wound thermbolysis. Generally, EPCs are help in the blood flow in retort to the injury and it present into the bones (Asahara et al. 1997; Liu and Velazquez 2008; Leone et al. 2009). Consequently, EPCs embed into the refurbished microvasculature, pleasing place adjoined to endothelial cells and contiguous to the injured area.

Mobilization of endothelial progenitor cell is transmitted by nitric oxide, metalloproteinases (MMP), VEGF, and matrix, specifically MMP-9. EPC engraftment and perhaps distinguishing happen in reaction to stromal cell-derived factor and has turn into probable freshly, insulin-like growth factor (IGF) (Leone et al. 2009; Maeng et al. 2009).

2.3   Matrix Remodeling and Scar Formation

Wound reconstruction and re-epithelialization are vital steps at this stage. This occurs due to epidermal and dermal cell migration and proliferation resulting from usual blood flow at the wound site (Fig. 1c). Fibroblasts multiply the surrounding by the extracellular matrix (ECM), and wound constitutes granulation tissue insertion with recently developed blood vessels. At the same time, a provisional matrix containing fibronectin, hyaluronic acid, collagen III, and fibrin gradually substitute ECM primarily containing collagen I. Then occurs shrinking of wound and matrix modification (Fig. 1c) (Wang et al. 2012). Wound contraction is principally attained by distinguishing myofibroblasts or fibroblasts that produce in respond to TGF-A. Tissue tension is induced in the influence of a few matrix proteins (e.g., tenascin C, α-smooth muscle actin (α-SMA), ED-A fibronectin) (Hinz 2007). Fibroblast-produced contraction power is subsequently put out to ECM through ECM receptor-dependent and cytoskeleton-associated, mechanic coupling focal adhesion complexes, i.e., integrin receptors (Hinz 2007). Further mechanism escorted shrinking of the wound is fibroblast motility in significant matrix rearrangement. This energetic and reciprocated procedure engages a delay sequence of ECM production and deterioration in a fibroblastic or stromal cell.

At this juncture, matrix modified enzymes, specifically MMPs, take part in the modification of the matrix microenvironment in the favor of the recovery process. Lastly, apoptosis of fibroblastic cells resulting in the development of a new cellular granulating tissue with tensile strength related to normal skin (Fig. 1c) (Ehrlich et al. 1999). Even though the significance of apoptosis in granulation tissue modification and lesion production is generally admitted, the stimulation responsible for apoptosis is not completely defined. It is indicated that FGF-2, TGF-A, and TNF can cause a rise in the total apoptotic cells during the final process of healing (Desmouliere et al. 1995; Akasaka et al. 2004).

3   Strategies for Wound Healing

Conventional treatment of wound involves the use of antimicrobial agents as ointment, cream, and other semi-solid dosage forms. In developing countries, people still rely on these conventional medications (Nimia et al. 2019). However, they are not preferred by the majority of the patients. These medical preparations are messy, stick and stain clothes; their applications by finger or applicator produce pain to the affected area and incur the possibility of further microbial contamination. Woven gauze impregnated with antimicrobial agents is preferred by hospital staff and clinicians. It requires force to remove during subsequent applications and lead to patient discomfort, pain, and mechanical debridement. This negatively impacts sleep, standard of life, and movement of daily living of the patients. Patients need to take pain killers orally or intravenously, which is inconvenient to older people and children. The post-burn injury leaves hypertrophic scars up to 67% of the affected persons, and they suffer from physical impairments, social stigma, and many a time a distinct syndrome called dysmorphophobia (Hawkins et al. 2018).

It is very clear from the previous study that wounds should be healed by protecting them from contamination and loss of moisture, which is achieved with the help of wound dressing (Hanna and Giacopelli 1997). However, in doing so, wound dressing delays or impairs wound healing. A good and effective wound dressing should have an impression of biological and structural features of the skin extracellular matrix and should deliver systematized oxygen permeability (Okur et al. 2020). According to the nature of the action, wound dressings are divided into three different classes: bioactive, inert/passive, and interactive (Okur et al. 2020). Various conventionally available wound dressings and their characteristics features are highlighted in Table 1. These dressings, the majority of them being inert or passive, bioactive and interactive in nature, have various drawbacks like stickiness, cytotoxicity, dryness issues, and not used in third-degree burns. This triggers the search for alternative and superior wound healing agents devoid of above-mentioned patient non-compliance issues.

Table 1 Commonly available wound dressings and their characteristic features

Dressing type	Formulation	Disadvantages
Inert/ passive	Gauzes	• Adhere to wounds • Interrupt the wound pallet when removed • Acceptable mainly for minor wound • Not used for third-degree burns supervision • Wounds with desiccate escharing
	Hydrocolloids	• Applied in the wound with high exudates • Can be cytotoxic • Support an acid pH at the application site
Bioactive	Alginates	• Less use in no or less exudating wounds • Responsible for lack of moisture • Escharing • Change daily
	Collagens	• Not prescribe to administration in wounds with necrosis • Third-degree burns
	Hydrofibers	• Inflammation of the wound dressing causing to extension and reduction of adhesion
	Hydrogels	• May cause an excess amount of water • Weak mechanical properties
Interactive	Semi-permeable films	• Used an extra sheet for hydrogels and foams
	Semi-permeable foams	• Can cause dryness and scabbing

From the earlier times, it has been believed that the wounds heal when they are kept dry. But this concept was changed, and the idea of moist wound healing emerged after the famous claim and scientific explanation by George Winter in 1960. According to one of his landmark studies, he opined that epithelial cells require more energy during the dry wound healing and consuming much time, and if a moist wound occurs, they require less time as compared to dry wounds (Winter 1962). Various other studies say that moist wound healing can accelerate the inflammatory response and also helps to enhance cell proliferation and wound healing in deeper dermal wounds (Bryan 2004). Since then, the concept of advanced medicated dressings made of hydrogel, hydrocolloid, alginate, silicone sheets, etc. have been introduced as advanced strategies of wound healing (Ousey et al. 2016).

4   Probiotics as Wound Healing Booster

As per the definition of WHO, probiotics are living organism when they take in specific amount gives positive effects on the host (O'Toole et al. 2017). The most consistently used probiotic microorganisms are *Lactobacillus acidophilus*, *Bifidobacterium bifidum*, *Saccharomyces boulardii*, *Lactobacillus rhamnosus*, and *Bacillus coagulans*. Administered orally, they are effective for the treatment of gastroenteritis, enteral dysbacteriosis, and pediatric post-antibiotic-associated diarrhea in humans. Presently, a variety of probiotics are being explored for wound healing because of their immunomodulating activity (Nole et al. 2014). Strains of *Lactobacillus* and *Bifidobacterium* have exhibited in vitro efficacy for wound healing against commonly used pathogens like *Staphylococcus aureus* and *Staphylococcus epidermidis* (Sikorska and Smoragiewicz 2013; Misic et al. 2014). Topically probiotics have been investigated by various in vitro studies and animal models for chronic wound therapy in burn-related wounds, gastric ulcers, diabetic and non-diabetic foot ulcers (Peral et al. 2009).

As effective wound healing agents, probiotics can provide the following advantages (McFarland 2009):

- It can reduce the chances of systemic sepsis.
- It is considered safe from its established and safe oral use.
- It will lead to or promote a positive bacterial balance in the skin.
- It will improve in skin's innate immunity.
- It is free from any type of unwanted effects.
- It is very compatible, and formulation of topical use would be easy to prepare.
- It is not much expensive.

4.1   Mechanism of Probiotic in Wound Healing

Wound healing action of probiotics is based on their battling with the pathogens at the wound site for adhesion and utilization of nutrients and growth factors. These help probiotics in the prevention of microbial colony formation and modulation of host immune response (Oelschlaeger 2010). Additionally, probiotics generate low molecular weight substances that is lactic acid and bacteriocins (Sikorska and Smoragiewicz 2013; Sonal et al. 2014). Probiotics have immunomodulatory abilities by activating transcriptional pathways and T-cell activation that are linked to cytokines. Moreover, there are some elements, i.e., bacterial components or its byproducts which activate specific receptors of the human body that initiate the inflammatory process. The wound healing mechanism is based on the same theory of beneficial actions obtained by probiotics in the gastrointestinal tract where bacteria host interactions involve epithelial cells, regulatory T lymphocytes, and dendritic cells (Wong et al. 2013). Furthermore, one study by Peral et al. (2009) has shown that probiotics may act in wound healing through the competitive inhibition of pathogenic *P. aeruginosa* and some disturbances in communication pathways of bacteria (Wong et al. 2013). All these activities of probiotics mentioned above can reduce bacterial load at the wound site, regulate inflammatory cell infiltration and thus promote wound healing (Peral et al. 2009, 2010; Brachkova et al. 2011).

4.2   Efficacy of Probiotics for Different Types of Wounds

When the skin undergoes any kind of injury, it is contaminated with the airborne pathogen. The skin cohesion, as well as integrity, is compromised. The whole process of natural wound healing has been described in Sect. 2. These stages are further impacted by the development of oxidative stress (Knackstedt et al. 2020). In recent past, growing trend has been observed to study the efficacy of probiotic microorganisms to address wound infection and improve the speed of the healing process (Fardin and Keri 2021). Burn wound is a special type of wound caused by fire, chemicals, radiation, and electricity. An overall increase in the healing process was observed when burn wounds were treated with *Saccharomyces cerevisiae* (Oryan et al. 2018). Meticillin-resistant *Staphylococcus aureus* (MRSA) is one of the most widely stated disease pathogens able to infect burn wounds (Karska-Wysocki et al. 2010). Some studies have shown the ability of *L. acidophilus*, *L. casei* to be effective against MRSA and efficacious for burn wounds. In another study, Prince et al. (2012) tested the efficacy of *L. reuteri*, *L. rhamnosus*.

For burn-related wounds caused by *S. aureus*, the researchers proposed this efficacy as the pathogen's ability to induce keratinocyte cell death. In a recent study, Jones et al. (2012) reported the healing efficacy *L. fermentum* loaded patches for wounds caused by *S. aureus*. An increased wound closure concurrent with the development of probiotic-induced nitric oxide (gNO) was established.

Diabetic foot ulcer is another chronic type of complication whose treatment is quite challenging as this is prone to microbial infection, which takes further delay in the healing process. There are many reports where efficacies of probiotics like *Lactobacillus acidophilus*, *Bifidobacterium bifidum*, *Lactobacillus fermentum*, and *Lactobacillus casei* have been established against the treatment of diabetic foot ulcer (Sekhar et al. 2014; Mohseni et al. 2018; Amini et al. 2021). Gastric ulcer is another complicated disease where use probiotics like *Lactobacillus rhamnosus*, *Lactobacillus gasseri*, *Lactobacillus acidophilus*, *Lactobacillus rhamnosus*, *Saccharomyces boulardii*, *Lactobacillus acidophilus*, and *Bifidobacterium longum* have shown proven track record of good wound healing activity (Lam et al. 2007; Khoder et al. 2016; Dharmani et al. 2013).

5   Systemic and Topical Use of Probiotics for Wound Healing

Probiotics can be effective in two ways: Topical application will outcompete the growth of biofilm-forming pathogens and help immune systems too overall boosting the healing process. Oral supplementation of probiotics boosts the immune system and defense mechanisms that can work at distant places (where wounds are present). Probiotics can promote the process of healing by modulating the inflammatory response. In one study, Valdez et al. (2005) reported the efficacy of a topical probiotic containing *Lactobacillus plantarum* for the treatment of burn wounds infected with P*seudomonas aeruginosa.* The observation was reconfirmed by Peral et al. (2009) by a different study. In humans, *Lactobacillus plantarum*, while put on to second- and third-degree burns, was equally beneficial as silver sulfadiazine in declining the infection probability, promoting granulation tissue, and wound healing. Skin commensals, organisms that reside on the skin, have also shown beneficiary action for wound healing. In one study, Lopes et al. (2017) claimed that *Propioniferax innocua*, a skin commensal, has the ability to deteriorate established biofilms. *Staphylococcus caprae* is capable to provide antimicrobial activity in opposition to methicillin-resistant *S. aureus* and inhibit *S. aureus* colonization in a mouse model (Paharik et al. 2017). *Staphylococcus epidermidis* gives antimicrobial activity that particularly focuses on *S. aureus* and *S. pyogenes* and suppresses swelling via lipoteichoic acid (Lai et al. 2009; Periasamy et al. 2012; Christensen and Brüggemann 2014). In another study, Wang et al. (2014) proved the efficacy of probiotic bacteria *C. acnes* in the mouse model to treat wound infection caused by *S. aureus* through the creation of propionic acid. Few more studies with the proof of beneficiary actions of probiotics are summarized in Table 2 (Cinque et al. 2011).

6   Challenges in the Utilization of Probiotics for Wound Healing

In spite of the promising aspect of probiotics for wound treatment, it has not emerged as a medical product yet. There are few concerns that need to be addressed. The improvement in new technologies to preserve the probiotics viability for effective delivery as topical formulation is a challenging task. It will permit the delivery of probiotics as an intact entity and will reach their target site. Considering the issue of bacterial viability in stressful environments, such as lyophilization and high temperatures, present studies have attention on enhancing the viability of probiotics (Rokka and Rantamäki 2010). Another major concern is that the probiotic bacteria may translocate into the bloodstream, causing sepsis and bacteremia. The safety of topical use of probiotics as wound healing agents has not been investigated extensively. Their possible entry to the systemic circulation might be a major concern to the immune-compromised patients apart from the chance of additional infection of the existing wounds by the probiotics (Wilmink et al. 2020). There is an immunological concern about the capacity of skin microflora in the wound healing procedure. It is already established that the lack of microbiota can reduce healing time (Canesso et al. 2014). On the other side, cases of wound infections increase only because of exogenous bacteria control into the systemic and local elements of patient skin. Thus, it is possible when equilibrium is attained within bacteria and host that permits wound healing progression (Robson 1997). Microbial microflora can negatively influence the wound healing procedure.

Table 2 Published research works on wound healing activity of Probiotic

S. No	Probiotic	Route of administration	Animal/ human	Outcome	References
1	*Bacillus subtilis*	Topical	Male Wistar rats	The probiotic strain was found to be useful ulcer healing	Shahsafi (2017)
2	*Lactobacillus brevis*	Topical	Male Wistar rat	The study demonstrated a momentous reduction in swelling and an accelerating of wound healing	Zahedi et al. (2011)
3	*Lactobacillus plantarum*	Topical	Mice	*L. plantarum* showed potential effect for treatment of *P. aeruginosa* burn wound infection	Valdez et al. (2005)
4	Recombinant *Escherichia coli*	In vitro physically wounded monolayer model	–	The study proved to be a significant milestone against the clinical application of probiotic vehicle for wound healing processes	Choi et al. (2012)
5	*Lactobacillus plantarum*	Topically	Male Wistar rats	The study showed to decrease in swelling and promoting of wound healing in rats	Heydari et al. (2011)
6	*Lactobacillus plantarum*	Topical	Rabbit	*Lactobacillus plantarum* showed great capacity as a therapeutic agent in reducing scarring and burn wound infection	Satish et al. (2017)
7	*Lactobacillus paracasei* LPC-37, *Lactobacillus acidophilus* NCFM, *Lactobacillus rhamnosus* HN001, and *Bifidobacterium lactis* HN0019	Oral	Rat	The study showed a rapid contraction of wound size in rats by decreasing the proactive phase, speeding up fibrosis, and deposition of collagen	Tagliari et al. (2019)
8	*Lactobacillus reuteri*	Topical	Sprague Dawley male rats	*Lactobacillus reuteri* successfully stimulated the wound healing procedure and produced encouraging results	Khodaii et al. (2019)
9	*Lactobacillus acidophilus*	Orally	Human	This study indicates the antibacterial efficacy and	Jebur (2010)

(continued)

Table 2 (continued)

S. No	Probiotic	Route of administration	Animal/human	Outcome	References
				immunological properties of *Lactobacillus acidophilus* through investigation of burn wound pathogenic agents with susceptibility testing	
10	*Saccharomyces cerevisiae*	Topical	Male Sprague Dawley rats	The topically applied *S. cerevisiae* improved the restoration procedure of burned wounds with additional beneficial actions	Oryan et al. (2018)
11	*L. bulgaricus*, *L. plantarum*, and *L. acidophilus*	Topical	Female mice	The combination of these probiotics showed rapid wound healing	Al-Mathkhury and Al-Aubeidi (2008)
12	Kefirs[a] natural probiotic	Topical	Rat	The kefir natural probiotic gives better healing to serious burn as contrast to conventional silver sulfadiazine treatment	Huseini et al. (2012)

[a]Kefir is fermented milk drink like a thin yogurt that is prepared from kefir grains enriched with Probiotic bacteria like *Bifidobacterium bifidum*, *Streptococcus thermophilus*, *Lactobacillus acidophilus*, *Lactobacillus delbrueckii subsp. bulgaricus*, *Lactobacillus kefiranofaciens*, *Lactobacillus helveticus*, *Lactococcus lactis*, and *Leuconostoc* species

7 Conclusion

This chapter highlights the basics of wound healing mechanism and pathophysiology involving various steps such as coagulation/inflammatory phase, proliferation phase granulation tissue formation, and matrix remodeling. Apart from conventional and some emerging novel strategies, use of probiotic for effective wound healing has shown promising new areas. Various studies of probiotics in animal and human models have confirmed compact in various skin conditions. Topical applied probiotics for burn infections have shown to decrease the pathogen load. Therefore, the possible use of probiotics for wound infections stays worthy to explore in the future. As such, there is no probiotic-based formulation available in the market that claims directly to be used for the treatment and management of wound. But, given enough scientific evidence with further research and patient-friendly dosage forms, probiotics can prove effective alternative treatment options for various acute and chronic wounds.

References

Akasaka Y, Ono I, Yamashita T et al (2004) Basic fibroblast growth factor promotes apoptosis and suppresses granulation tissue formation in acute incisional wounds. J Pathol 203:710–720. https://doi.org/10.1002/path.1574

Al-Mathkhury HJ, Al-Aubeidi HJ (2008) Probiotic effect of lactobacilli on mice incisional wound infections. Al-Nahrain J Sci 11:111–116. https://doi.org/10.22401/JNUS.11.3.14

Amini MR, Aalaa M, Sanjari M, Mehrdad N, Mohajeri Tehrani MR, Ejtahed HS (2021) The role of probiotics in management of infected diabetic foot ulcers. J Mazandaran Univ Med Sci 31 (193):59–70

Asahara T, Murohara T, Sullivan A et al (1997) Isolation of putative progenitor endothelial cells for angiogenesis. Science 275:964–966. https://doi.org/10.1126/science.275.5302.964

Brachkova MI, Marques P, Rocha J et al (2011) Alginate films containing *Lactobacillus plantarum* as wound dressing for prevention of burn infection. J Hosp Infect 79:375–377

Bryan J (2004) Moist wound healing: a concept that changed our practice. J Wound Care 13 (6):227–228

Bunman S, Dumavibhat N, Chatthanawaree W et al (2017) Burn wound healing: pathophysiology and current management of burn injury. Bangkok Med J 13(2):91–99

Canesso MC, Vieira AT, Castro TB et al (2014) Skin wound healing is accelerated and scarless in the absence of commensal microbiota. J Immunol 193:5171–5180. https://doi.org/10.4049/jimmunol.1400625

Choi HJ, Ahn JH, Park SH et al (2012) Enhanced wound healing by recombinant *Escherichia coli* Nissle 1917 via human epidermal growth factor receptor in human intestinal epithelial cells: therapeutic implication using recombinant probiotics. Infect Immun 80:1079–1087. https://doi.org/10.1128/IAI.05820-11

Christensen GJ, Brüggemann H (2014) Bacterial skin commensals and their role as host guardians. Benefic Microbes 5:201–215. https://doi.org/10.3920/BM2012.0062

Cinque B, La Torre C, Melchiorre E et al (2011) Use of probiotics for dermal applications. In: Probiotics. Springer, Berlin, pp 221–241. https://doi.org/10.3920/BM2012.0062

Delavary BM, van der Veer WM, van Egmond M, Niessen FB, Beelen RHJ (2011) Macrophages in skin injury and repair. Immunobiology 216:753–762. https://doi.org/10.1016/j.imbio.2011.01.001

Desmouliere A, Redard M, Darby I, Gabbiani G (1995) Apoptosis mediates the decrease in cellularity during the transition between granulation tissue and scar. Am J Pathol 146:56

Dharmani P, De Simone C, Chadee K (2013) The probiotic mixture VSL# 3 accelerates gastric ulcer healing by stimulating vascular endothelial growth factor. PLoS One 8(3):58671

Ehrlich HP, Keefer K, Myers RL, Passaniti A (1999) Vanadate and the absence of myofibroblasts in wound contraction. Arch Surg 134:494–501. https://doi.org/10.1001/archsurg.134.5.494

Eming SA, Krieg T, Davidson JM (2007) Inflammation in wound repair: molecular and cellular mechanisms. J Investig Dermatol 127:514–525. https://doi.org/10.1038/sj.jid.5700701

Falanga V (2005) Wound healing and its impairment in the diabetic foot. Lancet 366:1736–1743. https://doi.org/10.1016/S0140-6736(05)67700-8

Fardin S, Keri J (2021) The microbiome, probiotics, and prebiotics. In: Integrative dermatology. Springer, Cham, pp 1–30

Folkman J, Klagsbrun M (1987) Angiogenic factors. Science 235:442–447. https://doi.org/10.1126/science.2432664

Gilroy DW, Lawrence T, Perretti M et al (2004) Inflammatory resolution: new opportunities for drug discovery. Nat Rev Drug Discov 3:401–416. https://doi.org/10.1038/nrd1383

Hanna JR, Giacopelli JA (1997) A review of wound healing and wound dressing products. J Foot Ankle Surg 36(1):2–14

Hawkins HK, Jay J, Finnerty CC (2018) Pathophysiology of the burn scar. In: Total burn care. Elsevier, New York, pp 466–475. https://doi.org/10.1016/B978-0-323-47661-4.00044-7

Heydari NM, Tajabadi EM, Dehghan S et al (2011) Study the probiotic effects of *Lactobacillus plantarum* on cutaneous wound healing on rats. New Cell Mol Biotechnol J 1:21–28

Hinz B (2007) Formation and function of the myofibroblast during tissue repair. J Investig Dermatol 127:526–537. https://doi.org/10.1038/sj.jid.5700613

Humar RO, Kiefer FN, Berns H (2002) Hypoxia enhances vascular cell proliferation and angiogenesis in vitro via rapamycin (mTOR)-dependent signaling. FASEB J 16:771–780. https://doi.org/10.1096/fj.01-0658com

Huseini HF, Rahimzadeh G, Fazeli MR et al (2012) Evaluation of wound healing activities of kefir products. Burns 38:719–723. https://doi.org/10.1016/j.burns.2011.12.005

Jebur MS (2010) Therapeutic efficacy of *Lactobacillus acidophilus* against bacterial isolates from burn wounds. N Am J Med Sci 2:586. https://doi.org/10.4297/najms.2010.2586

Jones M, Ganopolsky JG, Labbé A et al (2012) Novel nitric oxide producing probiotic wound healing patch: preparation and in vivo analysis in a New Zealand white rabbit model of ischaemic and infected wounds. Int Wound J 9(3):330–343

Karska-Wysocki B, Bazo M, Smoragiewicz W (2010) Antibacterial activity of *Lactobacillus acidophilus* and *Lactobacillus casei* against methicillin-resistant *Staphylococcus aureus* (MRSA). Microbiol Res 165(8):674–686

Khodaii Z, Afrasiabi S, Hashemi S et al (2019) Accelerated wound healing process in rat by probiotic *Lactobacillus reuteri* derived ointment. J Basic Clin Physiol Pharmacol 30. https://doi.org/10.1515/jbcpp-2018-0150

Khoder G, Al-Menhali AA, Al-Yassir F, Karam SM (2016) Potential role of probiotics in the management of gastric ulcer. Exp Ther Med 12(1):3–17

Knackstedt R, Knackstedt T, Gatherwright J (2020) The role of topical probiotics on wound healing: a review of animal and human studies. Int Wound J 17(6):1687–1694

Lai Y, Di Nardo A, Nakatsuji T et al (2009) Commensal bacteria regulate Toll-like receptor 3–dependent inflammation after skin injury. Nat Med 15:1377. https://doi.org/10.1038/nm.2062

Lam EK, Yu L, Wong HP et al (2007) Probiotic *Lactobacillus rhamnosus* GG enhances gastric ulcer healing in rats. Eur J Pharmacol 565(1-3):171–179

Leone AM, Valgimigli M, Giannico M et al (2009) From bone marrow to the arterial wall: the ongoing tale of endothelial progenitor cells. Eur Heart J 30:890–899. https://doi.org/10.1093/eurheartj/ehp078

Liu ZJ, Velazquez OC (2008) Hyperoxia, endothelial progenitor cell mobilization, and diabetic wound healing. Antioxid Redox Signal 10:1869–1882. https://doi.org/10.1089/ars.2008.2121

Lopes EG, Moreira DA, Gullón P et al (2017) Topical application of probiotics in skin: adhesion, antimicrobial and antibiofilm in vitro assays. J Appl Microbiol 122:450–461. https://doi.org/10.1111/jam.13349

Maeng YS, Choi HJ, Kwon JY et al (2009) Endothelial progenitor cell homing: prominent role of the IGF2-IGF2R-PLCβ2 axis. Blood 113:233–243. https://doi.org/10.1182/blood-2008-06-162891

McFarland LV (2009) Evidence-based review of probiotics for antibiotic-associated diarrhea and *Clostridium difficile* infections. Anaerobe 15:274–280. https://doi.org/10.1016/j.anaerobe.2009.09.002

Misic AM, Gardner SE, Grice EA (2014) The wound microbiome: modern approaches to examining the role of microorganisms in impaired chronic wound healing. Adv Wound Care 3:502–510. https://doi.org/10.1089/wound.2012.0397

Mohseni S, Bayani M, Bahmani F et al (2018) The beneficial effects of probiotic administration on wound healing and metabolic status in patients with diabetic foot ulcer: a randomized, double-blind, placebo-controlled trial. Diabetes Metab Res Rev 34(3):e2970

Nimia HH, Carvalho VF, Isaac C (2019) Comparative study of Silver Sulfadiazine with other materials for healing and infection prevention in burns: a systematic review and meta-analysis. Burns 45(2):282–292. https://doi.org/10.1016/j.burns.2018.05.014

Nole KL, Yim E, Keri JE (2014) Probiotics and prebiotics in dermatology. J Am Acad Dermatol 71 (4):814–821

O'Toole PW, Marchesi JR, Hill C (2017) Next-generation probiotics: the spectrum from probiotics to live biotherapeutics. Nat Microbiol 2:1–6. https://doi.org/10.1038/nmicrobiol.2017.57

Oelschlaeger TA (2010) Mechanisms of probiotic actions–A review. Int J Med Microbiol 300:57–62

Okur ME, Karantas ID, Şenyiğit Z et al (2020) Recent trends on wound management: new therapeutic choices based on polymeric carriers. Asian J Pharm Sci 15(6):661–684. https://doi.org/10.1016/j.ajps.2019.11.008

Oryan A, Jalili M, Kamali A, Nikahval B (2018) The concurrent use of probiotic microorganism and collagen hydrogel/scaffold enhances burn wound healing: an *in vivo* evaluation. Burns 44 (7):1775–1786. https://doi.org/10.1016/j.ajps.2019.11.008

Ousey K, Cutting KF, Rogers AA et al (2016) The importance of hydration in wound healing: reinvigorating the clinical perspective. J Wound Care 25(3):122–130

Paharik AE, Parlet CP, Chung N et al (2017) Coagulase-negative staphylococcal strain prevents *Staphylococcus aureus* colonization and skin infection by blocking quorum sensing. Cell Host Microbe 22:746–756. https://doi.org/10.1016/j.chom.2017.11.001

Peral MC, Huaman Martinez MA, Valdez JC et al (2009) Bacteriotherapy with *Lactobacillus plantarum* in burns. Int Wound J 6:73–81. https://doi.org/10.1111/j.1742-481X.2008.00577

Peral MC, Rachid MM, Gobbato NM et al (2010) Interleukin-8 production by polymorphonuclear leukocytes from patients with chronic infected leg ulcers treated with *Lactobacillus plantarum*. Clin Microbiol Infect 16:281–286. https://doi.org/10.1111/j.1469-0691.2009.0279

Periasamy S, Chatterjee SS, Cheung GY et al (2012) Phenol-soluble modulins in staphylococci: what are they originally for? Commun Integr Biol 5(3):275–277

Prince T, Mcbain AJ, O'Neill CA (2012) *Lactobacillus reuteri* protects epidermal keratinocytes from *Staphylococcus aureus*-induced cell death by competitive exclusion. Appl Environ Microbiol 78:5119–5126

Robson MC (1997) Wound infection: a failure of wound healing caused by an imbalance of bacteria. Surg Clin N Am 77:637–650

Rokka S, Rantamäki P (2010) Protecting probiotic bacteria by microencapsulation: challenges for industrial applications. Eur Food Res Technol 231:1–12. https://doi.org/10.1007/s00217-010-1246-2

Satish L, Gallo PH, Johnson S et al (2017) Local probiotic therapy with *Lactobacillus plantarum* mitigates scar formation in rabbits after burn injury and infection. Surg Infect 18:119–127

Sekhar MS, Unnikrishnan MK, Vijayanarayana K, Rodrigues GS, Mukhopadhyay C (2014) Topical application/formulation of probiotics: will it be a novel treatment approach for diabetic foot ulcer? Med Hypotheses 82(1):86–88

Sen CK (2019) Human wounds and its burden: an updated compendium of estimates. Adv Wound Care 8:39–48

Shahsafi M (2017) The effects of *Bacillus subtilis* probiotic on cutaneous wound healing in rats. Nov Biomed 5:43–47

Sikorska H, Smoragiewicz W (2013) Role of probiotics in the prevention and treatment of meticillin-resistant *Staphylococcus aureus* infections. Int J Antimicrob Agents 42:475–481

Singer A, Dagum AB (2008) Current management of acute cutaneous wounds. N Engl J Med 359:1037–1046

Sonal SM, Unnikrishnan MK, Vijayanarayana K et al (2014) Topical application/formulation of probiotics: will it be a novel treatment approach for diabetic foot ulcer? Med Hypotheses 82:86–88

Tagliari E, Campos LF, Campos AC et al (2019) Effect of probiotic oral administration on skin wound healing in rats. Arq Bras Cir Dig 32(3):1–6

Valdez JC, Peral MC, Rachid M et al (2005) Interference of *Lactobacillus plantarum* with *Pseudomonas aeruginosa* in vitro and in infected burns: the potential use of probiotics in wound treatment. Clin Microbiol Infect 11:472–479

Velnar T, Bailey T, Smrkolj V et al (2009) The wound healing process: an overview of the cellular and molecular mechanisms. J Int Med Res 37:1528–1542

Wang Y, Wang G, Luo X et al (2012) Substrate stiffness regulates the proliferation, migration, and differentiation of epidermal cells. Burns 38:414–420

Wang Y, Dai A, Huang S et al (2014) Propionic acid and its esterified derivative suppress the growth of methicillin-resistant *Staphylococcus aureus* USA300. Benefic Microbes 5:161–168

Weyrich AS, Zimmerman GA (2004) Platelets: signaling cells in the immune continuum. Trends Immunol 25:489–495

Wilmink JM, Ladefoged S, Jongbloets A et al (2020) The evaluation of the effect of probiotics on the healing of equine distal limb wounds. PLoS One 15(7):e0236761

Winter GD (1962) Formation of the scab and the rate of epithelization of superficial wounds in the skin of the young domestic pig. Nature 193(4812):293–294

Wong VW, Martindale RG, Longaker MT et al (2013) From germ theory to germ therapy: skin microbiota, chronic wounds, and probiotics. Plast Reconstr Surg 132:854e–861e

Zahedi F, Heydari NM, Tajabadi EM et al (2011) Study the effect of Lactobacillus brevis isolated from Iranian traditional cheese on cutaneous wound healing in male rats on days 3 and 14. Razi J Med Sci 18(88):16–23

Use of Probiotic Bacteria and Their Bioactive Compounds for Wound Care

Sarita Devi and Prasun Kumar

1 Introduction

Various commensal microbes start to inhabit the human body at birth and remain throughout one's existence. In comparison to harmful and infective microbes that can destroy the host barriers and cause disease pathogenesis, commensal microorganisms found in symbiotic communities are adapted for survival without sacrificing the integrity of the host (Thursby and Juge 2017; Lukic et al. 2017). Exploration of the human body and the functional integration and harmonization of microbiomes has shown that the microbiota has a vital influence on various biological functions like modulation of the immune system and fortification against infections (Lukic et al. 2017). The significance of commensal microbes in keeping up host well-being has been identified initially in gut microbiome investigation. The germ-free animals have been shown to be more susceptible to pathogen invasion (Kamada et al. 2012), have disrupted mucosal wound healing (Hernández-Chirlaque et al. 2016), and are more vulnerable to chemical poisoning (Breton et al. 2013). The methods that alter its composition to strengthen the physiological, immunological and metabolic functions of the host have become increasingly significant due to the major systemic and local consequences of the gut microbiome. This resulted in the discovery of advantageous microbial species (symbiotic) and increased host well-being. The lactic acid-producing microorganisms (*Lactobacillus* and *Bifidobacteria*) are among the most

S. Devi (✉)
Biotechnology Division, CSIR-Institute of Himalayan Bioresource Technology, Palampur, Himachal Pradesh, India
e-mail: sarita@ihbt.res.in

P. Kumar
Department of Chemical Engineering, Chungbuk National University, Cheongju, Chungbuk, Korea (Republic of)

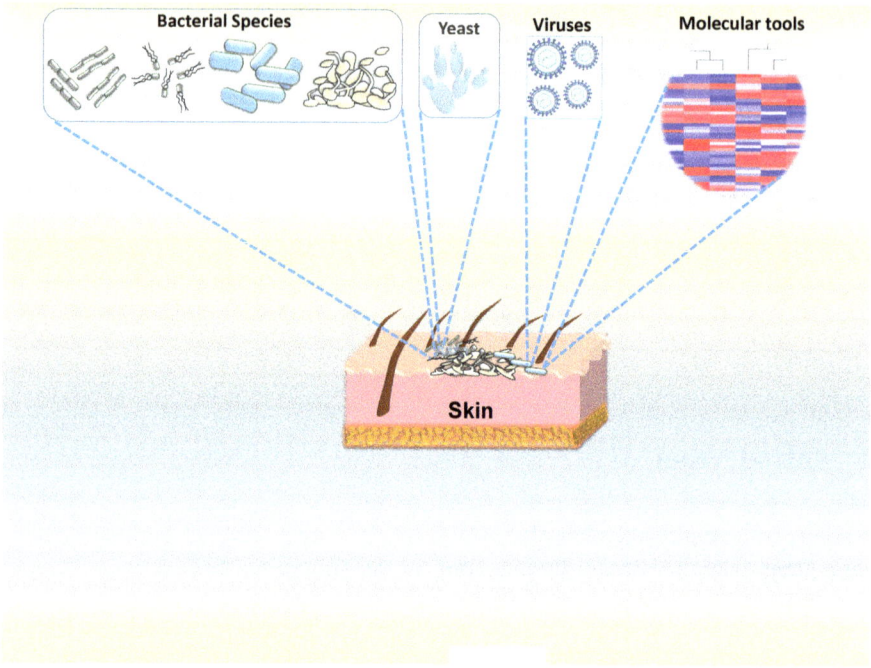

Fig. 1 Different inhabitants of skin microbiota

broadly investigated microbes, most often alluded to as probiotics, which have health benefits (Belkaid and Hand 2014).

The WHO (World Health Organization) has described the probiotic microbes as "live microbes that confer a health benefit on the host when administered in adequate amounts" (Mahajan and Singh 2014). These are live microbes (yeast or bacteria, predominantly, Fig. 1), the normal microbiome's members, which can equilibrate the human microflora and results in the predominance of advantageous microbes for the body (Tsiouris and Tsiouri 2017). These microorganisms help to reduce the level of low-density lipoproteins and contribute to down-regulate inflammation and the host's immune response (Hakansson and Molin 2011; Jones et al. 2012a; Wong et al. 2013). They are consumed as microbial food supplements. Different terms are used in the literature on nutrition for non-digestible fermented carbohydrates, which alter the gut microbiome (prebiotics) by endogenous bacteria, and combinations of probiotic microbes and prebiotics as synbiotics (Patel and Denning 2013). The probiotic microbes have been accounted for to be advantageous for treating or preventing various inflammatory cutaneous (skin) diseases (Hacini-Rachinel et al. 2009), respiratory tract infections (McFarland 2011), diabetes prevention and control (Rad et al. 2016), gastrointestinal (GI) disorders (Ringel-Kulka et al. 2011; Demers et al. 2014), ulcerative colitis (Abdin and Saeid 2008) and urogenital infections (Reid et al. 2001) to give some examples. Recent studies have also emphasized the use of non-viable compounds of probiotic microbes, known as postbiotics, as a more stable probiotics alternative (Tsilingiri et al. 2012; Volz et al. 2014; Christensen and Brüggemann 2014). Postbiotics have gained significant importance in treating various disorders related to inflammation, where the utilization of live microbes carries the perils linked with excessive immune system activation.

The mode of action of microbes (probiotics) is attributed to their capability towards the immune response improvement, contend with harmful microorganisms for adherence at specific locations, antagonize the pathogenic microbes, and antimicrobial substance fabrication (Mahajan and Singh 2014). The probiotics' health benefits include prevention and treatment of several diseases and conditions such as lactose intolerance, gastrointestinal disorders, necrotizing colitis, irritable bowel syndrome, inflammatory bowel disease, allergies, numerous cancers, Upper respiratory infections, urogenital infections, Arthritis, AIDS, different oral health diseases such as prevention of dental caries, halitosis and periodontal diseases and many other effects which are under exploration (Mahajan and Singh 2014). The outcome of numerous clinical studies suggests that the probiotic microbes may use their advantageous effect for the treatment and prevention of various disorders to accomplish human well-being. The typical foods containing probiotics include kefir, yogurt, miso, sauerkraut (non-pasteurized), tempeh, sourdough bread, kimchi, and pickles (in brine, not vinegar), e.g. the *L. acidophilus* containing yogurt, which gives the yogurt, its valuable gastrointestinal health-related properties (Rezac et al. 2018).

The studies further suggest that some microbial probiotic strains and the mixture of probiotics such as a milk drink kefir (fermented) can have a positive influence on the wound repair process either by per os administration or topical application (Rodrigues et al. 2005; Huseini et al. 2012; Bourrie et al. 2016). The main reason for mortality and morbidity is impaired wound healing for a substantial portion of the population (Menke et al. 2007). Moreover, a higher degree of focus and research investigation is required to assess novel pharmaceutical compounds that can enhance wound healing and reduce the occurrence of chronic wounds and ulcers due to the substantial financial encumbrance and social influence of wound demands (Frykberg and Banks 2015). This book chapter reviews current information about probiotics on both GI epithelium and skin associated with their therapeutic properties. It also addresses their antimicrobial potential and identifies molecular and cellular mechanisms of action, suggesting innovative approaches to treating wound healing disorders.

2   Effect of Probiotics on Skin Microflora

The inherent microbiota flourishing within the gut of any individual is known to play an important role in gut-health, but what about our skin? Many millions of microbes live there, and during wound healing, the probiotic microbes may have enormous ability to prevent infections. The skin can serve as a physical barrier with numerous functions, e.g. thermoregulation, fluid homeostasis, metabolic and neurosensory

Table 1 List of different probiotic strains

Lactobacillus	Bifidobacterium	Enterococcus	Lactococcus	Streptococcus
L. acidophilus	B. thermophilum	E. faecium	L. lactis	S. thermophilus
L. casei	B. animalis	E. faecalis		
L. brevis	B. breve			
L. fermentum	B. longum			
L. curvatus	B. infantis			
L. gasseri	B. adolescentis			
L. reuteri				
L. johnsonii				
L. rhamnosus				
L. salivarius				
L. plantarum				
Propionibacterium	**Saccharomyces**	**Kluyveromyces**	**Leuconostoc**	**Pediococcus**
P. jensenii	S. cerevisiae	K. lactis	L. mesenteroides	P. acidilactici
P. freudenreichii	S. boulardii			

Source: Lew and Liong (2013)

functions, immune responses and primary protection against infection as the skin's harsh environment prevents many microorganisms from inhabiting its surface (Romanovsky 2014; Sugiura et al. 2014). The skin contains two different kinds of microbe, i.e. resident and transient microbial strains (Table 1). The coagulase-negative *Staphylococci* (*S. epidermidis*), *Propionibacteria* (*P. avidum, P. acnes* and *P. granulosum*), *Bacillus* sp., *Acinetobacter*, *Micrococci*, and *Corynebacteria* are the most common resident species of skin. The transient microbial species include *E. coli, P. aeruginosa*, and *S. aureus*. The resident species are capable of establishing and reproducing the microbial colonies skin, thus offering an advantageous environment, while transient often refers to non-advantageous microbes which cannot produce colonies on the skin surface (Christensen and Brüggemann 2014).

In the competitive exclusion of antagonistic microbes that cause skin infection, skin processing proteins, sebum, and free fatty acids, the microbiota of the skin have a significant function. Intriguingly, the inhabitant microflora can be seen as "beneficial" to the healthy host but can be detrimental to the host with disrupted skin integrity (Cinque et al. 2011). The harmful microbes are preparing to advance into the body to colonize it at the point when the skin barrier is injured. This is especially perilous if the antibiotics-resistant harmful microbe in question causes significant harm to skin or other tissues (Cinque et al. 2011). Normally, *S. aureus* is present in the nose regions of about 30% of the population and generally does not cause skin damage. However, when the skin barrier is broken, *S. aureus* may result in serious infections. *S. aureus* is notorious for biofilms production, as soon as it occurs, the microbe appends to a surface, e.g. the sugar molecules and the skin create a matrix (protective) around the microbe. These films are generally antibiotic-resistant and are thus pose major health jeopardy (Kumar et al. 2020). *S. aureus* can cause sepsis when it spreads to the blood, the main cause of a child's death who has experienced serious burn injuries (Sakr et al. 2018). Another harmful microbe *P. aeruginosa*, also identified to develop biofilms, is frequently present in infected wounds caused by burns. Normally present in a gut, this pathogenic microbe attacks and colonizes the skin, accompanied by other body organs, e.g. the lungs and liver in immune-compromised persons such as blistered patients by burns (Church et al. 2006).

The microorganisms may even cause atopic dermatitis (AD), rosacea, eczema, acne, and psoriasis. Although there are insufficient investigations following the probiotic's approach for the treatment of microbiota associated cutaneous diseases, it is fascinating to believe that usage of probiotic (topical) may be helpful for the prevention and/or treatment of microorganisms-related skin disease (Simmering and Breves 2009; Krutmann 2009). Significant evidence is also available in the prior art showing that probiotic microbes are effective for atopic dermatitis prevention, mostly in children during the post- and pre-natal periods (Martinelli et al. 2020). The controversial evidence remains, however, that the probiotic strains are success-ful in treating atopic dermatitis, and for this, further research is required. While presently not acknowledged as standard dermatological clinical procedures, some of the investigational studies are also there, indicating promising outcomes in wound healing, acne vulgaris, photo-protection with probiotics, and eczema treatment. While such outcomes are encouraging, more large-scale trials must be carried out before the incorporation of such treatment modalities into clinical practice (Rather et al. 2016).

Although the precise "probiotics-action" mechanism on the skin is not clear, it was proposed that microbial strains may produce a shielding barrier that averts overlaying malicious microbes (known as bacterial interference) from being detected by skin cells. Such an incidence can hinder the communication of Langerhans cells and keratinocytes with the immune system, thus avoiding an immune response (Fijan et al. 2019). It has likewise been noticed that antimicrobial properties of probiotics, such as antibiotics, may be used as alternatives to traditional therapies. In addition, the probiotics' immune-modulating effects help to reduce immune responses such as inflammation, redness and irritation (Lukic et al. 2017).

Stokes and Pillsbury conjectured a correlation in 1930 between an individual's stress or emotional state and stomach health, which further affects the well-being of the skin. They were well ahead of their time with their observations and hypothesis, and recent studies provide convincing proof of the correlation between these three different anatomical regions (Bowe and Logan 2011). A somewhat different approach to improving the skin's microbiota is done in the field of cosmetic treatments by topical treatment. A variety of items are available in the market, the majority of them are in the form of probiotic skin care creams. Although most research on the probiotics beneficial effects on the health status of skin have been performed through the oral administration, there is substantial proof that direct application on the skin is also a viable mode of treatment (Guéniche et al. 2008; Huseini et al. 2012). Future research could help to elucidate variations in the effectiveness of probiotics administered topically and orally. While antibiotics are used in wound care, the resistance to multiple drugs is widespread and infections persist. Various alternatives are being searched by scientists. Can microbes have much-needed strategies to avert infections that are life-threatening?

3 Fermented Probiotic Supernatant/Extract and Wound Healing

According to the theory of the gut–brain–skin axis, the use of probiotics modulates the microbiome that may have significant benefits on skin inflammation and skin homeostasis (Arck et al. 2010).

Increasing evidence suggests that microbial compounds, such as fragments of the cell wall, intra- and/or extra- cellular metabolites and even dead bacterial cells, can evoke some skin immune responses and improve the function of the skin barrier. Antimicrobial and immunomodulatory activities have been confirmed in extracts (cell-free) of lactic acid bacteria that have probiotic potential, indicating the use of probiotics in non-viable forms (Iordache et al. 2008). The alternative option may be natural cell components and metabolites in cases where the delivery of live cells is not feasible. In addition, at room temperature, cell metabolites and components are more stable than viable cells and are thus more acceptable for topical applications. Human clinical trials have shown that probiotics exert not only dermal benefits via the gastrointestinal pathway, but also via topical applications. By means of in vitro studies, Iordache et al. (2008) showed that the expression of soluble virulence factors by opportunistic dermal pathogens such as *Pseudomonas aeruginosa* and *Staphylococcus aureus* was inhibited by cell-free extracts of lactic acid bacteria with probiotic potentials such as *Lactobacillus plantarum*, *L. casei* and *Enterococcus faecium* and decreased their adherence ability to the cellular substrate represented by HeLa cells. Meanwhile, Guéniche et al. (2010) observed a statistically significant change after the use of cell lysate from *Bifidobacterium longum* sp. versus placebo in different inflammation-related parameters, such as a reduction in vasodilation, oedema, TNF-alpha release, and mast cell degranulation, using human skin explants (ex vivo) model. Three nanogel formulations consisting of probiotic supernatants (*Bacillus subtilis* sp. natto, *Lactobacillus reuteri* and *L. fermentum*) loaded chitosan nanogels have been prepared from the corresponding culture (Iordache et al. 2008).

The characterization of the chitosan nanogels was done previously by Zetasizer, FTIR and TEM. The efficacy and dressing activity of the prepared formulations were evaluated by examining wound closure and histological trials in Sprague-Dawley rats. The findings showed that all formulations of probiotic lysate had advantages over the mechanism of wound healing. Nevertheless, *Bacillus subtilis* natto has an enhanced wound healing rate, which is well understood in pathology research. It is suggested as a promising candidate for wound healing purposes by the favourable effects of probiotic lysate nanogels, including rational wound closing rate, good wound appearance, and adequate histological observation through in vivo analysis (Ashoori et al. 2020). Tsiouris et al. (2017) suggest that as a pharmacological treatment of wounds, sterile kefir extracts (70% kefir gel, *L. fermentum*, *L. brevis*, *L. reuteri*, *L. plantarum*) are more effective than the probiotic treatment of yeast (*S. boulardii*). While several studies and patents on the use of probiotic extracts for topical application on the skin have been published, the underlying mechanisms or the specific compounds responsible for the benefits of bacterial extracts on the skin

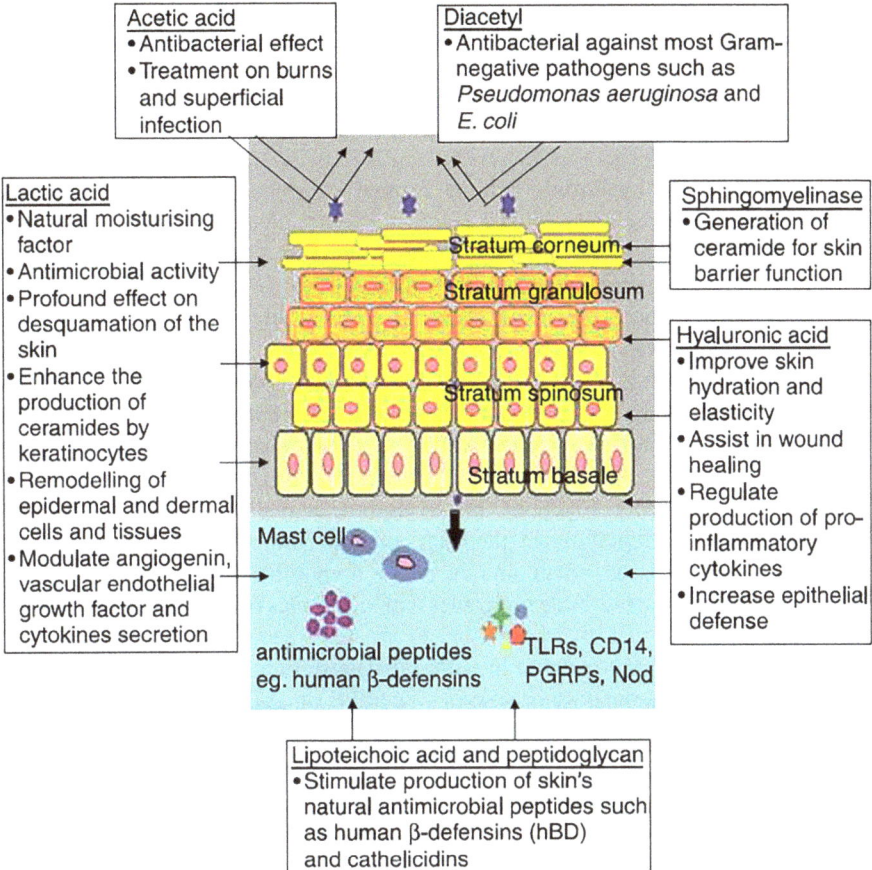

Fig. 2 Role of different bioactive compounds derived from probiotics in skincare (Source: Lew and Liong 2013)

remain unclear. The growing demand for probiotic dermal formulations further increases the need to understand the specific mechanisms of action. This chapter of the book is intended to report on the bacterial compounds that contribute to beneficial dermal effects and certain potential mechanisms of action of different bioactive compounds from probiotic supernatant or extract (Fig. 2).

4 Bioactive Compounds from Probiotics for Wound Healing

4.1 Hyaluronic Acid

The macromolecule hyaluronic acid is formed by polymerizing approximately 2000–25,000 repeating units of two sugar molecules, N-acetyl glucosamine and glucuronic acid (Chong et al. 2005). The hyaluronic acid's molecular weight, depending on the source, could range from 10^4 to 10^7 Da. In dermatology, hyaluronic acid has been used extensively as a biomaterial for stimulating wound healing and for bioengineering purposes. Besides, being utilized in cosmetic and dermatology goods, it is also broadly utilized in ophthalmology, drug delivery and pharmacology rheumatology (Kogan et al. 2007; Lew and Liong 2013). In most mammalian skin, hyaluronic acid is found to serve as a matrix. The hyaluronic acid is essential to preserve the structure of the standard stratum corneum and to maintain various epidermal barrier' functions. The hyaluronic acid also has a role in a number of other significant functions in the skin, e.g. in controlling cell proliferation, differentiation and tissue repair and in water immobilization in tissues. It also helps facilitate the water-soluble molecules and ion solutes transportation and retains the extracellular dermal matrix owing to its high-water binding ability. The hyaluronic acid is highly osmotic in nature that is significant for regulating tissue hydration during inflammatory processes (Weindl et al. 2004).

It has also been documented that by triggering β-defensin-2 via Toll-like receptors, the hyaluronic acid (low molecular weight) enhances epithelial protection (Gariboldi et al. 2008). β-Defensins, which are expressed in many body tissues, most remarkably epithelial surfaces and leukocytes, are the prevailing antimicrobial peptides involved in the host's response against bacterial infections (Menendez and Finlay 2007). Gariboldi et al. (2008) have stated that in all layers of the epidermal compartment, the low molecular weight hyaluronic acid treatment for murine skin enhanced mouse β-defensin-2 release.

It has been revealed by Taylor et al. (2004) that after injury, the fragments of hyaluronic acid are released, resulting in augmented chemokine IL-8 expression in the cells of the endothelium, thus stimulating the same to identify wound and initiate wound repair, while, during wound repair, the hyaluronic acid's antioxidant properties averted oxygen free radicals' damage on tissue granulation (Trabucchi et al. 2002). Moreover, the exogenous hyaluronic acid plays a supportive role in wound repairing due to its capability to retain moisture, thus promoting various physiological processes, e.g. provisional matrix's proteolytic degradation to facilitate epithelial migration, regeneration and remodelling (Chantre et al. 2019).

Commercially, Hyaluronic acid is obtained from rooster combs and few *Streptococcus* Group C (attenuated strains), which, as part of their capsule, naturally produce this compound. A detailed description has been documented of various sources from which hyaluronic acid can be extracted. It has been described that hyaluronic acid preparation from microbial sources contains very less contaminating nucleotides, endotoxins and proteins, than those from animal sources (Shiedlin et al. 2004). In order to produce hyaluronic acid, very few microbial strains are known to date, e.g. *Pasteurella multocida* (Gram negative) and group A and group C streptococci (Gram positive). It was described for the first time in 2009 that hyaluronic acid is produced by *S. thermophilus* YIT2084 (a putative probiotic strain) in milk broth through fermentation (Izawa et al. 2009). *S. zooepidemicus* fermentations have also been reported to produce hyaluronic acid (low molecular weight, <200 kDa) under optimized fermentation conditions (Lew and Liong 2013). Recently, an alternative has emerged to produce higher hyaluronic acid yield by fermenting recombinant microbial strains that are Generally Recognized As Safe (GRAS).

4.2 Sphingomyelinase

Sphingomyelinase enzyme, from glucosylceramide and precursors of sphingomyelin, produces phosphorylcholine and ceramides for the extracellular lipid bilayers development in the stratum corneum (Slotte 2013). Its activity has been shown to be significant for skin barrier function (Bocheńska and Gabig-Cimińska 2020). A drop in stratum corneum' ceramide results in epidermal barrier dysfunction and water loss (Mizutani et al. 2009), including a deprivation of protection against bacteria and antigens. In addition, the reduced stratum corneum's ceramide levels have been advised as a potential contact dermatitis's aetiological factor, atopic dermatitis, irritant dermatitis and psoriasis (Murata et al. 1996; Berardesca et al. 2001). It is present in the interstices of stratum corneum and epidermal lamellar bodies and has been graded as the basis of their pH optima as neutral, acidic and alkaline sphingomyelinase. The soluble glycoprotein with an optimum activity at acidic pH (pH 5.0) is identified as the acidic sphingomyelinase. The neurological disorder Niemann–Pick syndrome resulted from the absence of this enzyme in humans. It was identified that persons suffering from Niemann–Pick syndrome also exhibited an aberration in the homeostasis of the permeability barrier of skin with very slow recovery kinetics resulting in acute disruption of the barrier (Lew and Liong 2013). Taking into account that acid Sphingomyelinase is contained in the outer part of the epidermis, the production of ceramides, the acid Sphingomyelinase is therefore responsible and further for basal permeability barrier functions. Moreover, the skin ageing has been related to a reduction in inner epidermal acid Sphingomyelinase (Jensen et al. 2005). The neutral sphingomyelinase, on the other hand, is associated with the cell membrane and, during permeability barrier repair, is significant for cell signalling through increased ceramide accumulation (Kreder et al. 1999). In aged skin, the decreased neutral Sphingomyelinase activity in the outer and inner epidermal layers was found (Lew and Liong 2013), possibly due to decreased proliferation rates, resulting in decreased barrier repair capacity. Mice deficient in TNF-induced neutral Sphingomyelinase activation indicated a smaller increase in epidermal proliferation upon barrier disruption and abridged barrier repair capacity (Kreder et al. 1999). The neutral

Sphingomyelinase activities in lesional and non-lesional atopic dermatitis skin were also reported to be reduced, linked with impaired keratins expression and cornified envelope proteins, which are vital for skin barrier functions (Lew and Liong 2013).

Sphingomyelinase is found in mammalian cells and various microbes (bacteria and yeast), with large Sphingomyelinase activity variations among different microbial strains. The microbial Sphingomyelinase is a secretory protein released into the media from cells, whereas mammalian neutral Sphingomyelinase is a membrane-bound protein (Di Marzio et al. 2001). The alkaline sphingomyelinase can be extracted from probiotic microbes and is an enzyme located exclusively in the intestinal brush border and bile that hydrolyses sphingomyelin into sphingosine, sphingosine-1-phosphate and ceramide, contributing to apoptosis of epithelial cells. In premalignant and malignant intestinal epithelia and in ulcerative colitis tissues, decreased levels of alkaline sphingomyelinase have been identified (Soo et al. 2008). Reduced alkaline sphingomyelinase levels have been observed in premalignant and malignant epithelial and ulcerative colitis tissues (Soo et al. 2008).

4.3   Lipoteichoic Acid

One of the immune-stimulating structural constituents of both non-pathogenic and pathogenic Gram-positive bacterial cell walls is called lipoteichoic acid that has very critical role in bacterial growth and physiology (Villéger et al. 2014). Prior investigations revealed that Lipoteichoic acid could serve as a major pathogen-associated molecular pattern, resulting in nitric oxide (NO), activation of NF-κB (nuclear transcription factor), pro-inflammatory cytokines and other pro-inflammatory mediators' production (Kao et al. 2005; Lebeer et al. 2012). An infection or injury in the host's body is followed by the inflammatory reaction to restore and preserve homeostasis. The lipoteichoic acid from *S. aureus* (a pathogenic Gram-positive bacteria) may, however, induce chronic inflammation and resulting in septic shock, an example of systemic inflammatory response syndrome development (Lew and Liong 2013). The structure-activity correlation investigations of lipoteichoic acid revealed that important strain-specific variations may occur, although, most lipoteichoic acid molecules have a similar basic structure. Unlike Lipoteichoic acid from *S. aureus*, Lew and Liong (2013) isolated lipoteichoic acid from beneficial probiotics, e.g. *L. plantarum* that induced tolerance by protection against the pro-inflammatory cytokines production associated with TNF-α sepsis.

Lipoteichoic acid has been found to promote skin protection against microbial infections through toll-like-receptor induction upon topical application (Sumikawa et al. 2006). In the cutaneous pathogen recognition system, the toll-like receptor activation initiates the release of antimicrobial peptides (soluble effectors) that maintain dermis sterility (Lai et al. 2010). The most popular forms of antimicrobial peptides that contribute against skin bacterial infections, in the host response are human β-defensins and cathelicidins. Various *Lactobacilli* and *Bifidobacteria* species have adequate amounts of Lipoteichoic acid to upsurge dermal cellular defence against microbial infection (Lew and Liong 2013). Lipoteichoic acid also contributed to cutaneous wound healing by activating human β-defensins, in addition to the antimicrobial properties, and accomplished a number of immune-modulatory functions, performing not only as pro-inflammatory agents but also as a main connexion between the adaptive and the innate immune system (Diamond et al. 2009).

4.4   Peptidoglycan

The polymerization of N-acetylmuramic acids and $\beta(1–4)$-linked N-acetylglucosamine, cross-linked by short peptides containing alternating D- and L-amino acids produce peptidoglycan (PG) that is considered as the main structural constituent of microbial cell wall responsible for upholding the shape and to provide shield against osmotic lysis (Dziarski 2003). Peptidoglycans are particularly copious in Gram-positive bacterial strains, where it accounts for around 90% of the cell wall's weight and thickness up to 80 nm (Lew and Liong 2013). On the other hand, the cytoplasmic membrane under the lipopolysaccharide-containing outer membrane of Gram-negative bacteria is surrounded by a relatively thin layer of peptidoglycan (thickness $<$ 10 nm). Although the structure and development of the peptidoglycans are amazingly preserved across bacterial species, it has been found that the chain lengths depend on the bacterial species and various conditions of growth (Lew and Liong 2013).

By stimulating the innate immunity system through Toll-like receptor 2, the peptidoglycan plays a crucial role in the skin's protection against pathogenic microbes resulting in the secretion of numerous chemokines and cytokines that are involved in immune responses (Niebuhr et al. 2010). It has also been demonstrated that the peptidoglycan has the capability to activate NF-kB (nuclear factor) and triggers the interleukin-8 production abundantly from keratinocytes, indicating that peptidoglycan plays a vital role in the chemokines and cytokines production from keratinocytes (Matsubara et al. 2004). Numerous other peptidoglycan recognition molecules also recognize the peptidoglycan, including the nucleotide oligomerization domain-containing proteins (CD14), peptidoglycan lytic enzymes (lysozyme and amidase) and a family of Peptidoglycan recognition proteins (PGRPs, Dziarski 2003; Kumar et al. 2010). These molecules induce the responses of the host to microbes, mediate the antimicrobial peptides or degrade Peptidoglycan and chemokines release that results in recruitment of phagocytic cells to the site of infection (Dziarski and Gupta 2005; Lew and Liong 2013). The microbe-derived molecules like peptidoglycan have been reported to be able to induce or increase the expression of human β-defensins in whole skin keratinocytes of the humans, contributing to the stimulation of host's innate immunity (Sørensen et al. 2005). *Lactobacilli* peptidoglycan stimulates innate immune response through Toll-Like Receptor 2 and also to increases the IL-12 production and other regulatory factors by macrophages, which further results in skin protection (Paradis-Bleau et al. 2007; Lew and Liong 2013).

4.5 Lactic Acid

Lactic acid is an organic acid, classified as one of the α-hydroxy acids, with one hydroxyl group attached to the alpha position of the acid and produced by microbial fermentation or chemical synthesis. The lactic acid produced by chemical synthesis mostly consists of the racemic mixture (DL-lactic acid), while L(+)- or D(−)-lactic acid (optically pure) can be derived through fermentation using appropriate microorganisms (Wee et al. 2006; Tang and Yang 2018). At sufficient concentrations, *Lactobacilli* strains may produce lactic acid to show antibacterial activity against the majority of the pathogenic microbes on the skin (Lew et al. 2013). They metabolize carbohydrates with at least 50–85% lactic acid, either homo-fermentatively or hetero-fermentatively, for production of the main end product, i.e. lactic acid (Yeo and Liong 2010). Lactic acid has been extensively utilized for a long time in skin care products and cosmetic regiments, e.g. exfoliants, moisturizers and emollients (Smith 1996). One of the causes that lactic acid (α-hydroxy acids) is frequently used as a chemical peeling agent and an exfoliator is because of its profound effect on skin desquamation. The induction of skin desquamation is done by the dissociation of the cellular adhesions, which occurs via the chelating action of α-hydroxy acids as a result of reduced concentration of epidermal calcium ions. The reduced epidermal calcium ion level also tends to promote cell growth and delays cell differentiation, resulting in younger-looking skin (Soleymani et al. 2018). Moreover, due to its ability to boost the function of the stratum corneum barrier, it has the potentials for various skin applications and also improves the ceramide's production by keratinocytes. The improved ceramide 1-linoleate to oleate ratio has a significant role in enhancing the functions of the skin barrier (Yamamoto et al. 2006). Pasricha et al. (1979) have investigated lactic acid's antimicrobial activity against dermal pathogens, e.g. beta haemolytic *Streptococci*, *S. aureus*, *Proteus* species, *E. coli* and *P. aeruginosa*. Owing to its non-toxic and non-sensitizing properties, the long-lasting topical use of lactic acid cream has been suggested as a preventive remedy for acne vulgaris, in addition to its antimicrobial activity.

4.6 Acetic Acid

Acetic acid is produced both chemically and by microbial fermentation at industrial level. Heterofermentative lactic acid bacteria can produce acetic acid via the hexose monophosphate or pentose pathway (Yeo and Liong 2010). The acetic acid usage has been described from time to time as a topical agent in the treatment of microbial infections and also been used to treat superficial infections and burns. When several antibiotic-resistant strains cause infection and where therapeutic options are insufficient, it has been suggested as the best remedy (Nagoba et al. 2008). It has been shown that acetic acid exerts antibacterial effects on several microbes, including *S. aureus* and *P. aeruginosa* (Lew and Liong 2013).

4.7 Diacetyl

Some strains of the genera *Streptococcus, Leuconostoc, Lactobacillus* and *Pediococcus* can produce diacetyl, also referred to as 2,3-butanedione. *Lactobacilli* and *Bifidobacteria* strains might produce diacetyl (concentrations up to 30 mg ml^{-1} s), signifying their possibility for antimicrobial dermal activities with maximum sensitivity compared to Gram-positive bacteria against Gram-negative bacteria and fungi (Lew et al. 2013). Although the majority of the Gram-negative bacteria, e.g. *Bartonella* sp. *Borrelia burgdorferi, P. aeruginosa, Pasteurella multocida, Vibrio vulnificus, Klebsiella rhinoscleromatis, Helicobacter pylori* and *S. typhi*, are not typical skin microflora residents, it has been suggested that they cause cutaneous infections. Diacetyl, at a very low concentration of 100 ppm, has been proved to be bactericidal against *E. coli* and *S. aureus* (Lanciotti et al. 2003). A pathogen *S. aureus* has appeared as a major infectious microbe of skin and soft tissue, including cellulitis, folliculitis and impetigo (Miller and Cho 2011), and one of the most common skin pathogens identified is *E. coli* (Doern et al. 1999). The diacetyl's antimicrobial activity has been well acknowledged, but, there is very little research available on topical application of diacetyl and considerable investigations are needed to establish its effects on the skin and other tissues.

4.8 Antimicrobial Substances

The growing attention has been given to the possible topical application of probiotic microbial strains' ability to produce potent antimicrobial toxins (i.e. H_2O_2, organic acids, bacteriocins and bacteriocin-like substances) to effectively avert pathogen adhesion and outcompete undesired microorganisms (Fig. 3) (Gillor et al. 2008; Cinque et al. 2011). The compositions comprising probiotic microbes, spores and their products, have been described by Farmer (2005), apposite for topical usage on the skin, can be utilized to impede the growth of microbes and combinations thereof. Different treatment approaches and therapeutic systems to prevent the growth of pathogens and combinations thereof through topical application of pharmaceutical compositions, comprising of isolated species of *Bacillus*, spores or an extracellular product of *B. coagulans* (a supernatant or filtrate of a fermented *B. coagulans* culture) are also disclosed in the invention.

Spigelman and Ross (2008) have also given a composition and method for the probiotic microorganisms' application to skin surfaces to avert or constrain pathogenic microorganisms' contamination. The probiotic microorganisms include bacteria, yeast or fungi. The appropriate probiotics should be selected on the basis of one or more unique characteristics, the desired characteristics being the competitive exclusion of pathogenic microbes from the surface to which they are applied, antibiotic sensitivity, human tissue adherence, a high resistance to oxygen and acid and antimicrobial activity. More specifically, the procedure consists of multiple

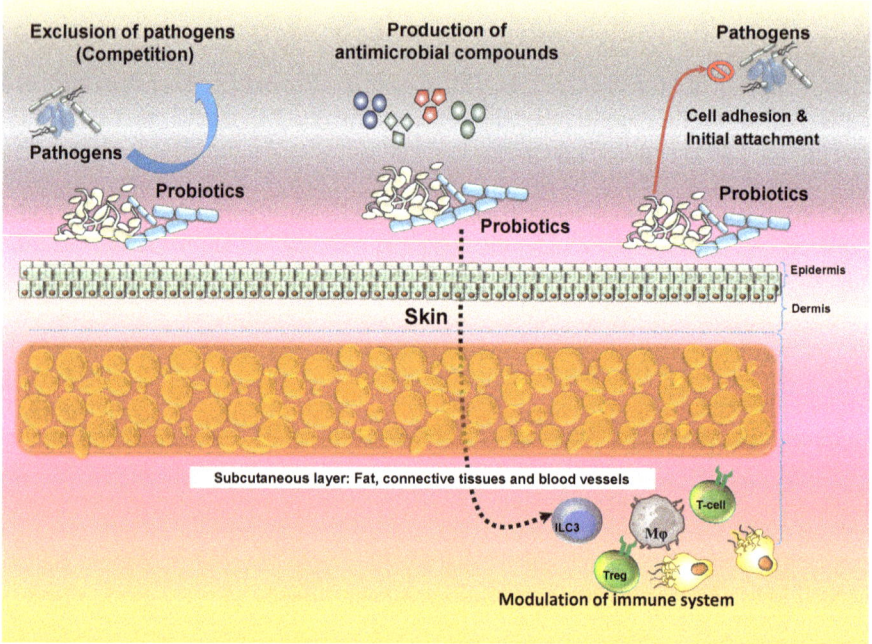

Fig. 3 Demonstration of potential mechanisms of action of probiotics' antagonistic effects

methods of application (e.g. wiping paper, spraying and lotions) of one or more probiotic microbes to a wide range of surfaces (e.g. hospital equipment, fixtures and human skin), effectively avoiding, at least partially, their infection, invasion, growth and cross-contamination by pathogenic microorganisms.

The technique depends on the probiotic's capability to produce isolated colonies to generate a protective layer that can prevent and eliminate pathogenic microbes unable to survive on top of other bacteria. Depending on various factors such as the therapeutically effective amount, type, probiotic application mode or the degree of contamination of the biological or non-biological surface, the probiotic application is recommended for a suitable period. The method, therefore, suggests the usage of a single or a multitude of diverse probiotic microbes, serially adding multiple layers of bacteria to combat single or many resistant kinds of pathogenic microorganisms. Numerous lactic acid bacterial species producing a number of bacteriocins, including *Lactobacillus*, *Pediococcus*, *Lactococcus*, *Leuconostoc*, *Carnobacterium* and *Propionibacterium* have been reported in relation to the potential use of bacteriocin-producing strains as probiotic and bio-protective agents (Mokoena 2017). *Lactococcus* sp. HY 449 bacteriocin was capable of preventing the growth of skin inflammatory bacteria, e.g. *S. epidermidis*, *S. aureus*, *P. acnes* and *Streptococcus pyogenes* (Oh et al. 2006). The bacteriocin's inhibitory effect used in the research work done by Oh et al. (2006) was triggered by the bacteriolytic activity on the cell membranes and cell walls of *P. acnes*. Acknowledgments to its antimicrobial properties, bacteriocin from *Lactococcus* can be used for many purposes in cosmetic applications.

The eubiotic product, according to Teodorescu (1999), is a mixture of three probiotic *Lactobacillus acidophilus* LR, LV and LD strains, the association in the mixture in equivalent parts before lyophilization, for the treatment and maintenance of tegument. In order to abolish pathogenic microflora and to be immune to cosmetic composition, this eubiotic substance is capable of preserving the skin pH at physiological values.

4.9 β-Defensins

Lactobacillus extracts can induce dose-dependent β-defensins production in skin cells, which may be useful in reducing or preventing the growth of skin microbial populaces (Sullivan et al. 2009). The effective quantity of *Lactobacillus* extract is applied to a skin wound or an open cut that may have been in direct contact with soil or unwanted microbes on a chronic basis, added to clean skin to uphold a hale and hearty skin flora. These extracts can also be beneficial in treating acne. Indeed, when applied consistently over a 2-month span, the *L. plantarum* extract containing topical compositions/lotions is shown to lessen the occurrence of both inflamed and non-inflamed acne lesions.

Furthermore, the extracts were used as a preservative in cosmetic or pharmaceutical items, especially *L. plantarum*, which has a wide range of activity against Gram-negative and Gram-positive bacteria. Acne vulgaris is a multifactorial disorder characterized by *P. acnes* hyper-colonization, inflammation and immune responses. In an in vivo study, the synbiotic capability of Konjac glucomannan hydrolysates and probiotic bacteria to inhibit *P. acnes* growth has recently been reported, indicating that it may be promising to develop a new probiotic therapy alternative to minimize the acne episodes (Cinque et al. 2011).

5 Probiotics, Infections and Intestinal Wound Healing

In antimicrobial treatment of pathogens, the close association of lactic acid bacteria and *Bifidobacteria* with epithelial cells makes them ideal probiotic candidates (Lukic et al. 2017). Lactic acid bacteria and *Bifidobacteria* use their interaction with gut epithelial cells to hinder pathogens' growth directly and capability to enhance tissue repair mechanisms and host mucosal defence systems. In combating overt and opportunistic pathogens, these properties are of paramount importance.

5.1   Probiotics: Direct Inhibition of Pathogens' Growth

The antimicrobials' production, pathogens displacement from mucus and epithelial cells, removal of pathogens by co-aggregation and quorum quenching are recommended direct probiotic action mechanisms against pathogens. The organic acids, H_2O_2, reuterin, diacetyl and bacteriocins are the antimicrobials produced by probiotic strains. The anti-pathogenic activity against Gram-negative pathogens is essentially due to the production of organic acids by numerous probiotic strains, from both lactic acid bacteria and *Bifidobacteria* (Piqué et al. 2019). Hydrogen peroxide producing *Lactobacilli* (*L. jensenii*, *L. fermentum* and *L. acidophilus*) have been associated with abridged count of Gram-positive bacteria (fastidious anaerobe), including *Prevotella, Bacteroides, Mycoplasma* sp. and *Gardnerella* (Atassi et al. 2006; Breshears et al. 2015). *L. reuteri* produced a well-known antimicrobial metabolite Reuterin (3-hydroxypropionaldehyde) and thought to exert its influence through the thiol groups' oxidation in the target pathogenic microorganisms (Schaefer et al. 2010). Importantly, without killing beneficial microorganisms, reuterine will precisely inhibit the growth of harmful gut bacteria, causing *L. Reuteri* to kill gut invaders while keeping the microbiota of the natural gut intact. Reuterin also exhibits antimicrobial activity against the common chronic wound pathogen, *Staphylococcus* (Arqués et al. 2008). Diacetyl is produced by *Lactobacilli*, another metabolic product that also shows a broad range of antimicrobial ability against Gram-negative and Gram-positive pathogens (Kang and Fung 1999; Langa et al. 2014).

Bacteriocins, the second class of metabolites from probiotic strains, are very small peptides by microbes that display a wide spectrum of antimicrobial activity both in vitro and in vivo (Minami et al. 2009). *S. salivarius*, (producing bacteriocin), a commensal of oral epithelium, is a potential inhibitor of *S. pyogenes* (a pathogen) triggers pharyngitis and cutaneous infections (Heng et al. 2011). Both in adults and children, *S. salivarius*'s prophylactic oral administration has shown beneficial effects in preventing recurrent infections of *S. pyogenes* (Di Pierro et al. 2013).

The potential of auto-aggregation by several microbial strains (probiotics), including *Bifidobacterium longum*, *L. delbrueckii* and *L. rhamnosus*, confer the antimicrobial ability to co-aggregate with other microorganisms that include the common wound pathogens *Candida albicans* and *S. aureus* (Barzegari et al. 2020). In addition to co-aggregation and antimicrobial metabolites' production, probiotic microbes can move intestinal pathogens from the epithelium of the gut or stomach. The specific surface molecules obtained from *Lactobacilli* (extracellular polysaccharides) have the ability of displacement, which further allow *L. paracasei* to competitively adhere to gut epithelial cells and displace the harmful microbes (Rutherford and Bassler 2012).

The inhibition of harmful microbes' quorum sensing (QS) system is another emerging antimicrobial mechanism of lactic acid-producing microbes. Quorum sensing refers to an intercellular communication mechanism that microorganisms use to modify cell-population density-based gene expression to form biofilm and confer virulence (Pastar et al. 2013; Kumar et al. 2020). Majority of the pathogens, including microbes usually found in chronic wound infections (e.g. *P. aeruginosa* and *S. aureus*), use quorum sensing for resistance to host defence, virulence and biofilm formation (Lukic et al. 2017). Nevertheless, probiotics can interfere with the quorum sensing of pathogens. Specifically, *L. plantarum* has been shown to prevent the quorum sensing signalling molecules (acyl-homoserine-lactone) production by *P. aeruginosa*, along with the decrease in the formation of biofilm (Valdez et al. 2005).

5.2 Epithelial Barrier and Probiotic Effects

The probiotics can enhance the epithelial barrier's function, thereby limiting pathogen invasion, besides direct antimicrobial effects on harmful microorganisms, (Ohland and MacNaughton 2010). By improving expression and controlling the localization of tight junction proteins both in vivo (Karczewski et al. 2010) and in vitro (Anderson et al. 2010), they have a well-defined function in strengthening the gastrointestinal barrier, e.g. increased occludin, claudin and zonula occludens 1 expression in the gut in newborn piglets resulting from oral administration of *L. reuteri* (Yang et al. 2015). In the same way, following oral administration of *L. plantarum*, occludins and zonula occludens 1 were recruited to the tight junction region (Karczewski et al. 2010). Additionally, a mixture of eight distinct probiotic bacterial strains stimulates the suppression of chronic inflammation by epithelial barrier function fortification, which have been shown in the murine models of chronic ileitis (Ewaschuk et al. 2008).

In addition to *Lactobacilli*, the probiotic strain belonging to genus *Bifidobacterium* has also shown similar effects. The expression and trans-epithelial resistance of tight junction proteins occluding and zonula occludens 1 were increased by *B. infantis* in human gut epithelia. Improved trans-epithelial resistance has also been correlated with enhanced cell signalling events significant for extracellular signal-regulated kinases phosphorylation, barrier formation and p38 (Fijan et al. 2019). In order to verify the effects of probiotics on wound healing in the gastrointestinal tract, various experimental models, including acetic acid-induced ulcers, full thickness wounds and intestinal anastomoses, have been extensively examined. The beneficial effects of *Lactobacilli* in these studies were largely mediated by stimulation and activation of fibroblast proliferation and/or migration by the epithelial cells (Lukic et al. 2017).

Aside from improving the epithelium repair, the presence of *L. plantarum* has been shown to linked with increased production of collagen in the intestine (Nasrabadi et al. 2011), and similar skin effects have been shown in hairless mouse model with UVB induced skin photo-ageing after oral administration of *L. acidophilus* (Lukic et al. 2017). Given that the chemokines, cytokines and growth factors have regulated the epithelial cells' and fibroblasts' functions (Pastar et al. 2014), the epidermal barrier fortification by probiotics is closely linked with their impact on immune components. Probiotics also have an effect on innate immune components of the intestinal barrier by inducing β-defensin, which is known to promote wound healing in addition to its function in fighting intestinal pathogens (Lukic et al. 2017).

In vivo murine studies have revealed that commercially available probiotic microbial mixture can be utilized for the stimulation of the vascular endothelial growth factor (VEGF) and transforming growth factor β (TGF β) expressions (Dharmani et al. 2013), while *Saccharomyces boulardii* containing probiotic formulation was shown to stimulate insulin-like growth factor (IGF), epidermal growth factor (EGF) and its receptor activity (EGFR) (Fordjour et al. 2010). Additionally, *L. rhamnosus* has been reported to stimulate hypoxia-inducible factor 2α (a master controller of progenitor stem cell recruitment during tissue repair) in vivo (Wang et al. 2011).

6 Chronic Wounds and Probiotic Therapy

Considering the potential role of skin microbiota and biofilms in skin-related diseases, the investigators have begun to investigate probiotics for the chronic wounds' treatment e.g. kefir extracts (natural probiotic compounds) containing topical gels have been applied to infected burn wounds in rats and have shown improved collagen formation and epithelialization compared to controls treated with silver sulfadiazine (Lukic et al. 2017). The probiotics administered to mice (orally) were able to modulate interleukin-10 levels and skin immune cell density after an injury caused by UV radiation, indicating that these beneficial microbes in cutaneous tissues can exert strong immune-modulatory effects (Guéniche et al. 2006). The investigators have also isolated microbes from burn wounds and demonstrated that most of the microbial strains were extremely vulnerable to *L. acidophilus* (Jebur 2010). Via competitive inhibition of pathogenic microbe *P. aeruginosa* and disruption of bacteria–bacteria communication pathways, i.e. quorum sensing, *L. plantarum* has a potential role in the topical treatment of wounds (Peral et al. 2009). Some of the useful tools for reconstructive surgery include bio-prostheses and implants, but they are at greater risk for formation of biofilm and chronic infection. Therapies based on probiotics can play a role in reducing these complications, e.g., surfactants obtained from probiotics have been shown in a voice prosthesis model to lessen pathogenic microbial colonization and extend graft function (Rodrigues et al. 2004). Some probiotic strains are capable of producing oxidative reactions that impede the growth of fungi and the formation of biofilms (Reid et al. 2006). These impacts may be due to pathogenic microbial adhesion changes and can also be used prophylactically in high-risk patients. However, in clinical environments such as infected (contaminated) mesh, contracture or extrusion for the breast implant, and other prosthetic complications, the use of probiotics remains unproven but is considered as a field of considerable research potential (Wong et al. 2013).

Some experiments have revealed the feasibility of manipulating microbial properties to facilitate wound healing through the application of principles in tissue engineering, e.g., investigators have produced a topical patch containing nitric oxide producing probiotics (e.g. lactic acid bacteria), a molecule well recognized to enhance the synthesis of fibroblast collagen and increase tissue blood flow (Isenberg et al. 2005). This bacteria-impregnated patch substantially improved wound closure in infected and ischemic wounds in rabbits measured after 3 weeks (Jones et al. 2012b). Further, investigations are needed in order to understand the probiotics' role and possible delivery mechanisms for non-healing wounds. In conjunction with traditional approaches to wound healing, probiotic-based therapies could be a significant adjunct for potential paradigms of wound treatment. The genetically modified microorganisms or engineered microbial by-products may play a role in regulating interactions of host-bacteria or bacteria–bacteria to facilitate the repair of cutaneous tissue in addition to the exogenously administered probiotic strains.

7 Probiotics and Cutaneous Wound Healing

Bifidobacteria and *Lactobacilli* are the most widely studied potential probiotics for numerous dermatological conditions, including non-healing wounds (Baquerizo Nole et al. 2014). The protective abilities of probiotic microbes against skin pathogens have been demonstrated in numerous in vitro experiments with human keratinocytes (Lukic et al. 2017). The probiotic strains, *L. rhamnosus* and *B. longum*, similar to their impact on the gut epithelium, have been shown to mend tight junction functions and expression of zonula occludens 1, claudin 1 and occludin in *S. aureus* infected keratinocytes (Sultana et al. 2013). Unlike *L. rhamnosus*, *B. longum* augmented the claudin 4 expression, another major tight junction protein (Sultana et al. 2013), proposing that *B. longum* can affect tight junction function through a substitute mechanism by lessening para-cellular permeability and therefore averting the pathogen invasion.

Furthermore, the Toll-like Receptor 2 activation increases the tight barrier function in keratinocytes as well as gut epithelial cells (Yuki et al. 2011). *B. longum*'s modulation related to functions of tight junction seems to be Toll-like Receptor 2 dependent as tight junction protein levels and trans-epithelial electrical resistance cease to upsurge when Toll-like Receptor 2 is neutralized or blocked, respectively (Sultana et al. 2013). The implications of the commonly used *L. rhamnosus*, on the other hand, on keratinocytes are Toll-like Receptor 2-independent, indicating that this probiotic species utilizes another method to augment tight barrier function (Sultana et al. 2013). The mitogen-activated protein kinase pathway, known to increase tight barrier function through modulation of extracellular signal-regulated kinases and p38 (Lukic et al. 2017), is a possible pathway involved in this process (Lukic et al. 2017).

The probiotic species (*L. plantarum* and *L. reuteri*) also possess the capability to upsurge tight barrier function in primary human keratinocytes (Sultana et al. 2013). *L. rhamnosus* and *L. reuteri* also improve re-epithelialization through enhanced keratinocyte migration and cellular proliferation (Mohammed et al. 2015). The probiotics are also able to cause re-epithelialization through chemokines induction, e.g. *L. rhamnosus* augmented the chemokine CXCL2 and its receptor CXCR2 expressions that stimulates proliferation and migration of keratinocyte during normal wound healing (Mohammed et al. 2015). Although the majority of probiotic microbes have been advantageous for function associated with keratinocyte, *L. fermentum* deceases viability of keratinocyte and re-epithelialization (Mohammed et al. 2015; Lukic et al. 2017), demonstrating strain-specific effects once again.

The antibacterial activities of already established probiotics, e.g. reduction of pathogen adhesion and inhibition of pathogen growth, are the methods of fortification against cutaneous wound infections. The protective effect exhibited by *L. rhamnosus* by *S. aureus* growth inhibition in infected keratinocytes is yet unknown mechanism (Mohammedsaeed et al. 2014). The probiotics like *L. rhamnosus* and *L. casei* Shirota exhibited antimicrobial activity that is not due to acid (Mohammedsaeed et al. 2014), hydrogen peroxide or bacteriocin production (Vesterlund et al. 2004), featuring a number of protective mechanisms by probiotics. Additionally, *L. plantarum* extract, by interfering with its quorum sensing system, disrupt *P. aeruginosa*'s pathogenic characteristics, a widespread chronic wound pathogen. Through inhibition of *P. aeruginosa* virulence factors pyocyanin, elastase and rhamnolipid, this extract without live probiotics was able to reduce biofilm growth and bacterial adhesions (Ramos et al. 2012).

Lactobacilli can also, by competitive exclusion, inhibit pathogen invasion into keratinocytes. *L. rhamnosus* and *L. reuteri* are capable to impede the initial *S. aureus* adhesion to keratinocytes and displace already attached *S. aureus* to human keratinocytes (Mohammedsaeed et al. 2014; Lukic et al. 2017). The pertinent molecules that contribute in *S. aureus* exclusion and displacement from keratinocytes derived from human are still unidentified, but they depend on moonlight proteins: a class of multi-functional bacterial adhesins that may, among many functions, bind to epithelial cells (Kainulainen and Korhonen 2014). The enolase from *L. crispatus*, an example of a moonlight protein, can bind to collagen-I and laminin (Antikainen et al. 2002), while *L. plantarum* enolase binds to fibronectin and avert *S. aureus* adhesion to epithelial cells (Castaldo et al. 2009). A mechanism for displacement, as illustrated by *L. rhamnosus*, will enable probiotics not only to protect keratinocytes from infection but also to rescue them, both of which are important features for potential clinical applications. The supernatants, lysates and metabolites from probiotic microorganisms have been extensively investigated in vivo and in vitro to address protection of utilizing live probiotic bacteria topically, showing beneficial effects similar to live microorganisms (Mohammedsaeed et al. 2014; Lukic et al. 2017).

In vivo wound repair investigations were mainly focused on topical application of probiotics that support in vitro data, showing enhanced wound repair via increased tissue repair and reduced bacterial load in rodent wound models (Rodrigues et al.

2005). *L. plantarum* (topical application) inhibited colonization of wounds caused by *P. aeruginosa* in a mouse model with burns by clearing *Pseudomonas* from the liver, spleen and skin, via lessening apoptosis and increasing phagocytosis and (Valdez et al. 2005). Even the use of kefir (mixture of lactic acid bacteria and yeasts) has resulted in improved healing with antifungal and antibacterial effects (Lukic et al. 2017).

L. plantarum (topical use) has interfered with pathogen colonization caused by *S. aureus*, *S. epidermidis* and *P. aeruginosa* in human burn wounds (Peral et al. 2009). Topically applied *L. plantarum* Treatment with lessened bacterial load and encouraged wound repair is found to be comparable with silver sulfadiazine treatment. One possible mechanism underlying *L. plantarum*'s antimicrobial/anti-pathogenic characteristics is that *P. aeruginosa* and *L. plantarum* stimulate reverse effects on the infection (Hessle et al. 2000). *L. plantarum* (Gram-positive bacteria) activates the secretion of interleukin-12, which activates natural killer cells and cytotoxic T cells to secrete IFNγ, while Gram-negative pathogen *P. aeruginosa* favourably stimulates interleukin -10, which prevents those functions (Hessle et al. 2000). The antagonistic inflammatory response regulation, however, does not account for antibacterial effects of *L. plantarum* on *S. aureus*, a Gram-positive pathogen. Topical use of *L. plantarum* also boosted wound healing in human chronic venous ulcers (Peral et al. 2010) infected predominantly with *P. aeruginosa* and *S. aureus*, stimulated a continuous process of healing that decreased microbial load and triggered the granulation tissue formation (Peral et al. 2010; Lukic et al. 2017). Polymorphonuclear cells screened from the ulcer bed showed augmented interleukin-8 production, decreased apoptosis percentage and necrosis upon *L. plantarum* treatment. Taking its antimicrobial and immunomodulatory effects in humans into account, *L. plantarum*, by controlling interleukin-8 levels and controlling the entry and activity of Polymorphonuclear cells travelling from peripheral blood to the ulcer, is thought to inhibit pathogen colonization (Peral et al. 2010).

8 Future Perspectives

The new insight and healing potential of advantageous microbial probiotics are illustrated in the book chapter as an alternative and healthy approach to treating patients with skin-related wounds/disorders. The examination of microbiota, including beneficial microorganisms by using high-throughput genomic technologies, will elucidate new pathways and molecular mechanisms that can improve our knowledge of how commensal microbes, including non-healing wounds, can cope with different diseases. Moreover, it is important to identify cross-communication among the beneficial microorganisms and the host's respective pathways. The selection of bacterial species is of particular significance, because the impacts of probiotic bacteria can be highly strain-specific. Their incorporation as an integrative therapy provides new possibilities to treat patients with wound healing disorders, taking into account the studies expended on probiotics and their important role in human health.

9 Conclusions

Based on recent in vitro and in vivo research, this book chapter documents the ability of cellular components or probiotic metabolites to promote skin health and dermatological advancement. Although numerous investigations have indicated encouraging potentials of probiotics and their supernatants for skin health, we predict that such an argument for skin health is still at its early stages, with the requirement of more comprehensive final topical applications and human trials (well-designed) to validate the exact doses required, safety and regulatory compliances, possible side effects, host dependency, and, essentially, the precise mechanisms for indirect and direct actions of the live cells and/or therapeutic compounds. The microbial colonization occurs immediately after injury, and the cross-talk between the innate immune response, pathogens and microflora almost simultaneously begins. Cutaneous microbiota has an advantageous impact on the wound curing process through several potential processes, both positively and harmfully, dependent on the microorganisms prevailing in the area of the wound. The identification of microorganisms derived from the nonspecific immune (innate) response is essential for triggering the process of wound healing, and particularly for the preliminary stage of severe (acute) inflammation. The microbe *S. aureus*, however, can cause infection, impaired healing of chronic and acute wounds. Besides the microflora on skin, the gastrointestinal microorganisms may similarly influence the wound curing process, by influencing, indirectly or directly, several features that control the therapeutic potential, e.g. blood pressure, tissue oxygenation levels, immune response and inflammations. Furthermore, several mechanisms have been recognized as to how the gut microbiome could affect the energy metabolism of the host and thus the incidence of indications of metabolic syndrome, e.g. Diabetes, hypertension, obesity and hyperlipidemia, which were also associated with very slow wound healing. For a substantial portion of the population, impaired wound healing is a major reason for morbidity and mortality. Taking the above seriously, the scientific community guides the investigation to a deeper understanding of the cross-talk amongst host immune response and microorganisms, with the objective of developing new methods for therapeutic wound care based on the therapeutic use of probiotic microbes. The probiotics are advantageous host microbes and, on the basis of evidence so far, may have a positive effect on the wound curing process. Probiotics administration has been associated with improvements in the topical and per os wound curing process. Is the management of the human microbiome, gastrointestinal as well as cutaneous, essentially crucial for the chronic wounds and ulcers treatment therapies that have eluded us so far, or is it just another factor that needs to be taken seriously for this therapeutic entity to be treated? Time is going to say.

Acknowledgments The author gratefully acknowledges The Director, CSIR-IHBT (Institute of Himalayan Bioresource Technology), Palampur, Himachal Pradesh and the Council of Scientific and Industrial Research (CSIR), New Delhi, for providing basic computational facilities for carrying out the present study.

Conflicts of Interest The author declares that there is no conflict of interests.

Funding No external funding was received for writing this book chapter.

References

Abdin AA, Saeid EM (2008) An experimental study on ulcerative colitis as a potential target for probiotic therapy by *Lactobacillus acidophilus* with or without "olsalazine". J Crohn's Colitis 2 (4):296–303

Anderson RC, Cookson AL, McNabb WC, Park Z, McCann MJ, Kelly WJ, Roy NC (2010) *Lactobacillus plantarum* MB452 enhances the function of the intestinal barrier by increasing the expression levels of genes involved in tight junction formation. BMC Microbiol 10(1):316

Antikainen J, Anton L, Sillanpää J, Korhonen TK (2002) Domains in the S-layer protein CbsA of *Lactobacillus crispatus* involved in adherence to collagens, laminin and lipoteichoic acids and in self-assembly. Mol Microbiol 46(2):381–394

Arck P, Handjiski B, Hagen E, Pincus M, Bruenahl C, Bienenstock J, Paus R (2010) Is there a 'gut–brain–skin axis'? Exp Dermatol 19(5):401–405

Arqués JL, Rodríguez E, Nuñez M, Medina M (2008) Antimicrobial activity of nisin, reuterin, and the lactoperoxidase system on *Listeria monocytogenes* and *Staphylococcus aureus* in cuajada, a semisolid dairy product manufactured in Spain. J Dairy Sci 91(1):70–75

Ashoori Y, Mohkam M, Heidari R, Abootalebi SN, Mousavi SM, Hashemi SA, Golkar N, Gholami A (2020) Development and *in vivo* characterization of probiotic lysate-treated chitosan nanogel as a novel biocompatible formulation for wound healing. Biomed Res Int 2020. https://doi.org/10.1155/2020/8868618

Atassi F, Brassart D, Grob P, Graf F, Servin AL (2006) *Lactobacillus* strains isolated from the vaginal microbiota of healthy women inhibit *Prevotella bivia* and *Gardnerella vaginalis* in coculture and cell culture. FEMS Immunol Med Microbiol 48(3):424–432

Baquerizo Nole KL, Yim E, Keri JE (2014) Probiotics and prebiotics in dermatology. J Am Acad Dermatol 71(4):814–821

Barzegari A, Kheyrolahzadeh K, Khatibi SM, Sharifi S, Memar MY, Vahed SZ (2020) The battle of probiotics and their derivatives against biofilms. Infect Drug Resist 13:659

Belkaid Y, Hand TW (2014) Role of the microbiota in immunity and inflammation. Cell 157 (1):121–141

Berardesca E, Barbareschi M, Veraldi S, Pimpinelli N (2001) Evaluation of efficacy of a skin lipid mixture in patients with irritant contact dermatitis, allergic contact dermatitis or atopic dermatitis: a multicenter study. Contact Dermatitis 45(5):280–285

Bocheńska K, Gabig-Cimińska M (2020) Unbalanced sphingolipid metabolism and its implications for the pathogenesis of psoriasis. Molecules 25(5):1130

Bourrie BC, Willing BP, Cotter PD (2016) The microbiota and health promoting characteristics of the fermented beverage kefir. Front Microbiol 7:647

Bowe WP, Logan AC (2011) Acne vulgaris, probiotics and the gut-brain-skin axis-back to the future? Gut Pathog 3(1):1–11

Breshears LM, Edwards VL, Ravel J, Peterson ML (2015) *Lactobacillus crispatus* inhibits growth of *Gardnerella vaginalis* and *Neisseria gonorrhoeae* on a porcine vaginal mucosa model. BMC Microbiol 15(1):1–12

Breton J, Daniel C, Dewulf J, Pothion S, Froux N, Sauty M et al (2013) Gut microbiota limits heavy metals burden caused by chronic oral exposure. Toxicol Lett 222(2):132–138

Castaldo C, Vastano V, Siciliano RA, Candela M, Vici M, Muscariello L, Marasco R, Sacco M (2009) Surface displaced alfa-enolase of *Lactobacillus plantarum* is a fibronectin binding protein. Microb Cell Factories 8(1):1–10

Chantre CO, Gonzalez GM, Ahn S, Cera L, Campbell PH, Hoerstrup SP, Parker KK (2019) Porous biomimetic hyaluronic acid and extracellular matrix protein Nanofiber scaffolds for accelerated cutaneous tissue repair. ACS Appl Mater Interfaces 11(49):45498–45510

Chong BF, Blank LM, Mclaughlin R, Nielsen LK (2005) Microbial hyaluronic acid production. Appl Microbiol Biotechnol 66(4):341–351

Christensen GJ, Brüggemann H (2014) Bacterial skin commensals and their role as host guardians. Benefic Microbes 5(2):201–215

Church D, Elsayed S, Reid O, Winston B, Lindsay R (2006) Burn wound infections. Clin Microbiol Rev 19(2):403–434

Cinque B, La Torre C, Melchiorre E, Marchesani G, Zoccali G, Palumbo P, Di Marzio L, Masci A, Mosca L, Mastromarino P, Giuliani M (2011) Use of probiotics for dermal applications. In: Probiotics. Springer, Berlin, pp 221–241

Demers M, Dagnault A, Desjardins J (2014) A randomized double-blind controlled trial: impact of probiotics on diarrhea in patients treated with pelvic radiation. Clin Nutr 33(5):761–767

Dharmani P, De Simone C, Chadee K (2013) The probiotic mixture VSL# 3 accelerates gastric ulcer healing by stimulating vascular endothelial growth factor. PLoS One 8(3):e58671

Di Marzio L, Paola Russo F, D'Alo S, Biordi L, Ulisse S, Amicosante G, De Simone C, Cifone MG (2001) Apoptotic effects of selected strains of lactic acid bacteria on a human T-leukemia cell line are associated with bacterial arginine deiminase and/or sphingomyelinase activities. Nutr Cancer 40(2):185–196

Di Pierro F, Adami T, Rapacioli G, Giardini N, Streitberger C (2013) Clinical evaluation of the oral probiotic *Streptococcus salivarius* K12 in the prevention of recurrent pharyngitis and/or tonsillitis caused by *Streptococcus pyogenes* in adults. Expert Opin Biol Ther 13(3):339–343

Diamond G, Beckloff N, Weinberg A, Kisich KO (2009) The roles of antimicrobial peptides in innate host defense. Curr Pharm Des 15(21):2377–2392

Doern GV, Jones RN, Pfaller MA, Kugler KC, Beach ML, SENTRY Study Group (1999) Bacterial pathogens isolated from patients with skin and soft tissue infections: frequency of occurrence and antimicrobial susceptibility patterns from the SENTRY Antimicrobial Surveillance Program (United States and Canada, 1997). Diagn Microbiol Infect Dis 34(1):65–72

Dziarski R (2003) Recognition of bacterial peptidoglycan by the innate immune system. Cell Mol Life Sci 60(9):1793–1804

Dziarski R, Gupta D (2005) Peptidoglycan recognition in innate immunity. J Endotoxin Res 11 (5):304–310

Ewaschuk JB, Diaz H, Meddings L, Diederichs B, Dmytrash A, Backer J, Looijer-van Langen M, Madsen KL (2008) Secreted bioactive factors from *Bifidobacterium infantis* enhance epithelial cell barrier function. Am J Physiol Gastrointest Liver Physiol 295(5):G1025–G1034

Farmer S (2005) Topical compositions containing probiotic bacillus bacteria, spores, and extracellular products and uses thereof. US Patent 6905692, 14 June 2005

Fijan S, Frauwallner A, Langerholc T, Krebs B, ter Haar née Younes JA, Heschl A, Mičetić Turk D, Rogelj I (2019) Efficacy of using probiotics with antagonistic activity against pathogens of wound infections: an integrative review of literature. BioMed Res Int 2019:7585486

Fordjour L, D'Souza A, Cai C, Ahmad A, Valencia G, Kumar D, Aranda JV, Beharry KD (2010) Comparative effects of probiotics, prebiotics, and synbiotics on growth factors in the large bowel in a rat model of formula-induced bowel inflammation. J Pediatr Gastroenterol Nutr 51 (4):507–513

Frykberg RG, Banks J (2015) Challenges in the treatment of chronic wounds. Adv Wound Care 4 (9):560–582

Gariboldi S, Palazzo M, Zanobbio L, Selleri S, Sommariva M, Sfondrini L, Cavicchini S, Balsari A, Rumio C (2008) Low molecular weight hyaluronic acid increases the self-defense of skin epithelium by induction of β-defensin 2 via TLR2 and TLR4. J Immunol 181(3):2103–2110

Gillor O, Etzion A, Riley MA (2008) The dual role of bacteriocins as anti-and probiotics. Appl Microbiol Biotechnol 81(4):591–606

Guéniche A, Benyacoub J, Buetler TM, Smola H, Blum S (2006) Supplementation with oral probiotic bacteria maintains cutaneous immune homeostasis after UV exposure. Eur J Dermatol 16(5):511–517

Guéniche A, Dahel K, Bastien P, Martin R, Nicolas JF, Breton L (2008) *Vitreoscilla filiformis* bacterial extract to improve the efficacy of emollient used in atopic dermatitis symptoms. J Eur Acad Dermatol Venereol 22(6):746–747

Guéniche A, Bastien P, Ovigne JM, Kermici M, Courchay G, Chevalier V, Breton L, Castiel-Higounenc I (2010) *Bifidobacterium longum* lysate, a new ingredient for reactive skin. Exp Dermatol 19(8):e1–e8

Hacini-Rachinel F, Gheit H, Le Luduec JB, Dif F, Nancey S, Kaiserlian D (2009) Oral probiotic control skin inflammation by acting on both effector and regulatory T cells. PLoS One 4(3): e4903

Hakansson A, Molin G (2011) Gut microbiota and inflammation. Nutrients 3(6):637–682

Heng NC, Haji-Ishak NS, Kalyan A, Wong AY, Lovrić M, Bridson JM, Artamonova J, Stanton JA, Wescombe PA, Burton JP, Cullinan MP (2011) Genome sequence of the bacteriocin-producing oral probiotic *Streptococcus salivarius* strain M18. Genome Announc 193:6402

Hernández-Chirlaque C, Aranda CJ, Ocón B, Capitán-Cañadas F, Ortega-González M, Carrero JJ (2016) Germ-free and antibiotic-treated mice are highly susceptible to epithelial injury in DSS colitis. J Crohn's Colitis 10(11):1324–1335

Hessle C, Andersson B, Wold AE (2000) Gram-positive bacteria are potent inducers of monocytic interleukin-12 (IL-12) while gram-negative bacteria preferentially stimulate IL-10 production. Infect Immun 68(6):3581–3586

Huseini HF, Rahimzadeh G, Fazeli MR, Mehrazma M, Salehi M (2012) Evaluation of wound healing activities of kefir products. Burns 38(5):719–723

Iordache F, Iordache C, Chifiriuc MC, Bleotu C, Pavel M, Smarandache D, Sasarman E, Laza V, Bucu M, Dracea O, Larion C (2008) Antimicrobial and immunomodulatory activity of some probiotic fractions with potential clinical application. Arch Zootech 11(3):41–51

Isenberg JS, Ridnour LA, Espey MG, Wink DA, Roberts DD (2005) Nitric oxide in wound-healing. Microsurgery 25:442–451

Izawa N, Hanamizu T, Iizuka R, Sone T, Mizukoshi H, Kimura K, Chiba K (2009) *Streptococcus thermophilus* produces exopolysaccharides including hyaluronic acid. J Biosci Bioeng 107 (2):119–123

Jebur MS (2010) Therapeutic efficacy of *Lactobacillus acidophilus* against bacterial isolates from burn wounds. N Am J Med Sci 2(12):586

Jensen JM, Förl M, Winoto-Morbach S, Seite S, Schunck M, Proksch E, Schütze S (2005) Acid and neutral sphingomyelinase, ceramide synthase, and acid ceramidase activities in cutaneous aging. Exp Dermatol 14(8):609–618

Jones ML, Martoni CJ, Prakash S (2012a) Cholesterol lowering and inhibition of sterol absorption by *Lactobacillus reuteri* NCIMB 30242: a randomized controlled trial. Eur J Clin Nutr 66 (11):1234–1241

Jones M, Ganopolsky JG, Labbé A, Gilardino M, Wahl C, Martoni C, Prakash S (2012b) Novel nitric oxide producing probiotic wound healing patch: preparation and *in vivo* analysis in a New Zealand white rabbit model of ischaemic and infected wounds. Int Wound J 9(3):330–343

Kainulainen V, Korhonen TK (2014) Dancing to another tune—adhesive moonlighting proteins in bacteria. Biology 3(1):178–204

Kamada N, Kim YG, Sham HP, Vallance BA, Puente JL, Martens EC, Núñez G (2012) Regulated virulence controls the ability of a pathogen to compete with the gut microbiota. Science 336 (6086):1325–1329

Kang DH, Fung DY (1999) Effect of diacetyl on controlling *Escherichia coli* O157:H7 and *Salmonella Typhimurium* in the presence of starter culture in a laboratory medium and during meat fermentation. J Food Prot 62(9):975–979

Kao SJ, Lei HC, Kuo CT, Chang MS, Chen BC, Chang YC, Chiu WT, Lin CH (2005) Lipoteichoic acid induces nuclear factor-κB activation and nitric oxide synthase expression via phosphatidylinositol 3-kinase, Akt, and p38 MAPK in RAW 264.7 macrophages. Immunology 115(3):366–374

Karczewski J, Troost FJ, Konings I, Dekker J, Kleerebezem M, Brummer RJ, Wells JM (2010) Regulation of human epithelial tight junction proteins by *Lactobacillus plantarum in vivo* and protective effects on the epithelial barrier. Am J Physiol Gastrointest Liver Physiol 298(6): G851–G859

Kogan G, Šoltés L, Stern R, Gemeiner P (2007) Hyaluronic acid: a natural biopolymer with a broad range of biomedical and industrial applications. Biotechnol Lett 29(1):17–25

Kreder D, Krut O, Adam-Klages S, Wiegmann K, Scherer G, Plitz T, Jensen JM, Proksch E, Steinmann J, Pfeffer K, Krönke M (1999) Impaired neutral sphingomyelinase activation and cutaneous barrier repair in FAN-deficient mice. EMBO J 18(9):2472–2479

Krutmann J (2009) Pre-and probiotics for human skin. J Dermatol Sci 54(1):1–5

Kumar M, Kumar A, Nagpal R, Mohania D, Behare P, Verma V, Kumar P, Poddar D, Aggarwal PK, Henry CJ, Jain S (2010) Cancer-preventing attributes of probiotics: an update. Int J Food Sci Nutr 61(5):473–496

Kumar P, Lee JH, Beyenal H, Lee J (2020) Fatty acids as antibiofilm and antivirulence agents. Trends Microbiol 28(9):753–768

Lai Y, Cogen AL, Radek KA, Park HJ, MacLeod DT, Leichtle A, Ryan AF, Di Nardo A, Gallo RL (2010) Activation of TLR2 by a small molecule produced by *Staphylococcus epidermidis* increases antimicrobial defense against bacterial skin infections. J Investig Dermatol 130 (9):2211–2221

Lanciotti R, Patrignani F, Bagnolini F, Guerzoni ME, Gardini F (2003) Evaluation of diacetyl antimicrobial activity against *Escherichia coli*, *Listeria monocytogenes* and *Staphylococcus aureus*. Food Microbiol 20(5):537–543

Langa S, Martín-Cabrejas I, Montiel R, Landete JM, Medina M, Arqués JL (2014) Combined antimicrobial activity of reuterin and diacetyl against foodborne pathogens. J Dairy Sci 97 (10):6116–6121

Lebeer S, Claes IJ, Vanderleyden J (2012) Anti-inflammatory potential of probiotics: lipoteichoic acid makes a difference. Trends Microbiol 20(1):5–10

Lew LC, Liong MT (2013) Bioactives from probiotics for dermal health: functions and benefits. J Appl Microbiol 114(5):1241–1253

Lew LC, Gan CY, Liong MT (2013) Dermal bioactives from *Lactobacilli* and *Bifidobacteria*. Ann Microbiol 63(3):1047–1055

Lukic J, Chen V, Strahinic I, Begovic J, Lev-Tov H, Davis SC, Tomic-Canic M, Pastar I (2017) Probiotics or pro-healers: the role of beneficial bacteria in tissue repair. Wound Repair Regen 25 (6):912–922

Mahajan B, Singh V (2014) Recent trends in probiotics and health management: a review. Int J Pharm Sci Res 5(5):1643

Martinelli M, Banderali G, Bobbio M, Civardi E, Chiara A, D'Elios S, Vecchio AL, Olivero M, Peroni D, Romano C, Stronati M (2020) Probiotics' efficacy in paediatric diseases: which is the evidence? A critical review on behalf of the Italian Society of Pediatrics. Ital J Pediatr 46(1):1–3

Matsubara M, Harada D, Manabe H, Hasegawa K (2004) *Staphylococcus aureus* peptidoglycan stimulates granulocyte macrophage colony-stimulating factor production from human epidermal keratinocytes via mitogen-activated protein kinases. FEBS Lett 566(1–3):195–200

McFarland LV (2011) *Lactobacillus* GG prevented nosocomial gastrointestinal and respiratory tract infections. Arch Dis Child Educ Pract Ed 96(6):238

Menendez A, Finlay BB (2007) Defensins in the immunology of bacterial infections. Curr Opin Immunol 19(4):385–391

Menke NB, Ward KR, Witten TM, Bonchev DG, Diegelmann RF (2007) Impaired wound healing. Clin Dermatol 25(1):19–25

Miller LS, Cho JS (2011) Immunity against *Staphylococcus aureus* cutaneous infections. Nat Rev Immunol 11(8):505–518

Minami M, Ohmori D, Tatsuno I, Isaka M, Kawamura Y, Ohta M, Hasegawa T (2009) The streptococcal inhibitor of complement (SIC) protects *Streptococcus pyogenes* from bacteriocin-like inhibitory substance (BLIS) from *Streptococcus salivarius*. FEMS Microbiol Lett 298(1):67–73

Mizutani Y, Mitsutake S, Tsuji K, Kihara A, Igarashi Y (2009) Ceramide biosynthesis in keratinocyte and its role in skin function. Biochimie 91(6):784–790

Mohammed SW, Cruickshank S, McBain AJ, O'Neill CA (2015) *Lactobacillus rhamnosus* GG lysate increases re-epithelialization of keratinocyte scratch assays by promoting migration. Sci Rep 5:16147

Mohammedsaeed W, McBain AJ, Cruickshank SM, O'Neill CA (2014) *Lactobacillus rhamnosus* GG inhibits the toxic effects of *Staphylococcus aureus* on epidermal keratinocytes. Appl Environ Microbiol 80(18):5773–5781

Mokoena MP (2017) Lactic acid bacteria and their bacteriocins: classification, biosynthesis and applications against uropathogens: a mini-review. Molecules 22(8):1255

Murata Y, Ogata J, Higaki Y, Kawashima M, Yada Y, Higuchi K, Tsuchiya T, Kawaminami S, Imokawa G (1996) Abnormal expression of sphingomyelin acylase in atopic dermatitis: an etiologic factor for ceramide deficiency? J Investig Dermatol 106(6):1242–1249

Nagoba B, Wadher B, Kulkarni P, Kolhe S (2008) Acetic acid treatment of pseudomonal wound infections. Eur J Gen Med 5(2):104–106

Nasrabadi MH, Aboutalebi H, Ebrahimi MT, Zahedi F (2011) The healing effect of *Lactobacillus plantarum* isolated from Iranian traditional cheese on gastric ulcer in rats. Afr J Pharm Pharmacol 5(12):1446–1451

Niebuhr M, Baumert K, Werfel T (2010) TLR-2-mediated cytokine and chemokine secretion in human keratinocytes. Exp Dermatol 19(10):873–877

Oh S, Kim SH, Ko Y, Sim JH, Kim KS, Lee SH, Park S, Kim YJ (2006) Effect of bacteriocin produced by *Lactococcus* sp. HY 449 on skin-inflammatory bacteria. Food Chem Toxicol 44 (4):552–559

Ohland CL, MacNaughton WK (2010) Probiotic bacteria and intestinal epithelial barrier function. Am J Physiol Gastrointest Liver Physiol 298(6):G807–G819

Paradis-Bleau C, Cloutier I, Lemieux L, Sanschagrin F, Laroche J, Auger M, Garnier A, Levesque RC (2007) Peptidoglycan lytic activity of the *Pseudomonas aeruginosa* phage φKZ gp144 lytic transglycosylase. FEMS Microbiol Lett 266(2):201–209

Pasricha A, Bhalla P, Sharma KB (1979) Evaluation of lactic acid as an antibacterial agent. Indian J Dermatol Venereol Leprol 45(3):159–161

Pastar I, Nusbaum AG, Gil J, Patel SB, Chen J, Valdes J, Stojadinovic O, Plano LR, Tomic-Canic M, Davis SC (2013) Interactions of methicillin resistant *Staphylococcus aureus* USA300 and *Pseudomonas aeruginosa* in polymicrobial wound infection. PLoS One 8(2):e56846

Pastar I, Stojadinovic O, Yin NC, Ramirez H, Nusbaum AG, Sawaya A, Patel SB, Khalid L, Isseroff RR, Tomic-Canic M (2014) Epithelialization in wound healing: a comprehensive review. Adv Wound Care 3(7):445–464

Patel RM, Denning PW (2013) Therapeutic use of prebiotics, probiotics, and postbiotics to prevent necrotizing enterocolitis: what is the current evidence? Clin Perinatol 40(1):11–25

Peral MC, Huaman Martinez MA, Valdez JC (2009) Bacteriotherapy with *Lactobacillus plantarum* in burns. Int Wound J 6(1):73–81

Peral MC, Rachid MM, Gobbato NM, Martinez MH, Valdez JC (2010) Interleukin-8 production by polymorphonuclear leukocytes from patients with chronic infected leg ulcers treated with *Lactobacillus plantarum*. Clin Microbiol Infect 16(3):281–286

Piqué N, Berlanga M, Miñana-Galbis D (2019) Health benefits of heat-killed (Tyndallized) probiotics: an overview. Int J Mol Sci 20(10):2534

Rad AH, Sahhaf F, Hassanalilou T, Ejtahed HS, Motayagheni N, Soroush AR, Javadi M, Mortazavian AM, Khalili L (2016) Diabetes management by probiotics: current knowledge and future perspectives. Curr Diab Rev 86:215–217

Ramos AN, Sesto Cabral ME, Noseda D, Bosch A, Yantorno OM, Valdez JC (2012) Antipathogenic properties of *Lactobacillus plantarum* on *Pseudomonas aeruginosa*: the potential use of its supernatants in the treatment of infected chronic wounds. Wound Repair Regen 20 (4):552–562

Rather IA, Bajpai VK, Kumar S, Lim J, Paek WK, Park YH (2016) Probiotics and atopic dermatitis: an overview. Front Microbiol 7:507

Reid G, Bruce AW, Fraser N, Heinemann C, Owen J, Henning B (2001) Oral probiotics can resolve urogenital infections. FEMS Immunol Med Microbiol 30(1):49–52

Reid G, Kim SO, Köhler GA (2006) Selecting, testing and understanding probiotic microorganisms. FEMS Immunol Med Microbiol 46(2):149–157

Rezac S, Kok CR, Heermann M, Hutkins R (2018) Fermented foods as a dietary source of live organisms. Front Microbiol 9:1785

Ringel-Kulka T, Palsson OS, Maier D, Carroll I, Galanko JA, Leyer G, Ringel Y (2011) Probiotic bacteria *Lactobacillus acidophilus* NCFM and *Bifidobacterium lactis* Bi-07 versus placebo for the symptoms of bloating in patients with functional bowel disorders: a double-blind study. J Clin Gastroenterol 45(6):518–525

Rodrigues L, Van der Mei HC, Teixeira J, Oliveira R (2004) Influence of biosurfactants from probiotic bacteria on formation of biofilms on voice prostheses. Appl Environ Microbiol 70 (7):4408–4410

Rodrigues KL, Caputo LR, Carvalho JC, Evangelista J, Schneedorf JM (2005) Antimicrobial and healing activity of kefir and kefiran extract. Int J Antimicrob Agents 25(5):404–408

Romanovsky AA (2014) Skin temperature: its role in thermoregulation. Acta Physiol 210 (3):498–507

Rutherford ST, Bassler BL (2012) Bacterial quorum sensing: its role in virulence and possibilities for its control. Cold Spring Harb Perspect Med 2(11):a012427

Sakr A, Brégeon F, Mège JL, Rolain JM, Blin O (2018) *Staphylococcus aureus* nasal colonization: an update on mechanisms, epidemiology, risk factors, and subsequent infections. Front Microbiol 9:2419

Schaefer L, Auchtung TA, Hermans KE, Whitehead D, Borhan B, Britton RA (2010) The antimicrobial compound reuterin (3-hydroxypropionaldehyde) induces oxidative stress via interaction with thiol groups. Microbiology 156(6):1589–1599

Shiedlin A, Bigelow R, Christopher W, Arbabi S, Yang L, Maier RV, Wainwright N, Childs A, Miller RJ (2004) Evaluation of hyaluronan from different sources: *Streptococcus zooepidemicus*, rooster comb, bovine vitreous, and human umbilical cord. Biomacromolecules 5(6):2122–2127

Simmering R, Breves R (2009) Pre-and probiotic cosmetics. Hautarzt 60(10):809–814

Slotte JP (2013) Biological functions of sphingomyelins. Prog Lipid Res 52(4):424–437

Smith WP (1996) Epidermal and dermal effects of topical lactic acid. J Am Acad Dermatol 35 (3):388–391

Soleymani T, Lanoue J, Rahman Z (2018) A practical approach to chemical peels: a review of fundamentals and step-by-step algorithmic protocol for treatment. J Clin Aesthet Dermatol 11 (8):21

Soo I, Madsen KL, Tejpar Q, Sydora BC, Sherbaniuk R, Cinque B, Di Marzio L, Cifone MG, Desimone C, Fedorak RN (2008) VSL# 3 probiotic upregulates intestinal mucosal alkaline sphingomyelinase and reduces inflammation. Can J Gastroenterol 22(3):237–242

Sørensen OE, Thapa DR, Rosenthal A, Liu L, Roberts AA, Ganz T (2005) Differential regulation of β-defensin expression in human skin by microbial stimuli. J Immunol 174(8):4870–4879

Spigelman M, Ross M (2008) Method of using topical probiotics for the inhibition of surface contamination by a pathogenic microorganisms and composition therefor. US Patent 0107699 A1, 8 May 2008

Sugiura A, Nomura T, Mizuno A, Imokawa G (2014) Reevaluation of the non-lesional dry skin in atopic dermatitis by acute barrier disruption: an abnormal permeability barrier homeostasis with defective processing to generate ceramide. Arch Dermatol Res 306(5):427–440

Sullivan M, Schnittger SF, Mammone T, Goyarts EC (2009) Skin treatment method with Lacto-bacillus extract. US Patent 7,510,734 B2, 31 Mar 2009

Sultana R, McBain AJ, O'Neill CA (2013) Lysates of *Lactobacillus* and *Bifidobacterium* augment tight junction barrier function in human primary epidermal keratinocytes in a strain-dependent manner. Appl Environ Microbiol 79:4887–4894

Sumikawa Y, Asada H, Hoshino K, Azukizawa H, Katayama I, Akira S, Itami S (2006) Induction of β-defensin 3 in keratinocytes stimulated by bacterial lipopeptides through toll-like receptor 2. Microbes Infect 8(6):1513–1521

Tang SC, Yang JH (2018) Dual effects of alpha-hydroxy acids on the skin. Molecules 23(4):863

Taylor KR, Trowbridge JM, Rudisill JA, Termeer CC, Simon JC, Gallo RL (2004) Hyaluronan fragments stimulate endothelial recognition of injury through TLR4. J Biol Chem 279 (17):17079–17084

Teodorescu R (1999) A natural eubiotic product for maintenance and treatment of teguments. WO Patent 007332, 18 Feb 1999

Thursby E, Juge N (2017) Introduction to the human gut microbiota. Biochem J 474 (11):1823–1836

Trabucchi E, Pallotta S, Morini M, Corsi F, Franceschini R, Casiraghi A, Pravettoni A, Foschi D, Minghetti P (2002) Low molecular weight hyaluronic acid prevents oxygen free radical damage to granulation tissue during wound healing. Int J Tissue React 24(2):65–71

Tsilingiri K, Barbosa T, Penna G, Caprioli F, Sonzogni A, Viale G, Rescigno M (2012) Probiotic and postbiotic activity in health and disease: comparison on a novel polarised ex-vivo organ culture model. Gut 61(7):1007–1015

Tsiouris CG, Tsiouri MG (2017) Human microflora, probiotics and wound healing. Wound Med 19:33–38

Tsiouris CG, Kelesi M, Vasilopoulos G, Kalemikerakis I, Papageorgiou EG (2017) The efficacy of probiotics as pharmacological treatment of cutaneous wounds: meta-analysis of animal studies. Eur J Pharm Sci 104:230–239

Valdez JC, Peral MC, Rachid M, Santana M, Perdigon G (2005) Interference of *Lactobacillus plantarum* with *Pseudomonas aeruginosa* in vitro and in infected burns: the potential use of probiotics in wound treatment. Clin Microbiol Infect 11(6):472–479

Vesterlund S, Paltta J, Lauková A, Karp M, Ouwehand AC (2004) Rapid screening method for the detection of antimicrobial substances. J Microbiol Methods 57(1):23–31

Villéger R, Saad N, Grenier K, Falourd X, Foucat L, Urdaci MC, Bressollier P, Ouk TS (2014) Characterization of lipoteichoic acid structures from three probiotic *Bacillus* strains: involvement of D-alanine in their biological activity. Antonie Van Leeuwenhoek 106(4):693–706

Volz T, Skabytska Y, Guenova E, Chen KM, Frick JS, Kirschning CJ, Kaesler S, Röcken M, Biedermann T (2014) Nonpathogenic bacteria alleviating atopic dermatitis inflammation induce IL-10-producing dendritic cells and regulatory Tr1 cells. J Investig Dermatol 134(1):96–104

Wang Y, Kirpich I, Liu Y, Ma Z, Barve S, McClain CJ, Feng W (2011) *Lactobacillus rhamnosus* GG treatment potentiates intestinal hypoxia-inducible factor, promotes intestinal integrity and ameliorates alcohol-induced liver injury. Am J Pathol 179(6):2866–2875

Wee YJ, Kim JN, Ryu HW (2006) Biotechnological production of lactic acid and its recent applications. Food Technol Biotechnol 44(2):163–172

Weindl G, Schaller M, Schäfer-Korting M, Korting HC (2004) Hyaluronic acid in the treatment and prevention of skin diseases: molecular biological, pharmaceutical and clinical aspects. Skin Pharmacol Physiol 17(5):207–213

Wong VW, Martindale RG, Longaker MT, Gurtner GC (2013) From germ theory to germ therapy: skin microbiota, chronic wounds, and probiotics. Plast Reconstr Surg 132(5):854e–861e

Yamamoto Y, Uede K, Yonei N, Kishioka A, Ohtani T, Furukawa F (2006) Effects of alpha-hydroxy acids on the human skin of Japanese subjects: the rationale for chemical peeling. J Dermatol 33(1):16–22

Yang F, Wang A, Zeng X, Hou C, Liu H, Qiao S (2015) *Lactobacillus reuteri* I5007 modulates tight junction protein expression in IPEC-J2 cells with LPS stimulation and in newborn piglets under normal conditions. BMC Microbiol 15(1):1–11

Yeo SK, Liong MT (2010) Effect of prebiotics on viability and growth characteristics of probiotics in soymilk. J Sci Food Agric 90(2):267–275

Yuki T, Yoshida H, Akazawa Y, Komiya A, Sugiyama Y, Inoue S (2011) Activation of TLR2 enhances tight junction barrier in epidermal keratinocytes. J Immunol 187(6):3230–3237

Development of Novel Anti-infective Formulations for Wound Disinfection

Regalin Rout

1 Introduction

A wound infection is the localization of pathogenic microorganisms which have invaded into viable tissue surrounding the wound. This leads to inflammation and damage of tissues as well as delays the healing process. Though there are some infections which are self-contained and heal on their own, some infections if not treated can be life-threatening and need medical attention (White et al. 2006; Cutting and White 2005). Nowadays, wound care has acquired global attention in the healthcare sector to improve the research, and to fasten the healing process, the wound needs to be protected from infection from microorganisms (Maxson et al. 2012). Basically wound healing entails a three-step process according to some authors such as inflammation, proliferation, and maturation/remodeling, whereas nowadays it is described as a four-step process such as hemostasis, inflammation, proliferation, and maturation (Broughton et al. 2006).

The functional aim of these processes is to heal the wound through different stages such as prevention of loss of blood, cleaning the area of infection, and finally repairing the wound site. This hemostasis phase is included in the inflammation process in the three-phase approach. These processes are programmed and time bound to heal the wound successfully and appear in a programmed manner. Certain factors such as oxygenation, infection, age and sex hormones, stress or anxiety, diabetes, obesity, medications, alcohol, smoking, and nutrition obstruct the process of wound healing, which demands further research to improve the healing process as well as protecting the wound before it gets fatal. More research on information relating to these factors could lead to improvement of healing of wounds. The different stages of wound healing are described as follows and in Fig. 1.

R. Rout (✉)
School of Biotechnology, KIIT University, Bhubaneswar, Odisha, India

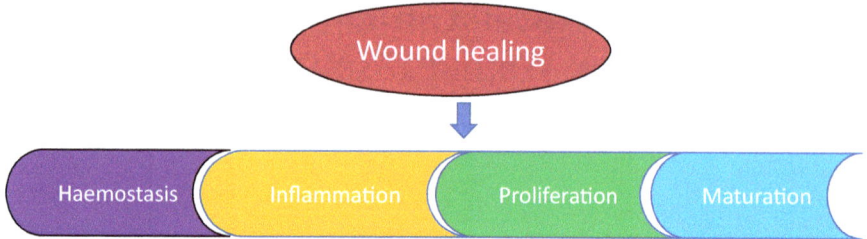

Fig. 1 Sequential order of different stages of wound healing

1.1 Hemostasis

Hemostasis begins when an injury occurs in our body. It is the first and foremost stage of healing process. It starts with vascular constriction followed by blood clotting. When the blood vessel shrinks, it eventually slows down the blood flow platelets and adheres together to close the gap in the blood vessels. Then with the process of clotting the platelet binds with threads of fibrin. Then the platelets attach to the sub-endothelium surface of the ruptured blood vessel's epithelial wall followed by the adherence of the strands of fibrin. Once the fibrin mesh starts, the blood is converted to a gel form with the release of the protein prothrombin. The release of prothrombin initiates clot formation which retains the platelets and blood cells in the site of wound. Though the clot or thrombus is an important factor in the healing stages of wound, sometimes it may lead to strokes or heart attacks if it enters through the circulatory system after detaching from the vessel walls.

1.2 Inflammation

Hemostasis which is immediately followed by inflammation is the second stage that happens in wound healing. It begins right after the bleeding is controlled, and the inflammatory cells are migrated to wound infection area (Gosain and DiPietro 2004; Broughton et al. 2006; Campos et al. 2008). This phase is basically characterized by removal of damaged cells or debris, pathogens, and bacteria from the site of infection. During the stage of healing process, the neutrophils, growth factors, and enzymes cause swelling with redness and pain at the wound site. Though inflammation is a natural process of wound recovery, it could be challenging if it is prolonged or excessive. This happens mostly in immunocompromised individuals (Midwood et al. 2004). Therefore, reduction of this phase is often the main goal in therapeutics. As this induces proliferative phase of healing, it is a necessary part in the complete process.

1.3 Proliferation

The proliferative phase which is also known as rebuilding phase is the third step to wound healing and overlaps with the inflammatory phase. The wound is remodeled with novel tissues that are made up of collagen and extracellular matrix when it comes to the step of contraction. In this stage, the macrophage recruits fibroblasts which create collagen fibers. In the presence of adequate oxygen and vitamin C, granulation of tissue forms at the site of the wound bed. Oxygen along with two amino acids proline and lysine is required for the collagen synthesis, whereas vitamin C is needed for the hydroxylation of proline to hydroxyproline, an amino acid present in collagen. Fibroblasts form a collagen bed during the process of granulation to fill the defect and produce new capillaries. The edges of the wound are pulled together to reduce the size of infection. Finally, the process of epithelialization occurs in which the epithelial cells enter the site of injury and resurface it to provide protection from the outer environment. This is a fast process when wounds are kept moist and hydrated. This humidity can be retained if the wounds are dressed within 48 hours after the injury to optimize the final process of proliferation, i.e., epithelialization (Broughton et al. 2006; Campos et al.).

1.4 Maturation

Maturation which is also called as remodeling stage occurs when the wound is completely healed. It is also the concluding phase of wound healing. This phase includes transformation of collagen from type III to type II. The phase starts to work in the third stage of infection and continues up to 2 years (Mancl et al. 2013). In this stage, the cells that are involved in repairing the wound are eliminated by the process of apoptosis. Collagen fibers are aligned along the tension lines and cross link occurs. This reduces the thickness of the scar and makes the skin stronger than before. The final resulting scar will never have the original strength of the wound, and only about 80% of the tensile strength (Broughton et al. 2006; Campos et al. 2008; Midwood et al. 2004).

In the abovementioned phases, it is found that wound healing is a much complex and fragile process. Certain exogenous and endogenous factors can regulate such events as well as influence the process of healing. It can also be affected by some systemic disorders, such as diabetes, immunosuppression, venous disease, and metabolic deficiencies in the elderly. Apart from this, smoking and cortico-therapy also hinder the process of early healing of wound.

It is already found that wounds are also susceptible to certain pathogenic bacteria such as *Pseudomonas aeruginosa* (Gram-positive) and *Staphylococcus aureus* (Gram-positive) which colonize in skin wounds, thereby producing biofilms. Once the biofilms are produced, bacterial cells are aggregated and immobilized in an adhesive matrix of extracellular polymeric substances (Kumar et al. 2013, 2020;

Wolcott et al. 2010). This leads to the weak penetration of antibiotics and subsequently makes the elimination of bacteria difficult due to the host clearance mechanisms, i.e., antibodies and phagocytes through the microbial biofilm (Kumar et al. 2013, 2020; Breidenstein et al. 2011; Mah et al. 2003; Taylor et al. 2014; Ovington 2003). In addition, toxins produced from bacteria lead to an excessive, detrimental inflammatory response such as development of antibiotic resistance and delayed wound healing followed by prolonged hospitalization (Rajpaul 2015; Wilson 2003). Therefore, wound infections and its healing have emerged as a big cause of death and burden toward the healthcare system (Kish et al. 2010; Hannigan et al. 2015; Howell-Jones et al. 2005; Rajanbabu and Chen 2011). Based on the above scenario, different anti-infective therapies and formulations were suggested which will be described in this chapter.

2 Anti-Infective Formulations for Wound Disinfection

2.1 Antimicrobial Therapy for Treating Chronic Wounds

Due to the emergence of multidrug resistance, treatment of wound infection leads to the invention of different anti-infective formulations. Among those antimicrobial peptides (AMPs) are short-amino-acid chains that are involved in resolving the wound infections (Kalia et al. 2014; Rajanbabu and Chen 2011; Zasloff 2002). They also play a vital role in defending against pathogenic microorganisms and also take part in modulation of immune response. They are believed to be evolutionarily conserved in all living kingdoms from prokaryotes to eukaryotes (Giuliani et al. 2007). Based on their secondary structure, they are divided into beta-sheet, alpha-helical loop, and extended peptides. A lot of research has already been done on beta-sheet and alpha-helical groups as they are the two most frequently occurring groups in nature. Each AMPs consist of 100 amino acid residues with a cationic charge from +2 to +9, lysine, arginine, and hydrophobic residues (Martin et al. 1995: Huang et al. 2010; Dorica-Mirela and Ionel 2009; Huang et al. 2011; Halstead et al. 2015). It has a broad spectrum of antimicrobial activity toward the microbes that also include drug resistant bacteria which is not hindered by any biological fluids, exudates, or biofilms. AMPs act faster at numerous sites of the wound infected area reducing the chance of resistance as well as any further microbial infection (Gottrup et al. 2014; Lai and Gallo 2009; Felgueiras and Amorim 2017a, b; Malmsten et al. 2007).

In the process of wound healing, biomolecules are invaded at the site of infection and induce the consecutive phases of healing. Several antimicrobial agents including AMPs being part of our immunity system provide protection against foreign invaders. Once the infection appears, proteases are cleaved at specific sites to release heparin-binding epidermal growth factors (HB-EGF) and amphiregulin. These HB-EGF induce the expression of epidermal AMPs to fasten wound healing due to the presence of their antimicrobial properties (Pasupuleti et al. 2007). In the hemostasis phase, the cleavage of proteins like fibrinogen or thrombin occurs via the complement and coagulation cascades which leads to production of many AMPs with antimicrobial activity from the fragments of those proteins (Felgueiras and Amorim 2017a, b).

During the inflammatory phase, neutrophils entered the site of wound infection which are known to be one of the most significant producers of AMPs. Monocytes and lymphocytes enter the site of infection once the neutrophils are already in. Neutrophils also contain α-defensins which are known as human neutrophil peptides (HNPs) in azurophilic granules, cathelicidins when stored in specific granules, and calgranulins in the cytosol (Borregaard and Cowland 1997; Ganz 1987). Despite the fact that defensins are commonly used in neutrophil phagolysosome because of their antiviral and antibacterial activity, they also promote bacterial phagocytosis. Besides phagocytosis, macrophages also play an important role in inflammation and chemotactic activity toward monocytes, T cells, and immature dendritic cells (Ganz et al. 1985; Edwards and Harding 2004). Like other AMPs, cathelicidins also possess a broad spectrum of antimicrobial activity against bacteria, fungi, and viruses. They are trimmed by proteinase 3 into the AMP, hCAP-18, once they are released from the granules (Rodríguez-Martínez et al. 2008). This hCAP-18 is also known as LL37. These cathelicidins are liable for the expression of the vascular endothelial growth factor (VEGF) and by producing monocytes it also transactivates the epidermal growth factor (EGFR). This leads to the migration of keratinocytes to the site of wound infection (Tokumaru et al. 2000; Sørensen 2016).

In the proliferation stage, it is found that the epidermal keratinocytes, like hBD-2, hBD-3, RNase7, and psoriasin, possess most of the AMPs that are involved in healing. During this phase, the LL37 and calgranulins such as S100A8 (calgranulin A)/S100A9 (calgranulin B) reach their peak of expression. It is found that many defensins belong to the same ancestral gene due to which the neutrophils (inflammation) and keratinocytes (proliferation) share the same AMPs and antimicrobial proteins. The expression of these defensins is basically dependent on the phases of wound healing (Roupé et al. 2010; Nurjadi et al. 2012). Upon a cut or infection, the EGFR activation in epidermal keratinocytes induces the expression of beta defensin-3 (hBD-3). Unlike beta defensins the expression of calgranulins, S100A8, or S100A9 is induced by the activation of growth factors or by pro-inflammatory cytokines. This leads to linking growth and tissue regeneration with expression by AMPs (Nurjadi et al. 2012; Steinbakk et al. 1990). Certain epidermal AMPs such as nBD-3 and RNase7 showed antibacterial activity against *Staphylococcus aureus*, whereas psoriasin is effective against *Escherichia coli* and calgranulins against the strain *Candida albicans* (Gläser et al. 2005; Zanger et al. 2010; Niyonsaba et al. 2007). This shows that epidermal AMPs have a broad spectrum of antibiotic activity that helps in wound healing. Apart from protecting the wound from foreign attackers, the AMPs are also involved in other non-antimicrobial functions.

It is also found that hBD-2, hBD3, and psoriasin have chemoattractant properties. The former activates dendritic cells via TLR-4 because of its chemoattractant properties toward immature dendritic cells and memory T cells. This is also involved in inducing the expression of keratinocytes as well as in its proliferation, migration, and cytokine production (Röhrl et al. 2010). hBD-3 takes part in activating the mast cells and increases its vascular permeability. This also induces the second-level expression of keratinocyte differentiation markers, thereby simultaneously promoting the proliferation of endothelial cells (Roupé et al. 2010; Abdillahi et al. 2012). During the remodeling phase, the highly antimicrobial collagen type VI gives protection to the connective tissue of the skin (Ye et al. 2018.).

2.2 Antibacterial Agents in Wound Dressings

2.2.1 Antibiotics

Though it is found that many bacteriostatic or bactericidal antibiotics contribute in wound healing, their influence on wound repair is still unnoticed. Among several effective antibiotics against microorganisms, only quinolones, aminoglycosides, tetracyclines, and cephalosporins have been proved to be useful as antimicrobial wound dressings for effective wound healing which is later described in Table 1.

The above table is precise information regarding the method of action of the antibiotics involved in wound dressings as it is already known that antibiotics interfere with certain bacterial functions and metabolic pathways. They inhibit some of the bacterial pathways such as inhibition of synthesis of bacterial cell wall and interfere with the synthesis of proteins. They also obstruct in blockage of certain metabolic pathways and inhibit the synthesis of nucleic acids (Kohanski et al. 2010; Rai et al. 2016). Though antibiotics have already proved to be useful for the wound treatment, their irregular usage may lead to antibiotic resistance (Kalia et al. 2014; Pîrvănescu et al. 2014). Researchers already found that some microorganisms like *S. aureus* and *P. aeruginosa* considerably developed resistance against antibiotics (Friedman et al. 2016). It is already found that 70% of bacteria causing wound infections are resilient to some regularly used antibiotics (Chávez-González et al. 2016). This leads to an urgent demand for finding different substitutes which can be used for wound healing such as essential oils (Shrestha et al. 2014) and nanoparticles (Pîrvănescu et al. 2014).

2.2.2 Natural Antimicrobials for Wound Infections

Because of the increasing antibiotic resistance against contagious strains, researchers are now focused on the huge collection of bioresources. Most of them in the collection are naturally available herbs. There are certain natural agents such as garlic, curcumin, ginger, clove, goldenseal, and oregano with antimicrobial properties. It has also been reported that some wounds have complex polymicrobial infections that are caused due to activities of certain natural agents (Seow et al.

Table 1 List of antibiotics contained within wound dressings

Antibiotic	Material used during wound dressing	Susceptible species	References
Amoxicillin (class: quinolones)	Sponges collected from bacterial cellulose	*Escherichia coli, Candida albicans, Staphylococcus aureus*	Liu et al. (2018)
Ciprofloxacin (class: quinolones)	Calcium alginate films	*E. coli, S. aureus, Pseudomonas aeruginosa*	Contardi et al. (2017)
	Films and nanofiber mats of Povidone	*E. coli, Bacillus subtilis*	Li et al. (2017)
	Electrospun fibers based on thermoresponsive polymer poly(*N*-isopropylacrylamide), poly(l–lactic acid–co-ε-caprolactone)	*E. coli, S. aureus*	Pamfil et al. (2017)
	Hydrogels from 2-hydroxyethyl methacrylate/citraconic anhydride-modified collagen	*S. aureus*	Anjum et al. (2016)
Tetracycline (class: tetracycline)	Cotton fabric enclosed with chitosan-poly(vinyl pyrrolidone)–PEG	*E. coli, S. aureus*	Khampieng et al. (2014)
Doxycycline (class: tetracycline)	Poly(acrylic acid) nanofiber mats	*S. aureus, Strepto-coccus agalactiae*	Michalska-Sionkowska et al. (2018)
Gentamicin (class: aminoglycosides)	Thin films made from collagen, chitosan and hyaluronic acid	*E. coli, S. aureus, P. aeruginosa*	Ahire et al. (2017)
	Sodium carboxymethyl cellulose loaded with antibiofilm agents (xylitol and ethylenediaminetetraacetic acid)	*S. aureus, Bacillus subtilis, P. aeruginosa, E. coli*	Khampieng et al. (2014)
Kanamycin (class: aminoglycosides)	Nanofibers made with a combination of polyethylene oxide and hyaluronic acid	*Listeria monocytogenes, P. aeruginosa*	Rădulescu et al. (2016)
Cefuroxime and cefepime (class: cephalosporins)	Biocompatible nanostructured composite based on naturally derived biopolymers (chitin and sodium alginate)	*E. coli, S. aureus*	Etebu and Arikekpar (2016)

2014). The antimicrobial properties of essential oils and honey have already been confirmed by studies.

2.2.3 Essential Oils

Essential oils also abbreviated as "EOs" are plant secondary metabolites and are a derivative product of leaves, twigs, seeds, barks, and roots of plants. These isolated metabolites from plants consist of wound repairing properties such as antioxidant, anti-inflammatory, anti-allergic, antiviral, antimicrobial, and regenerative (Agyare et al. 2016). The antimicrobial properties of plant-based metabolite for wound healing lie in different chemical constituents such as cinnamaldehyde, thymol, geraniol, menthol, and carvacrol (Semeniuc et al. 2017; Scagnelli 2016; Aumeeruddy-Elalfi and Mahomoodally 2016). The presence of these constituents is based on their isolation processes such as hydro-distillation, microwave-assisted extraction, microwave-generated hydro-distillation, steam distillation, microwave steam diffusion, and ultrasound-assisted extraction (Aumeeruddy-Elalfi et al. 2016; Kavoosi et al. 2013). Numerous publications explained that antimicrobial properties of essential oils are mostly dependent on their constituents. The constituents found are mostly thymol, cinnamaldehyde, geraniol, and carvacrol. EOs attack lipids and phospholipids of the cell membrane (Altiok et al. 2010). In the bacterial cell wall, EO is involved in the cytoplasm outflow and decrease in pH, whereas its role in impairment of certain cellular processes such as ATP biosynthesis, DNA transcription, and protein synthesis is also seen. The role of interference of EOs in the cytoplasmic membrane is also described by other groups. This occurs by interfering with the transportation of nutrients through the cell membrane and coagulation of bacterial cell matter (Walsh et al. 2003).

After so many in vitro tests, usage of EO toward multidrug resilient microorganisms proved to be beneficial because of its no or very minute effects on the resistance development in comparison to antibiotics (Sienkiewicz et al. 2014; Zenati et al. 2014; Liakos et al. 2014). EOs of other plant-based products such as thyme, peppermint, lavender, cinnamon, rosemary, eucalyptus, and lemon grass have been found to possess antimicrobial properties. Over and above their antibiotic and antiseptic properties, essential oils have antimicrobial properties as they been used in wound dressings (Liakos et al. 2015; Nogueira et al. 2014). Sometimes during the treatment of wound infections, EOs may be required in high concentrations or a repetitive application which may arise in some adverse effects (Table 2).

Helichrysum Essential Oil

Helichrysum essential oil is useful in healing wound infections because of its anti-inflammatory, antifungal, and antibacterial properties. It contains arzanol a substance that plays an important role in healing because of its anti-inflammatory properties (Popoola et al. 2015). Helichrysum oil helps to prevent scarring by regenerating new cells because of its antioxidant properties (Ammon 2006). This oil increases collagen production and thereby reduces skin rashes and infections. This oil may cause sunburn as it is sensitive to sun. It is not advisable to use in case of any recent surgery or internal hemorrhaging.

Table 2 Description about essential oils that play a major role in healing scars

Essential Oils	Effective time period	Benefits	References
Helichrysum essential oil	More than 3 months	Anti-inflammatory, antifungal, and antibacterial properties	Popoola et al. (2015), Ammon (2006)
Frankincense essential oil	1 month	Improves skin tone and kills bacteria	Boukhatem et al. (2013)
Geranium essential oil	1 month or more	Antibacterial and anti-inflammatory properties	Ferdaous et al. (2016)
Lavender essential oil	1 week	Antibiotic, antioxidant, and antiseptic properties	Mori et al. (2016), Bai et al. (2014), Imane et al. (2017), Kwakman et al. (2010)
Carrot seed essential oil	1 month	Antibacterial and antifungal properties	Shinde et al. (1999)
Cedar wood essential oil	1 month	Anti-inflammatory and analgesic properties; may treat acne	Lauren et al. (2016), Marino et al. (2001)
Hyssop essential oil	3 months or more	Heals wounds, prevents infections, and reduces the appearance of wrinkles and acne scars	Fatemeh et al. (2011), Pariya et al. (2015)
Vitamin E oil	3 weeks or more	Boosts collagen production	Nesrine et al. (2013)
Tea tree oil	1 month	Antiviral, antibacterial, and antifungal nature	Appendino et al. (2007), Edmondson et al. (2011), Evandri et al. (2005)
Neroli essential oil	1 month	Reduces pain and inflammation	Ammar et al. (2012), Valerón-Almazán et al. (2015)
Rosehip seed oil	6 weeks	Improves wrinkles and acne	Kim et al. (2017)
Coconut oil	10 days	Softens the skin and reduces inflammation	Varma et al. (2019), Nevin and Rajamohan (2010)
Almond oil	1 month	Comprises vitamin E; soothes and moisturizes the skin	Ahmad (2010)

Frankincense Essential Oil

Frankincense essential oil is also known to be beneficial for the skin by soothing the skin and evening out the complexion. It is also useful for the treatment of scars as it kills microorganisms. This oil enhances the growth of new skin cells, tightens the skin, and diminishes the scars. Because of the presence of the active ingredient boswellic acid, it shows anti-inflammatory benefits (Boukhatem et al. 2013). This essential oil sometimes causes skin irritation and abdominal problems. This oil is not recommended to use if you take blood thinners.

Geranium Essential Oil

Geranium oil is used for the treatment of skin burns or scars by soothing inflammation and improving the skin tone. It also stimulates new cell growth and helps in making the scars less visible due to its antiseptic, antimicrobial, and antibacterial nature. It is found that rose geranium oil is very effective for the prevention and treatment of inflammatory skin conditions which can be visible within a month (Mori et al. 2016). Precautions should be taken in case of patients having high blood pressure or cardiovascular diseases.

Lavender Essential Oil

Lavender commonly known as *Lavandula angustifolia* is a well-known traditional medicine used all over the world. It inhibits the growth of infection caused by microorganisms because of its antibiotic, antioxidant, and antiseptic properties (Imane et al. 2017). The antibacterial activity of lavender oil is mainly due to the presence of two components, i.e., linalool and linalyl. Their EO contains excess amounts of constituents like linalool, linalyl acetate, 1,8-cineole, and camphor. In vitro studies with lavender EO showed strong antibacterial activity and inhibition of the growth of microbial strains such as *E. coli*, *S. aureus*, and *P. aeruginosa* (Kwakman et al. 2010). For surface disinfection, lavender EO was used in the form of a prophylactic or topical application. This oil encouraged wound healing in the primary phase which includes stimulation of collagen production, differentiation of fibroblasts, and also fastening the establishment of granulation tissue (Bai et al. 2014).

This oil helps to prevent scars by promoting cell and tissue growth. It was found in 2016 that this oil shows potential in wound healing as well as is a natural remedy for repairing damaged skin tissues (Mori et al. 2016). It also increases collagen production and regeneration of tissue, thereby encouraging wound healing or repairing especially in the early phase of treatment. It fastens wound healing by shrinking its size by topical application.

Carrot Seed Essential Oil

Carrot seed essential oil is also having beneficial effects toward wound scars. This is due to its antibacterial and antifungal properties. It is also useful in treating older scars and the results can be visible in a month. This is an inexpensive oil to be used for wound healing (Shinde et al. 1999).

Cedar Wood Essential Oil

Cedar wood oil has the potential in treating the skin scars and acne. It has the property to lower inflammation and helps in pain relieving. The results can be seen after a month of consistent use. It also causes allergic reactions.

Hyssop Essential Oil

Hyssop essential oil assists in repairing damaged skin like the abovementioned oils. It also possesses antiseptic, antifungal, antimicrobial, and antibacterial properties. It has the potential to heal wound infections and reduce the wrinkles and scars produced from acne (Fatemeh et al. 2011). A publication in 2011 shows that this essential oil can be used for medicine (Pariya et al. 2015). The results with the use of hyssop oil can be visible within few months. Hyssop oil is not recommended to be used in case of epilepsy and hypertension patients. It should also not be used in higher doses.

Vitamin E Oil

Vitamin E oil prevents scars and retains the moisture of the skin while boosting the production of collagen. It protects the skin from damage caused by UV radiation. The results can be visible after a few weeks of use. It may cause allergic reactions, skin rash, and itchiness. Studies have shown that it improves the cosmetic appearance of scars significantly (Nesrine et al. 2013).

Tea Tree Oil

Melaleuca alternifolia which is commonly known as tea tree oil is a very well-known EO for its use as traditional remedies. Extraction of this oil is mainly from the leaves and terminal branches of the *Melaleuca alternifolia*. The final product after extraction is a combination of 100 different components such as monoterpenes and sesquiterpenes (Appendino et al. 2007). Among these the mixture of terpinen-4-ol and 1,8-cineole is the highly active component having analgesic, antiviral, antibacterial, antifungal, antiprotozoal, and anti-inflammatory properties. Because of its healing properties in history, it has importance in modern medicine. Nowadays, its useful components are used in dermatological creams and ointments.

Numerous data show that wound infections upon treatment with tea tree oil integrated into functional dressings and were investigated against microorganisms. For example, an electrospun polycaprolactone nonwoven mat covered with a layer of chitosan and containing tea tree EO was studied against *S. aureus* in vitro. This data showed inhibition of growth of *S. aureus when exposed to tea tree oil* (Edmondson et al. 2011). Likewise, it is also found that chitosan loaded with tea tree oil exhibits antimicrobial activity and inhibits the growth of strains such as *S. aureus*, *E. coli,* and *C. albicans* (Evandri et al. 2005).

Neroli Essential Oil

Neroli essential oil is also known to be used for the treatment of scars and a variety of skincare conditions. This is mostly used for reducing pain and inflammation at the wound infection site. It is known as a good healer because of its antimicrobial, antifungal, and antioxidant properties. The effects can be visible after a month of prolonged use. It also causes skin irritation and allergic reactions. It is one of the expensive essential oils used for wound healing (Ammar et al. 2012; Kim et al. 2017).

Rosehip Seed Oil

Rosehip seed oil is a relatively less expensive essential oil. This oil is used for the improvement of skin condition, treatment of scars, wrinkles, and acne. In 2015, researchers found that rosehip seed oil is used for the healing of the appearance of the scars after surgery (Kim et al. 2017). It is also seen that using the oil twice a day for a period of 12 weeks found significantly less discoloration, atrophy, and redness. The results can be seen after 6 weeks of use. This oil causes irritation and allergic reaction and is not recommended to use in case of diabetes, kidney stones, or anemia. It can be used as a carrier oil for essential oils.

Coconut Oil

Coconut oil helps to reverse skin damage, moisturize skin, and heal skin disorders because of the presence of fatty acids and micronutrients in it. Owing to its antioxidant and anti-inflammatory properties, it also improves in healing skin conditions (Varma et al. 2019.). It also increases collagen production and also moisturizes and softens the skin. In 2019, an ongoing research found that coconut oil also protects skin cells from inflammation (Nevin and Rajamohan 2010). The research to find out the real mechanism of action is still continuing. Sometimes it causes irritation or allergic reaction after consistent use of 10 days (Nevin and Rajamohan 2010).

Almond Oil

Almond oil has the potential to reduce the appearance of scars. This oil has numerous benefits toward skin cells as it contains vitamin E. Due to the presence of vitamin E, it hydrates, soothes, and moisturizes dry or damaged skin. It also helps in rejuvenating the skin cells and improves the skin complexion and tone. Owing to its anti-inflammatory properties, it improves the skin tone soon. It also helps in reducing the stretch marks and itching. The results can be visible after a month of use. It is not advisable to use this oil if you have diabetes and any recent surgery (Zeeshan 2010).

2.2.4   Honey

Honey is derived from the collection of nectar by the honeybee, *Apis mellifera*, and is then modified to honey. Honey has been used as a traditional medication since decades, whereas the science behind its efficacy has been recently discovered. Besides its use as a natural healer for controlling diseases such as cardiovascular, gastrointestinal tract ailments, and upper respiratory tract infections, it is also used in healing wound infections (Kwakman et al. 2010). It is not just a sugary syrup that helps in wound dressings, but it helps in healing the wounds because of the presence of certain bioactive components. These components help in reducing inflammation, granulation, angiogenesis stimulation, and wound epithelialization (Aumeeruddy-Elalfi et al. 2016). Numerous publications have ascribed several factors about honey's bacteriostatic and bactericidal activity which are as follows:

- Its acidic nature (pH:3.4–6.1) encourages macrophages to eliminate bacteria and helps in microbial biofilm inhibition (Molan 2006).
- The high osmolality in its chemical composition obstructs microbial development (Israili 2014).
- Owing to certain components such as hydrogen peroxide, lysozyme, antioxidants, flavonoids, phenolic acids, methylglyoxal, and bee peptides, it possesses antibacterial activity (Boateng and Diunase 2015; Simon et al. 2009). Particularly, inhibition of bacterial growth occurs via hydrogen peroxide, thereby triggering oxidative damage to pathogen macromolecules. It prevents the growth of microorganisms upon reacting with the bacterial cell wall, as well as with intracellular lipids, proteins, and nucleic acids (Kuś et al. 2016).
- There are 14 different varieties of honey that are responsible for its high efficacy toward microorganisms that take part in infection in wounds. Investigations published by Kus et al. have shown that cornflower, buckwheat, and thyme honey were the most effective ones among the different types of honey and inhibit the growth of bacterium *S. aureus* (Sherlock et al. 2010). Honey from Ulmo tree (Ng and Lim 2015), melaleuca (Jantakee and Tragoolpua 2015), and longan flower (Cooper 2014) inhibits the growth of MRSA microorganisms.
- Floral source, species of bee, and geographical distribution basically determine the composition of honey. To overcome certain limitations that exist in the traditional honey such as deactivation of glucose oxidase by spores, only honey with certified activities is recommended. This introduces medical honey in the market; e.g., chestnut, manuka, thyme, and revamil (Packer et al. 2012) are compared with "traditional honey" with good certainty and quality. Nowadays, few companies produce dressings containing Manuka honey such as Actilite®,

Algivon®, MediHoney®, and Activon Tulle® (Aumeeruddy-Elalfi and Mahomoodally 2016).

- Manuka honey is extracted from Manuka tree which comprises antibacterial activity in biological fluids (Aumeeruddy-Elalfi and Mahomoodally 2016). The antibacterial properties are not only because of the hydrogen peroxide but also due to the high amounts of the antibacterial compound (methylglyoxal) (Bulman et al. 2017). Manuka honey also obstructs the growth of MRSA microorganisms and *S. pyogenes*, along with gram-negative strains such as *E. coli* and *P. aeruginosa* (Lu et al. 2014). It also restricts establishment of microbial biofilm at the infection sites of wounds (Yang et al. 2017a, b). Manuka honey is used in wound dressings as it possesses anti-inflammatory and antibacterial activity along with wound repair efficacy. From researchers such as Yang and coworkers, Manuka honey is found to be a functional antibacterial agent that has antibacterial activity against MRSA as well as other bacteria such as *E. coli,* and *P. aeruginosa* (Saikaly and Khachemoune 2017).

- As per the clinical beneficial effects of honey, based on randomized controlled trials a recent report was published (Rai et al. 2009). Several publications even show varied effects depending on the type of wound, and it has also been found that honey has some detrimental effects. Therefore, more research and clinical trials are essential for establishment of clinical benefits of honey to reduce the wound infections.

2.3   Nanoparticles

With the discovery of nanoparticles the field of nanoscience emerged in 1959 as one of the fastest growing developments in medicine in the twenty-first century. Ever since its application in medicine, nanoparticles have the potential to be used for the improvement of current therapies and diagnostics as we have already mentioned about the occurrence of multidrug resistance as a major problem in wound disinfection. Treatment of wounds with such multidrug-resistant bacteria has become a major task due to the failure of antibiotics in controlling the infection (Yang et al. 2017a, b). In this regard, nanoparticles are considered to be a promising replacement as they exhibit bactericidal activity against several pathogens compared to the conventional antibiotics. Their potential in wound disinfection cannot be ignored with their diminished side effects and no microbial resistance (Kumar et al. 2018).

A nanoparticle upon invading a bacterial cell wall discharges toxic metal ions or reactive oxygen species (ROS), thereby achieving its bactericidal effect. Upon contact with bacterial cell wall, the negatively charged groups attract positively charged nanoparticles at bacterial surfaces by establishing van der Waals forces, receptor–ligand, and hydrophobic interactions (Kandi and Kandi 2015; Simões et al. 2018; Baek and An 2011). By entering the cell wall of bacteria, nanoparticles also interfere with metabolic path and cellular organelle such as mitochondria. Besides bacterial cell wall, certain extra factors can also affect the tolerance of bacteria toward nanoparticles.

Depending on the mechanism of nanoparticle toxicity, susceptibility of microorganisms is determined. For example, *S. aureus* and *Bacillus subtilis* are less susceptible, whereas *E. coli* is very vulnerable to CuO and ZnO NPs (Ashkarran et al. 2012). Its toxicity depends on the composition, modification, and intrinsic properties of nanoparticles. Silver nanoparticles are higher effective toward strains of *E. coli* and *S. aureus* than Cu nanoparticles (Lu et al. 2009; Pramanik et al. 2012). CuO nanoparticles exhibit higher toxicity against *E. coli*, *B. subtilis*, and *S. aureus* than zinc nanoparticles. The toxicity of Cu nanoparticles is directed by several factors such as high temperature, aeration, low pH, and bacteria concentration (Zewde et al. 2016).

Silver nanoparticles are found to be extensively used against microbes as they possess inhibitory action toward more than 650 microbial species and antibiotic resistant bacteria (Volkan et al. 2016).

Wounds that are non-healing due to infection are still a challenge to consider. Therefore, designing of advanced novel materials that could be used at wound site for dressings is required. Recent publications showed that the use of nanoparticles in wound dressings has the potential to become ideal candidates for the delivery of beneficial molecules and drugs to the wound site to improve the healing. In this chapter, we present the incorporation of several antibiotics, essential oils, honey, and inorganic nanoparticles, allowing the creation of composite materials for multi-pharmacological goals during the process of wound healing.

References

Abdillahi SM, Balvanović S, Baumgarten M, Mörgelin M (2012) Collagen VI encodes antimicrobial activity: novel innate host defense properties of the extracellular matrix. J Innate Immun 4:371–376

Agyare C, Duah Y, Oppong E, Hensel A, Oteng S, Appiah T (2016) Review: African medicinal plants with wound healing properties. J Ethnopharmacol 177:85–100

Ahire JJ, Robertson DD, van Reenen AJ, Dicks LMT (2017) Polyethylene oxide (PEO)-hyaluronic acid (HA) nanofibers with kanamycin inhibits the growth of listeria monocytogenes. Biomed Pharmacother 86:143–148

Ahmad Z (2010) The uses and properties of almond oil. Complement Ther Clin Pract 16(1):10–12

Altiok D, Altiok E, Tihminlioglu F (2010) Physical, antibacterial and antioxidant properties of chitosan films incorporated with thyme oil for potential wound healing applications. J Mater Sci Mater Med 21:2227–2236

Ammar AH, Bouajila J, Lebrihi A, Mathieu F, Romdhane M, Zagrouba F (2012) Chemical composition and in vitro antimicrobial and antioxidant activities of *Citrus aurantium* l. flowers essential oil (Neroli oil). Pak J Biol Sci 15(21):1034–1040

Ammon HPT (2006 Oct) Boswellic acids in chronic inflammatory diseases. Planta Med 72 (12):1100–1116

Anjum S, Arora A, Alam MS, Gupta B (2016) Development of antimicrobial and scar preventive chitosan hydrogel wound dressings. Int J Pharm 508:92–101

Appendino G, Ottino M, Marquez N, Bianchi F, Giana A, Ballero M, Sterner O, Fiebich BL, Munoz E (2007) Arzanol, an Anti-inflammatory and Anti-HIV-1 Phloroglucinol α-Pyrone from *Helichrysum italicum* ssp. *Microphyllum*. Nat. Prod 70(4):608–612

Ashkarran AA, Ghavami M, Aghaverdi H, Stroeve P, Mahmoudi M (2012) Bacterial effects and protein corona evaluations: crucial ignored factors for prediction of bio-efficacy of various forms of silver nanoparticles. Chem Res Toxicol 25:1231–1242

Aumeeruddy-Elalfi Z, Mahomoodally M (2016) Chapter: Extraction techniques and pharmacological potential of essential oils from medicinal and aromatic plants of Mauritius. In: Peters M (ed) Essential oils: historical significance, chemical composition and medicinal uses and benefits. Nova Publisher, Hauppauge, NY, pp 51–80. isbn:978-1-63484-367-6

Aumeeruddy-Elalfi Z, Gurib-Fakim A, Mahomoodally M (2016) Chemical composition, antimicrobial and antibiotic potentiating activity of essential oils from 10 tropical medicinal plants from Mauritius. J Herb Med 6:88–95

Baek YW, An YJ (2011) Microbial toxicity of metal oxide nanoparticles (CuO, NiO, ZnO, and Sb_2O_3) to *Escherichia coli*, *Bacillus subtilis*, and *Streptococcus aureus*. Sci Total Environ 409:1603–1608

Bai M-Y, Chou T-C, Tsai J-C, Yu W-C (2014) The effect of active ingredient-containing chitosan/polycaprolactone nonwoven mat on wound healing: in vitro and in vivo studies. J Biomed Mater Res Part A 102:2324–2333

Boateng J, Diunase KN (2015) Comparing the antibacterial and functional properties of Cameroonian and Manuka honeys for potential wound healing—have we come full cycle in dealing with antibiotic resistance? Molecules 20:16068–16084

Borregaard N, Cowland JB (1997) Granules of the human neutrophilic polymorphonuclear leukocyte. Blood 89:3503–3521

Boukhatem MN, Kameli A, Ferhat MA, Saidi F, Mekarnia M (2013) Rose geranium essential oil as a source of new and safe anti-inflammatory drugs. Libyan J Med 8(1):22520

Breidenstein EB, de la Fuente-Nunez C, Hancock RE (2011) Pseudomonas aeruginosa: all roads lead to resistance. Trends Microbiol 19:419–426

Broughton G 2nd, Janis JE, Attinger CE (2006) The basic science of wound healing (retraction of Witte M., Barbul A. In: Surg Clin North Am 1997; 77:509-528). Plast Reconstr Surg 117 (7 Suppl):12S–34S

Bulman SE, Tronci G, Goswami P, Carr C, Russell SJ (2017) Antibacterial properties of non-woven wound dressings coated with Manuka honey or methylglyoxal. Materials 10:954

Campos AC, Groth AK, Branco AB (2008) Assessment and nutritional aspects of wound healing. Curr Opin Clin Nutr Metab Care 11:281–288

Chávez-González ML, Rodríguez-Herrera R, Aguilar CN (2016) Chapter 11—essential oils: a natural alternative to combat antibiotics resistance. In: Antibiotic resistance. Mechanisms and new antimicrobial approaches. Elsevier Science, New York, NY, pp 227–237

Contardi M, Heredia-Guerrero JA, Perotto G, Valentini P, Pompa PP, Spanò R, Goldonic L, Bertorelli R, Athanassiou A, Bayera IS (2017) Transparent ciprofloxacin-povidone antibiotic films and nanofiber mats as potential skin and wound care dressings. Eur J Pharm Sci 104:133–144. https://doi.org/10.1016/j.ejps.2017.03.044

Cooper R (2014) Honey as an effective antimicrobial treatment for chronic wounds: Is there a place for it in modern medicine? Chronic Wound Care Manag Res 1:15–22

Cutting KF, White RJ (2005) Criteria for identifying wound infection—revisited. Ostomy Wound Manage 51:28–34. https://www.prnewswire.com/news-releases/advanced-wound-care-products-market-global-industry-analysis-trends-market-size-and-forecasts-up-to-2023-300558761.html

Dorica-Mirela S, Ionel J (2009) Biologically active natural peptides. J Agroaliment Processes Technol 15:484–499

Edmondson M, Newall N, Carville K, Smith J, Riley TV, Carson CF (2011) Uncontrolled, open-label, pilot study of tea tree (*Melaleuca alternifolia*) oil solution in the decolonisation of methicillin-resistant *Staphylococcus aureus* positive wounds and its influence on wound healing. Int Wound J 8:375–384

Edwards R, Harding KG (2004) Bacteria and wound healing. Curr Opin Infect Dis 17:91–96

Etebu E, Arikekpar I (2016) Antibiotics: classification and mechanisms of action with emphasis on molecular perspectives. Int J Appl Microbiol Biotechnol Res 4:90–101

Evandri MG, Battinelli L, Daniele C, Mastrangelo S, Bolle P, Mazzanti G (2005) The antimutagenic activity of *Lavandula angustifolia* (lavender) essential oil in the bacterial reverse mutation assay. Food Chem Toxicol 43:1381–1387

Fatemeh F, Masoumeh M, Sanaz H (2011) Phytochemical analysis and antioxidant activity of Hyssopus officinalis L from Iran. Adv Pharm Bull 1(2):63–67

Felgueiras HP, Amorim MT (2017a) Electrospun polymeric dressings functionalized with antimicrobial peptides and collagen type I for enhanced wound healing. IOP Conf Series Mater Sci Eng 254:062004

Felgueiras HP, Amorim MT (2017b) Functionalization of electrospun polymeric wound dressings with antimicrobial peptides. Colloids Surf B: Biointerfaces 156:133–148

Friedman ND, Temkin E, Carmeli Y (2016) The negative impact of antibiotic resistance. Clin Microbiol Infect 22:416–422

Ganz T (1987) Extracellular release of antimicrobial defensins by human polymorphonuclear leukocytes. Infect Immun 55:568–571

Ganz T, Selsted ME, Szklarek D, Harwig SS, Daher K et al (1985) Defensins. Natural peptide antibiotics of human neutrophils. J Clin Invest 76:1427–1435

Ghrab F, Djemaa B, Bellassoued K, Zouari S, El Feki A, Ammar E (2016 Nov) Antioxidant and wound healing activity of Lavandula aspic L. ointment. J Tissue Viability 25(4):193–200

Giuliani A, Pirri G, Nicoletto SF (2007) Antimicrobial peptides: an overview of a promising class of therapeutics. Central Eur J Biol 2:1–33

Gläser R, Harder J, Lange H, Bartels J, Christophers E et al (2005) Antimicrobial psoriasin (S100A7) protects human skin from *Escherichia coli* infection. Nat Immunol 6:57

Gosain A, DiPietro LA (2004) Aging and wound healing. World J Surg 28:321–326

Gottrup F, Apelqvist J, Bjarnsholt T, Cooper R, Moore Z et al (2014) Antimicrobials and non-healing wounds. Evidence, controversies and suggestions-key messages. J Wound Care 23:477–482

Halstead FD, Rauf M, Bamford A, Wearn CM, Bishop JR et al (2015) Antimicrobial dressings: comparison of the ability of a panel of dressings to prevent biofilm formation by key burn wound pathogens. Burns 41:1683–1694

Hannigan GD, Pulos N, Grice EA et al (2015) Adv Wound Care (New Rochelle) 4:59–74

Howell-Jones RS, Wilson MJ, Hill KE et al (2005) A review of the microbiology, antibiotic usage and resistance in chronic skin wounds. J Antimicrob Chemother 55:143–149

Huang Y, Huang J, Chen Y (2010) Alpha-helical cationic antimicrobial peptides: relationships of structure and function. Protein Cell 1:143–152

Huang HN, Pan CY, Rajanbabu V, Chan YL, Wu CJ, Chen JY (2011) Modulation of immune responses by the antimicrobial peptide, epinecidin (Epi)-1, and establishment of an Epi-1-based inactivated vaccine. Biomaterials 32:3627–3636. https://doi.org/10.1016/j.biomaterials.2011.01.061

Imane MM, Houda F, Amal AHS, Kaotar N, Mohammed T, Imane R, Farid H (2017) Phytochemical composition and antibacterial activity of Moroccan *Lavandula angustifolia* mill. J Essent Oil Bear Plants 20:1074–1082

Israili ZH (2014) Antimicrobial properties of honey. Am J Ther 21:304–423

Jantakee K, Tragoolpua Y (2015) Activities of different types of Thai honey on pathogenic bacteria causing skin diseases, tyrosinase enzyme and generating free radicals. Biol Res 48:4

Kalia C, Wood TK, Kumar P (2014) Evolution of resistance to quorum sensing inhibitors. Microbiol Ecol 68(1):13–23

Kandi V, Kandi S (2015) Antimicrobial properties of nanomolecules: potential candidates as antibiotics in the era of multi-drug resistance. Epidemiol Health 37:e2015020

Kavoosi G, Dadfar SMM, Purfard AM, Mehrabi R (2013) Antioxidant and antibacterial properties of gelatin films incorporated with Carvacrol. J Food Saf 33:423–432

Khampieng T, Wnek GE, Supaphol P (2014) Electrospun DOXY-h loaded-poly(acrylic acid) nanofiber mats: in vitro drug release and antibacterial properties investigation. J Biomater Sci-Polym Ed 25:1292–1305

Kim S, Jang JE, Kim J, In Lee Y, Lee DW, Song SY, Lee JH (2017) Enhanced barrier functions and anti-inflammatory effect of cultured coconut extract on human skin. Food Chem Toxicol 106 (Part A):367–375

Kish TD, Chang MH, Fung HB (2010) Treatment of skin and soft tissue infections in the elderly: a review. Am J Geriatr Pharmacother 8:485–513

Kohanski MA, Dwyer DJ, Collins JJ (2010) How antibiotics kill bacteria: from targets to networks. Nat Rev Microbiol 8:423–435

Kumar P, Patel SKS, Lee JK, Kalia CV (2013) Extending the limits of *Bacillus* for novel biotechnological application. Biotechnol Adv 31(8):1543–1561

Kumar M, Curtis A, Hoskins C (2018) Application of nanoparticle Technologies in the Combat against anti-microbial resistance. Pharmaceutics 10:11

Kumar P, Lee JH, Beyenal H, Lee J (2020) Fatty acids as Antibiofilm and Antivirulence agents. Trends Microbiol 28(9):753–768

Kuś PM, Szweda P, Jerković I, Tuberoso CI (2016) Activity of Polish unifloral honeys against pathogenic bacteria and its correlation with colour, phenolic content, antioxidant capacity and other parameters. Lett Appl Microbiol 62:269–276

Kwakman PH, te Velde AA, de Boer L, Speijer D, Vandenbroucke-Grauls CM, Zaat SA (2010) How honey kills bacteria. FASEB J 24:2576–2582

Lai Y, Gallo RL (2009) AMPed up immunity: How antimicrobial peptides have multiple roles in immune defense. Trends Immunol 30:131–141

Lauren AH, Jennifer NO, Raja KS (2016) Cedar wood oil as complementary treatment in refractory acne. J Alternat Complement Med 22(3):252–253. https://doi.org/10.1089/acm.2015.0208

Li H, Williams GR, Wang JWH, Sun X, Zhu LM (2017) Poly(*N*-isopropyl acrylamide)/poly(l-lactic acid-*co*-ε-caprolactone) fibers loaded with ciprofloxacin as wound dressing materials. Mater Sci Eng C Mater Biol Appl 79:245–254. https://doi.org/10.1016/j.msec.2017.04.058

Liakos I, Rizzello L, Scurr DJ, Pompa PP, Bayer IS, Athanassiou A (2014) All-natural composite wound dressing films of essential oils encapsulated in sodium alginate with antimicrobial properties. Int J Pharm 463:137–145

Liakos I, Rizzello L, Hajiali H, Brunetti V, Carzino R, Pompa P, Athanassiou A, Mele E (2015) Fibrous wound dressings encapsulating essential oils as natural antimicrobial agents. J Mater Chem B 3:1583–1589

Liu X, Nielsen LH, Kłodzińska SN, Nielsen HM, Quc H, Christensen LP, Rantanen J, Yangad M (2018) Ciprofloxacin-loaded sodium alginate/poly(lactic-*co*-glycolic acid) electrospun fibrous mats for wound healing. Eur J Pharm Biopharm 123:42–49. https://doi.org/10.1016/j.ejpb.2017.11.004

Lu C, Brauer MJ, Botstein D (2009) Slow growth induces heat-shock resistance in normal and respiratory-deficient yeast. Mol Biol Cell 20:891–903

Lu J, Turnbull L, Burke CM, Liu M, Carter DA, Schlothauer RC, Whitchurch CB, Harry EJ (2014) Manuka-type honeys can eradicate biofilms produced by *Staphylococcus aureus* strains with different biofilm-forming abilities. PeerJ 2:e326

Mah TF, Pitts B, Pellock B et al (2003) A genetic basis for Pseudomonas aeruginosa biofilm antibiotic resistance. Nature 426:306–310

Malmsten M, Davoudi M, Walse B, Rydengard V, Pasupuleti M et al (2007) Antimicrobial peptides derived from growth factors. Growth Factors 25:60–70

Mancl KA, Kirsner RS, Ajdic D (2013) Wound biofilms: lessons learned from oral biofilms. Wound Repair Regen 21:352–362

Marino M, Bersani C, Comi G (2001) Impedance measurements to study the antimicrobial activity of essential oils from Lamiaceae and Compositae. Microbiol 67(3):187–195

Martin E, Ganz T, Lehrer RI (1995) Defensins and other endogenous peptide antibiotics of vertebrates. J Leukoc Biol 58:128–136

Maxson S, Lopez EA, Yoo D, Danilkovitch-Miagkova A, Lerox MA (2012) Concise Review: Role of mesenchymal stem cells in wound repair. Stem Cells Transl 1(2):142–149. (Medline)

Michalska-Sionkowska M, Kaczmarek B, Walczak M, Sionkowska A (2018) Antimicrobial activity of new materials based on the blends of collagen/chitosan/hyaluronic acid with gentamicin sulfate addition. Mater Sci Eng C Mater Biol Appl 86:103–108

Midwood KS, Williams LV, Schwarzbauer JE (June 2004) Tissue repair and the dynamics of the extracellular matrix. Int J Biochem Cell Biol 36(6):1031–1037

Molan PC (2006) The evidence supporting the use of honey as a wound dressing. Int J Lower Extrem Wounds 5:40–54

Mori H, Kawanami H, Kawahata H, Aoki M (2016) Wound healing potential of lavender oil by acceleration of granulation and wound contraction through induction of TGF-β in a rat model. BMC Complement Altern Med 16:144

Nesrine R, Yassine M, Salma D, Hedia C, Marwa J, Xavier F, Abdennacer B (2013) Variation of the chemical composition and antimicrobial activity of the essential oils of natural populations of Tunisian *Daucus carota* L. (Apiaceae). Chem Divers 10(12):2278–2290

Nevin KG, Rajamohan T (2010) Effect of topical application of virgin coconut oil on skin components and antioxidant status during dermal wound healing in young rats. Skin Pharmacol Physiol 23:290–297

Ng WJ, Lim MS (2015) Anti-staphylococcal activity of melaleuca honey. Southeast Asian J Trop Med Public Health 46:472–479

Niyonsaba F, Ushio H, Nakano N, Ng W, Sayama K et al (2007) Antimicrobial peptides human β-defensins stimulate epidermal keratinocyte migration, proliferation and production of proinflammatory cytokines and chemokines. J Investig Dermatol 127:594–604

Nogueira MNM, Aquino SG, Rossa Junior C, Spolidorio DMP (2014) Terpinen-4-ol and alpha-terpineol (tea tree oil components) inhibit the production of IL-1b, IL-6 and IL-10 on human macrophages. Inflamm Res 63:769–778

Nurjadi D, Herrmann E, Hinderberger I, Zanger P (2012) Impaired β-defensin expression in human skin links DEFB1 promoter polymorphisms with persistent Staphylococcus aureus nasal carriage. J Infect Dis 207:666–674

Ovington L (2003) Ostomy Wound Manage 49:8–12

Packer JM, Irish J, Herbert BR, Hill C, Padula M, Blair SE, Carter DA, Harry EJ (2012) Specific non-peroxide antibacterial effect of manuka honey on the *Staphylococcus aureus* proteome. Int J Antimicrob Agents 40:43–50

Pamfil D, Vasile C, Tarţău L, Vereştiuc L, Poiată A (2017) pH-responsive 2-hydroxyethyl methacrylate/citraconic anhydride–modified collagen hydrogels as ciprofloxacin carriers for wound dressings. J Bioact Compat Polym 32:355–381. https://doi.org/10.1177/0883911516684653

Pariya K, Hamed S, Jinous A (2015) Analgesic and anti-inflammatory activities of *Citrus aurantium* L. blossoms essential oil (neroli): involvement of the nitric oxide/cyclic-guanosine monophosphate pathway. J Nat Med 69:324–331

Pasupuleti M, Walse B, Nordahl EA, Morgelin M, Malmsten M et al (2007) Preservation of antimicrobial properties of complement peptide C3a, from invertebrates to humans. J Biol Chem 282:2520–2528

Pîrvănescu H, Bălăşoiu M, Ciurea ME, Bălăşoiu AT, Mănescu R (2014) Wound infections with multi-drug resistant bacteria. Chirurgia 109:73–79

Popoola OK, Marnewick JL, Rautenbach F, Ameer F, Iwuoha EI, Hussein AA (2015) Inhibition of oxidative stress and skin aging-related enzymes by Prenylated Chalcones and other flavonoids from *Helichrysum teretifolium*. Molecules 20(4):7143–7155

Pramanik A, Laha D, Bhattacharya D, Pramanik P, Karmakar P (2012) A novel study of antibacterial activity of copper iodide nanoparticle mediated by DNA and membrane damage. Colloids Surf B Biointerfaces 96:50–55

Rădulescu M, Holban AM, Mogoantă L, Bălșeanu TA, Mogoșanu GD, Savu D, Popescu RC, Fufă O, Grumezescu AM, Bezirtzoglou E et al (2016) Fabrication, characterization, and evaluation of bionanocomposites based on natural polymers and antibiotics for wound healing applications. Molecules 21:761

Rai M, Yadav A, Gade A (2009) Silver nanoparticles as a new generation of antimicrobials. Biotechnol Adv 27:76–83

Rai M, Kon K, Gade A, Ingle A, Nagaonkar D, Paralikar P, da Silva SS (2016) Chapter 6— antibiotic resistance: can nanoparticles tackle the problem? In: Antibiotic resistance. Mechanisms and new antimicrobial approaches. Elsevier Science, New York, NY, pp 121–143

Rajanbabu V, Chen JY (2011) The antimicrobial peptide, tilapia hepcidin 2-3, and PMA differentially regulate the protein kinase C isoforms, TNF-α and COX-2, in mouse RAW264.7 macrophages. Peptides 32:333–341. https://doi.org/10.1016/j.peptides.2010.11.004

Rajpaul K (2015) Biofilm in wound care. Br J Community Nurs 20:S6

Rodríguez-Martínez S, Cancino-Diaz JC, Vargas-Zuñiga LM, Cancino-Diaz ME (2008) LL-37 regulates the overexpression of vascular endothelial growth factor (VEGF) and c-IAP-2 in human keratinocytes. Int J Dermatol 47:457–462. https://doi.org/10.1111/j.1365-4632.2008.03340.x

Röhrl J, Yang D, Oppenheim JJ, Hehlgans T (2010) Human β-defensin 2 and 3 and their mouse orthologs induce chemotaxis through interaction with CCR2. J Immunol 184:6688–6694

Roupé KM, Nybo M, Sjöbring U, Alberius P, Schmidtchen A et al (2010) Injury is a major inducer of epidermal innate immune responses during wound healing. J Investig Dermatol 130:1167–1177

Saikaly SK, Khachemoune A (2017) Honey and wound healing: an update. Am J Clin Dermatol 18:237–251

Scagnelli AM (2016) Therapeutic review: Manuka honey. J Exot Pet Med 25:168–171

Semeniuc CA, Popa CR, Rotar AM (2017) Antibacterial activity and interactions of plant essential oil combinations against Gram-positive and Gram-negative bacteria. J Food Drug Anal 25:403–408

Seow YX, Yeo CR, Chung HL, Yuk H-G (2014) Plantessentialoilsasactiveantimicrobialagents. Crit Rev Food Sci Nutr 54:625–644

Sherlock O, Dolan A, Athman R, Power A, Gethin G, Cowman S, Humphreys H (2010) Comparison of the antimicrobial activity of Ulmo honey from Chile and Manuka honey against methicillin-resistant *Staphylococcus aureus, Escherichia coli* and *Pseudomonas aeruginosa*. BMC Complement Altern Med 10:47

Shinde UA, Phadke AS, Nair AM, Mungantiwar AA, Dikshit VJ, Saraf MN (1999) Studies on the anti-inflammatory and analgesic activity of *Cedrus deodara* (Roxb.). loud. Wood Oil J Et Hnopharmacol 65(1):21–27

Shrestha G, Raphael J, Leavitt SD, St Clair LL (2014) In vitro evaluation of the antibacterial activity of extracts from 34 species of North American lichens. Pharm Biol 52:1262–1266

Sienkiewicz M, Głowacka A, Kowalczyk E, Wiktorowska-Owczarek A, Jóźwiak-Bębenista M, Łysakowska M (2014) The biological activities of cinnamon, Geranium and lavender essential oils. Molecules 19:20929–20940

Simões D, Miguel SP, Ribeiro MP, Coutinho P, Mendonça AG, Correia IJ (2018) Recent advances on antimicrobial wound dressing: a review. Eur J Pharm Biopharm 127:130–141

Simon A, Traynor K, Santos K, Blaser G, Bode U, Molan P (2009) Medical honey for wound care—still the 'latest resort'? Evid-Based Complement Altern Med 6:165–173

Sørensen OE (2016) Antimicrobial peptides in cutaneous wound healing. In: Antimicrobial peptides. Springer, Cham, pp 1–15

Steinbakk M, Naess-Andresen C, Fagerhol M, Lingaas E, Dale I et al (1990) Antimicrobial actions of calcium binding leucocyte L1 protein, calprotectin. Lancet 336:763–765

Taylor PK, Yeung AT, Hancock RE (2014) Antibiotic resistance in Pseudomonas aeruginosa biofilms: towards the development of novel anti-biofilm therapies. J Biotechnol 191:121–130

Tokumaru S, Higashiyama S, Endo T, Nakagawa T, Miyagawa JI et al (2000) Ectodomain shedding of epidermal growth factor receptor ligands is required for keratinocyte migration in cutaneous wound healing. J Cell Biol 151:209–220

Valerón-Almazán P, Gómez-Duaso AJ, Santana-Molina N, García-Bello MA, Carretero G (2015) Evolution of post-surgical scars treated with pure rosehip seed oil. J Cosmet Dermatol Sci Appl 5:161–167

Varma SR, Sivaprakasam TO, Ilavarasu A, Dilip N, Raghuraman M, Pavan KB, Rafiq M, Paramesh R (2019) In vitro anti-inflammatory and skin protective properties of virgin coconut oil. J Tradit Complement Med 9(1):5–14

Volkan T, Jurek C, Masoud M, René van der H, Berend van der L (2016) The role of topical. Vitamin E in scar management: a systematic review. Aesthet Surg J 36(8):959–965

Walsh SE, Maillard J-Y, Russell AD, Catrenich CE, Charbonneau DL, Bartolo RG (2003) Development of bacterial resistance to several biocides and effects on antibiotic susceptibility. J Hosp Infect 55:98–107

White RJ, Cutting K, Kingsley A (2006) Topical of wound bioburden. Ostomy Wound Manage 52:26–58

Wilson MA (2003) Skin and soft-tissue infections: impact of resistant gram-positive bacteria. Am J Surg 186:35S–41S

Wolcott RD, Rhoads DD, Bennett ME et al (2010) Chronic wounds and the medical biofilm paradigm. J Wound Care 19:45–46

Yang X, Fan L, Ma L, Wang Y, Lin S, Yu F, Pan X, Luo G, Zhang D, Wang H (2017a) Green electrospun Manuka honey/silk fibroin fibrous matrices as potential wound dressing. Mater Des 119:76–84

Yang Y, Qin Z, Zeng W, Yang T, Cao Y, Mei C, Kuang Y (2017b) Toxicity assessment of nanoparticles in various systems and organs. Nanotechnol Rev 6:279–289

Ye S, Jiang L, Wu J, Su C, Huang C, Liu X, Shao W (2018) Flexible amoxicillin-grafted bacterial cellulose sponges for wound dressing: in vitro and in vivo evaluation. ACS Appl Mater Interfaces 10:5862–5870. https://doi.org/10.1021/acsami.7b16680

Zanger P, Holzer J, Schleucher R, Scherbaum H, Schittek B et al (2010) Severity of Staphylococcus aureus infection of the skin is associated with inducibility of human β-defensin 3 but not human β-defensin 2. Infect Immun 78:3112–3117

Zasloff M (2002) Antimicrobial peptides of multicellular organisms. Nature 415:389–395

Zenati F, Benbelaid F, Khadir A, Bellahsene C, Bendahou M (2014) Antimicrobial effects of three essential oils on multidrug resistant bacteria responsible for urinary infections. J Appl Pharm Sci 4:15–18

Zewde B, Ambaye A, Stubbs J III, Raghavan D (2016) A review of stabilized silver nanoparticles—synthesis, biological properties, characterization, and potential areas of applications. JSM Nanotechnol Nanomed 4:1043

Part III
Interdisciplinary Approach to Wound Care

SilverSol® a Nano-Silver Preparation: A Multidimensional Approach to Advanced Wound Healing

A. de Souza, A. H. Vora, A. D. Mehta, K. Moeller, C. Moeller, A. J. M. Willoughby, and C. S. Godse

Abbreviations

ABL	American Biotech Labs
ATCC	American Type Culture Collection
CFU	Colony forming units
DNA	Deoxy ribose nucleic acid
DTA	Differential thermal analysis
EPA	Environmental Protection Agency
ESBL	Extended-spectrum β-lactamase
FDA	Food and Drug Administration
FHSA	Federal Hazardous Substances Act
FTIR	Fourier-transform infrared spectroscopy
ICP-MS	Inductive coupled plasma-mass spectrophotometry
IL	Interleukin
MCP	Monocyte chemoattractant protein
MDR	Multi-drug resistant
MIC	Minimum inhibitory concentration
MRI	Magnetic resonance imaging
MRSA	Methicillin resistant *Staphylococcus aureus*
NADH	Nicotinamide adenine dinucleotide hydrogen
NP(s)	NP(s)

A. de Souza · A. H. Vora · A. D. Mehta · C. S. Godse (✉)
Viridis BioPharma Pvt Ltd, Mumbai, India
e-mail: chhaya@viridisbiopharma.com

K. Moeller · C. Moeller
American Biotech Labs, LLC, American Fork, USA
e-mail: keith@ablsilver.com; cam@ablsilver.com

A. J. M. Willoughby
Direct Dental Source (DD Source) Manufacturing Inc, Surrey, Canada
e-mail: andrew.willoughby@ddsource.com

NQO	NADH quinone oxidoreductase
PPM (ppm)	Parts per million
RNA	Ribonucleic acid
SEM	Scanning electron microscopy
TGA	Thermogravimetric analysis
TEM	Transmission electron microscopy
USP	United States Pharmacopoeia
UV-VIS	Ultraviolet visible spectrophotometry
VRE	Vancomycin resistant enterococci
XRD	X-ray diffraction

1   Introduction: Silver in Medicine

1.1   Ancient Use: Silver Compounds

Silver, a lustrous metal, has been known globally since ancient times. Metallic silver was known to the Chaldeans since 4000 BC, especially for making valuable goods. Traditionally the royal families used silver in tableware. It is interesting to note that silver came into medicinal use during the first millennium AD. Primeval Egyptians, Romans, Phoenicians, Greeks and others used silver to preserve freshness to prevent spoilage of water, milk, food (Alexander 2009). North Americans used to drop a silver coin in water for its preservation and for a long-distance transportation prior to the discovery of refrigerators. Persian King (600 to 530 BC) would drink water only if it is brought in silver vessels.

These ancient simple practices of preservation of water and food got connected to antimicrobial properties of silver much later when microbes were discovered (1665) and their role in infections was established (1884). However, the question arises whether silver is soluble enough in the water to impart a protective effect for health? As silver is sparingly soluble in water, just a few parts per trillion, it is believed that the potential ligands of goblet forming silver salts give rise to its antimicrobial potential. Most silver salts viz. silver sulfadiazine, silver halides and silver nitrate being water soluble have antimicrobial properties and oligodynamic effects.

Silver nitrate was discovered in the thirteenth century. However, its medicinal use came into existence in 1614, when it was used internally by Angelo Sala as a counterirritant, purgative, and for the treatment of brain infections. Carl Siegmund Franz Crede, a German obstetrician, in the 1880s effectively used silver nitrate eye drops for treating gonorrhoeal ophthalmia (*ophthalmia neonatorium*) in infants. The practice of using 1% solution of silver nitrate, reduced the incidence of gonorrhoeal ophthalmia to 0.13% from 7.8% (Schneider 1984). Systematic studies of silver ions were conducted by Vonnaegele. He reported the effect of silver against 650 species of unicellular organisms, proposing its promising bactericidal effects (Searle 1920). Colloidal silver also has been shown to be effective in puerperal sepsis, staphylococcal sepsis, tonsillitis, acute epididymitis, and other infectious diseases (Duhamel 1912; Sanderson-Wells 1916; Brown 1916).

It is only by the twentieth century, the use of metallic silver for water purification was scientifically established. Later it was expanded for the treatment of wound and eye infections and for dental hygiene including the prevention and correction of pyorrhoea, gingivitis, and bad breath. Traditionally silver was used as a blood purifier, for the prevention of palpitation of the heart and for the reatment of offensive breath has been reported in 980 AD. Swallowing silver was also found to be useful in stopping epileptic seizures. In India, the Ayurvedic system of medicine describes the use of processed metals viz. gold, silver, lead, mercury for various disease conditions. The use of silver in Ayurvedic therapeutics dates back to the period of *Charaka (300 BC)* (Galib 2011).

1.2   Modern Trends: Ionic Versus Metallic Silver

Historically silver has been used for a medical purpose in different forms— as an additive in ceramic (zeolite) or in glass matrix, as silver salts (chlorides, nitrates, sulphides) or as elemental silver, though the nomenclature emerged much later. The Microbicidal capacity of different silver forms was mainly dependent on their capacity to release silver ions (Das et al. 2005). Silver nitrate has been mentioned in the Pharmacopeia in Rome in 69 B.C. (Hill and Pillsbury 1939), whereas, record of its medicinal use has been found in 702–705 AD. Colloidal silver was first used in 1891 (vide infra). Consumption of silver by humans over millennia bespeaks its safety. In India, 275 tons of silver is consumed every year in metallic form (Silver foil) as an additive in foods and sweets (Das et al. 2005).

With the advancement of nanomedicine in the twentieth century, several products containing metallic silver in nano form (<100 nm particle size) have been developed. The advantage of nano-silver over traditional forms is mainly due to its nano size with large effective surface area. This allows it to be used at extremely low concentrations without side effects compared to silver ions or silver nitrate. In the early twentieth century, medicinal nanoscale silver colloids became available commercially under trade names Collargol, Argyrol, and Protargol and thereafter their use became widespread within 50 years (Nowack et al. 2011). Collargol, a silver preparation with 10 nm particle size was used as early as in 1897 (Nowack et al. 2011). The estimated worldwide production and use of nano-silver by 2011 was 320 tons per year (Nowack et al. 2011). Among the 1300 nanotechnology-based marketed metallic products one fourth comprise nano-silver (Munger et al. 2014). Compared to ionic silver (chemical forms), nano-silver has distinct physicochemical properties which lead to their efficacy as an antibacterial, anti-viral, and anti-inflammatory agent (Yardley 1998).

1.3   SilverSol®: Coated Nano Silver

SilverSol®, developed by American Biotech Labs (ABL), USA using a patented technology,—(US6743348B2 United States) is a uniquely engineered colloidal silver preparation having microbicidal, wound healing and several other activities (Holladay et al. 2001). A decade of efforts by ABL has resulted in the development of SilverSol®. The term "sol" specifies the chemical nature of the silver preparation as 'a pure mineral permanently suspended in the water where the mineral's charge is transferred to the entire body of the water'. The elemental form of zero-valent metallic silver particles is coated with silver oxide, with particle size ranging between 5 and 50 nm. The nano size confers multidimensional bioactivity and high stability to the product. SilverSol® is the only patented engineered product in the world, containing nano-silver particles with proven safety and multidimensional efficacy at extremely low concentrations. A wide variety of SilverSol® formulations available both in liquid and gel forms are colourless, odourless, and tasteless.

1.4   The Multidimensional SilverSol®

The nano-silver particles in SilverSol® have a unique structural arrangement, which confers a wide range of activities to SilverSol® at extremely low concentrations. SilverSol® has significant antimicrobial activity, effective against several microbes, including fungi, bacteria, protozoa and viruses, which makes it efficacious in various infectious conditions such as Malaria, Influenza, human immunodeficient viral (HIV), and Hepatitis B viral infections, vaginal infections, oral and urinary tract infections. Severe infectious conditions caused by antibiotic-resistant bacteria—MRSA and VRE can also be treated successfully with SilverSol®. Besides antimicrobial activity, SilverSol® has been shown to possess activity in cancer cell lines. Several in vitro and in vivo studies; conducted to confirm its potential, have yielded positive results with clinical evidence in Malaria, MRSA infections, HIV, pain, inflammation of various origins as well as in wound healing of various aetiology. SilverSol® oral products have also been used as immune enhancers.

The current review focuses on the promising effects of various SilverSol® products in the treatment of acute and chronic wounds. It also gives its detailed attributes regarding safety, efficacy and pharmacology. Physicochemical properties and regulatory status of the product are also described. Clinical cases with complex, infected wounds of varied aetiology and severity, treated with SilverSol® have been described in a separate section. Case histories of over 22,000 patients undergoing various dental procedures treated with SilverSol® products have been summarized. Quicker healing time with relief from post-surgical pain and swelling was found to be prominent in these patients.

2 Silversol® Technology

2.1 History of SilverSol® Discovery

Dr. William D. Moeller (Photograph 1), an entrepreneur; transitioned in the mid-1970s from a highly successful insurance career to become part owner and CEO of multiple mining companies, including American Consolidated Mining Company and Clifton Mining Company. Along with his sons, he developed property in and around Gold Hill, Utah, USA. A former mining boomtown during the turn of the twentieth century, Gold Hill is riddled with valuable mineral deposits. Its original operations ceased after World War II as the demand for metals waned (https://en.wikipedia.org/wiki/Gold_Hill_Utah). William and his team picked up where the old miners left off, assaying, drilling, and mining the veins for their famously high levels of copper, silver, gold, lead, and tungsten.

William had a great talent and tenacity when it came to solving problems. He was able to focus and engineer a way to solve complex problems. He acquired the mining property from the scattered owners with small plots, gradually building a block of mining claims that incorporated about 13 square miles of land. During the 1990s, the prices of precious metals dropped to historic lows. William overcame this hurdle by diversifying into the application of silver, which had been a mainstay in medicine prior to the advent of antibiotics. He teamed up with Robert Holladay—an electrician and chemist, and Herbert Christensen—an engineer, and together they began to research colloidal silver products and the manufacturing methods for them. Their efforts resulted in the formation of nano-silver with antimicrobial and other properties that are far better than those of the usual colloidal silver, due to their innovative tetrahedral structure coating. William was not only successful in obtaining patents to protect the invention but was also successful in getting FDA approvals for several versions of the product.

Photograph 1: Dr. William D. Moeller (1936–2014)

Today ABL, the company founded by Williams, has continued the original spirit of advancing the manufacturing and providing highly effective and very stable forms of medicinal silver. Along with universities and government labs, and in close collaboration with other research groups, such as Viridis Biopharma Pvt. Ltd. (VBPL) from Mumbai (India), they spent the next decade researching the silver particles they had created and testing them against a large number of varied microorganisms. The vast experimental data that was generated proved that SilverSol® was capable of killing most pathogenic bacteria including *Yersinia pestis* causing bubonic plague and also many yeasts, fungi and viruses (Roy et al. 2007; Pedersen et al. 2008; Pedersen and Moeller 2009; Revelli et al. 2011). They also proved it was safe for use on and in humans and animals (Munger et al. 2014). They showed that if taken orally it did not harm the probiotics and when used topically on a wound, it protected the wound from infection, expedited the healing, reduced the inflammation and pain (Revelli et al. 2011), and reduced the formation of scar tissue (Pedersen and Moeller 2009). This unmatched silver technology created by William and his team is known worldwide as SilverSol® Technology. Over 400 major studies conducted including published clinical studies demonstrate the capability of the technology. SilverSol® besides safe was found to be highly effective against pathogens that are difficult to manage. It is making waves on many disease fronts, including treatments for ailments like Malaria (Pedersen and Hedge 2010), MRSA, and HIV (Pedersen et al. 2008), to name but a few. The studies done in various government laboratories in US, showed the killing of infectious viruses viz. SARS and H5N1 by the technology (Pedersen et al. 2008).

William's diligent work led to the technology obtaining several US FDA clearances, as well as clearances and approvals by other international governments. It has garnered more than 70 US and international patents, with numerous new patents currently pending. Before he died, William's goal was to change the medicinal history of the world. Having sold more than 22,000,000 units of SilverSol Technology products worldwide and having helped improve the lives of hundreds of thousands of patients, it is safe to say he achieved his goal.

2.2 Characterization of SilverSol®

SilverSol®, a colloidal solution of silver is a 2-phase stable solution of metallic silver. The crystalline solid phase of ultrahomogenous silver in the liquid phase of water forms solid state epitaxy—an amazing phenomenon of metallic aquasols in which the crystalline templet imparts its structure to the amorphous solid phase, imposing the amorphous phase to crystallize as per the crystalline template. In SilverSol®, silver in Ag_4O_4 forms a thin coating around metallic silver suspended in water. This confers a unique tetrahedral structure to SilverSol®, wherein metallic nano-silver particles are surrounded by 4 AgO and 2 water molecules (Fig. 1). This structure, while being stable, also imparts multidimensional biological properties to

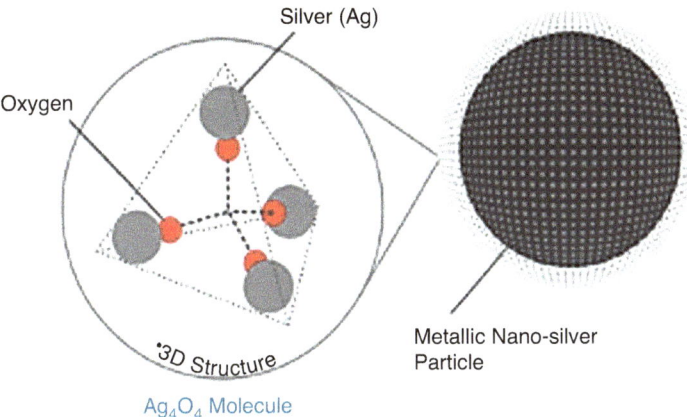

Fig. 1 Tetrahedral structure of SilverSol®

SilverSol® nanoparticles (NPs) without any side effects (Sect. 2.3 safety studies). This is of key importance, as the safety of the NPs is a major concern in nanomedicine.

The phase analysis of SilverSol®, using advanced material science has been reported by Roy et al. They conducted phase equilibrium studies of the biphasic SilverSol®. Various techniques viz. Differential thermal analysis (DTA), Thermogravimetric analysis (TGA), X-ray diffraction (XRD), Scanning electron microscopy (SEM), and Transmission electron microscopy (TEM) were used for the solid phase analysis. On the other hand by using Fourier-transform infrared spectroscopy (FTIR), UV–VIS, and Raman spectroscopy the liquid water phase was analysed. The readers may refer to the paper published by the group for the detailed methodology and the data (Roy et al. 2007). Some salient features highlighting the peculiar distinctive characteristics of SilverSol® as enlisted below:

- The water phase of the SilverSol® was reported to be pure and the purity was maintained in the presence SilverSol® nano particles.
- Transmission electron microscopy revealed that the particle size of nano-silver ranged between 10–50 nm with a median around 30 nm. It was observed that the bigger particles of 20–30 nm size comprised of a group of 5–7 nm particles, seized together by relatively weak bonds (vide infra). These 5 nm, metallic, silver-containing 'mobile' units, with oxide cover (in the form of layers or 'skins) around them, confer stability and unique bio-activity to SilverSol®.
- Roy et al. also compared SilverSol® to several silver colloids that were commercially available. Using Raman Spectroscopy, the SilverSol® could clearly be distinguished from any of the silver colloids, as well as from the deionized water and the HPLC water. This led to the conclusion that while the water was pure in the presence of metallic silver, its structural arrangement was different than that of HPLC and deionized water (Roy et al. 2007).

Later in 2011, an interesting study was conducted by Revelli and the group. They observed using TEM and photoelectron spectroscopic imaging (PSI), various microorganisms treated with SilverSol® and demonstrated the localization of silver NPs within the microbes. Using TEM, they observed the localisation of clustered silver NPs both within *Escherichia coli* and *Staphylococcus aureus* but not in several other bacterial strains studied. However, when the PSI technique was used, all treated strains showed dispersed silver particles localized within all the bacterial cells (Revelli et al. 2011).

2.3 Safety Studies of SilverSol®

Recently, Metallic NPs with a wide range of composition and size, have been introduced for the treatment and diagnosis of several ailments and diseases. This has, however, led to an increase in human exposure to several metals including silver. The safety of metallic NPs is an ongoing concern due to their accumulation in the body which is likely to cause some side effects (Magaye et al. 2012). It has been shown that NPs can have varied degree of cellular uptake and toxicity through their interactions (influenced by their size, shape, charge, and the constituent material) with biological systems (Albanese et al. 2012). Almost a quarter of currently available commercial nanoproducts used for the medicinal purpose are made up of nanoscale silver (Munger et al. 2014) which have extensive medicinal and surgical applications (Rai et al. 2014; Lee and Jun 2019). It is therefore the responsibility of the manufacturer to ensure the safety of their silver NP product.

2.3.1 Safety of SilverSol®: Unique Structure

The unique arrangement of metallic silver and its oxide in water (vide supra) ensures the essential non-toxicity and the safety of SilverSol®. Safety of SilverSol® has been established both for topical application and oral administration. It is widely known that excessive oral intake of ionic (soluble form) silver, may cause irreversible blackening of the skin termed as Argyria. Trop et al. reported raised liver enzymes and argyria when silver-based dressing was used in burn patients (Trop et al. 2006).

2.3.2 Safety of SilverSol®: Ultra-Low Effective Dose

SilverSol® containing 10–40 ppm silver is more efficient, about 1000 times than other forms of colloidal or ionic silver, in destroying pathogens. On the contrary, some silver products contain up to 300,000 ppm silver (Sellman 2010). Consumption of at least 900 mg of silver in a year (2.4 mg/day for 1 year) is needed to develop Argyria. The recommended dosage of 10 ppm SilverSol®—½ to 1 spoon (15–30 μg in 1.5–3 ml) once or thrice a day is 50 times less than the dose needed to develop argyria. As per the Environmental Protection Agency (EPA) standards, 5 µg/kg body weight silver per day (350 µg/day for a 70 kg adult) can safely be taken by an adult (https://www.govinfo.gov/content/pkg/FR-1999-08-17/html/99-21253.htm). Considering this cut-off, consumption of 6 tablespoon 10 ppm ASAP solution every day by an average weight person would not cross the safety limits even if continued up to 72 years. Besides these safety calculations, a complete safety profile of SilverSol®, has been established by ABL by conducting more than 30 studies. Some of them are enlisted below.

2.3.3 Cytotoxicity Studies In Vitro

ASAP 10 ppm silver solution was studied on Murine fibroblast cell line—L929 by Nelson Laboratories USA using agar overlay test method. In brief, in a 6 well plate, L929 cells were grown to 80% confluency. Over the cell mat, a layer of 1% agar was put, on which the discs containing 100 µl of 10 ppm ASAP solution were placed. The plates were then incubated. The effect of silver that diffuses through the agar layer was studied on cell growth. The results were assessed in terms of the zone around the disc as per the criteria stated in the United States Pharmacopeia & National Formulary. The score of cytotoxicity due to ASAP was 1 as compared to the positive control—latex natural rubber, which gave a score of 4 (Nelson Laboratories Report 2013).

2.3.4 Animal Toxicity Studies

Acute oral toxicity ASAP 22 ppm solution was tested in rats as per the guidelines of the Federal Hazardous Substances Act (FHSA) regulations, 16 CFR 1500 (NAMSA, California report 1999). ASAP 22 ppm, 5 gm/kg (a dose 50 times higher than that of the human dose) was given to rats. There was no significant toxicity or mortality observed in rats monitored over 14 days. Another study was conducted at Shri CB Patel Research Centre, Mumbai, India, using 10 and 32 ppm ASAP solutions. A dose of 50 ml/kg body weight was injected into the peritoneal cavity of Swiss albino mice. There was no mortality and no organ toxicity seen at the end of 72 h, which implied that the safety criteria as per the USP requirements were met.

2.3.5 Selective Inaction on Probiotics

Colloidal silver has emerged as an effective antimicrobial agent, acting across a wide spectrum of the microbial population. Chemical compounds and antibiotics are effective over a smaller range of microbes, whereas colloidal silver can kill over 600 types of microorganisms. Interestingly, it has been reported by VBPL that SilverSol® 10 and 22 ppm does not kill bacteria used in probiotics. In an in vitro study, SilverSol® 10 and 22 ppm was studied against two marketed probiotics—

Lactisyn (containing *Lactobacillus lactis, Lactobacillus acidophilus, Streptococcus lactis and Streptococcus thermophilus*) and Kyo-Dophilus® (containing *Lactobacillus acidophilus, Bifidobacterium bifidum and Bifidobacterium longum*). There was no inhibition of bacteria in the formulation when grown in presence of ASAP (Data on file at ABL). Another more extensive study conducted at Dr. Ron W Leavitt's laboratory, Brigham Young University, in 2004 showed similar results. They showed ASAP to be bactericidal against the pathogens tested at varying degrees but not against the various bacteria used as probiotics viz. *Lactobacillus, Bifidobacterium and Streptococcus* species. They used several concentrations of ASAP 32 ppm (0.13 ppm to 16.0 ppm) in broth microdilution assay against standard ATCC bacterial cultures. Based on Minimum inhibitory concentration (MIC) results, they reported that the growth of *Bifidobacterium* was not affected at all at the highest concentration (16 ppm) and growth of *Lactobacillus* was marginally affected at 4 and 8 ppm. Whereas all pathogens tested were inhibited at MIC of 2 ppm (Data on file at the ABL). The study concluded that the consumption of probiotics in conjunction with ASAP would be beneficial.

2.3.6 Clinical Safety: Effect on Haematology and Metabolic Markers

Clinical safety of silver and other metallic NP has been a continuous concern. Several in vitro and in vivo studies have been shown the safety of SilverSol® as described in Sect. 2.1.3 to 2.1.5. However, it is important to ensure the clinical safety of SilverSol®. Munger et al. conducted a Phase 1 clinical study. Healthy volunteers were given 10 and 32 ppm SilverSol® for 14 days and the effect on biological systems was evaluated.

The first prospective double-blind placebo-controlled crossover study in the 60 healthy volunteers was conducted by Munger et al. This systematic study was conducted as par the 'International Conference on Harmonisation of Technical Requirements for Registration of Pharmaceuticals for Human Use' Guidelines for Good Clinical Practice and the Declaration of Helsinki, with the approval from the Institutional Review Board—University of Utah (Munger et al. 2014). The study registration was done with Clinical-Trials.gov (identifier: NCT01243320). Total 36 subjects received orally placebo as 15 ml sterile water or 15 ml 10 ppm oral SilverSol® solution and 24 subjects received 32 ppm oral silver particle daily for 14 days. At the end of 14 days, there was 3 days wash out period, after which the volunteers received a crossover dose of sterile water or dose of respective silver particles. All the subjects were investigated at baseline and at the end of 3, 7 and 14 days for physical examination, medical and drug history, a panel of metabolic markers, hematology and urine analysis. There were no significant changes in clinical or physical findings and in the metabolic, hematologic or urine analysis.

In the above-mentioned study, eighteen subjects receiving 10 ppm and eleven subjects receiving 32 ppm SilverSol®, also underwent cardiac and abdominal MRIs post 3–14 days of treatment. There were no morphological or structural changes noted. The markers of oxidative damage and inflammation—hydrogen peroxide production or peroxiredoxin protein expression and pro-inflammatory cytokine RNA expression analysed in these volunteers were unchanged. IL-8, IL-1α, IL-1β, MCP1 (Monocyte chemoattractant protein-1) and NQO1 (NADH quinone oxidoreductase-1) showed no statistical difference between the subjects treated with active silver and placebo solutions.

2.4   Efficacy of SilverSol®

The activity of SilverSol® has been shown in vitro and in vivo studies. Studies conducted in human suffering from several illnesses too have indicated its efficacy. ABL so far has conducted more than 400 studies at 60 different private, U.S. government, university and military labs across the world. All the data collectively confirm the activity, efficacy and safety of SilverSol® (Data on file at the ABL).

2.4.1   Microbicidal Activity In Vitro

Traditionally use of ionic silver (silver nitrate) as an antimicrobial was most popular. It was used as both bacteriostatic and bactericidal (Ricketts et al. 1970; Berger et al. 1976b; Tilton and Rosenberg 1978; Ritchie and Jones 1990), antifungal (Miller and McCallan 1957; Brown and Smith 1976; Berger et al. 1976a), protozoicidal, (Wysor and Zollinhofer 1972) and antiviral (Coleman et al. 1973) agent. However, it was not so effective against bacterial spores, cysts of *Entamoeba histolytica* and Mycobacteria (Zanger et al. 2008). In case of *Pseudomonas aeruginosa* a non-linear order of death was observed with silver ion (Brown and Anderson 1968). Whereas a rapid bactericidal action of silver ion (silver nitrate 0.5 and 1 µg/ml) was observed in water but not in broth (Ricketts et al. 1970). Several other forms of silver viz. silver citrate, lactate and proteinate, and silver sulfadiazine have also been developed.

Ravelin in 1869 (Ravelin 1869) was the first to report antimicrobial effect at extremely low concentrations of metallic silver and other metal derivatives. Von Naegeli found that metallic silver at 0.0000001% (1 ppm) concentration would kill the common fresh-water *Spirogyra* (Von Naegelli 1893). Germination of *Aspergillus niger* spores was prevented by metallic silver at 60 ppm (0.00006%) (Russell and Hugo 1994). Interestingly, SilverSol® too kills nearly all microorganisms at similar concentrations 10–50 ppm including the ones resistant to antibiotics.

In the studies conducted with standard ATCC culture and resistant strains; SilverSol® showed antimicrobial potential. An in vitro study was conducted using SilverSol®- Silver Water Dispersion™ Solution (De Souza et al. 2006). Eight microorganisms viz. *Shigella flexneri, Salmonella typhi, S. aureus* 6538 P. *Bacillus subtilis, Candida albicans,* MDR (multiple-drug resistant) strains of *E. coli and P. aeruginosa,* methicillin-resistant *S. aureus* were treated individually with Silver

Water Dispersion™ and nineteen commonly used antibiotics. Further synergistic activity of Silver Water Dispersion™ was studied by the group using it in a combination with individual antibiotics. MICs of Silver Water Dispersion™ and the antibiotic solutions were determined using the macro-dilution test. The agar cup method was used for zone of inhibition studies using individual antibiotics and synergistic combinations. MIC of Silver water dispersion™ Solution was found to be in the range of 2–17 ppm, the lowest inhibitory concentration was against *S. typhi*. The highest inhibitory concentration was found to be against *B. subtilis*. Total 96 tests were conducted to determine the synergistic activity of Silver Water Dispersion™ in combination with other standard antibiotics. Five combinations were found to be synergistic- with amakacin, cefoperazone, ceftizidine (against MDR—*E. coli* and *P. aeruginosa*) and kanamycine, 89 were additive, and two were antagonistic—with amoxicillin and oxacillin.

Microbicidal activity of ASAP nano-silver solution was reported by Bhat et al. against drug-resistant pathogens—bacteria and Candida. The resistant bacteria studied included Methicillin-resistant *Staphylococcus aureus* (MRSA), Vancomycin-resistant *Enterococcus faecalis*, drug-resistant *Escherichia coli*, ESBL (extended-spectrum β-lactamase) producing *Klebsiella pneumoniae*, drug-resistant *Pseudomonas aeruginosa*, *Salmonella typhi*, and *Shigella flexneri* were isolated from clinical samples. The inhibitory and microbicidal effects of ASAP were determined by broth dilution and suspension test. *S. aureus* ATCC 25923 and *E. coli* ATCC 25922 were used as controls. Bacteria were found to be more susceptible than Candida. It was observed that an exposure time of around 30–60 min would kill bacteria while *C. albicans* was killed after 120 min of exposure to ASAP (Bhat et al. 2009).

In another study by Revelli et al., the activity of 10 ppm SilverSol® was compared with 5 antibiotics—Erythromycin, Ofloxacin, Tetracycline, Penicillin and Cefaperazone. MIC was also determined against *Streptococcus pyogenes* (ATCC 19615), *Streptococcus gordonii* (ATCC 10558), *Escherichia coli* O157: H7 (ATCC 43895), *Streptococcus mutans* (ATCC 25175), *Streptococcus pneumoniae* (ATCC 6303), *E. coli* (S.E. 163 Luria Strain B ATCC 11303), *Klebsiella pneumoniae* (ATCC 13883), *S. typhimurium* (ATCC 14028), *Enterobacter aerogenes* (ATCC 13048), *P. aeruginosa* (ATCC 27853), *Streptococcus faecalis*, *Shigella boydii*, *Staphylococcus aureus*, *Klebsiella oxytoca*, *Salmonella enterica* subsp. *arizonae* and, *Enterobacter cloacae*. MIC for the majority of gram-negative organisms tested was found to be 2.5 ppm and for *P. aeruginosa*, *Shigella boydii*, and *K. oxytoca showed* lower MICs (1.67 ppm, 2.19 ppm, 1.25 ppm, respectively) (Revelli et al. 2011).

Recent collaborative studies conducted at Texas Tech University have investigated the antibacterial activity of SilverSol® Gel (Ag-gel). The group investigated the inhibitory activity of Ag-gel against bacteria causing tooth decay and plaque formation (Tran et al. 2019). The activity of Ag-gel was first tested against individual bacteria and colony forming units (CFU) were monitored. the effect of Ag-gel on biofilm formation was determined by placing suspensions of these bacteria (approximately 4×10^2 CFU—colony forming units of each) on a 6 mm paper disc. Ag-gel

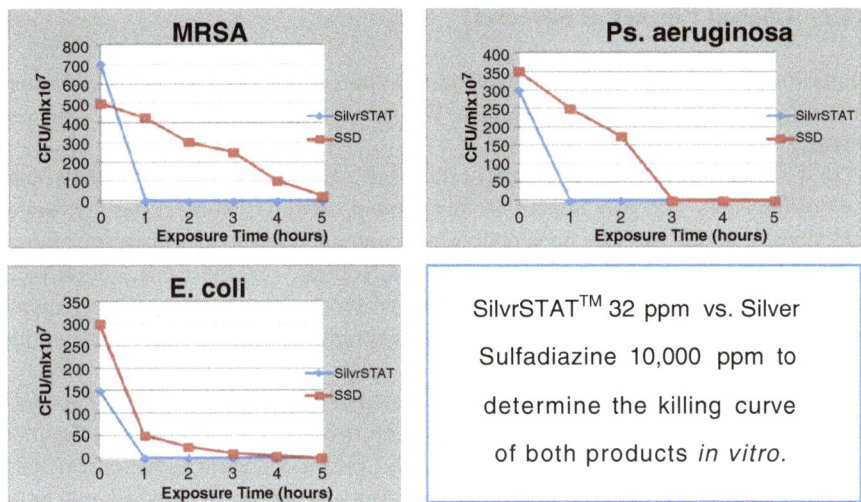

Fig. 2 Kill Time Curve for SilverSTAT® versus Silver Sulfadiazine Cream

0.5 gm was applied on the inoculated discs which were then placed on the growth medium in plates. Micro-aerobic conditions were generated using EZ GasPak in a jar in which the plates were incubated at 37 °C for 24 h. Biofilm formation was monitored by scanning electron microscopy. In the CFU assay, Ag-gel inhibited 100% growth of all the test bacteria, as against in the placebo gel and in the blank cellulose disc wherein bacterial growth was seen to be as high as 6 logs. In the biofilm formation assay, *S. salivarius*, *S. sanguis*, *S. mutans* and the mixture of all three strains developed microcolonies on the cellulose discs indicating the formation of typical biofilms. In presence of Ag-gel, no such biofilm formation was seen.

In another study done by the group, synergistic action of silver colloidal gel (Ag-gel) with 5% Betadine solutions was determined against both Gram negative and Gram-positive bacteria. The experiments were carried out using colony forming unit assay, and confocal laser scanning microscopy. Ag-gel alone inhibited the growth of all the bacteria, except *Klebsiella pneumoniae* CI strain. On the contrary, bactericidal activity was not seen when 5% Betadine was used alone. Interestingly, Ag-gel in combination with 5% Betadine solution, completely eliminated even *K. pneumoniae* (Tran et al. 2020).

SilvrSTAT® 32 PPM hydrogel—one of the products of SilverSol® used as an antibacterial wound dressing gel has shown significant bactericidal effect as seen in the comparative kill curves with currently available silver products approved for use in wound management. In-vitro comparison of kill curves for SilvrSTAT® (32 ppm silver) exemplify clinical relevance of SilvrSTAT® over silver sulfadiazine cream (10,000 ppm), for eradication of MDR microbes in infected wounds (Fig. 2)—Data on file at ABL.

2.4.2 Clinical Efficacy of SilverSol®

Multidimensional SilverSol® products have demonstrated clinical effects in numerous ailments, besides wound healing. This section gives an overview of these studies.

In 4 various hospitals, about 120 studies have been conducted in patients with various illnesses viz. eye infections, upper respiratory tract infections, retroviral infections, malaria, external cuts, abdominal pain and diarrhoea, urinary tract infections, sore throats, bronchitis, vaginal yeast infections, ear infections, gonorrhoea, pelvic inflammatory disease, various mouth problems, etc. Almost all the patients treated orally with 10 ppm SilverSol® recovered within 1 week (Data on file at ABL).

In Ghana, West Africa studies were conducted at three hospitals by Dr. Kwabiah, Dr. Sackey, Dr. Abraham at Air Force Station Hospital, Korie-Bu Teaching Hospital and at the Justab Clinic/Maternity respectively. Fifty-eight patients were treated by their attending physicians in respective hospitals using ASAP 10 ppm Solution as an alternative to antibiotics both as an oral and topical application depending upon the illness. Orally the dosages given were 5–10 ml twice or thrice a day for infections including malaria, HIV, fungal skin infections, upper respiratory and urinary tract infections, peptic ulcers, infectious abscesses, sore throat and pelvic inflammation. In case of halitosis and gingivitis it was used as a mouth wash, and in vaginal yeast infection as a douche. ASAP drops were used for eye infections such as conjunctivitis, ear infections viz. otitis media and upper respiratory tract infections such as sinusitis and rhinitis. Within 7 days all the patients showed a complete recovery. The details of all these studies are deposited as ABL's proprietary records and we are not describing these studies here in detail as the focus of the article is wound healing.

SilverSol® products, 10 ppm and 30 ppm SilverSol® liquid as well as the 32 ppm SilvrSTAT™ Hydrogel, have been used effectively in dental surgeries (e.g. extractions, bone grafts, Guided Tissue Regeneration, Periodontal, Laser, Dental Implant and Endodontic surgeries). A profound ability of these products has been demonstrated by Willoughby AJM in treating oral infections and speeding up wound healing without negatively impacting the oral microbiome (and probiotic bacteria) or gut health. The readers can visit www.ddsource.com to for specific dental protocols and learn more. Some of the data has been given in a separate Sect. 3.6.8 below.

2.5 Bioavailability and Pharmacodynamics of SilverSol®

After an oral administration of SilverSol®, its concentration in urine and blood was measured by Munger et al. in a study conducted in healthy volunteers (Munger et al. 2014). In this study, healthy volunteers were given 15 ml of 10 ppm and 32 ppm SilverSol® for 14 days. Blood and urine were collected on the 14 day at ≤2 h of ingestion of the last dose. In the group receiving a 10 ppm dose, blood samples were also collected at ≥ 24 h. after ingestion of the last dose on the third and seventh day of administration. Silver concentration was measured by using ICP-MS. In the 10 ppm dosing group, 42% of the subjects showed peak silver concentration (mean concentration 1.6 ± 0.4 µg/L) in the serum at the end of 14 days. In the group receiving the higher dose −32 ppm, 92% of the subjects showed peak silver levels in the serum with the mean value 6.8 ± 4.5 µg/L. There was no silver detected in the urine irrespective of time and dose.

As to the drug interaction, SilverSol® does not interfere with most of the other pharmaceutical compounds but works synergistically with medications. If taken in conjunction with antibiotics, it will work synergistically to increase the antibiotic effectiveness by tenfold (Sellman 2010). Importantly, its antibacterial action does not affect the beneficial gut flora. The nano-silver particles in SilverSol® do not interfere with the hydrochloric acid production in the stomach. They do not fall out of suspension and hence do not accumulate in the tissues or the skin reducing risk of developing Argyria, (i.e., blue man syndrome) (Roy et al. 2007).

2.6   SilverSol®: Regulatory Status and Products

Different forms of silver nano products viz. silver citrate, lactate, proteinate and nitrate have been listed in pharmacopoeias and formularies around the world. The old British National Formulary included a silver nitrate lotion, but it was removed from the 1993 issue (BNF 1993). and only silver sulphadiazine came thereafter. Cream of silver sulphadiazine and ophthalmic solution of silver nitrate have been included in The USP XXII (USP 1990). Other silver products such as silver metal, silver protein, silver acetate, silver nitrate, and silver sulphadiazine have been mentioned in Martindale, The Extra Pharmacopoeia (Martindale 1993).

In the case of SilverSol® products, several approvals from various regulatory agencies in the US, India and Canada have been received by ABL and VBPL. Product licence for Silver Biotics Antimicrobial Wound Cream and Silver Biotics Antimicrobial Hand & Body Wound Lotion has been given in Canada. Similar products for dental applications have been available under the brand names OraSIL™ and CuraSIL®. US FDA has given marketing approvals to various SilverSol® products viz. ASAP OTC® Wound Dressing Gel, SilvrSTAT® Antibacterial Wound Dressing Gel, AGRX Wound Wash Antibacterial Silver Skin and Wound Cleanser (Prescription), and AGX Wound Wash Antibacterial Skin and Wound Cleanser (Over the Counter). The FDA has recently approved the SilverSol® gel as a prescription medicine for its use in the management of diverse wounds viz. caused by first and second degree burns, abrasions, lacerations, diabetic ulcers, skin, tears, and surgical wounds (Sellman 2010). In India, VBPL has been granted product licenses by FDA; for Colloidal Silver Solution 10 ppm, Amorphous hydrogel with 32 ppm colloidal silver, Amorphous hydrogel Wound dressing with 32 ppm colloidal silver and Colloidal Silver Solution 40 ppm. In 1991, the USEPA established an oral reference dose of 0.005 mg/kg/day for silver. According to this recommendation, in the case of SilverSol® the daily intake limit of silver, for an average size adult, would be about an ounce/day of a 10-ppm product.

Considering the wider application of SilverSol®, a large number of products in gel or liquid form with different strengths are being produced and marketed by ABL. The main treatment categories where SilverSol® can be used, are immune support, skin and wound care, and oral care. These products are prepared without any artificial ingredients, dyes or flavours, preservatives or additives and are gluten free, 100% vegetarian and probiotic friendly.

3 Wound Healing

3.1 Global Challenge of Wound Healing

Would healing, controlled by several biological and molecular events, is a complex physiological cascaded process of damage repair involving three major phases viz. (1) cell migration and proliferation (2) extracellular matrix deposition and (3) remodeling. This normal course of healing may get impaired or delayed in certain pathophysiological and metabolic conditions viz. uncontrolled diabetes, diabetic neuropathy and vascular diseases resulting in the formation of chronic non-healing wounds (Mustoe et al. 2006). Infection of such wounds; especially with drug resistant microorganisms further complicates the management. Such situations often pose a therapeutic challenge to a medical practitioner.

Chronic wounds, due to prolonged morbidity pose a humanitarian and economic burden both at the individual and at the national level. Foot ulcers are common in diabetes (around 15–25%) (Marston 2006) which can become chronic and nonhealing due to complications of diabetes leading to a need of amputation of lower extremities in 12% of them (Greer et al. 2013). The costs involved with the healing of an ulcer can be up to US$45,000 (Paquette et al. 2002). In addition, the detrimental consequences on the patient's quality of life because of diminished mobility and significant loss of productivity will have a socioeconomic impact. In addition, acute and emergency wound care resulted through trauma, surgery and burns requires several procedures and high cost. In developed countries, the treatment cost of chronic wounds has been estimated to be 1–3% or even more of the total health care expenditure (Olsson et al. 2019). The prevalence of chronic wounds was reported to be 6% in 2016 in Wales, UK, which resulted in to about 5.5% cost to the National Health Services (NHS) (Phillips et al. 2016). Worldwide, the annual average cost for wound care has been reported to be around $2.8 billion in 2014 and it is estimated to rise up to $3.5 billion by 2021 (Settipalli 2015). Skin scarring is an additional burden implicated in wound healing, which can bring about an annual cost of US$12 billion.

Overall, wound healing is a great challenge both in case of handling acute emergencies and managing chronic non-healing wounds. Healthcare professionals and patients look for novel medicine, medical devices, and newer treatment modalities, that can offer better treatment options to improve healing rates, minimizing complications and reduce hospital stays. This has created great attention for both the scientific fraternity and the commercial enterprises. The current market of wound healing products surpasses US$15 billion annually and the amount spent per year for handling wound scarring is about US$12 billion (Sen et al. 2009). Innovative biotechnology-based treatment procedures and medical devices are being developed for improved wound management. The SilverSol® products for wound healing, developed using advanced technology are at the top of the list, scientifically addressing various aspects of wound healing viz. safety, efficacy, faster healing and reducing morbidity.

3.2 Historical Aspect of Wound Healing

Wounds and their management have been a part of human existence since the time man first arrived on earth. It is interesting to examine the methods and preparations used to heal wounds over the centuries. One of the oldest records of wound management—the Smith Papyrus, was discovered by Dr. Edwin Smith—a well-known scholar in 1862. The writings (discovered by Smith) date back to around 2600 BC and cover many aspects of patients' care. It describes the cleaning and suturing of wounds, the use of antiseptics, adhesives such as acacia gum and resins, and bandaging. The resin coated bandages that were used to wrap mummies were used for wound dressing. Over thousands of years, a variety of materials have been used for wound healing such as spider webs, dung, various species of animal and insects, vinegar, beer, wine, honey, leaves and tree bark. The first description of 'three healing gestures"—washing the wounds, making the plasters, and bandaging the wound was found to be recorded on mud tablets during 2200 BC (Ackerknech 1982; Richard 1991; Yardley 1998). An interesting description of using a bandage for wound healing was found in Mesopotamian culture. It states that "mix in milk and beer (the bandage) in a small copper pan; spread on the skin; bind on him (on patient's wound), and he shall recover" (Farrar and Krosnick 1991). Egyptians used adhesive bandages which contained honey and grease for the protection of wounds from infection and vegetable fibres' lint to aid drainage of the wound. Honey has been used for thousands of years and is still a part of many advanced wound dressings. Even in India, a long before the birth of Christ, honey was used for wound care. Greek Physician Hippocrates practiced 3 measures for wound healing— (1) cleaning and drying the wound edges (2) bringing wound edges as close as possible to accelerate healing and (3) applying warm or cold wine as an antiseptic (Farrar and Krosnick 1991; Brown 1992; Yardley 1998). This description interestingly covers the modern concept of 'The TIME' in wound management (vide infra).

3.3 Challenges in Wound Healing: Current Practices and Novel Approaches

A wound, whether minor or major causes a lot of suffering to the patient due to pain, swelling and inflammation. It can cause temporary disability to prolong immobility and affect the overall quality of life. However, inflammation of the wound, the first response of the body to an injury, induces migration of various polymorphic mononuclear blood cells and monocytes to the site of injury to remove cell debris and bacteria by phagocytosis. This initial inflammation is needed for initiating the proliferative phase of the damage-repair cascade. Under normal physiological conditions, the body follows the damage repair cascade—(1) cell migration and proliferation (2) extracellular matrix deposition and (3) remodeling, which leads to wound healing. Under such normal physiological condition, therapeutic management of wounds includes supportive drug/non-drug modalities to enhance the natural process. Various wound dressings and treatments have been evolved considerably to handle such conditions. However, in case of impaired physiological condition as in diabetes, or for severe wounds, their management is a foremost challenge. Wounds tend to become chronic, infectious and non-healing in patients with co-morbidities (vide supra). Such complex wounds that include lacerations, diabetic-, pressure-, and venous- ulcers, infectious third-degree burns require a systematic management strategy, simultaneously addressing inflammation, infection and impaired physiological process.

The inflammation phase is a part of a wound healing process at the initial stage of an injury (vide supra). However, prolonged inflammation can lead to tissue damage and hamper the natural healing process. Infection of wounds is a major concern that complicates wound management. Especially wounds that take longer to heal are more prone to infections. Skin is a natural habitat for common bacteria such as *Staphylococcus epidermidis,* and various other species viz. *Staphylococcus* and *Corynebacterium, Brevibacterium, Proprionibacterium acnes, Pityrosporum,* hence, serves as a potential source of wound contamination (Bowler et al. 2001b; Broughton et al. 2006; Schreml et al. 2010). Normally wounds can heal in presence of these bacteria, but colonization of bacteria in slow healing wounds may hamper the healing process. Cell debris and local hypoxia at the site of the wound promote bacterial colonization and subsequent chronic infection. Colonization by drug-resistant bacteria viz. MRSA and VRE further complicate wound healing. Proliferating bacteria at the site of the wound penetrate deeper healthy tissue resulting in tissue damage and uncontrolled inflammation, leading to severe wound. β-haemolytic *Streptococcus pyogenes* and *Streptococcus agalactiae, Staphylococcus aureus, Proteus, Klebsiella, Pseudomonas, Escherichia coli, Stenotrophomonas, Acinetobacter, a*nd *Xanthomonas* are common pathogens that can infect wounds (Bowler et al. 2001a; Ovington 2003).

Prolonged inflammation and severe infection of wounds affect the normal wound healing process. Infections reduce the essential growth factors and degrade fibrin that required for natural healing. Moreover, prolonged, and uncontrolled inflammation induces tissue damage, supports bacterial growth and collectively hampers the healing process. Biofilms formed by colonizing bacteria further create a hypoxic environment that damages tissues, supports bacterial growth and hampers fibroblast proliferation and collagen production required for the natural wound healing process. These factors collectively cause wound complications (Anderson and Hamm 2012; Okur et al. 2020). Besides these factors, other co-morbidities such as immune suppression and smoking have a negative influence on the wound healing process.

With the advancement of science and technology wound repair has become more organized through a holistic approach involving wound factors, local tissue factors, patient factors, and environmental factors. Increasing knowledge of the mechanisms of wound healing and advancement of technology has led to the expansion of superior wound healing modalities, such as hyperbaric oxygen therapy, bioengineered skin and tissue equivalents and negative pressure wound therapy (Wu et al. 2010). The details of such advancements are covered by others in this book. However, the authors would like to mention here that the 'The TIME' concept is now considered to be an essential Wound Care Process. It covers 4 important measures of wound healing namely **T**—Tissue, Removal of devitalized tissue, **I**—Inflammation/Infection and its prevention & control, **M**—Moisture Management and **E**—Edge Protection. SilverSol® wound healing products made from silver NPs address all these aspects of 'The TIME' concept.

3.4 Silver in Wound Healing

Hippocrates (400 BC) mentioned in his medical writing, beneficial effects of silver in healing and in disease-alleviating properties. He highlighted an ability of silver in tissue repair and wound healing and applied silver preparations for the treatment of ulcers. Marion Sims; an American physician in 1852, who was a pioneer in the field of surgery and known as the "father of modern gynaecology" used fine silver wires to close the fistulas after surgical repair of vesico-vaginal fistula. He also employed silver catheters for urinary diversion until complete healing of repairs (Sims 1884). Later, Halsted, an American surgeon, who stressed upon the strict sterile practices during surgical methods, treated wound infections using silver foil (Hill and Pillsbury 1939). In 1520 the Swiss physician Paracelsus used Silver nitrate as a medication for wounds,' both for an internal therapy and for a topical application (Alexander 2009). Later during the period of 1800–1900 silver nitrate was effectively used to treat complicated wounds like skin ulcers, compound fractures, and oozing wounds. Crusius treated burn injuries with silver nitrate in the 1890s (Alexander 2009). The first record of applying colloidal silver on the wound as an antiseptic was found in 1891 (Grier 1968). Roe in 1915, an ophthalmic surgeon used successfully colloidal silver to treat infected corneal ulcers (Roe 1915).

Though the knowledge of using silver in wound healing is age-old, efforts to develop newer forms of silver for optimizing delivery, efficacy and safety are still on even today. SilverSol® qualifies all these aspects to be a successful medicinal product.

3.5 SilverSol® in Wound Healing

SilverSol® is a breakthrough among the currently available advanced wound-healing technologies. It offers a next generation therapy for the treatment and management of severe, chronic and infectious wounds that are difficult to manage. The clinical efficacy of SilverSol® products has been proven in wounds of varied etiology such as acute and traumatic wounds, lacerations, diabetic-, pressure-, and venous- ulcers, infectious wounds, third degree burns, MRSA and VRE infected wounds. It is among the few nano-silver technologies which have received FDA approvals for various formulations for wound healing. Various formulations manufactured using the SilverSol® technology are available under brand names such as Armor Gel™, ASAP OTC™ Wound dressing gel, AGRX Wound Wash, Antibacterial Silver Skin and Wound Cleanser (Prescription), and AGX Wound Wash, Antibacterial Skin and Wound Cleanser (Over the Counter). SilvrStat®, Megaheal, Hyrdoheal, and Silverex Heal are some of the approved prescription varieties of the SilverSol® gel available for the treatment of lacerations, first- and second-degree burns, diabetic ulcers, skin tears, abrasions, various surgical wounds, and MRSA and VRE infected wounds.

3.6 SilverSol®: Mechanisms of Action

SilverSol® consists of metalic nano-silver with a silver oxide coating. Being metallic in nature, it acts through quite a different mechanism as compared to that of ionic silver. Various silver products irrespective of the form of silver present in them, require ionization of the metallic silver for their antimicrobial activity. The highly reactive Positively charged silver ions (Ag+) are highly reactive and kill microbes by binding to proteins, DNA, RNA, and chloride ions of microbes which are negatively charged. These ionic silvers can steal 1 electron, however, the SilverSol® metallic nano-silver has the ability to steal multiple electrons.

Conventional silver products hence (ionic form) impart their effect through the direct contact with microbes—chemical reaction. SilverSol® technology works by catalytic action, which allows the silver NP to first destroy the pathogens and then instantly recharge and "kill" continually—like a rapid-fire machine gun (Sellman 2010). This makes SilverSol® incredibly powerful, destroying pathogens thousands of times more effectively than a simple colloidal or ionic silver. This explains why other silver solutions/suspensions need to be used at concentrations up to 300,000 ppm of silver, while SilverSol® performs effectively even at 5–30 ppm. In addition, ionic silver can also bind to negatively charged particles viz. proteins and chlorides in the wound bed fluid, reducing the bioavailability of ionic silver. Hence the high concentration of silver will have to be used for maintaining prolonged activity.

It is also known that microbes develop resistance to ionic silver but not to metallic silver by sequestering silver in its more innocuous state—ionic or sulphide form and thereby detoxify the silver. However, as reported by Revelii et al., silver present in SilverSol® being metallic, accumulates in most bacterial cells in a quite different form—as small particles seen by PSI technique (vide supra) (Revelli et al. 2011). Secondly, each silver NP in SilverSol® remains always embedded with a resonant frequency, which allows the particles to have a continuous impact on things, without direct contact with them. Moreover, the particles also have an electrostatic charge that adds to its effect.

The unique wound healing action of SilverSol® has also been attributed to its stimulatory effect on stem cell regeneration (Sellman 2010). The two properties of metallic silver viz. high conductivity and bactericidal action impart the overall efficacy to SilverSol®. In addition, SilverSol® has been shown to be non-toxic to healthy cells in cytotoxicity experiments in vitro. This has an added advantage as local application of SilverSol® as the wound site will not have any damaging effect on healthy tissue.

SilverSol® is effective in wound healing. It exerts effects of silver at 3 levels namely, prevention or clearing of infections, improving healing process, and controlling inflammation. This ultimately speeds up the wound healing and reduces pain leading to a positive clinical outcome. SilverSol® has shown promising clinical efficacy in infectious wounds through its remarkable bactericidal activity even against MRSA and VRE. Several case studies demonstrating the effect of SilverSol® in the treatment of wounds of varied etiology and severity are discussed in the following section.

3.7 Effect of SilverSol® in Wound Healing: Case Studies

This section describes various case studies to illustrate the efficacy of various SilverSol® branded products (ARMOR GEL™, MEGAHEAL, ASAP OTC, SilvrSTAT®, Silver Biotics Pet Vet Veterinary Gel) used for the treatment of various skin infections, and wounds of varied etiology including mild cuts, lacerations, first and second degree burns, pressure ulcers, traumatic wounds, and chronic wounds. In addition, SilverSol® in a gel form, wound wash and dressing has been used in tens of thousands of patients undergoing dental procedures. The wound healing potential of these SilverSol® products viz. wash solution ASAP™ (10 ppm) and a gel SilvrSTAT™ (32 ppm) was evident through the remarkable recovery in these patients. A separate subsection is dedicated to summarizing effects in a large number of cases undergoing dental procedures. It opens another avenue for wound healing in oral care and focal infections (vide infra).

3.7.1 Use of SilverSol® in Mild to Moderate Wounds

Case 1 A 11-year-old female got a large abrasion on the elbow and hip due to a scooter accident. After thorough washing of wounds with water, ARMOR GEL™ was applied to the affected area once for the first 2 days, and every 2 days after that. Wounds were covered with bandages, which were changed and each time the ARMOR GEL™ was reapplied. The bandages were discontinued after day 7, but the ARMOR GEL™ was still applied once daily up to day 10. A complete recovery was seen by day 12 (Figs. 3 and 4).

Case 2 A 22-month-old male child got a deep injury on the forehead. The wound was approx. 2 cm. long and 0.75 cm deep. It was cleaned and sutured at the hospital. Bacitracin was applied to the wound and was protected with a bandage. After returning home, ARMOR GEL™ was applied instead of bacitracin. The gel was then reapplied with the bandage change—1–4 times in a day (a repeated application was needed as being small, the child tended to remove the bandage). A bandage and ARMOR GEL™ were continued till day 9 after removing the sutures by a physician on day 5. Complete healing was achieved by day 9 (Fig. 5).

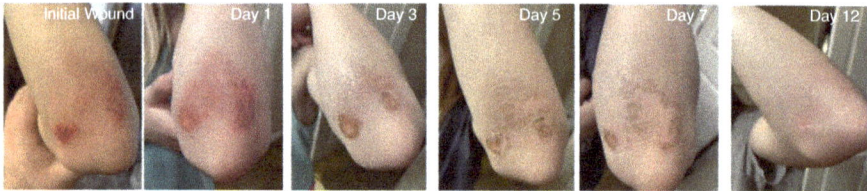

Fig. 3 CASE 1

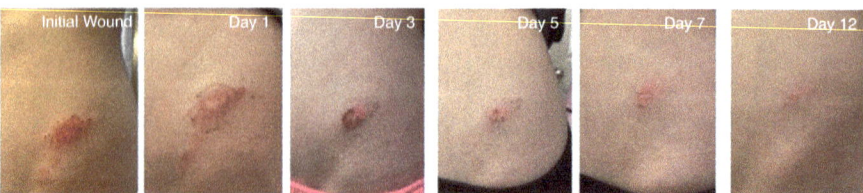

Fig. 4 CASE 1

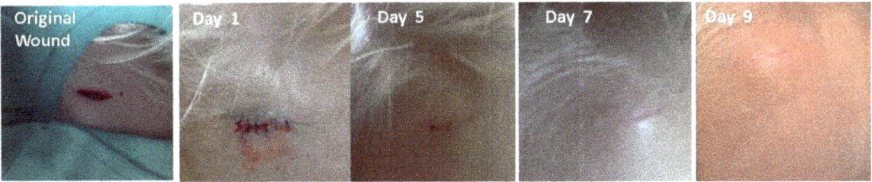

Fig. 5 CASE 2

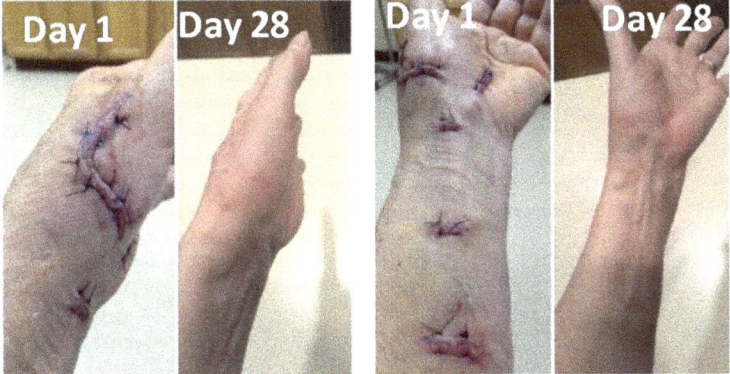

Fig. 6 CASE 4

3.7.2 Use of SilverSol® in Superficial Pellet Wounds

Case 3 (Published) SilverSol® product Megaheal amorphous hydrogel was used for multiple pellet wounds. A 35-year-old male patient was injured accidentaly by shotgun. The wounds caused by multiple pellets were superficial and had not damaged bones and distal neurovascular status been intact. Musculoskeletal injuries were significant. Supportive treatment was given along with broad spectrum antibiotics and anti-tetanus therapy. After washing the wounds with saline, Megaheal was applied thrice a day which was followed by dressing. Patient followed every fifth day, showed gradual healing of the wounds and painless shading of remnants pellet. A complete healing was achieved by 2 weeks. The detailed case report is available as a published paper (Dharmshaktu et al. 2016).

3.7.3 Use of SilverSol® in Surgical Wounds

Case 4 A 63-year-old female had multiple surgical wounds. ARMOR GEL™ was applied to the wounds and were covered with bandages. It was reapplied 5 times daily every time with the new dressing. No other treatment or products were used for wound healing. Pictures of the wounds were taken on day 1 and day 28 (Fig. 6).

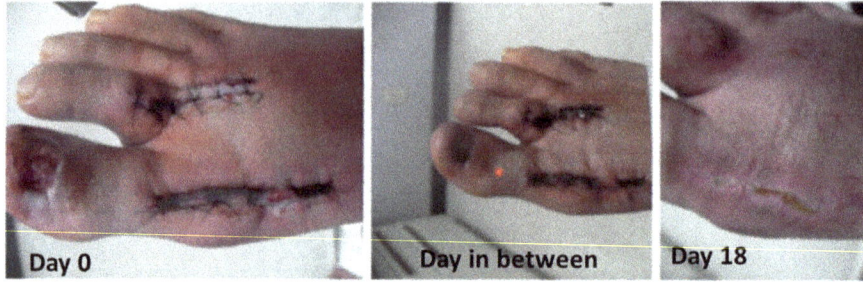

Fig. 7 CASE 5

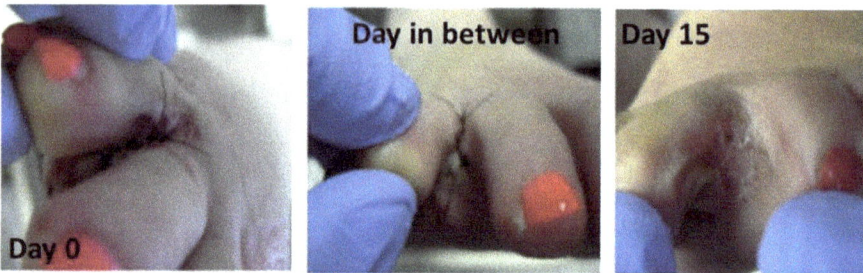

Fig. 8 CASE 6

Case 5 A 52-year-old diabetic male undertook the voluntary corrective repair of hallux valgus and pre-dislocated second metatarsophalangeal joint (MPJ) right foot. The Silver hydrogel was applied to sutures immediately after surgery and every third day throughout the post-surgical care. The patient's sutures were removed on day 18. There was no indication of dehiscence/pull out or any sign of infection (Fig. 7).

Case 6 A 56-year-old female patient underwent plantar digital neuroplastic surgery to her third interspace right foot. The Silver hydrogel was used after surgery with dressing which was changed every third day till the 15th day. On day 15 sutures were removed. There was no pull-out, dehiscence or infection seen (Fig. 8).

Case 7 A 72-year-old male diabetic patient had to go for surgical resection of his fourth metatarsal head which was secondary to acute osteomyelitis. After surgery on the day 3 silver hydrogel was applied to the surgical wound with the change of dressing. Thereafter on every third day repeated dressing with hydrogel application was continued for 21 days. The wound healed completely by day 21 (Fig. 9).

Case 8 (Published) A 34-year-old female is a case of drug abused (IV Dilaudid injection) who developed an abscess to her right foot. Incision and drainage procedure (I&D) was urgently needed. After I&D, SilvrStat® was applied for 4 days. The dressing was changed weekly thereafter till the coaptation of tissue. Sutures were removed on day 21 without signs of infection, dehiscence or pull out (Lullove and Bernstein 2015).

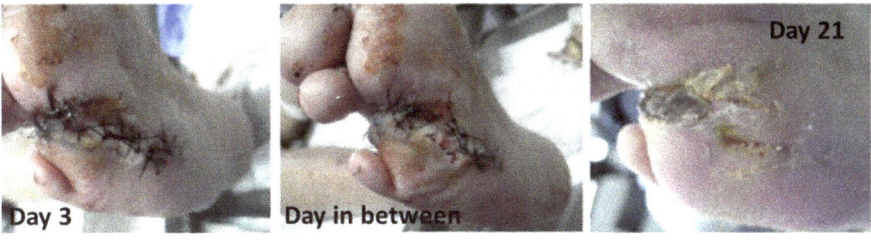

Fig. 9 CASE 7

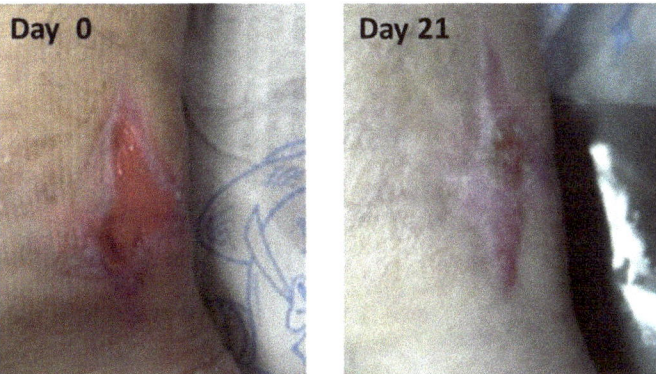

Fig. 10 CASE 9

Case 9 A 38-year-old female had a brown-recluse spider bite at the right medial lower leg. The wound so caused was irrigated and an ovine forestomach dermal template was used to obtained wound closure along with silver hydrogel application and collagen dressing. At the end of 6 weeks, there was no evidence of cytotoxicity with the collagen dressing or recurrence of infection. The wound healed completely thereafter (Fig. 10).

Case 10 Surgical wound with MRSA infection—A female patient after mastectomy got her wound infected with MRSA. Silver Sol gel was applied 4 times a day for 5 weeks. A series of photographs reveal gradual healing and complete resolution in 5 weeks (Fig. 11).

3.7.4 Use of SilverSol® for Skin Diseases

Case 11 A 5-year-old child suffering from eczema involving the bottom sides, and toes of the right foot. He was treated with the repeated application ARMOR GEL™ 24 ppm 2–3 times a day. On each application of ARMOR gel, after it could absorb, the Silver Biotics Skin Cream was applied over. After the open cracks and wounds

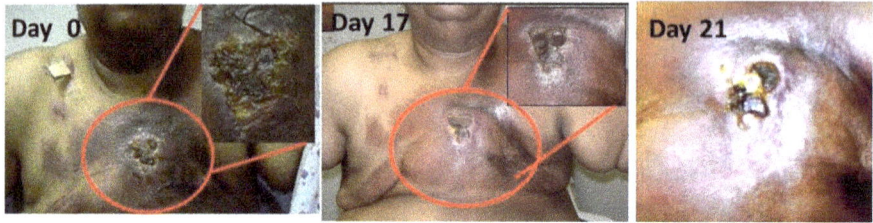

Fig. 11 CASE 10

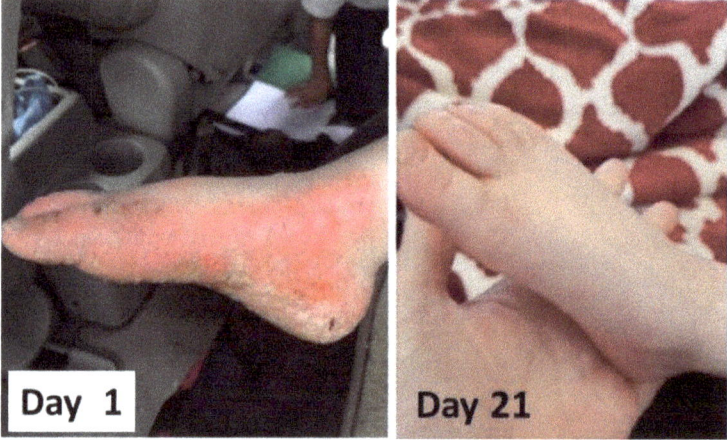

Fig. 12 CASE 11

healed, only the skin cream was used daily. There was complete healing of eczema without any sacring by day 21 (Fig. 12).

Case 12 A 10-year-old female suffering from ringworm infection with inflamed and painful skin was treated with ARMOR GEL™ local application twice a day and the affected area was left uncovered. The infection got completely cured by day 10 (Fig. 13).

3.7.5   Use of SilverSol® in Burns

Case 13 A 62-year-old male got a wound because of the spillage of hot grease onto the foot. The patient had diabetes and had undergone a liver and kidney transplant. The wound was covered with a bandage after the application of ASAP OTC and ARMOR GEL™ (24 ppm) hydrogel-wound dressing at the hospital. The ASAP OTC was then reapplied every day at the time of bandage change for 30 days. Thereafter only ASAP OTC was used on the wound which was completely healed by day 128 (Fig. 14).

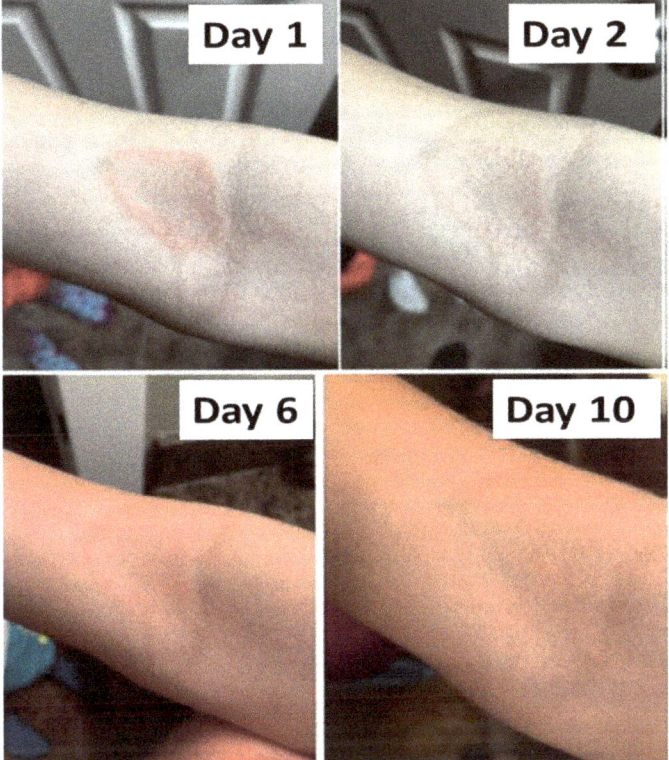

Fig. 13 CASE 12

Case 14 A 1.5-year-old female child burnt her hand on a stove burner. The patient was hospitalized for 3 days and treated by a physician for severe second degree burns. After being discharged, ARMOR GEL™ (24 ppm hydrogel wound dressing) was used topically whenever the dressings were changed. Skin grafts were suggested by the physician but were put on hold. At the physician's direction, the product used was switched from ARMOR GEL™ to the prescription version SilvrSTAT® from around day 17. The wounds were healed completely by day 105 (Fig. 15).

Case 15 A 62-year-old male underwent a laser procedure that was done for hyperpigmentation. His hand got burnt during the procedure. The patient's wound was cleaned after the examination by a physician. The wound was covered with a bandage after the application of ASAP OTC/ARMOR GEL™ (24 ppm hydrogel wound dressing). The bandage was changed once a day with the application of ASAP every time until day 8. After day 8, only ASAP OTC was used on the wound, and it was applied 3–4 times per day. Complete healing was seen by day 27 (Fig. 16).

Case 16 An 88-year-old woman had complex wounds due to burns which was needed skin implantation. But the skin grafting was not successful due to her age and

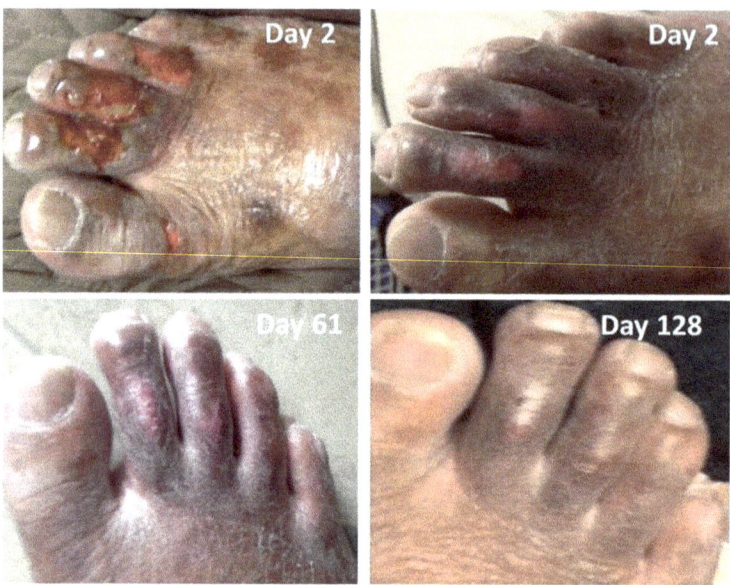

Fig. 14 CASE 13

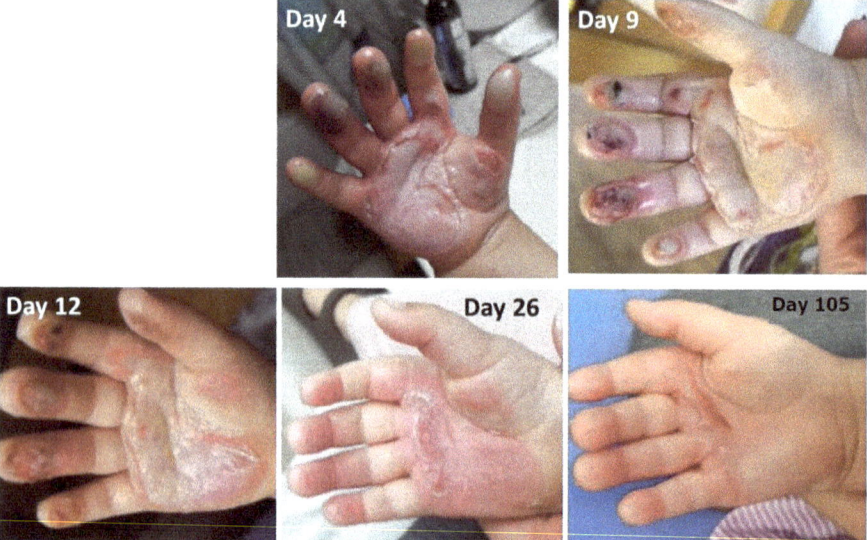

Fig. 15 CASE 14

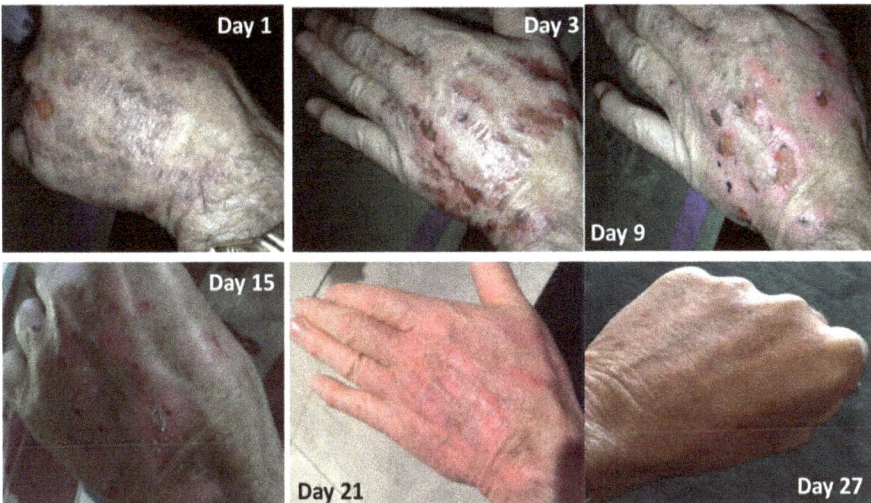

Fig. 16 CASE 15

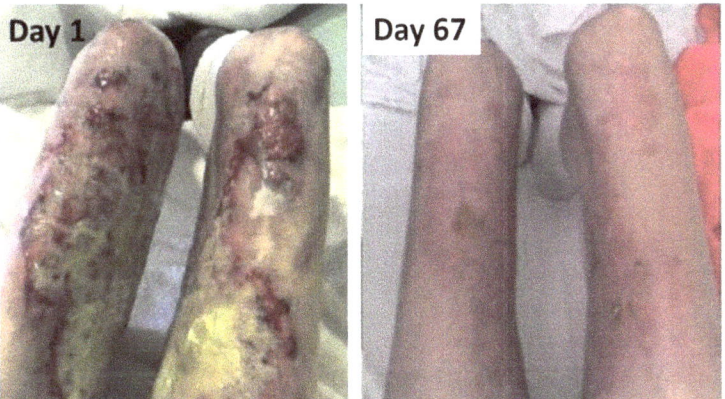

Fig. 17 CASE 16

compromised immunity. The patient recovered completely as the wound healed by 67 days of treatment with SivrSTAT® (Fig. 17).

3.7.6 Use of SilverSol® for Complicated Wounds

This section covers several complicated cases with traumatic laceration, diabetic wounds, infectious wounds (MRSA), various ulcers due to diabetes or vasculopathies, pressure ulcers (bed sores) etc. Several cases have shown recovery

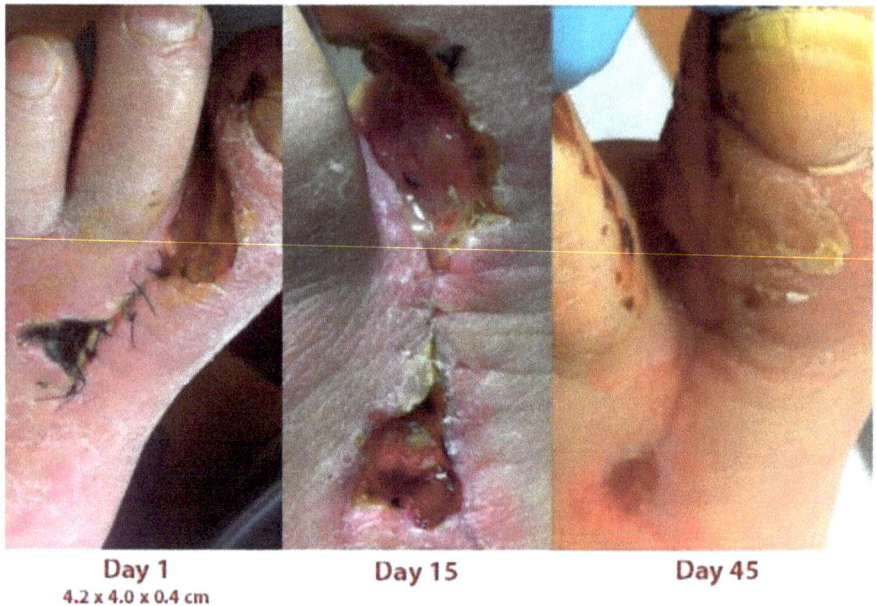

Day 1
4.2 x 4.0 x 0.4 cm
                    Day 15   **Day 45**

Fig. 18 CASE 17

from ulcers with various severity when treated with SilverSol® products. Some representative cases have been shown below.

Case 17 Diabetic wound infected with MRSA—A 71-year-old male diabetic patient with peripheral artery disease and past medical record of having sensory neuropathy, hypertension and dyslipidaemia. He was suffering from a complex wound located at the lateral left hallux extending to the dorsal left foot, secondary to his co-morbid conditions. Wound cultures were done which showed intense growth of MRSA along with *Proteus vulgaris* and *Enterobacter cloacae*. The wound was cleaned with debridement weekly and SilverSTAT® was applied daily with Adaptec dressing. A complete wound resolution was seen at the final evaluation on day 45 (Fig. 18).

Case 18 Diabetic wound infected with MRSA—A 70-year-old diabetic patient with an amputated limb suffered from a chronic wound. After amputation, the prosthetic device could not be fitted for 1 year due to the non-healing wound that further became complicated with MRSA infection. On the use of SilvrSTAT®, the wound healed completely within 4 months, and the patient became well enough for prosthetic fitting (Fig. 19).

Case 19 Traumatic serious laceration—A 47-year-old healthy male got a traumatic laceration about 2.5-inch-long to the eye and forehead. The orbital bone was broken, and a serious hematoma developed on both eyelids and the bridge of the nose. The wound was cleaned and closed by suturing that required eighteen stitches over the

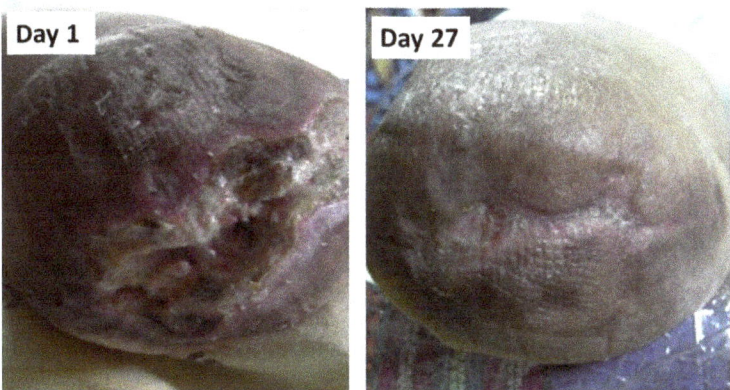

Fig. 19 CASE 18

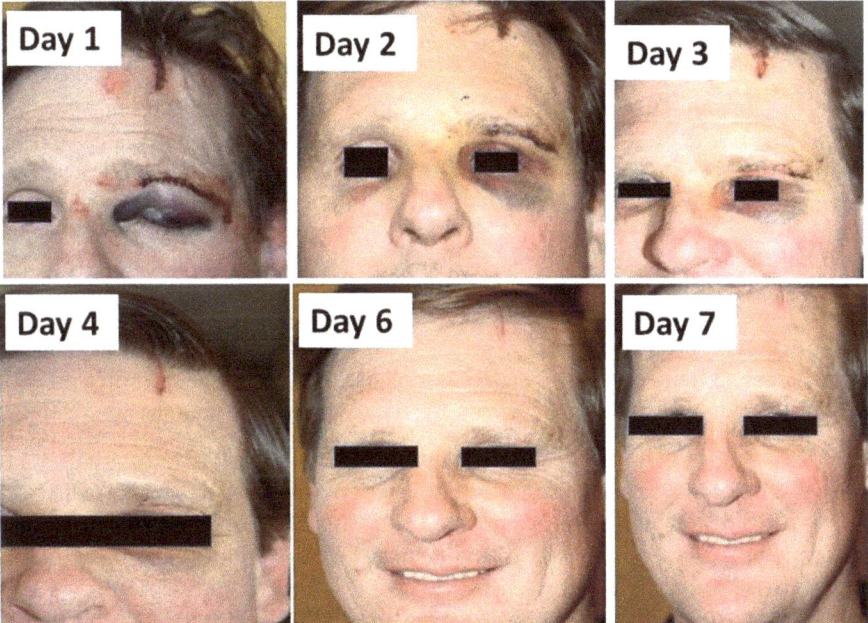

Fig. 20 CASE 19

eye and ten at the forehead. SilverSol® liquid was given orally two teaspoons twice a day and SilverSol® gel was applied topically 4 times a day. Pictures were taken at the hospital immediately after suturing and every day thereafter. Complete healing occurred by day 7 (Fig. 20).

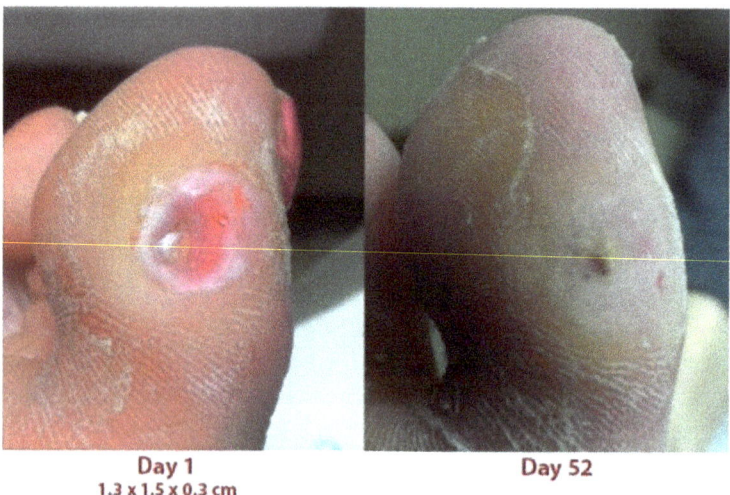

Day 1
1.3 x 1.5 x 0.3 cm   **Day 52**

Fig. 21 CASE 20

Case 20 Chronic Diabetic Ulcer A 70-year-old female patient presented with a wound that had not healed for 13 months. The wound was located at plantar medial right foot hallux. Medical history included diabetes; rheumatoid arthritis; hypertension and end stage renal disease. Treatment by debridement every 2 weeks and daily application of SilverSTAT® with Adaptic dressing brought about complete resolution of the wound by the final evaluation on day 52 (Fig. 21).

Case 21 and 22: Chronic Diabetic and Pressure Ulcers
ASAP wound dressing was used in 2 cases one having diabetic foot ulcers and the other having chronic pressure ulcer for 6 months. Both the cases were treated with SilverSol® and they showed recovery by 65 days and 3½ months of treatment respectively (Fig. 22).

Case 23: Bullous Pemphigoid MRSA Infected Wound
A 48-year-old woman having MRSA infection with Bollous Pemphigoid (autoimmune) complication was treated with SilverSol®. Oral administration of two table spoons twice a day of SilverSol® liquid was prescribed. The SilverSol® gel was given for topical application once a day at the time of bandage change. The treatment led to a reduction in MRSA infection and autoimmune attacks on tissues. Wound epithelialization was seen after 10 days of treatment (Fig. 23).

Case 24: Diabetic Ulcer
A 70-year-old diabetic female patient with sensory neuropathy developed ulceration. Earlier history indicated suffering from rigid bunion deformity, end stage renal disease, hypertension, and dyslipidaemia.

The patient was diagnosed having chronic neuropathic clinical infection Wagner grade 2, IDSA-no (Infectious Diseases Society of America) associated with the ulcer located at plantar medial right foot first metatarsophalangeal joint. Wound cultures report indicated moderate infection with Oxacillin-susceptible *S. aureus* and light

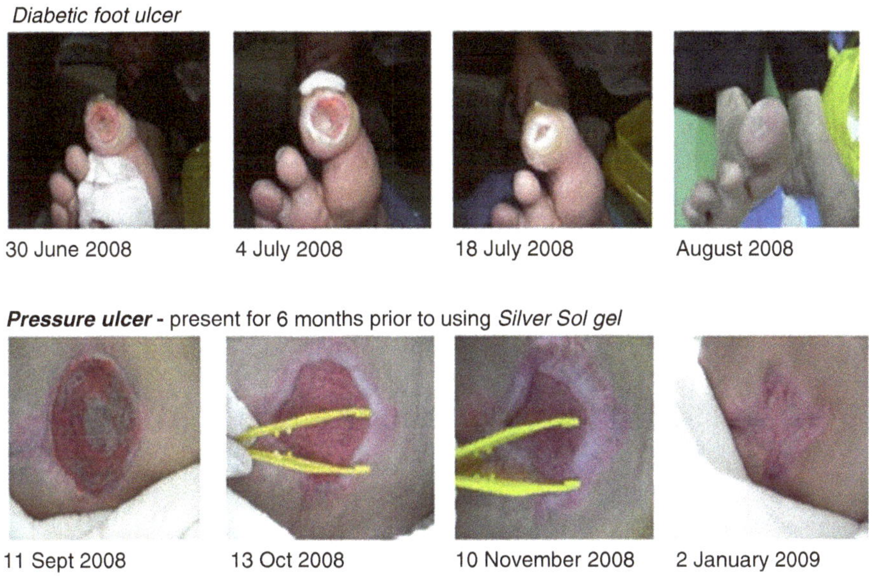

Diabetic foot ulcer

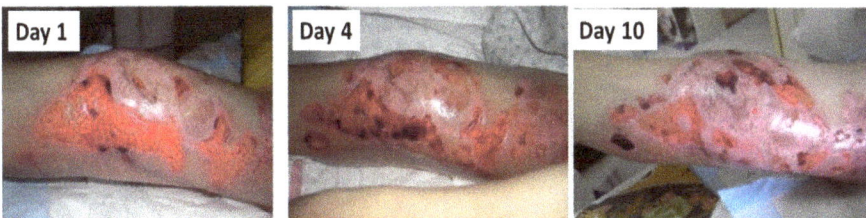

| 30 June 2008 | 4 July 2008 | 18 July 2008 | August 2008 |

Pressure ulcer - present for 6 months prior to using *Silver Sol gel*

| 11 Sept 2008 | 13 Oct 2008 | 10 November 2008 | 2 January 2009 |

Fig. 22 CASE 21, 22

Day 1   Day 4   Day 10

Fig. 23 CASE 23

growth of gram-negative rods. The patient's treatment included debridement and cleaning of the wound every 2 weeks. SilverSTAT® application with adaptic dressing was done daily. A Complete wound resolution occurred by the final evaluation on day 55 (Fig. 24).

Case 25 (published) A 33-year-old patient was seen in the hospital for an infected right great toe ulceration. The patient had the previous history of diabetic foot infections and this was the third occurrence on the same foot. The Patient encountered extreme pain of his right foot when he was brought to the hospital. Investigations revealed a 2.0 cm diameter ulcer to the medial aspect of the right great toe and a 4 cm tunnel from the proximal plantar first MPJ to distal plantar right great toe. No probing to the bone was identified. Past medical history included IDDM (for 26 years); depression; asthma; left great toe amputation (in 2012) The patient was allergic to Erythromycin and iodine. The patient was given operative treatment

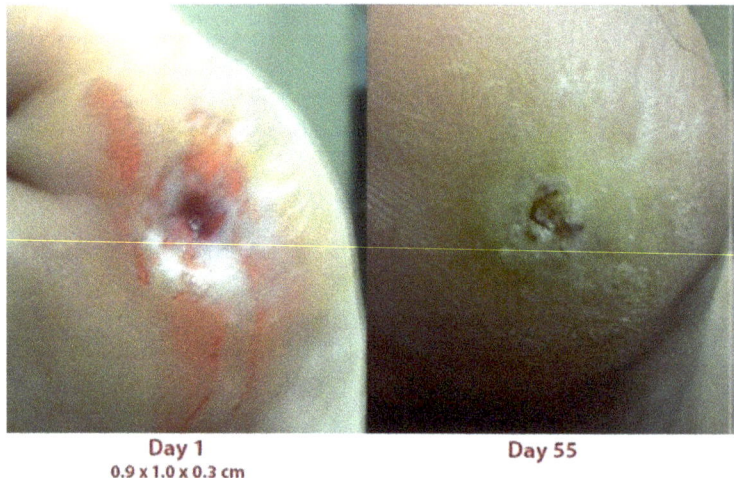

Day 1
0.9 x 1.0 x 0.3 cm

Day 55

Fig. 24 CASE 24

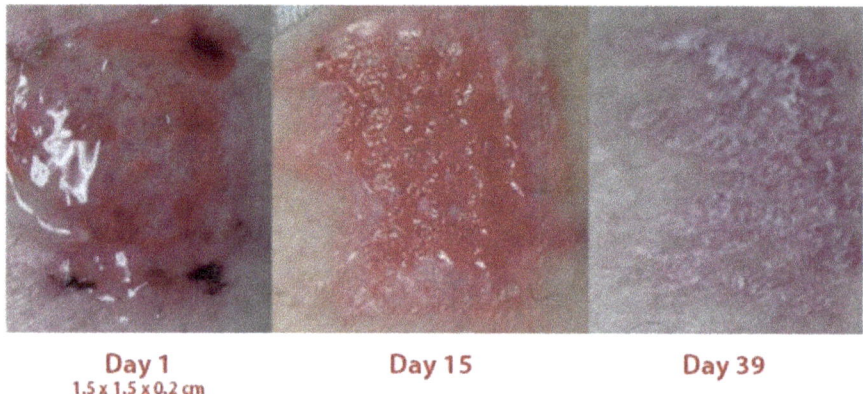

Day 1
1.5 x 1.5 x 0.2 cm

Day 15

Day 39

Fig. 25 CASE 26

which included debridement and incision/drainage of the abscess. Post-operative treatment included the application of SilverSTAT® every 3 days (Lullove and Bernstein 2015).

Case 26 A 58-year-old patient presented with an open wound to the left foot. The patient underwent a scheduled split thickness skin graft from the left thigh as a donor site. Past medical history included NIDDM and hypertension. Post-surgery, the patient was treated with SilverSTAT® to protect the donor site, viz. the left anterior thigh, and the application was changed every 3 days (Fig. 25).

Case 27 A 68-year-old non-diabetic patient came to the hospital for an initial consultation for a non-healing wound to the right foot. The patient was unable to

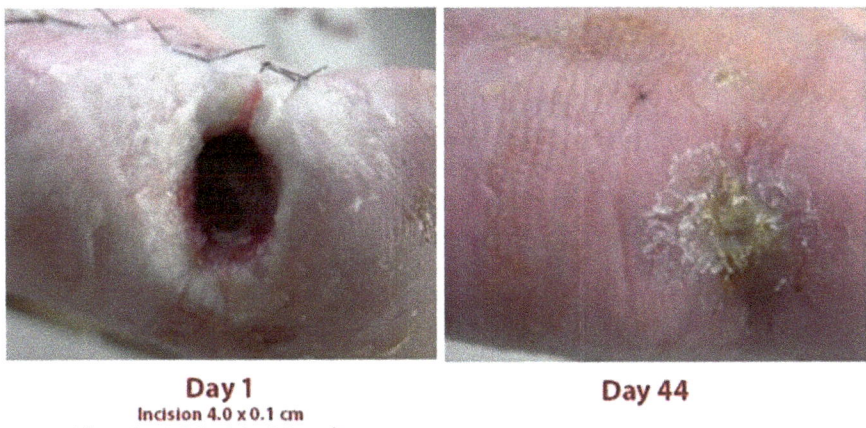

Day 1
Incision 4.0 x 0.1 cm
Ulceration is 2.0 x 2.0 x 1.2 cm deep

Day 44

Fig. 26 CASE 27

relate further information regarding the wound, however, he stated that he had problems walking and was also complaining of pain. The patient underwent interventional angioplasty, which showed occlusion of the peroneal artery of the right leg.

The patient was operated for osteomyelitis of the fifth metatarsal. The patient was placed on negative pressure wound therapy for 2 weeks. Also, SilverSTAT® was placed into the wound and the incision on the foot was protected with a collagen dermal template with ECM. The wound completely epithelialized by post op day 44. The integrity of the incision dorsally without large evidence of adhesion scarring or fibrosis was easily seen (Fig. 26).

3.7.7 Horse Leg Wound Healing Using PetVet Gel

Case 28 The cause of the wound was unknown. The horse was kept in an outdoor corral, and the owner thought it may have been caused by a cougar or possibly by another horse. There was a large flap of skin missing from the wound and it was infected. The attending vet suggested that putting the horse down was probably the best option due to the infection, the wound's large surface area, and the inability to cover the wound due to its location. The owner was familiar with the Silver Biotics Pet Vet gel and opted to try and heal the wound first. The Silver Biotics Pet Vet gel was applied to the wound once daily by the owner and left uncovered. No other wound healing products were used. The tissue regrew and the wound completely closed after about 6 months. The infection quickly subsided and did not reoccur. It continued to heal to almost unnoticeable conditions, with both the skin and hair growing back, and almost no scar tissue formation (Fig. 27).

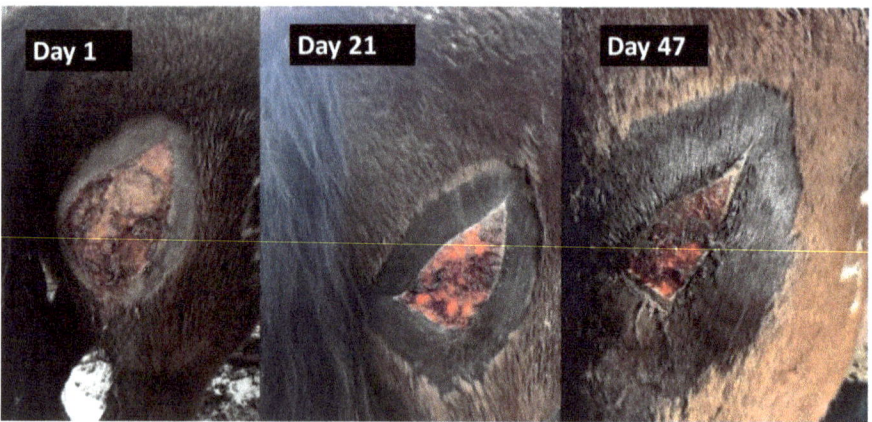

Fig. 27 CASE 28

3.7.8 SilverSol® in Dental Pain Infection and Inflammation: Case Studies

Andrew Willoughby, Reconstructive General Dentist from Canada and one of the co-authors of this chapter, used over 14 years SilverSol® products on his patients as a topical application during various dental surgeries *viz.* extractions, bone grafts, Guided Tissue Regeneration, Periodontal, Laser, Dental Implant and Endodontic surgeries. He has developed dozens of novel clinical protocols for specific surgical, endodontic and periodontal treatments utilizing these nano-silver products for various dental procedures. He has evaluated and assessed the efficacy and performance of the 10 ppm and 30 ppm SilverSol® liquid as well as the 32 ppm SilvrSTAT™ Hydrogel. Willoughby has confirmed the effect of these products in reducing the twelve most common oral pathogens and the formation of biofilms by them. He monitored DNA of these pathogens in saliva applying the polymerase chain reaction (PCR) test.

His exhaustive dental research with SilverSol® has demonstrated its profound ability to treat oral infections and speed up wound healing without negatively impacting the oral microbiome (and probiotic bacteria) or gut health. When utilized as a part of an integrated clinical protocol, Willoughby found that these Nano-silver products effectively eradicated bacterial infections associated with gum disease, tooth decay and dental infections, as well as accelerated wound healing and reduced inflammation. The fact that these products prevented post-operative/surgical infections and ultimately contributed to better patient outcomes far more effectively than other commonly used chemical disinfectants and antiseptics. He reported his findings and experiences using a combination of direct visual observations, intra-oral pictures, digital x-rays, computed tomography scans, ortho-pantographs, oral DNA tests and periodontal probing on over 22,000 patients having undergone almost 39,000 procedures for 20 different dental conditions (Fig. 28). He used different

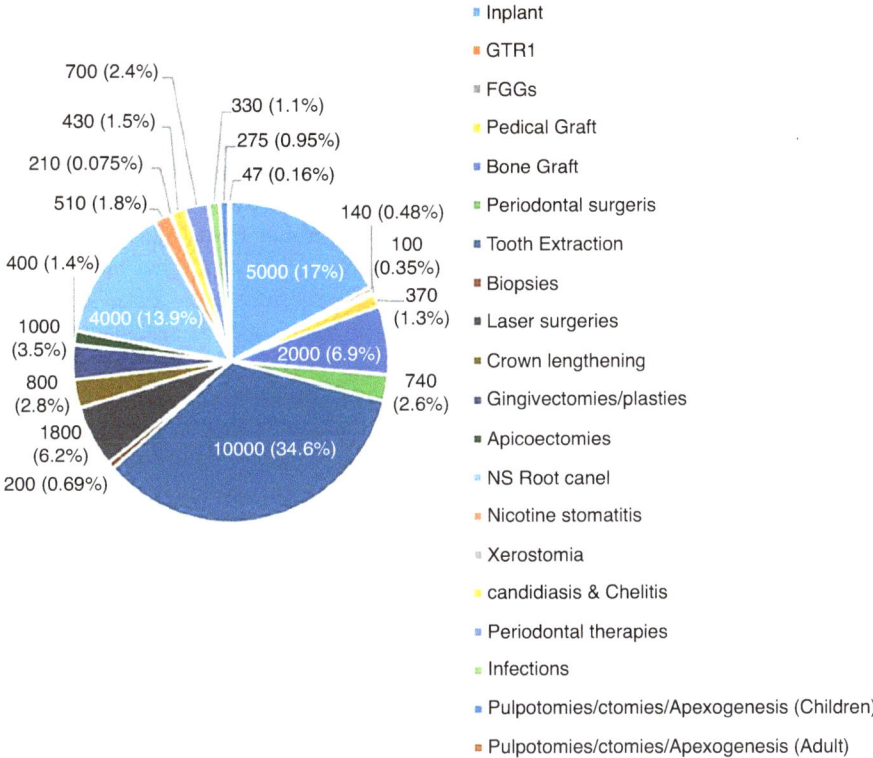

Legend:
- Inplant
- GTR1
- FGGs
- Pedical Graft
- Bone Graft
- Periodontal surgeris
- Tooth Extraction
- Biopsies
- Laser surgeries
- Crown lengthening
- Gingivectomies/plasties
- Apicoectomies
- NS Root canel
- Nicotine stomatitis
- Xerostomia
- candidiasis & Chelitis
- Periodontal therapies
- Infections
- Pulpotomies/ctomies/Apexogenesis (Children)
- Pulpotomies/ctomies/Apexogenesis (Adult)

Fig. 28 Various dental procedures for which SilverSol® products were used

Table 1 SilverSol® products used for dental procedures

1	Silverstat® Antimicrobial gel during the procedure followed by a 10-day regimen of ASAP 10 ppm + 3% diluted H_2O_2 mouth rinse (3×daily)
2	ASAP 10 ppm + 3% diluted H_2O_2 rinsing followed by ASAP 10PPM for final rinse during the procedure
3	ASAP 10 ppm during the procedure
4	ASAP 10 ppm + 3% diluted H_2O_2 rinsing followed by SilverStat®

Nano-silver products alone or in combination with 1–3% dilute H_2O_2 (Table 1). Willoughby states his experience in his report "...*After the use of these Nano-silver products (both during and after surgery) on thousands of my dental patients, the antimicrobial/wound healing benefits of this technology are truly impressive....the products help create a noticeable reduction in post-operative infection rates, accelerate healing times along with a clinically significant reduction in pain and swelling*..." Corroborating Dr. Willoughby's clinical research, other leading Dental Implant surgeons have experienced very similar clinical outcomes with these Nano-silver products.

3.7.9   Oral Care and Focal Infection: (Proposed?) Role of SilverSol®

The oral cavity and teeth are one of the major sites that provide favourable environment for the growth of microbes. Poor oral care can lead to biofilm formation. Adental plaque containing such biofilms can have higher than 10^{11} microbes/mg (Li et al. 2000). Such oral microflora comprisings of mainly anaerobic gram-negative rods of diverse species. Apical periodontitic teeth can harbour around 200 species and marginal periodontitic teeth can harbour more than 500 species (Moore and Moore 1994; Tronstad 1992). It is now well accepted that such a focal infection in the oral cavity can spread to distant sites in the body—'Focal infection theory' (Miller 1891). The theory proposed by W Miller in 1891 speculated diffusion of microbes and their toxins from the focal infection to a distant body site. Immunocompromised hosts suffering from chronic diseases viz. cancer, diabetes arthritis or patients receiving immunosuppressive treatment are more prone to focal infection. Moreover, marginal and apical periodontitis can be a potential risk factor for the development of systemic diseases. Various dental conditions viz. alveolar abscesses, pyorrhoea alveolaris (periodontitis), and apical periodontitis, cellulitis, general oral sepsis and endodontically treated teeth, pulppless teeth, with the infection caused by viridans group streptococci are the major cause of focal infection (Easlick 1951; Pallasch and Wahl 2000; Murray and Saunders 2000). Various dental procedures including endodontic treatment, periodontal surgery or even tooth extraction and root scaling to treat these conditions facilitate the dissemination of bacteria into the systemic circulation. This dissemination leading to bacteraemia may occur within a minute after the oral procedure. The displaced microbes can reach the peripheral blood capillary system, lungs and heart causing injury through local infection, microbial toxins and inflammation (Kilian 1982).

The SilverSol® product—Ag-gel has been shown to have remarkable activity by Tran et al. against bacteria contributing to tooth decay and plaque formation—Sect. 2.4.1 (Tran et al. 2019). It was found to be bactericidal and was able to prevent biofilm formation. The above-mentioned clinical experience of Andrew Willoughby during dental procedures enables authors to extrapolate that the effect of SilverSol® may have the potential to control focal infections and the resulting systemic consequences described above. However, experimental evidence have to be generated through further studies.

Conflict of Interest   The authors, A de Souza, Managing Director; Vora AH, Director; Mehta AD, director; and Godse CS Assistant Medical Director are from Viridis BioPharma Pvt. Ltd., whereas, Moeller K, is the Managing Director and Chief Executive Officer and Moeller C is the Director of Communication at ABL, LLC. Both, Viridis Biopharma and American Biotech manufacture and market SilverSol® products. Willoughby AJM is associated with DDSource, which is promoting SilverSol® products OraSIL™ and CuraSIL for dental applications.

References

Ackerknech E (1982) A short history of medicine. John Hopkins University Press, Baltimore, MD

Albanese A, Tang PS, Chan WC (2012) The effect of NP size, shape, and surface chemistry on biological systems. Annu Rev Biomed Eng 14:1–16. https://doi.org/10.1146/annurev-bioeng-071811-150124

Alexander J (2009) History of the medical use of silver. Surg Infect 10(3):289–292

Anderson K, Hamm RL (2012) Factors that impair wound healing. J Am Coll Clin Wound Spec 4 (4):84–91. https://doi.org/10.1016/j.jccw.2014.03.001

Berger TJ, Spadaro JA, Bierman R, Chapin SE, Becker RO (1976a) Antifungal properties of electrically generated metallic ions. Antimicrob Agents Chemother 10(5):856–860. https://doi.org/10.1128/aac.10.5.856

Berger TJ, Spadaro JA, Chapin SE, Becker RO (1976b) Electrically generated silver ions: quantitative effects on bacterial and mammalian cells. Antimicrob Agents Chemother 9(2):357–358. https://doi.org/10.1128/aac.9.2.357

Bhat GK, Suman E, Shetty A, Hegde BM (2009) A study on the ASAP nano-silver solution on pathogenic bacteria and candida. J Ind Acad Clin Med 10(1–2):15–17

BNF (1993) British national formulary no. 26. British Medical Association and The Pharmaceutical Press, London

Bowler P, Duerden B, Armstrong D (2001a) Wound microbiology and associated approaches to wound management. Clin Microbiol Rev 14(2):244–269. https://doi.org/10.1128/cmr.14.2.244-269.2001

Bowler PG, Duerden BI, Armstrong DG (2001b) Wound microbiology and associated approaches to wound management. Clin Microbiol Rev 14(2):244–269. https://doi.org/10.1128/cmr.14.2.244-269.2001

Broughton G 2nd, Janis JE, Attinger CE (2006) The basic science of wound healing. Plast Reconstr Surg 117(7 Suppl):12s–34s. https://doi.org/10.1097/01.prs.0000225430.42531.c2

Brown V (1916) Colloidal silver in sepsis. Am J Obstet Dis Women Childr 20:136–143

Brown H (1992) Wound healing research through the age. WB Saunders, Philadelphia, PA

Brown M, Anderson R (1968) The bactericidal effect of silver ions on *Pseudomonas aeruginosa*. J Pharm Pharmacol 20(S1):1S–3S. https://doi.org/10.1111/j.2042-7158.1968.tb09850.x

Brown T, Smith D (1976) The effects of silver nitrate on the growth and ultrastructure of the yeast Cryptococcus albidus. Microbios Lett 3:155–162

Coleman VR, Wilkie J, Levinson WE, Stevens T, Jawetz E (1973) Inactivation of herpesvirus hominis types 1 and 2 by silver nitrate in vitro and in vivo. Antimicrob Agents Chemother 4 (3):259–262. https://doi.org/10.1128/aac.4.3.259

Das M, Dixit S, Khanna S (2005) Justifying the need to prescribe limits for contaminants in food grade silver foil. Food Addit Contam 22(12):1219–1223

De Souza A, Mehta D, Leavitt R (2006) Bactericidal activity of combinations of silver--water dispersion with 19 antibiotics against seven microbial strains. Curr Sci 91(7):926–929

Dharmshaktu GS, Aanshu S, Tanuja P (2016) Colloidal silver-based nanogel as nonocclusive dressing for multiple superficial pellet wounds. J Family Med Prim Care 5(1):175–177. https://doi.org/10.4103/2249-4863.184659

Duhamel B (1912) Electric metallic colloids and their therapeutic applications. Lancet 1:89–90

Easlick KA (1951) An evaluation of the effect of dental foci of infection on health. J Am Dent Assoc 42(6):615–697

Farrar GEJ, Krosnick (1991) A: wound healing. Clin Ther 13:430–434

Greer N, Foman N, MacDonald R, Dorrian J, Fitzgerald P, Rutks I, Wilt TJ (2013) Advanced wound care therapies for nonhealing diabetic, venous, and arterial ulcers: a systematic review. Ann Intern Med 159(8):532–542. https://doi.org/10.7326/0003-4819-159-8-201310150-00006

Grier N (1968) Silver and its compounds. In: Block SS (ed) Disinfection, sterilization and preservation. Lea & Febiger, Philadelphia

Hill WR, Pillsbury DM (1939) Argyria–the pharmacology of silver. Williams and Williams, Baltimore

Holladay RJ, Herbert C, Moeller WD (2001) Apparatus and method for producing antimicrobal silver solution. US Patent Patent, US6743348B2 United States

Kilian M (1982) Systemic disease: manifestations of oral bacteria. In: JR MG, Michalek SM, Cassell GH (eds) Dental microbiology. Philadelphia, PA, Harpers & Row, pp 832–838

Lee S, Jun B (2019) Silver NPs: synthesis and application for nanomedicine. Int J Mol Sci 20 (4):865. https://doi.org/10.3390/ijms20040865

Li X, Kolltveit KM, Tronstad L, Olsen I (2000) Systemic diseases caused by oral infection. Clin Microbiol Rev 13(4):547–558. https://doi.org/10.1128/cmr.13.4.547-558.2000

link W Gold Hill, Utah Wikipedia. https://en.wikipedia.org/wiki/Gold_Hill_Utah

Lullove E, Bernstein B (2015) Use of SilvrSTAT® in lower extremity wounds: a two center case series. J Diabetic Foot Complications 7:13–16

Magaye R, Zhao J, Bowman L, Ding M (2012) Genotoxicity and carcinogenicity of cobalt-, nickel- and copper-based NPs. Exp Ther Med 4(4):551–561. https://doi.org/10.3892/etm.2012.656

Marston W (2006) Dermagraft diabetic foot ulcer study group. Risk factors associated with healing chronic diabetic foot ulcers: the importance of hyperglycemia. Ostomy Wound Manage 52 (3):26–28. 30, 32 passim

Martindale (1993) The Extra Pharmacopoeia, 30th edn. Pharmaceutical Press, London

Miller W (1891) The humanmouth as a focus of infection. Dental Cosmos 33:689–713

Miller LP, McCallan SEA (1957) Fungicides, toxic action of metal ions to fungus spores. J Agric Food Chem 5(2):116–122. https://doi.org/10.1021/jf60072a003

Moore WE, Moore LV (1994) The bacteria of periodontal diseases. Periodontology 2000 5:66–77. https://doi.org/10.1111/j.1600-0757.1994.tb00019.x

Munger MA, Radwanski P, Hadlock GC, Stoddard G, Shaaban A, Falconer J, Grainger DW, Deering-Rice CE (2014) In vivo human time-exposure study of orally dosed commercial silver NPs. Nanomed Nanotechnol Biol Med 10(1):1–9. https://doi.org/10.1016/j.nano.2013.06.010

Murray CA, Saunders WP (2000) Root canal treatment and general health: a review of the literature. Int Endod J 33(1):1–18. https://doi.org/10.1046/j.1365-2591.2000.00293.x

Mustoe T, O'Shaughnessy K, Kloeters O (2006) Chronic wound pathogenesis and current treatment strategies: a unifying hypothesis. Plast Reconstr Surg 117(7 Suppl):35S–41S. https://doi.org/10.1097/01.prs.0000225431.63010.1b

Nowack B, Krug HF, Height M (2011) 120 years of nanosilver history: implications for policy makers. Environ Sci Technol 45(4):1177–1183. https://doi.org/10.1021/es103316q

Okur ME, Karantas ID, Şenyiğit Z, Üstündağ Okur N, Siafaka PI (2020) Recent trends on wound management: new therapeutic choices based on polymeric carriers. Asian J Pharm Sci. https://doi.org/10.1016/j.ajps.2019.11.008

Olsson M, Järbrink K, Divakar U, Bajpai R, Upton Z, Schmidtchen A, Car J (2019) The humanistic and economic burden of chronic wounds: a systematic review. Wound repair and regeneration : official publication of the Wound Healing Society [and] the European Tissue Repair Society 27 (1):114–125. https://doi.org/10.1111/wrr.12683

Ovington L (2003) Bacterial toxins and wound healing. Ostomy Wound Manage 49 (7A Suppl):8–12

Pallasch TJ, Wahl MJ (2000) The focal infection theory: appraisal and reappraisal. J Calif Dent Assoc 28(3):194–200

Paquette D, Falanga V, Falanga V (2002) Leg ulcers. Clin Geriatr Med 18(1):77–88., vi. https://doi.org/10.1016/s0749-0690(03)00035-1

Pedersen G, Hedge B (2010) Silver sol completely removes malaria parasites from the blood of human subjects infected with malaria in an average of five days: a review of four randomized, multi-centered, clinical studies performed in Africa. Indian Pract 63(9):567–574

Pedersen G, Moeller K (2009) Silver sol improves wound healing. J Sci Healing Out 1(4)

Pedersen G, Sidwell R, Moloff A, Saum R (2008) Effect of prophylactic treatment with ASAP-AGX-32 and ASAP solutions on an avian influenza a (H5N1) virus infection in mice. J Sci Healing Out 1(1)

Phillips CJ, Humphreys I, Fletcher J, Harding K, Chamberlain G, Macey S (2016) Estimating the costs associated with the management of patients with chronic wounds using linked routine data. Int Wound J 13(6):1193–1197. https://doi.org/10.1111/iwj.12443

Rai M, Kon K, Ingle A, Duran N, Galdiero S, Galdiero M (2014) Broad-spectrum bioactivities of silver NPs: the emerging trends and future prospects. Appl Microbiol Biotechnol 98 (5):1951–1961. https://doi.org/10.1007/s00253-013-5473-x

Ravelin J (1869) Chemistry of vegetation. Sci Nat 11:93–102

Revelli DA, Lydiksen CG, Smith JD, Leavitt RW (2011) A unique SILVERSOL with surprisingly 1 broad antimicrobial properties. Antimicrobial 3(11):5–16

Richard M (1991) In: Carmichael AG, Richard RM (eds) Medicine: a Treasury of art and literature. Hugh Lauter Levin, New York, NY

Ricketts CR, Lowbury EJ, Lawrence JC, Hall M, Wilkins MD (1970) Mechanism of prophylaxis by silver compounds against infection of burns. Br Med J 1(5707):444–446. https://doi.org/10.1136/bmj.2.5707.444

Ritchie JA, Jones CL (1990) Antibacterial testing of metal ions using a chemically defined medium. Lett Appl Microbiol 11(3):152–154. https://doi.org/10.1111/j.1472-765X.1990.tb00147.x

Roe AL (1915) Collosol Argentum and its opthalmic use. Br Med J 16:104

Roy R, Hoover MR, Bhalla AS, Slawecki T, Dey S, Cao W, Li J, Bhaskar S (2007) Ultradilute Ag-aquasols with extraordinary bactericidal properties: role of the system Ag–O–H2O. Mater Res Innov 11(1):3–18. https://doi.org/10.1179/143307507X196167

Russell AD, Hugo WB (1994) Antimicrobial activity and action of silver. Prog Med Chem 31:351–370. https://doi.org/10.1016/s0079-6468(08)70024-9

Sanderson-Wells TH (1916) A case of puerperal septicaemia successfully treated with intravenous injections of collosol argentum. Lancet 1:258–259

Schneider G (1984) Silver nitrate prophylaxis. Can Med Assoc J 131:193–196

Schreml S, Szeimies RM, Prantl L, Landthaler M, Babilas P (2010) Wound healing in the 21st century. J Am Acad Dermatol 63(5):866–881. https://doi.org/10.1016/j.jaad.2009.10.048

Searle AB (1920) Colloids as germicides and disinfectants. In: The use of colloids in health and disease. Constable & Co., London

Sellman S (2010) A silver lining for Women's health. Total Health 30(5):24–27

Sen CK, Gordillo GM, Roy S, Kirsner R, Lambert L, Hunt TK, Gottrup F, Gurtner GC, Longaker MT (2009) Human skin wounds: a major and snowballing threat to public health and the economy. Wound repair and regeneration: official publication of the Wound Healing Society [and] the European Tissue Repair Society 17(6):763–771. https://doi.org/10.1111/j.1524-475X.2009.00543.x

Settipalli S (2015) A robust market rich with opportunities: advanced wound dressings.

Sims M (1884) The story of my life. D. Appleton & Co., New York

Tilton RC, Rosenberg B (1978) Reversal of the silver inhibition of microorganisms by agar. Appl Environ Microbiol 35(6):1116–1120

Tran PL, Luth K, Wang J, Ray C, de Souza A, Mehta D, Moeller KW, Moeller CD, Reid TW (2019) Efficacy of a silver colloidal gel against selected oral bacteria in vitro. F1000Research 8:267. https://doi.org/10.12688/f1000research.17707.1

Tran P, Luth K, Dong H, Dev A, Mehta D, Mitchell K, Moeller K, Moeller C, Reid T (2020) The in vitro efficacy of betadine antiseptic solution and colloidal silver gel combination in inhibiting the growth of bacterial biofilms. Preprints 2020:2020020419. https://doi.org/10.20944/preprints202002.0419.v1

Tronstad L (1992) Recent development in endodontic research. Scand J Dent Res 100(1):52–59. https://doi.org/10.1111/j.1600-0722.1992.tb01809.x

Trop M, Novak M, Rodl S, Hellbom B, Kroell W, Goessler W (2006) Silver-coated dressing acticoat caused raised liver enzymes and argyria-like symptoms in burn patient. J Trauma 60 (3):648–652. https://doi.org/10.1097/01.ta.0000208126.22089.b6

USP (1990) United States Pharmacopoeia. In, vol XXII. US Pharmacopoeial Convention, Rockville, Maryland.

Von Naegelli V (1893) Silver nitrate: a very effective antimicrobial agent. Deut schr Schweiz Naturforsch Ges 33:174–182

Wu SC, Marston W, Armstrong DG (2010) Wound care: the role of advanced wound healing technologies. J Vasc Surg 52(3 Suppl):59S–66S. https://doi.org/10.1016/j.jvs.2010.06.009

Wysor MS, Zollinhofer RE (1972) Antibacterial properties of silver chelates of uracil and uracil derivatives in vitro. Chemotherapy 17(3):188–199. https://doi.org/10.1159/000220853

Yardley PA (1998) A brief history of wound healing. Oxford clinical communications. Ortho McNeil Pharmaceuticals and Janssen-Cilag

Zanger UM, Turpeinen M, Klein K, Schwab M (2008) Functional pharmacogenetics/genomics of human cytochromes P450 involved in drug biotransformation. Anal Bioanal Chem 392 (6):1093–1108. https://doi.org/10.1007/s00216-008-2291-6

Preclinical Models for Wound-Healing and Repair Studies

Subramani Parasuraman

1   Introduction

The skin is a physical barrier and protects against microbial invasions and maintains temperature and fluid homeostasis. Destruction of skin barrier causes wound and this wound is major causes of gangrene and septic death (Öhnstedt et al. 2019). In addition, wound environment, including bacterial infection, oxidative stress, hypoxia, and ischemia are playing a major role in wound progression (Whittam et al. 2016). In recent years, many of the natural, synthetic, and semisynthetic compounds are explored for their pharmacological properties including wound care.

A wound is a form of injury that occurs relatively rapidly when the skin is torn, sliced, or punctured or when a contusion is created by blunt force trauma and therefore compromises its protective function; as a result, the wound can be contaminated by microorganism (Kenneth 2017). Healing is a process where the normal structure and functions of the injured cells are restored by two distinct processes such as regeneration and repair (Mohan and Mohan 2011). Acute and chronic wounds are the different forms of the wound (Demidova-Rice et al. 2012). In acute wounds, the normal processes of regeneration, tissue development, and remodeling occur in a timely manner. In a chronic wound, the wound fails to heal due to various pathological and physiological reasons that contribute to impaired healing (Clark. 2014; Krzyszczyk et al. 2018).

S. Parasuraman (✉)
Department of Pharmacology, Faculty of Pharmacy, AIMST University, Bedong, Kedah, Malaysia
e-mail: parasuraman@aimst.edu.my

© The Author(s), under exclusive license to Springer Nature Singapore Pte Ltd. 2021   397
P. Kumar, V. Kothari (eds.), *Wound Healing Research*,
https://doi.org/10.1007/978-981-16-2677-7_13

2   Wound Healing

Wound healing is a normal biological process that consists of four phases: hemostasis, inflammation, proliferation, and tissue remodeling or resolution. The process of wound healing depends on the patient factor (age, underlying illnesses/disease, and the effect of the injury on healing), wound factor (organ or tissue injured, extent of injury, nature of injury, contamination or infection), and other local factors such as hemostasis and debridement, and timing of closure (Wound Management).

Wounds that show impaired healing have not progressed through the normal stages of the wound healing process and become a condition of pathologic inflammation. In the United States, nonhealing wounds affect around 3–6 million people and result in an immense annual expense of >$3 billion per year (Zarubova et al. 2020). The percentage of nonhealing wounds is increasing globally; hence, the search for newer agents has also increased. To study the therapeutic efficacy of newer agents, preclinical models are used. These models are used to study the complex biochemical and cellular process of wound healing and repair and to study the safety, efficacy, and potency of any investigational compounds. In this chapter, the animal models used in wound healing and repair studies are briefly described.

3   Preclinical Models

The preclinical models are used to study the pharmacological and toxicological properties of investigational drugs/compounds. In pharmacological studies, drug release patterns (pharmacokinetic) and efficacy are evaluated using cell lines or animal models. In toxicological studies, local and systemic toxicities are studies. The pharmacological models used in wound-healing and repair studies are summarized in Table 1 and different types of rodent wound models are summarized in Fig. 1.

Mice, rats, pigs, and rabbits are most commonly used as experimental animals in wound-healing research. In that, pig models have a higher correlation to human healing. In wound-healing research, in vivo models are most commonly used, because these models allow investigating the biochemical, molecular, and histological changes in wound and healing parameters. In vivo models are simulating the real body condition and the experiment is clinically relevant but it is a time-consuming process. All the in vivo experiments should be carried out with the prior approval of the Institute ethical committee.

In recent decades, in vitro experiments are recommended in preclinical research to avoid the utilization of animals in research. In vitro experiments also reduce the duration of the experiment and help the researchers to screen the larger number of investigational compounds in a short duration. The in vitro experiment has few limitations which include, (1) not suitable for the long-term pharmacological and

Table 1 Pharmacological models used in wound healing and repair studies

Type of study	Model
In vitro	Cell proliferation assay
Ex vivo	Skin explant
In vivo (acute wound models)	Excision wound
	Incision wound
	Burn wound
	Dead-space
	Wound chamber
In vivo	Ischemic wound (ear wound)
	Pressure ulcer
	Pressure-induced deep tissue injury
	Biofilm-infected wound
	Diabetic wound
	Flap surgery
	Chemically impaired wound healing
	Parabiosis
	Denervated wound
	Tape stripping
	Xeno-grafts
In silico	Molecular docking studies

toxicological experiment; (2) may give false-positive results because of low specificity; (3) in general, the culture methods are more expensive.

3.1 Cell Proliferation Assay

Coagulation, inflammation, migration-proliferation, and remodeling are four different phases of wound healing. After tissue injury, the wound margins may exhibit a higher proliferative activity. The proliferative activities may be influenced by the various growth factors and cytokines, keratinocytes. The effect of the investigational compounds on cell migration, cell proliferation, and wound healing is studied using scratch assays.

Requirement: Human keratinocyte cell or Human peritoneal mesothelial cells, mouse keratinocyte cell, culture media, EDTA-trypsin solution, methylthiazol tetrazolium (MTT), colorimetric assay, culture flasks/microtiter plate, contrast microscopy.

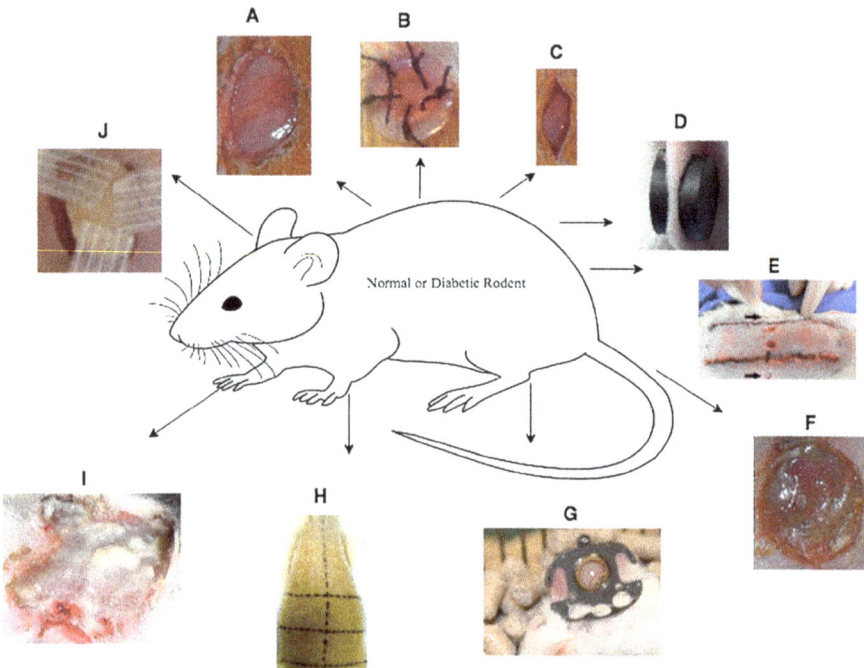

Fig. 1 Different types of rodent wounds models A: Excisional wound; B: Excisional splinted wound; C: Incisional wound; D: Ischemia/Reperfusion ulcer; E: Ischemic wound; F: Open ulcer; G: Skin fold chamber; H: Burn model; I: Infected wound and J: Xenograft. (Reprinted with permission from Elsevier (License Number 4886771206862; Dated 12 Aug 2020); Reference: Sami et al. 2019)

3.1.1 Procedure

- Cutaneous wound healing is a multistep process involving different types of cells, including epidermal keratinocytes, endothelial cells, fibroblasts, and peripheral nerve cells. During wound healing, re-epithelialization could be a critical step involving the relocation and proliferation of keratinocytes from the encompassing epidermis and appendages. Keratinocytes at the wound margin continue to proliferate behind actively moving cells after an injury, resulting in a dense, hyperproliferative epithelium seen at the wound margin as migrating cell sheets. (Hara-Chikuma and Verkman 2008).
- *Human keratinocyte cell cultures:* Neonatal human keratinocytes (cultured within keratinocyte basal medium [KBM]) or Human keratinocytes from adult skin (cultured within Dulbecco's Modified Eagle Medium [DMEM] with 10% heat-inactivated fetal bovine serum [FBS] and 1% L-glutamine) are cultured in suitable culture medium. EDTA-trypsin solution is used for detaching cells from flasks/microtiter plates (Hara-Chikuma and Verkman 2008). In vitro wound-healing assay is used to measure cell proliferation and cell migration.

- *Cell proliferation assay:* Cell proliferation activity is evaluated by using the MTT colorimetric assay. The assay is based on the reduction of a yellow tetrazolium salt to purple formazan crystals. The cells are seeded at a density of 1×10^4 cells/well in a 96-well microtiter plate, incubated at 37 °C under 5% CO_2 for 24 h prior to treatment. Later, 100 μL of test compound (different concentration) is added to the well and incubated at 37 °C under 5% CO_2 for 48 h. After the exposure, cell viability is determined by the MTT assay. An aliquot of MTT solution (25 μL, 5 mg/mL) is added into each well and incubated at 37 °C in a 5% CO_2 incubator. After 3 hours (h) incubation, 200 μL of Dimethyl sulfoxide (DMSO) is added to each well and shaken for 10 min to dissolve the purple formazan crystals formed. The Optical Density (OD) of the resulting purple dye is measured at 570 nm using an ELISA plate reader (Wang et al. 2011).
- *Wound scratch assay:* The cells are seeded at a density of 5×10^4 cells per well are seeded in six-well cell culture plates. The cells are allowed to grow to 70–80% confluence as a monolayer. Later the monolayer confluent cells are scrapped over the horizontally with a sterile pipette tip. Another scratch is made in a perpendicular way to the first, making a cross in each well. After scratching, the medium is removed by washing with phosphate-buffered saline (PBS). Later, a fresh medium containing 5% V/V of heat-inactivated FBS and treatments is added to each well, and cells are grown for 24 h (Governa et al. 2019). The scratch induced that represented wound is photographed at 0 h (before the treatment) and 24 h (after the treatment) using phase-contrast microscopy at × 40 magnifications. Wound region and total area are measured using suitable software. The rate (percentage) of wound closure is calculated mathematically.

$$\text{Percentage of wound closure} = \frac{(\text{Wound area 0 h} - \text{Wound area 24 h})}{\text{Wound area 0 h}} \times 100$$

- *Rat/Human peritoneal mesothelial cells:* Rat peritoneal mesothelial cells are obtained from the peritoneal walls of male Wistar rats and identified by immunocytochemical assay techniques (Matsumoto et al. 2012). Human peritoneal mesothelial cells are collected from the omental tissue of the patients who are experiencing abdominal surgery (Ryu et al. 2012). The peritoneal mesothelial cells are maintained in DMEM containing glucose (5.6 mM), FBS (10%), penicillin (100 units/ml), and streptomycin (100 μg/ml) in 5% CO_2 at 37 °C (Matsumoto et al. 2012). In vitro wound-healing assay is used to measure cell migration.

– *Wound Scratch Assay:* The cells are seeded in a 24-well culture plate at a density of 2×10^5 cells/well in DMEM medium, incubated at 37 °C and 5% CO_2. The cells are allowed to grow to 70–80% confluence as a monolayer. Later the monolayer confluent cells are scrapped over the horizontally with a sterile pipette tip and debris is removed by washing with PBS. The cells are treated at various concentrations of investigational compound by diluting with serum-free DMEM. The scratch induced that represented the wound is photographed at 0 h (before the treatment) and 24 h (after the treatment) using phase-contrast microscopy at $\times$ 40 magnifications. The migration rate is calculated using suitable software and wound closure is calculated mathematically (Muniandy et al. 2018).

$$\text{Percentage of wound closure} = \frac{(\text{Measurement at 0 h} - \text{Measurment at 24 h})}{\text{Measurement at 0 h}} \times 100$$

– *Cell spreading assay*: Increased cell spreading can enhance epithelial wound closure. The intestinal epithelial cells are seeded on 13-mm collagen-coated coverslips at ~40 confluencies, and allowed to adhere overnight. Later, the cells are stimulated with an investigational compound for 4 h. Prior to the experiment and after the stimulation of cells, the cell surface area is measured and compared to nonactivated cells. E-Cadherin, a junctional marker is to strain the cell surface area and individual cell surface area is quantified using the images obtained from confocal microscopy with the help of suitable software (Sumagin et al. 2013).

3.2   Skin Explant

Skin explant is an ex vivo model. In this model, explant culture is used for quantitative assessment of the wound epithelialization. It is also used to study the marker of interest in the epithelialization process at specific time points (Mazzalupo et al. 2002). Human and Swine skin are used as a model in skin explant wound.

Requirement: Hair remover, 70% ethanol, anesthetic agent, rodent surgical table, surgical scalpel handle, scalpel blade, surgical scissors, surgical sutures, rodent cage with fasting grills, cotton, polyethylene bottle, hydroxyproline.

3.2.1   Procedure

• This model requires anesthesia. The animals are anesthetized with a suitable anesthetic agent.

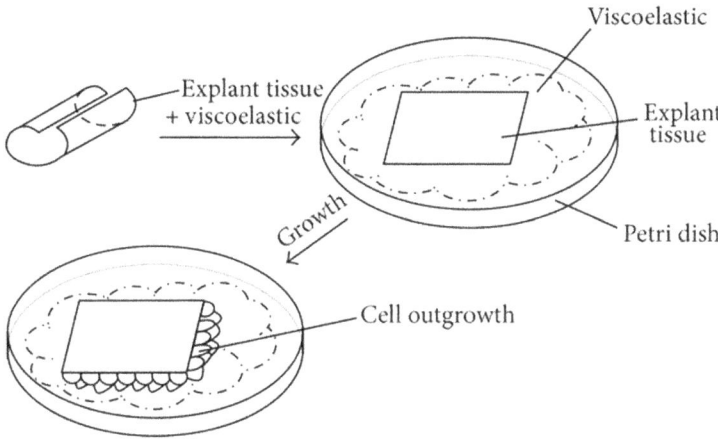

Fig. 2 Method for adherent ex vivo cultivation (adapted from Andjelic et al. 2014)

- *Human skin:* Human abdominal skin is obtained from healthy donors. Dermal fragments are obtained by cutting fresh skin samples in sterile conditions in 1 mm^3 fragment (Isnard et al. 2001).
- *Swine skin:* Skin specimen is excised from the paravertebral region of a young swine (2–4 months old). The skin is excised with an electrokeratome at a depth of 0.2–0.3 mm and explants are prepared using the sterile technique (Hebda. 1988).
- Explants are incubated in a culture medium (contains culture medium, FBS, antibiotics, and growth factors) and examined daily under a phase-contrast microscope.
- The medium is replenished with a DMSO vehicle or investigational compound daily. Equal numbers of cells are seeded in the culture plate. The control cultures are maintained in order to determine the increase in cell number during the incubation period (Isnard et al. 2001).
- Explants are examined daily for initiation of epidermal outgrowth under phase contrast microscope. Representative outgrowths are photographed and the region of outgrowth is approximated as the result of the width and height measurements. (Mazzalupo et al. 2002). Normally, the percentage growth of explants is measured on days 1, 2, and 3, and the radius of outgrowth is measured on days 2, 4, and 7 of the experiment (Fig. 2).

3.3 Excision Wound

The rodent excision wound model is one of the widely accepted models to study the effect of the investigational compound for its wound-healing activity because of the simplicity and reproducibility of the method. The rodent excisional wound model is an acute model and widely used to study the healing index of any investigational compound. This model is considered to resemble acute clinical wounds that required healing by second intention. Wound healing in rodents is primarily by contraction whereas a human heals by re-epithelialization. Mice, rats, rabbits, and pigs are commonly used as experimental animals in the excision wound model.

Requirement: Hair remover, 70% ethanol, anesthetic agent, rodent surgical table, surgical scalpel handle, scalpel blade, surgical scissors, rodent cage with fasting grills, cotton, graph paper.

3.3.1   Procedure

- This model requires anesthesia. The animals are anesthetized with a suitable anesthetic agent.
- The animal is placed on a rodent surgical table.
- The back of the animals is shaved. If requires, a hair remover can be applied to remove the hair on the dorsal surface. Preferably on the thoracic region (1 cm away from the vertebral column and 5 cm away from the ear of the rats).
- After shaving the skin surface is cleaned with 70% ethanol.
- Later, an area of 7×7 mm or about ≈ 500 mm^2 is marked to make an excision.
- An excision wound is created by a surgical blade and the wound is left undressed (Fig. 3).
- The wounded animals are house individually on rodent cages that contain fasting grills to avoid coprophagia and other infections.
- After the creation of the wound, the investigational compounds are applied once/ twice daily for 21/24 consecutive days.
- The wound area is measured at regular intervals after the creation of the wound with the help of a transparent sheet using millimeter-scale graph paper. Generally, wound contraction is measured at every 4 days interval until completion of the study.

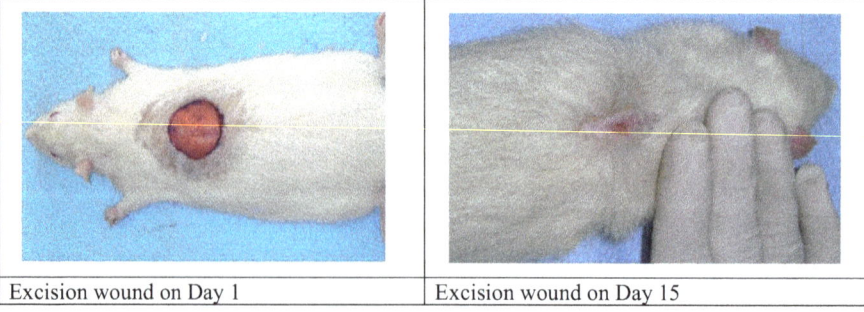

| Excision wound on Day 1 | Excision wound on Day 15 |

Fig. 3 Excision wound model (adapted from Nayak et al. 2011)

- The wound-healing effect of the investigational compound is compared with the untreated group and the percentage of wound healing is calculated using the following formula (James and Victoria. 2010; Murthy et al. 2013):

$$\text{Percentage of wound healing} = \frac{\text{Healed area}}{\text{Total wound area}} \times 100$$

- At the end of the study, tissue of the excision wound is collected and used for histopathological analysis to study the microscopic structure of inflammatory cells, necrotic cells, fibroblast cells, collagen fibers, and blood vessels.

3.4 Incision Wound

The rodent incisional wound model is an acute model. Incisional is classified as primary (first intention) or secondary closure (second intention) and sutured immediately after wound infliction or not respectively. Primary and secondary closure is the model to study the biomechanical investigation of wound strength and examine scarring at late time points, respectively (Masson-Meyers et al. 2020). Mice, rats, rabbits, and porcine are commonly used as experimental animals in the incision wound model.

Requirement: Hair remover, 70% ethanol, anesthetic agent, rodent surgical table, surgical scalpel handle, scalpel blade, surgical scissors, surgical sutures, rodent cage with fasting grills, cotton, polyethylene bottle, hydroxyproline.

3.4.1 Procedure

- This model requires anesthesia. The animals are anesthetized with a suitable anesthetic agent.
- The animal is placed on a rodent surgical table.
- The back of the animals is shaved. If requires, hair remover can be applied to remove the hair on the dorsal surface of the paravertebral area.
- About 6-cm long, two linear-paravertebral incisions are made using a sterile surgical blade through the full thickness of the shaved skin at a distance of 1.5 cm (for rats) from the dorsal midline of each side of the vertebral column.
- Later, the wound is closed with three surgical interrupted sutures (non-absorbable braided non-capillary and siliconized) of 1 cm apart (Fig. 4).
- The wounded animals are house individually on rodent cages that contain fasting grills to avoid coprophagia and other infections.
- After the creation of the wound, the investigational compounds are applied once/ twice daily for 8 consecutive days.

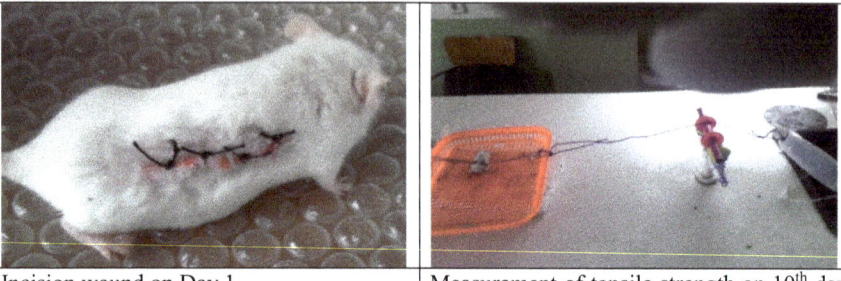

| Incision wound on Day 1 | Measurement of tensile strength on 10th day using water flow technique |

Fig. 4 Incision wound model (adapted from Demilew et al. 2018)

- The sutures are removed eighth post wound day and skin breaking strength, protein content and collagen content are measured on the tenth day.
- *Measurement of breaking strength:* On day 10, the animals are sacrificed using a suitable euthanasia method. Later, the sutures are gently removed and wound stripes of equal size (width) are cut and removed. Both the ends of the wound stripes are fixed at a fixed distance using steel clips. One side of the wound stripe is connected with a polyethylene bottle and it is filled with water gradually till the wound strip is broken at the site of the wound. The volume of water requires to break the wound is noted and expressed as the tensile strength of the wound in gram (Fig. 4) (Patil et al. 2012).
- *Estimation of protein content:* The total protein content of the skin tissue is measured by the Lowry method using bovine serum albumin as a standard.
- *Estimation of collagen (hydroxyproline content):* The collagen content of wound tissue is measured using a hydroxyproline assay. The sample is incubated at 60 °C for 15 min to get the dry weight. Then the sample is homogenized with 6 N hydrochloric acid (HCl) and incubated for 4 h at 130 °C to promote acid hydrolysis. Later, the pH is adjusted to 7.0 at room temperature. The concentration of hydroxyproline in wound tissue is measured by the interpolation method (using the absorbance of hydroxyproline standard solutions). The hydroxyproline standards are prepared at a concentration of 1.0–100 µg/mL. A 10 µL of standard and samples are transferred to a 96-well microplate. Then, 90 µL of 0.056 M chloramine-T solution and 100 µL of Ehrlich reagent are added to each well. The absorbance of standards and sample is measured at 550 nm. The concentrations of hydroxyproline in tissue homogenates are calculated by the amount of HCl used and finally per milligram of dry tissue, based on a standard curve (Caetano et al. 2016).
- *Immunohistochemistry and Western blotting:* During the study, skin sample from wound area is collected from different time points and used for immunohisto-chemistry and Western blotting analysis to study the protein expression rate.

3.5   Burn Wound

A burn wound is created by heat (thermal), chemical, or radiation to the skin or tissues. Burn is one of the common injuries that can lead to serious morbidities. Healing response in burn wound is due to the effects of their impact on the viability of cells and tissue. A burn wound by thermal injury is used to measure angiogenesis, contraction, re-epithelialization, granulation tissue formation, scarring, and wound tissue biochemistry (Masson-Meyers et al. 2020). The rate of wound healing is depended on the depth of the burn. Thermal burns create an extensive zone of frank necrosis including dead cells. Rats are commonly used as experimental animals.

Requirement: Hair remover, 70% ethanol, anesthetic agent, rodent surgical table, hot plate, rodent cage with fasting grills, cotton, graph paper.

3.5.1   Procedure

- This model requires anesthesia. The animals are anesthetized with a suitable anesthetic agent.
- The animal is placed on a rodent surgical table.
- In the dorsal surface of the animal between the lower parts of both scapulas are shaved and hair is removed.
- Later, a partial thickness burn is made by placing a hot plate (size may vary depending on animal model) on the prepared area for 10 s at the temperature of 75 °C.
- The burnt area is about 10% of the total body surface area.
- The wounded animals are housed individually on rodent cages that contain fasting grills to avoid coprophagia and other infections.
- After the creation of the wound, the investigational compound is applied once/ twice daily for 14 consecutive days.
- Immediately after the burn and on days 3, 7, 10, and 14 after burn injury, the burnt area is measured using millimeter-scale graph paper.
- The lesion/injury of the wounds of animals is evaluated using color, exudates, swelling of the wound surface, wound bed, and the consistency of tissues encompassing the wound.
- The wound-healing effect of the investigational compound is compared with the untreated group and the percentage of wound healing is calculated using the following formula.

Percentage of wound healing = 1

$$- \left(\frac{\text{Wound area on the corresponding day (cm}^2)}{\text{Wound area on day zero (cm}^2)} \right)$$
$$\times 100$$

- At the end of the experiment, the animals are sacrificed using a suitable euthanasia method, and tissue of the healed wound is collected for histological examination (Somboonwong et al. 2012).

3.6   Dead-Space

This model is used to study the physical changes in granuloma tissues and ideal for biochemical assessment. Rats are commonly used as experimental animals in dead-space wound model.

Requirement: Hair remover, 70% ethanol, anesthetic agent, rodent surgical table, polypropylene tube/sterile cotton pellets, rodent cage with fasting grills, cotton, and other chemicals for biochemical/histopathological analysis.

3.6.1   Procedure

- This model requires anesthesia. The animals are anesthetized with a suitable anesthetic agent.
- The animal is placed on a rodent surgical table.
- The dorsal surface of the animal (in the regions of dorsal and lumbar vertebrae) is shaved and hair is removed.
- Later, 1 cm incision is made on dorsolumbar part (two sides) and a polypropylene tube (0.5×2.5 cm^2) or sterile cotton pellets (5 mg each) is placed on the dead-space of the lumbar region on the dorsal surface of the rat on each side and wounds are closed with suture material.
- In this model, the investigational compounds are administered orally.
- On the tenth post-wounding day, the experimental animals are sacrificed and granulation tissue formed around the implanted tubes/sterile cotton pellets are carefully dissected out.
- The wet weight of the tissue is measured and tensile strength (force required to open a healing skin wound) is determined using a tensiometer. Collected granulation tissue is processed for the estimation of free radicals, antioxidants, and collagen (hydroxyproline, a major component of the protein collagen) tissue parameters. The tissue sample is also used to study histology.
- The collected granulation tissue samples are dried at 60 °C for 12 h and used for the determination of dry granulation tissue weight. The dried tissue is added with

5 ml 6 N HCl and kept at 110 °C for 24 h. The neutralized acid hydrolysate of the dry tissue is used for the determination of hydroxyproline (Murthy et al. 2013; Nayak et al. 2009; Agarwal et al. 2009; Gautam et al. 2014).

3.7 Wound Chambers

This model is useful to observe the wound healing process, such as angiogenesis and extravasation of blood-borne inflammatory cells into the wound site (Davidson et al. 2013). Rats and pigs are commonly used as experimental animals in the wound chambers model.

Requirement: Hair remover, 70% ethanol, anesthetic agent, rodent surgical table, wound chambers, animal/rodent cage with fasting grills, cotton, and other chemicals for biochemical/histopathological analysis.

3.7.1 Procedure

- This model requires anesthesia. The animals are anesthetized with a suitable anesthetic agent.
- The animal is placed on a rodent surgical table.
- The dorsal surface of the animal (from the scapula to pelvis) is shaved and hair is removed and sterilized with polyvidone iodine.
- A full-thickness skin incision is made vertical to the spine through the panniculus carnosus to the fascial plane. A space that is around the same estimate as the chamber is opened under the dermis and the sterile wound chambers (wound chambers are made of stainless steel wire mesh measuring 1 cm × 2.5 cm long and they are closed at both ends by Teflon caps/titanium) with caps are slipped below the skin. The incisions are closed through individual 4.0 nylon sutures (Fig. 5).
- Two/four champers are implanted on the dorsum of each animal and the investigational compounds are injected directly into the chamber.
- Evaluation plan 1:
  - *Wound in a skin island:* In this model, wound healing is monitored for 4 and 8 days. This model comprises two concentric punch (4 mm) biopsies, which results in a central full-thickness wound encompassed by a skin island. A 4 mm punch is used to create a full-thickness wound extending through the panniculus carnosus muscle. The encompassing skin island is then concentric on the original full-thickness wound using a 10 mm punch biopsy needle.
  - *Wound without skin:* In this model, wound healing is monitored for 10 days. In this model, a 10-mm punch is used to make a full-thickness wound and the outer rim of the wound encompasses the edge of the wound chamber. In both

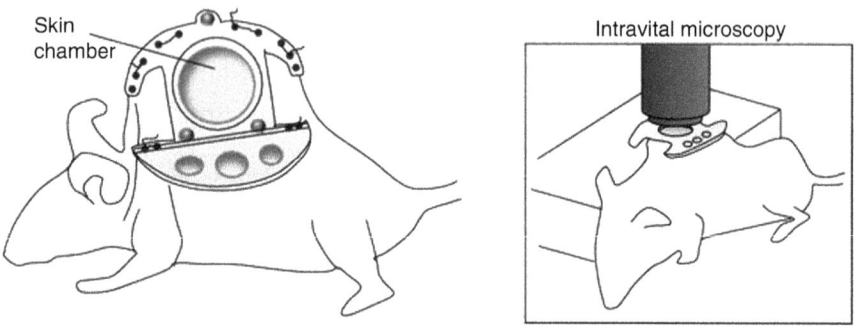

Fig. 5 Dorsal skin fold chamber (adapted from Wong et al. 2010)

wound models, the wound chamber is inserted with the flange buried under the skin edges.

– In this model wound contraction, wound closure, and re-epithelialization are evaluated.

– Wound contraction and wound closure are evaluated by measuring tattooed margins from macroscopic wound images and comparing the digital photographs of the wounds on different time points respectively.

– At the end of the experiment, wounds are biopsied and the tissue samples are fixed in 4% neutral buffered formalin for histopathological examination. The tissue sample is embedded in paraffin and sectioned for staining with hematoxylin-eosin (H&E). The stained tissue sections are examined for re-epithelialization under a light microscope (Nuutila et al. 2016).

• Evaluation plan 2:

– After implantation, the rats are sacrificed on days 3, 7, 12, 18, or 22 after chamber implantation and wound fluid is collected by aspiration with the help of a syringe and discarded. The substance deposited in the chambers is collected with a scalpel and stored at −80 °C until analysis. From the sample total protein, collagen (hydroxyproline content), and glycosaminoglycans (GAG) leaves are estimated (Siméon et al. 2000).

3.8   Ear Wound

The ear wound model is used to measure epithelialization and neovascularization of the wound in living animals. In this model, mice and rabbits are used as experimental animals.

Requirement: Hair remover, 70% ethanol, anesthetic agent, rodent surgical table, wound chambers, animal/ rodent cage with fasting grills, cotton, and other chemicals for biochemical/histopathological analysis.

3.8.1 Procedure

- The ear wound model is an ischemic wound model and has been used in reperfusion injuries, burn studies, and flap necrosis.
- This model requires anesthesia. The animals are anesthetized with a suitable anesthetic agent.
- The animal is placed on a rodent surgical table.
- In mice, 2-mm full-thickness hole punched through the center of each ear is made to create a wound (Buckley et al. 2011).
- In rabbits, 6 mm full-thickness dermal punches (4 numbers) are made on the inner surface of both ears down to bare cartilage by removing the epidermis, dermis, and perichondrium. The wound in one ear is served as control and a wound in the other ear is treated with the investigational compound (Jia et al. 2011).
- In the ear wound model, the animals are treated for a week's time. The animals are sacrificed on the eighth post-wounding day and tissue samples are collected in two parts. One part is used for histological examination and the other part of the tissue is used for biochemical or protein analysis.
- The major limitations of the ear wound model are the process of healing over an avascular cartilage base on materials testing and thickness of the skin as compared to the trunk (Davidson et al. 2013).
- *Ischemic wound model:*

  - This model requires anesthesia. Ears of the experimental animals are shaved and injected with pre-anesthetic medication. Later, the animal is anesthetized and placed on the surgical table. One ear of the animal is rendered ischemic and the other ear serves non-ischemic control. The dorsal surface of the animal (rabbit) ears is saved and the surgical site is cleaned with betadine solution. Using a #15 blade, a small incision is made to the level of bare cartilage at the base of the ear. The caudal, central arteries, and circumferential circulation are identified and ligated so the ear is perfused only rostral artery with preservation of the caudal, central, and rostral veins to render the rabbit ear ischemia. This method results in ischemia in 7–10 days.
  - The incision is closed with 4–0 or 5–0 polypropylene suture and covered with sterile gauze.
  - Six-millimeter full-thickness dermal punches (4 numbers) are made on the inner surface of both ears down to bare cartilage by removing the epidermis, dermis, and perichondrium. The distance between the wounds is a minimum of 20–30 mm. The wound in one ear is served as control and a wound in the other ear is treated with investigational compound. To keep the wound from being desiccated, an occlusive dressing is used to cover the wound site (Fig. 6).
  - The use of aged rabbits may delay the healing process and healing may take up to 26 days (Jia et al. 2011; Chien and Wilhelmi 2012).

Fig. 6 Ischemic wound
model in rabbit ears

6 mm full-
thickness
dermal punches

3.9   Pressure Ulcer

This model is a resemblance of localized skin/underlying tissue damage due to bone
prominence due to pressure. Rats and mice are used as experimental animals in
pressure ulcer models.

Requirement: Hair remover, 70% ethanol, anesthetic agent, rodent surgical table,
wound chambers, animal/rodent cage with fasting grills, cotton, and other chemicals
for biochemical/histopathological analysis.

3.9.1   Procedure

- This model is also referred to as "Vessel ligation".
- This model requires anesthesia. The animals are anesthetized with a suitable
  anesthetic agent.
- The animal is placed on a rodent surgical table.
- The dorsal surface of the animal is shaved and hair is removed and cleaned with
  70% isopropanol.
- A prototype is used to mark the location of the magnetic plates in animals. The
  dorsal surface of the skin is carefully pulled up and positioned between two,
  circular ceramic magnetic plates that are 12 mm diameter and 5.0 mm thick, with
  a mean weight of 2.4 g and 1000 G magnetic force. This technique is designed to
  leave a 5.0-mm skin bridge between the two magnets (Fig. 7) (Lanzafame et al.
  2007).
- During ischemia–reperfusion cycles, animals are not immobilized, anesthetized
  and animals are allowed food and water ad libitum.
- The use of the magnet compressed the skin and reduced blood flow causes
  ischemia, and removal of the magnet allows blood to reperfuse into the ischemic
  region of the skin. During the ischemic phase of the ischemia–reperfusion cycle,
  magnets are held to the skin purely by magnetic attraction (Peirce et al. 2000).
- One ischemia–reperfusion cycle consisted of 2 h of ischemia and 0.5 h of
  reperfusion. A maximum of five compression cycles (five ischemia–reperfusion
  cycles) is administered per day followed by a period of 11.5 h of reperfusion

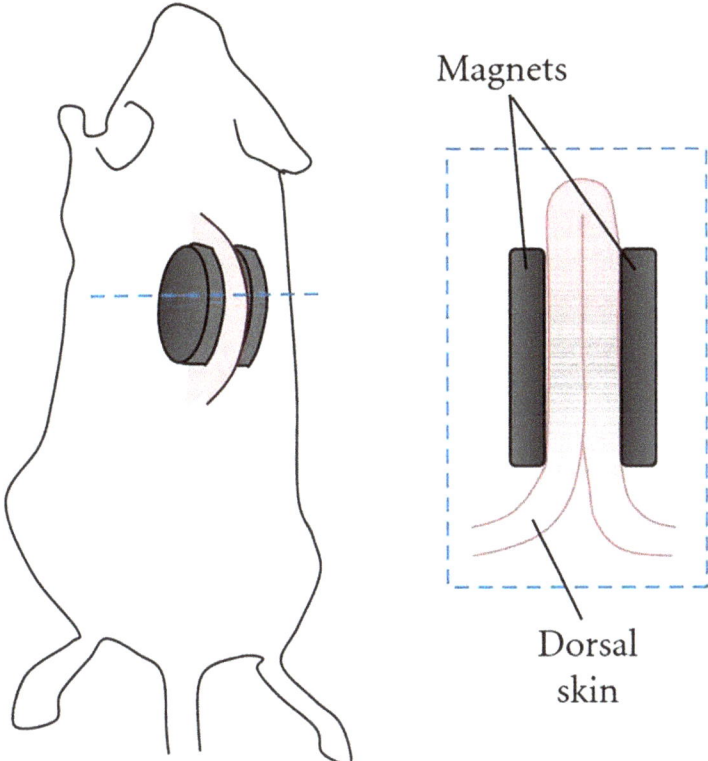

Fig. 7 Pressure ulcer models (adapted from Wong et al. 2010)

conducted for 2 or 3 days. After completion of the specified total number of compression cycles, the magnets are removed for a final 11.5 h of reperfusion, and the animals are euthanized and their treatment sites analyzed. Or 3 cycles of 12 h of ischemia and 12 h of reperfusion used to decubitus ulcer formation.

- The animals resume normal activity within a few minutes of magnet placement.
- The wounded animals are house individually on rodent cages and treated with the investigational drug.
- Animals are observed over a 21-day period following the ischemia–reperfusion cycles. The ulcer stage is graded using a standardized grading scale (Ulcer stage 0: Intact, normal, skin with normal capillary refill; Ulcer stage 1: Intact skin, nonblanchable erythema; Ulcer stage 2: Superficial/partial skin loss involving epidermis and dermis; Ulcer stage 3: Full-thickness loss with damage and necrosis of subcutaneous tissue; Ulcer stage 4: Full-thickness loss with extensive destruction, tissue necrosis, and with exposure of muscle and bone) (Stadler et al. 2004).
- Ulcer stages and skin temperature alterations are recorded at regular intervals. Few animals in a group are euthanized on days 5, 10, 15, and 21 and the tissue sample is collected for histologic evaluation.

3.10 Pressure-Induced Deep Tissue Injury

It is a form of pressure ulcer and is induced by moderate compression. This model mimics the pathology of pressure-induced deep tissue injury without producing any morphological damage to the skin layer. In this model, rats are used as experimental animals.

Requirement: Anesthesia, surgical table, povidone-iodine, isopropyl alcohol, 3 M™ Steri-Strip™ Reinforced Adhesive Skin Closures, others.

3.10.1 Procedure

- This model requires anesthesia. The animals are anesthetized with a suitable anesthetic agent.
- The animal is placed on a rodent surgical table throughout the experimental period and sustainability of the anesthetic level is determined by the whisker movement.
- The fur on the compression site of the rat is removed and static pressure of 100 mmHg (13.3 kPa equivalent) is applied to an area of 1.5 cm^2 in the tibialis region of the right limb of the rats. The compression force applied to the rat is continuously monitored by an electronic balance under the indenter (Teng et al. 2011). A laser Doppler flowmeter is used to monitor the blood flow at the compression site.
- The duration of compression is 6 h on each of the two consecutive days. The animals are allowed to recover from anesthesia and monitored for their ambulatory activities (ambulatory activity is restored ~0.5 h after each compression procedure).
- The investigational compound is administered to the animals for 14 or 21 days and the wound healing effect is assessed by measuring wound area (Shi et al. 2016).
- At the end of the study, rats are euthanized and tibialis anterior muscles are dissected and stored for histopathological analysis (Nelissen et al. 2018).

3.11 Biofilm-Infected Wound

Pathogenic biofilms represent a critical component of nonhealing wounds, employing many specific strategies to suppress endogenous inflammatory responses and to resistance to traditional therapeutics (Seth et al. 2012a). Mice, rats, and rabbits are used as experimental animals in biofilm-infected wound models.

Requirement: Hair remover, 70% ethanol, anesthetic agent, local anesthetics, rodent surgical table, wound chambers, wound-infective bacteria (example:

Pseudomonas aeruginosa) culture, animal/rodent cage with fasting grills, cotton, and other chemicals for biochemical/histopathological analysis.

3.11.1 Procedure

Rabbit ear model
- This model requires anesthesia. The animals are anesthetized with a suitable anesthetic agent.
- The animal is placed on a rodent surgical table.
- The ears of the rabbits are shaved and hair is removed and cleaned with 70% ethanol and injected 1% lidocaine/1:100,000 epinephrine, intradermally at the planned wound sites.
- The wound is created down to the perichondrium on the ventral surface of the ear and dressed with semi-occlusive transparent film (Tegaderm dressing). Individual wounds are inoculated with bacteria (*Pseudomonas aeruginosa*) (10^6 colony-forming units (CFU)/mL of bacteria at a volume of 10 μL) on postoperative day 3 (Seth et al. 2012b).
- The wound is treated with an investigational compound or redressed on postoperative days 4–12 days or until harvesting of wounds.
- At the end of the study, the animals are euthanized; wounds are harvested and used to test for the presence of bacteria in the wound bed and bacterial counts. Also, confirm biofilm formation scanning electron microscopy (SEM) analysis is performed to visually confirm biofilm formation within the wound bed. Part of the tissue is used for the histological analysis using H&E stain. The tissue sample is also used for biochemical and molecular biological studies to determine the protein levels (Gurjala et al. 2011; Seth et al. 2012c).

Rodent Model
- In the rodent model, a full-thickness, 1.5×1.5 cm surgical excision wound is created on the dorsal of the body and covered with semi-occlusive/semipermeable polyurethane transparent film. Approximately 2×10^6 CFU/mL of *Pseudomonas aeruginosa* strain applied topically to the wound of each rodent. The day when the wound is created is designated day 0. Wounds are harvested at 8 h, and 1, 3, and 7 days post wounding and then processed for biochemical, histological, and immunohistochemical examinations (Watters et al. 2013; Kanno et al. 2010).

3.12 Diabetic Wound

The diabetic wound does not follow the normal wound healing pattern of events and microbial infections further disrupt this process. Diabetic wound models are an example of impaired wound healing. The animal models of diabetic wound healing are used in combination with additional inducing factors that delay healing (Elliot et al. 2018). Rats are used as experimental animals in the diabetic wound model.

Requirement: Streptozotocin, tuberculin syringe (1 cc) with needle, glucometer, others.

3.12.1 Procedure

• *Induction of diabetics:* In overnight-fasted rats, diabetes mellitus is induced by administering freshly prepared, single intraperitoneal injection of streptozotocin (55 mg/kg) in distilled water, pH 4 or cold 0.1 M citrate buffer, pH 4.5. Diabetes mellitus is confirmed by measurement of fasting blood glucose level after 48 h of induction. Rats with fasting blood glucose of >200 mg/dL are considered as diabetics and used for the experiment (Parasuraman et al. 2019; Nayak et al. 2007).
• Diabetic animals used for further wound healing studies using incision wound/ excision wound/dead-space or wound chambers models.

3.13 Flap Surgery

Flap surgery is a type of chronic wound model and represents partial skin flap necrosis in humans. This model is used to repair wounds caused by congenital abnormalities, trauma, tumor excision, or other causes (Hsueh et al. 2016). Rats are used as experimental animals in flap surgery wound models.

Requirement: Hair remover, 70% ethanol, anesthetic agent, local anesthetics, rodent surgical table, wound chambers, animal/rodent cage with fasting grills, cotton, and other chemicals for biochemical/histopathological analysis.

3.13.1 Procedure

• This model requires anesthesia. The animals are anesthetized with a suitable anesthetic agent.
• The animal is placed on a rodent surgical table in the prone position and their limbs are immobilized with adhesive tape.
• The dorsal thoracic region of the animal is shaved and hair is removed and cleaned with 70% isopropanol or 0.5% topical alcoholic chlorhexidine solution.
• A skin flap measuring 10 × 4 cm (approx.) is raised with a cranial base on the dorsal surface of a rat (Fig. 8) (Cury et al. 2013).
• The skin flap is elevated using an incision with a number 15 scalpel blade in the delimited area and adjacent muscles are removed by using blunt scissors. The skin flap is released from the encompassing tissues, brought up to the relevant

Fig. 8 Ischemic flap model (adapted from Wong et al. 2010)

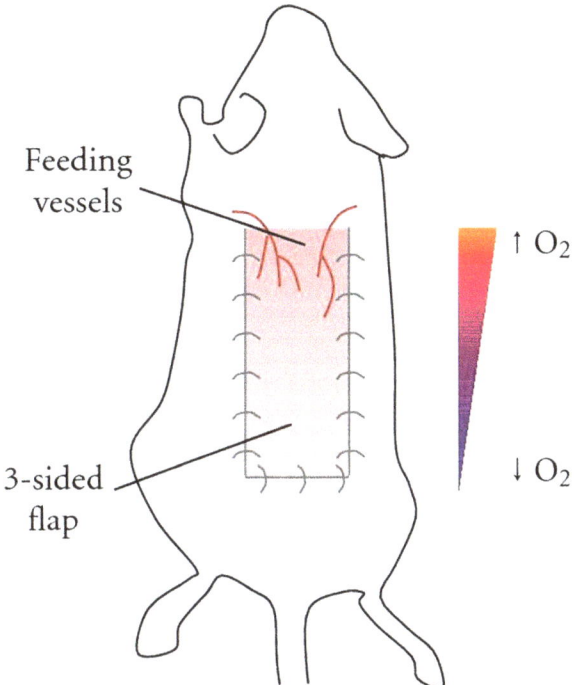

area, and sutured with a simple interrupted suture at 4.0 monofilament nylon line. The skin flap comprised superficial fascia, a fleshy panicle, subcutaneous tissue, and skin (Estevão et al. 2013).

- Immediately, after the surgical procedure, the rats are treated at least once for 7 consecutive days. At the end of the study, the animals are anesthetized and photographed with a digital camera with the demarcation of the necrotic area and total area for calculating the percentage of the necrotic area (Estevão et al. 2013).

$$\text{Percentage of flap necrosis} = \frac{\text{Area equivalent of necrotic tissue}}{\text{Total area to the flaps}} \times 100$$

- The tissue fragments are collected from cranial (containing healthy tissue), median (between the cranial and caudal areas), and the caudal (including the area of necrosis) areas for histological analysis.

3.14 Chemically Impaired Wound

It is an impaired wound healing model. Premature rabbits are used as animal models.

Requirement: Hair remover, anesthetic agent, rodent surgical table, concentrated HCl (80%), cotton, graph paper.

3.14.1 Procedure

- This model requires anesthesia. The animals are anesthetized with a suitable anesthetic agent.
- The animal is placed on a rodent surgical table.
- The back of the animals is shaved. If required, a hair remover can be used to remove the hair on the dorsal surface and left for 24 h.
- Few (two or three) drops of concentrated HCl (80%) are topically applied carefully on the shaved skin. The skin-burned rabbits are housed separately under sterile conditions in an isolated room (Abu-Zinadah 2009).
- After the creation of the wound, the investigational compounds are applied once/twice daily for 5 weeks.
- The wound area is measured at regular intervals after creation of the wound using millimeter-scale graph paper.
- The wound-healing effect of the investigational compound is compared with the untreated group and the percentage of wound healing is calculated using the following formula (James and Victoria 2010; Murthy et al. 2013).

$$\text{Percentage of wound healing} = \frac{\text{Healed area}}{\text{Total wound area}} \times 100$$

3.15 Parabiosis

Parabiosis is the method of joining two animals so that they can share blood supply with each other (Conese et al. 2017). In the parabiosis model, two animals are surgically joined to establish common blood circulation to study the circulatory physiology, immunology, metabolic diseases, cancer metastasis, and various biological processes including the hematopoietic cells migrating to the place of injury from circulating blood in tissue remodeling and repair and roles of nonresident progenitors/stem (Frozoni et al. 2012; Wong et al. 2010). Transgenic mice are used in the parabiosis model.

Requirement: Anesthesia, surgical table, surgical instruments, povidine-iodine solution, isopropyl alcohol, silk suture, painkiller, others.

3.15.1 Procedure

- This model requires anesthesia. The animals are anesthetized with a suitable anesthetic agent.
- Age-matched, adult transgenic mice are used for the experiment.

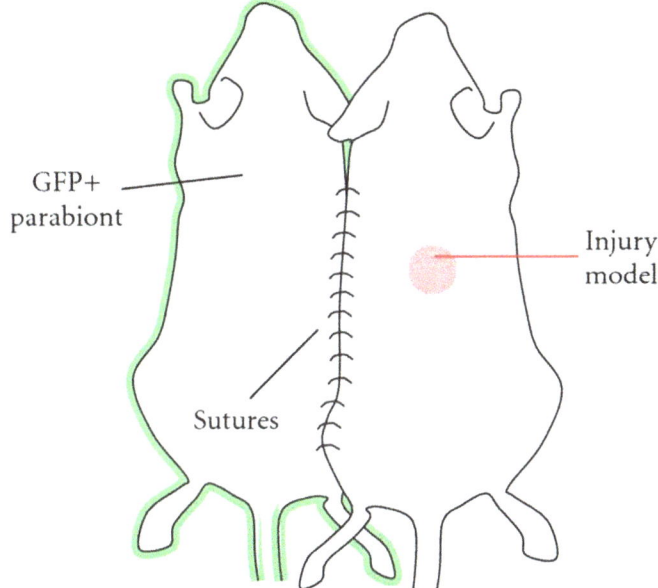

Fig. 9 Parabiotic models permit (adapted from Wong et al. 2010)

- The animals (two mice) are anesthetized and placed on a rodent surgical table.
- The surgical site of animals (both the animals) is shaved and sterilized with 5% povidine-iodine solution. For both the animals, on the dorsal side, a skin incision is made from the hip to the angle of the mandible. Animals are aligned, and the connective tissue and the skin edges are sutured together with absorbable silk suture material and joined with 7 or 9-mm wound clips at the dorsal side (Fig. 9) (Frozoni et al. 2012; Sung et al. 2019).
- After surgery, each parabiont is administered with a suitable painkiller to relieve both acute and chronic pain for the first 2 days. The parabiont is also administered with sterile normal saline to prevent dehydration.
- The wound clips are removed after 2 weeks.
- Reparative dentinogenesis is stimulated by the administration of test compound for the next 4 weeks (maybe varies based on study protocol).
- During/at the end of the experiment, changes in body weight, weight loss, levels of pro-inflammatory cytokines are measured. This model is used to study the importance of circulating cells in normal and diseased animals (Wong et al. 2010; Sung et al. 2019).

3.16   Denervated Wound

This model results in impaired healing of wounds and has lower inflammatory cells, particularly macrophages. This model demonstrates decreased microvascular responses due to denervation-induced desensitization of vascular smooth muscle. Reduction in microvascular responses may delay wound healing (Shu et al. 2015). Rats and mice are used in the denervated wound model.

Requirement: Anesthesia, a surgical instrument for rodent surgery, 70% ethanol, rodent surgical table, surgical scalpel handle, scalpel blade, surgical scissors, rodent cage with fasting grills, cotton, graph paper, others.

3.16.1   Procedure

- This model requires anesthesia.
- The animals are fasted overnight and anesthetized with a suitable anesthetic agent. If a nude mouse is used no need to remove the fur. The normal mice or rats are used as experimental animals fur should be removed on the surgical surface.
- A midline skin incision (approx. 5 cm for rats/3 cm for mice) is made at the 13th costovertebral angle (identified by palpation). The skin and subcutaneous tissues are separated and the T9 to L1 vertebrae are exposed. The nerve roots are exposed bilaterally, the nerve is transected distal to the point of trifurcation, and the wound is closed. After recovery from anesthesia, the animal is checked for response to stimuli on the wound site (if no response to the needle stimuli indicates successful nerve resection).
- After 2 days of the nerve resection, the animal is re-anesthetized and two 5 mm full-thickness excisional skin wounds are created on each side of the midline in the denervated skins.
- The experimental animals are housed individually after surgery. The investigational compound/material is administered intradermally around the wound.
- The wound area is measured on 7, 14, and 21 days of the experiment. Wound biopsies are also performed on 7, 14, and 21 days of the experiment including encompassing unwounded tissues. The collected tissue sample is used for enzyme-linked immunosorbent assay, histological, and immunohistochemical analyses (Fukai et al. 2005; Shu et al. 2015).

3.17   Tape Stripping

In this model, the stratum corneum is removed with the help of adhesive tape. Generally, the epidermal compartment is left intact in this model of wounding. This model depends on various factors including pressure exerted when applying the tape onto the skin, adhesiveness of the tape, velocity and direction of tape removal and the number of tape strips (Wilhelm et al. 2017). Mice are used in the tape stripping model.

Requirement: Anesthesia, elastic adhesive bandage, homogenizer, others.

3.17.1   Procedure

- The tape stripping model is used to evaluate a topical antibiotic for *Staphylococcus aureus* infections. This model is also known as the superficial skin bacterial colonization and infection models.
- This model requires anesthesia. The animals are anesthetized with a suitable anesthetic agent.
- The fur is stripped (area of 2 cm^2) from the anesthetized animal with an elastic adhesive bandage.
- An elastic adhesive bandage is stripping the back of the animal 7–10 times in succession, the transepidermal water loss (TEWL) reached approximately 70 g/ m^2 h.
- Following this procedure, the animal skin becomes visibly damage and no frank bleeding. The damage is characterized by glistening and reddening.
- After stripping off the animal skin, 5 μL droplets containing 10^7 *Staphylococcus aureus* cells and allowed overnight to initiate bacterial infection (Dai et al. 2011).
- After 24 h of infection, animals are treated with respective assigned treatment. During the study period, animals are sacrificed at different time points and skin samples (area of 2 cm^2) are collected and homogenated with water. The diluted homogenate is used for the determination of colony-forming units (CFU) (Hu et al. 2010).

3.18   Xeno-Grafts

Xeno-grafts are tissues transplanted from one species to another species. Xeno-grafts protect wounds from bacterial and physical trauma, reduce pain, and increase moisture and heat retention (Buchbinder and Buchbinder 2007). Athymic (nude) mice and nude rats are used for the experiments.

Requirement: Anesthesia, surgical table, povidone-iodine, isopropyl alcohol, 3 M™ Steri-Strip™ Reinforced Adhesive Skin Closures, others.

3.18.1   Procedure

- *Human skin harvest:* Normal human skin is harvested from unidentified individuals by elective abdominoplasty procedures. About 2 cm punch biopsy is used to harvest uniform samples of skin for engraftment. The dermis is dissected from the underlying fat tissue; grafted tissue included the stratum germinativum. The punch biopsy specimens are dissected and xenografted within 1 h of surgical harvest (Shanmugam et al. 2015).

- *Xeno-graft:*

  - This model requires anesthesia. The animals are anesthetized with a suitable anesthetic agent.
  - The animal is placed on a rodent surgical table and all the surgical procedures are performed in sterile conditions.
  - The dorsal region of the animal is sterilized using povidone-iodine followed by isopropyl alcohol.
  - On the flank region (both sides) two graft beds of 2 cm in diameter are removed. Full-thickness human skin xenografts are placed on each mouse wound bed and secured using 3 M™ Steri-Strip™ Reinforced Adhesive Skin Closures.
  - On postoperative days 7 and 14 dressings are changed at which time 3 M™ Steri-Strip™ Reinforced Adhesive Skin Closures are removed. Xenografts remained dressed until day 30. Graft viability is assessed at 2 months.
  - The animals are housed separately under sterile conditions and allowed to engraft.
  - At the end of the experiment imaging analysis, histochemical analysis, and immunohistochemistry are performed (Shanmugam et al. 2015).

3.19   In Silico Models

In silico models, the experiment is performed on the computer or via computer simulation. It was first introduced to the public in 1989 in the workshop "Cellular Automata: Theory and Applications" in Los Alamos, New Mexico (Vanjari et al. 2012). Drug discovery is a time-consuming multi-step process comprising synthesis, quality control, preclinical testing, toxicity testing, and clinical trials. With the help of in silico approach, the biological activity and toxicity profile of any investigational compounds are predicted using computer programs and this may help the researchers to find promising compounds with better therapeutic efficacy. These promising compounds can be taken for the further drug discovery process. In silico approaches are reducing the time gap between synthesis and clinical trials, and also reduce the cost.

The ligand-based or target-based virtual screening method is used to predict the binding affinity between investigational compounds and target receptors. Restoration of injured tissue/skin is mediated through blood cells, cytokines, and growth factors activity. Regulation of growth factors and inflammatory cytokine and inhibition of Glycogen synthase kinase 3-β (GSK3-β) enzyme are essential to enhance the wound-healing process (Tatke 2020). Many of in silico studies are reported on the prediction of the interaction between the investigational compound and GSK3-β, which minimize the cost and time duration of the experiment (Raja Naika et al. 2015; Harish et al. 2008).

4 Summary

Humans and other mammals are complex organisms that involve a complex network of cells, circulating factors, and hormones. The use of animals in preclinical research may give the lead for the therapeutic effect and toxicological properties of the investigational drugs/compounds. In preclinical research, in vitro models are also used to screen the therapeutic effect and toxicological properties, but the test is carried out with specific cells or organs and it may not give the effect of investigational drugs on the whole organism. In vitro studies may reduce the number of animals used for the experiment. In recent years, the disease models are emerging from in vitro studies but they must be validated in a whole organism; otherwise, they remain speculative (Barré-Sinoussi and Montagutelli 2015).

Wound healing is a complex process and various animal models are used to study the effect of drugs on the healing process and the factors affecting the healing process. In vitro models are relatively inexpensive, convenient, and fast for the researcher and helpful to obtain the results in a short time. In vitro studies are also used to screen multiple samples simultaneously using specific cell lines. However, in vitro models are difficult to simulate a "real world" application (Perez and Davis 2008). The in vivo animal models are used to study the safety and efficacy of the investigational compound. The rodents and small mammals are commonly used as experimental animals in wound healing research but time-consuming process. The animal experiment should be performed with prior approval from Institute Animal Ethical Committee and followed the three Rs (Reduction, Refinement, and Replacement) principles to minimize the discomfort to the animals and reduce the number of animals in the experiment. Features of general wound models are represented in Fig. 10.

Wound healing models involved surgical procedures, which can increase the pain and experimental stress to the animals. Whenever possible, the wound healing models can be refining to reduce the pain and experimental stress to the animals. Also, the appropriate measures should be taken on postoperative or post-experimental care. Another major issue with the experiment is the disposal of biological waste and animal carcasses. Once the experiment is over, or any animal died during the experiment it should be disposed of as per Institutional biosafety guidelines to prevent environmental contamination.

In recent years, the wound healing research focus has moved to in silico and in vitro models to avoid the use of a large number of animals in the research which helps to understand the cell growth and phases of wound healing (Sami et al. 2019). A simulation model (mathematical/computational modeling) is also available to understand the healing mechanism, which is also called a non-animal model. This

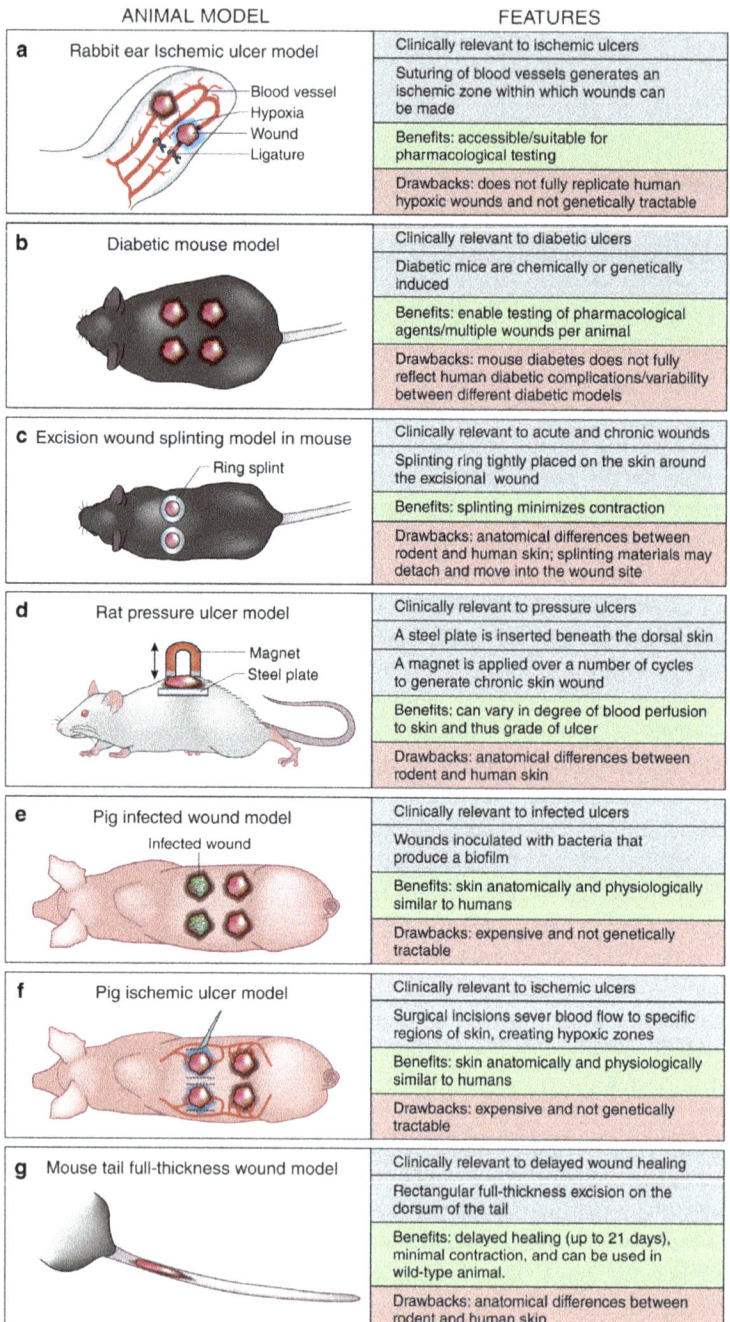

ANIMAL MODEL	FEATURES
a Rabbit ear Ischemic ulcer model	Clinically relevant to ischemic ulcers
	Suturing of blood vessels generates an ischemic zone within which wounds can be made
	Benefits: accessible/suitable for pharmacological testing
	Drawbacks: does not fully replicate human hypoxic wounds and not genetically tractable
b Diabetic mouse model	Clinically relevant to diabetic ulcers
	Diabetic mice are chemically or genetically induced
	Benefits: enable testing of pharmacological agents/multiple wounds per animal
	Drawbacks: mouse diabetes does not fully reflect human diabetic complications/variability between different diabetic models
c Excision wound splinting model in mouse	Clinically relevant to acute and chronic wounds
	Splinting ring tightly placed on the skin around the excisional wound
	Benefits: splinting minimizes contraction
	Drawbacks: anatomical differences between rodent and human skin; splinting materials may detach and move into the wound site
d Rat pressure ulcer model	Clinically relevant to pressure ulcers
	A steel plate is inserted beneath the dorsal skin
	A magnet is applied over a number of cycles to generate chronic skin wound
	Benefits: can vary in degree of blood perfusion to skin and thus grade of ulcer
	Drawbacks: anatomical differences between rodent and human skin
e Pig infected wound model	Clinically relevant to infected ulcers
	Wounds inoculated with bacteria that produce a biofilm
	Benefits: skin anatomically and physiologically similar to humans
	Drawbacks: expensive and not genetically tractable
f Pig ischemic ulcer model	Clinically relevant to ischemic ulcers
	Surgical incisions sever blood flow to specific regions of skin, creating hypoxic zones
	Benefits: skin anatomically and physiologically similar to humans
	Drawbacks: expensive and not genetically tractable
g Mouse tail full-thickness wound model	Clinically relevant to delayed wound healing
	Rectangular full-thickness excision on the dorsum of the tail
	Benefits: delayed healing (up to 21 days), minimal contraction, and can be used in wild-type animal.
	Drawbacks: anatomical differences between rodent and human skin

Fig. 10 Features of general wound models. Reprinted with permission from Elsevier (License Number 4886770742994; Dated 12 Aug 2020); Reference: Grada et al. 2018)

model helps to predict the treatment of wounds with novel therapies and reduces the cost and duration of the experiment (Flegg et al. 2015).

References

Abu-Zinadah OA (2009) Using *Nigella sativa* oil to treat and heal chemical induced wound of rabbit skin. JKAU: Sci 21(2):335–346. https://doi.org/10.4197/Sci.21-2.11

Agarwal PK, Singh A, Gaurav K, Goel S, Khanna HD, Goel RK (2009) Evaluation of wound healing activity of extracts of plantain banana (*Musa sapientum* var. *paradisiaca*) in rats. Indian J Exp Biol 47:32–40

Andjelic S, Lumi X, Veréb Z, Josifovska N, Facskó A, Hawlina M, Petrovski G (2014) A simple method for establishing adherent *ex vivo* explant cultures from human eye pathologies for use in subsequent calcium imaging and inflammatory studies. J Immunol Res 2014:232659. https://doi.org/10.1155/2014/232659

Barré-Sinoussi F, Montagutelli X (2015) Animal models are essential to biological research: issues and perspectives. Future Sci OA 1(4):FSO63. https://doi.org/10.4155/fso.15.63

Buchbinder D, Buchbinder SB (2007) Wound healing: adjuvant therapy and treatment adherence. In: Venous ulcers. Academic Press, New York, pp 91–103

Buckley G, Metcalfe AD, Ferguson MW (2011) Peripheral nerve regeneration in the MRL/MpJ ear wound model. J Anat 218(2):163–172. https://doi.org/10.1111/j.1469-7580.2010.01313.x

Caetano GF, Fronza M, Leite MN, Gomes A, Frade MA (2016) Comparison of collagen content in skin wounds evaluated by biochemical assay and by computer-aided histomorphometric analysis. Pharm Biol 54(11):2555–2559. https://doi.org/10.3109/13880209.2016.1170861

Chien S, Wilhelmi BJ (2012) A simplified technique for producing an ischemic wound model. J Vis Exp 63:e3341. https://doi.org/10.3791/3341

Clark RAF (2014) Chapter 76 – wound repair: basic biology to tissue engineering. In: Lanza R, Langer R, Vacanti J (eds) Principles of tissue engineering, 4th edn. Academic Press, San Deigo, CA, pp 1595–1617. https://doi.org/10.1016/B978-0-12-398358-9.00076-8

Conese M, Carbone A, Beccia E, Angiolillo A (2017) The fountain of youth: a tale of Parabiosis, stem cells, and rejuvenation. Open Med (Wars) 12:376–383. https://doi.org/10.1515/med-2017-0053

Cury V, Moretti AI, Assis L, Bossini P, de Souza CJ, Neto CB et al (2013) Low level laser therapy increases angiogenesis in a model of ischemic skin flap in rats mediated by VEGF, HIF-1α and MMP-2. J Photochem Photobiol B 125:164–170. https://doi.org/10.1016/j.jphotobiol.2013.06.004

Dai T, Kharkwal GB, Tanaka M, Huang YY, Bil de Arce VJ, Hamblin MR (2011) Animal models of external traumatic wound infections. Virulence 2(4):296–315. https://doi.org/10.4161/viru.2.4.16840

Davidson JM, Yu F, Opalenik SR (2013) Splinting strategies to overcome confounding wound contraction in experimental animal models. Adv Wound Care 2(4):142–148. https://doi.org/10.1089/wound.2012.0424

Demidova-Rice TN, Durham JT, Herman IM (2012) Wound healing angiogenesis: innovations and challenges in acute and chronic wound healing. Adv Wound Care (New Rochelle) 1(1):17–22. https://doi.org/10.1089/wound.2011.0308

Demilew W, Adinew GM, Asrade S (2018) Evaluation of the wound healing activity of the crude extract of leaves of *Acanthus polystachyus* Delile (Acanthaceae). Evid Based Complement Alternat Med 2018:2047896. https://doi.org/10.1155/2018/2047896

Elliot S, Wikramanayake TC, Jozic I, Tomic-Canic M (2018) A modeling conundrum: murine models for cutaneous wound healing. J Investig Dermatol 138(4):736–740. https://doi.org/10.1016/j.jid.2017.12.001

Estevão LR, Medeiros JP, Baratella-Evêncio L, Simões RS, Mendonça FD, Evêncio-Neto J (2013) Effects of the topical administration of copaiba oil ointment (*Copaifera langsdorffii*) in skin flaps viability of rats. Acta Cir Bras 28(12):863–869. https://doi.org/10.1590/s0102-86502013001200009

Flegg JA, Menon SN, Maini PK, McElwain DL (2015) On the mathematical modeling of wound healing angiogenesis in skin as a reaction-transport process. Front Physiol 6:262. https://doi.org/10.3389/fphys.2015.00262

Frozoni M, Zaia AA, Line SR, Mina M (2012) Analysis of the contribution of nonresident progenitor cells and hematopoietic cells to reparative dentinogenesis using parabiosis model in mice. J Endod 38(9):1214–1219. https://doi.org/10.1016/j.joen.2012.05.016

Fukai T, Takeda A, Uchinuma E (2005) Wound healing in denervated rat skin. Wound Repair Regen 13(2):175–180. https://doi.org/10.1111/j.1067-1927.2005.130208.x

Gautam MK, Purohit V, Agarwal M, Singh A, Goel RK (2014) *In vivo* healing potential of *Aegle marmelos* in excision, incision, and dead space wound models. Sci World J 2014:740107. https://doi.org/10.1155/2014/740107

Governa P, Carullo G, Biagi M, Rago V, Aiello F (2019) Evaluation of the *In vitro* wound-healing activity of Calabrian honeys. Antioxidants (Basel) 8(2):36. https://doi.org/10.3390/antiox8020036

Grada A, Mervis J, Falanga V (2018) Research techniques made simple: animal models of wound healing. J Investig Dermatol 138(10):2095–2105. https://doi.org/10.1016/j.jid.2018.08.005

Gurjala AN, Geringer MR, Seth AK, Hong SJ, Smeltzer MS, Galiano RD et al (2011) Development of a novel, highly quantitative *in vivo* model for the study of biofilm-impaired cutaneous wound healing. Wound Repair Regen 19(3):400–410. https://doi.org/10.1111/j.1524-475X.2011.00690.x

Hara-Chikuma M, Verkman AS (2008) Aquaporin-3 facilitates epidermal cell migration and proliferation during wound healing. J Mol Med (Berl) 86(2):221–231. https://doi.org/10.1007/s00109-007-0272-4

Harish BG, Krishna V, Kumar HS, Ahamed BK, Sharath R, Swamy HK (2008) Wound healing activity and docking of glycogen-synthase-kinase-3-β-protein with isolated triterpenoid lupeol in rats. Phytomedicine 15(9):763–767

Hebda PA (1988) Stimulatory effects of transforming growth factor-beta and epidermal growth factor on epidermal cell outgrowth from porcine skin explant cultures. J Investig Dermatol 91(5):440–445. https://doi.org/10.1111/1523-1747.ep12476480

Hsueh YY, Wang DH, Huang TC, Chang YJ, Shao WC, Tuan TL et al (2016 May 17) Novel skin chamber for rat ischemic flap studies in regenerative wound repair. Stem Cell Res Ther 7(1):72. https://doi.org/10.1186/s13287-016-0333-0

Hu Y, Shamaei-Tousi A, Liu Y, Coates A (2010) A new approach for the discovery of antibiotics by targeting non-multiplying bacteria: a novel topical antibiotic for staphylococcal infections. PLoS One 5(7):e11818. https://doi.org/10.1371/journal.pone.0011818

Isnard N, Legeais JM, Renard G, Robert L (2001) Effect of hyaluronan on MMP expression and activation. Cell Biol Int 25(8):735–739. https://doi.org/10.1006/cbir.2001.0759

James O, Victoria IA (2010) Excision and incision wound healing potential of *Saba florida* (Benth) leaf extract in *Rattus novergicus*. Inter J Pharm Biomed Res 1(4):101–107

Jia S, Zhao Y, Law M, Galiano RD, Mustoe TA (2011) The effect of collagenase on ischemic wound healing: results of an *in vivo* study. Ostomy Wound Manage 57(8):20–26

Kanno E, Toriyabe S, Zhang L, Imai Y, Tachi M (2010) Biofilm formation on rat skin wounds by *Pseudomonas aeruginosa* carrying the green fluorescent protein gene. Exp Dermatol 19(2):154–156. https://doi.org/10.1111/j.1600-0625.2009.00931.x

Kenneth IE (2017) Identification of bacteria associated with wounds in Wukari and environs, north-east. Nigeria AASCIT J Health 4(5):63–67

Krzyszczyk P, Schloss R, Palmer A, Berthiaume F (2018) The role of macrophages in acute and chronic wound healing and interventions to promote pro-wound healing phenotypes. Front Physiol 9:419. https://doi.org/10.3389/fphys.2018.00419

Lanzafame RJ, Stadler I, Kurtz AF, Connelly R, Brondon P, Olson D (2007) Reciprocity of exposure time and irradiance on energy density during photoradiation on wound healing in a murine pressure ulcer model. Lasers Surg Med 39(6):534–542. https://doi.org/10.1002/lsm.20519

Masson-Meyers DS, Andrade TA, Caetano GF, Guimaraes FR, Leite MN, Leite SN, Frade MA (2020 Feb) Experimental models and methods for cutaneous wound healing assessment. Int J Exp Pathol 101(1–2):21–37. https://doi.org/10.1111/iep.12346

Matsumoto M, Tamura M, Miyamoto T, Furuno Y, Kabashima N, Serino R et al (2012) Impacts of icodextrin on integrin-mediated wound healing of peritoneal mesothelial cells. Life Sci 90 (23–24):917–923. https://doi.org/10.1016/j.lfs.2012.04.036

Mazzalupo S, Wawersik MJ, Coulombe PA (2002) An *ex vivo* assay to assess the potential of skin keratinocytes for wound epithelialization. J Investig Dermatol 118(5):866–870. https://doi.org/10.1046/j.1523-1747.2002.01736.x

Mohan H, Mohan S (2011) Essential pathology for dental students. JP Medical Ltd, New Delhi, p 138

Muniandy K, Gothai S, Tan WS, Kumar SS, Mohd Esa N, Chandramohan G et al (2018) *In vitro* wound healing potential of stem extract of *Alternanthera sessilis*. Evid Based Complement Alternat Med 2018:3142073. https://doi.org/10.1155/2018/3142073

Murthy S, Gautam MK, Goel S, Purohit V, Sharma H, Goel RK (2013) Evaluation of *in vivo* wound healing activity of *Bacopa monniera* on different wound model in rats. Biomed Res Int 2013:972028. https://doi.org/10.1155/2013/972028

Nayak SB, Pinto Pereira L, Maharaj D (2007) Wound healing activity of *Carica papaya* L. in experimentally induced diabetic rats. Indian J Exp Biol 45(8):739–743

Nayak BS, Sandiford S, Maxwell A (2009) Evaluation of the wound-healing activity of Ethanolic extract of *Morinda citrifolia* L. Leaf. Evid Based Complement Alternat Med 6(3):351–356. https://doi.org/10.1093/ecam/nem127

Nayak BS, Kanhai J, Milne DM, Pereira LP, Swanston WH (2011) Experimental evaluation of ethanolic extract of *Carapa guianensis* L. leaf for its wound healing activity using three wound models. Evid Based Complement Alternat Med 2011:419612. https://doi.org/10.1093/ecam/nep160

Nelissen JL, Traa WA, de Boer HH, de Graaf L, Mazzoli V, Savci-Heijink CD et al (2018) An advanced magnetic resonance imaging perspective on the etiology of deep tissue injury. J Appl Physiol (1985) 124(6):1580–1596. https://doi.org/10.1152/japplphysiol.00891.2017

Nuutila K, Singh M, Kruse C, Philip J, Caterson EJ, Eriksson E (2016) Titanium wound chambers for wound healing research. Wound Repair Regen 24(6):1097–1102. https://doi.org/10.1111/wrr.12472

Öhnstedt E, Lofton Tomenius H, Vågesjö E, Phillipson M (2019) The discovery and development of topical medicines for wound healing. Expert Opin Drug Discov 14(5):485–497. https://doi.org/10.1080/17460441.2019.1588879

Parasuraman S, Ching TH, Leong CH, Banik U (2019) Antidiabetic and antihyperlipidemic effects of a methanolic extract of *Mimosa pudica* (Fabaceae) in diabetic rats. EJBAS 6(1):137–148. https://doi.org/10.1080/2314808X.2019.1681660

Patil MV, Kandhare AD, Bhise SD (2012) Pharmacological evaluation of ethanolic extract of *Daucus carota* Linn root formulated cream on wound healing using excision and incision wound model. Asian Pac J Trop Biomed 2(2):S646–S655. https://doi.org/10.1016/S2221-1691(12)60290-1

Peirce SM, Skalak TC, Rodeheaver GT (2000) Ischemia-reperfusion injury in chronic pressure ulcer formation: a skin model in the rat. Wound Repair Regen 8(1):68–76. https://doi.org/10.1046/j.1524-475x.2000.00068.x

Perez R, Davis SC (2008) Relevance of animal models for wound healing. Wounds: a compendium of clinical research and practice. 20(1):3–8. Available in https://www.woundsresearch.com/article/8200 Last Assessed 27 December 2020

Raja Naika H, Krishna V, Lingaraju K, Chandramohan V, Dammalli M, Navya PN et al (2015) Molecular docking and dynamic studies of bioactive compounds from *Naravelia zeylanica* (L.) DC against glycogen synthase kinase-3β protein. J Taibah Univ Sci 9(1):41–49

Ryu HM, Oh EJ, Park SH, Kim CD, Choi JY, Cho JH et al (2012) Aquaporin 3 expression is up-regulated by TGF-β1 in rat peritoneal mesothelial cells and plays a role in wound healing. Am J Pathol 181(6):2047–2057. https://doi.org/10.1016/j.ajpath.2012.08.018

Sami DG, Heiba HH, Abdellatif A (2019) Wound healing models: a systematic review of animal and non-animal models. Wound Med 24(1):8–17. https://doi.org/10.1016/j.wndm.2018.12.001

Seth AK, Geringer MR, Hong SJ, Leung KP, Mustoe TA, Galiano RD (2012a) *In vivo* modeling of biofilm-infected wounds: a review. J Surg Res 178(1):330–338. https://doi.org/10.1016/j.jss.2012.06.048

Seth AK, Geringer MR, Hong SJ, Leung KP, Galiano RD, Mustoe TA (2012b) Comparative analysis of single-species and polybacterial wound biofilms using a quantitative, *in vivo*, rabbit ear model. PLoS One 7(8):e42897. https://doi.org/10.1371/journal.pone.0042897

Seth AK, Geringer MR, Gurjala AN, Hong SJ, Galiano RD, Leung KP et al (2012c) Treatment of *Pseudomonas aeruginosa* biofilm-infected wounds with clinical wound care strategies: a quantitative study using an *in vivo* rabbit ear model. Plast Reconstr Surg 129(2):262e–274e. https://doi.org/10.1097/PRS.0b013e31823aeb3b

Shanmugam VK, Tassi E, Schmidt MO, McNish S, Baker S, Attinger C et al (2015) Utility of a human-mouse xenograft model and *in vivo* near-infrared fluorescent imaging for studying wound healing. Int Wound J 12(6):699–705. https://doi.org/10.1111/iwj.12205

Shi H, Xie H, Zhao Y, Lin C, Cui F, Pan Y et al (2016) Myoprotective effects of bFGF on skeletal muscle injury in pressure-related deep tissue injury in rats. Burns Trauma 4(1):26. https://doi.org/10.1186/s41038-016-0051-y

Shu B, Xie JL, Xu YB, Lai W, Huang Y, Mao RX et al (2015) Effects of skin-derived precursors on wound healing of denervated skin in a nude mouse model. Int J Clin Exp Pathol 8(3):2660–2669

Siméon A, Wegrowski Y, Bontemps Y, Maquart FX (2000) Expression of glycosaminoglycans and small proteoglycans in wounds: modulation by the tripeptide–copper complex glycyl-l-histidyl-l-lysine-Cu^{2+}. J Invest Dermatol 115(6):962–968. https://doi.org/10.1046/j.1523-1747.2000.00166.x

Somboonwong J, Kankaisre M, Tantisira B, Tantisira MH (2012) Wound healing activities of different extracts of *Centella asiatica* in incision and burn wound models: an experimental animal study. BMC Complement Altern Med 12(1):103. https://doi.org/10.1186/1472-6882-12-103

Stadler I, Zhang RY, Oskoui P, Whittaker MB, Lanzafame RJ (2004) Development of a simple, noninvasive, clinically relevant model of pressure ulcers in the mouse. J Investig Surg 17 (4):221–227. https://doi.org/10.1080/08941930490472046

Sumagin R, Robin AZ, Nusrat A, Parkos CA (2013) Activation of PKCβII by PMA facilitates enhanced epithelial wound repair through increased cell spreading and migration. PLoS One 8 (2):e55775. https://doi.org/10.1371/journal.pone.0055775

Sung J, Sodhi CP, Voltaggio L, Hou X, Jia H, Zhou Q et al (2019) The recruitment of extra-intestinal cells to the injured mucosa promotes healing in radiation enteritis and chemical colitis in a mouse parabiosis model. Mucosal Immunol 12(2):503–517. https://doi.org/10.1038/s41385-018-0123-3

Tatke P (2020) Wound healing activity and in silico binding studies with Gsk 3-β receptor of bioactive extract of *Anacardium Occidentale* leaves. Available at SSRN 3536846. https://papers.ssrn.com/sol3/papers.cfm?abstract_id=3536846. Last Assessed 13 February 2021

Teng BT, Tam EW, Benzie IF, Siu PM (2011) Protective effect of caspase inhibition on compression-induced muscle damage. J Physiol 589(Pt 13):3349–3369. https://doi.org/10.1113/jphysiol.2011.209619

Vanjari S, Chimandare N, Gandhi S (2012) A review on *in silico* approach in pharmacology. Adv Res Pharm Biol 2(2):129–141

Wang JP, Ruan JL, Cai YL, Luo Q, Xu HX, Wu YX (2011) *In vitro* and *in vivo* evaluation of the wound healing properties of *Siegesbeckia pubescens*. J Ethnopharmacol 134(3):1033–1038. https://doi.org/10.1016/j.jep.2011.02.010

Watters C, DeLeon K, Trivedi U, Griswold JA, Lyte M, Hampel KJ et al (2013) *Pseudomonas aeruginosa* biofilms perturb wound resolution and antibiotic tolerance in diabetic mice. Med Microbiol Immunol 202(2):131–141. https://doi.org/10.1007/s00430-012-0277-7

Whittam AJ, Maan ZN, Duscher D, Wong VW, Barrera JA, Januszyk M et al (2016) Challenges and opportunities in drug delivery for wound healing. Adv Wound Care (New Rochelle). 5 (2):79–88. https://doi.org/10.1089/wound.2014.0600

Wilhelm KP, Wilhelm D, Bielfeldt S (2017) Models of wound healing: an emphasis on clinical studies. Skin Res Technol 23(1):3–12. https://doi.org/10.1111/srt.12317

Wong VW, Sorkin M, Glotzbach JP, Longaker MT, Gurtner GC (2010) Surgical approaches to create murine models of human wound healing. J Biomed Biotechnol 2011:969618. https://doi.org/10.1155/2011/969618

Wound Management. In: Best practice guidelines in disaster situations [WHO/EHT/CPR 2005, formatted 2009]. Available in https://www.who.int/surgery/publications/WoundManagement.pdf. Last Assessed on 28 April 2020

Zarubova J, Hasani-Sadrabadi MM, Bacakova L, Li S (2020) Nano-in-micro dual delivery platform for chronic wound healing applications. Micromachines 11(2):158. https://doi.org/10.3390/mi11020158

Chronic Wounds: An Overview of Wound Healing and Experimental Models for Wound Studies

Diana G. Sami and Ahmed Abdellatif

1 Introduction

The skin is the first-line defense against the environment. Other functions of the skin include thermoregulation, fluid homeostasis and immune surveillance (Clark 2014). Delayed wound healing is a serious problem affecting millions of patients and costs healthcare systems billions of dollars annually (Steiner et al. 2006; Badia et al. 2017).

Wound management is a challenge for healthcare providers (Sibbald et al. 2012). It is expected that the annual wound care products market will reach $15–22 billion by 2024 (Sen 2019). Chronic infections and resistant wounds increase the length of hospital stay and cost billions of dollars of patient care (Gainza et al. 2015; Sen 2019; Han and Ceilley 2017). Chronic wounds also cause high mortality rates (Escandon et al. 2011; Sen 2019).

Wound healing models are necessary to understand the pathophysiology of wound healing, as well as to test new therapeutic approaches (Ud-Din and Bayat 2017). Wound models can be designed In silico, In vitro, Ex vivo, and In vivo using computational, cell culture, wound biopsies, and animal models (Andrade et al. 2015; Ud-Din and Bayat 2017).

In this chapter, we will provide an overview of skin anatomy, types of wounds, and the mechanism of healing in acute and chronic wounds. Also, we will discuss the wound microbiome in different wounds and experimental methods to assess wound healing.

D. G. Sami · A. Abdellatif (✉)
Department of Biology, School of Sciences and Engineering, The American University in Cairo, New Cairo, Egypt
e-mail: ahmed.abdellatif@aucegypt.edu

© The Author(s), under exclusive license to Springer Nature Singapore Pte Ltd. 2021   431
P. Kumar, V. Kothari (eds.), *Wound Healing Research*,
https://doi.org/10.1007/978-981-16-2677-7_14

2 Skin Anatomy

2.1 Skin Histology

The epidermis is the superficial layer of the skin (Carmichael 2014; Yousef et al. 2020). The epidermis is composed of different cell types organized in various layers, e.g., Stratum spinosum, Stratum granulosum, Stratum lucidum, and Stratum corneum as the outermost layer, made of keratin and dead keratinocytes (Fig. 1) (Murphrey et al. 2020; Schlüter et al. 2014; Yousef et al. 2020).

    The dermis is the connective tissue layer underneath the epidermis. It consists of the papillary dermis and the deeper reticular dermis. The dermis contains sweat glands, blood vessels, hair follicles, muscles, and sensory neurons (Brown and Krishnamurthy 2020). **The hypodermis or subcutaneous tissue** is deep to the

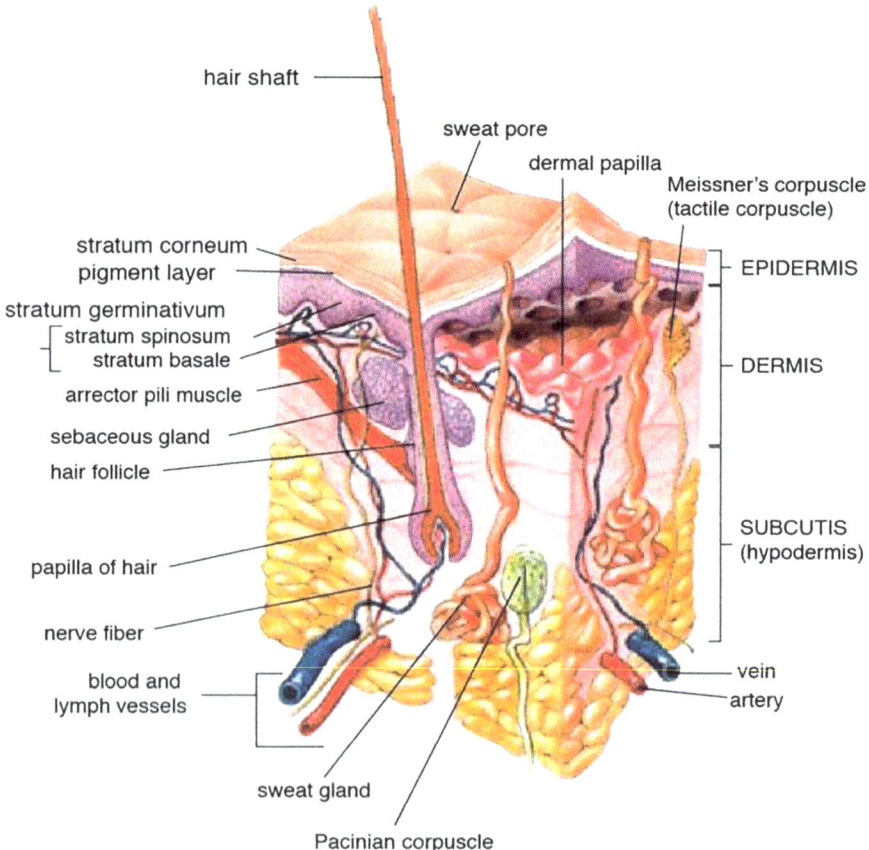

Fig. 1 Anatomy of the human skin. Wikimedia Commons, USGOV (Public Domain)

dermis and contains adipose lobules that insulate and protects the skin (Yousef et al. 2020).

2.2   Skin Proteomics

Tissues have distinctive architecture and functional characteristics. The extracellular matrix and its proteins contribute to the structure and function of the skin. Proteins such as glycoproteins, collagen, and regulators of collagen assembly are known as small leucine-rich proteoglycans (SLRPs) (Mikesh et al. 2013).

The basement membrane is rich in collagen type IV, laminin, nidogen/entactin, and heparin sulfate proteoglycans (Yurchenco and Schittny 1990). Collagen VII and anchoring fibrils connect the basal and reticular lamina in the basement membrane, thus promoting dermal to epidermal adhesion (Burgeson et al. 1990).

The dermis contains collagen types I and III (Watt and Fujiwara 2011). Fibril-associated collagens such as collagen types XII and XIV are distributed throughout the matrix (Kielty and Shuttleworth 1997). Collagen type XII plays a part in the stromal architecture, while collagen type XIV is important for the regulation of fibrillogenesis (Ansorge et al. 2009; Young et al. 2002). Tenascin-X is an extracellular matrix protein that is also found in the dermis (Egging et al. 2007).

2.3   Skin Immunological System

Communication between immune and non-immune cells and the skin microbiota contributes to the produce immune responses. Some anatomical sites are considered immunologically important such as hair follicles and sweat glands (Kabashima et al. 2019). In addition to their function as a physical barrier, Keratinocytes are the main component of the epidermis and are considered as part of the innate immune system. On inflammation, Keratinocytes express antigen identification receptors and secrete cytokines, such as tumor necrosis factor (TNF), and Interleukins (IL-33 and IL-1) (Carmi-Levy et al. 2011). These cytokines, in turn, activate and recruit other immune cells (Nestle et al. 2009). The hair follicle is classified as a site of immune privilege (Paus et al. 2003). Hair follicles also house Tissue-resident memory T cells (TRM cells), which are responsible for long-term skin immunity (Adachi et al. 2015).

Sebaceous glands produce lipids that contribute to forming the skin barrier and its function. Also, they produce antimicrobial peptides, cytokines, and chemokines that regulate skin immunity (Mattii et al. 2018). In vitro analysis showed that *Propionibacterium acnes* induced production of IL-6, Transforming growth factor β) TGFβ, and IL-1β, which activates dermal dendritic cells to prime TH17 cells, which promote neutrophil recruitment and inflammation. For further details, see (Kabashima et al. 2019).

The skin is covered with numerous microorganisms that significantly affect the immune system (Byrd et al. 2018). Microbiota composition vary between the surface and deeper areas of the skin (Grice et al. 2008). The metabolites and the structural components of microorganisms might affect both innate and adaptive immunity. *Staphylococcus epidermidis,* is the most common commensal bacteria of the skin, produces lipoteichoic acid, which inhibits the release of inflammatory cytokines from keratinocytes (Lai et al. 2009). Pathogenic microorganisms stimulate the production of inflammatory cytokines and chemokines, therefore recruiting neutrophils and monocytes, leading to infections such as folliculitis and hidradenitis. On the other hand, viruses invade the skin and produce latent infections and form their nest in skin appendages, probably due to their unique immune nature (Egawa et al. 2015).

The dermis is characterized by extracellular matrix (ECM), which provides a scaffold for immune cell migration (Wolf et al. 2003). A network of blood and lymphatic vessels and neurons is distributed through the ECM. This vascular system constantly recruits immune cells to the skin, including skin-resident dendritic cells (DCs), macrophages, neutrophils, and T cells, which detect foreign pathogens. Dendritic cells that have caught antigens go to the draining lymph nodes through the lymphatic system (Tomura et al. 2010). Neurons provide the sensory function to the skin and can communicate directly with the immune system (Riol-Blanco et al. 2014).

Adipose tissue has a group of immune cells, including T cells, B cells, and macrophages. It is essential for the defense against *S. aureus* infection as they produce cathelicidin, which kills bacteria (Zhang et al. 2015).

3 Chronic Wounds

The healing of acute wounds depends on the extent of the injury. Clinical assessment of acute wounds takes into consideration the method of injury and damage to the soft tissues and bony structures to optimize wound care and ensure sound healing (Nagle et al. 2020). Wounds that do not heal within 2–3 weeks are considered chronic. These wounds negatively impact the quality of life of patients (Rahman et al. 2010; Han and Ceilley 2017). Chronic wounds are caused mainly by the following causes; vascular, traumatic, malignant, pressure, and diabetic wounds (McCosker et al. 2019; Iyun et al. 2016).

3.1 Diabetic Wounds

Diabetic wounds and foot infections are a major problem affecting about 20% of diabetic patients worldwide (Ogurtsova et al. 2017). Diabetic foot management costs nearly USD10 billion annually (Raghav et al. 2018). Foot complications include infection, ulceration, and gangrene (Ray et al. 2005).

Metabolic complications due to diabetes, hyperglycemia, peripheral neuropathy, and impaired circulation disrupt the wound healing process (Popov 2010; Zhao et al. 2016b; Baltzis et al. 2014). Poor oxygenation and circulation are contributing factors to poor wound healing in diabetic patients (Berlanga-Acosta et al. 2013).

3.2   Pressure Ulcers

Pressure ulcers or bedsores are the most common example of tissue necrosis (Roaf 2006), developing mainly in elderly and bedridden patients (Tubaishat et al. 2018; Grey et al. 2006). Pressure ulcers affect about 2.5 million patients per year in the USA and cost over 10 billion USD per year. Skin surfaces over the bony prominences (e.g., hips, ankles, heels, coccyx, scapulae) are the most vulnerable areas (Baron et al. 2016; Mendoza-Garcia et al. 2015).

Pressure, shear, friction, and moisture are the main factors involved in the pathogenesis of pressure ulcers (Grey et al. 2006). Due to continuous pressure, the blood supply to the skin is obstructed, leading to poor circulation resulting in tissue death and ulcer development (Anders et al. 2010). Pressure ulcers may be prevented by changing patient positions frequently (Nageswaran et al. 2015).

3.3   Other Types of Ulcers

Venous Leg Ulcers occur due to chronic venous insufficiency, especially in the lower limbs (Guest et al. 2018; Comerota and Lurie 2015), due to the local rise in blood pressure and leakage of macromolecules into the extra-vascular space. Tissue edema and fibrosis impair oxygen diffusion, therefore, causing tissue ischemia and death (Guenin-Macé et al. 2014; Morton and Phillips 2016).

Arterial Ulcers are a less common type of ulcers. They occur because of arterial insufficiency and poor perfusion, leading to insufficient skin oxygenation and tissue breakdown (Guenin-Macé et al. 2014).

Traumatic wounds usually occur after road traffic injuries, gunshot wounds, and bone fractures, these wounds are, in some cases, life-threatening, depending on the site of injury. In most cases, traumatic wounds are usually contaminated by skin flora and to a certain level by environmental organisms (Robson 1997). Infected traumatic wounds cause heavy bacterial burden, which impacts wound healing negatively by increasing the metabolic requirements of the patient, and by stimulating pro-inflammatory cytokines and by the effects of cytokines secreted by bacteria (Devriendt and de Rooster 2017). The presence of necrotic tissue at the wound site increases the severity of infection, complicates healing, and prolongs care.

Malignant wounds usually occur as a complication of cancer spreading to the subcutaneous tissues. They are commonly seen on the breast and chest wall. Malignant wounds are usually complicated with the poor general condition and the associated pain, exudate, and sometimes hemorrhage. Treatment of malignant wounds is primarily palliative to relieve pain and control infection (Ramasubbu et al. 2017).

Other types of skin ulceration include burn and immune dysfunction (Tomioka et al. 2018).

4 Mechanism of Wound Healing

4.1 Normal Healing Process

The normal healing process of the skin is characterized by coagulation, acute inflammation, proliferation, and remodeling (Khodaeian et al. 2015).

After the acute injury, the coagulation cascade is activated, leading to the formation of a clot to prevent bleeding (Gilbert et al. 2016), and to protect the wound from infection (Bielefeld et al. 2013). The release of pro-inflammatory cytokines and growth factors from activated plates recruits inflammatory cells to the wound and initiates an inflammatory phase (Gilbert et al. 2016; Hameedaldeen et al. 2014). Neutrophils and macrophages are key players in this phase. Neutrophils remove bacteria, foreign objects from the wound and produce proteolytic enzymes such as matrix metalloproteinase (MMP) to break down dead tissue (Hameedaldeen et al. 2014). Monocytes later differentiate to macrophages and phagocytose foreign organisms and dead neutrophils, and also release transforming growth factor ß (TGF-ß) and other cytokines, therefore, enhance fibroblasts and epithelial cells migration into the wound area (Bielefeld et al. 2013; Hameedaldeen et al. 2014).

The proliferation phase is characterized by angiogenesis, extracellular matrix (ECM) synthesis, and re-epithelialization (Gilbert et al. 2016; Emanuelli et al. 2016). Macrophages shift to an anti-inflammatory phenotype expressing anti-inflammatory mediators, proteases, and growth factors, such as vascular endothelial growth factor (VEGF) and TGF- ß to encourage cell proliferation and protein synthesis. Endothelial cells and fibroblasts then enhance new blood vessels and fibrous tissue formation to form granulation tissue (Tsourdi et al. 2013).

The remodeling phase is the last step leading to skin recovery. Immature ECM and collagen type III are degraded by MMPs and replaced with collagen type I (Zhao et al. 2016a; Gilbert et al. 2016; Bielefeld et al. 2013). Subsequently, collagen fibers rearrange across tension lines, facilitating cross-linking, and increasing the tensile strength of the wound (Emanuelli et al. 2016; Baltzis et al. 2014).

4.2 *Chronic Wound Healing Process*

In chronic wounds, reactive oxygen species (ROS) are the key players. Elevated levels of ROS cause oxidative damage in DNA, proteins, and lipids leading to tissue damage (Donato-Trancoso et al. 2016). They induce inflammation, which in turn leads to epithelial dysfunction, decreased reperfusion, impaired angiogenesis resulting in poor ulcer healing (Blakytny and Jude 2006). Chronic skin ulcers are characterized by reduced levels of tissue inhibitors of matrix metalloproteinases, which result in elevated levels of matrix metalloproteinases to accelerate tissue degradation (Baltzis et al. 2014; Amin and Doupis 2016). MMPs destroy growth factors involved in healing, such as TGF- β1, insulin growth factor (IGF-I), and platelet-derived growth factor (PDGF), which are crucial for the healing process, there inhibiting re-epithelization (Falanga 2005). High concentrations of ROS and low TGF- β1 expression level increase macrophage chemoattractant protein-1 (MCP-1) levels, which attracts greater numbers of macrophage, leading to sustained inflammation (Blakytny and Jude 2009). Impaired angiogenesis is seen in patients with chronic ulcers. Hypoxia-inducible factor-1α (HIF-1α), which iduces angiogenesis is induced in response to hypoxia resulting in the transcription of growth factors like VEGF, which is important for angiogenesis. In chronic ulcers, HIF-1α is downregulated, leading to low expression of VEGF as a result of poor angiogenesis and impaired wound healing (Catrina et al. 2004; Larouche et al. 2018).

5 Wound Management

Pressure ulcer prevention requires risk assessment, patient mobility and nutrition, skincare, and regular pressure redistribution (Langemo et al. 2015). Treatment plans may include the use of **local (topical) treatment,** such as antimicrobials, antioxidants, growth factors, and analgesics (Gupta et al. 2017). Although antibiotics are the first line of treatment of wounds to prevent infection, the extensive use of antibiotics leads to antimicrobial resistance (Norman et al. 2016; Ayukekbong et al. 2017).

Wound Dressings have been used to protect wounds and accelerate healing. The choice of dressing varies depending on the type of the wound, its location, and the amount of exudate (Gupta et al. 2017). Non-healing or chronic and infected wounds may require **debridement** or the excision of necrotic tissue to clean the wound, and decrease infection (Leaper et al. 2011; Burtis and Dobbs 2009). Debridement is done surgically or biologically (Woo et al. 2015; Falabella 2006). Biological debridement employs the use of enzymes, which may lead to inflammation, while surgical debridement removes necrotic and healthy tissues non-selectively (Falabella 2006).

Other plans include the use of **Vacuum-Assisted Closure (VAC), or negative pressure** to accelerate chronic wound healing (Nain et al. 2011; Han and Ceilley 2017). The vacuum enhances oxygenation, blood flow, and tissue repair (Schreiber

2016; Huang et al. 2014). Alternative plans include the use of **Hyperbaric Oxygen therapy (HBOT),** aiming to increase the oxygen concentration in the patient's blood. There are still some doubts that high oxygen pressure might harm the brain (Tuk et al. 2014). Finally, some cases require surgical treatment and the use of autologous, full, or partial thickness skin grafts (Serena 2015).

Artificial Intelligence and Machine Learning can lead to a revolution in wound care practice. In the past few years, applications for wound photography and measurement of the wound area were developed. Such electronic tools can be integrated easily with the use of telemedicine and the use of electronic medical record systems (Queen 2019). Machine learning tools can analyze the wound area and recommend treatment for effective wound management (Queen 2019).

6 Wound Microbiome

Skin commensal microorganisms are essential for the development of the host immune response and protection against pathogenic microorganisms (Tr et al. 2018). In the following part, we will discuss the microbiome of healthy skin and how this microbiome is disturbed in acute and chronic wounds. Understanding the wound microbiome may improve the understanding of wound healing and lead to advances in treatment strategies.

6.1 The Healthy Skin Microbiome

Over 1000 bacterial species belonging to 19 phyla are commensals of the superficial layers of the skin (Grice et al. 2008, 2009). The bacterial composition differs according to the anatomical location of the skin and its moisture content. *Propionibacterium* and *Staphylococci* species dominate sites with many sebaceous glands (Tr et al. 2018), while sites with high moisture are dominated by *Corynebacteria* and *Staphylococci* spp. (Grice et al. 2009). Whereas dry sites are highly abundant in *β-proteobacteria*, *Flavobacteriales*, and other Gram-Negative organisms (Grice and Segre 2011). Recent studies proved that the healthy skin microbiome extends into the deeper parts of the dermis with higher proportions of *Proteobacteria* and *Actinobacteria* (Tr et al. 2018). In contrast to the bacterial microbiome, the mycobiome (fungal microbiome) component differs by anatomical location rather than moisture or sebaceous content (Findley et al. 2013). The fungal skin community is mainly composed of the *Malassezia* genus (Paulino et al. 2008). Feet as an exception are characterized by high fungal diversity and lower stability. This explains why diseases of the feet are common sites of recurrent fungal infections. Most recent, high throughput metagenomic sequencing has identified the Human Polyomavirus and Circoviruses are the main constituents of the skin virome (Tr et al. 2018). Moreover, bacteriophages (*Staphylococcus* phages) are also major components of the skin virome (Landini et al. 2015). Recent studies have shown that environmental and genetic factors help define normal skin flora (Chen and Tsao 2013).

6.2 Acute Wound Microbiome

Understanding the microbiome of acute wounds helps in the understanding of the healing process (Xu and Hsia 2018). One study investigated the microbiota of open fractures. There were significant differences in the bacterial composition of wounds at initial admission in the hospital and after discharge. On admission, Pseudomonas, Corynebacterium, and Anaerococcus were more abundant in wounds if compared to normal skin. After the wound was healed, the skin was rich in Staphylococcus and decreased in Pseudomonas (Hannigan et al. 2014). Another study investigated the microbiome of burn wound patients. Burn wounds in these patients show an abundance of thermophilic organisms such as *Aeribacillus* and *Nesterenkonia* and a decrease of *Corynebacterium* (Plichta et al. 2017). There was a significant relationship between different bacterial species present and postburn complications, as wound infection, sepsis & pneumonia. *Propionibacterium acnes* was associated with a greater risk of pneumonia and wound infection (Xu and Hsia 2018).

The microbes available in acute wounds do not obstruct the inflammatory response. *Staphylococcus epidermidis* produces lipoteichoic acid, which decreases inflammation. Keratinocytes express antimicrobial peptides (AMPs) that provide protection against pathogenic bacteria. Probiotic supplements such as *Lactobacillus reuteri* may accelerate wound healing. Bacteria such as *Staphylococcus aureus* produce superantigens that reduce interleukin (IL-17) and consequently enhance the wound healing process. *Pseudomonas* enhances epithelization and angiogenesis through transforming growth factor beta-activated kinase 1 (TAK1) signaling. (Tr et al. 2018).

6.3 Chronic Wound Microbiome

Polymicrobial biofilms play a critical role in impaired wound healing (James et al. 2008). Non-healing wounds have high numbers of microorganisms, including anaerobic bacteria, which may hinder the healing process. These bacterial biofilms increase the expression of cytokines IL-1B, IL-8, IL-6, chemokine ligands 1 and 8, and TNF-α. IL-8 is a potent neutrophil chemoattractant. The biofilm decreases matrix metalloproteinase -3, and vascular endothelial growth factor (VEGF) expression. Thus contributing to poor vascularization of the wound bed further prevents the delivery of exogenous therapeutics. Although chronic wounds have inflammation phase, the total number of macrophages and activated fibroblasts are low preventing healing (Tr et al. 2018).

Modern tools such as next-generation sequencing shed more light on the diversity of the wound microbiome. For example, in pressure ulcers, the most dominant phyla observed were Firmicutes, Proteobacteria, and Actinobacteria (Ammons et al. 2015) There is a clear difference in the microbial composition of pressure ulcers in diabetic and non-diabetic patients (Xu and Hsia 2018). In diabetic foot ulcers, the most abundant bacteria found were Corynebacterium, Bacteroides, Peptoniphilus, Finegoldia, Anaerococcus, Streptococcus, and *Serratia* spp. (den Reijer et al. 2016). Another study showed that *Streptococcus* and Clostridiales Family XI, a family of anaerobic bacteria, were more abundant in diabetic rather than non-diabetic patients (Clark 1989). Deeper non-healing ulcers have high levels of anaerobes and Gram-negative bacteria. Shallow healed ulcers showed a high abundance of *Staphylococcus aureus*. The most dominating fungal species found in diabetic foot ulcers (DFU) are Candida (parapsilosis, albicans, and tropicalis) and Trichophyton mentagrophytes as (Uberoi et al. 2020). Finally, in venous ulcer, the most dominant bacteria include Bacteroides species, S. aureus, Pseudomonas, Corynebacterium species, various anaerobes, and Serratia (Thomsen et al. 2010). A study used debridement samples to demonstrate the association between bacteria and ulcer healing. They found that non-healed wounds at 6 months showed higher bacterial abundance and diversity. Those wounds had significantly high Actinomycetales and low Pseudomonas if compared with healed ulcers (Tuttle et al. 2011).

7 Experimental Wound Models

7.1 In Silico Models

Computational and mathematical models or in silico models utilize data analysis and bioinformatics to study wound healing. They may be used to test potential therapeutics and to design synthetic tissues for skin regeneration. Their main disadvantage is that they rely on the parameters that are chosen by the researcher and lack the biological complexity of the human skin. The outcome of such studies should be applied in a biological model to confirm the results (Menke et al. 2010).

7.2 In Vitro

The pathogenesis of wound healing may be studied in vitro; such models may help understand scar formation and healing processes (van den Broek et al. 2014). Examples of these models include single-layer cell cultures and co-cultures. Skin explants are also used as or three-dimensional cultures (Fig. 2).

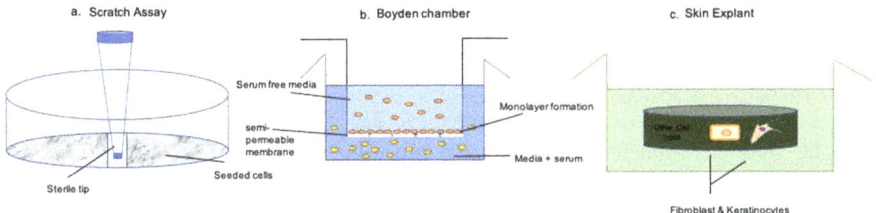

Fig. 2 Different types of In vitro wound healing assays. (**a**) Scratch Assay in mono-layer cell culture. (**b**) Boyden co-culture Chamber. (**c**) Skin Explant. (with permission from Sami et al. 2019)

8 Single or Mono-layer Cell Culture

Conventional cell culture is a relatively easy and fast approach to wound healing study (van den Broek et al. 2014). Single or Mono-layers of cells, e.g., human or animal fibroblasts or keratinocytes, are commonly grown in culture. The layer of cells is disrupted using a scratch tool (sterile glass or plastic instrument) (Fig. 2). this scratch assay is commonly used to study the migration of cells in vitro. This method is usually utilized when studying various factors or drugs that affect cell migration and proliferation. The cell culture environment is controlled, which may be seen as an advantage for this technique, where factors may be added or removed depending on the purpose of the study (Henemyre-Harris et al. 2008). Human skin is a complex organ with many interacting cell types, which makes this model of limited value (Henemyre-Harris et al. 2008).

8.1 Co-Cultured Cell Cultures

To overcome the problems of single-layer cell culture, different systems are developed to allow for the co-culture of different cell types to study (van den Broek et al. 2014). The trans-well system employs a chamber (Boyden Chamber) with a filter membrane separating it into two. The pore size is used to control the migration of cells (Fig. 2) (Boyden 1962).

Co-cultures provide insight into the factors affecting migration and cell–cell interaction. (Werner et al. 2007). However, this system is a two-dimensional system with the availability to study only two cell types at a time, which limits its use (Ud-Din and Bayat 2017).

8.2   Ex Vivo Models

Skin explant is a type of culture in which the top layers of the skin are grown in vitro after removing fat and subcutaneous tissue (Fig. 2). Healthy and diseased skin samples may be collected from surgical procedures. This model is used for the study of wound repair and inflammation and to test the effects of different drugs (Cho et al. 2013; Ud-Din and Bayat 2017; Reus et al. 2012). This system has the advantage of being a three-dimensional structure with multiple cell types in which the micro-environment can be studied (Cho et al. 2013; Nayak et al. 2013; Karamichos et al. 2009). Other factors can be controlled and changed to simulate in vivo conditions such as pH, nutrients, and temperature (Nayak et al. 2013).

A major disadvantage is the lack of nerve supply, which is thought to be essential in skin repair (Cho et al. 2013). This model is also highly variable depending on the age, sex, and health condition of the donor.

Other examples of Ex vivo models are used to study scars and evaluate therapeutics affecting keloids and hypertrophic scars (Iqbal et al. 2010; Zhang et al. 2009; Syed et al. 2013).

8.3   In Vivo Models

Models utilizing living animals are the most clinically relevant models in wound healing studies. Humans, small and large animals, are often used. They provide a complex biological environment involving multiple aspects of the pathophysiology of wound healing. Animal models are essential for testing new therapeutic techniques before moving to human clinical trials (Stephens et al. 2013). For a full review of the different wound models see Sami et al. 2019.

8.3.1   Human Wound Models

Using humans as a model is the ideal situation for wound research since the pathophysiology of healing, especially in acute wounds, is almost identical among all human beings. They are clinically important, especially in clinical trials (Wilhelm et al. 2017). A summary of some of the acute wound models is shown in Table 1. Unfortunately, such studies have legal, ethical, and practical concerns (Trøstrup et al. 2016).

Chronic wounds include venous leg ulcers, diabetic ulcers, and pressure ulcers, and others (Frykberg and Banks 2015). Human chronic wounds provide an opportunity to study the delayed healing process. They are accessible, and scientists are able to get wound tissue samples post debridement surgery. Wound swabs and tissue specimens from chronic ulcers also can be obtained during outpatient visits (Pastar et al. 2018). They thus provide a better opportunity to study the underlying factors of

Table 1 Acute wound models in human

Model	Technique and depth	Comments	Applications	References
Partial thickness model	Skin stripping with adhesive tape. Limited to stratum corneum, without affecting the dermal blood vessels	Simple and non-invasive. Variable wound, depending on the number of strips, pressure, and adhesive power.	Used to assess new wound dressing, therapeutic agents & skincare products.	Sobiepanek et al. (2019), Gao et al. (2013)
	Suction blister model involves the epidermis.	The negative pressure causes splitting in the basal membrane and separation between epidermis and dermis.	To assess healing and collagen synthesis. Measurements of pharmacological agents.	Koivukangas and Oikarinen (2003)
	Abrasive wound model involves the epidermis.	A semi-invasive technique using surgical brush to scrub skin.	Useful for differentiating wound dressing properties	Wigger-Alberti et al. (2009)
	Laser wounds. Variable depth	A laser used to induce superficial wounds affecting epidermis or deeper layers. High wound reproducibility.	Useful for topical pharmacological studies	Marquardt et al. (2015)
	Microdermabrasion (MDA). Variable depth	A minimally invasive technique using abrasive crystals to cause mechanical abrasion and removing the stratum corneum. Less expensive than lasers.	A nonsurgical cosmetic procedures to treat scars, photoaging. Improves the transdermal drug delivery	Shah and Crane (2020)
	Split-thickness wounds. Variable depth	It involves the removal of a 100–1500 micron-thick layer of the epidermis/upper dermis using a sharp blade. The epidermal appendages are left to determine re-epithelization.	It is used to evaluate the effect of age on wound healing and to test the topical application of therapeutics and growth factors.	Wilhelm et al. (2017)
Full-thickness wound model	A full-thickness wound and complete removal of epidermis and dermis.	A scalpel or dermatome is used to create standardized wounds.	It is used to test new treatments.	Misic, et al. (2014)

wound healing on the molecular and histopathological levels (Pastar et al. 2018). Another point is that wound fluid is rich in information on metabolomics, proteomics, and wound microbiome environment, which plays an important role in the impaired healing process. Such studies will lead to personalized and targeted treatments that result in improvements in the outcome (Dowd et al. 2011).

Many difficulties prevent the widespread use of patients with chronic wounds as study subjects. For example, histological studies require multiple biopsies at multiple time points, which is not practical as it disturbs the healing process and is counterintuitive to the study's purpose of improving healing. Patients compliance and cooperation is also a major factor in establishing a well-controlled and statistically significant study (Nuutila et al. 2014). Moreover, ethical considerations prevent infecting the human wound or the use of an untreated control subject (Ud-Din and Bayat 2017; Ito and Cotsarelis 2008).

8.3.2 Rodents

The skin anatomy of the rodents is different from that of humans, as rodents have a thin epidermis and dense hair, as well as the presence of subcutaneous muscle (Panniculus carnosus), which accelerate wound healing (Dorsett-Martin 2004; Ito and Cotsarelis 2008). Also, rodents lack apocrine and eccrine glands (Dorsett-Martin 2004 and have endogenous vitamin C, which is crucial in all phases of healing (Wong et al. 2011; Moores 2013; Dorsett-Martin 2004). Finally, the rodents' immune system is stronger compared to humans (Seaton et al. 2015).

Despite all these differences, rodents remain extensively used in wound healing studies due to their low cost, availability, which makes them suitable for large studies (Trøstrup et al. 2016). Different models of wounds were developed to study acute and chronic wounds (Fig. 3).

8.3.3 Rabbits

The rabbit ear is commonly used as a model for wound healing. The cartilage of the rabbit ear heals by re-epithelialization and granulation formation. It is characterized by high vascularization, which is similar to the human dermis. The high breeding cost is the main disadvantage of using rabbits (Rittié 2016). Table 2 summarizes previous research using rabbits as a model for wound healing.

8.3.4 Pigs

Porcine (pigs) have shared anatomical and physiological similarities to humans that make them promising models for wound healing. Pigs have thick skin with apocrine glands, sparse hair, and rely on exogenous vitamin C. Wounds heal by re-epithelization. In contrast to humans, pigs have poor dermis vasculature

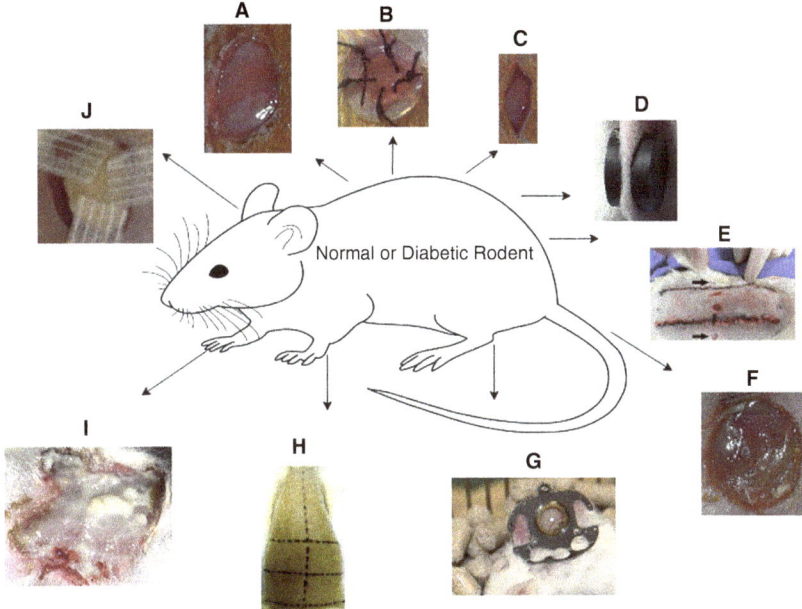

Fig. 3 Different types of wound models in rodents. (with permission from Sami et al. 2019)

(Summerfield et al. 2015). Pigs are not an ideal model due to the high cost and the requirement for large experimental setups, which limits their use, especially in large-scale experiments where a large number of animals is needed (Parnell and Volk 2019). Table 3 is a summary of previous research that used pigs as a wound model.

8.3.5 Zebrafish

Zebrafish (*Danio rerio*) are small freshwater fish that are commonly used in biological research (Li et al. 2011; Trompouki et al. 2011). Among other models, zebrafish have several advantages. They have a high degree of genetic conservation with humans, a large number of offspring, and their maintenance cost is cheap (Trompouki et al. 2011). Wounds in the epidermis of zebrafish and mammalian embryos close in the same way both of them re-epithelialize without inflammation, granulation tissue formation, and scarring (Brockes et al. 2004). The caudal fin of the zebrafish is unique since it can fully regenerate within 10–14 days (Alvarez et al. 2018). A previous study used a laser to introduce full-thickness wounds on the flank of adult zebrafish (Richardson et al. 2013). They showed that wound re-epithelialized rapidly without inflammation and blood clot formation (Richardson et al. 2013). From our point of view, this model is not suitable to study acute wounds that occur in humans as they do not go through the phases of wound healing (inflammation, granulation, and scaring).

Table 2 The use of rabbit ear as wound Models

Wound type	Method	Comment	References
Excisional wound model	A full-thickness excision on the backside of the ears.	This model imitates the acute healing process in humans.	Qian et al. (2017)
Diabetic model	A full-thickness wound on the ear in diabetic animals. Diabetes is induced by Alloxan, or Streptozotocin (STZ), or by a high-fat diet.	**Application** Useful to study diabetic wounds. **Limitations** High mortality rate of diabetic rabbits.	Wang et al. (2010)
Cutaneous ischemia-reperfusion ulcer	Ischemia/reperfusion injury was induced by applying and removing magnets over the ear between the central artery and marginal veins.	**Application** This model is used to study type 1 pressure ulcers. **Limitations** A small sample size (four animals) was used in this study.	Schivo et al. (2017)
Ischemic skin wound model	Ligation of two arteries at the base of the ear. A punch biopsy was used to create deep ulcers.	**Application** Used to study pressure ulcers as it resembles the healing process of the human.	Ahn and Mustoe (1990)
Skinfold chamber model	A chamber is placed on top of holes punched through the cartilage.	**Application** To study the vascular changes in the presence of shear stress.	Ichioka et al. (1997)
Burn wound model	Heated rod (90 °C) to induce burns.	**Application** It is considered a good model example to study hypertrophic scar.	Friedrich et al. (2017)
Infected model	Bacteria were inoculated in full-thickness wounds.	**Application** It is used to study the antibacterial effect of drugs in chronic and infected wounds.	Dai et al. (2011), Seth et al. (2012)
Parabiosis model	Two rabbits are joined surgically at the ears.	**Application** It is used to study immunology, cancer metastasis, and metabolic disorders. **Limitations** High risk of mortality.	Andresen et al. (1957), Dai et al. (2011)
Denervated wound model	Surgical excision of the sensory nerves of one ear.	**Limitations** Wound completely healed by 15 days.	Yagmur et al. (2011)

8.3.6   *Caenorhabditis elegans*

The epidermis of *Caenorhabditis elegans* consists of a simple epithelium on top of a basal lamina. Epithelial cells secrete flexible collagenous cuticles (Chisholm and Hsiao 2012). *C. elegans* has been used as a model to study the different biological

Table 3 The use of pigs as wound Models (modified from Sami et al. 2019)

Wound Type	Method	Comment	References
Excisional wound model	Full-thickness wounds are created on the back of the animal in normal or diabetic pigs.	**Limitations** High cost and small sample size. Wounds in diabetic pigs heal faster (18 days) compared to humans.	Singer and McClain (2003)
Radiation-induced wound	The leg was irradiated, and a full-thickness excision is made.	**Limitations** Skin became normal after 2 weeks later. High cost	Bernatchez et al. (1998)
Cutaneous ischemia-reperfusion ulcer	Two magnets are implanted subcutaneously and external magnets are applied to cause compression and ischemia of the skin.	**Application** Used to study vascular supply and the circulation of the skin. **Limitations** The small sample size used in the study.	Vaena et al. (2017)
Ischemic skin wound model	A skin flap is created and a silicone sheet to avoid reattachment. Circular excisional wounds are made in the flap.	Decreased blood reperfusion and delayed macrophages occurred in the ischemic wounds. **Limitations** High surgical skills are required Pocket filled with fluid form under the flap. Small sample size and small wounds are hard to analyze.	Patil et al. (2017), Roy et al. (2009), Seaton, et al. (2015)
Burn wound model	A heated metal bar (80–110 °C) is used to create burn wound.	**Application** A simple, non-invasive model for burn wounds. It is important to study the therapeutic effect of new drugs and study hypertrophic scarring. May be used to study wound infection. **Limitations** Small burn area. A modification uses a steel bar to control the depth of the burn.	Rapp et al. (2015)
Infected wound model	An incisional/excisional wound is created and inoculated with microorganisms, or foreign bodies, in normal and diabetic pigs.	**Application** Used to study antibacterial drugs. Used to study bacterial interaction in the wound	Dai et al. (2011); Hirsch et al. (2008), Kalan et al. (2016), Seaton et al. (2015)

(continued)

Table 3 (continued)

Wound Type	Method	Comment	References
		environment. Useful to study infection in chronic diabetic wound. **Limitations** Diabeteic wound healing did not resemble human wound healing pattern.	
Dead space wound model	Skin incision and removing part of the back muscles.	**Application** The model was useful to study the effect of negative pressure wound therapy.	Huang et al. (2014), Murthy et al. (2013), Nayak et al. (1999)
Denervated wound model	Nerve is cut followed by skin compression.	**Application** Suitable to study bedsores. **Limitations** Requires high surgical skills and is complicated by muscle atrophy. Wound heals within 3 weeks, which does not allow for studying chronic wounds.	Seaton et al. (2015)

processes involved in wound healing. Its development is well understood, which enables researchers to trace the lineage of different cells involved in the wound repair process (Bao et al. 2006).

9 Conclusion

The wound healing process is complex and requires extensive research to be understood and for the development of new therapeutic strategies. To the best of our knowledge, there are no animal models that demonstrate identical wound healing criteria to human wound pathophysiology. Since human models are of limited value to research, various animal models have been developed to enhance our understanding of the healing process. Multiple models should be used to study different types of wounds, either sequentially or in combination, e.g., in vitro and in silico models could be used before moving to animal models. Despite the ethical concerns and their associated disadvantages, animal models remain necessary for the study of the complex pathophysiology of wound healing and for testing different treatment approaches.

Rodent models are the most cost-effective, and therefore, large numbers can be used, leading to statistically significant studies. Rats are, in our view, are preferred for wound studies despite the presence of the subcutaneous muscle (Panniculus carnosus). Recent studies (Maldonado et al. 2014; Sami and Abdellatif 2020) report wound models that overcome the contraction element of the subcutaneous muscle and create ulcers very similar to human pressure ulcers, which provide an excellent tool for future wound research.

Many reliable wound assessment tools have been developed (Lima et al. 2018) to assess wound healing in small animal models with high reliability, therefore making these models even more significant for translational wound research.

Conflict of Interest The authors declare no competing financial interest.

References

Adachi T, Kobayashi T, Sugihara E, Yamada T, Ikuta K, Pittaluga S, Saya H, Amagai M, Nagao K (2015) Hair follicle-derived IL-7 and IL-15 mediate skin-resident memory T cell homeostasis and lymphoma. Nat Med 21(11):1272–1279. https://doi.org/10.1038/nm.3962

Ahn ST, Mustoe TA (1990) Effects of ischemia on ulcer wound healing: a new model in the rabbit ear. Ann Plast Surg 24(1):17–23. https://doi.org/10.1097/00000637-199001000-00004

Alvarez M, Chávez MN, Miranda M, Aedo G, Allende ML, Egaña JT (2018) A novel in vivo model to study impaired tissue regeneration mediated by cigarette smoke. Sci Rep 8(1):10926. https://doi.org/10.1038/s41598-018-28687-1

Amin N, Doupis J (2016) Diabetic foot disease: from the evaluation of the 'foot at risk' to the novel diabetic ulcer treatment modalities. World J Diabetes 7(7):153–164. https://doi.org/10.4239/wjd.v7.i7.153

Ammons MCB, Morrissey K, Tripet BP, Van Leuven JT, Han A, Lazarus GS, Zenilman JM, Stewart PS, James GA, Copié V (2015) Biochemical association of metabolic profile and microbiome in chronic pressure ulcer wounds. PLoS One 10(5):e0126735. https://doi.org/10.1371/journal.pone.0126735

Anders J, Heinemann A, Leffmann C, Leutenegger M, Pröfener F, von Renteln-Kruse W (2010) Decubitus ulcers: pathophysiology and primary prevention. Dtsch Arztebl Int 107(21):371–382. https://doi.org/10.3238/arztebl.2010.0371

Andrade TA, Aguiar AF, Guedes FA, Leite MN, Caetano GF, Coelho EB, Das PK, Frade MA (2015) Ex vivo model of human skin (HOSEC) as alternative to animal use for cosmetic tests. Procedia Engineering, 4th International Conference on Tissue Engineering, ICTE2015, An ECCOMAS Thematic Conference 110(January):67–73. https://doi.org/10.1016/j.proeng.2015.07.011

Andresen RH, Hass GM, Madden DA, Monroe CW (1957) Postparabiotic tissue reactions of rabbits to Musculofascial cross-transplants. J Exp Med 105(1):85–92. https://doi.org/10.1084/jem.105.1.85

Ansorge HL, Meng X, Zhang G, Veit G, Sun M, Klement JF, Beason DP, Soslowsky LJ, Koch M, Birk DE (2009) Type XIV collagen regulates Fibrillogenesis: premature collagen fibril growth and tissue dysfunction in null mice. J Biol Chem 284(13):8427–8438. https://doi.org/10.1074/jbc.M805582200

Ayukekbong JA, Ntemgwa M, Atabe AN (2017) The threat of antimicrobial resistance in developing countries: causes and control strategies. Antimicrob Resist Infect Control 6(May). https://doi.org/10.1186/s13756-017-0208-x

Badia JM, Casey AL, Petrosillo N, Hudson PM, Mitchell SA, Crosby C (2017) Impact of surgical site infection on healthcare costs and patient outcomes: a systematic review in six European countries. J Hosp Infect 96(1):1–15. https://doi.org/10.1016/j.jhin.2017.03.004

Baltzis D, Eleftheriadou I, Veves A (2014) Pathogenesis and treatment of impaired wound healing in diabetes mellitus: new insights. Adv Ther 31(8):817–836. https://doi.org/10.1007/s12325-014-0140-x

Bao Z, Murray JI, Boyle T, Ooi SL, Sandel MJ, Waterston RH (2006) Automated cell lineage tracing in *Caenorhabditis Elegans*. Proc Natl Acad Sci 103(8):2707–2712. https://doi.org/10.1073/pnas.0511111103

Baron J, Jillian S, Presseau J, Aspinall A, Jaglal S, White B, Wolfe D, Grimshaw J (2016) Self-management interventions to improve skin Care for Pressure Ulcer Prevention in people with spinal cord injuries: a systematic review protocol. Syst Rev 5(September):150. https://doi.org/10.1186/s13643-016-0323-4

Berlanga-Acosta J, Schultz GS, López-Mola E, Guillen-Nieto G, García-Siverio M, Herrera-Martínez L (2013) Glucose toxic effects on granulation tissue productive cells: the diabetics' impaired healing. BioMed Res Int 2013:256043. https://doi.org/10.1155/2013/256043

Bernatchez SF, Parks PJ, Grussing DM, Matalas SL, Nelson GS (1998) Histological characterization of a delayed wound healing model in pig. Wound Repair Regen 6(3):223–233. https://doi.org/10.1046/j.1524-475X.1998.60308.x

Bielefeld KA, Amini-Nik S, Alman BA (2013) Cutaneous wound healing: recruiting developmental pathways for regeneration. Cell Mol Life Sci 70(12):2059–2081. https://doi.org/10.1007/s00018-012-1152-9

Blakytny R, Jude E (2006) The molecular biology of chronic wounds and delayed healing in diabetes. Diabet Med 23(6):594–608. https://doi.org/10.1111/j.1464-5491.2006.01773.x

Blakytny R, Jude EB (2009) Altered molecular mechanisms of diabetic foot ulcers. Int J Low Extrem Wounds 8(2):95–104. https://doi.org/10.1177/1534734609337151

Boyden S (1962) The chemotactic effect of mixtures of antibody and antigen on polymorphonuclear leucocytes. J Exp Med 115(3):453–466

Brockes JP, Martin P, Redd MJ, Cooper L, Wood W, Stramer B, Martin P (2004) Wound healing and inflammation: embryos reveal the way to perfect repair. Philos Trans R Soc Lond Ser B Biol Sci 359(1445):777–784. https://doi.org/10.1098/rstb.2004.1466

Brown TM, Krishnamurthy K (2020) Histology, dermis. In: StatPearls. StatPearls Publishing, Treasure Island, FL. http://www.ncbi.nlm.nih.gov/books/NBK535346/

Burgeson RE, Lunstrum GP, Rokosova B, Rimberg CS, Rosenbaum LM, Keene DR (1990) The structure and function of type VII collagen. Ann N Y Acad Sci 580:32–43. https://doi.org/10.1111/j.1749-6632.1990.tb17915.x

Burtis DB, Dobbs MR (2009) Chapter 38 – tetanus toxin. In: Clinical neurotoxicology. W.B. Saunders, Philadelphia, pp 427–435. https://doi.org/10.1016/B978-032305260-3.50044-7

Byrd AL, Belkaid Y, Segre JA (2018) The human skin microbiome. Nat Rev Microbiol 16 (3):143–155. https://doi.org/10.1038/nrmicro.2017.157

Carmichael SW (2014) The tangled web of langer's lines. Clinical Anatomy (New York, N.Y.) 27 (2):162–168. https://doi.org/10.1002/ca.22278

Carmi-Levy I, Homey B, Soumelis V (2011) A modular view of cytokine networks in atopic dermatitis. Clin Rev Allergy Immunol 41(3):245–253. https://doi.org/10.1007/s12016-010-8239-6

Catrina S-B, Okamoto K, Pereira T, Brismar K, Poellinger L (2004) Hyperglycemia regulates hypoxia-inducible factor-1alpha protein stability and function. Diabetes 53(12):3226–3232

Chen YE, Tsao H (2013) The skin microbiome: current perspectives and future challenges. J Am Acad Dermatol 69(1):143–155.e3. https://doi.org/10.1016/j.jaad.2013.01.016

Chisholm AD, Hsiao TI (2012) The C. Elegans epidermis as a model skin. I: development, patterning, and growth. Wiley Interdiscip Rev Dev Biol 1(6):861–878. https://doi.org/10.1002/wdev.79

Cho H-i, Won CH, Chang SE, Lee MW, Park G-H (2013) Usefulness and limitations of skin explants to assess laser treatment. Medical Lasers; Engineering, Basic Research, and Clinical Application 2(2):58–63. https://doi.org/10.25289/ML.2013.2.2.58

Clark RAF (1989) Wound repair. Curr Opin Cell Biol 1(5):1000–1008. https://doi.org/10.1016/0955-0674(89)90072-0

Clark RAF (2014) Chapter 76 – wound repair: basic biology to tissue engineering. In: Lanza R, Langer R, Vacanti J (eds) Principles of tissue engineering, 4th edn. Academic Press, Boston, pp 1595–1617. https://doi.org/10.1016/B978-0-12-398358-9.00076-8

Comerota A, Lurie F (2015) Pathogenesis of venous ulcer. Semin Vasc Surg 28(1):6–14. SI: Contemporary Management of Lower Extremity Venous Ulceration. https://doi.org/10.1053/j.semvascsurg.2015.07.003

Dai T, Kharkwal GB, Tanaka M, Huang Y-Y, Bil de Arce VJ, Hamblin MR (2011) Animal models of external traumatic wound infections. Virulence 2(4):296–315. https://doi.org/10.4161/viru.2.4.16840

den Reijer PM, Haisma EM, Lemmens-den Toom NA, Willemse J, Koning RA, Demmers JAA, Dekkers DHW et al (2016) Detection of alpha-toxin and other virulence factors in biofilms of *Staphylococcus Aureus* on polystyrene and a human epidermal model. PLoS One 11(1): e0145722. https://doi.org/10.1371/journal.pone.0145722

Devriendt N, de Rooster H (2017) Initial management of traumatic wounds. Veterinary Clinics of North America: Small Animal Practice, Wound Management 47(6):1123–1134. https://doi.org/10.1016/j.cvsm.2017.06.001

Donato-Trancoso A, Monte-Alto-Costa A, Romana-Souza B (2016) Olive oil-induced reduction of oxidative damage and inflammation promotes wound healing of pressure ulcers in Mice. J Dermatol Sci 83(1):60–69. https://doi.org/10.1016/j.jdermsci.2016.03.012

Dorsett-Martin WA (2004) Rat models of skin wound healing: a review. Wound Repair and Regeneration: Official Publication of the Wound Healing Society [and] the European Tissue Repair Society 12(6):591–599. https://doi.org/10.1111/j.1067-1927.2004.12601.x

Dowd SE, Wolcott RD, Kennedy J, Jones C, Cox SB (2011) Molecular diagnostics and personalised medicine in wound care: assessment of outcomes. J Wound Care 20(5):232., 234–39. https://doi.org/10.12968/jowc.2011.20.5.232

Egawa N, Egawa K, Griffin H, Doorbar J (2015) Human papillomaviruses; epithelial tropisms, and the development of neoplasia. Viruses 7(7):3863–3890. https://doi.org/10.3390/v7072802

Egging D, van den Berkmortel F, Taylor G, Bristow J, Schalkwijk J (2007) Interactions of human tenascin-X domains with dermal extracellular matrix molecules. Arch Dermatol Res 298 (8):389–396. https://doi.org/10.1007/s00403-006-0706-9

Emanuelli T, Burgeiro A, Carvalho E (2016) Effects of insulin on the skin: possible healing benefits for diabetic foot ulcers. Arch Dermatol Res 308(10):677–694. https://doi.org/10.1007/s00403-016-1686-z

Escandon J, Vivas AC, Tang J, Rowland KJ, Kirsner RS (2011) High mortality in patients with chronic wounds. Wound Repair and Regeneration: Official Publication of the Wound Healing Society [and] the European Tissue Repair Society 19(4):526–528. https://doi.org/10.1111/j.1524-475X.2011.00699.x

Falabella AF (2006) Debridement and wound bed preparation. Dermatol Ther 19(6):317–325. https://doi.org/10.1111/j.1529-8019.2006.00090.x

Falanga V (2005) Wound healing and its impairment in the diabetic foot. Lancet 366 (9498):1736–1743. https://doi.org/10.1016/S0140-6736(05)67700-8

Findley K, Julia O, Yang J, Conlan S, Deming C, Meyer JA, Schoenfeld D et al (2013) Human skin fungal diversity. Nature 498(7454):367–370. https://doi.org/10.1038/nature12171

Friedrich EE, Niknam-Bienia S, Xie P, Jia S-X, Hong SJ, Mustoe TA, Galiano RD (2017) Thermal injury model in the rabbit ear with quantifiable burn progression and hypertrophic scar. Wound Repair and Regeneration: Official Publication of the Wound Healing Society [and] the European Tissue Repair Society 25(2):327–337. https://doi.org/10.1111/wrr.12518

Frykberg RG, Banks J (2015) Challenges in the treatment of chronic wounds. Adv Wound Care 4 (9):560. https://doi.org/10.1089/wound.2015.0635

Gainza G, Villullas S, Pedraz JL, Hernandez RM, Igartua M (2015) Advances in drug delivery systems (DDSs) to release growth factors for wound healing and skin regeneration. Nanomedicine 11(6):1551–1573. https://doi.org/10.1016/j.nano.2015.03.002

Gao Y, Wang X, Chen S, Li S, Liu X (2013) Acute skin barrier disruption with repeated tape stripping: an in vivo model for damage skin barrier. Skin Research and Technology: Official Journal of International Society for Bioengineering and the Skin (ISBS) [and] International Society for Digital Imaging of Skin (ISDIS) [and] International Society for Skin Imaging (ISSI) 19(2):162–168. https://doi.org/10.1111/srt.12028

Gilbert RWD, Vickaryous MK, Viloria-Petit AM (2016) Signalling by transforming growth factor Beta isoforms in wound healing and tissue regeneration. J Dev Biol 4(2):21. https://doi.org/10.3390/jdb4020021

Grey JE, Harding KG, Enoch S (2006) Pressure Ulcers. BMJ: Br Med J 332(7539):472–475

Grice EA, Segre JA (2011) The skin microbiome. Nat Rev Microbiol 9(4):244–253. https://doi.org/10.1038/nrmicro2537

Grice EA, Kong HH, Renaud G, Young AC, Comparative Sequencing Program NISC, Bouffard GG, Blakesley RW, Wolfsberg TG, Turner ML, Segre JA (2008) A diversity profile of the human skin microbiota. Genome Res 18(7):1043–1050. https://doi.org/10.1101/gr.075549.107

Grice EA, Kong HH, Conlan S, Deming CB, Davis J, Young AC, Bouffard GG et al (2009) Topographical and temporal diversity of the human skin microbiome. Science (New York, N. Y.) 324(5931):1190–1192. https://doi.org/10.1126/science.1171700

Guenin-Macé L, Oldenburg R, Chrétien F, Demangel C (2014) Pathogenesis of skin ulcers: lessons from the Mycobacterium Ulcerans and Leishmania Spp. pathogens. Cell Mol Life Sci 71 (13):2443–2450. https://doi.org/10.1007/s00018-014-1561-z

Guest JF, Fuller GW, Vowden P (2018) Venous leg ulcer Management in Clinical Practice in the UK: costs and outcomes. Int Wound J 15(1):29–37. https://doi.org/10.1111/iwj.12814

Gupta S, Andersen C, Black J, de Leon J, Fife C, Lantis Ii JC, Niezgoda J, Snyder R, Sumpio B, Tettelbach W, Treadwell T, Weir D, Silverman RP (2017) Management of chronic wounds: diagnosis, preparation, treatment, and follow-up. Wounds 29(9):S19–S36. PMID: 28862980

Hameedaldeen A, Liu J, Batres A, Graves GS, Graves DT (2014) FOXO1, TGF-β regulation and wound healing. Int J Mol Sci 15(9):16257–16269. https://doi.org/10.3390/ijms150916257

Han G, Ceilley R (2017) Chronic wound healing: a review of current management and treatments. Adv Ther 34(3):599–610. https://doi.org/10.1007/s12325-017-0478-y

Hannigan GD, Hodkinson BP, McGinnis K, Tyldsley AS, Anari JB, Horan AD, Grice EA, Mehta S (2014) Culture-independent pilot study of microbiota colonizing open fractures and association with severity, mechanism, location, and complication from presentation to early outpatient follow-up. J Orthop Res: Official Publication of the Orthopaedic Research Society 32 (4):597–605. https://doi.org/10.1002/jor.22578

Henemyre-Harris CL, Adkins AL, Chuang AH, Graham JS (2008) Addition of epidermal growth factor improves the rate of sulfur mustard wound healing in an in vitro model. Eplasty 8(March): e16

Hirsch T, Spielmann M, Zuhaili B, Koehler T, Fossum M, Steinau H-U, Yao F, Steinstraesser L, Onderdonk AB, Eriksson E (2008) Enhanced susceptibility to infections in a diabetic wound healing model. BMC Surg 8(February):5. https://doi.org/10.1186/1471-2482-8-5

Huang C, Leavitt T, Bayer LR, Orgill DP (2014) Effect of negative pressure wound therapy on wound healing. Curr Probl Surg 51(7):301–331. https://doi.org/10.1067/j.cpsurg.2014.04.001

Ichioka S, Shibata M, Kosaki K, Sato Y, Harii K, Kamiya A (1997) Effects of shear stress on wound-healing angiogenesis in the rabbit ear chamber. J Surg Res 72(1):29–35. https://doi.org/10.1006/jsre.1997.5170

Iqbal SA, Syed F, McGrouther DA, Paus R, Bayat A (2010) Differential distribution of Haematopoietic and Nonhaematopoietic progenitor cells in Intralesional and Extralesional keloid: do keloid scars provide a niche for Nonhaematopoietic mesenchymal stem cells? Br J Dermatol 162(6):1377–1383. https://doi.org/10.1111/j.1365-2133.2010.09738.x

Ito M, Cotsarelis G (2008) Is the hair follicle necessary for Normal wound healing? J Invest Dermatol 128(5):1059–1061. https://doi.org/10.1038/jid.2008.86

Iyun AO, Ademola SA, Olawoye OA, Michael AI, Oluwatosin OM (2016) Point prevalence of chronic wounds at a tertiary Hospital in Nigeria. Wounds Compendium Clin Res Practice 28 (2):57–62

James GA, Swogger E, Wolcott R, deLancey Pulcini E, Secor P, Sestrich J, Costerton JW, Stewart PS (2008) Biofilms in chronic wounds. Wound Repair Regen 16(1):37–44. https://doi.org/10.1111/j.1524-475X.2007.00321.x

Kabashima K, Honda T, Ginhoux F, Egawa G (2019) The immunological anatomy of the skin. Nat Rev Immunol 19(1):19–30. https://doi.org/10.1038/s41577-018-0084-5

Kalan L, Loesche M, Hodkinson BP, Heilmann K, Ruthel G, Gardner SE, Grice EA (2016) Redefining the chronic-wound microbiome: fungal communities are prevalent, dynamic, and associated with delayed healing. MBio 7(5). https://doi.org/10.1128/mBio.01058-16

Karamichos D, Lakshman N, Matthew Petroll W (2009) An experimental model for assessing fibroblast migration in 3-D collagen matrices. Cell Motil Cytoskeleton 66(1):1–9. https://doi.org/10.1002/cm.20326

Khodaeian M, Enayati S, Tabatabaei-Malazy O, Amoli MM (2015) Association between genetic variants and diabetes mellitus in Iranian populations: a systematic review of observational studies. J Diabetes Res 2015:585917. https://doi.org/10.1155/2015/585917

Kielty CM, Shuttleworth CA (1997) Microfibrillar elements of the dermal matrix. Microsc Res Tech 38(4):413–427. https://doi.org/10.1002/(SICI)1097-0029(19970815)38:4<413::AID-JEMT9>3.0.CO;2-J

Koivukangas V, Oikarinen A (2003) Suction blister model of wound healing. In: DiPietro LA, Burns AL (eds) Wound healing: methods and protocols, Methods in molecular medicine™. Humana Press, Totowa, NJ, pp 255–261. https://doi.org/10.1385/1-59259-332-1:255

Lai Y, Di Nardo A, Nakatsuji T, Leichtle A, Yang Y, Cogen AL, Wu Z-R et al (2009) Commensal Bacteria regulate toll-like receptor 3-dependent inflammation after skin injury. Nat Med 15 (12):1377–1382. https://doi.org/10.1038/nm.2062

Landini MM, Borgogna C, Peretti A, Doorbar J, Griffin H, Mignone F, Lai A et al (2015) Identification of the skin Virome in a boy with widespread human Papillomavirus-2-positive warts that completely regressed after administration of tetravalent human papillomavirus vaccine. Br J Dermatol 173(2):597–600. https://doi.org/10.1111/bjd.13707

Langemo D, Haesler E, Naylor W, Tippett A, Young T (2015) Evidence-based guidelines for pressure ulcer Management at the end of life. Int J Palliat Nurs 21(5):225–232. https://doi.org/10.12968/ijpn.2015.21.5.225

Larouche J, Sheoran S, Maruyama K, Martino MM (2018) Immune regulation of skin wound healing: mechanisms and novel therapeutic targets. Adv Wound Care 7(7):209–231. https://doi.org/10.1089/wound.2017.0761

Leaper DJ, Meaume S, Apelqvist J, Teot L, Gottrup F (2011) 24 – debridement methods of non-viable tissue in wounds. In: Farrar D (ed) Advanced wound repair therapies, Series in Biomaterials. Woodhead Publishing, Sawston, pp 606–632. https://doi.org/10.1533/9780857093301.5.606

Li Q, Frank M, Thisse CI, Thisse BV, Uitto J (2011) Zebrafish: a model system to study heritable skin diseases. J Investig Dermatol 131(3):565–571. https://doi.org/10.1038/jid.2010.388

Lima RO, Fechine FV, Lisboa MR, Leitão FK, Vale ML (2018) Development and validation of the experimental wound assessment tool (EWAT) for pressure ulcer in laboratory animals. J Pharmacol Toxicol Methods 90(April):13–18. https://doi.org/10.1016/j.vascn.2017.10.011

Maldonado AA, Cristóbal L, Martín-López J, Mallén M, García-Honduvilla N, Buján J (2014) A novel model of human skin pressure ulcers in Mice. PLoS One 9(10):e109003. https://doi.org/10.1371/journal.pone.0109003

Marquardt Y, Amann PM, Heise R, Czaja K, Steiner T, Merk HF, Skazik-Voogt C, Baron JM (2015) Characterization of a novel standardized human three-dimensional skin wound healing model using non-sequential fractional Ultrapulsed CO2 laser treatments. Lasers Surg Med 47 (3):257–265. https://doi.org/10.1002/lsm.22341

Mattii M, Lovászi M, Garzorz N, Atenhan A, Quaranta M, Lauffer F, Konstantinow A et al (2018) Sebocytes contribute to skin inflammation by promoting the differentiation of T helper 17 cells. Br J Dermatol 178(3):722–730. https://doi.org/10.1111/bjd.15879

McCosker L, Tulleners R, Cheng Q, Rohmer S, Pacella T, Graves N, Pacella R (2019) Chronic wounds in Australia: a systematic review of key epidemiological and clinical parameters. Int Wound J 16(1):84–95. https://doi.org/10.1111/iwj.12996

Mendoza-Garcia J, Sebastian A, Alonso-Rasgado T, Bayat A (2015) Ex vivo evaluation of the effect of photodynamic therapy on skin scars and Striae Distensae. Photodermatol Photoimmunol Photomed 31(5):239–251. https://doi.org/10.1111/phpp.12180

Menke NB, Cain JW, Reynolds A, Chan DM, Segal RA, Witten TM, Bonchev DG, Diegelmann RF, Ward KR, Virginia Commonwealth University Reanimation, Engineering Shock Center, The Wound Healing Group (2010) An in silico approach to the analysis of acute wound healing. Wound Repair and Regeneration: Official Publication of the Wound Healing Society [and] the European Tissue Repair Society 18(1):105–113. https://doi.org/10.1111/j.1524-475X.2009.00549.x

Mikesh LM, Aramadhaka LR, Moskaluk C, Zigrino P, Mauch C, Fox JW (2013) Proteomic anatomy of human skin. J Proteome 84(June):190–200. https://doi.org/10.1016/j.jprot.2013.03.019

Misic AM, Gardner SE, Grice EA (2014) The wound microbiome: modern approaches to examining the role of microorganisms in impaired chronic wound healing. Adv Wound Care 3 (7):502–510. https://doi.org/10.1089/wound.2012.0397

Moores J (2013) Vitamin C: a wound healing perspective. British J Community Nurs Suppl (December): S6, S8–11

Morton LM, Phillips TJ (2016) Wound healing and treating wounds: differential diagnosis and evaluation of chronic wounds. J Am Acad Dermatol 74(4):589–605.; quiz 605–6. https://doi.org/10.1016/j.jaad.2015.08.068

Murphrey MB, Miao JH, Zito PM (2020) Histology, stratum Corneum. In: StatPearls. StatPearls Publishing, Treasure Island, FL. http://www.ncbi.nlm.nih.gov/books/NBK513299/

Murthy S, Gautam MK, Goel S, Purohit V, Sharma H, Goel RK (2013) Evaluation of in vivo wound healing activity of bacopa monniera on different wound model in rat. BioMed Res Int 2013 (6):972028. https://doi.org/10.1155/2013/972028

Nageswaran S, Vijayakumar R, Sivarasu S (2015) Design of Mechanical Interface to re-distribute excess pressure to prevent the formation of decubitus ulcers in bed ridden patients. In: 2015 37th Annual International Conference of the IEEE Engineering in Medicine and Biology Society (EMBC). IEEE, New York, pp 1021–1024. https://doi.org/10.1109/EMBC.2015.7318538

Nagle SM, Waheed A, Wilbraham SC (2020) Wound assessment. In: StatPearls. StatPearls Publishing, Treasure Island, FL. http://www.ncbi.nlm.nih.gov/books/NBK482198/

Nain PS, Uppal SK, Garg R, Bajaj K, Garg S (2011) Role of negative pressure wound therapy in healing of diabetic foot ulcers. J Surg Technique Case Rep 3(1):17–22. https://doi.org/10.4103/2006-8808.78466

Nayak, B. S., A. L. Udupa, and S. L. Udupa. 1999. "Effect of *Ixora Coccinea* flowers on dead space wound healing in rats." Fitoterapia 70 (3): 233–236. https://doi.org/10.1016/S0367-326X(99)00025-8

Nayak S, Dey S, Kundu SC (2013) Skin equivalent tissue-engineered construct: co-cultured fibroblasts/keratinocytes on 3D matrices of Sericin Hope cocoons. PLoS One 8(9):e74779. https://doi.org/10.1371/journal.pone.0074779

Nestle FO, Di Meglio P, Qin J-Z, Nickoloff BJ (2009) Skin immune sentinels in health and disease. Nat Rev Immunol 9(10):679–691. https://doi.org/10.1038/nri2622

Norman G, Dumville JC, Moore ZEH, Tanner J, Christie J, Goto S (2016) Antibiotics and antiseptics for pressure ulcers. In: The cochrane library. John Wiley & Sons, Ltd., New York. https://doi.org/10.1002/14651858.CD011586.pub2

Nuutila K, Katayama S, Vuola J, Kankuri E (2014) Human wound-healing research: issues and perspectives for studies using wide-scale analytic platforms. Adv Wound Care 3(3):264–271. https://doi.org/10.1089/wound.2013.0502

Ogurtsova K, da Rocha Fernandes JD, Huang Y, Linnenkamp U, Guariguata L, Cho NH, Cavan D, Shaw JE, Makaroff LE (2017) IDF diabetes atlas: global estimates for the prevalence of diabetes for 2015 and 2040. Diabetes Res Clin Pract 128(June):40–50. https://doi.org/10.1016/j.diabres.2017.03.024

Parnell LKS, Volk SW (2019) The evolution of animal models in wound healing research: 1993–2017. Adv Wound Care 8(12):692–702. https://doi.org/10.1089/wound.2019.1098

Pastar I, Wong LL, Egger AN, Tomic-Canic M (2018) Descriptive vs mechanistic scientific approach to study wound healing and its inhibition: is there a value of translational research involving human subjects? Exp Dermatol 27(5):551–562. https://doi.org/10.1111/exd.13663

Patil P, Martin JR, Sarett SM, Pollins AC, Cardwell NL, Davidson JM, Guelcher SA, Nanney LB, Duvall CL (2017) Porcine ischemic wound-healing model for preclinical testing of degradable biomaterials. Tissue Eng Part C, Methods 23(11):754–762. https://doi.org/10.1089/ten.TEC.2017.0202

Paulino LC, Tseng C-H, Blaser MJ (2008) Analysis of Malassezia microbiota in healthy superficial human skin and in psoriatic lesions by multiplex real-time PCR. FEMS Yeast Res 8 (3):460–471. https://doi.org/10.1111/j.1567-1364.2008.00359.x

Paus R, Ito N, Takigawa M, Ito T (2003) The hair follicle and immune privilege. J Investig Dermatol Symp Pro 8(2):188–194. https://doi.org/10.1046/j.1087-0024.2003.00807.x

Plichta JK, Gao X, Lin H, Dong Q, Toh E, Nelson DE, Gamelli RL, Grice EA, Radek KA (2017) Cutaneous burn injury promotes shifts in the bacterial microbiome in autologous donor skin: implications for skin grafting outcomes. Shock (Augusta, Ga.) 48(4):441–448. https://doi.org/10.1097/SHK.0000000000000874

Popov D (2010) Endothelial cell dysfunction in hyperglycemia: phenotypic change, intracellular signaling modification, ultrastructural alteration, and potential clinical outcomes. Int J Diabetes Mellitus 2(3):189–195. https://doi.org/10.1016/j.ijdm.2010.09.002

Qian L-W, Fourcaudot AB, Leung KP (2017) Silver sulfadiazine retards wound healing and increases hypertrophic scarring in a rabbit ear excisional wound model. J Burn Care Res 38 (1):e418–e422. https://doi.org/10.1097/BCR.0000000000000406

Queen D (2019) Artificial intelligence and machine learning in wound care—the wounded machine! Int Wound J 16(2):311–311. https://doi.org/10.1111/iwj.13108

Raghav A, Khan ZA, Labala RK, Ahmad J, Noor S, Mishra BK (2018) Financial burden of diabetic foot ulcers to world: a progressive topic to discuss always. Ther Adv Endocrinol Metab 9 (1):29–31. https://doi.org/10.1177/2042018817744513

Rahman GA, Adigun IA, Fadeyi A (2010) Epidemiology, etiology, and treatment of chronic leg ulcer: experience with sixty patients. Ann Afr Med 9(1):1. https://doi.org/10.4103/1596-3519.62615

Ramasubbu DA, Smith V, Hayden F, Cronin P (2017) Systemic antibiotics for treating malignant wounds. Cochrane Database Syst Rev 8:CD011609. https://doi.org/10.1002/14651858.CD011609.pub2

Rapp SJ, Rumberg A, Visscher M, Billmire DA, Schwentker AS, Pan BS (2015) Establishing a reproducible hypertrophic scar following thermal injury: a porcine model. Plast Reconstr Surg Glob Open 3(2):e309. https://doi.org/10.1097/GOX.0000000000000277

Ray JA, Valentine WJ, Secnik K, Oglesby AK, Cordony A, Gordois A, Davey P, Palmer AJ (2005) Review of the cost of diabetes complications in Australia, Canada, France, Germany, Italy and Spain. Curr Med Res Opin 21(10):1617–1629. https://doi.org/10.1185/030079905X65349

Reus AA, Usta M, Krul CAM (2012) The use of ex vivo human skin tissue for genotoxicity testing. Toxicol Appl Pharmacol 261(2):154–163. https://doi.org/10.1016/j.taap.2012.03.019

Richardson R, Slanchev K, Kraus C, Knyphausen P, Eming S, Hammerschmidt M (2013) Adult zebrafish as a model system for cutaneous wound-healing research. J Investig Dermatol 133 (6):1655–1665. https://doi.org/10.1038/jid.2013.16

Riol-Blanco L, Ordovas-Montanes J, Perro M, Naval E, Thiriot A, Alvarez D, Paust S, Wood JN, von Andrian UH (2014) Nociceptive sensory neurons drive Interleukin-23-mediated Psoriasiform skin inflammation. Nature 510(7503):157–161. https://doi.org/10.1038/nature13199

Rittié L (2016) Cellular mechanisms of skin repair in humans and other mammals. J Cell Commun Signal 10(2):103–120. https://doi.org/10.1007/s12079-016-0330-1

Roaf R (2006) The causation and prevention of bed sores. J Tissue Viability 16(2):6–8. https://doi.org/10.1016/S0965-206X(06)62002-0

Robson, M. C. 1997. "Wound infection. A failure of wound healing caused by an imbalance of Bacteria." Surg Clin North Am 77 (3): 637–650. https://doi.org/10.1016/s0039-6109(05)70572-7

Roy S, Biswas S, Khanna S, Gordillo G, Bergdall V, Green J, Marsh CB, Gould LJ, Sen CK (2009) Characterization of a preclinical model of chronic ischemic wound. Physiol Genomics 37 (3):211–224. https://doi.org/10.1152/physiolgenomics.90362.2008

Sami DG, Abdellatif A (2020) Histological and clinical evaluation of wound healing in pressure ulcers: a novel animal model. J Wound Care 29(11):632–641. https://doi.org/10.12968/jowc.2020.29.11.632

Sami DG, Heiba HH, Abdellatif A (2019) Wound healing models: a systematic review of animal and non-animal models. Wound Med 24(1):8–17. https://doi.org/10.1016/j.wndm.2018.12.001

Schivo M, Aksenov AA, Pasamontes A, Cumeras R, Weisker S, Oberbauer AM, Davis CE (2017) A rabbit model for assessment of volatile metabolite changes observed from skin: a pressure ulcer case study. J Breath Res 11(1):016007

Schlüter H, Upjohn E, Varigos G, Kaur P (2014) Chapter 34 – burns and skin ulcers. In: Lanza R, Atala A (eds) Essentials of stem cell biology, 3rd edn. Academic Press, Boston, pp 501–513. https://doi.org/10.1016/B978-0-12-409503-8.00034-2

Schreiber ML (2016) Evidence-based practice. Negative pressure wound therapy. Medsurg Nurs 25 (6):425–428

Seaton M, Hocking A, Gibran NS (2015) Porcine models of cutaneous wound healing. ILAR J 56 (1):127–138. https://doi.org/10.1093/ilar/ilv016

Sen CK (2019) Human wounds and its burden: an updated compendium of estimates. Adv Wound Care 8(2):39–48. https://doi.org/10.1089/wound.2019.0946

Serena TE (2015) The increasing role of epidermal grafting utilizing a novel harvesting system in chronic wounds. Wounds Compendium Clin Res Pract 27(2):26–30

Seth AK, Geringer MR, Gurjala AN, Hong SJ, Galiano RD, Leung KP, Mustoe TA (2012) Treatment of *Pseudomonas Aeruginosa* biofilm–infected wounds with clinical wound care strategies: a quantitative study using an in vivo rabbit ear model. Plast Reconstr Surg (2):129, 262e. https://doi.org/10.1097/PRS.0b013e31823aeb3b

Shah M, Crane JS (2020) Microdermabrasion. In: StatPearls. StatPearls Publishing, Treasure Island, FL. http://www.ncbi.nlm.nih.gov/books/NBK535383/

Sibbald RG, Goodman L, Norton L, Krasner DL, Ayello EA (2012) Prevention and treatment of pressure ulcers. Skin Therapy Lett 17(8):4–7

Singer AJ, McClain SA (2003) Development of a porcine excisional wound model. Acad Emerg Med 10(10):1029–1033. https://doi.org/10.1111/j.1553-2712.2003.tb00570.x

Sobiepanek A, Galus R, Kobiela T (2019 December) Application of the tape stripping method in the research on the skin condition and its diseases. Rev Res Cancer Treat 5:2–14

Steiner CA, Karaca Z, Moore BJ, Imshaug MC, Pickens G (2006) Surgeries in hospital-based ambulatory surgery and hospital inpatient settings, 2014: statistical brief #223. In: Healthcare Cost and Utilization Project (HCUP) statistical briefs. Agency for Healthcare Research and Quality (US), Rockville, MD. http://www.ncbi.nlm.nih.gov/books/NBK442035/

Stephens P, Caley M, Peake M (2013) Alternatives for animal wound model systems. In: Wound regeneration and repair, Methods in molecular biology. Humana Press, Totowa, NJ, pp 177–201. https://doi.org/10.1007/978-1-62703-505-7_10

Summerfield A, Meurens F, Ricklin ME (2015) The immunology of the porcine skin and its value as a model for human skin. Mol Immunol 66(1):14–21. https://doi.org/10.1016/j.molimm.2014.10.023

Syed F, Bagabir RA, Paus R, Bayat A (2013) Ex vivo evaluation of Antifibrotic compounds in skin scarring: EGCG and silencing of PAI-1 independently inhibit growth and induce keloid shrinkage. Lab Invest; J Technical Methods Pathol 93(8):946–960. https://doi.org/10.1038/labinvest.2013.82

Thomsen TR, Aasholm MS, Rudkjøbing VB, Saunders AM, Bjarnsholt T, Givskov M, Kirketerp-Møller K, Nielsen PH (2010) The bacteriology of chronic venous leg ulcer examined by culture-independent molecular methods. Wound Repair Regen 18(1):38–49. https://doi.org/10.1111/j.1524-475X.2009.00561.x

Tomioka T, Soma K, Sato Y, Miura K, Endo A (2018) S. Gupta. Auris Nasus Larynx 45 (5):1130–1134. https://doi.org/10.1016/j.anl.2018.04.004

Tomura M, Honda T, Tanizaki H, Otsuka A, Egawa G, Tokura Y, Waldmann H et al (2010) Activated regulatory T cells are the major T cell type emigrating from the skin during a cutaneous immune response in Mice. J Clin Invest 120(3):883–893. https://doi.org/10.1172/JCI40926

Tr J, Bi G, Mk MI, Ma D, Rj C, Se N, Dm B (2018, September 11) The cutaneous microbiome and wounds: new molecular targets to promote wound healing. Int J Mol Sci 19:2699. https://doi.org/10.3390/ijms19092699

Trompouki E, Bowman TV, DiBiase A, Zhou Y, Zon LI (2011) Chapter 19 – Chromatin immunoprecipitation in adult zebrafish red cell. In: Detrich HW, Westerfield M, Zon LI (eds) Methods in cell biology, The Zebrafish: Genetics, Genomics and Informatics, vol 104. Academic Press, Amsterdam, pp 341–352. https://doi.org/10.1016/B978-0-12-374814-0.00019-7

Trøstrup H, Thomsen K, Calum H, Høiby N, Moser C (2016 November 1) Animal models of chronic wound care: the application of biofilms in clinical research. Chronic Wound Care Manage Res 3:123–132. https://www.dovepress.com/animal-models-of-chronic-wound-care-the-application-of-biofilms-in-cli-peer-reviewed-fulltext-article-CWCMR

Tsourdi E, Barthel A, Rietzsch H, Reichel A, Bornstein SR (2013) Current aspects in the pathophysiology and treatment of chronic wounds in diabetes mellitus. BioMed Res Int 2013:385641. https://doi.org/10.1155/2013/385641

Tubaishat A, Papanikolaou P, Anthony D, Habiballah L (2018) Pressure ulcers prevalence in the acute care setting: a systematic review, 2000–2015. Clin Nurs Res 27(6):643–659. https://doi.org/10.1177/1054773817705541

Tuk B, Tong M, Fijneman EMG, van Neck JW (2014) Hyperbaric oxygen therapy to treat diabetes impaired wound healing in rats. PLoS One 9(10):1–8. https://doi.org/10.1371/journal.pone.0108533

Tuttle MS, Mostow E, Mukherjee P, Hu FZ, Melton-Kreft R, Ehrlich GD, Dowd SE, Ghannoum MA (2011) Characterization of bacterial communities in venous insufficiency wounds by use of conventional culture and molecular diagnostic methods. J Clin Microbiol 49(11):3812–3819. https://doi.org/10.1128/JCM.00847-11

Uberoi A, Campbell A, Grice EA (2020) Chapter 12 – the wound microbiome. In: Bagchi D, Das A, Roy S (eds) Wound healing, tissue repair, and regeneration in diabetes. Academic Press, London, pp 237–258. https://doi.org/10.1016/B978-0-12-816413-6.00012-5

Ud-Din S, Bayat A (2017) Non-animal models of wound healing in cutaneous repair: in silico, in vitro, ex vivo, and in vivo models of wounds and scars in human skin. Wound Repair Regen 25(2):164–176. https://doi.org/10.1111/wrr.12513

Vaena MLHT, Sinnecker JP, Pinto BB, Neves MFT, Serra-Guimarães F, Marques RG, Vaena MLHT et al (2017) Effects of local pressure on cutaneous blood flow in pigs. Rev Col Bras Cir 44(5):498–504. https://doi.org/10.1590/0100-69912017005012

van den Broek LJ, Limandjaja GC, Niessen FB, Gibbs S (2014) Human hypertrophic and keloid scar models: principles, limitations and future challenges from a tissue engineering perspective. Exp Dermatol 23(6):382–386. https://doi.org/10.1111/exd.12419

Wang J, Wan R, Mo Y, Zhang Q, Sherwood LC, Chien S (2010) Creating a long-term diabetic rabbit model. Exp Diabetes Res 2010. https://doi.org/10.1155/2010/289614

Watt FM, Fujiwara H (2011) Cell-extracellular matrix interactions in normal and diseased skin. Cold Spring Harb Perspect Biol 3(4). https://doi.org/10.1101/cshperspect.a005124

Werner S, Krieg T, Smola H (2007) Keratinocyte–fibroblast interactions in wound healing. J Investig Dermatol 127(5):998–1008. https://doi.org/10.1038/sj.jid.5700786

Wigger-Alberti W, Kuhlmann M, Ekanayake S, Wilhelm D (2009) Using a novel wound model to investigate the healing properties of products for superficial wounds. J Wound Care 18 (3):123–128., 131. https://doi.org/10.12968/jowc.2009.18.3.39813

Wilhelm K-P, Wilhelm D, Bielfeldt S (2017) Models of wound healing: an emphasis on clinical studies. Skin Research and Technology: Official Journal of International Society for Bioengineering and the Skin (ISBS) [and] International Society for Digital Imaging of Skin (ISDIS) [and] International Society for Skin Imaging (ISSI) 23(1):3–12. https://doi.org/10.1111/srt.12317

Wolf K, Müller R, Borgmann S, Bröcker E-B, Friedl P (2003) Amoeboid shape change and contact guidance: T-lymphocyte crawling through Fibrillar collagen is independent of matrix remodeling by MMPs and other proteases. Blood 102(9):3262–3269. https://doi.org/10.1182/blood-2002-12-3791

Wong VW, Sorkin M, Glotzbach JP, Longaker MT, Gurtner GC (2011) Surgical approaches to create murine models of human wound healing. BioMed Res Int 2011:969618. https://doi.org/10.1155/2011/969618

Woo KY, Keast D, Nancy P, Gary Sibbald R, Mittmann N (2015) The cost of wound debridement: a Canadian perspective. Int Wound J 12(4):402–407. https://doi.org/10.1111/iwj.12122

Xu Z, Hsia HC (2018) The impact of microbial communities on wound healing: a review. Ann Plast Surg 81(1):113–123. https://doi.org/10.1097/SAP.0000000000001450

Yagmur C, Guneren E, Kefeli M, Ogawa R (2011) The effect of surgical denervation on prevention of excessive dermal scarring: a study on rabbit ear hypertrophic scar model. J Plast Reconstr Aesthet Surg: JPRAS 64(10):1359–1365. https://doi.org/10.1016/j.bjps.2011.04.028

Young BB, Zhang G, Koch M, Birk DE (2002) The roles of types XII and XIV collagen in Fibrillogenesis and matrix assembly in the developing cornea. J Cell Biochem 87(2):208–220. https://doi.org/10.1002/jcb.10290

Yousef H, Alhajj M, Sharma S (2020) Anatomy, skin (integument), epidermis. In: StatPearls. StatPearls Publishing, Treasure Island, FL. http://www.ncbi.nlm.nih.gov/books/NBK470464/

Yurchenco PD, Schittny JC (1990) Molecular architecture of basement membranes. FASEB Journal: Official Publication of the Federation of American Societies for Experimental Biology 4(6):1577–1590. https://doi.org/10.1096/fasebj.4.6.2180767

Zhang Q, Takayoshi Y, Paul Kelly A, Shi S, Wang S, Brown J, Wang L, French SW, Shi S, Anh DL (2009) Tumor-like stem cells derived from human keloid are governed by the inflammatory niche driven by IL-17/IL-6 Axis. PLoS One 4(11):e7798. https://doi.org/10.1371/journal.pone.0007798

Zhang L-j, Guerrero-Juarez CF, Hata T, Bapat SP, Ramos R, Plikus MV, Gallo RL (2015) Innate immunity. Dermal adipocytes protect against invasive *Staphylococcus Aureus* skin infection. Science (New York, N.Y.) 347(6217):67–71. https://doi.org/10.1126/science.1260972

Zhao R, Liang H, Clarke E, Jackson C, Xue M (2016a) "Inflammation in chronic wounds." edited by Terrence Piva. Int J Mol Sci 17(12):2085. https://doi.org/10.3390/ijms17122085

Zhao Y, Crimmins EM, Hu P, Yang S, Smith JP, Strauss J, Wang Y, Zhang Y (2016b) Prevalence, diagnosis, and management of diabetes mellitus among older Chinese: results from the China health and retirement longitudinal study. Int J Public Health 61(3):347–356. https://doi.org/10.1007/s00038-015-0780-x

Experimental Wound-Care Models: In Vitro/In Vivo Models and Recent Advances Based on Skin-on-a-Chip Models

Sónia P. Miguel, Maximiano P. Ribeiro, and Paula Coutinho

1 Introduction

The ageing of the EU population, generally associated with comorbidities, is a considerable challenge for the healthcare systems. Chronic wounds are one of the most frequent comorbidities related to aged populations (Frykberg and Banks 2015).

The treatment of complicated skin injuries continues to be an emergent problem for the scientific and medical community. Although many available therapeutic options in the clinical environment to treat skin lesions, like conventional (e.g. gauzes, adhesive strips), advanced dressings (e.g. hydrocolloids, hydrogels, creams), implants, and skin substitutes, they do not present the "ideal" properties (Sood et al. 2014). So, the researchers have been concentrating on the production of wound dressings, able to reduce the healing process time, as well as re-establish the quality of patients' life (Dreifke et al. 2015; Rezvani Ghomi et al. 2019).

Further, the clinical translation of wound devices is greatly correlated to the physicochemical characterisation and preclinical evaluation that assess the formulations' quality, safety, and efficacy (Ruggeri et al. 2020). Moreover, the wound-healing therapies' commercialisation process is laborious, expensive, and prolonged involving the in vitro, in vivo, and clinical assays. The preclinical evaluation of most therapies is performed through in vitro and in vivo animal models, that present low predictability when tested in humans (Dellambra et al. 2019). Apart from these facts, the EU's seventh Amendment to the Cosmetics Directive banned animal models for testing cosmetic products, and different agencies have supported the 3R principle (Replacement, Reduction, Refinement) (Ranganatha and Kuppast 2012; Doke and

S. P. Miguel · M. P. Ribeiro · P. Coutinho (✉)
CPIRN-IPG Center of Potential and Innovation of Natural Resources, Polytechnic Institute of Guarda, Guarda, Portugal

CICS-UBI Health Sciences Research Centre, University of Beira Interior, Covilhã, Portugal
e-mail: coutinho@ipg.pt

Dhawale 2015). Also, the European Medicines Agency (EMA) described the guidelines to implement the 3Rs policy in the testing of medicinal products, referring to the requirements to test medical products for human use, as reviewed by Dal Negro et al. (2018)

In this way, different efforts made by companies (e.g. GlaxoSmithKline, Johnson & Johnson, L'Oréal, Novartis, Proctor & Gamble, and Unilever) and researchers have resulted in the use of non-animal alternative strategies to estimate the biological performance of therapies through the development of in vitro 3D models. Several industries already developed skin models, skin tissue engineered and protocols to evaluate skin features, such as SkinEthic®, epiCS®, EpiDerm®, and EpiSkin®. These are examples of reconstructed skin models validated, available in the market, and able to be used as an alternative to the in vivo assays (European Commission, 2020).

In this way, it was possible to obtain in vitro assays with reproducibility, by using skin constructs to characterise genetic damage from topical therapeutic formulations; full-thickness skin models to study the drug penetration into damaged skin; and skin structures to ascertain the effect of the LED light in the acne' therapy (Yu et al. 2019; Wei et al. 2020).

However, most of these 3D skin models are static and cannot reproduce all skin functional and morphological features, as well as cell interactions, and dynamic microenvironment found on native human skin (Wufuer et al. 2016; Sriram et al. 2018; Wang et al. 2017).

Taking this into account, the organ-on-a-chip platforms arise as revolutionary systems that require a small quantity of reagents, cells, and samples to replicate the structure, morphology, and dynamic conditions of a specific tissue. Furthermore, this preclinical system affords vast possibilities to replicate different disease conditions by perfusing the chip with diverse molecules (Low and Tagle 2017; van den Berg et al. 2019; Wu et al. 2020).

This chapter describes the leading preclinical platforms used for the biological evaluation of wound healing therapies, evidencing their main advantages and disadvantages. Special attention is dedicated to the new promising in vitro skin models, reporting the main production techniques and their use as a platform for testing wound dressings, topical formulations, and cosmetic products or studying pathological and physiological alterations associated with skin diseases. We also address the main future perspectives enrolling in the development of the skin models accelerating the clinical translation of different innovative wound healing therapies.

2   Biological Assessment of Wound Healing Therapies: Actual Preclinical Platforms

The constant development of wound healing therapies has revolutionised the pre-clinical research area. The evaluation of safety, toxicity, and efficacy in a dose-response trend are some parameters that need to be determined before the clinical trials, following the regulatory guidelines, as reviewed by Bernard et al. (2018)

In the first instance, the wound healing process is assessed through in vitro assays (e.g. wound scratch or transwell migration assays) (Grada et al. 2017; Liang et al. 2007; Van Meerloo et al. 2011). These are easily executed, and are approachable assays that can be used to evaluate the primary skin cells' migration ability (keratinocytes and fibroblasts). These in vitro models afford an effective first approach to discover and screen the drug's effects (Abd et al. 2016).

However, to reproduce the biological conditions (immune response, cell–cell and cell–matrix interactions) on wound healing mechanisms, researchers have created the organotypic cultures. These are composed of two layers: dermis and the epider-mis, being used to characterise the keratinocyte–fibroblast crosstalk during the wound healing mechanism and evaluate the therapeutic effects of a topical treatment applied in 3D cultures (Oh et al. 2013; Weinmüllner et al. 2020).

Similarly, the ex vivo human wound samples have been employed to study the mechanism of wound epithelialization.

Even so, animal models remain the standard preclinical platform that gives a complete knowledge of the biological response of therapies at the local and systemic level (Barre-Sinoussi and Montagutelli 2015).

However, the use of alternatives to animal experimentation has been a topic very explored by cosmetic and pharmaceutical companies, due to ethical and economic concerns.

Herein, we depict the most current in vivo, ex vivo, and in vitro models in the skin regeneration area, describing their advantages and disadvantages. It is also discussed the recent innovations performed in the development of the skin-on-a-chip, which are classified as micro-engineered models that replicate the morphology and func-tions of the human skin.

2.1   In vivo Models

Actually, the preclinical testing is essentially conducted through in vivo models. In general, in vivo wound models have several potential features since they enable: (1) the study of multiple interactions between diverse cell populations that normally occur during the wound repair process; (2) the search of multiple players enrolled in the healing process (e.g. cytokines, hormones, growth factors); (3) the specific reduction/silencing of target genes to evaluate their impact on healing process; (4) the investigation of an efficient immune response; (5) the induction of various wounds in the same animal, and (6) the induction of different types of wounds (burns, surgical injuries, crushing, etc.) (Barre-Sinoussi and Montagutelli 2015; Masson-Meyers et al. 2020; Sami et al. 2019).

The utilisation of in vivo animal models for studying the pathophysiology of cutaneous wound repair and testing/optimising novel therapeutic approaches enables the evaluation of biological performance under systemic conditions, making possible the promotion of cell communications and interactions between local cells, and immune cells recruitment (Barre-Sinoussi and Montagutelli 2015; Sami et al. 2019).

Thus, different animal models have been employed for wound healing assay, including rodents (e.g. mice and rats), pigs, rabbits, and zebrafish, which share some anatomic and physiologic properties with human skin, as illustrated in Fig. 1.

Among the different animal models, rodents are the most utilised in wound-healing studies since they are easily available, to handle and low cost to acquire and preserve. Further, its small body size avoids high quantities of chemical use, and the duration of wound healing assay is shorter ($\approx$ 21 days).

Besides, the mice and/or rats used in in vivo assays are genetically modified organisms (transgenic or various knockouts/knockins) that simplifies the study of the molecular pathways, and they can be easily controlled to reproduce the particular wound conditions (e.g. diabetes, skin infections, hypoxia, and ischemia) (Dellambra et al. 2019; Pastar et al. 2018; Wilhelm et al. 2017).

During the wound healing assay, the histology analysis, and the mechanical properties of the regenerated skin tissue, biochemistry as well as the expression of genes provide essential parameters to estimate the progression of the wound healing process. Furthermore, non-invasive methods like bioluminescence and laser Doppler imaging can also examine blood flow and other wound healing parameters.

The incisional wound, obtained by cutting the skin with a sharp blade, is a model suitable to study the scar tissue formation over long periods, but not appropriate for the characterisation of the epithelialisation process. Excisional wound models are mostly used for the characterisation of the contraction rate and epithelialization in murine models, which allow the identification of the essential signalling pathways related to the wound healing progress, such as Wnt/β-catenin, transforming growth factor-β (TGF-β), Notch, and bone morphogenetic proteins (BMPs) (Pastar et al. 2018). Additionally, to evaluate the biological processes dependent on epithelialisation, cell proliferation, and angiogenesis, a silicon splint wound murine model is employed (Park et al. 2015). Further, the parabiosis model has been exploited to evidence the role of cytokines and cells (inflammatory and mesenchymal cells) on delayed wounding in diabetic mice (Song et al. 2010; Wong et al. 2010).

The dorsal skinfold chamber enables real-time high-resolution microscopic imaging, and it is widely applied in the study of vasculature physiology (Wong et al. 2010). Finally, the least invasive murine wound model is the tape-striping wound model, which consists of disruption of the epidermal layer by withdrawing just the *stratum corneum* and *stratum granulosum*. However, to reproduce the partial-thickness wounds, the murine skin is not suitable since it possesses a thin epidermis and large hair follicles, compromising the induction of superficial wounds

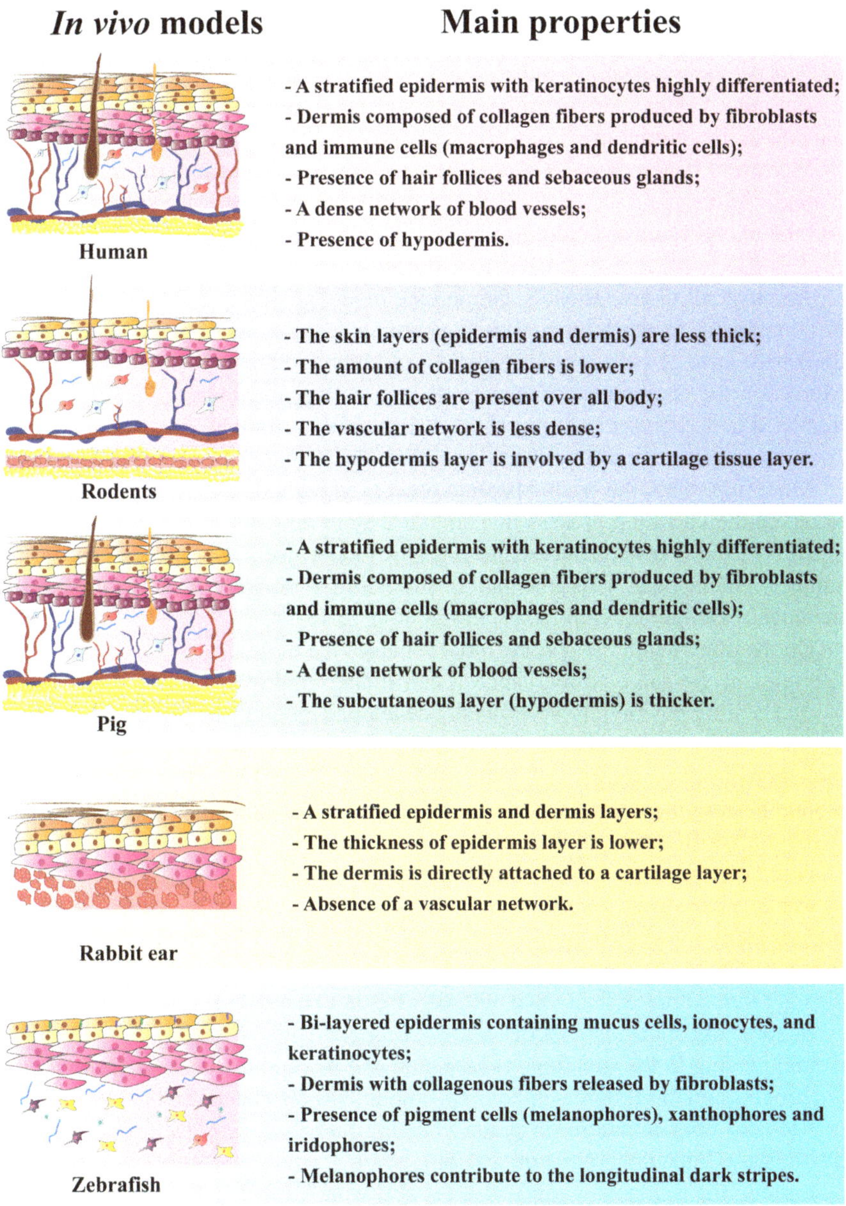

In vivo models   ## Main properties

Human

- A stratified epidermis with keratinocytes highly differentiated;
- Dermis composed of collagen fibers produced by fibroblasts and immune cells (macrophages and dendritic cells);
- Presence of hair follicles and sebaceous glands;
- A dense network of blood vessels;
- Presence of hypodermis.

Rodents

- The skin layers (epidermis and dermis) are less thick;
- The amount of collagen fibers is lower;
- The hair follicles are present over all body;
- The vascular network is less dense;
- The hypodermis layer is involved by a cartilage tissue layer.

Pig

- A stratified epidermis with keratinocytes highly differentiated;
- Dermis composed of collagen fibers produced by fibroblasts and immune cells (macrophages and dendritic cells);
- Presence of hair follicles and sebaceous glands;
- A dense network of blood vessels;
- The subcutaneous layer (hypodermis) is thicker.

Rabbit ear

- A stratified epidermis and dermis layers;
- The thickness of epidermis layer is lower;
- The dermis is directly attached to a cartilage layer;
- Absence of a vascular network.

Zebrafish

- Bi-layered epidermis containing mucus cells, ionocytes, and keratinocytes;
- Dermis with collagenous fibers released by fibroblasts;
- Presence of pigment cells (melanophores), xanthophores and iridophores;
- Melanophores contribute to the longitudinal dark stripes.

Fig. 1 Representative illustration and description of the "skin" structure and composition of the in vivo animal models commonly used in wound healing assays and the human skin

which assure the consistency and reproducibility (Dellambra et al. 2019; Sami et al. 2019).

On the other hand, considering the clinical relevance and prevalence of chronic wounds in the human population, significant efforts were made to create diabetic models for wound healing studies, namely the diabetic db/db mice, which exhibited a pronounced impaired wound healing (excisional and incisional wounds) in comparison to the normal in vivo models (Mendes et al. 2012).

Mertz and Eaglstein in 1978 created a porcine wound-healing model to study the wound-healing processes, namely the formation of scar tissue, the skin regeneration, and the microorganisms colonisation at the wound site (Meyer et al. 1978; Sullivan et al. 2001).

Amongst all in vivo models, the porcine model possesses a similar skin structure in comparison to the human skin, concerning features like epidermal and dermal thickness, arrangements of hair follicles, and blood vessels (Dellambra et al. 2019). Moreover, the compounds of the dermis (e.g. collagen and elastin) are analogous to human skin and present a comparable distribution of immune cells and melanocyte. Also, the pigskin sticks to underlying tissue like human skin.

More important, the wound healing process in pig animals occurs essentially due to the epithelialization process in a similar way to humans.

Due to these resemblances, the pig is a model used in several dermatological subjects (melanoma, depigmentation and vitiligo disorders, and wound healing therapies) (Dellambra et al. 2019; Pastar et al. 2018).

On the other hand, the porcine model enables the biological evaluation of distinct therapies in the same animal since it can induce multiple wounds. Indeed, a few wound healing pig models were already developed namely, the incisional, excisional injuries, skin infection, burns, UV radiation injury, and skin hypoxia. The porcine models have been validated for the assessment of the biological effect of diverse wound healing therapies, including various common wound care products available in the market (Rittié et al. 2013).

However, these models are expensive and have some limitations such as the lack of the eccrine sweat glands, their larger size and complexity which implies the management by skilled professionals, as well as the challenging induction and maintenance of diabetes in pigs. On the other hand, the development of transgenic pigs is more complex than mice and rats (Pastar et al. 2018).

In turn, the rabbit ear model has been applied in the research of deregulated wound healing in the situations as ischemia and infection, which are determinant in the progression of chronic wounds. This model has also been used to test the effects of bacterial contamination on wound healing and discover/develop new potential treatments (Chien and Wilhelmi 2012).

Further, the ischemic wound model is also widely used in the rabbit ear, which is generated by suturing off the ear's arterial blood supply. This model is relatively inexpensive to maintain and create multiple wounds in a single animal. Despite this, the rabbit ear model is incapable of fully recapitulating the human setting since the dermis is firmly connected to the cartilage and it is avascular (Chien and Wilhelmi 2012; Pastar et al. 2014).

Finally, zebrafish is an animal model generally used, since this model presents a genome analogous to humans, embryonic transparency, rapid embryonic development, many offspring, and easily manipulated genetic content (Spitsbergen and Kent 2003; Chávez et al. 2016).

The "zebrafish" skin is composed of an epidermis separated by a basal membrane from a collagen-based dermis, becoming this model proper for wound healing studies, since it possesses a structure and primary mechanisms similar to the human wound healing process (Li and Uitto 2014; Richardson et al. 2013).

On the other hand, zebrafish are also used in studies regarding melanogenesis, pigment disorders, and melanoma (Choi et al. 2007; van Rooijen et al. 2017). Indeed, the zebrafish skin possesses cells with black pigment, which present a similar behaviour to the human melanocytes (Li and Uitto 2014; Li et al. 2011). However, the zebrafish did not possess an epidermal barrier and skin appendages.

Despite the potential of using animal models in the testing/screening of wound care products, recently, the researchers/industries have been centred on discovering non-animal alternative approaches. The animal experiments are expensive, time-consuming, require ethical concerns, and the results are frequently non-predictive when tested in humans (Stirland et al. 2013). According to the Humane Society International Organization, about 90% of the medicines that displayed promising results in animal models were unsuccessful when tested on humans. Also, different agencies (European Union Reference Laboratory-European Centre for the Validation of Alternative Methods (ECVAM) and Organisation for Economic Cooperation and Development (OECD)) are enforcing the 3 Rs strategy, which defends the reduction, refinement, and replacement of animal use in experimentation. The ex vivo and in vitro models are detailed in the following sub-sections, evidencing the primary potentialities.

2.2 Ex vivo Models

The ex vivo skin models are widely used to determine topical or injected therapies' effectiveness due to their time and cost-effectiveness. Compared to the in vitro skin models, ex vivo models can replicate the structure of extracellular matrix (ECM), cell signalling mechanisms and metabolism, incorporation of skin appendages, and the effects of dermal absorption (Flaten et al. 2015).

These models' skin samples are typically acquired from surgical operations such as abdominoplasty (Xu et al. 2012a). The skin is separated from underlying fat to enable the appropriate attachment of the dermis layer to the culture plate. Different types of wounds can be induced, like an excisional wound with a punch (3–4 mm) and a full-thickness skin injury by using biopsy punch (8–10 mm). After that, the therapeutic approaches can be directly applied to the wound site, and then the reepithelisation mechanism can be monitored through histomorphometric analysis and keratin immunostaining (Planz et al. 2016).

Ex vivo skin models are more appropriate to determine the effect of drug formulations, transdermal delivery, topical penetration, and percutaneous absorption. As potential advantages, the human ex vivo wound models present the total thickness of epidermis and dermis, Langerhans cells, melanocyte cells, and nerve endings, allowing the establishment of standardised wound depths (Pastar et al. 2014).

However, ex vivo models fail in the blood supply, restricting the assays including immune cell infiltrates, and the culture of the human skin ex vivo during at least 2 weeks is too laborious, as highlighted in the literature (Pastar et al. 2014; Gordon et al. 2015).

The NativeSkin® is an ex vivo skin model currently available in the market. It consists of skin biopsies obtained from patients submitted to the surgery, in which skin samples are placed within the supportive matrix that provides a suitable moist environment for skin dermis. This model has been utilised as a standardised ex vivo skin model before clinical trials since it possesses a protective and physiological barrier and *stratum corneum* layer appropriated for absorption assays. However, NativeSkin® is not able to replicate a pathological condition.

Another wound model example is the murine skin explant utilised to characterise the keratinocytes' migration. In this case, the fascia is separated from the mouse skin, and the removed skin sample is immersed in the cell culture medium. After that, the influence of pharmacological therapies on keratinocyte migration can be assessed. This model can study the interaction between keratinocytes and fibroblasts. However, it does not possess immune cells, blood supply, and the ECM (Pastar et al. 2014).

Also, porcine skin samples can be used as ex vivo models for wound infection and screening the activity of antimicrobial therapies. In the first instance, the porcine skin sample is sterilised, and the partial-thickness wounds can be induced with a dermatome. In contrast, the chronic wound environment can be mimicked by the growth of bacterial biofilms (e.g. *Pseudomonas aeruginosa* (*P. aeruginosa*) and *Staphylococcus aureus* (*S. aureus*)) on wounds. This type of model was already used to evaluate the therapeutic effect of formulations containing iodine and silver, surfactant-based dressings, and honey ointments. This model presents similar cutaneous physiology to the human skin, and the tissue can be easily achieved from meat processing. However, the porcine ex vivo skin models exhibit a limited host immune response, and the commensal microbiome is not present (due to the sterilisation process) (Gordon et al. 2015).

Having in mind, the limitations of material used in ex vivo explants, bioengineered skin substitutes have emerged. Different improvements have been performed in tissue engineering, allowing the development of diverse wound reepithelialization models.

2.3   In vitro Skin Models

The in vitro skin models arise as an alternative more accessible, rapid, no presenting ethical concerns neither immune responses. In general, in vitro models to test/screen the wound healing therapies are subdivided into: (1) 2D skin models; (2) 3D static

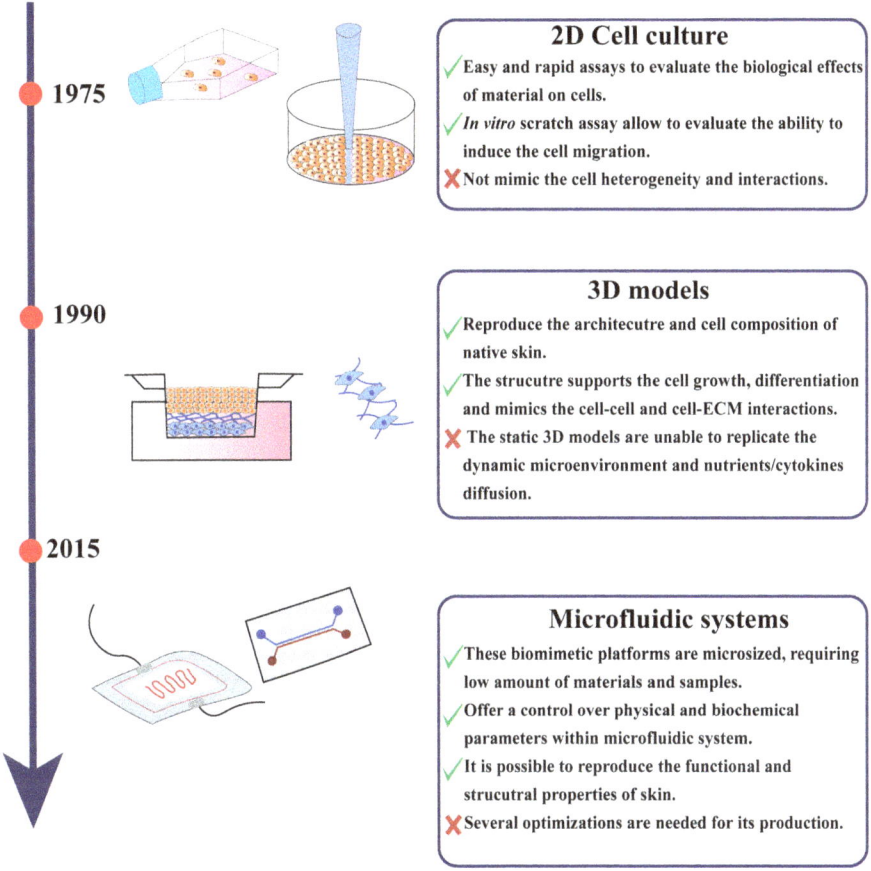

2D Cell culture

✓ Easy and rapid assays to evaluate the biological effects of material on cells.

✓ *In vitro* scratch assay allow to evaluate the ability to induce the cell migration.

✗ Not mimic the cell heterogeneity and interactions.

3D models

✓ Reproduce the architecutre and cell composition of native skin.

✓ The strucutre supports the cell growth, differentiation and mimics the cell-cell and cell-ECM interactions.

✗ The static 3D models are unable to replicate the dynamic microenvironment and nutrients/cytokines diffusion.

Microfluidic systems

✓ These biomimetic platforms are microsized, requiring low amount of materials and samples.

✓ Offer a control over physical and biochemical parameters within microfluidic system.

✓ It is possible to reproduce the functional and strucutral properties of skin.

✗ Several optimizations are needed for its production.

Fig. 2 Representation of the main in vitro models used for testing/screening wound healing therapies, describing their advantages and disadvantages and their temporal evolution

skin models and more recent; (3) Skin-on-a-chip models based on microfluidic systems.

Figure 2 summarises the main strong points and limitations of these in vitro skin models.

In the initial phase of the drug screening process, the 2D in vitro assays are the most used to investigate the formulations' biocompatibility and ability to promote cell activities. These conventional assays are commonly used since they are easy, cheap, and relatively fast, providing a quick assessment of a large number of samples under controlled conditions (Abd et al. 2016).

Cell 2D monolayers are extensively used to screen compounds for initial assessment of the toxicity profile of therapeutic formulations. Such assays act as quick and pre-selective screening tools (Planz et al. 2016).

The skin irritancy was the first assay performed by evaluating the products' effect in normal human keratinocytes (Rheinwald 1989).

In turn, the cell proliferation assay is done to ascertain the material/drug compound's ability to promote cell growth, whereas the cell cytotoxicity assay evaluates the cytotoxic effect of the therapies on cell viability (Wilhelm et al. 2017).

In general, the cytotoxicity is evaluated following the guidelines described in ISO 10993-5 standard, which was updated in 2009. There are available in vitro cytotoxic assays such as 3-(4,5-dimethylthiazol-2-yl)-2,5-diphenyltetrazolium bromide (MTT) that ascertain the percentage of viable cells through the quantification of the formazan converted by the mitochondrial enzymes. The quantity of formazan converted by the cells' metabolic activity is correlated to the percentage of viable cells (Van Meerloo et al. 2011). Other cytotoxicity assays include the neutral red uptake (NRU) test, colony formation test, and (2,3-bis-(2-methoxy-4-nitro-5-sulfophenyl)-2H-tetrazolium-5-carboxanilide) (XTT) test. In these assays, the cell cultures should be selected according to the intended in vivo application.

The cytotoxicity assays afford qualitative and quantitative approximations about the cytotoxic potential of a therapeutic formulation since it is possible to examine the cell morphology (through optic microscopy) and the cells' activity (Bernard et al. 2018). The continuous (immortalised) cell lines such as HeLa, L929, WI-38, 3 T3 or CHO are most used in the screening phase (Pizzoferrato et al. 1994). In the evaluation of materials for wound healing purposes, the fibroblasts are the most used cell type, since they are the main cells enrolled on the different mechanisms of the healing process, namely in the production of the ECM compounds. Further, the keratinocytes can also be used considering their role in maintaining the epidermal layer and other mechanisms of the wound healing process (Bernard et al. 2018).

Another in vitro experiment widely done by researchers is the scratch wound-healing assays that evaluate the "materials" ability to induce cell migration and proliferation (Liang et al. 2007). This assay enables the measurement of the gap "closure" rates and/or the cell number that migrate to the "wound gap". However, this assay does not mimic the intricate relationship between cell migration and proliferation (Buenzli et al. 2020; Jin et al. 2020).

On the other hand, 2D co-cultures have also been used, since they give more insight regarding cell–cell interaction, namely the keratinocytes and fibroblasts interaction which is relevant during the wound healing and scar formation processes. However, the 2D cell culture model does not precisely imitate the in vivo cell microenvironment. These models are unable to simulate the complex interaction between the epidermal layer, skin and immune cells, as well as are insufficient to the accurate representation of what occurs in vivo (Sami et al. 2019).

In this way, 3D in vitro models arise since they faithfully recapitulate the human skin structure, cell–cell interactions and mimic the drug metabolism in a specific environment. In 3D skin models, the structure encourages cell adhesion, favouring tissue growth and development. So, the 3D skin models are produced in vitro by using diverse cell types, and materials aim to provide: (1) an epidermal barrier with keratinocytes, that prevent "microorganisms" infiltration and dehydration; and (2) a dermal layer containing mature fibroblast or stem cells able to stimulate the wound healing mechanism (Flaten et al. 2015).

Researchers have used culture systems based on static 3D matrices that support the epidermal differentiation and maturation, and the air–liquid interface that enables the stratification process, allowing a more accurate evaluation of cell proliferation into a microenvironment more similar to found in vivo conditions (Löwa et al. 2018; Dhiman et al. 2005).

Similarly, to ex vivo wound models, the 3D in vitro models can also evaluate the effects of diverse molecules on cells' migration and keratinocyte–fibroblasts interactions during the healing process. Generally, the human keratinocytes are seeded on a matrix composed of type I collagen incorporating fibroblasts or devitalised human dermis (Xu et al. 2012b; Mendoza-Garcia et al. 2015).

Furthermore, these models can be employed to examine the pathophysiology of the healing process by inducing a wound through a biopsy punch. The wound can also be induced through the CO_2 laser, and needle punctures Then, the reepithelialisation process is evaluated by histomorphometry analysis or immunostaining (Pastar et al. 2014; Planz et al. 2016).

The most used 3D in vitro skin models are EpiDerm®, EpiSkin®, and SkinEthic®, which are essentially composed of human fibroblasts and keratinocytes layers inserted on the 3D supportive matrix. All the models present the proper cell metabolic and mitotic activity, expressing the epidermal differentiation markers (Flaten et al. 2015; Planz et al. 2016; Hänel et al. 2013).

However, this type of model presents some shortcomings, namely the basic organisation and few cell types compared to the in vivo native skin. In this way, the complex cellular interactions that occur during inflammation and angiogenesis processes are not replicated. Further, the functional dermal–epidermal junction is not mimicked, leading to incomplete differentiation and instability of cell interactions. Also, these models presented other limitations like unability of: (1) replicating all structural and functional properties of native skin; (2) promoting a controlled microenvironment, (3) imitating the cell–cell interactions, nutrients and cytokines diffusion, and (4) conducting the unreliable results in dermal penetration evaluation, due to its weak barrier function (Wufuer et al. 2016; Sriram et al. 2018; Wang et al. 2017).

In this sense, a critical and actual trend is developing microfluidic devices for better mimicking skin function and acting as wound healing research platforms.

The skin-on-a-chip technologies are the most recent in vitro models reported in the literature, which have been developing to screen drugs and treatments. These biomimetic platforms are characterised by possessing a small size and increased complexity, requiring low quantities of samples (from nanoliters to picoliters) (van den Berg et al. 2019; Wu et al. 2020; Lee et al. 2017). Skin-on-a-chip is a system that allows the culture of skin tissues within a microfluidic platform, provides management of physical and biochemical issues, as well as mimics the 3D microenvironment of the native skin. It is also possible to generate the model with three layers, sensory organs, appendages, and a full vascular network (van den Berg et al. 2019; Kilic et al. 2018; Mori et al. 2017).

These devices are typically composed of glass and silicone functionalised with bioactive materials that aim to reproduce the specific organ microenvironment. The multi-channel merged with a micro-perfusion system creates dynamic culture conditions, enabling the delivery of nutrients for long periods of time, and simultaneously promote the transport of wastes and the maintenance of oxygen levels (Planz et al. 2016; Stark et al. 2004).

After the creation of the chip, it can be incorporated into a multiorgan chips system, simulating a full "organism" native environment, offering wide opportunities to develop more potent therapeutic approaches for wound healing disorders.

3 The Emergent Strategies to Fabricate In Vitro Skin Models

Most skin models constitute a 3D structure that affords a guide for cell adhesion and proliferation during wound healing and neovascularization processes. The morphology, structural integrity, and elasticity are strongly influenced by the production methodology. In this way, the researchers have been used different techniques that allow obtaining 3D scaffolds with peculiar structural properties (as represented in Fig. 3).

3.1 Freeze-Drying Technique

The freeze-drying method commonly uses water as a porogen to generate porous hydrogels. In brief, after freezing the polymer solution, the lyophilisation process is performed to promote solvent evaporation, and hence the pores are created within the structure. The composition of a hydrogel, the freezing temperature, and the duration of freeze-drying cycles are critical factors that influence the pore size and scaffold morphology (O'Brien et al. 2005). This is an easy but time-consuming and laborious method, that does not use organic solvents, requiring rigorous temperature control to avoid the collapse of the internal pores hydrogel (O'Brien et al. 2005; Yun et al. 2018; Nicholas et al. 2016).

Hilmi et al. produced chitosan scaffolds composed of a single layer for dermal fibroblasts' culture, through freeze-drying technique (Hilmi et al. 2013). They presented a highly porous architecture ($93 \pm 12.57\%$) with pore sizes between 40 and 140µm. The fibroblast cells were cultured into chitosan scaffolds, and cell–chitosan interaction was enhanced, confirming structures' suitability for acting as a template for cell proliferation.

In turn, Ma et al. developed a bilayer skin model based on chitosan, which was composed of a chitosan film (casting method) and a chitosan sponge layer

Techniques	Advantages	Disadvantages
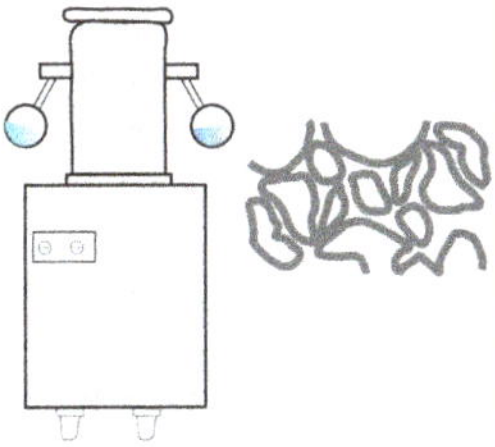 Freeze-drying	- Easy and acessible; - Did not require solvent organics; - Promotes the obtainment of highly porous structures.	- Requires a strict control over the temperature of freezing; - The time of lyophilization process also influences the pore sizes; - The mechanical integrity is weak.
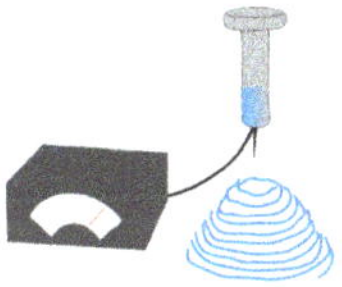 Electrospinning	- Fibrous structure similar to the skin extracellular matrix; - The membranes present a high surface-to-volume ratio and porosity; - Different types of polymers can be used.	- Poor tensile strength; - Limited control over pore structures; - Poor cellular infiltration; - Unable to create complex 3D architecture.
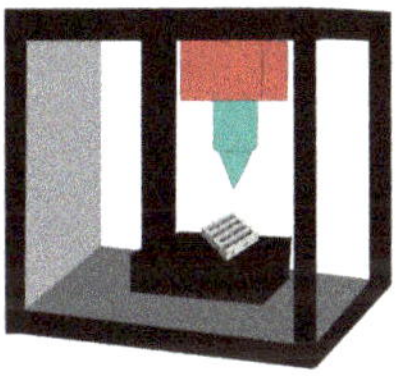 3D Printing	- 3D structures with a precise control on overall structural architecture; - The materials can be selected according to desired mechanical properties and biodegradation profile; - High reproducibility and throughput.	- Post-processing steps increase the time and cost of production; - The crosslinking agents can be non-compatible with the cell incorporation.
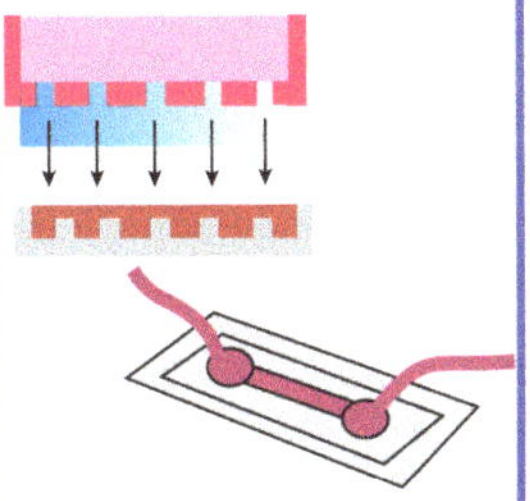 Micropatterning Micromolding	- 3D structures with specific geometry and size; - Precise control over dynamic fluid behaviours and external physical factors; - Requires a small amount of materials and reagents; - High reproducibility and throughput.	- The optimization production process can be slow; - High complexity; - The UV-crosslinkable polymers are the most used materials.

Fig. 3 List of the main advantages and disadvantages of the techniques mostly used in the production of skin models

(freeze-drying technique) (Ma et al. 2001). The stability of the bilayer model structure promoted the fibroblasts' proliferation during 4 weeks.

3.2 Electrospinning

The electrospinning technique uses outside the applied electric field to produce thin fibres from a polymeric solution injected through a syringe. During the ejection process of polymeric solution, the solvent evaporates, and the fibres are deposited in a collector (Yun et al. 2018; Nicholas et al. 2016; Miguel et al. 2018).

The electrospinning enables the production of networks of interconnected fibres with dimensions and morphology similar to the skin ECM's collagen fibres. Further, the electrospun membranes are also characterised by possessing high porosity and surface-to-volume ratio, which afford a surface area suitable for cell activity (Yun et al. 2018; Nicholas et al. 2016; Miguel et al. 2018).

Different polymers, natural and synthetic, or blends can be employed to fabricate electrospun membranes. However, electrospinning still has some challenges, including the low tensile strength, restricted control of pore sizes, and incapacity to produce complex 3D morphology. Besides, the low cellular infiltration into electrospun scaffolds is a significant issue, which has been overcome by managing the diameter of fibre and pores as well as combining biological signs that propel the cell migration and proliferation (Miguel et al. 2018; Heydarkhan-Hagvall et al. 2008).

Recently, Miguel et al. created a bilayer skin substitute by using PCL/silk fibroin to produce the top layer, mimicking the epidermis features. In contrast, the silk fibroin/hyaluronic acid/thymol combination was used to obtain a dermis bottom layer to absorb the exudate excess and promote cell proliferation (Miguel et al. 2019a). The results evidenced that the top layer presented a higher fibre diameter (471.4 ± 151.6 nm), low porosity ($64.28 \pm 2.59\%$), high water contact angle ($103.10 \pm 6.57°$), and avoid the *S. aureus* and *P. aeruginosa* infiltration. On the other side, the bottom layer exhibited a lower fibre diameter (295.4 ± 88.4 nm), high porosity ($85.24 \pm 2.47\%$), high swelling ratio (≈ 45), low water contact angle ($38.77 \pm 5.32°$), which promoted the fibroblasts attachment and proliferation.

The same authors also fabricated a bilayer skin substitute combining electrospinning and 3D printing techniques (Miguel et al. 2019b). To accomplish that, the authors electrospun PCL/silk sericin to obtain the top layer and 3D printed a chitosan/sodium alginate to achieve a hydrogel-based bottom layer. The authors verified that the epidermis layer was mimicked by a dense polymeric electrospun membrane. In contrast, a 3D printed hydrogel reproduced the dermis layer with suitable hydrophilic character, porosity, and biological properties for promoting cell activity.

3.3   3D Printing

3D printing is a newer technology that enables the production of 3D structures with accurate control structure and composition of the 3D matrix (Yun et al. 2018; Nicholas et al. 2016). This technique also displays other strong points, namely the custom design, high reproducibility, and high throughput (Koch et al. 2012; Koch et al. 2010).

Inkjet printing, laser-assisted printing, fused deposition model (FDM), and stereolithography are some examples of the 3D printing techniques developed. The laser printing technique was optimised to arrange the keratinocytes and fibroblasts impregnated into a 3D collagen matrix (Koch et al. 2012).

To accomplish the production of a multicellular skin graft, the bio-ink was composed of: (1) NIH-3 T3 fibroblasts; (2) HaCaT keratinocytes; (3) a collagen hydrogel; and (4) a sheet of Matriderm®, that act as a supportive matrix.

Lee et al. reported the layer-by-layer assembly of the collagen and keratinocytes and fibroblasts cells (Lee et al. 2014). The authors verified that the cell viability on 3D skin construct was high (>94%), and the skin tissue displayed 3–7 different cell layers, after 14 days of air–liquid interface culture.

However, these 3D-printed skin models cannot fully reproduce the human skin due to the lack of the vascular network, which is determinant for the cellular supply of nutrients and growth factors.

In this way, the potential of 3D printing has been explored in microfluidic systems production. This approach will allow obtaining structures with tailored geometry and size containing a spatial distribution of stem cells and/or bioactive molecules, essential in the healing process.

In work performed by Leng et al. a skin microfluidic device was produced by 3D printing (Leng et al. 2013). A mixture of alginate and collagen was employed as a supportive matrix for cell proliferation. The human fibroblast cells were integrated within the biopolymer sheet, which was continuously extruded, under spatiotemporal control in terms of cellular localisation and density. In vivo experiments were done on immunodeficient mice, and the printed biopolymer sheets were placed on the excision position, revealing that the skin substitute promoted the keratinisation and wound healing process (demonstrated through the trichrome and keratin 14 staining). Additionally, the printing process of this device (1 m^2) lasts only 48 min.

Apart from producing microfluidic devices, 3D printing can promote cell incorporation into structures, since it can deposit different cell lines in a precise location, creating diverse cell patterns in 3D structures. In general, the cell patterning in skin models consists essentially of incorporating just skin cells in bilayer substitutes recapitulating the native skin structure. However, to match the human skin constitution, other skin' compounds like melanocytes, stem cells, and glands should also be incorporated (Koch et al. 2012; Lee et al. 2014).

Another challenge is the post-processing steps customarily used in 3D printing techniques. Some examples of post-processing steps that complicate the workflow and increase the duration and the cost are the thermal treatment and filtration and the photopolymerisation process (Yun et al. 2018; Nicholas et al. 2016).

More recently, significant research is centred on producing novel biomaterials for 3D printing—referred to as 4D printing. The 3D printing of these stimuli-responsive materials enables to modify/functionalise the constructs' shape after the printing process, by exposing it to an external stimulus. Lastly, the issues regarding regulatory aspects for printed skin models need to be established to allow the use of these materials in clinical trials (Ashammakhi et al. 2018).

3.4 Micropatterning and Micromolding

Advanced techniques like micropatterning and micromolding enable to produce 3D structures with specific geometry and size. Nowadays, these technologies are the most used in the production of microfluidic platforms (Nicholas et al. 2016; Verhulsel et al. 2014).

The microfluidics-based platforms have been received significant attention since it allows accurate adjustment of skin-on-a-chip models' cellular microenvironment, controlling, at the microscale, the fluid flow and operation factors (e.g. gaseous exchanges, temperature, mechanical force).

Soft lithography is one example of how to create these patterns, which consisted of the micromolding of hydrogel solution using a stamp composed of polydimethylsiloxane (PDMS) (Song et al. 2017; Lee and Sung 2018).

Through this strategy, it is possible to obtain several channels running throughout a hydrogel, which can be personalised according to the stamp's design. Furthermore, live cells can be incorporated inside the hydrogel solution, obtaining a structure with a homogeneous cell distribution. Stereolithography and laser microstructuration are other micropatterning methods that can be used to produce such structures (Nicholas et al. 2016; Verhulsel et al. 2014).

Also, the skin-on-a-chip devices can recapitulate the highly organised skin tissue architecture and confer a microenvironment that simulates the dynamic forces exerted during the healing process (Yun et al. 2018; Lee and Sung 2018).

In general, the microfluidic technology requires small fluid' volumes (10^{-9} to 10^{-18} L) in hollow microchannels, which exhibit an appropriated laminar flow within the microfluidic systems (Mohammadi et al. 2016; Hasan et al. 2014). Further, the samples' required small amount overcome the limitations associated with the low availability of the patient-derived samples. The use of low amounts of reagents and drugs to be tested enables the detection of biomarkers with more sensitivity (Lee et al. 2017; Mohammadi et al. 2016; Hasan et al. 2014).

Wufuer et al. created a microfluidic-based skin-on-a-chip system, which was composed of three layers: keratinocytes (epidermal), fibroblasts (dermal), and endothelial cells (vascular), composed of PDMS and separated by transparent and porous membranes (Wufuer et al. 2016). To simulate the inflammatory condition into the device, the tumour necrosis factor-α (TNF-α) was perfused and then was evaluated the effect of dexamethasone. The results indicated that the skin-on-chip model can be used for screening and testing the drugs' therapeutic effect, replacing animal experimentation. On the other hand, Lee et al. engineered a 3D in vitro skin chip by using PDMS and collagen hydrogels with vascular structure. After that, the dermal fibroblasts and keratinocytes were perfused into the microfluidic device, and the cells remained viable for 10 days (Lee et al. 2017).

In turn, Mah et al. produced a microfluidic system to be used in in vitro assays for skin permeation studies. It requires low medium volumes (70–200 μL) and a small skin tissue sample (0.283 cm^3), which makes it a device widely employed in the pre-formulation assays for testing expensive and low-available drugs (Mah et al. 2013).

The combination of conventional microfluidic and tissue engineering technologies has enabled the fabrication of the more complex skin microsystems, offering helpful and effective platforms for evaluation of the drug screening (Wang et al. 2017).

Besides, in situ biosensors could be incorporated into the chip to provide real-time readouts about wound healing parameters and simultaneously detect the therapeutic responses to the drugs (Kilic et al. 2018). Such emerging technology is up-and-coming and can constitute a precious device for animal-free testing, namely in the cosmetic and pharmaceutical industries.

4 Applications of the In vitro Skin Models

In general, the main purposes of developing in vitro skin models are the testing of the efficacy and penetration of drugs on the skin and the study/evaluation of the mechanisms and therapeutic effects on skin disease conditions (e.g. wounding, burns, psoriasis, or melanoma). The main in vitro skin models developed for drug screening and skin disorders study purposes will be described in the following subsections.

4.1 Evaluation of Drugs Toxicity and Penetration

The majority of the currently available in vitro skin models are intended to mimic the main skin layers (epidermis, dermis). EpiSkin® and EpiDerm® were the first skin models validated, in 1998, by ECVAM as a predictive skin corrosion model, accepted in 2004, with the publication of the OECD test guideline (Fentem and Botham 2002).

Afterward, the EpiDerm® SIT and the SkinEthic® Reconstructed Human Epidermis (RHE) were approved as skin irritation model in 2008 (Alepee et al. 2010; Kandarova et al. 2009), and SkinEthic®, epiCS®, EpiDerm®, and EpiSkin® as in vitro artificial skin models for skin irradiation, corrosion and phototoxicity evaluation.

SkinEthic®, epiCS®, EpiDerm® are composed of keratinocytes seeded on polycarbonate membrane, whereas EpiSkin® consists of a stratified differentiated human keratinocytes cell layer seeded on a type I bovine collagen matrix and type IV human collagen film.

The LabCyte EPI-MODEL 24 was produced in Japan as a reconstructed human epidermis model and validated for hazard prediction. This model is formed by a human keratinocytes layer cultured under favourable conditions at an air–liquid interface, creating the stratified structure characteristic to the "skin" epidermis after 14 days.

Apart from these skin models that replicate the epidermis and dermis layers, the skin models incorporating additional skin compounds can reproduce the physiological and functional features of the native human skin, which will contribute to obtain results more predictive.

Concerning the drug screening/testing purpose, different skin models were already developed as extensively reviewed in (Planz et al. 2016; Mathes et al. 2014).

Among different works reported, we highlighted the work conducted by Abaci et al. who prepared a human skin-on-a-chip system comprised of epidermal and dermal compounds and then incorporated into a pumpless microfluidics system (Abaci et al. 2015). The chip was created to display a stable air–liquid interface enabling the blood circulation rate similar to that found in human skin tissues. The size of the chip decreases the volume of the culture medium and the cells needed. The medium's recirculation at the desired flow rate was achieved without using a pump or external tube connection, and this system was suitable to maintain the barrier function for 3 weeks. Further, the immunohistochemistry assays indicated that the keratinocytes' differentiation and location were achieved, forming the epidermis' sub-layers, after 1 week of the culture. The authors also validated the skin-on-a-chip potential to act as a preclinical platform for drug testing purposes. The haematoxylin/eosin staining and immunostaining assays showed that the doxorubicin promotes a spatial detachment of the basal layer.

In turn, Mori and co-workers described an alternative strategy to fabricate a perfusable vasculature in human skin equivalents (Mori et al. 2017). The evaluation of cell distribution within the perfused and non-perfused skin-equivalents enabled to evidence the important role of the vascular channels in the transport of the nutrients essentials to maintain the cell proliferation and viability (as shown in Fig. 4).

Afterward, the diffusion of different drug molecules from the epidermal layer to the vascular channels was measured under perfusion. All data gathered demonstrated that the skin-on-a-chip system exhibits promising properties to be used in drug testing, evidencing the importance of vascular perfusion.

In another work, Wagner and co-workers reported a multiorgan microfluidic device production able to maintain 3D tissues derived from human primary hepatocytes and human skin biopsies (Wagner et al. 2013). This two-organ system maintains a long-term functionality for at least 28 days, promoting the tissues' molecular

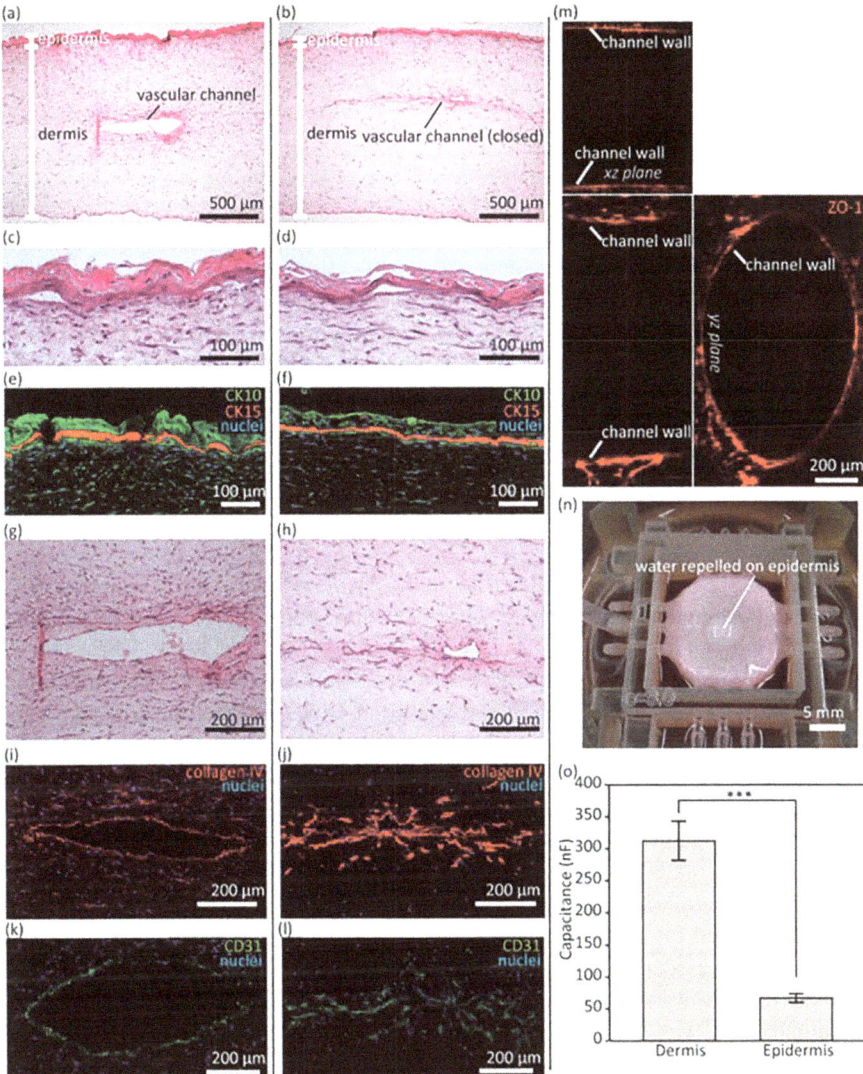

Fig. 4 Histological analysis relating to the assessment of the barrier function of in vitro skin model: Perfused and non-perfused skin equivalents stained with haematoxylin and eosin staining (**a**, **b**); Histological and immunostained images of epidermal layers (**c–f**) and vascular channels (**g–l**); Confocal images of vascular channel immunostained with ZO-1 antibody (**m**); Evaluation of barrier function through the "water" repellency by the epidermal layer (**n**); Assessment of barrier function by measurement of capacitance (**o**). Reprinted from Biomaterials, vol. 116, Mori et al. Skin integrated with perfusable vascular channels on a chip, 48–56, with permission from Elsevier (Mori et al. 2017)

interplay. Also, this platform revealed an outstanding performance in the testing of troglitazone at different molecular levels.

4.2   Skin Diseases Models

As previously mentioned, the skin models that mimic specific skin disorders are equally relevant in discovering new and more effective therapies. In this way, cutaneous wounds, inflammatory skin diseases, and skin cancer are examples of the pathologies that were already replicated in in vitro skin models, as reviewed by (Randall et al. 2018; Semlin et al. 2011; Sarkiri et al. 2019). Some disease models are also commercially available, such as MelanoDerm®, Melanoma®, and Psoriasis® (Amelian et al. 2017).

For example, the full-thickness skin models utilised as a photodermatitis model enabling the evaluation of UV light exposition. On the other hand, the skin inflammation response to UV stimulus can be induced by incorporating into the epidermal layer dendritic cells.

Additionally, several approaches to induce wounds on skin model, including abrading, burning, and scratching. The most common injury model is the burn induced by a brass string heat to 150 °C (Emanuelsson and Kratz 1997). As an alternative, Vaughan et al. used the laser to induce wounds (6 mm × 1 mm × 400μm) for assessing the reepithelialisation and ageing in vitro (Vaughan et al. 2004).

Wufuer et al. replicated the inflammation and edema mechanisms on an in vitro human skin-on-a-chip device (Wufuer et al. 2016). In order to accomplish such a purpose, the chip was composed of three PDMS layers interspersed by porous membranes that allow the co-culture of three cell layers (keratinocytes, fibroblasts, and endothelial cells) mimicking the skin structure. Then, TNF-α was perfused through the microfluidic channels to mimic an inflammatory condition. The drug' efficacy is based on dexamethasone's perfusion within the microfluidic platform, and the results evidenced that the levels of IL-1β, IL-6, and IL-8 were decreased. Further, the results also highlighted that the permeability of chips treated with dexamethasone was lower than the chips treated just with TNF-α (Fig. 5). The authors could conclude that the perfusion of culture medium and other bioactive compounds into microfluidic channels of the skin-on-a-chip device allowed to recreate the microenvironment peculiar to the native human tissue.

Further, the skin disorders such as atopic dermatitis and psoriasis have been replicated in artificial skin models.

The researchers started to generate in vitro skin substitutes by utilising skin cells isolated from psoriasis' patients. As an example, the TESTSKIN™ model was employed to study psoriasis, in which it was verified that the psoriatic "donors fibroblasts induce the healthy donors" keratinocytes hyperproliferation (Saiag et al. 1985).

Besides, the inflammation was already induced (through the perfusion of the CD4 + T cells) in the human primary keratinocytes' culture after being seeded on the

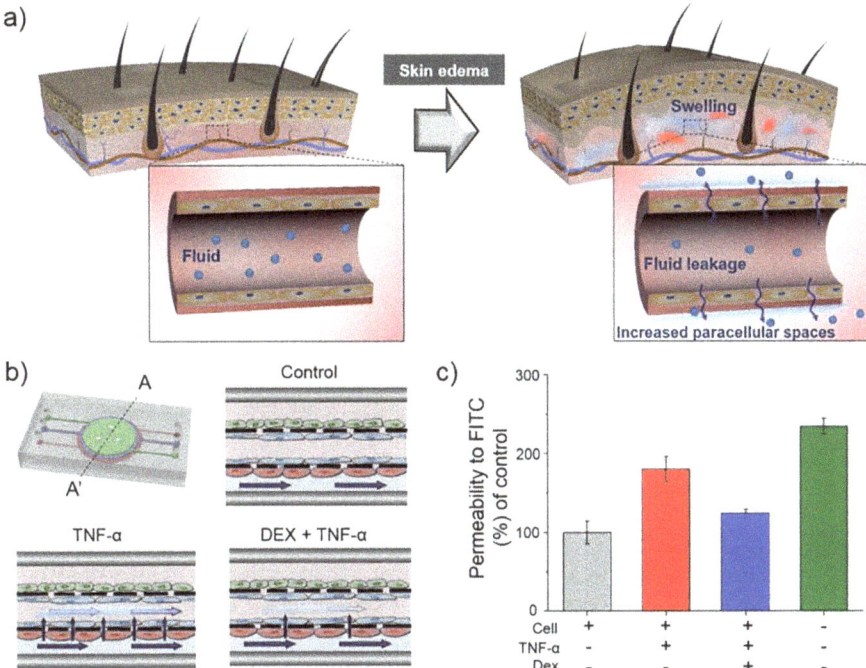

Fig. 5 Analysis of the permeability and therapeutic effect on skin-on-a-chip edema model. Illustrative representation of the edema skin model induced by TNF-α, resulting in vascular leakage (**a**); Schematic representation of the skin edema model in the microfluidic device with TNF-α exposure after the treatment with Dexamethasone (**b**); The chip treated with TNF-α presented improved permeability to FITC-dextran in comparison to the non-treated chips (**c**). Reprinted from Scientific Reports, vol. 6, Wufuer et al. Skin-on-a-chip model simulating inflammation, edema and drug-based treatment, 37,471, open-access article with permission from Springer Nature (Wufuer et al. 2016)

decellularised dermis for 7 days at the air–liquid interface (Lorthois et al. 2019; Shin et al. 2020).

On the other side, the psoriatic phenotype was postulated into in vitro skin models through the addition of IL-22. The secretion of IL-22 by Th17 cells is the most potent keratinocyte activation (Bernard et al. 2012; Boniface et al. 2005; Boniface et al. 2007). The gene expression results revealed that the IL-22 treatment induced a decline in the expression of filaggrin, loricrin, and involucrin, the main keratinocyte differentiation markers.

Considering this, BIOalternatives, Dermatest GmbH, MatTek Corporation, or Straticell construct organotypic psoriatic models for compounds screening (Bernard et al. 2012).

Besides that, the skin cancer models were also developed by incorporating the tumour cells into a 3D matrix. Among different skin carcinoma models, Commandeur et al. proved the suitability of a cell carcinoma for drug development (Commandeur et al. 2012).

Fig. 6 Production of an organotypic melanoma-spheroid skin model. The melanoma cells were seeded on a non-adhesive petri dish with PBS and incubated for 15 days, which promote the formation of the melanoma spheroids with 500μm of diameter. The live/dead assay was then performed on spheroids, staining the live and dead cells with green and red colour, respectively (**a**); The spheroids were collected and mixed with fibroblasts and collagen I to constitute the dermal section of the full-thickness skin substitute (**b**). Reprinted from Cell death & disease, vol. 11, Vorsmann et al. Development of a human three-dimensional organotypic skin-melanoma spheroid model for in vitro drug testing, open-access article with permission from Springer Nature (Vorsmann et al. 2013)

An organotypic hybrid system replicating the metastatic melanoma was also developed (Vorsmann et al. 2013). In this case, the melanoma spheroids with standard sizes were incorporated into the dermal matrix to mimic the 3D structure and multicellular complexity of tumour in vivo (Fig. 6). The authors verified that the TRAIL (tumour necrosis factor-related apoptosis-inducing ligand) cell sensitisation and co-application of sublethal doses of UV-B cisplatin was distinguished between 2D and 3D models. In 2D culture, both treatments affected equally the cancer cell line. However, the melanoma spheroids embedded into skin equivalent were potently compromised by TRAIL+cisplatin treatment, whereas the TRAIL+UVB combination did not affect the melanoma spheroids. Through this work, the authors could establish a human skin model that recreates the melanoma conditions, which will help improve the search for tailored therapies.

5   Conclusions and Future Perspectives

The preclinical development of innovative skin therapies requires the realisation of assays aims to assess their absorption, efficacy, and safety.

In this way, the human skin excised and animal models are still the "gold-standard" methodologies that better mimic the structural and physiological features.

Nevertheless, due to the restricted accessibility, differences in anatomy and physiological processes, and ethical concerns, these strategies have been replaced by using non-animal alternatives. Thus, different in vitro 3D models have been created and applied to rapidly screen and optimise skin formulations. In general, these 3D models mimic the skin "layers'" epidermis and dermis, and they are cultured in air–liquid interface conditions. However, most skin 3D models available are static, incapable of recapitulating the cell interactions and dynamic environment found on native skin tissue.

In this way, with the recent advances in the wound care area, the combination of 3D printing and microfluidic devices arises as an attractive future direction for the production of the microengineered skin-on-a-chip platforms capable of incorporating different skin cells and compounds as well as providing control over dynamic conditions of the cell culture within the microfluidic system (blood and nutrients flow, mechanical forces, etc.).

In these systems, the "cells" growth occurs in a microenvironment, in which dynamic perfusion of the nutrients, gases are possible, promoting the epidermal morphogenesis and differentiation. Such strategy constitutes the starting point for developing preclinical models that more precisely represent the human conditions.

However, the economic and technical constraints related to the microfluidic devices require further investigations to adjust these procedures. The production of accessible, readily and reproducible, with elevated robustness and at a low price, has not yet been achieved. Furthermore, future developments should include the addition of other cells (immune and endothelial cells) and skin appendages (sweat glands or hair follicles), providing a more realistic in vivo situation.

Future efforts for implementing these models have also to be performed, as for (1) establish systemic methodologies to improve the predictability; (2) correlate in vivo data and systematic investigation; and (3) adjust the cultivation procedures to obtain tissue structures similar to the native conditions.

Overall, the future trend depends on the advancement of multi-organ-chips aiming to generate human-on-a-chip systems able to replace animal experimentation during preclinical assessment, and consequently speed up the commercialisation process of the new therapeutic formulations. We believe that the 3D skin models will become a crucial tool for developing more effective therapeutic formulation for skin disorders.

References

Abaci HE, Gledhill K, Guo Z, Christiano AM, Shuler ML (2015) Pumpless microfluidic platform for drug testing on human skin equivalents. Lab Chip 15(3):882–888

Abd E, Yousef SA, Pastore MN, Telaprolu K, Mohammed YH, Namjoshi S et al (2016) Skin models for the testing of transdermal drugs. Clin Pharmacol Adv Appl 8:163

Alepee N, Tornier C, Robert C, Amsellem C, Roux MH, Doucet O et al (2010) A catch-up validation study on reconstructed human epidermis (SkinEthic RHE) for full replacement of the Draize skin irritation test. Toxicol In Vitro 24(1):257–266

Amelian A, Wasilewska K, Megias D, Winnicka K (2017) Application of standard cell cultures and 3D in vitro tissue models as an effective tool in drug design and development. Pharmacol Rep 69 (5):861–870

Ashammakhi N, Ahadian S, Zengjie F, Suthiwanich K, Lorestani F, Orive G et al (2018) Advances and future perspectives in 4D bioprinting. Biotechnol J 13(12):e1800148

Barre-Sinoussi F, Montagutelli X (2015) Animal models are essential to biological research: issues and perspectives. Future Sci OA 1(4):FSO63

Bernard FX, Morel F, Camus M, Pedretti N, Barrault C, Garnier J et al (2012) Keratinocytes under fire of proinflammatory cytokines: bona fide innate immune cells involved in the physiopathology of chronic atopic dermatitis and psoriasis. J Allergy (Cairo) 2012:718725

Bernard M, Jubeli E, Pungente MD, Yagoubi N (2018) Biocompatibility of polymer-based biomaterials and medical devices–regulations, in vitro screening and risk-management. Biomater Sci 6(8):2025–2053

Boniface K, Bernard FX, Garcia M, Gurney AL, Lecron JC, Morel F (2005) IL-22 inhibits epidermal differentiation and induces proinflammatory gene expression and migration of human keratinocytes. J Immunol 174(6):3695–3702

Boniface K, Diveu C, Morel F, Pedretti N, Froger J, Ravon E et al (2007) Oncostatin M secreted by skin infiltrating T lymphocytes is a potent keratinocyte activator involved in skin inflammation. J Immunol 178(7):4615–4622

Buenzli PR, Lanaro M, Wong CS, McLaughlin MP, Allenby MC, Woodruff MA et al (2020) Cell proliferation and migration explain pore bridging dynamics in 3D printed scaffolds of different pore size. Acta Biomater 114:285–295

Chávez MN, Aedo G, Fierro FA, Allende ML, Egaña JT (2016) Zebrafish as an emerging model organism to study angiogenesis in development and regeneration. Front Physiol 7:56

Chien S, Wilhelmi BJ (2012) A simplified technique for producing an ischemic wound model. JoVE (Journal of Visualized Experiments) 63:e3341

Choi TY, Kim JH, Ko DH, Kim CH, Hwang JS, Ahn S et al (2007) Zebrafish as a new model for phenotype-based screening of melanogenic regulatory compounds. Pigment Cell Res 20 (2):120–127

Commandeur S, van Drongelen V, de Gruijl FR, El Ghalbzouri A (2012) Epidermal growth factor receptor activation and inhibition in 3D in vitro models of normal skin and human cutaneous squamous cell carcinoma. Cancer Sci 103(12):2120–2126

European Commision (2020) Alternative methods for toxicity testing. https://ec.europa.eu/jrc/en/eurl/ecvam/alternative-methods-toxicity-testing. Accessed 16 Sept, 2020

Dal Negro G, Eskes C, Belz S, Bertein C, Chlebus M, Corvaro M et al (2018) One science-driven approach for the regulatory implementation of alternative methods: a multi-sector perspective. Regul Toxicol Pharmacol 99:33–49

Dellambra E, Odorisio T, D'Arcangelo D, Failla CM, Facchiano A (2019) Non-animal models in dermatological research. ALTEX-Altern Anim Exp 36(2):177–202

Dhiman HK, Ray AR, Panda AK (2005) Three-dimensional chitosan scaffold-based MCF-7 cell culture for the determination of the cytotoxicity of tamoxifen. Biomaterials 26(9):979–986

Doke SK, Dhawale SC (2015) Alternatives to animal testing: a review. Saudi Pharm J 23 (3):223–229

Dreifke MB, Jayasuriya AA, Jayasuriya AC (2015) Current wound healing procedures and potential care. Mater Sci Eng C Mater Biol Appl 48:651–662

Emanuelsson P, Kratz G (1997) Characterization of a new in vitro burn wound model. Burns 23 (1):32–36

Fentem JH, Botham PA (2002) ECVAM's activities in validating alternative tests for skin corrosion and irritation. Altern Lab Anim 30(Suppl 2):61–67

Flaten GE, Palac Z, Engesland A, Filipović-Grčić J, Vanić Ž, Škalko-Basnet N (2015) In vitro skin models as a tool in optimization of drug formulation. Eur J Pharm Sci 75:10–24

Frykberg RG, Banks J (2015) Challenges in the treatment of chronic wounds. Adv Wound Care (New Rochelle). 4(9):560–582

Gordon S, Daneshian M, Bouwstra J, Caloni F, Constant S, Davies DE et al (2015) Non-animal models of epithelial barriers (skin, intestine and lung) in research, industrial applications and regulatory toxicology. ALTEX 32(4):327–378

Grada A, Otero-Vinas M, Prieto-Castrillo F, Obagi Z, Falanga V (2017) Research techniques made simple: analysis of collective cell migration using the wound healing assay. J Investig Dermatol 137(2):e11–ee6

Hänel KH, Cornelissen C, Lüscher B, Baron JM (2013) Cytokines and the skin barrier. Int J Mol Sci 14(4):6720–6745

Hasan A, Paul A, Vrana NE, Zhao X, Memic A, Hwang Y-S et al (2014) Microfluidic techniques for development of 3D vascularized tissue. Biomaterials 35(26):7308–7325

Heydarkhan-Hagvall S, Schenke-Layland K, Dhanasopon AP, Rofail F, Smith H, Wu BM et al (2008) Three-dimensional electrospun ECM-based hybrid scaffolds for cardiovascular tissue engineering. Biomaterials 29(19):2907–2914

Hilmi AB, Halim AS, Hassan A, Lim CK, Noorsal K, Zainol I (2013) In vitro characterization of a chitosan skin regenerating template as a scaffold for cells cultivation. Springerplus 2(1):79

Jin W, Lo K-Y, Sun Y-S, Ting Y-H, Simpson MJ (2020) Quantifying the role of different surface coatings in experimental models of wound healing. Chem Eng Sci 20, 115609

Kandarova H, Hayden P, Klausner M, Kubilus J, Kearney P, Sheasgreen J (2009) In vitro skin irritation testing: improving the sensitivity of the EpiDerm skin irritation test protocol. Altern Lab Anim 37(6):671–689

Kilic T, Navaee F, Stradolini F, Renaud P, Carrara S (2018) Organs-on-chip monitoring: sensors and other strategies. Microphysiol Syst 2:5

Koch L, Kuhn S, Sorg H, Gruene M, Schlie S, Gaebel R et al (2010) Laser printing of skin cells and human stem cells. Tissue Eng Part C Methods 16(5):847–854

Koch L, Deiwick A, Schlie S, Michael S, Gruene M, Coger V et al (2012) Skin tissue generation by laser cell printing. Biotechnol Bioeng 109(7):1855–1863

Lee SH, Sung JH (2018) Organ-on-a-chip technology for reproducing multiorgan physiology. Adv Healthc Mater 7(2):1700419

Lee V, Singh G, Trasatti JP, Bjornsson C, Xu X, Tran TN et al (2014) Design and fabrication of human skin by three-dimensional bioprinting. Tissue Eng Part C Methods 20(6):473–484

Lee S, Jin S-P, Kim YK, Sung GY, Chung JH, Sung JH (2017) Construction of 3D multicellular microfluidic chip for an in vitro skin model. Biomed Microdevices 19(2):22

Leng L, Amini-Nik S, Jeschke M, Guenther A (2013) Skin printer: microfluidic approach for skin regeneration and wound dressing. US Prov Patent Application 61817860

Li Q, Uitto J (2014) Zebrafish as a model system to study skin biology and pathology. J Invest Dermatol 134(6):e21

Li Q, Frank M, Akiyama M, Shimizu H, Ho S-Y, Thisse C et al (2011) Abca12-mediated lipid transport and Snap29-dependent trafficking of lamellar granules are crucial for epidermal morphogenesis in a zebrafish model of ichthyosis. Dis Model Mech 4(6):777–785

Liang C-C, Park AY, Guan J-L (2007) In vitro scratch assay: a convenient and inexpensive method for analysis of cell migration in vitro. Nat Protoc 2(2):329

Lorthois I, Simard M, Morin S, Pouliot R (2019) Infiltration of T cells into a three-dimensional psoriatic skin model mimics pathological key features. Int J Mol Sci 20(7):1670

Low LA, Tagle DA (2017) Tissue chips – innovative tools for drug development and disease modeling. Lab Chip 17(18):3026–3036

Löwa A, Jevtić M, Gorreja F, Hedtrich S (2018) Alternatives to animal testing in basic and preclinical research of atopic dermatitis. Exp Dermatol 27(5):476–483

Ma J, Wang H, He B, Chen J (2001) A preliminary in vitro study on the fabrication and tissue engineering applications of a novel chitosan bilayer material as a scaffold of human neofetal dermal fibroblasts. Biomaterials 22(4):331–336

Mah CS, Kochhar JS, Ong PS, Kang L (2013) A miniaturized flow-through cell to evaluate skin permeation of endoxifen. Int J Pharm 441(1–2):433–440

Masson-Meyers DS, Andrade TAM, Caetano GF, Guimaraes FR, Leite MN, Leite SN et al (2020) Experimental models and methods for cutaneous wound healing assessment. Int J Exp Pathol 101(1–2):21–37

Mathes SH, Ruffner H, Graf-Hausner U (2014) The use of skin models in drug development. Adv Drug Deliv Rev 69:81–102

Mendes JJ, Leandro CI, Bonaparte DP, Pinto AL (2012) A rat model of diabetic wound infection for the evaluation of topical antimicrobial therapies. Comp Med 62(1):37–48

Mendoza-Garcia J, Sebastian A, Alonso-Rasgado T, Bayat A (2015) Optimization of an ex vivo wound healing model in the adult human skin: functional evaluation using photodynamic therapy. Wound Repair Regen 23(5):685–702

Meyer W, Schwarz R, Neurand K (1978) The skin of domestic mammals as a model for the human skin, with special reference to the domestic pig1. Skin-drug application and evaluation of environmental hazards. Karger Publishers, Berlin, pp 39–52

Miguel SP, Figueira DR, Simões D, Ribeiro MP, Coutinho P, Ferreira P et al (2018) Electrospun polymeric nanofibres as wound dressings: a review. Colloids Surf B: Biointerfaces 169:60–71

Miguel SP, Simoes D, Moreira AF, Sequeira RS, Correia IJ (2019a) Production and characterization of electrospun silk fibroin based asymmetric membranes for wound dressing applications. Int J Biol Macromol 121:524–535

Miguel SP, Cabral CSD, Moreira AF, Correia IJ (2019b) Production and characterization of a novel asymmetric 3D printed construct aimed for skin tissue regeneration. Colloids Surf B Biointerfaces 181:994–1003

Mohammadi MH, Heidary Araghi B, Beydaghi V, Geraili A, Moradi F, Jafari P et al (2016) Skin diseases modeling using combined tissue engineering and microfluidic technologies. Adv Healthc Mater 5(19):2459–2480

Mori N, Morimoto Y, Takeuchi S (2017) Skin integrated with perfusable vascular channels on a chip. Biomaterials 116:48–56

Nicholas MN, Jeschke MG, Amini-Nik S (2016) Methodologies in creating skin substitutes. Cell Mol Life Sci 73(18):3453–3472

O'Brien FJ, Harley BA, Yannas IV, Gibson LJ (2005) The effect of pore size on cell adhesion in collagen-GAG scaffolds. Biomaterials 26(4):433–441

Oh JW, Hsi T-C, Guerrero-Juarez CF, Ramos R, Plikus MV (2013) Organotypic skin culture. J Invest Dermatol 133(11):e14

Park SA, Covert J, Teixeira L, Motta MJ, DeRemer SL, Abbott NL et al (2015) Importance of defining experimental conditions in a mouse excisional wound model. Wound Repair Regen 23 (2):251–261

Pastar I, Stojadinovic O, Yin NC, Ramirez H, Nusbaum AG, Sawaya A et al (2014) Epithelialization in wound healing: a comprehensive review. Adv Wound Care (New Rochelle) 3 (7):445–464

Pastar I, Liang L, Sawaya AP, Wikramanayake TC, Glinos GD, Drakulich S et al (2018) Preclinical models for wound-healing studies. Skin tissue models. Elsevier, San Diego, pp 223–253

Pizzoferrato A, Ciapetti G, Stea S, Cenni E, Arciola CR, Granchi D (1994) Cell culture methods for testing biocompatibility. Clin Mater 15(3):173–190

Planz V, Lehr C-M, Windbergs M (2016) In vitro models for evaluating safety and efficacy of novel technologies for skin drug delivery. J Control Release 242:89–104

Randall MJ, Jüngel A, Rimann M, Wuertz-Kozak K (2018) Advances in the biofabrication of 3D skin in vitro: healthy and pathological models. Front Bioeng Biotechnol 6:154

Ranganatha N, Kuppast I (2012) A review on alternatives to animal testing methods in drug development. Int J Pharm Pharm Sci 4(SUPPL 5):28–32

Rezvani Ghomi E, Khalili S, Nouri Khorasani S, Esmaeely Neisiany R, Ramakrishna S (2019) Wound dressings: current advances and future directions. J Appl Polym Sci 136(27):47738

Rheinwald J (1989) Methods for clonal growth and serial cultivation of normal human epidermal keratinocytes and mesothelial cells. In: Cell growth and division: a practical approach. Oxford University Press, New York, pp 81–94

Richardson R, Slanchev K, Kraus C, Knyphausen P, Eming S, Hammerschmidt M (2013) Adult zebrafish as a model system for cutaneous wound-healing research. J Investig Dermatol 133 (6):1655–1665

Rittié L, Sachs DL, Orringer JS, Voorhees JJ, Fisher GJ (2013) Eccrine sweat glands are major contributors to reepithelialization of human wounds. Am J Pathol 182(1):163–171

Ruggeri M, Bianchi E, Rossi S, Vigani B, Bonferoni MC, Caramella C et al (2020) Nanotechnology-based medical devices for the treatment of chronic skin lesions: from research to the clinic. Pharmaceutics 12(9):185

Saiag P, Coulomb B, Lebreton C, Bell E, Dubertret L (1985) Psoriatic fibroblasts induce hyperproliferation of normal keratinocytes in a skin equivalent model in vitro. Science 230 (4726):669–672

Sami DG, Heiba HH, Abdellatif A (2019) Wound healing models: a systematic review of animal and non-animal models. Wound Med 24(1):8–17

Sarkiri M, Fox SC, Fratila-Apachitei LE, Zadpoor AA (2019) Bioengineered skin intended for skin disease modeling. Int J Mol Sci 20(6):1407

Semlin L, Schäfer-Korting M, Borelli C, Korting HC (2011) In vitro models for human skin disease. Drug Discov Today 16(3–4):132–139

Shin JU, Abaci HE, Herron L, Guo Z, Sallee B, Pappalardo A et al (2020) Recapitulating T cell infiltration in 3D psoriatic skin models for patient-specific drug testing. Sci Rep 10(1):1–12

Song G, Nguyen DT, Pietramaggiori G, Scherer S, Chen B, Zhan Q et al (2010) Use of the parabiotic model in studies of cutaneous wound healing to define the participation of circulating cells. Wound Repair Regen 18(4):426–432

Song HJ, Lim HY, Chun W, Choi KC, Sung JH, Sung GY (2017) Fabrication of a pumpless, microfluidic skin chip from different collagen sources. J Ind Eng Chem 56:375–381

Sood A, Granick MS, Tomaselli NL (2014) Wound dressings and comparative effectiveness data. Adv Wound Care (New Rochelle). 3(8):511–529

Spitsbergen JM, Kent ML (2003) The state of the art of the zebrafish model for toxicology and toxicologic pathology research—advantages and current limitations. Toxicol Pathol 31 (1_suppl):62–87

Sriram G, Alberti M, Dancik Y, Wu B, Wu R, Feng Z et al (2018) Full-thickness human skin-on-chip with enhanced epidermal morphogenesis and barrier function. Mater Today 21(4):326–340

Stark H-J, Willhauck MJ, Mirancea N, Boehnke K, Nord I, Breitkreutz D et al (2004) Authentic fibroblast matrix in dermal equivalents normalises epidermal histogenesis and dermo-epidermal junction in organotypic co-culture. Eur J Cell Biol 83(11–12):631–645

Stirland DL, Nichols JW, Miura S, Bae YH (2013) Mind the gap: a survey of how cancer drug carriers are susceptible to the gap between research and practice. J Control Release 172 (3):1045–1064

Sullivan TP, Eaglstein WH, Davis SC, Mertz P (2001) The pig as a model for human wound healing. Wound Repair Regen 9(2):66–76

van den Berg A, Mummery CL, Passier R, van der Meer AD (2019) Personalised organs-on-chips: functional testing for precision medicine. Lab Chip 19(2):198–205

Van Meerloo J, Kaspers GJ, Cloos J (2011) Cell sensitivity assays: the MTT assay. In: Cancer cell culture. Springer, Berlin, pp 237–245

van Rooijen E, Fazio M, Zon LI (2017) From fish bowl to bedside: the power of zebrafish to unravel melanoma pathogenesis and discover new therapeutics. Pigment Cell Melanoma Res 30 (4):402–412

Vaughan MB, Ramirez RD, Brown SA, Yang JC, Wright WE, Shay JW (2004) A reproducible laser-wounded skin equivalent model to study the effects of aging in vitro. Rejuvenation Res 7 (2):99–110

Verhulsel M, Vignes M, Descroix S, Malaquin L, Vignjevic DM, Viovy JL (2014) A review of microfabrication and hydrogel engineering for micro-organs on chips. Biomaterials 35 (6):1816–1832

Vorsmann H, Groeber F, Walles H, Busch S, Beissert S, Walczak H et al (2013) Development of a human three-dimensional organotypic skin-melanoma spheroid model for in vitro drug testing. Cell Death Dis 4:e719

Wagner I, Materne EM, Brincker S, Sussbier U, Fradrich C, Busek M et al (2013) A dynamic multi-organ-chip for long-term cultivation and substance testing proven by 3D human liver and skin tissue co-culture. Lab Chip 13(18):3538–3547

Wang YI, Oleaga C, Long CJ, Esch MB, McAleer CW, Miller PG et al (2017) Self-contained, low-cost body-on-a-Chip systems for drug development. Exp Biol Med (Maywood) 242 (17):1701–1713

Wei Z, Liu X, Ooka M, Zhang L, Song MJ, Huang R et al (2020) Two-dimensional cellular and three-dimensional bio-printed skin models to screen topical-use compounds for irritation potential. Front Bioeng Biotechnol 8:109

Weinmüllner R, Zbiral B, Becirovic A, Stelzer EM, Nagelreiter F, Schosserer M et al (2020) Organotypic human skin culture models constructed with senescent fibroblasts show hallmarks of skin aging. NPJ Aging Mech Dis 6(1):1–7

Wilhelm KP, Wilhelm D, Bielfeldt S (2017) Models of wound healing: an emphasis on clinical studies. Skin Res Technol 23(1):3–12

Wong VW, Sorkin M, Glotzbach JP, Longaker MT, Gurtner GC (2010) Surgical approaches to create murine models of human wound healing. J Biomed Biotechnol 2011:969618

Wu Q, Liu J, Wang X, Feng L, Wu J, Zhu X et al (2020) Organ-on-a-chip: recent breakthroughs and future prospects. Biomed Eng Online 19(1):9

Wufuer M, Lee G, Hur W, Jeon B, Kim BJ, Choi TH et al (2016) Skin-on-a-chip model simulating inflammation, edema and drug-based treatment. Sci Rep 6:37471

Xu W, Jong Hong S, Jia S, Zhao Y, Galiano RD, Mustoe TA (2012a) Application of a partial-thickness human ex vivo skin culture model in cutaneous wound healing study. Lab Investig 92 (4):584–599

Xu W, Hong SJ, Jia S, Zhao Y, Galiano RD, Mustoe TA (2012b) Application of a partial-thickness human ex vivo skin culture model in cutaneous wound healing study. Lab Investig 92 (4):584–599

Yu JR, Navarro J, Coburn JC, Mahadik B, Molnar J, Holmes JH et al (2019) Current and future perspectives on skin tissue engineering: key features of biomedical research, translational assessment, and clinical application. Adv Healthc Mater 8(5):e1801471

Yun YE, Jung YJ, Choi YJ, Choi JS, Cho YW (2018) Artificial skin models for animal-free testing. J Pharm Invest 48(2):215–223

Potential Biomedical Applications of Marine Sponge-Derived Chitosan: Current Breakthroughs in Drug Delivery for Wound Care

Harekrishna Roy, Asha Gummadi, and Sisir Nandi

1 Introduction

Chitosan is a polymer of natural origin linear polysaccharide obtained from marine shrimp, crabs, and cell walls of yeasts and fungi. Commercially various grades of chitosan are produced from chitin. It is available in the market in various forms ranging from dry flakes to fine powder. Some techniques including capillary viscometers, size exclusion chromatography with light scattering, and ultracentrifugation are used to estimate the average molecular weight of chitosan. Based on the process and degree of deacetylation (DDA), the molecular weight varies from about 3800 to 2,000,000 Da (Barbosa et al. 2011). Chitosan has unique features of the amino-polysaccharide group which possesses a high binding capacity and unique functionality and makes it available for wide applicability (Salehi et al. 2016).

Chitosan solubility primarily depends upon the quantity of protonated amino groups within the polymeric chains, i.e., the proportion of acetylated and non-acetylated D-glucosamine units. It is soluble in organic and inorganic acids like perchloric, nitric, acetic, hydrochloric, and phosphoric acids (El-Gamal et al. 2016). The solubility of chitosan in water is pH dependent which allows processing ability under several conditions, which makes it available to a good range of applications particularly within the field of the cosmetic industry (Croisier and Jérôme 2013). Prompt differences between chitin and chitosan are remarked by their solubility in a different solvent system. The chitin is soluble in N, N-dimethylacetamide (DMAc) in the presence of 5–10% w/v lithium chloride and

H. Roy (✉) · A. Gummadi
Department of Pharmaceutics, Nirmala College of Pharmacy, Affiliated to Acharya Nagarjuna University, Guntur, Andhra Pradesh, India

S. Nandi (✉)
Department of Pharmaceutical Chemistry, Global Institute of Pharmaceutical Education and Research, Affiliated to Uttarakhand Technical University, Kashipur, India

insoluble in dilute acid solutions while the reverse is for chitosan (El-Gamal et al. 2016).

According to Cheung et al. (2015), chitosan-related drug delivery systems are extensively used in gels, tablets, films, and particulate dosage forms in the form of a novel carrier of delivering the active ingredient. Few techniques are being widely used, especially chemical crosslinking, ionic gelation, microemulsion, and spray drying method. Though they are beneficial, having some limitations including time-consuming, less profitability, and more processing cost makes chitosan to be used less on a commercial scale (Badhe et al. 2015). Chitosan has been extensively studied for various applications because of its nontoxicity, biocompatibility, less immunogenicity, biodegradability, easy availability, derivability, and inexpensiveness. It shows wound healing actions due to hemostasis and bactericidal and antimicrobial properties and also acts as a biomedical drug carrier in the pharmaceutical product (Abdul Khalil et al. 2016). Other than the pharmaceutical field, chitosan is being used in the process of food product manufacturing, food preservative, and biocompatible film and recycling of products during waste material management in the food industry. Chitosan has also been used as an absorbent for the treatment of wastewater including pollutants, heavy metals, dyes, as well as the complexation process (Al-Manhel et al. 2018).

2 Degree of Deacetylation of Chitin for the Synthesis of Chitosan

The degree of deacetylation defines the molar ratio of the D-glucosamine units to the sum of both portions of N-acetyl-D-glucosamine and D-glucosamine units (Fig. 1). The enzymes like lipases, lysozyme, and glucosaminidases cause depolymerization of chitosan and yield active chitooligosaccharides which are having greater antimicrobial activity, and its monomeric products like glucosamine are quickly metabolized and eliminated from the human body. Hence chitosan is regarded as mostly biodegradable and biocompatible with the human body organs (Logith Kumar et al. 2016). Successfully it has been used in the treatment of Alzheimer's disease, along with few more brain diseases such as anxiety, and glaucoma by Xu et al. The study also reported significant increases in the permeability of the hydrophobic drug compounds (Xu et al. 2017). According to Chiappisi and Gradzielski (2015), chitosan helps to self-associate the hydrophobic polymers and origin. This inherent tendency can be further improved by suitable hydrophobic modification of chitosan such as N-alkylation as seen in other hydrophilic polymers. Several research papers were published on the biocompatibility issue of chitosan, but it was extensively studied and reported with a degree of deacetylation by Giri and coworkers. The study reported that the entrapment of biological cells in hydrogel made by chitosan did not produce incompatibility with normal biological function (Giri et al. 2012).

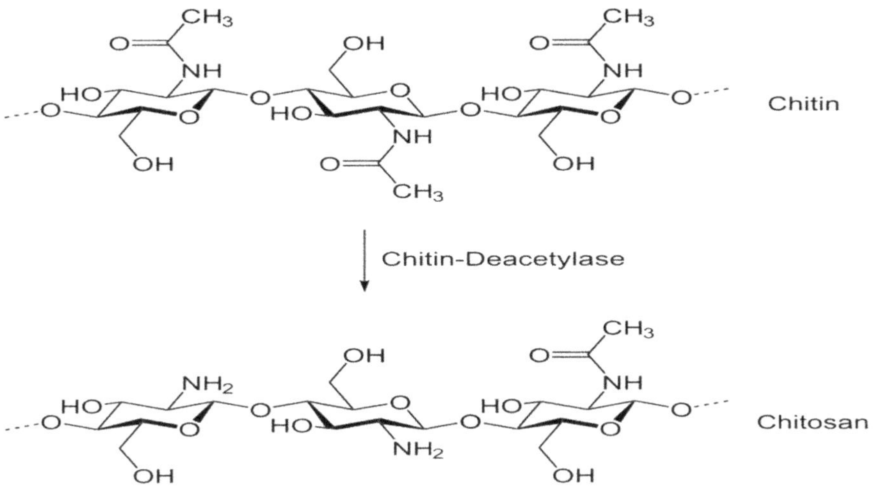

Fig. 1 Conversion of chitin to chitosan on deacetylation

3 Biomedical Applications

The biocompatible chitosan molecule has many amino moieties, due to the protonation of amino groups; it may produce many positive heads and can bind with the negative charged microbial cell membrane and ultimately damage the membrane, which causes leakage of intracellular microbial components residues. The very simple mechanism of antimicrobial actions of chitosan may have great attention in the generation of various biomedical formulations (Fig. 2).

3.1 Chitosan in Cosmetic Technology

Libio et al. (2016) studied and suggested chitosan to be suitable for cosmetic applications. The study reported biocompatible film made up of chitosan utilizing citrate and acetate as a salt neutralizer (Libio et al. 2016). The cosmetic film was prepared with an appropriate concentration of excipients. During the study, the plasticizer such as glycerol was added and further evaluated for moisture content, the thickness of the biofilm, swelling index, acid-soluble substances, X-ray diffraction, and differential scanning calorimetry (DSC). The citrate-neutralized chitosan film exhibited greater stability along with acceptable thickness value, swelling property, low moisture value, and reduced solubility in the acidic buffer. The DSC study revealed possible interaction with water molecules, a critical factor in the cosmetic industry. From the relevant study, it was supposed to provide a more stable biofilm in citrate medium without the presence of glycerol; hence, the release study of hyaluronic acid (HA) was ascertained in the skin model. The in vivo model

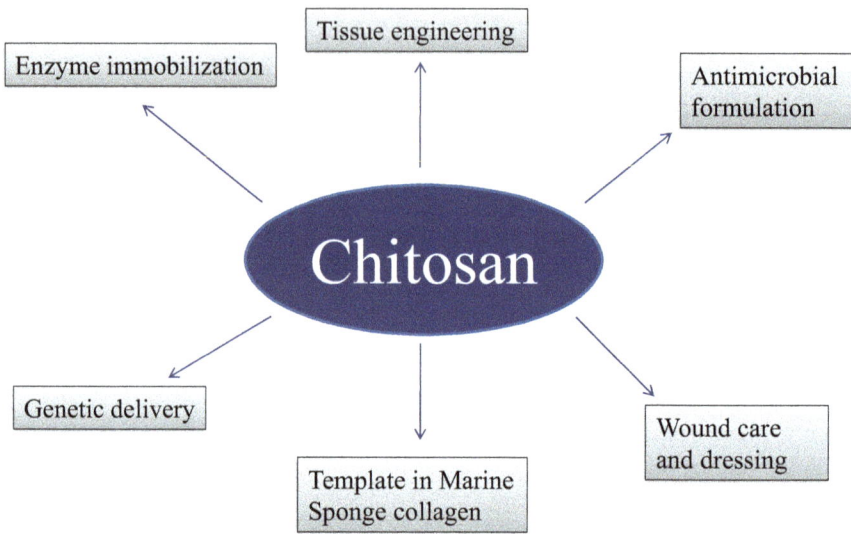

Fig. 2 Chitosan in biomedical application

showed the desquamation of stratum corneum with or without the presence of HA on the application of film and resulted in hydration. The study also suggested that the citrate-based chitosan film in the absence of glycerol produced exfoliation of the skin needed in cosmetic preparation.

Chitosan and its derivatives being highly soluble in water make them most preferable in the cosmetic industry. Carboxymethylated chitin and chitosan have their similarity characteristic in the three-dimensional arrangement of polysaccharides, which makes them hydrophilic, improved viscosity, cationic binding property, satisfied osmotic value, gelling property, and biodegradability. Moreover, modified chitosan and chitin could function as a biological membrane permeability barrier, as seen in hyaluronate, which formed a protective layer surrounding the human ovum. Similarly in another review, the carboxymethyl chitosan in the cosmetic industry for potential applications is being used and discussed in terms of antioxidant, drug delivery, and antibacterial property, the stability of the emulsion, and moisture retention property (Jimtaisong and Saewan 2016).

Chitosan is one of the few popular cationic polymers of natural origin that has been used in the cosmetic industry for the preparation of lotions, skincare products, nail paints, creams, and other skincare and hair treatment products. Chitosan on application as topical preparation forms an elastic film that protects and moisturizes the skin surface. In cosmetic formulation, chitosan is being popularly used as moisturizers such as in sunscreen preparation, lotion, etc. (Chen 2008). It protects and smoothens the skin from harsh environmental conditions and topical exposure to chemicals. It also showed that chitosan drastically increases the hydration property of HA. The study already proved that it protects the skin and lips from UV A and UV B radiation from sunlight and moisturizes in the winter season (Chen 2008). In lip care preparation, the addition of chitosan makes the lips softer and also retains the color for a long period with excellent water repellant property. In another study, Tzaneva and coworkers studied the antimicrobial property of chitosan in deodorant preparation. The result revealed the microbial enzyme inhibition character of chitosan which resulted in the antimicrobial property (Tzaneva et al. 2017). The anti-odor property, adhesion property, and compatibility with the biological membrane of chitosan formulation were found superior to triclosan (Tzaneva et al. 2017).

3.2 Antimicrobial Activity of Chitosan

The antimicrobial property of chitosan and its related derivatives have proved to be most effective against the majority of viruses, fungus, bacteria, and yeast (Rabea et al. 2003). It is reported to have added advantages and a significantly broad range of antimicrobial properties, as well as chances of transmitting the pathogenic antigen toward human cells are the least (Campaniello et al. 2008; Campaniello and Corbo 2010). Hence, there has been a tremendous use of chitosan in most food and packaging industries for a broad range of antimicrobial properties (Ganguly 2013). Chitosan can be fabricated and molded to biofilms which form a protective barrier and can effectively protect against disease-causing pathogens. A literature study revealed an excellent solubility of chitosan under acidic conditions which facilitated easy incorporation into solid and liquid dosage products. Friedman and coworkers found effective use of low-molecular-weight chitosan at acidic medium preferably below pH 6.0 for its antioxidant, antimicrobial, and preservative activity in solid and liquid content (Friedman and Juneja 2010; Goy et al. 2009). There has been market availability of few chitosan-related products especially for wound treatment, such as Kyoto cell™ and Chitoderm™ (Jardine and Sayed 2014; Cheung et al. 2015), and fabrics with antimicrobial properties, such as CRABYON™ and Bac-Shield™ (Lim and Hudson 2003). In the recent era, there is a big challenge for the formulators to develop an antimicrobial drug against antimicrobial resistance. Few articles were published with the key focus of chitosan and derived products for its unique antimicrobial activity for the quest of new antimicrobials (Tomayko et al. 2014). In brief, it can be stated that the prominent antimicrobial activity, as well as the least precipitation of toxicity in the human body, made it favorable in the wide application of food and packaging, in topical preparations, and also in the seed industry. However, the exact underlying mechanism of chitosan interaction with a microbial strand for antimicrobial effect is still to be unveiled. Few scientists argue for the minimum inhibitory concentration (MIC) below 20 ppm for chitosan to be reliably used in systemic infection and novel techniques of medicine preservation. However, chemical modification of the chitosan principal structure could make it possible to achieve the above stated, whereas such modification may result in desired polymeric character, biocompatibility, and so on.

3.3   Chitosan as Anti-Infective

The drug molecule and the excipients such as chitosan derived from natural resources are the most preferable sources by the pharmaceutical industry. Natural products including chitosan have added advantages including biocompatibility, biosafety, green source, biodegradability, and expected broad range of antimicrobial activity (Atay 2019). Dai and coworkers reported its significant advantage over other antimicrobials in terms of low toxicity, prominent antimicrobial activity, higher drug diffusion through polymer mass, and less chance to develop microbial resistance (Dai et al. 2011). It was observed that chitosan targets through the outer cell membrane of bacteria and interacts mainly with lipopolysaccharide and peptidoglycan along with teichoic acid for gram-negative and gram-positive bacteria, respectively. The groups such as N-acetyl glucosamine and muramic acid, carboxy, and most importantly phosphate groups interact tightly with polycations of chitosan (Sashiwa and Aiba 2004). Chitosan also forms a complex with covalent ions present on the surface of the microbial wall and ultimately damages the outer cell membrane. Similarly, the antibacterial activity of chitosan was found at a concentration of 200 mg/l; it found that polycationic chitosan binds to a charged phosphate layer of bacteria and causes agglutination. At higher concentrations, the higher amount of cations most easily maximizes the net positive charge on the bacterial surface, ultimately increases density, and makes the bacterial colony in suspension (Sashiwa and Aiba 2004). Chitosan profoundly acts on the microbial cell membrane and changes cellular permeability. In a UV-spectrophotometric absorption study, it ascertained the leakage of intracellular components by chitosan binding. The subsequent disruption of the cellular membrane could be the result of cellular depolarization. Furthermore, it is well described as toxins to microbes and their growth (Raafat et al. 2008; Badawy and Rabea 2011). There are few more antimicrobial mechanisms well described including inhibition of the microbial enzyme, solubilization of the bacterial toxins, and excretion as a soluble mass from the host cell. Another major mechanism is the inhibition of mRNA and protein formation by interacting cationic chitosan with microbial including bacteria, fungus, and yeast deoxyribonucleic acid (DNA) (de Britto et al. 2011). It also reported dose-dependent microbial inhibitory action in the in vitro study (Bakshi et al. 2018).

Rabea and coworkers provided extended information on the antimicrobial effect of chitosan, generally extraneous and intrinsic factors such as degree of deacetylation (DDA), temperature, pH of the medium, molecular weight of chitosan, polymerization, solubility in the solvent, organisms, etc. The wider antimicrobial activity of chitosan was reported to be further increased by chemical modifications such as the incorporation of a new functional group and complexation by polyelectrolyte between microbes and polymers (Rabea et al. 2003). The chitosan-derived mass does possess the same principle of the antibacterial mechanism as that of chitosan (Kong et al. 2010). Kim and coworkers prepared quaternary chitosan and observed the antimicrobial property. They observed that combined with pentameric alkyl compound the said derived chitosan possessed an additive effect on microbes (Kim et al. 1997). Similarly, it reported that the quaternary salt of chitosan was superior to parent chitosan and gradually the antimicrobial activity increased with an increase in alkyl chain length, which could be attributed to the hydrophobicity of alkylated chitosan (Xu et al. 2010). In another study, the quaternary compound of chitosan was successfully evaluated for antimicrobial action against a broad range of bacteria, fungus, and yeast. It assumed that the strong interaction between hydrophobic alkylated chitosan and microbial phospholipid attenuated antimicrobial property (Rúnarsson et al. 2010; Fu et al. 2011). Likewise, Jia and co-researchers developed several moieties of quaternary chitosan and tested effectiveness on *Escherichia coli*. It was observed that those salts exhibited superior activity in an acidic condition against *E. coli* as compared to chitosan. Streptococcus species are commonly responsible for major dental caries in humans. It reported that chitosan-oligosaccharide along with glycidyl trimethylammonium chloride linked via covalent bond was found to be effective against Streptococcus species (Badawy 2010; Jia and Xu 2001; Kim and Choi 2002). Similarly, Belalia and coworkers successfully found and reported the antimicrobial effect of quaternary chitosan salt against *Listeria monocytogenes* (96%) and *Salmonella typhimurium* (100%) (Belalia et al. 2008). Likewise, trimethoxy chitosan (TMC) was tested for antimicrobial effectiveness in wound healing; the experiment provided evidence of electrospun mats of TMC/polyvinyl alcohol and mats of TMC/polyvinyl pyrrolidone for antimicrobial activity against *E.coli* and *S. aureus* (Avadi et al. 2004). Trimethyl chitosan is preferred as one of the most studied quaternary salts of chitosan for its MIC at 0.125 µg/ml against *E. coli* and 0.0625 µg/ml against S. aureus (Sajomsang et al. 2008, 2009). Few other bacteria such as *Enterococcus feacalis* and *Pseudomonas aeruginosa* are also found to be susceptible to TMC, with an MIC of chitosan as 128 µg/ml and 256 µg/ml, respectively (Kim et al. 2003). Similarly, bactericidal cotton fabrics are also designed and impregnated with alkylated chitosan and further quaternary substitution and methylation. From the relevant research, it proved that the incorporation of quaternary ammonium salts in chitosan structure increased antimicrobial property in wound healing including hydrophilic character (Belalia et al. 2008).

3.4 Lyophilized Chitosan Sponges for Wound Dressing

Biomedical sponges fabricated with chitosan were reported to be used in many topical drug delivery devices (Freier et al. 2005). The chitosan-loaded sponges with enormous micropores have their unique benefit, which includes ease in soaking the medicament, lightweight, and absorption of excess amount of medicament. To fabricate and develop chitosan sponges, there have been reported many processes including leaching, gas foaming, and lyophilization (Lim and Hudson 2003; Pezeshki-Modaress et al. 2014; Gupta and Shivakumar 2010; Ji et al. 2011), but the most accessible and popular technique to design the sponges is the process of lyophilization. In the case of wound recovery, the process to accelerate the natural healing mechanism and compatibility with host cells by chitosan proved to be a major advantage in sponge fabrication consideration (Tucci et al. 2001; Khor and Lim 2003; Harkins et al. 2014). It was reported that chitosan-based sponges are preferably hydrophilic as the presence of the amino group keeps the contact tissue area moist and prevents the loss of moisture in dosage delivery consideration (Harkins et al. 2014; Ozkaynak et al. 2005). There is a significant finding in severe musculoskeletal wound treatment about chitosan sponge. It has been observed that there is a drawback of systematic antibiotics which are unable to reach the injury site and the poor blood flow to the infected site diminishes the natural healing process and thus immunity. Hence to overcome the difficulty, antibiotics loaded in chitosan sponge are considered as a better alternative in local delivery (Dai et al. 2011; Hanssen 2005). On application, over time chitosan sponges facilitate the drug release to the wound area without compromising the biocompatibility, frequency of administration, and secondary topical treatment.

3.5 Chitosan-Based Coatings Formed by Chemical Bonding Methods

The technique such as the phase inversion method was successfully employed to fix chitosan on fabric materials, e.g., cellulose and cotton by chemical and mechanical bonding method (nonwoven) (Wang et al. 2016), for effective treatment in wound healing. It was noticed that fabrics bonded with chitosan exhibited a decrease in water absorption property as well as antimicrobial property (Wang et al. 2016). Fibroblasts are the key point during the process of tissue repair in wound healing and migration to the broken tissue is needed during the healing process (Rahmati et al. 2020). A study was done by Lou (2008) on a mice model that compared chemically bonded fabrics and an untreated one. The result indicated a superior healing process and there was no significant cytotoxic effect even after 24 h of chitosan impregnated fabricated material (Lou 2008).

3.6 Marine Sponge Collagen as a Template for Wound Treatment

Collagen is termed as a matrix system of a three-dimensional network enclosing connective tissue and forms a porous network in a dense network. Generally, they are arranged as a layer of a network and also termed as a sponge. The presence of porosity and unique surface character of the network like mesh is an essential and important characteristic of collagen sponge (Lim et al. 2019). In several biomedical applications including wound repair and bone tissue engineering, the antigenic property of sponge collagen of the marine source must be considered (Nakamura et al. 2019; Meyer 2019). The structural finding of marine collagen reveals a lack of tyrosine which is responsible for the antigenic property (Bardakova et al. 2018), whereas the antigenic telopeptide region comprised of helical C and N terminal easily cleaved by pepsin and other enzymes ultimately loses the antigenic property of the collagen sponge. The unique clinical approach of collagen sponge is in wound care, cartilage and bone repair, intravaginal contraception, implants, cosmetics, and drug delivery because of its excellent moisture-retaining property (Chvapil 1977, 1982). In a study carried by Lin and coworkers, *Callyspongiidae* is a family of marine sponges collected from the coast of Western Australia and assessed for the characteristic by using scanning electron microscopy (SEM) and eosin. The result found the excellent bioabsorbable character, cell attachment, and most ideal property for ideal biomedical collagen (Lin et al. 2011). As discussed earlier in this chapter, chitosan and its carboxymethylated derivatives have excellent antimicrobial properties. Hence there is a quest for the suitability of chitosan along with marine collagen sponge in biomedical applications including wound care and bandage, bone and tissue regeneration, and novel drug delivery system (Khor and Lim 2003).

It already reported the use of chitosan in prosthetics, orthopedic, hernia treatment, implants, etc. (Khor and Lim 2003). Fibrin glue is a biodegradable agent used in abdominal operations for the stoppage of bleeding and wound healing. In a study, it prepared bioadhesive gel of photo cross-linkable chitosan including azide and lactose group and exhibited superior adhesive property than fibrin glue (Kucharska et al. 2010). Similarly, chitosan does possess the required characteristics for ideal contact lenses such as gas permeability, clarity, stability, wettability, and optical corrections (Ishihara et al. 2001).

3.7 Role of Chitosan in Gene Delivery

Popularly chitosan is used as a vector in a nonviral genetic delivery strategy. It strongly binds with deoxyribonucleic acid by electrostatic interaction and is followed by cellular endocytosis without disrupting the chitosan–DNA complex (Grossman and Nwabunma 2013). It also plays a significant role in both membrane attachment and lysosomal degradation and removal of the encapsulated complex for efficient cellular genetic transmission (Bergmann and Stumpf 2013). It was reported to develop a novel hybrid DNA–chitosan complex and fabricated to a nanosphere system. The complex is further conjugated with a ligand for effecting targeting and to accelerate the process of macromolecular endocytosis. During the study, a suitable amount of lysosomotropic agents was incorporated for protection against lysosomes. It reported that the prepared complex resisted serum nuclease and could help encapsulate other biomedical agents without compromising stability on long-term storage and several procedures including lyophilization (Kumar et al. 2004; Tehrani et al. 2012).

The major drawback associated with chitosan-fabricated DNA nanoparticles reported poor solubility in water and low passage across the cell. Few strategies were adopted for improving transfection and aqueous solubility in the case of chitosan and its derivatives at different pH labels. Successfully chitosan-anchored DNA nanoparticles were administered through oral, peroral, topical, and pulmonary routes. In most cases, loaded chitosan nanoparticles were generally prepared by ion gelation technique with genetic material optimized by altering as well as a selection of suitable excipients. In most cases, chitosan-tripolyphosphate (TPP)-impregnated nanoparticles showed promised results.

3.8   Chitosan in Tissue Engineering

In the case of tissue engineering, chitosan has been successfully used as a bio-scaffold for skin and tissue regeneration (Hoemann et al. 2007). Chitosan and its derivatives gradually biodegrade as new tissues formed at the application site. Moreover, it does not develop any kind of inflammatory response and toxicity, making it most ideal for a scaffold in tissue engineering. Likewise, microporous chitosan structures are successfully being employed as scaffolds for bone osteoblasts as well as cutaneous repair and regeneration process (Sarkar et al. 2013; Romanova et al. 2015; Cañas et al. 2016). The three-dimensional arrangement of chitosan and scaffold makes it more similar in relating extracellular matrix components; hence, it provides a stimuli response for regeneration, proliferation, and adhesiveness to tissues (Keong and Halim 2009). Reports already published related to biocompatibility, broad antimicrobial activity, and wound repair property make it more preferable for bone and tissue engineering. The marketed formulations comprising chitosan and fabricated scaffolds for biomedical applications such as sponges, biofilms, topical, and cutaneous gels were successfully developed (Arca and Senel 2008). Similarly, a review reported by Hsieh and coworkers in 2015 reported chitosan applicability in soft and adipose tissue, hepatic target delivery, central nervous system, optical delivery, and blood capillary reengineering (Hsieh et al. 2015).

3.9   Clinical Studies on Chitosan-Based Medical Textile Products for Wound Care

The manufacturing of fabrics made by either natural or synthetic fibers needs a deep study on charge development characteristics. Successfully it studied the prevention of induced static charge on polysaccharide fiber surface by a chitosan or derived chitosan coating (Lim and Hudson 2003). In the medical-related textile field, it has profoundly been used in surgical wound threads, nanocomposite fibers, nonwoven fibers, antimicrobial sponges, and sutures. Lim and Hudson in 2003 published a manuscript on antimicrobial cotton impregnated by derivatized chitosan in wound care (Lim and Hudson 2003). On a severe battlefield, there is always a need for emergency treatment for bleeding. In this connection, HemCon® chitosan-coated hemostatic bandage was developed for the treatment of emergency hemorrhage on the battlefield. Likewise, Celox™ and ChitoGauze® invented by the US military exactly work in a similar way to that of surgical gauze. The underlying mechanism reveals swelling and the formation of a gel-like network on exposure to blood. The unique mechanism says that it does not interfere with the normal clotting mechanism even also best work on heparinized blood (Bennet et al. 2014). Recently, similar to that mentioned above ChitoGauze®, ChitoFlex®, and GuardaCare® are few surgical dressings with antimicrobial property developed by US military services. A comparison was made in a clinical trial between Celox™ and traditional pressure bandage. It showed extreme effectiveness in hemostasis in limb injury by Celox™ (Hatamabadi et al. 2015). Likewise, Chito-Seal™ and Clo-Sur^PLUS PAD are a few topical hemostasis chitosan-fabricated products developed for improving hemostasis in the vascular and percutaneous site. The mechanism reveals that positive charge chitosan binds to negatively charged red blood cell and shortens the clotting time and also hemostasis time. Chitosan is equally effective in binding the platelets and shortens the hemostasis time (Nguyen et al. 2007). Few of the wound dressing marketed products such as Tegasorb™ and Tegaderm™ are effectively being used in the prevention of infection in severe burns, deep wounds, dermal ulcers, and limb cosmetic surgery. Chitosan generally swells and develops three-dimensional gel on the wound surface and promotes the healing process (Weng 2008). Similarly, few more marketed products available for wound care are ChiGel, Chitopack C®, and TraumaStat™.

4 Drug Delivery Systems Utilizing Chitosan Based Sponges

The topic discusses and provides a summary of chitosan along with a sponge-related scaffold in biomedical applications (Murray et al. 2019; Bramhachari et al. 2016). Generally, sponges are the porous structure comprising of all evolutionary organisms of the lowest multicellular animal *Phylum Porifera*, pore bearing species. Generally, they comprise several layers in the body cells having layers of porous channels that allow water and water-soluble substances to circulate through. Ideally, these contain a thin layer of specialized cells that represent a sandwich-like structure consisting of jelly but spongy like mesophyll (internal ground tissue). Unlike other animals, they are asymmetrical (round, cylindrical, and sac-like) with unique two-layer (outer dermal and gastral innermost layer). Most sponges comprise an internal skeleton of silicon dioxide or calcium carbonate which allows the maximum amount of water to flow and leave through the osculum. Sessile aquatic organisms

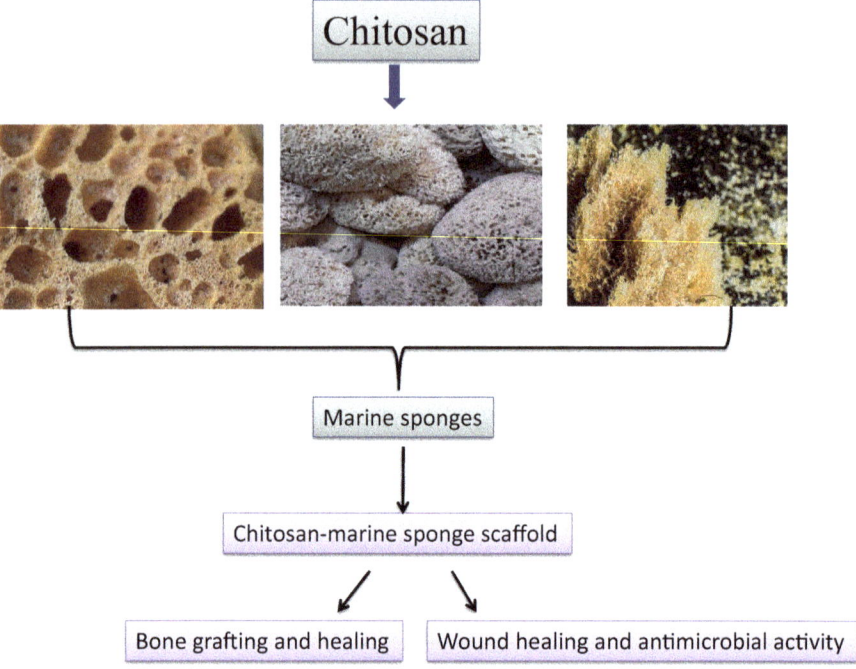

Fig. 3 Chitosan-marine sponge scaffold in biomedical application

including sponges and corals were found under the wall of deep water (Pisani et al. 2015; Blanquer and Uriz 2011). Diverse marine microbial dense sponges which include the yellow tube sponge, *Aplysina fistularis*, the purple vase sponge, *Niphates digitalis*, the red encrusting sponge, *Spiratrella coccinea*, and the gray rope sponge, *Callyspongia* sponge, have attracted the attention of the global community including drug delivery, organ-specific delivery, and regeneration of tissue. Recently, scientists from the global community including oncology have been devoted to the application of sponge in biomaterial study, bone and tissue engineering, antimicrobial study, wound care and regenerative medicine, and advanced methodologies used for the bioprospection of marine microorganisms (Bramhachari et al. 2016). There has been a tremendous discovery of marine-derived sources and their biopharmaceutical applications including chitosan. This study also gave a brief account of the potent antimicrobial bioactivity of chitosan and fabricated sponge (Matica et al. 2019) along with their biomedical applications in drug delivery systems and targeting (Fig. 3).

In biomedical and tissue engineering in bone and cartilage, the scaffold systems by sources from natural origin have gained much more attention from scientists. In this scenario, a tri-scaffold system was successfully prepared by lyophilization and freeze-drying technique. The mentioned scaffold system (Chi-HAp-MSCol) comprises chitosan (Chi), hydroxyapatite (HAp), and marine sponge collagen (MSCol) of *Ircinia fusca*. The system was successfully characterized and tested for significance in bone grafting and healing. During the study, the individual components such as Chi, Chi-HAp, and tri-scaffold were compared with each other (Pallela et al. 2012). The report says, Chi and HAp obtained from marine sponge collagen together are considered for successful novel drug delivery systems including nanoparticles, nanosphere, and microparticulate dosage form. Successfully a nanocomposite system comprised chitosan/gelatin/nanohydroxyapatite (nHA) prepared by Bagheri-Khoulenjani and coworkers in 2013. While preparing the nanosphere, few variables and their effect on nanocomposite were considered such as chitosan to gelatin ratio, gelatin to nanohydroxyapatite ratio, the concentration of chitosan, and stirring rate studied. The result showed the excellent value in particulate size, morphology, and rate of healing in cartilage tissue injury (Bagheri-Khoulenjani et al. 2013). Similarly, a localized drug delivery system for the treatment of severe gum injury was formulated utilizing a suitable combination of chitosan and polyvinyl alcohol (PVA). The emulsification technique (water in oil) was employed for fabricating the core active carboxymethylated chitosan and tested for antimicrobial and healing efficacy in gum disease (Wang et al. 2009). There have been suitable considerations of chitosan–nHA nanocomposite in cell regeneration and proliferation, healing, increased expression of the protein, and cartilage formation (Liu et al. 2013). Similarly, a chitosan with nHA was successfully used in the scaffold system to release prenylated flavonol glycoside (icariin) for bone regeneration and healing (Fan et al. 2012). Venkatesan and coworkers designed a multiwalled carbon nanotube (f-MWCNT) scaffold system utilizing marine sponge-derived natural hydroxyapatite along with chitosan employing the freeze-drying technique. Later on, the multiwalled nanotube was characterized and tested for efficacy in drug delivery related to bone repair and graft substitute (Venkatesan et al. 2011).

In the recent century, the novel drug delivery system utilizes natural origin especially chitosan as an active carrier (Dash et al. 2011; Sonia and Sharma 2011). Earlier stated in the chapter, chitosan along with other bioactive material has profoundly been used in controlled drug delivery devices because of its nontoxic and biocompatible properties (Sharma et al. 2015). Chitosan as an excipient as well as active constituent is used in many pharmaceutical products such as in hydrogel, biodegradable implants, polyelectrolyte complex, and bioconjugates. Similarly, it has successfully been used in protein and peptide delivery, gene therapy, integration with growth factors, vaccine delivery, antimicrobial agent delivery, and bioimaging procedures (Habibi and Lucia 2012; Grossman and Nwabunma 2013; Kumar et al. 2004). Chitosan is a kind of excipient, which is successfully used in almost all routes of drug delivery which includes oral, ocular, topical, and targeted drug delivery (Hagenaars et al. 2009; Verheul 2010; Subbiah et al. 2012). Few reports have already been published on chitosan as a carrier along with bioactive materials including marine-derived sponges in nanospheres, nanogels, biofilms, etc. (Subbiah et al. 2012; Benediktsdóttir et al. 2014). There has been a tremendous survey on the mucoadhesive character of chitosan, as its positive charge can easily undergo ionic interaction with a negatively charged mucus layer. The negative charge to mucus is attributed to sialic and sulfonic acid only (Jayakumar et al. 2010; Abd Elgadir et al.

2015). Few scientists also claimed its penetration enhancement property as well as in nasal drug delivery (Kim 2013; Pardeshi and Belgamwar 2016; Rassu et al. 2016). Several chitosan formulations have been proposed and commercially available for the brain targeting the delivery of drugs (solutions, suspension, gels, microemulsion, and powders). As discussed in this section, chitosan and its derivatives along with marine sponges and other suitable excipients are also effective in dosage design, targeting, and delivering drugs (Anitha et al. 2011a, b, Anitha et al. 2012).

5   Chitosan in Biotechonology

Cytolytic T-lymphocytes are the class of lymphocytes termed as killer T cell which kills cancerous cells as well as the cells infected with viruses. It has been reported that chitosan possesses antiviral property as it can successfully kill infected cells as well as cancerous cells. Hence it has already been reported as successful in various biomedical systems (Snima et al. 2012; Kim and Rajapakse 2005). The unique cationic charges of amino groups of chitosan could possess additional functional property to activate both immune and defense systems. On the lab scale, it was observed that chitosan and its derivatives can stimulate the extracellular lysosomal activity in mammalian cells followed by the formation of connective tissue. It also found chitosan to act as a self-defense mechanism against microbial infection in almost all biological cells. The free radical scavenging activity and antioxidant property of chitosan were evaluated by Klotzbach in 2008. The study demonstrated chitooligosaccharides had potential scavenging activity on lab-induced free radicals in the cellular system (Klotzbach et al. 2008). In biotechnology, there has been a potential role of enzyme immobilization technique (Urrutia et al. 2018). Successfully it has been involved in the immobilization technique to preserve the property of enzyme and protein. The preservation helps in structurally redesigning enzymes, improvement of stability, three-dimensional configuration, the introduction of new functional group, and improvement of selectivity toward macromolecular structure, which is highly essential for biotransformation (Coma 2013).

6   Recent Patents on Chitosan Based Sponges Promoting
Wound Healing

Chitosan-based biofilm was successfully developed and studied for bactericidal efficacy along with tricrotic acid (Hoggarth and Hardy 2015). Villaneuva et al. (2016) successfully invented an edible bioactive film of chitosan with extracted quinoa protein (Villaneuva et al. 2016). Similarly, an invention was made on water-soluble modified chitosan carried out by Michael addition with 2-methacryloyloxyethyl choline phosphate. The antibacterial film coating was further studied and characterized by the antimicrobial property (Jie et al. 2019). Liu et al. (2007) prepared marine-derived chitosan/carboxymethyl chitosan-based rapid hemostatic sponge having long hemostasis time and limited hemostasis capability toward the wounds with a large amount of bleeding during use (Liu et al. 2007). The hemostatic material of the invention has good toughness, can be bent and folded at will, has short hemostasis time, can be used for treating the wounds with a larger amount of bleeding, achieves a more rapid hemostasis effect in clinical application, and can be used for hemostasis in clinical surgeries and promoting wound healing. Yu et al. (2010) and coworkers formulated a water-soluble chitosan-based hemostatic sponge and described its preparation method. This formulation is used for stanching and boosting wound healing of clinical operation (Xiao-Juan et al. 2010). In an attempt made by Arthur and coworkers, a highly dense chitosan structure utilizing simultaneous vacuum and high compression technique was prepared. The technique resulted in denied chitosan membrane which possessed satisfied characterization along with significant clinical activity in mammals (Decarlo et al. 2014). Further, in 2012, patent number CN102526795A has been published. This invention relates to chitosan-based styptic sponge with a thrombin immobilization effect, and the chitosan-based styptic sponge is a porous sponge made from chitosan with the thrombin immobilization effect and hemostatic. The preparation method of the chitosan-based styptic sponge provided by the invention comprises the following steps: immobilizing thrombin with chitosan or carboxymethyl chitosan; adding other styptics, cryoprotectants, and crosslinking agents to prepare the porous styptic sponge, wherein the weight ratio of the chitosan or carboxymethyl chitosan to the thrombin is 100:(0.1–20); and pre-freezing, lyophilizing in a vacuum, casting, cutting, encapsulating, and sterilizing to prepare the chitosan-based styptic sponge. According to the chitosan-based styptic sponge, the thrombin is immobilized through the chitosan or carboxymethyl chitosan, so that the stability and procoagulant activity of the thrombin can be improved, and the property of the prepared chitosan-based styptic sponge is more stable, the procoagulant and wound healing effects are remarkably improved, and the chitosan-based styptic sponge can be widely applied to wound or surgery hemostasis (Guo et al. 2012). Chitosan microfibrils and chitosan-alginate microfibrils with the addition of calcium were used to construct dressing material. The microfibrils are fabricated in a sponge form and successfully evaluated in the wound healing procedure. The prepared microfibrils were evaluated for mechanical strength, uptake of biological fluid, compatibility to the cell, and hemostatic property, which showed that the prepared sponge meets the basic criteria of the physicomechanical and biological activity. This demonstrated the use of chitosan-based microfibril as an effective dressings material for the treatment of wounds (Polish patent application no. P 385031 and P 385032, 2008).

7 Conclusion

The wound is damage and breakdown of cellular components of tissues. Growth and regeneration of cellular moieties initiate the healing process which consists of a series of biological reactions to regenerate and reestablish the cellular components in broken tissue. Therefore, the wound healing process should be carried out in a normal biological way with the help of a drug formulation having zero toxicity. If it is delayed due to abnormal reactions or side effects, it may produce cancer. Such a situation can be controlled by the activation of the caspase-mediated apoptotic program with the application of natural formulations having zero side effects. One of the important natural polymers is marine-sourced chitosan which can be incorporated to produce many formulations to achieve this effect. Therefore, it could be an attempt to develop marine sponge-derived chitosan-incorporated formulations, may stimulate the apoptotic system, as well as impart wound healing actions.

References

Abd Elgadir M, Uddin MS, Ferdosh S et al (2015) Impact of chitosan composites and chitosan nanoparticle composites on various drug delivery systems: a review. J Food Drug Anal 23:619–629

Abdul Khalil HPS, Saurabh CK, Adnan AS et al (2016) A review on chitosan-cellulose blends and nanocellulose reinforced chitosan biocomposites: properties and their applications. Carbohydr Polym 150:216–226

Al-Manhel AJ, Al-Hilphy AR, Niamah AK (2018) Extraction of chitosan, characterisation and its use for water purification. J Saudi Soc Agric Sci 17:186–190

Anitha A, Deepa N, Chennazhi KP et al (2011a) Development of mucoadhesive thiolated chitosan nanoparticles for biomedical applications. Carbohydr Polym 83:66–73

Anitha A, Deepagan VG, Rani VD et al (2011b) Preparation, characterization, in vitro drug release and biological studies of curcumin loaded dextran sulphate–chitosan nanoparticles. Carbohydr Polym 84:1158–1164

Anitha A, Maya S, Deepa N et al (2012) Curcumin-loaded N, O-carboxymethyl chitosan nanoparticles for cancer drug delivery. J Biomater Sci Polym Ed 23:1381–1400

Arca HÇ, Senel S (2008) Chitosan based systems for tissue engineering part II: soft tissues. Fabad J Pharm Sci 33:211

Atay HY (2019) Antibacterial activity of chitosan-based systems. In: Functional chitosan. Springer, Singapore, pp 457–489

Avadi MR, Sadeghi AM, Tahzibi A et al (2004) Diethylmethyl chitosan as an antimicrobial agent: synthesis, characterization and antibacterial effects. Eur Polym J 40:1355–1361

Badawy ME (2010) Structure and antimicrobial activity relationship of quaternary N-alkyl chitosan derivatives against some plant pathogens. J Appl Polym Sci 117:960–969

Badawy ME, Rabea EI (2011) A biopolymer chitosan and its derivatives as promising antimicrobial agents against plant pathogens and their applications in crop protection. Int J Carbohydr Chem 2011:29

Badhe RV, Nanda RK, Chejara DR et al (2015) Microwave-assisted facile synthesis of a new tri-block chitosan conjugate with improved mucoadhesion. Carbohydr Polym 130:213–221

Bagheri-Khoulenjani S, Mirzadeh H, Etrati-Khosroshahi M et al (2013) Particle size modeling and morphology study of chitosan/gelatin/nanohydroxyapatite nanocomposite microspheres for bone tissue engineering. J Biomed Mater Res A 101:1758–1767

Bakshi PS, Selvakumar D, Kadirvelu K et al (2018) Comparative study on antimicrobial activity and biocompatibility of N-selective chitosan derivatives. React Funct Polym 124:149–155

Barbosa MA, Pêgo AP, Amaral IF (2011) Comprehensive biomaterials. In: Ducheyne P (ed) Chitosan, 1st edn. Elsevier, Oxford, pp 221–237

Bardakova KN, Grebenik EA, Istranova EV et al (2018) Reinforced hybrid collagen sponges for tissue engineering. Bull Exp Biol Med 165:142–147

Belalia R, Grelier S, Benaissa M et al (2008) New bioactive biomaterials based on quaternized chitosan. J Agric Food Chem 56:1582–1588

Benediktsdóttir BE, Baldursson Ó, Másson M (2014) Challenges in evaluation of chitosan and trimethylated chitosan (TMC) as mucosal permeation enhancers: from synthesis to in vitro application. J Control Release 173:18–31

Bennet BL, Littlejohn LF, Kheirabadi BS et al (2014) Management of external hemorrhage in tactical combat casualty care: chitosan-based hemostatic gauze dressings—TCCC guidelines-change 13-05. J Spec Oper Med 14:40–57

Bergmann CP, Stumpf A (2013) Biomaterials. In: Bergmann CP (ed) Dental ceramics. Topics in mining, metallurgy and materials engineering. Springer, Berlin, pp 9–13

Blanquer A, Uriz MJ (2011) "Living together apart": the hidden genetic diversity of sponge populations. Mol Biol Evol 28:2435–2438

Bramhachari PV, Ehrlich H, Pallela R (2016) Introduction to the global scenario of marine sponge research. In: Ehrlich H (ed) Marine sponges: Chemicobiological and biomedical applications. Springer, New Delhi, pp 1–23

Campaniello D, Corbo MR (2010) Chitosan: a polysaccharide with antimicrobial action. In: Bevilacqua A, Corbo MR, Sinigaglia M (eds) Application of alternative food-preservation technologies to enhance food safety and stability, 1st edn. Bentham Publisher, Emirate of Sharjah, pp 92–113

Campaniello D, Bevilacqua A, Sinigaglia M et al (2008) Chitosan: antimicrobial activity and potential applications for preserving minimally processed strawberries. Food Microbiol 25 (8):992–1000

Cañas AI, Delgado JP, Gartner C (2016) Biocompatible scaffolds composed of chemically crosslinked chitosan and gelatin for tissue engineering. J Appl Polym Sci 133(33):43814

Chen YL (2008) Preparation and characterization of water-soluble chitosan gel for skin hydration. https://core.ac.uk/download/pdf/11932521.pdf. Accessed 12 July 2020

Cheung RC, Ng TB, Wong JH et al (2015) Chitosan: an update on potential biomedical and pharmaceutical applications. Mar Drugs 13:5156–5186

Chiappisi L, Gradzielski M (2015) Co-assembly in chitosan–surfactant mixtures: thermodynamics, structures, interfacial properties and applications. Adv Colloid Interf Sci 220:92–107

Chvapil M (1977) Collagen sponge: theory and practice of medical applications. J Biomed Mater Res 11:721–741

Chvapil M (1982) Considerations on manufacturing principles of a synthetic burn dressing: a review. J Biomed Mater Res 16:245–263

Coma V (2013) Polysaccharide-based biomaterials with antimicrobial and antioxidant properties. Polímeros 23:287–297

Croisier F, Jérôme C (2013) Chitosan-based biomaterials for tissue engineering. Eur Polym J 49:780–792

Dai T, Tanaka M, Huang YY et al (2011) Chitosan preparations for wounds and burns: antimicrobial and wound-healing effects. Expert Rev Anti-Infe 9:857–879

Dash M, Chiellini F, Ottenbrite RM et al (2011) Chitosan—a versatile semi-synthetic polymer in biomedical applications. Prog Polym Sci 36:981–1014

de Britto D, Celi Goy R, Campana Filho SP et al (2011) Quaternary salts of chitosan: history, antimicrobial features, and prospects. Int J Carbohydr Chem 2011:1–12

Decarlo AA, Ellis A, Dooley TP et al. (2014) Composition, preparation, and use of dense chitosan membrane materials US Patent 20130164311 A1, 27 May 2014

El-Gamal R, Nikolaivits E, Zervakis GI et al (2016) The use of chitosan in protecting wooden artifacts from damage by mold fungi. Electron J Biotechnol 24:70–78

Fan J, Bi L, Wu T et al (2012) A combined chitosan/nano-size hydroxyapatite system for the controlled release of icariin. J Mater Sci Mater Med 23:399–407

Freier T, Koh HS, Kazazian K et al (2005) Controlling cell adhesion and degradation of chitosan films by N-acetylation. Biomaterials 26:5872–5878

Friedman M, Juneja VK (2010) Review of antimicrobial and antioxidative activities of chitosans in food. J Food Prot 73:1737–1761

Fu X, Shen Y, Jiang X et al (2011) Chitosan derivatives with dual-antibacterial functional groups for antimicrobial finishing of cotton fabrics. Carbohydr Polym 85:221–227

Ganguly S (2013) Antimicrobial properties from naturally derived substances useful for food preservation and shelf-life extension—a review. Int J Bioassays 2:929–931

Giri TK, Thakur A, Alexander A (2012) Modified chitosan hydrogels as drug delivery and tissue engineering systems: present status and applications. Acta Pharm Sin B 2:439–449

Goy RC, Britto DD, Assis OB (2009) A review of the antimicrobial activity of chitosan. Polímeros 19:241–247

Grossman RF, Nwabunma D (2013) Biopolymer nanocomposites: processing, properties, and applications. John Wiley & Sons, New York

Guo M-M, Liu H, Jin-Hui P et al (2012) Chitosan-based styptic sponge and preparation method thereof. China patent number CN102526795A, 04 July 2012

Gupta NV, Shivakumar HG (2010) Preparation and characterization of superporous hydrogels as gastroretentive drug delivery system for rosiglitazone maleate. DARU J Pharm Sci 18:200

Habibi Y, Lucia LA (2012) Polysaccharide building blocks: a sustainable approach to the development of renewable biomaterials. John Wiley & Sons, New York

Hagenaars N, Verheul RJ, Mooren I et al (2009) Relationship between structure and adjuvanticity of N, N, N-trimethyl chitosan (TMC) structural variants in a nasal influenza vaccine. J Control Release 140:126–133

Hanssen AD (2005) Local antibiotic delivery vehicles in the treatment of musculoskeletal infection. Clin Orthop Relat Res 437:91–96

Harkins AL, Duri S, Kloth LC et al (2014) Chitosan–cellulose composite for wound dressing material. Part 2. Antimicrobial activity, blood absorption ability, and biocompatibility. J Biomed Mater Res B 102:1199–1206

Hatamabadi HR, Asayesh Zarchi F, Kariman H et al (2015) Celox-coated gauze for the treatment of civilian penetrating trauma: a randomized clinical trial. Trauma Mon 20:e23862

Hoemann CD, Sun J, McKee MD et al (2007) Chitosan–glycerol phosphate/blood implants elicit hyaline cartilage repair integrated with porous subchondral bone in microdrilled rabbit defects. Osteoarthr Cartilage 15:78–89

Hoggarth A, Hardy C (2015) Composition for a wound dressing. US Patent US20180008742A1, 27 Jan 2015

Hsieh FY, Tseng TC, Hsu SH (2015) Self-healing hydrogel for tissue repair in the central nervous system. Neural Regen Res 10:1922

Ishihara M, Ono K, Saito Y et al (2001) Photocrosslinkable chitosan: an effective adhesive with surgical applications. In: International congress series. Elsevier, Amsterdam, pp 251–257

Jardine A, Sayed S (2014) Chitosan as an advanced healthcare material. In: Tiwari A, Nordin AN (eds) Advanced biomaterials and biodevices. Wiley, Hoboken, pp 147–182

Jayakumar R, Menon D, Manzoor K et al (2010) Biomedical applications of chitin and chitosan based nanomaterials—a short review. Carbohydr Polym 82:227–232

Ji C, Annabi N, Khademhosseini A et al (2011) Fabrication of porous chitosan scaffolds for soft tissue engineering using dense gas CO_2. Acta Biomater 7:1653–1664

Jia Z, Xu W (2001) Synthesis and antibacterial activities of quaternary ammonium salt of chitosan. Carbohydr Res 333:1–6

Jie W, Min L, Xingyu C (2019) Water-soluble modified chitosan and preparation method and application thereof. CN110551235A, 12 Sept 2019

Jimtaisong A, Saewan N (2016) Utilization of carboxymethyl chitosan in cosmetics. Int J Cosmet Sci 36:12–21

Keong LC, Halim AS (2009) In vitro models in biocompatibility assessment for biomedical-grade chitosan derivatives in wound management. Int J Mol Sci 10:1300–1313

Khor E, Lim LY (2003) Implantable applications of chitin and chitosan. Biomaterials 24:2339–2349

Kim SK (2013) Chitin and chitosan derivatives: advances in drug discovery and developments. CRC Press, Boca Raton

Kim CH, Choi KS (2002) Synthesis and antibacterial activity of quaternized chitosan derivatives having different methylene spacers. J Ind Eng Chem 8:71–76

Kim SK, Rajapakse N (2005) Enzymatic production and biological activities of chitosan oligosaccharides (COS): a review. Carbohydr Polym 62:357–368

Kim CH, Choi JW, Chun HJ et al (1997) Synthesis of chitosan derivatives with quaternary ammonium salt and their antibacterial activity. Polym Bull 38:387–393

Kim JY, Lee JK, Lee TS (2003) Synthesis of chitooligosaccharide derivative with quaternary ammonium group and its antimicrobial activity against Streptococcus mutans. Int J Biol Macromol 32:23–27

Klotzbach TL, Watt M, Ansari Y et al (2008) Improving the microenvironment for enzyme immobilization at electrodes by hydrophobically modifying chitosan and Nafion® polymers. J Membrane Sci 311:81–88

Kong M, Chen XG, Xing K et al (2010) Antimicrobial properties of chitosan and mode of action: a state of the art review. Int J Food Microbiol 144:51–63

Kucharska M, Ciechańska D, Niekraszewicz A et al (2010) Potential use of chitosan–based materiale in medicine. Prog Chem Appl Chitin Deriv 15:169–176

Kumar MR, Muzzarelli RA, Muzzarelli C et al (2004) Chitosan chemistry and pharmaceutical perspectives. Chem Rev 104:6017–6084

Libio IC, Demori R, Ferrão MF (2016) Films based on neutralized chitosan citrate as innovative composition for cosmetic application. Mater Sci Eng C 67:115–124

Lim SH, Hudson SM (2003) Review of chitosan and its derivatives as antimicrobial agents and their uses as textile chemicals. J Macromol Sci Polymer Rev 43:223–269

Lim YS, Ok YJ, Hwang SY et al (2019) Marine collagen as a promising biomaterial for biomedical applications. Mar Drugs 17:467

Lin Z, Solomon KL, Zhang X et al (2011) In vitro evaluation of natural marine sponge collagen as a scaffold for bone tissue engineering. Int J Biol Sci 7:968

Liu W, Han B, Gu Q (2007) A chitosan based fiber material, preparing method and application thereof. China patent application number WO2008138202A1, 20 Nov 2011

Liu H, Peng H, Wu Y et al (2013) The promotion of bone regeneration by nanofibrous hydroxyapatite/chitosan scaffolds by effects on integrin-BMP/Smad signaling pathway in BMSCs. Biomaterials 34:4404–4417

Logith Kumar R, Keshav Narayan A, Dhivya S et al (2016) A review of chitosan and its derivatives in bone tissue engineering. Carbohydr Polym 151:172–188

Lou CW (2008) Process technology and properties evaluation of a chitosan-coated Tencel/cotton nonwoven fabric as a wound dressing. Fiber Polym 9:286–292

Matica MA, Aachmann FL, Tøndervik A et al (2019) Chitosan as a wound dressing starting material: antimicrobial properties and mode of action. Int J Mol Sci 20:5889

Meyer M (2019) Processing of collagen based biomaterials and the resulting materials properties. Biomed Eng Online 18:24

Murray RZ, West ZE, Cowin AJ et al (2019) Development and use of biomaterials as wound healing therapies. Burns Trauma 7:s41038–s41018

Nakamura R, Katsuno T, Kitamura M et al (2019) Collagen sponge scaffolds containing growth factors for the functional regeneration of tracheal epithelium. J Tissue Eng Regen M 13:835–845

Nguyen N, Hasan S, Caufield L et al (2007) Randomized controlled trial of topical hemostasis pad use for achieving vascular hemostasis following percutaneous coronary intervention. Catheter Cardiovasc Interv 69:801–807

Ozkaynak MU, Atalay-Oral C, Tantekin-Ersolmaz SB et al (2005) Polyurethane films for wound dressing applications. Macromol Symp 228:177–184

Pallela R, Venkatesan J, Janapala VR et al (2012) Biophysicochemical evaluation of chitosan-hydroxyapatite-marine sponge collagen composite for bone tissue engineering. J Biomed Mater Res A 100:486–495

Pardeshi CV, Belgamwar VS (2016) Controlled synthesis of N, N, N-trimethyl chitosan for modulated bioadhesion and nasal membrane permeability. Int J Biol Macromol 82:933–944

Pezeshki-Modaress M, Rajabi-Zeleti S, Zandi M et al (2014) Cell-loaded gelatin/chitosan scaffolds fabricated by salt-leaching/lyophilization for skin tissue engineering: in vitro and in vivo study. J Biomed Mater Res A 102:3908–3917

Pisani D, Pett W, Dohrmann M et al (2015) Genomic data do not support comb jellies as the sister group to all other animals. Proc Natl Acad Sci 112:15402–15407

Raafat D, Von Bargen K, Haas A et al (2008) Insights into the mode of action of chitosan as an antibacterial compound. Appl Environ Microbiol 74:3764–3773

Rabea EI, Badawy ME, Stevens CV et al (2003) Chitosan as antimicrobial agent: applications and mode of action. Biomacromolecules 4:1457–1465

Rahmati M, Blaker JJ, Lyngstadaas SP et al (2020) Designing multigradient biomaterials for skin regeneration. Materials Today Advances 5:100051

Rassu G, Soddu E, Cossu M et al (2016) Particulate formulations based on chitosan for nose-to-brain delivery of drugs. A review. J Drug Deliv Sci Tec 32:77–87

Romanova OA, Grigor'ev TE, Goncharov ME et al (2015) Chitosan as a modifying component of artificial scaffold for human skin tissue engineering. B Exp Biol Med+ 159:557–566

Rúnarsson ÖV, Holappa J, Malainer C et al (2010) Antibacterial activity of N-quaternary chitosan derivatives: synthesis, characterization and structure activity relationship (SAR) investigations. Eur Polym J 46:1251–1267

Sajomsang W, Tantayanon S, Tangpasuthadol V et al (2008) Synthesis of methylated chitosan containing aromatic moieties: Chemoselectivity and effect on molecular weight. Carbohydr Polym 72:740–750

Sajomsang W, Gonil P, Saesoo S (2009) Synthesis and antibacterial activity of methylated N-(4-N, N-dimethylaminocinnamyl) chitosan chloride. Eur Polym J 45:2319–2328

Salehi E, Daraei P, Shamsabadi AA (2016) A review on chitosan-based adsorptive membranes. Carbohydr Polym 152:419–432

Sarkar SD, Farrugia BL, Dargaville TR et al (2013) Chitosan–collagen scaffolds with nano/microfibrous architecture for skin tissue engineering. J Biomed Mater Res A 101:3482–3492

Sashiwa H, Aiba SI (2004) Chemically modified chitin and chitosan as biomaterials. Prog Polym Sci 29:887–908

Sharma RK, Lalita D, Singh AP (2015) Synthesis and characterization of chitosan based graft copolymers for drug release applications. J Chem Pharm Res 7:612–621

Snima KS, Jayakumar R, Unnikrishnan AG et al (2012) O-Carboxymethyl chitosan nanoparticles for metformin delivery to pancreatic cancer cells. Carbohydr Polym 89:1003–1007

Sonia TA, Sharma CP (2011) Chitosan and its derivatives for drug delivery perspective. In: Chitosan for biomaterials I. Springer, Berlin, Heidelberg, pp 23–53

Subbiah R, Ramalingam P, Ramasundaram S et al (2012) N, N, N-Trimethyl chitosan nanoparticles for controlled intranasal delivery of HBV surface antigen. Carbohydr Polym 89:1289–1297

Tehrani MR, Safari S, Jafari S et al (2012) Gene delivery using N-diethylmethyl chitosan to Hela cell line. J Control Release 1:34–38

Tomayko JF, Rex JH, Tenero DM et al (2014) The challenge of antimicrobial resistance: new regulatory tools to support product development. Clin Pharmacol Ther 96:166–168

Tucci MG, Ricotti G, Mattioli-Belmonte M et al (2001) Chitosan and gelatin as engineered dressing for wound repair. J Bioact Compat Pol 16:145–157

Tzaneva D, Simitchiev A, Petkova N et al (2017) Synthesis of carboxymethyl chitosan and its rheological behaviour in pharmaceutical and cosmetic emulsions. J Appl Pharm Sci 7:70–78

Urrutia P, Bernal C, Wilson L et al (2018) Use of chitosan heterofunctionality for enzyme immobilization: β-galactosidase immobilization for galacto-oligosaccharide synthesis. Int J Biol Macromol 116:182–193

Venkatesan J, Qian ZJ, Ryu B et al (2011) Preparation and characterization of carbon nanotube-grafted-chitosan–natural hydroxyapatite composite for bone tissue engineering. Carbohydr Polym 83:569–577

Verheul RJ (2010) Tailorable Trimethyl chitosans as adjuvant for intranasal immunization doctoral dissertation, Utrecht University

Villaneuva CT, James LA, Fuentes NC (2016) Edible bio-active films based on chitosan or a mixture of quinoa protein-chitosan; sheets having chitosan-tripolyphoshate-thymol nanoparticles; production method; bio-packaging comprising same; and use thereof in fresh fruit with a low ph. WO2017132777A1, 01 Feb 2016

Wang LC, Wu H, Chen XG et al (2009) Biological evaluation of a novel chitosan-PVA-based local delivery system for treatment of periodontitis. J Biomed Mater Res A 91:1065–1076

Wang B, Wu X, Li J et al (2016) Thermosensitive behavior and antibacterial activity of cotton fabric modified with a chitosan-poly (N-isopropylacrylamide) interpenetrating polymer network hydrogel. Polymers 8:110

Weng MH (2008) The effect of protective treatment in reducing pressure ulcers for non-invasive ventilation patients. Intensive Crit Care Nurs 24:295–259

Ding Xiao-Juan, Liu Wan-Shun, Peng Yan-Fei et al (2010) A water-soluble chitosan-based hemostatic sponge and its preparation method and application. China Patent Number CN101053669 B, 26 May 2010

Xu T, Xin M, Li M et al (2010) Synthesis, characteristic and antibacterial activity of N, N, N-trimethyl chitosan and its carboxymethyl derivatives. Carbohydr Polym 81:931–936

Xu Y, Asghar S, Yang L et al (2017) Lactoferrin-coated polysaccharide nanoparticles based on chitosan hydrochloride/hyaluronic acid/PEG for treating brain glioma. Carbohydr Polym 157:419–428

Zhou Yu, Sun Zhan, Zhang Yu-Xi, et al (2010) Chitosan/Carboxymethyl chitosan rapid hemostatic sponge and preparation method thereof. China Patent Number CN101927027 A, 29 Dec 2010

Biomedical Applications of Biodegradable Polymers in Wound Care

Sónia P. Miguel, Maximiano P. Ribeiro, and Paula Coutinho

1 Introduction

Until nowadays, the health care professionals continue to concern with the pain/ complications induced by the different skin injuries, namely to the wounds difficult to treat and heal, which can evolve to a chronic state (Frykberg and Banks 2015). The treatment implies to cover the wound site with a dressing that confers protection to the injury against harmful external agents and dehydration. Furthermore, an active dressing can provide a moist environment adequate for stimulating the healing process (Mogoşanu and Grumezescu 2014). Indeed, several biomaterials such as foams, gauzes, hydrocolloids, hydrogels, and films are recommended for wound coverage. These dressings possess unique features such as (i) protect the peri-wound skin, (ii) sustain suitable moisture at the wound and prevent the dehydration, (iii) restrict microbial biofilms, (iv) clean the damaged tissues, (v) reduce the pain, (vi) remove dead and nonviable tissues, and (vii) limit the odors (Mogoşanu and Grumezescu 2014; Song et al. 2018).

Apart from these properties, essential requirements for a wound dressing are the biocompatibility and biodegradability. When an 'ideal' wound dressing is implanted on human body, it must not elicit any immunological response, and it should be degraded at a controlled rate, presenting non-toxic degradation products easily metabolized and eliminated (Mogoşanu and Grumezescu 2014; Song et al. 2018). The wound dressing should act as a transitory supportive matrix for cell migration, adhesion and proliferation until tissue regeneration process is attained. So, the biodegradable materials are preferable because the non-degradable materials require

S. P. Miguel · M. P. Ribeiro · P. Coutinho (✉)
CPIRN-IPG- Center of Potential and Innovation of Natural Resources, Polytechnic Institute of Guarda, Guarda, Portugal

CICS-UBI- Health Sciences Research Centre, University of Beira Interior, Covilhã, Portugal
e-mail: coutinho@ipg.pt

© The Author(s), under exclusive license to Springer Nature Singapore Pte Ltd. 2021   509
P. Kumar, V. Kothari (eds.), *Wound Healing Research*,
https://doi.org/10.1007/978-981-16-2677-7_17

removing/replacement, increasing the pain/discomfort to the patient and risk of infection (Song et al. 2018; Mogoşanu et al. 2012).

Among different types of biodegradable materials, the natural polymers such as polysaccharides (e.g., alginates, chitosan, heparin, hyaluronic acid), proteins (e.g., collagen, gelatin, keratin, silk fibroin) and proteoglycans are broadly used in wound care, due to their biocompatibility, biodegradability and similarity to macromolecules recognized by the human body (Song et al. 2018; Mogoşanu et al. 2012).

Furthermore, synthetic biodegradable polymers like Poly(lactide-co-glycolide) (PLGA), polyglycolic acid (PGA), polylactic acid (PLA), poly-ε-caprolactone (PCL), polyurethane (PU) and polyethylene glycol (PEG) have also been explored in the wound dressings field (Mir et al. 2018).

Despite the excellent biological properties exhibited by natural polymers, they had weak stability and mechanical properties, whereas synthetic polymers present remarkable mechanical performance but fail in the cell interaction. In this way, the strategy most adopted by the researchers passes through the blending of natural and synthetic polymers targeting the attractive properties of both types of biodegradable polymers.

Here, we describe the biodegradable natural and synthetic polymers, presenting their degradation mechanisms and application on wound dressings development and respective biological effects in vitro and/or in vivo.

2 Biodegradable Polymers in Wound Care

2.1 Protein-Based Polymers

2.1.1 Collagen

Collagen remains the major extracellular matrix (ECM) proteic compound that exhibits excellent biological properties for wound healing applications. Collagen is mainly obtained from animal tissues. However, the collagen extracted from fish skin, fins, scales, bones, or swim bladders has recently gained special attention (Venkatesan et al. 2017; Nagai and Suzuki 2000; Zhang et al. 2014; Chandika et al. 2015; Felician et al. 2018).

In general, the collagen has been widely employed in wound dressings production due to their biological features, namely haemostatic, biocompatible, low antigenicity, controlled biodegradability, and ability to stimulate the cell attachment and growth (Song et al. 2018; Chattopadhyay and Raines 2014; Lee et al. 2001).

Furthermore, the collagen dressings are flexible and can absorb high amounts of exudate, acting as a competitive substrate for collagenase, which can reduce enzymatic degradation of tissue. Simultaneously, their low pH minimizes the risk of bacterial colonization (Song et al. 2018; Chattopadhyay and Raines 2014).

The in vitro/in vivo degradation of collagen is essentially performed by enzymes available on the human body like matrix metalloproteinases and collagenases, resulting in amino acids fragments that exhibit chemotactic effect for cells enrolled in the wound healing process (Song et al. 2018; Chattopadhyay and Raines 2014; Lee et al. 2001). In general, the collagen's biodegradability is manipulated by using chemical crosslinker agents (e.g., glutaraldehyde, carbodiimides and succinimides), which promotes the establishment of covalent bonds between amino and carboxyl groups. Collagen can also be blended/mixed with other polymers (natural or synthetic) or modified chemically to achieve the improvement of mechanical properties and biodegradation profile.

In the literature, it is possible to find different collagen-based wound dressings from powders, gels, films, scaffolds to electrospun membranes. Several works reported the biocompatibility of collagen dressings in vitro (Parenteau-Bareil et al. 2010; Craciunescu et al. 2014) and in vivo (Chen et al. 2019; Oancea et al. 2000). Furthermore, Promogran®, Biobrane®, Helitene®, ActiFoam®, and SkinTemp® are examples of collagen-based wound dressings available in the market. In this sub-section, different collagen-based wound dressings reported in the literature will be described and listed in Table 1.

In Werner's work, the collagen-based matrices' performance was evaluated to treat full-thickness wounds and compared with Matriderm® (a bioengineered construct for dermal regeneration composed of bovine collagen coated with α-elastin hydrolysate) (Petersen et al. 2016). The animal assays were performed using minipigs as an animal model due to its anatomical and physiological similarities to human skin. The authors noticed that the developed biomaterials promoted a significant acceleration and improvement of dermal wound repair since the interconnected porous structure of scaffold enables keratinocytes, fibroblasts, and endothelial cells migration.

In turn, Serdar and their co-workers used the electrospinning technique to obtain three-layered doxycycline (DOX) collagen-loaded nanofibrous membranes (Tort et al. 2017a). To accomplish that, the authors produced the first layer, which was composed of coaxial nanofibers containing 1% PCL and 4.5% collagen in the core, while the shell was composed of 2.5% DOX and 2.5% polyethylene oxide. Then, the chitosan (CS) was used to produce the second layer, and the sodium alginate to fabricate the third layer.

Through this layered structure, the authors intend that the first and second layers would contact the wound to reduce/control the inflammation, whereas the collagen–DOX coaxial nanofibers (third layer) will improve the wound healing process. The incorporation of DOX into the shell allowed its rapid release, which is crucial to inhibit matrix metalloproteinase (MMP-2). On the other hand, the core part of coaxial nanofibers composed of collagen will create an appropriate environment to promote cell proliferation in the production of an ECM-like structure at the wound site. The main findings revealed that this layered nanofiber wound dressings presented good bioadhesion, mechanical and wettability properties.

Alternatively, Mehta et al. compared the performance of collagen dressings impregnating silver sulfadiazine with conventional dressings in second-degree burns (Mehta et al. 2019). The authors applied silver-sulphadiazine-impregnated collagen and conventional dressings in 25 patients with similar burn wounds to

Table 1 Description of the several collagen-based wound dressings reported in the literature

Polymeric combinations	Incorporated bioactive compounds	Type of wound dressing	Main findings	Refs.
Collagen	None	Electrochemically deposited collagen wound matrix	– A novel tissue-engineered was successfully produced by deposition of collagen matrix and human adipose-derived stem cells using electrochemical procedure. – The collagen matrix possessed a good physical–chemical and biological properties namely tensile strength, porosity and biocompatibility – The collagen material stimulated the healing and tissue regeneration process, attaining an increase of granulation tissue and epidermal thickness	(Edwards et al. 2018)
Collagen	Ampicillin	Collagen matrix discs	– The collagen concentration influences the matrix stiffness and the material resistance against collagenase action – The collagen matrices can release ampicillin which was crucial to inhibit the bacterial growth of *S. aureus* over 3 days – The collagen matrices revealed high stability without any degradation, and no inflammation response, after 15 days of subcutaneous implantation in rats	(Helary et al. 2015)
Collagen–gelatin	None	Collagen–gelatin fleece	– The biological performance of the collagen–gelatin fleece was evaluated in full-thickness skin injuries, induced on minipigs and compared with Matriderm® – The results of wound closure and histological analysis showed that both treatments improved the skin regeneration process. The collagen–gelatin fleece's interconnected porous structure allowed the keratinocytes, fibroblasts, and endothelial cells migration	(Petersen et al. 2016)

Collagen/PCL PEO	DOX	Coaxial electrospun nanofibers	– The core of nanofibers was composed of PCL and collagen, whereas the shell comprised DOX and PEO – The complete drug release was obtained after 15 min of incubation – The DOX was rapidly released from the outer shell of coaxial nanofibers, and then collagen presented showed its ECM remodelling capacity – The nanofibrous wound dressing was biocompatible	(Tort et al. 2017b)
Sharkskin collagen/PU	None	Films	– The collagen isolated from blue sharkskin (SSCS) is collagen type I – The SSCS/PU dressings promoted wound healing – The dressings promoted the TGF-β and CD31 expression at initial phases of the wound healing process	(Shen et al. 2017)
Collagen/alginate	None	3D structure matrix	– The fibroblasts proliferated on the collagen matrix – The collagen structure was able to control the metalloproteinases expression – The collagen-based wound dressing also stabilized the PDGF-BB in vitro	(Wiegand et al. 2016)
Collagen/oxidized regenerated cellulose (ORC)	Silver	Membrane	– The dressing was applied in chronic leg ulcers of patients – Total reepithelialization was obtained between the 10th and 34th days of treatment – The wounds treated with collagen-based dressings incorporating silver did not present wound infection – The composite dressing allowed fast healing with minimal postoperative pain and bleeding	(Konstantinow et al. 2017)
Collagen	None	Matrix	– The collagen dressing promoted a decrease in wound surface area of 29% – The wound pain was also reduced by 66.66% when treated with collagen matrices	(Ricci and Cutting 2016)

(continued)

Table 1 (continued)

Polymeric combinations	Incorporated bioactive compounds	Type of wound dressing	Main findings	Refs.
Collagen	Silver	Membrane	– The combination of pulsed-current electric therapy with silver-collagen membrane decreased patients' wound surface area with chronic full-thickness wound	(Zhou et al. 2016)
Collagen	Probiotic microorganism (*S. cerevisiae*)	Hydrogel	– The biological effect of collagen hydrogels was characterized on burn wounds in rat, with or without topical use of *S. cerevisiae* – Collagen hydrogel modulated the inflammation, encouraged the epithelialization and reduced the scar formation – The combination with collagen hydrogel and *S. cerevisiae* increased the collagen fibres production	(Oryan et al. 2018)
Collagen	None	Sponge, electrospun membrane	– The electrospun soluble collagen sponge, acid-soluble collagen, bovine collagen electrospun membranes were successfully produced and applied on the animal wounds – All the types of collagen-based wound dressings increased wound contraction and epithelization, promoting the healing, and reducing the inflammatory cells infiltration – The tilapia skin and bovine collagen presented significant bioactivity and accelerated the rat model's wound healing	(Chen et al. 2019)
Collagen	Wheat grass	Aerogel	– The reinforced 3D sponge-like aerogel revealed capacity for water retention, which enables the exchange of nutrients and gaseous compounds – The wheat grass and the collagen augmented the angiogenic ability and in situ production of collagen – The collagen-based aerogel was biocompatible, biodegradable and nonadhesive	(Govindarajan et al. 2017)

Collagen/PVA	Indomethacin	Sponge	– The sponges were successfully obtained through freeze-drying technique – Collagen sponges presented a porous architecture as well as a pore size of 20–200μm – The indomethacin release was compatible with the targeted skin wound healing, namely the control of inflammation mechanism and local pain – The treatment of burns, induced on animals, with the collagen sponges incorporating indomethacin was improved compared to the other groups	(Marin et al. 2018)
Collagen	None	Cream	– The collagen was obtained from bovine tissue and used for the preparation of a cream formulation – The presence of the atelocollagen confirmed the purity of collagen – The collagen-based cream was biocompatible when incubated in contact with fibroblasts cells – The therapeutic formulation also exhibited inhibitory effect against *Bacillus subtilis*, *S. aureus* and *E. coli* – Collagen-based formulation promoted the wound healing process in in vivo rat models	(Udhayakumar et al. 2017)
Collagen	Silver sulphadiazine	Membrane	– The collagen-based dressings incorporating silver sulphadiazine (SIC) were applied on 25 patients with second-degree burn wounds – The patients treated with SIC presented an improved wound healing process after 7 days, whereas the patients treated with conventional dressings required 14 days – The SIC enhanced healing process controlled the infection and reduced the pain	(Mehta et al. 2019)

(continued)

Table 1 (continued)

Polymeric combinations	Incorporated bioactive compounds	Type of wound dressing	Main findings	Refs.
Collagen/PVA	Graphene oxide	Electrospun membrane	– The combination between collagen was obtained from fish bones, PVA and graphene oxide enabled the synthesis of nanofibrous membranes with suitable physicochemical and mechanical properties – The membranes showed biocompatibility for keratinocytes in vitro assays – In vivo assays revealed that the membranes accelerated the wound healing process, by stimulating the cell proliferation	(Senthil et al. 2018)
Collagen/gelatin	Bifunctional peptide: QKCMP	Film	– The polymeric films presented a fibrous micro-structure similar to the native extracellular structure – The collagen–gelatin films modified with QKCMP supported the angiogenesis process – The QKCMP presented high affinity to the collagens, inducing the endothelial cell morphogenesis	(Chan et al. 2015)
Collagen/CS/ Alginate	None	Cushion	– The composite wound dressing was produced by paintcoat and freeze-drying technique – The polymeric cushion presented favourable water absorption capacity and mechanical properties – In vivo assays showed that the composite dressings promoted the healing process, presenting higher values of healing rate (48.49% ± 1.07%) – The expression of EGF, bFGF, TGF-β and CD31 were augmented on the rats treated with composite dressing	(Xie et al. 2018)

Collagen/alginate/HA	AMP Tet213	Sponge	– An antimicrobial peptide (AMP Tet213) was loaded on the surfaces of the collagen-based sponges to provide antimicrobial properties – The wound dressings presented good properties such as biodegradability, mechanical, porosity, swelling capacity – The AMP Tet213-loaded in dressings displayed antimicrobial effect – The in vivo assays evidenced the promotion of wound healing process	(Lin et al. 2019)
Collagen/gelatin	None	Sponge	– The potential of the collagen–gelatin sponges was evaluated in wound closure in a minipig model – The frequent application of sponges improved the wound healing with a better quality of epidermis – The epidermal thickness increased on treated wounds with multiple applications	(Schiefer et al. 2016)
Collagen/CS	Nanoparticles TiO_2	Sponge	– Nano-TiO_2 was introduced into a polymeric scaffold to improve the antibacterial properties – The collagen-based scaffold conferred a moist environment for wound repairing and presented good stability along 4 weeks – Scaffolds displayed an evident inhibitory effect against S. aureus	(Fan et al. 2016)

accomplish such a purpose. The results evidenced that the patients treated with silver-sulphadiazine-impregnated collagen presented an improved wound healing after 7 days of treatment. Contrarily, for the group treated with conventional dressings, the wound healing process was enhanced in just 14 patients. Such evidence reinforces the silver-sulphadiazine-impregnated collagen's ability to treat second-degree burn wounds since it controlled infection, reduced the pain and did not induce any severe complications.

As already mentioned, the most source of collagen used is the bovine and/or porcine animal. These sources can present some limitations associated with immune rejection, viral transmission, and others. In this way, the researchers identified the marine collagen as a new promising viable alternative. For example, Zhang et al. showed that collagen extracted from Tilapia skin presented biocompatibility, and it is absorbed and degraded by tissues (Zhang et al. 2016).

Further, the same team evaluated wound healing properties of tilapia collagen sponges (pepsin soluble collagen sponge (PCS) and acid-soluble collagen sponge (ACS) compared with bovine collagen sponge (BCS) and bovine collagen electrospun (BCE I and BCE II) in in vivo assays using rats full-thickness wounds (Chen et al. 2019). The results revealed improved wound healing in the PCS, ACS, BCE I and BCE II groups, as well as an increase in hydroxyproline and protein contents. Also the fibroblasts proliferation and the synthesis of collagen was enhanced in these groups when compared to control group and the groups treated with BCS and Woundplast. Furthermore, histopathological and immunohistochemical examinations confirmed the positive effect of PCS, ACS, BCE I, BCE II on neovascularization, collagen synthesis, fibroblasts' proliferation, re-epithelialization and restoration of skin appendages. Then, the collagen-based materials' ability to induce the angiogenesis was evaluated through the staining of the vascular endothelial marker CD31 (Fig. 1). The results confirmed that the CD31 expression was significantly encouraged by PCS, ACS, BCE I and BCE II, indicating the vascularization process as fundamental in the progression of the healing.

Regarding the augment of wound dressings' healing potential, researchers have been combined the collagen with other natural polymers, including CS, alginate, gelatin and others. For example, Xie et al. prepared a composite dressing blending collagen, CS and alginate (Xie et al. 2018). The CS–collagen–alginate composite (CCA) produced by paintcoat and lyophilization presented swelling ability, porosity, degradation and mechanical properties proper for wound dressing applications. Moreover, the in vivo assays revealed that CCA composite dressing, gauze and CS healing rates of $48.49 \pm 1.07\%$, $28.02 \pm 6.4\%$ and $38.97 \pm 8.53\%$, respectively, after 5 days, as shown in Fig. 2. Further, the CCA composite dressing exhibited, in the early stage, a stimulatory effect on the epidermal growth factor (EGF), basic fibroblast growth factor (bFGF), transforming growth factor β (TGF-β) and CD31 expression.

In turn, Lin and their collaborators immobilized antimicrobial peptides (AMP) on top of the substrate of alginate, collagen and hyaluronic acid (HA) to develop a new antimicrobial wound dressing (Lin et al. 2019). After that, the chemical crosslinking between compounds was achieved, obtaining the ALG/HA/COL–AMP composite

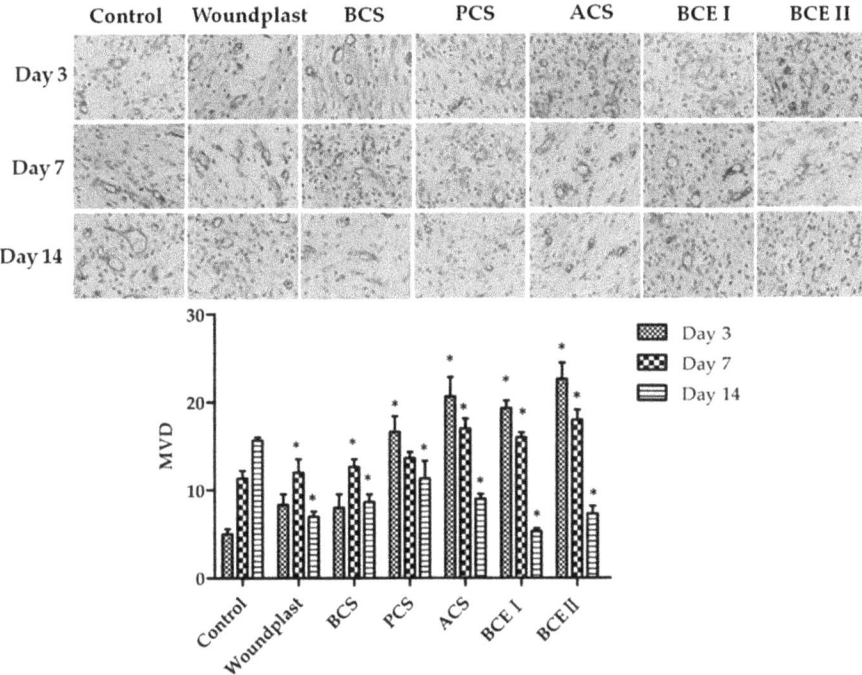

Fig. 1 Immunohistochemical analysis of CD31 expression and the microvessel density (MVD) in wounded tissues. Reprinted from Marine Drugs, vol. 17, Chen et al., Fish Collagen Surgical Compress Repairing Characteristics on Wound Healing Process In vivo, 33, open-access article with permission from MDPI (Chen et al. 2019)

wound dressing. The chemical bonding approach allowed the modulation of the porosity, elongation at break, tensile strength, swelling profile, and biodegradability, as demonstrated in vitro. The weight loss of dressings increased when the incubation time also increases, i.e., the dressings degraded 25%–30%, 55%–60% and 80%–90% after one, two and four weeks, respectively. Furthermore, the AMP-loaded wound dressing promoted a sustained release of AMP, had good biocompatibility, and displayed antimicrobial activity. Finally, the authors also verified that the ALG/HA/COL–AMP wound dressing improved the recovery process of infected full-thickness wounds in animal models, inhibiting or killing bacteria in these infected wounds (*Escherichia coli* and *Staphylococcus aureus*) (as shown in Fig. 3).

2.1.2 Gelatin

Gelatin is derived from collagen through thermal, acid or alkaline treatment. It is broadly used in regenerative medicine due to its abundance, low cost and biological properties, namely biocompatibility, biodegradability and low antigenicity (Mogoşanu and Grumezescu 2014). The gelatin is known as an excellent biological

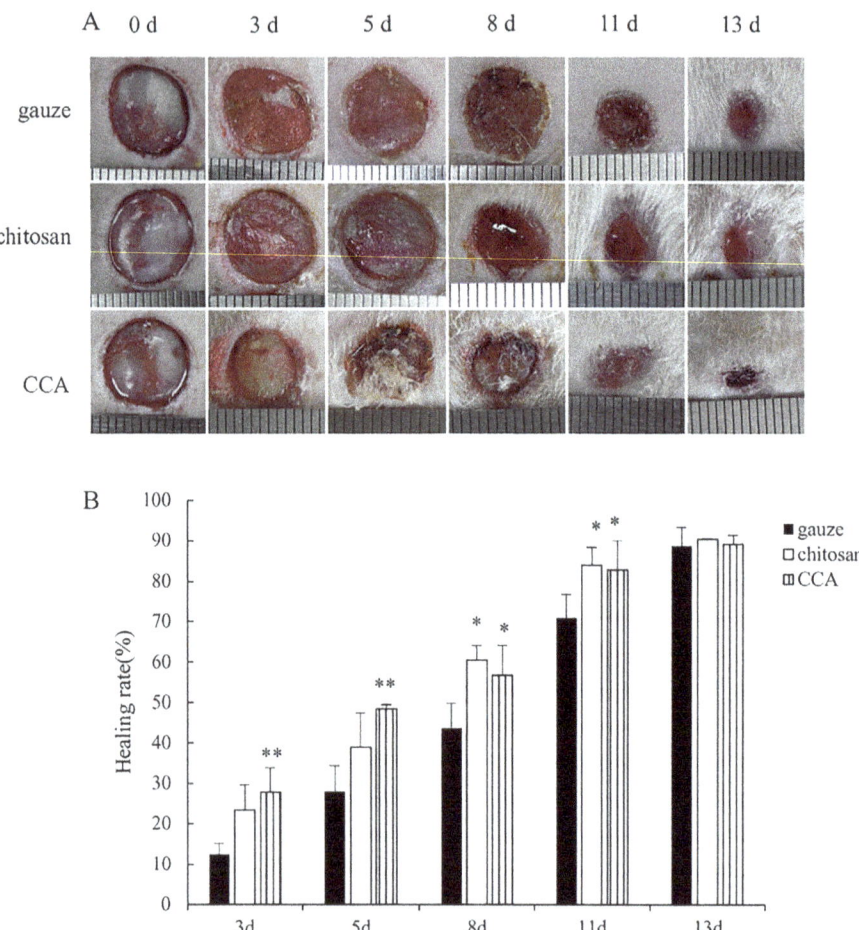

Fig. 2 Evaluation of in vivo performance of the chitosan–collagen–alginate (CCA) composite dressings on the treatment of wounds induced on rats: Macroscopic images of the wound surface at different timepoints (**a**) and the wound healing ratio of the dressings and the gauze control group (**b**). Reprinted from *International Journal of Biological Macromolecules*, vol. 107, Xie et al., Preparation of chitosan-collagen-alginate composite dressing and its promoting effects on wound healing, 93–104, with permission from Elsevier (Xie et al. 2018)

material since it contains adhesive peptide sequences (arginine–glycine–aspartic acid (RGD) domains) recognized by integrin cell receptors, playing a crucial role in cell adhesion mechanism (Ulubayram et al. 2002).

In the wound dressing's field, the more attractive characteristics of gelatin are biodegradability and biocompatibility. It promotes cell adhesion and proliferation and presents reduced immunogenicity and risk of pathogen transmission in relation to the collagen.

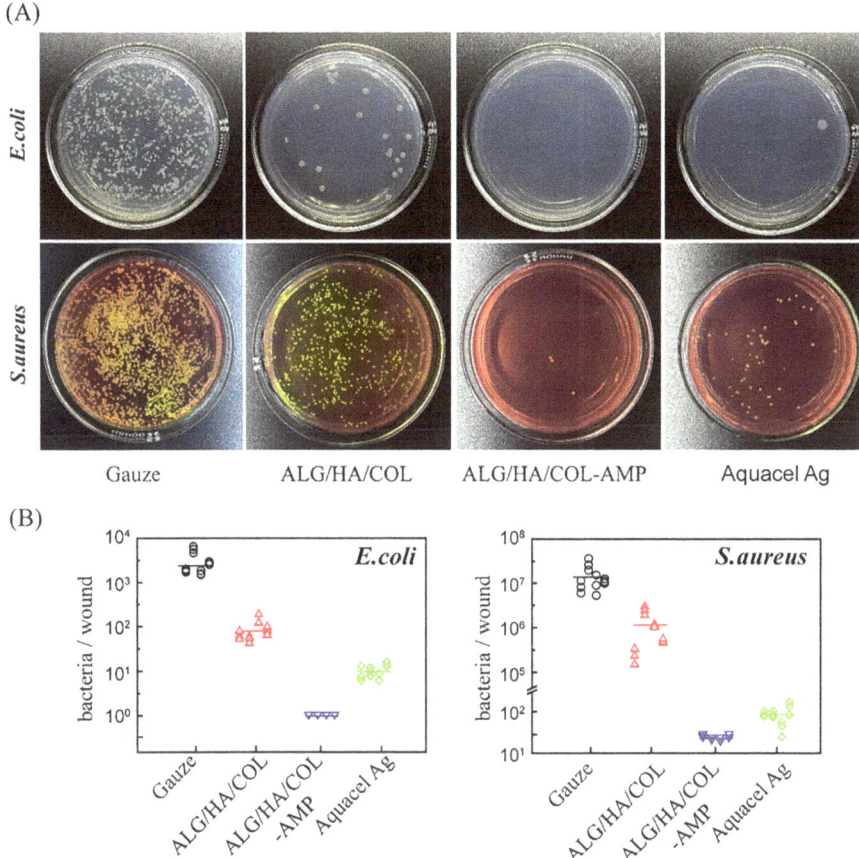

Fig. 3 Determination of the antibacterial properties of the wound dressings: *E. coli* and *S. aureus* colonies obtained from the wound tissues treated with gauze, alginate/hyaluronic acid/collagen (ALG/HA/COL), ALG/HA/COL-antimicrobial peptides (AMP), Aquacel Ag and formed on agar plates (**a**); Quantification of the number of bacteria remaining in wound area at day 4 (**b**). Reprinted from *International Journal of Biological Macromolecules*, vol. 140, Lin et al., Biofunctions of antimicrobial peptide-conjugated alginate/hyaluronic acid/collagen wound dressings promote wound healing of a mixed-bacteria-infected wound, 330–342, with permission from Elsevier (Lin et al. 2019)

However, the weak mechanical properties and water stability of gelatin are the main complications in wound dressings production. For example, gelatin has been chemically functionalized with unsaturated methacrylic and thiol groups, and cross-linked by genipin, glutaraldehyde or periodate-oxidized alginate (Song et al. 2018). Besides, the gelatin can be used in blends with other polymers (natural and synthetic) to augment gelatin-based wound dressings' structural integrity.

Table 2 Description of the most recent work reporting the production and characterization of several gelatin-based wound dressings

Polymeric combinations	Incorporated bioactive compounds	Types of wound dressing	Main findings	Refs.
Gelatin/oxidized starch	*Lawsonia inermis*	Electrospun membrane	– The addition of henna into membrane promoted the adhesion and proliferation of fibroblasts – The presence of henna in nanofibers' composition confers antibacterial activity to the membranes – In vivo assays evidenced the excellent biological properties of the nanofibrous membranes, promoting the re-epithelialization, angiogenesis and collagen production, and avoiding an exuberant inflammatory response	(Hadisi et al. 2018)
Gelatin/cellulose acetate/ nanohydroxyapatite (nHA)	None	Electrospun membrane	– The concentration of nHA influenced the porosity as well as water contact angle, water uptake and water transmission rate – In vitro assays revealed that nanofibers containing 25 mg of nHA had the greatest proliferation ratio – In vivo assays confirmed that this formulation promoted the highest wound closure ($93.5 \pm 1.6\%$), encouraging the collagen synthesis, reepithelialization and neovascularization process	(Samadian et al. 2018)
Gelatin/cellulose acetate	Berberin	Electrospun membrane	– The nanofibrous membranes exhibited good properties for wound dressing applications such as contact angle, porosity, tensile strength, water vapour permeability and water uptake ratio – The membranes presented antibacterial effect front to the Gram-positive and Gram-negative bacteria, without affecting the cells' viability – The in vivo studies on diabetic rats revealed that the membranes improved the healing process	(Samadian et al. 2020)

Gelatin/PVA	Carica papaya	Electrospun membrane	– The nanofibers presented an average diameter between 140 and 160 nm – The nanofibers presented an antimicrobial effect against S. aureus and E. coli, due to incorporation of Carica papaya – The nanofibrous membranes also were haemocompatible and biocompatible	(Ahlawat et al. 2019)
Gelatin	None	Electrospun membrane	– The gelatin nanofibrous membranes were UV crosslinked at different times (5, 10 and 20 min) – The degradation assay demonstrated that the membranes, crosslinked at 5 and 10 min, remained in medium culture for 14 days – All membranes promoted the adhesion and proliferation of keratinocytes – In vitro wound scratch assays demonstrated that the gelatin nanofibers crosslinked after 5 min, induced a rapid cell migration (79% of wound closure after 24 h)	(Beishenaliev et al. 2019)
Gelatin/methoxyl pectin/ Carboxymethyl cellulose	Povidone iodine	Film	– The films were obtained by using glycerin as a plasticizer, while calcium chloride and glutaraldehyde were employed as crosslinking agents – The F-Glu-Ca-G30 film presented a high percentage of elongation at break (32.80%), fluid uptake ability (88.45% at 2 h), and water retention capacity (81.70% at 2 h) – The povidone-iodine was incorporated into films, which enabled the inhibition of microbial growth	(Jantrawut et al. 2019)
Gelatin/salmon	None	Sponge	– The in vivo performance of dressing was evaluated by using pigs as skin wound models – The wound size decreased ≈10% in groups treated with gelatin-based dressings, indicating a faster recovery than the commercial control group	(Acevedo et al. 2019)

(continued)

Table 2 (continued)

Polymeric combinations	Incorporated bioactive compounds	Types of wound dressing	Main findings	Refs.
Gelatin/bacterial cellulose	None	Sponge	– The produced sponges had a high porosity (94–95%), and a great swelling ability (3000–3150%); – The release of ampicillin from sponges enabled to avoid the bacterial growth of *E. coli*, *C. albicans* and *S. aureus* – The sponges did not induce any cytotoxic effect on keratinocytes cells	(Ye et al. 2019)
Gelatin/CS	Nanocrystalline cellulose (NCC) and calcium peroxide particles (CP)	Film	– The combination of NCC and CP improved the mechanical properties of films – The addition of the CP into films' composition conferred antimicrobial properties to the films against *E. coli* – The growth of fibroblasts cells was promoted when they were seeded in contact with films for 7 days	(Akhavan-Kharazian and Izadi-Vasafi 2019)
Gelatin/PCL	None	Electrospun membrane	– The gelatin was chemically modified with methacrylate groups to allow its chemical crosslinking with UV irradiation – PCL/GelMA nanofibers provided a wettability suitable for the bottom layer of an asymmetric membrane – The membrane was haemocompatible and bio-compatible in contact with fibroblast cells	(Alves et al. 2019)
Gelatin/PCL	*Gymnena sylvestre*	Electrospun membrane	– A natural herbal extract was incorporated into gelatin-based electrospun membranes to prevent bacterial colonization – The membranes presented good wettability and retained its mechanical properties when immersed in aqueous solutions for 30 days – The electrospun membranes encouraged the	(Ramalingam et al. 2019)

Material	Additive	Form	Description	Reference
Gelatin/collagen/PCL	None	Sponge	cellular behaviour namely the attachment, spreading and proliferation of fibroblasts – The biocomposite dressing was prepared by impregnation of gelatin/collagen sponges in PCL solutions – The biocomposites presented biocompatibility in the presence of epidermal keratinocytes and dermal fibroblasts – The materials promoted the rapid closure of the skin wounds in nude mice	(Wei et al. 2019)
Gelatin/cellulose acetate	*Zataria multiflora*	Electrospun membrane	– The nanofibers incorporating higher amounts of gelatin promoted an enhanced fibroblasts adhesion and proliferation – The nanofibers impregnated with nanoemulsion promoted the wound healing process, after 22 days	(Farahani et al. 2020)
Gelatin/CS/HA acid	None	Electrospun membrane	– The gelatin/CS and CS/HA nanofibers were simultaneously electrospun – The gelatin/CS remained their fibrous structure, while the CS/HA produced a get state – The nanofibrous scaffold promoted cell adhesion and proliferation and improved the in vivo healing process	(Bazmandeh et al. 2020)
Gelatin/HA	β-Cyclodextrin (βCD)-functionalized graphene oxide (GO)	Hydrogel	– The methacrylate-modified gelatin was mixed with HA graft dopamine to obtain a hydrogel – The GO-βCD provided a photothermal activity to the system – The combination of different compounds on hydrogels' composition resulted in ideal dressing with antibacterial activity and able to promote the collagen deposition and angiogenesis in an in vivo assay	(Huang et al. 2020)
Gelatin/PVA	Silver nanoparticles	Electrospun membrane	– The incorporation of silver nanoparticles into nanofibers conferred antimicrobial properties to	(Amer et al. 2020)

(continued)

Table 2 (continued)

Polymeric combinations	Incorporated bioactive compounds	Types of wound dressing	Main findings	Refs.
			the mats – The electrospun mats accelerated the healing rate when applied as a full-thickness wound dressing	
Gelatin	Exosomes derived from HUVECs	Hydrogel	– The gelatin methcryloyl hydrogel was developed incorporating HUVECs-exosomes to promote fibroblasts and keratinocytes proliferation and migration – When applied in full-thickness cutaneous wounds, the hydrogel repaired the wound defect and achieved the controlled release of exosomes – The in vivo results confirmed that the hydrogels accelerated the re-epithelialization, collagen maturity and angiogenesis	(Zhao et al. 2020b)
Gelatin/PCL	Dopamine; antimicrobial peptide	Nanosheet	– A flexible nanosheet made of two layers was produced: An antimicrobial and haemostatic gelatin/dopamine/antimicrobial peptide layer and another is a mechanical layer composed of PCL – The nanosheet presented excellent mechanical strength and a high platelet adhesion capacity – The nanosheet promoted the reduced clotting time of 4 min and a high bactericidal rate ($\approx$100%) – In vivo assays performed in murine dorsal skin and liver models demonstrated that the nanosheet efficiently controlled the haemorrhaging and exerted antibacterial effect for two weeks	(Xuan et al. 2020)
Gelatin/CS	Tannic acid or bacterial nanocellulose	Film	– The incorporation of nanocellulose and/or tannic acid into films improved the mechanical properties	(Taheri et al. 2020)

Polymer	Additive	Form	Findings	References
Gelatin/CS	Allantoin	Film	of films – In vivo assays on rats revealed that the wound contraction was higher for the wounds treated with films – The biocomposite films displayed antioxidant and anti-inflammatory and antibacterial activities – The blood compatibility tests demonstrated that the films were non-haemolytic – The fibroblasts adhesion and proliferation were noticed in biocomposite films	(Sakthiguru and Sithique 2020)
Gelatin	Zinc oxide/ graphene oxide	Electrospun membrane	– The zinc oxide/graphene oxide nanocomposites were incorporated into gelatin nanofibres to provide antibacterial properties – The decrease of *E. coli* and *S. aureus* viability (<90%) was observed in nanofibers incorporating the nanocomposites – The fibres were degraded entirely within 7 days of incubation	(Li et al. 2020)
Gelatin/sodium alginate	Silver nanoparticles	Hydrogel	– The hydrogel incorporating silver nanoparticles presented a significant bactericidal activity on *P. aeruginosa* and *S. aureus* – The in vitro assays demonstrated that the incorporation of silver nanoparticles did not affect fibroblasts' viability – The in vivo assays confirmed the hydrogels' ability to reduce the wound size, promoting the development and maturation of granulation tissue	(Diniz et al. 2020)
Gelatin/PCL/PU	Propolis extract	Electrospun membrane	– The gelatin/PCL electrospun membranes acted as a sublayer of bilayer scaffold aims to support adhesion and proliferation of cells – The gelatin/PCL nanofibers possessed average fibres' diameter of 237.3 ± 65.1 nm – The bilayer wound dressing presented high hydrophilicity, biodegradability and	(Eskandarinia et al. 2020b)

(continued)

Table 2 (continued)

Polymeric combinations	Incorporated bioactive compounds	Types of wound dressing	Main findings	Refs.
			biocompatibility – In vivo assays in rats' skin wound model showed that the bilayer accelerated the wound closure and collagen deposition	
Gelatin/Konjac	Gold nanoparticles/Gentamicin sulphate	Sponge	– The gold nanoparticles amplified the antibacterial activity, eliminating efficiently the bacteria – The sponges incorporating bioactive compounds were biocompatible in contact with fibroblasts cells – The sponges presented good swelling ratio and mechanical properties, supplying a moist environment at the wound site	(Zou et al. 2020)
Gelatin	Reduced graphene oxide (rGO)	Hydrogel	– The hydrogels were produced by UV crosslinking – The hydrogel presented porosity and swelling capacity to be applied in wound healing – The hydrogel was biocompatible, and the rGO incorporation improved the cell proliferation and migration and the angiogenesis process	(Rehman et al. 2019)
Gelatin/PCL	Clove essential oil	Electrospun membrane	– The addition of the clove essential oil into gelatin-based nanofibers resulted in an increase of the fibre diameter (from 241 ± 96 to 305 ± 82 nm) and wettability – The electrospun membranes loaded with essential oil exhibited antibacterial effect against *S. aureus* and *E. coli*, without comprising human dermal fibroblasts' viability	(Unalan et al. 2019)

Gelatin	Clinoptilolite zeolite; silver ions	Film	– The gelatin-based films presented suitable properties for the desired application – The silver ions were slowly released from the films, which was crucial to avoid the microorganism growth	(Hubner et al. 2020)
Gelatin	Polydopamine-coated carbon nanotubes (CNT-PDA), DOX	Hydrogel	– The gelatin-grafted dopamine (GT-DA), CS and CNT-PDA were used to produce a hydrogel with antibacterial, adhesive, antioxidant and conductive activity – The antibiotic (DOX) allows the use of hydrogel in the treatment of infected full-thickness defect wounds – The polydopamine improved the hydrogel capacity namely the tissue adhesiveness, haemostatic and antioxidant activity – The in vitro and in vivo assays demonstrated that hydrogels were biocompatible and promoted wound closure and collagen deposition	(Liang et al. 2019b)
Gelatin/CS	Cinnamaldehyde	Membrane	– The bacterial inhibition ability of the membranes augmented with increasing cinnamaldehyde content – Membranes also presented blood compatibility, biodegradability and biocompatibility	(Kenawy et al. 2019)
Gelatin	None	Hydrofilm	– The gelatin was crosslinked with lactose to obtain a resistant upper layer which was also non-degradable – The lower layer was obtained by the gelatin crosslinking with citric acid enabling the porosity and swelling ability – The hydrofilm presented excellent biocompatibility in an ex vivo wound healing assay	(Garcia-Orue et al. 2019)

(continued)

Table 2 (continued)

Polymeric combinations	Incorporated bioactive compounds	Types of wound dressing	Main findings	Refs.
Gelatin/alginate	None	Film	– The photocrosslinkable functionalities of gelatin and alginate were combined to produce films – The inclusion of alginate promoted films water uptake capacity – The films with high gelatin amount presented better cell adhesion and mechanical properties	(Stubbe et al. 2019)
Gelatin	Poly([2-(methacryloyloxy)ethyl] trimethylammonium chloride) (PMETAC),	Electrospun membrane	– The gelatin nanofibers were loaded with PMETAC to provide it with antimicrobial properties – After 14 days of incubation, the total weight loss of nanofibers was >90% – The PMETAC nanofibers exhibited great antibacterial activity – The in vitro assays demonstrated the promotion of cell adhesion and proliferation on the surface	(Inal and Mulazimoglu 2019)
Gelatin/PCL	Lawsone	Electrospun membrane	– Coaxial nanofibers were produced aims to prolong the release of lawsone over 20 days – The nanofibers increased cell attachment and proliferation, promoting the expression of TGF-β1 and COL1 – The in vivo assays demonstrated that the nanofibers carried with lawsone augmented the wounds' epithelization after 14 days	(Adeli-Sardou et al. 2019)
Gelatin/alginate/ Hydroxyapatite	Tetracycline hydrochloride	Film	– The alginate and hydroxyapatite content on the films' composition promoted a decreased on swelling ratio and weight loss – The amount of tetracycline released from the films was dependent on the amounts of alginate and hydroxyapatite – The films containing tetracycline presented antibacterial activity	(Türe 2019)

Thus, gelatin has been used to develop the different biomaterials (hydrogels, sponges, hydrofilms, membranes etc.) as listed in Table 2. In the clinic, the gelatin is used as a haemostatic sponge (Hu et al. 2012).

Ramalingam et al. reported the in vitro effect of integration gelatin into the PCL/*Gymnema sylvestre* extract on the physical and biological properties (Ramalingam et al. 2019). The authors produced electrospun nanofibers by combining PCL (8%), gelatin (4%) and natural herbal extracts. The different assays demonstrated that gelatin's inclusion into nanofibers resulted in increased wettability, which encourages fibroblasts and keratinocytes attachment, spreading and proliferation (Fig. 4). Additionally, the initial burst release of extracts from electrospun membranes is crucial to avoid bacterial colonization. Overall, *Gymnema sylvestre* loaded PCL/Gel hybrid mats are suggested to be used as an anti-infective nanofibrous wound dressing.

Posteriorly, Zhao et al., chemically modified gelatin with UV-crosslinkable monomers, yielding to a gelatin methacrylate hydrogel (GelMA) (Zhao et al. 2020b). The authors incorporated into hydrogel isolated exosomes derived from human umbilical vein endothelial cells (HUVECs). The in vitro assays showed the internalization of HUVECs-exosomes by keratinocytes and fibroblast cells, as well as the promotion of cells' proliferation and migration. In turn, in vivo assays demonstrated the controlled release of HUVECs-exosomes from GelMA hydrogel promoted the re-epithelialization, collagen deposition and angiogenesis, which have beneficial effects on the wound healing progression.

More recently, Rehman et al. reported GelMA-based hydrogels' production incorporating different concentrations of reduced graphene oxide (rGO) to be used in the chronic wound care (Rehman et al. 2019). To accomplish such purpose, the authors initially modified gelatin with methacrylate groups and then mixed directly with rGO (at 0.001, 0.002 and 0.004%) and photoinitiator (Irgacure 2959), enabling the production of porous hydrogels (pore size of 50μm) after 10 seconds under UV irradiation. Furthermore, the authors observed a slight decrease on the degradation rate by the incorporation of rGO into GelMA hydrogel, presenting a degradation rate of ≈40% in GelMA_rGO in comparison to GelMA hydrogel, which presented ≈70% of degradation rate, after incubation in phosphate-buffered saline (PBS) solution for 28 days. The in vitro assays with fibroblasts, keratinocytes and endothelial cells evidenced the hydrogels' biocompatibility, highlighting its ability to promote cell migration. The authors also performed the in vivo chicken embryo angiogenesis assay, and they verified that GelMA hydrogel inclosing 0.002% w/w rGO improved the production of a highly branched capillary network compared to the blank GelMA hydrogel (as shown in Fig. 5).

In another study, the gelatin was grafted with dopamine to improve the adhesiveness, haemostatic and antioxidant assets (Liang et al. 2019b). The authors also added polydopamine-coated carbon nanotubes (CNT-PDA) with excellent photothermal effect and antibiotic DOX to treat the infected wounds. All results suggested that CNT-PDA concentration influenced the porosity, conductivity, mechanical and rheological properties, swelling and biodegradability. Moreover, hemo- and biocompatibility of these hydrogels were confirmed by haemolysis and culture of

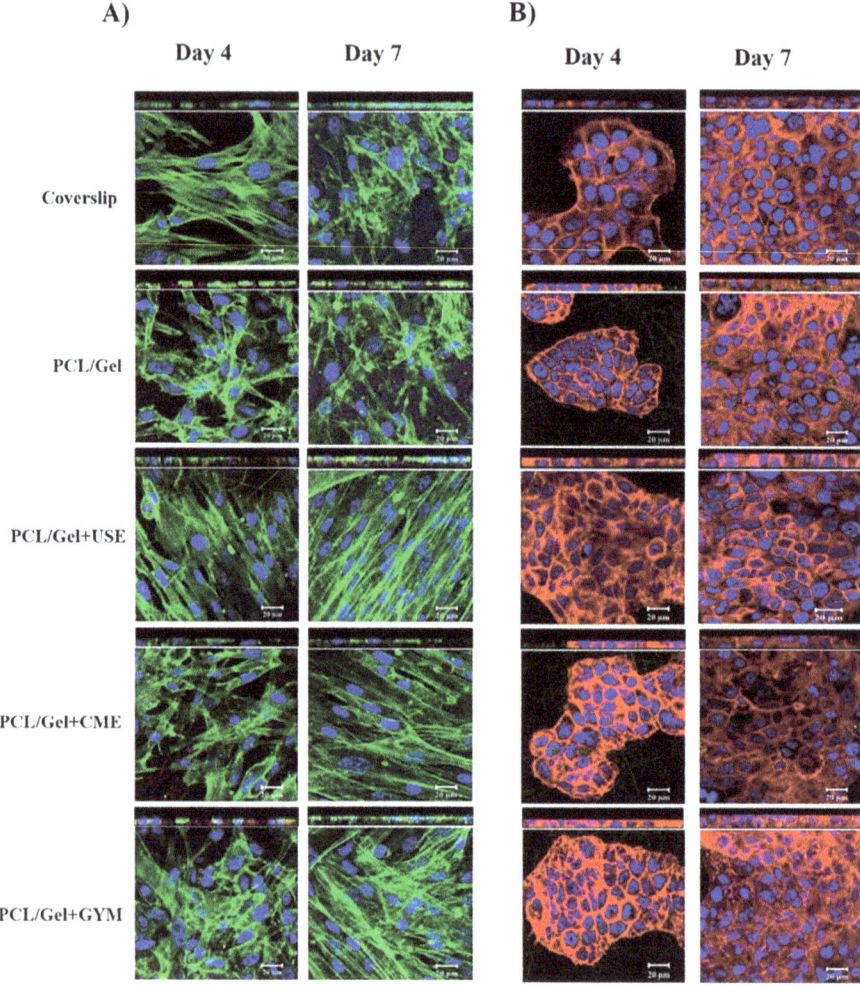

Fig. 4 Fluorescence microscopic images showing that the human primary dermal fibroblasts (**a**) and keratinocytes (**d**) adhered and proliferated at the surface of electrospun mats. F-Actin of fibroblasts and keratinocytes cells was stained green and red, respectively; the nuclei were stained blue. Scale bar corresponds to 20μm. Reprinted from Nanomaterials, vol. 9, Ramalingam et al., Poly-ε-caprolactone/gelatin hybrid electrospun composite nanofibrous mats containing ultrasound-assisted herbal extract: Antimicrobial and cell proliferation study, 462, open-access article with permission from MDPI (Ramalingam et al. 2019)

mouse fibroblast cells. In contrast, the hydrogels' in vivo performance in an infected wound was demonstrated by wound closure rate, collagen metabolism, granulation tissue thickness, epidermis regeneration and immunofluorescence staining of TGF-β3 and CD31.

Further, to obtain a bilayer wound dressing, Garcia-Orue and their co-workers succeeded in a bilayer wound dressing with gelatin and different crosslinkers

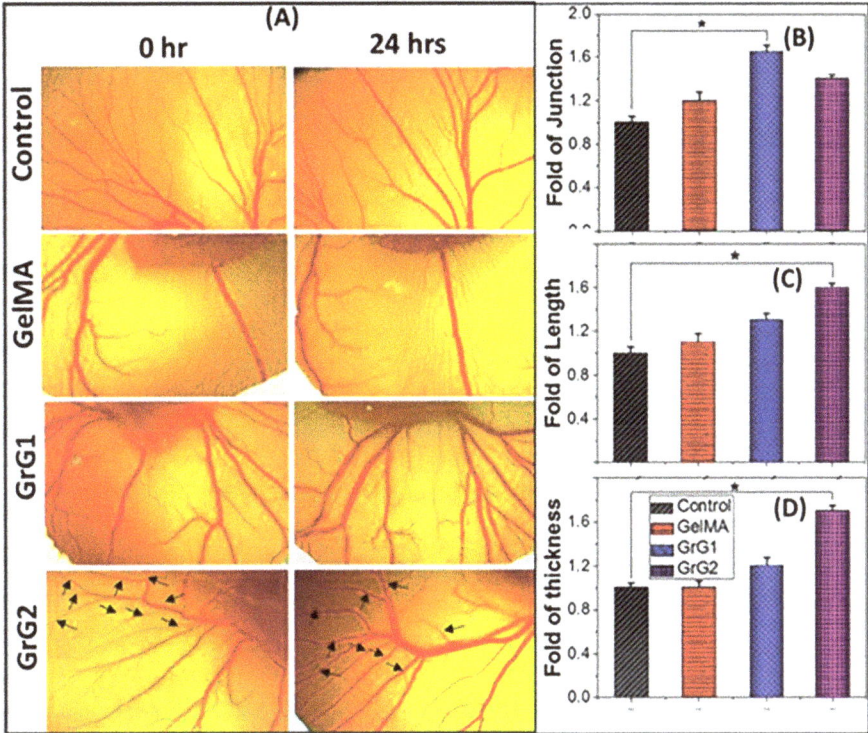

Fig. 5 In vivo Chicken Embryo Angiogenesis (CEO) assay to evaluate the angiogenic potential of the GelMA hydrogels: the formation of the matured blood vessels (marked with black arrows) was increased on group treated with 0.002 wt% rGO nanoparticles (GrG2) (**a**); Quantification of the different angiogenic parameters (blood vessel junction, length and thickness) (**b–d**). Reprinted from *International Journal of Nanomedicine*, vol. 14, Rehman et al., Reduced Graphene Oxide Incorporated GelMA Hydrogel Promotes Angiogenesis For Wound Healing Applications, 9603, open-access article with permission from Dove Medical Press Limited (Rehman et al. 2019)

(Garcia-Orue et al. 2019). To accomplish that, they developed a resistant and non-degradable upper layer by lactose-mediated crosslinking of gelatin and a porous lower layer with a great swelling ability through the crosslinking of gelatin with citric acid. Additionally, to improve its wound healing ability CS was incorporated into the lower layer. Further, the authors evaluated the stability of the hydrofilms, and the results showed that all hydrofilms, when immersed in PBS for 72 h, maintained about 96% of their dry weight. In general, the dressings presented good swelling and occlusivity features, and they did not show cytotoxicity in vitro. The authors also performed an ex vivo assay, where the biological performance of hydrofilms was determined by incubating them with skin explants samples. The results proved the biocompatibility of the dressings and their excellent beneficial effect on the wound healing process.

In turn, Adeli-Sardou et al. evaluated the role of electrospun polycaprolactone-gelatin nanofibres containing lawsone in wound healing (Adeli-Sardou et al. 2019). Initially, the authors incorporated lawsone (at different concentrations: 0.5%, 1% and 1.5%) into PCL/gelatin nanofibers, through coaxial electrospinning. After nanofibers' production, authors confirmed by electronic microscopy the nanofibres core shell structure, bead less morphology, with a mean diameter of ≈200 nm. Then, the determination of nanofibers' weight loss in PBS revealed that the PCL/gelatin mats had the lowest weight loss (≈12%) in comparison to PCL/gelatin/0.5% lawsone (≈20%), PCL/gelatin/1% lawsone (≈ 30%) and PCL/gelatin/1.5% lawsone (≈ 40%) after 14 days of incubation. The coaxial PCL/gelatin nanofibers prolonged the lawsone release for 20 days, which increased the cell attachment and proliferation. The gene expression of TGF-β1, collagen (COL1) and EGF were also quantified in vitro, evidencing that the PCL/gelatin incorporating 0.5% and 1% of lawsone significantly promoted the gene expression of TGF-β1 and COL1. About in vivo assays, the data revealed the PCL/gelatin/1% lawsone have the highest impact on healing, with an increase of wound reepithelization after 14 days.

2.1.3 Silk Fibroin

Silk is a natural protein extracted from the silkworm *Bombyx mori*. Silk fibroin (SF) is sorted out the mulberry silk after removing silk sericin (Altman et al. 2003). It is a semi-crystalline structure that exhibits remarkable mechanical properties, with superior tensile strength, exceptional elasticity and flexibility (Jao et al. 2016).

Furthermore, SF can also be obtained from spiders, such as *Antheraea mylitta*, which produce silks with better cell adhesion and mechanical properties, and reduced solubility in acidic solvents (Jao et al. 2016).

Besides, SF possesses excellent biocompatibility, controllable biodegradability, low immunogenicity and inflammatory potential, and water vapour and oxygen permeability, making them an attractive option for skin regeneration applications (Kundu et al. 2013). More importantly, it has been reported that SF promotes human keratinocytes and fibroblasts adhesion and enhances the type I collagen deposition in vitro.

Likewise, SF possesses a diverse variety of amino acids on its backbone structure that enable the functionalization through the attachment of biomolecules or antibodies (Jao et al. 2016; Kundu et al. 2013).

On the other hand, SF can also be blended with other polymers and crosslinked to achieve stable and biocompatible wound dressings. Finally, SF presents a responsive intrinsic capacity to pH changes, allowing the control of the biodegradation mechanism, which occurs through the proteolytic body enzymes (Jao et al. 2016).

It is easy to modify its structure, which controls the degradability, tensile strength, elasticity and flexibility of SF-based wound dressings. Furthermore, SF enables the introduction of physical crosslinker, haemostatic and self-healing agents and their processing into different constructs, such as films, hydrogels and sponges (as listed in Table 3).

Table 3 List of recent works reporting the production and characterization of silk fibroin-based wound dressings

Polymeric combinations	Incorporated bioactive compounds	Type of wound dressing	Main findings	Refs.
SF/soy protein	None	Electrospun membrane	– The incorporation of soy protein enhanced the hydrophilicity and water retention ability of electrospun membranes – The scaffolds were biocompatible and promoted the full-thickness wounds healing, in rat models	(Varshney et al. 2020)
SF	Zinc oxide nanoparticles	Hydrogel	– The SF hydrogels were coated with zinc oxide nanoparticles to confer antimicrobial properties to the dressings – The fibroblasts cells adhered and proliferated at the surface of SF hydrogels coated with zinc oxide nanoparticles – The hydrogel presented antibacterial effect against *E. coli*	(Majumder et al. 2020)
SF/PCL/ quaternized chitin	None	Nanofiber film	– The SF/PCL nanofibers were obtained through electrospinning technique, and then were functionalized with quaternized chitin by layer-by-layer self-assembly approach – The dressings presented suitable flexibility, tensile strength and antibacterial activity – In vivo assays demonstrated the acceleration of the in vivo vascular reconstruction within 15 days	(Hu et al. 2020)
SF	None	Film and sponge	– The SF films induced higher cell migration in comparison with collagen films – The cells on the surface of SF films exhibited higher gene expression associated with wound repair than on the collagen films	(Hashimoto et al. 2020)

(continued)

Table 3 (continued)

Polymeric combinations	Incorporated bioactive compounds	Type of wound dressing	Main findings	Refs.
			– The ECM-production-related were upregulated in cells seeded on SF sponges	
SF/HA	Zinc oxide nanoparticles	Coaxial electrospun nanofibers	– The core-shell configuration of nanofibers improved the sustained release of the antimicrobial agent (zinc oxide nanoparticles) – The addition of zinc oxide nanoparticles conferred antibacterial activity to the dressings in a dose-dependent fashion – High concentrations of zinc oxide (>3 wt%) induced cytotoxic effects on the cells – The incorporation of zinc oxide nanoparticles significantly reduced the in vivo inflammatory response	(Hadisi et al. 2020)
SF/CS/alginate	Diclofenac sodium	Membrane	– The membranes combined the mechanical properties of SF, the antimicrobial action of CS and the ideal exudate absorption of alginate – The diclofenac sodium was released from the membranes in 7 h when incubated in a simulated solution of wound exudate – The biocompatibility of the membranes was demonstrated in contact with fibroblasts	(Pacheco et al. 2020)
SF	Poly (hexamethylene biguanide) hydrochloride (PHMB)	Sponge	– The PHMB-loaded sponges presented pore sizes of 80–300μm – The PHMB was continuously released from sponges for up to 20 days – The PHMB-loaded sponges promoted the growth inhibition of *E. coli* and *S. aureus*	(Liang et al. 2020)

(continued)

Table 3 (continued)

Polymeric combinations	Incorporated bioactive compounds	Type of wound dressing	Main findings	Refs.
SF/ polydopamine (PDA)	None	Electrospun membrane	– The PDA coating enhanced the wettability and protein adsorption capacity of SF nanofibers – The adhesion as well as cell proliferation were improved by SF nanofibers coated with PDA – The membranes enhanced wound healing in animal model compared to the commercial dressing (3 M™ Tegaderm™)	(Zhang et al. 2019b)
SF/paramylon	None	Film	– Paramylon possessed suitable biological properties that improve the immune response – The films presented high thermal stability and high stiffness values – The SF/paramylon films showed blood and cells' compatibility	(Arthe et al. 2020)
SF	Curcumin	Film	– The films presented a prolonged release of curcumin, flexibility and gas permeability – The prepared films provided a significant protective barrier against bacterial penetration, without inducing cytotoxic effects on cells	(Zhang et al. 2019a)
SF/PDA	None	Film	– The SF was obtained through *Antheraea perny*, which is rich in Arg-Gly-Asp adhesive sequences – The PDA presence induced the roughness and hydrophilicity of films – The histological analysis demonstrates that the PDA-coated SF films minimized the wound inflammation and improved the epithelialization and collagen deposition	(Wang et al. 2019a)

(continued)

Table 3 (continued)

Polymeric combinations	Incorporated bioactive compounds	Type of wound dressing	Main findings	Refs.
SF/CS	Polydopamine nanoparticles (PDA-NPs)	Cryogel	– The cryogel possessed an ECM-like macroporous structure, enabling the cell adhesion and tissue ingrowth – The cryogel presented antioxidative activity during inflammatory responses – The cryogel containing PDA-NPs exhibited photothermally assisted antibacterial activity – The combination of photobiostimulation of infrared light enabled to the cryogel perform the bio-chemo-photothermal synergistic therapy	(Han et al. 2019)
Sf/PVA	None	Electrospun membrane	– The methanol-treated SF/PVA membranes mimicked the structure of the endogenous ECM – The in vivo assays demonstrated that the dressings seeded with epidermal stem cells promoted granulation tissue regeneration on the third day	(Huang et al. 2019)
SF/Konjac glucomannan	None	Sponge	– The blend ratio of SF/konjac glucomannan sponges influenced the pore structure – The konjac glucomannan improved the water absorption and compression strength of sponges – The dermal fibroblast cells adhered and proliferated on sponges	(Feng et al. 2019)
SF	Gelatin microspheres loaded with neurotensin	Membrane	– Neurotensin is an inflammatory modulator in wound healing – The SF-based film promoted the best healing performance in the treatment of wounds on rat diabetic models, promoting the fibroblast proliferation and the formation of collagen fibres	(Liu et al. 2019)

(continued)

Table 3 (continued)

Polymeric combinations	Incorporated bioactive compounds	Type of wound dressing	Main findings	Refs.
			– The gelatin microspheres presented suitable properties to acts as a carrier for controlled release drugs	
SF/HA/alginate	None	Sponge	– The scaffold presented soft and elastic properties with a mean pore diameter of 93μm – The scaffold exhibited good physical stability – The fibroblasts adhered and proliferated on scaffolds – The scaffold showed an improved reepithelialization and ECM remodelling	(Yang et al. 2019)
SF	Silver nanoparticles	Electrospun membrane	– The SF nanofibrous mats were coated by silver nanoparticles in situ using dandelion leaf extract – The silver nanoparticles-coated SF mats showed good mechanical strength, water absorption and adequate porosity – The nanofibrous mat could be used as a therapeutic device, drug delivery vehicle and tissue-engineered constructs	(Srivastava et al. 2019)
SF/ carboxymethyl chitosan (CMCS)	None	Electrospun membrane	– After the production of SF electrospun nanofibers, CMCS was adsorbed at its surface, through electrostatic layer-by-layer self-assembly techniques – The mats presented remarkable hydrophilicity and robust mechanical properties – The modified mats were biocompatible and presented enhanced antibacterial activity	(Tu et al. 2019)
SF/PCL SF/HA	Thymol	Electrospun membrane	– SF was blended with PCL to obtain a dense and protective upper layer of an asymmetric membrane – The combination of SF	(Miguel et al. 2019)

(continued)

Table 3 (continued)

Polymeric combinations	Incorporated bioactive compounds	Type of wound dressing	Main findings	Refs.
			with HA enabled to obtain a porous and bioactive bottom layer that avoids the exudate accumulation and promote cell migration – The electrospun asymmetric wound dressings presented excellent biological properties and structural similarities to the native skin structure	
SF	EGF	Electrospun membrane	– The SF nanofibers were functionalized with EGF aims to improve the biological performance in the wound healing process treatment of chronic wounds – The EGF-functionalized membranes promoted the adhesion of fibroblasts until 2.5-fold – The controlled release of EGF from membranes reduced the wound area in an in vitro wound model until 15%	(Woltje et al. 2018)
SF/alginate	Amniotic fluid	Hydrogel and electrospun fibres	– The electrospun SF fibre was combined with alginate hydrogel loading amniotic fluid to produce a bioactive wound dressing – The amniotic fluid is composed of multiple bioactive molecules that promote the wound healing – An increase in amniotic fluid improved the cell adhesion, proliferation and collagen deposition	(Ghalei et al. 2018)
SF	Recombinant spider silk protein (4RepCT)	Electrospun membrane	– The SF nanofibrous scaffold was coated with 4RepCT to promote the cell binding, growth and confer antimicrobial features – The recombinant spider silk protein was linkage to the SF surface of the scaffold by self-assembly – The functionalized SF	(Chouhan et al. 2018a)

Table 3 (continued)

Polymeric combinations	Incorporated bioactive compounds	Type of wound dressing	Main findings	Refs.
			scaffolds allowed the cocultivation of different cell types	
SF/collagen	Fenugreek	Electrospun membrane	– The SF content improved the fibre diameter and tensile strength of membranes – The nanofibres showed excellent antioxidant properties – The presence of collagen enhanced the migration of fibroblasts	(Selvaraj et al. 2018)
SF/chitin	Silver nanoparticles	Sponge	– The 3D porous scaffolds inhibited the growth of *E. coli*, *S. aureus* and *C. albicans* – The scaffolds were biocompatible and promoted cell attachment	(Mehrabani et al. 2018)
SF/PVA/aloe vera	Vitamin E	Electrospun membrane	– Starch nanoparticles loaded with vitamin E were incorporated into SF/PVA/ aloe vera nanofibres – The VE release was controlled, which improved the antioxidant activity and promoted the wound healing process – The electrospun membranes promoted the fibroblast adhesion, proliferation and collagen secretion	(Kheradvar et al. 2018)
SF/glucose	None	Film	– The addition of glucose to the SF films improved the flexibility and absorption capacity, without affecting the biocompatibility of the dressing – The films were biocompatible and supported the wound closure	(Panico et al. 2019)
SF/PVA	EGF, bFGF, LL-37	Electrospun membrane	– After 14 days of treatment, the SF-based dressings accelerated the wound closure – The dressings promoted	(Chouhan et al. 2018b)

(continued)

Table 3 (continued)

Polymeric combinations	Incorporated bioactive compounds	Type of wound dressing	Main findings	Refs.
			the development of angio-genesis, granulation tissue, and wound reepithelialization – A more organized and resistant ECM were noticed for SF-based dressings	
SF	None	Electrospun membrane	– SF nanomatrix promoted the decrease of wound size effectively and promoted the epithelialization – The expression of collagen in the dermis was aug-mented on wound area cov-ered with SF nanomatrix – The expression of inflam-matory cytokines (IL-1α and TGF-β1) were also con-trolled by SF membranes	(Ju et al. 2016)
SF	None	Film	– The SF film demonstrated to be waterproofness, acting as a barrier against bacterial penetration – The SF film effectively reduced the average wound healing time in rabbit full-thickness skin defect – The biological perfor-mance and long-term safety of SF films were demon-strated in porcine models and clinical trials	(Zhang et al. 2017)
SF	Cys-KR12	Electrospun membrane	– The antimicrobial peptide motif (Cys-KR12) was immobilized on SF electrospun membranes – The functionalized SF membranes exhibited anti-microbial activity – The membranes also supported skin cells prolif-eration, suppressing the expression of the TNF-α	(Song et al. 2016)

To the best of our knowledge, to date only three SF-based medical products have been approved for clinical use: SeriScaffold® (Allergan Medical, Inc.) from the U.S. Food and Drug Administration (FDA), TymPaSil® (CG Bio Inc.) from the Ministry of Food and Drug Safety of South Korea and Sidaiyi® (Suzhou Soho Biomaterial Science and Technology Co., Ltd) from the China Food and Drug Administration (CFDA). Among these, only Sidaiyi® has application on skin wound healing (Zhang et al. 2017).

Regarding the literature, Zhang and their collaborators developed a SF film clinically oriented for skin repair (Zhang et al. 2017). In the first instance, the morphology of SF film was compared with the Suprathel® and Sidaiyi® films. SF films revealed fluid handling capacity, gaseous permeability, good transmittance, waterproofness, as well as biocompatibility and acting as a bacterial barrier. After that, SF films' effect on the full-thickness skin defects healing was assessed in animal models, rabbits and porcine. Concerning the rabbit full-thickness skin defects, the groups treated with the SF films healed after 14 days post-surgery, whereas the groups treated with Suprathel or Sidaiyi healed after 17 and 21 days, respectively. Further, the wounds covered with the SF films presented a faster re-epithelialization, better angiogenesis, and more hair follicles compared to other groups. With respect to the porcine full-thickness skin defects, the long-term bio-compatibility evaluation of the SF films (30 and 90 days post-implantation) has demonstrated that the regenerating skin attained the thickness of normal skin after 30 and 90 days, which revealed improved remodelling than other groups. Notably, the data obtained highlighted the superior healing time performance of SF film compared to other spongy wound dressings. Besides, the SF film promoted a fast and remarkable skin regeneration process without antimicrobial agents' addition. Moreover, the efficacy and safety of the SF film were assessed in a randomized, single-blind parallel controlled clinical study with 71 patients. The results confirmed the SF's excellent biological properties, promoting faster wound healing than Sidaiyi® with a median healing time of (9.86 ± 1.79 vs 11.35 ± 3.03 days). Such evidence of a short healing time is of great importance to reduce hospitalization stay and health care costs.

SF's promising properties on skin defects treatment have been motivating several researchers to develop different types of SF-based biomaterials. For example, Ju et al. developed an SF nanomatrix through the electrospinning technique and then, evaluated their performance in a burn rat model (Ju et al. 2016). To accomplish that, the authors mixed SF with polyethylene oxide and electrospun the solution onto a rotating drum collector, in which sodium chloride crystals were dispensed. The sodium chloride crystals act as porogens that increase the thickness and pore size of SF nanomatrix. After the electrospinning process, SF nanomatrix was immersed in ethanol 100%, for 1 h for SF's re-crystallization, after be drying. In turn, the in vivo assays showed that the SF nanomatrix accelerated the burn wounds healing in rats, which was substantiated by the wound size reduction, collagen, epithelialization and PCNA expression. At the molecular level, the data obtained noticed that the SF nanomatrix suppressed the pro-inflammatory cytokines (IL-1α, IL-6, IL-10 and TGF-β), stimulating the re-epithelialization and reducing the duration of wound healing process and formation of scar.

Furthermore, Han et al. produced a mussel-inspired CS/SF cryogel functionalized with polydopamine nanoparticles (PDA-NPs) responsive to near-infrared light (Han et al. 2019). The cryogel was obtained through the cryo-gelation of CS, SF and PDA-NPs at sub-zero temperatures. CS/SF cryogel combines the excellent biocompatibility and inherent antibacterial properties of CS with high air permeability, breathability, flexibility, tensile strength and SF's moisture retention. In turn, the PDA-NPs promote both cell activity and skin tissue regeneration process, preventing the microbial infection and controlling ROS's accumulation at the wound site.

After the cryogel production, the water absorption ability was characterized by monitoring the adsorbed water mass after immersion in deionized water for 24 h. The results revealed that the PDA-NPs-CS/SF cryogel can maintain a moist environment at wound bed. In this work, the authors also immobilized EGF into cryogels, and they determined that the EGF release profile was gradual along time. Then, the photothermal effect of the PDA-NPs-CS/SF cryogels was confirmed in the antibacterial and antioxidant assays, where the NIR irradiation of cryogels promoted an enhanced antibacterial effect and a decrease on the level of ROS. Similar findings were observed in in vivo assays, where NIR irradiation groups showed improved tissue regeneration and wound healing.

On the other hand, the SF obtained from wild silkworm *Antheraea pernyi* is rich in RGD sequences and was used by Wang et al. to develop a wound dressing (Wang et al. 2019a). Firstly, the authors coated the SF film with polydopamine (PDA)-(PAF) to improve cell adhesion and wound healing. Indeed, roughness and hydrophilicity were increased by the PDA coating (water contact angle values decreased from $61.17 \pm 0.80°$ to $40.30 \pm 1.28°$). In this way, the attachment and spreading of rat marrow mesenchymal stem cells (rMSCs) was improved on PAF films, where it is possible to see the cell lamellipodia and filopodia (Fig. 6a). Furthermore, the cell activity in PAF presence was characterized (as shown in Fig. 6b). The authors also evaluated the films' ability to induce the migration of rMSCs, and the PAF films showed better results (Fig. 6c and d). In turn, the in vivo assays proved that the PAF films improved wound healing in rats, promoting the formation of the hair follicle, new epithelial tissue, collagen deposition and reduced inflammatory cell infiltration.

In turn, Hashimoto et al. investigated the application SF biomaterials as wound dressing, compared with a commercial collagen material (Hashimoto et al. 2020). So, the SF films and sponges were produced, and then fibroblasts were cultured on these SF-based biomaterials. In the SF films, the cell migration assays were done using time-lapse imaging, and the results evidenced that the migration rate of fibroblasts on SF film was more than triple about collagen film and glass surfaces. On the other hand, the authors also characterized the gene expression profiles of MMP3, FGF2, IL-1β, Col3a1 and TGF-β1 for fibroblasts seeded on SF sponges. In general, the results showed that SF sponges could accelerate skin epithelialization, reconstruction, wound repair and minimize scar tissue formation.

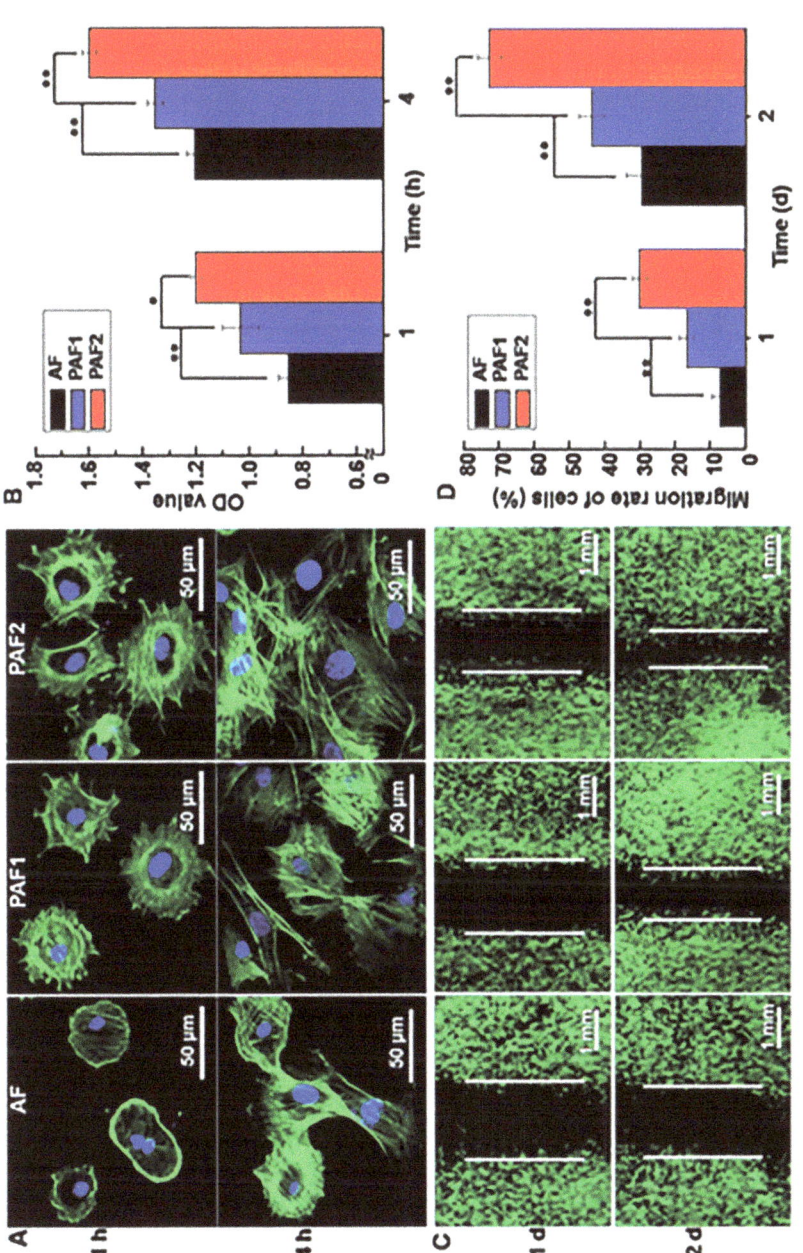

Fig. 6 Evaluation of the in vitro biological properties of the unfunctionalized SF films (AF) and SF films coated with PDA (PAF): Cell adhesion and migration on films after 1 and 4 h of incubation (**a**); Cell activity in contact with films determined by MTS assay (**b**); Cell migration on different films after 1 and 2 days of incubation (**c**); Quantification of the migration rate of cells (**d**). Reprinted from ACS applied materials & interfaces, vol. 11, Wang et al., Polydopamine-Coated Antheraea pernyi (A. pernyi) Silk Fibroin Films Promote Cell Adhesion and Wound Healing in Skin Tissue Repair, 34,736–34,743, with permission from ACS (Wang et al. 2019a). Copyright (2019) American Chemical Society

2.1.4 Keratin

Keratin, a major fibrous protein, constitutes the body's outer coverings such as hair, wool, nail, etc. Recently, keratin due to its outstanding biocompatibility and biodegradability has attracted enough attention. The good mechanical durability of keratin is assured by the presence of disulphide bonds (Rouse and Van Dyke 2010). It also contains various amino acids, namely cystine, lysine, proline and serine, and it can be handily processed into various types of wound dressings (Rouse and Van Dyke 2010).

The degradation of the keratin complex supramolecular organization is mainly accomplished by the synergic action of microbial keratinases and other keratinolytic enzymes. The resulting predominant product is the cysteine, which is readily available to other microbial hydrolytic enzymes (Jin et al. 2017).

Additionally, the keratin-based wound dressings are suitable for accelerating wound healing mainly in chronic wounds due to the interaction with the proteolytic environment at the wound site. Moreover, keratin extracted from hair contains RGD and leucine–aspartic acid–valine (LDV) cell adhesion sequences, noticed also in different ECM proteins namely fibronectin (Hamasaki et al. 2008). This contributes to the role of keratin on the support of cell adhesion and proliferation. Indeed, Yamauchi et al. verified improved fibroblasts adhesion and proliferation in the presence of keratin than type I collagen (Yamauchi et al. 1998).

Table 4 describes different keratin-based biomaterials reported in the literature employed in skin regeneration applications.

Wang et al. extracted through the action of urea, sodium dodecyl sulphate (SDS) and 2-mercaptoethanol, keratin from human hair. The authors then combined keratin with PU and silver nanoparticles (AgNPs) to produce nanofibrous mats (Wang et al. 2016). The presence of keratin on mats' composition improved the hydrophilic character (water contact angle reached $39 \pm 3.5°$ at 30% of keratin), promoting an enhanced cell interaction, adhesion and spreading. The cells cultured on PU/keratin mats displayed cytoplasmic extensions, evidencing the bioadhesive character of the keratin. On the other hand, the PU/keratin/AgNPs mats presented a pronounced antibacterial effect against *S. aureus* and *E. coli*. In turn, the in vivo assays proved that the composite nanofibrous mats did not elicit TNF-α secretion and promoted the skin regeneration process.

Despite the advantages of the keratin, hair keratin possesses fragility and brittleness, limiting its biomedical application. In this way, the researchers overcome such limitation through the blend or crosslink keratin with other materials. Hartrianti et al. produced a sponge of human hair keratin with alginate crosslinked by 1-ethyl-3-dimethylaminopropyl carbodiimide hydrochloride (EDC) (Hartrianti et al. 2017). The chemical interaction between keratin and alginate was promoted by EDC, which resulted in amide bonds between carboxylic and amine groups found on compounds. After preparing the alginate–EDC–keratin mixture, it was subjected to a freeze-drying cycle to attain a sponge. The mechanical assays revealed a higher value of compression and tensile modulus for the crosslinked keratin–alginate sponges. The

Table 4 Description of the most recent works reporting the use of keratin in the production of wound dressings

Polymeric combinations	Incorporated bioactive compounds	Types of wound dressing	Main findings	Refs.
Keratin	Halofuginone	3D printed hydrogel	– The 3D printed keratin hydrogels displayed water uptake ability and promoted the release of Halofuginone – The incorporation of Halofuginone induced a more organized dermal healing after burn, improving the healing process – The 3D keratin scaffolds promoted the healing process, reducing the scar tissue formation	(Navarro et al. 2020)
Keratin	Silver nanoparticles	Powder	– The keratin-derived powder was coated with silver nanoparticles aimed to be used as a wound dressing in a diabetic full-thickness skin – In vitro tests revealed that the dressing did not induce an inhibitory effect on fibroblast growth and haemolysis – The keratin-derived dressings accelerated wound closure and epithelization when compared with controls groups	(Konop et al. 2020)
Keratin/PAM	None	Sponge	– An expandable keratin sponge was obtained due to the intrinsic properties of the keratin and PAM – The sponge exhibited adequate haemostasis on rat penetrating liver haemorrhage – In vivo tests indicated the keratin-based sponge's effectiveness for haemostasis in a swine femoral artery transection haemorrhage model	(Wang et al. 2019b)
Keratin/bacterial cellulose	Hydrogel nanoparticles	Electrospun membrane	– The hydrogel particles (500 nm–2μm) were uniformly embedded into a fibrous network, without affecting its porous structure and the fibre diameter – The hydrogel nanoparticles improved the hydrophilicity, elasticity module, tensile strength and ductility – The fibrous composite supported the fibroblasts' adhesion and proliferation	(Azarniya et al. 2019)

(continued)

Table 4 (continued)

Polymeric combinations	Incorporated bioactive compounds	Types of wound dressing	Main findings	Refs.
Keratin	Diclofenac loaded hydrotalcites (HTD)	Electrospun membrane	– The incorporation of nanosized HTD into keratin nanofibres induced a decreased swelling ratio and a slower degradation profile – The cell viability assays confirmed the biocompatibility of keratin/HTD electrospun mats – A controlled diclofenac release was obtained within the first 24 h of incubation	(Giuri et al. 2019)
Keratin	None	Membrane	– The keratin biomaterial was used in a full-thickness surgical skin wound model – In vitro assays demonstrated that the membrane is biocompatible for murine fibroblasts and inhibited the bacterial growth – The dressings accelerated healing wound healing in mice with iatrogenically induced diabetes, no inducing any inflammatory response signs	(Konop et al. 2018)
Keratin/CS	None	Membrane	– The composite membrane combined the keratin and chitosan properties such as biodegradability, biocompatibility, improved cell adhesion and proliferation – The CS-azide/keratin combination enabled to produce UV-crosslinked membranes – The biocompatibility and biodegradability of membranes demonstrated the mild inflammation in the experiment of subcutaneous implantation	(Lin et al. 2018)
Keratin/CS	ZnO	Bandage	– The nanocomposite presented an increased swelling and bactericidal activity – In vivo assessments evidenced that the nanocomposite bandages improved the wound healing process through increasing of skin cell remodelling and collagen production	(Zhai et al. 2018)

Keratin/konjac glucomannan	Extract of *Avena sativa*	Hydrogel	– The porous hydrogels presented a remarkable swelling ability, when immersed in PBS solution – The hydrogels were biocompatible, promoting the cell attachment and infiltration – The in vivo assays disclosed the promotion of wound healing process in a diabetic rat excision wound model	(Veerasubramanian et al. 2018)
Keratin/gelatin_PU	None	Electrospun membrane	– A bilayer membrane comprised an outer layer (PU) and an inner layer (keratin/gelatin) was produced – The keratin/gelatin composite nanofibrous mats promoted cell adhesion and proliferation – The bilayer membrane promoted an enhanced angiogenesis and significant reduction in wound closure rate, at 4 days than other groups (gauze and Comfeel ®)	(Yao et al. 2017)
Keratin/alginate Keratin/agar Keratin/gellan	Silver nanoparticles Papain Glucose oxidase Trolox ® bFGF	Patch	– The patches with green synthesized silver nanoparticles displayed antimicrobial activity – The patches presented suitable mechanical properties, porosity and water absorption capacity to the desired application – The patches displayed the promising potential to act as a drug delivery system	(Nayak and Gupta 2017)
Keratin	None	Membrane	– The keratin-based dressing presented a high surface porosity appropriated for cell migration and proliferation – In vivo studies showed that keratin was biocompatible and accelerated wound closure and epithelialization on day 5 – Keratin dressings also enhanced the cosmetic effect (e.g., scar formation and appearance)	(Konop et al. 2017)
Keratin	Ciprofloxacin	Hydrogel	– The keratin hydrogels inhibited the growth of *P. aeruginosa* and *S. aureus*, in a porcine model of an infected partial-thickness burn, by >99% – The keratin hydrogels presented promising properties to deliver antibiotics and sustain the healing of partial-thickness burns	(Roy et al. 2016)

(continued)

Table 4 (continued)

Polymeric combinations	Incorporated bioactive compounds	Types of wound dressing	Main findings	Refs.
Keratin/alginate	None	Sponge	– The sponges displayed an increased tensile strength and compression modulus when the alginate content increased – The keratin content decreased the swelling abilities and increased the degradation rates by proteinase K – The sponges promoted the adhesion and proliferation of cells, as well as supported the in vivo cellular infiltration, neo-tissue formation and vascularization	(Hartrianti et al. 2017)
Keratin/fibrin/gelatin	Mupirocin	Sponge	– The sponges exhibited an interconnected porous structure with suitable mechanical properties – The biocompatibility of scaffolds were demonstrated by the encouraged adhesion and proliferation of fibroblasts and keratinocytes	(Singaravelu et al. 2016)
Keratin	None	Hydrogel	– The keratin-based hydrogels were formulated by electron beam irradiation and presented adequate tensile strength and elongation percentage values – After 7 days and 14 days of excision wound induction, the wounds treated with hydrogels presented a reduced wound size – The wound treated with keratin-based hydrogel presented a total epithelial regeneration	(Park et al. 2015)
Keratin/PU	Silver nanoparticles	Electrospun membrane	– The keratin obtained from human hair was modified with iodoacetic acid to afford S-(carboxymethyl) keratin – The silver nanoparticles were produced in situ to confer antibacterial properties to the electrospun mats – In vitro and in vivo assays showed the improvement of the fibroblast cell proliferation by keratin, accelerating the wound recovery in comparison to the conventional dressings based on gauze sponges	(Wang et al. 2016)

enzymatic degradation assays were also done by incubating the sponges with proteinase K and chymotrypsin for 30 days. The data obtained revealed that the sponges with the highest keratin content presented a higher weight loss of their original weight (74.5 ± 4.5%) in comparison to the sponges with highest alginate content (17.5 ± 3.7%) when incubated with proteinase K. Unlike, when the sponges were immersed in contact with chymotrypsin, the keratin-alginate sponges were more resistant. Moreover, in vitro and in vivo assays highlighted the high keratin content in sponges promoted the adhesion and proliferation of fibroblasts, controlled the immunological reaction and supported the cellular infiltration, neotissue formation and neovascularization.

In turn, Lin et al. developed a composite membrane composed of keratin and CS via UV irradiation (Lin et al. 2018). To perform that, the keratin was prepared from human hair, mixed with CS-azide and exposed for 15 min to UV radiation. The SEM analysis noticed that the samples had a compact structure, presenting great structural integrity. On the other hand, the composite membranes with more keratin amount presented a lower water contact angles, i.e., keratin: CS at 0.25:1, 0.5:1 and 1:1 ratios had water contact angles values of 101 ± 3°, 95 ± 4° and 86 ± 3°, respectively. Afterwards, the composite membranes' ability to induce cell migration was evaluated, and the results revealed that the keratin had a crucial role in cell migration (as shown in Fig. 7). Then, the subcutaneous implantation of membranes in animals highlighted the biocompatibility and biodegradability of the wound dressings.

More recently, Konop and their co-workers evaluated the role of insoluble fur (from mouse) keratin-derived powder loading silver nanoparticles (Konop et al. 2020). After preparing keratin-derived wound dressing, it was coated with the colloidal silver suspension by soaking for 60 min. The keratin dressings containing AgNPs exerted excellent antimicrobial properties against *E. coli* and *S. aureus*. Concerning the in vivo biological evaluation, the dressings promoted a faster epithelialization compared to the control wounds, encouraging the epithelial cellular migration and proliferation. Besides, the keratin dressings provided a suitable moist environment for wound healing mechanisms.

In turn, Navarro et al. produced a 3D keratin-based construct by UV crosslinking in a lithography-based printer (Navarro et al. 2020). In the first instance, the keratin was obtained from human hair and then combined with a photosensitive initiator-catalyst-inhibitor (riboflavin-SPS-hydroquinone) solution 0.001% wt/vol hydroquinone (the small molecule that reduces the scarring of severe burn wounds). After the printing process with the lithography-based printer, the keratin hydrogels were exposed to the UV radiation. The in vivo efficacy of Halofuginone-loaded hydrogels on dermal wound healing using a porcine burn model showed that the printed hydrogels provided a low inflammation environment, low late-stage vascular proliferation, with the improvement in collagen deposition over 70 days. Further, keratin and Halofuginone's combination displayed significant healing enhancement from 30 to 70 day.

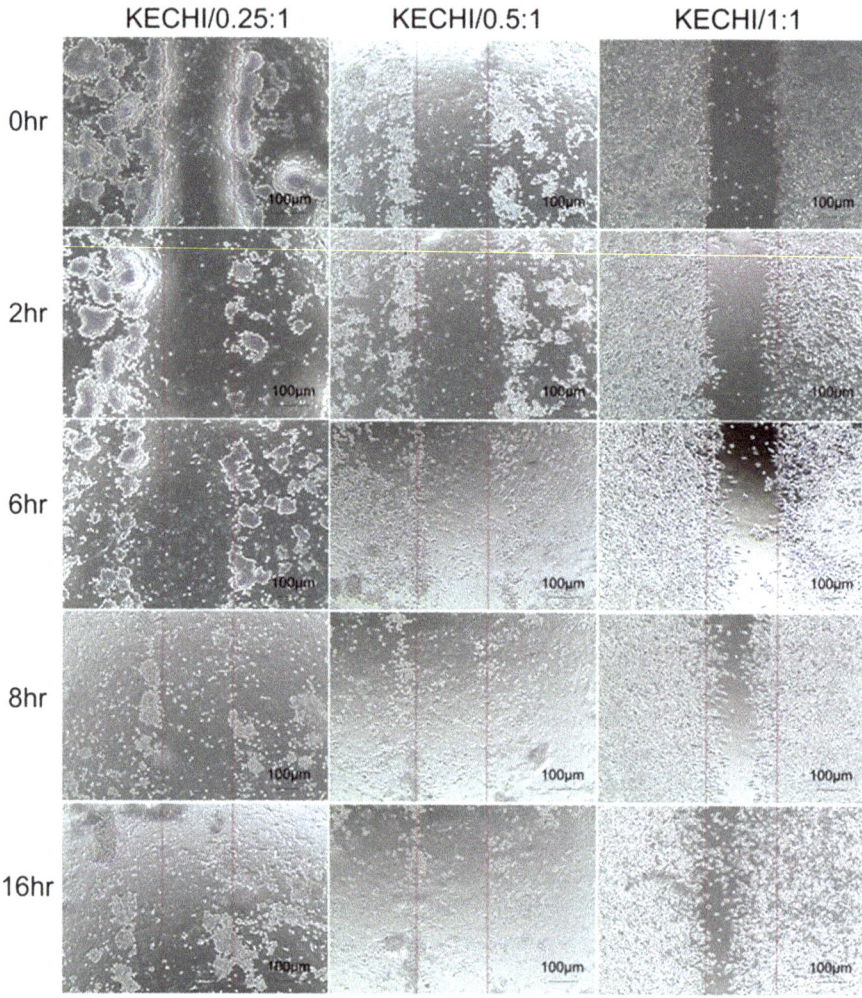

Fig. 7 Evaluation of keratin-chitosan membranes' ability to induce the migration of the fibroblast cells. The microscopic images were acquired at 0 h, 2 h, 6 h, 8 h, and 16 h after removing the culture-insert well mould. Reprinted from Polymers, vol. 10, Lin et al., Photo-crosslinked keratin/ chitosan membranes as potential wound dressing materials, 987, open-access article with permission from MDPI (Lin et al. 2018)

2.2 Naturally-Derived Polysaccharides

Apart from the natural biodegradable proteins, the natural polysaccharides are also considered an attractive therapeutic option for wound dressings development. Among different natural polysaccharides described in the literature, this section will only focus on the natural polysaccharides-based wound dressings most reported in the literature, namely HA, CS and alginate.

2.2.1 Hyaluronic Acid

Hyaluronic acid (HA) is a natural polymer included in heteropolysaccharides group known as glycosaminoglycans (GAGs), and can be found in the human vitreous humour, joints, rooster comb, umbilical cord, skin and connective tissues. HA is comprised of repeated disaccharide units of β-d-glucuronic acid and N-acetyl-d-glucosamine, alternately linked by β-1,3 and β-1,4 glycosidic bonds (Graça et al. 2020). HA possesses unique viscoelastic properties, good biocompatibility and biodegradability, making it a promising material for regenerative medicine. In wound healing applications, HA offers a high potential due to its high-water uptake capacity, preventing the wound dehydration, providing moist environment and promoting the healing process (Voigt and Driver 2012). HA is rapidly metabolized by hyaluronidases, with half-lives going from hours to days.

At initial phases of wound healing, HA acts a transitory matrix that helps the nutrients supply and remove waste products from wounds. Further, HA promotes the keratinocytes proliferation and migration. Also, HA induces the activity of endothelial cells and macrophages, which can contribute to blood vessels formation and regulate the collagen fibres production (Voigt and Driver 2012). HA can interact with endothelial cell receptors (CD44), promoting cell proliferation and angiogenesis, enhancing collagen deposition, and increasing the re-epithelialization in skin regeneration (Seol et al. 2018; Silva et al. 2016).

However, the weak mechanical properties and biodegradation profile of HA can be modulated through the blending of other polymers or producing HA chemical derivatives. Many carboxyl and hydroxyl groups, available within the HA structure, enable easy chemical modification and chemical crosslinking (Graça et al. 2020).

Different HA-based wound dressings were already developed from the clinical perspective, such as HylaSponge® System, Hyalomatrix® and Hyalosafe®, and others (Graça et al. 2020). However, the researchers have been developing HA-based wound dressings to overcome shortcomings of commercial HA-based wound dressings (e.g., high production costs and low mechanical stability). Table 5 described the more recent works dedicated to the production and characterization of HA-based wound dressings.

Wu et al. produced a hydrogel of gelatin and HA crosslinked by EDC (Wu et al. 2017). Through this strategy, the composite gelatin and HA hydrogel will meet both components' excellent biological properties. The crosslinked hydrogels presented a porosity between 40% and 70% with a pore size of 100–400μm. Further, in vitro assays confirmed hydrogels biocompatibility and ability to induce fibroblasts migration, while in vivo assays indicated that the hydrogels promoted the wound healing process on animal wound models.

Hong et al. produced two kinds of HA-based hydrogels: one is made through physical crosslinking of HA solution (HA1) and other by chemical crosslinking of HA (HA2) (Hong et al. 2018). The authors then evaluated the healing effect of HA hydrogels in the repair of full-thickness skin defects on rabbits and they noticed that the HA2 hydrogel presented a great promotion in wound reduction and healing

Table 5 Description of the most recent work reporting the production and characterization of several hyaluronic acid-based wound dressings

Polymeric combinations	Incorporated bioactive compounds	Types of wound dressing	Main findings	Refs.
HA/ε-polylysine (EPL)	None	Hydrogel	– The sol-gel transition occurred due to horse-radish peroxidase enzymatic cross-linking and Schiff base reaction – The hydrogel promoted the death of bacteria on the surface of wounds effectively – The rats treated with hydrogel dressings presented high thickness of the newborn skin and density of granulation tissue	(Liu et al. 2020)
HA/ Polygalacturonic acid	Silver nanoparticles	Electrospun membrane	– The nanofibrous membrane-embedded silver nanoparticles displayed antimicrobial properties both Gram-positive and Gram-negative bacteria – The in vivo assay in albino rat revealed that wound epithelization and collagen fibres formation reached maximum values after 14 days of treatment with electrospun membranes	(El-Aassar et al. 2020)
HA/pullulan	Curcumin	Film	– The films containing curcumin had a higher swelling ratio, enhanced cell proliferation and exhibited bactericidal activity – The materials also showed antioxidant activity – The films promoted an effective wound healing process	(Duan et al. 2020)
HA/CS	None	Film	– The HA incorporation into CS matrix reduced the film transparency and homogeneity – The HA presence	(Silvestro et al. 2020)

(continued)

Table 5 (continued)

Polymeric combinations	Incorporated bioactive compounds	Types of wound dressing	Main findings	Refs.
			improved the swelling and wettability of the film – The concentration of HA $\geq$ 5% avoided the adhesion of *S. epidermis*	
HA/lysozyme	None	Gel	– The gel presented suitable viscoelasticity and excellent adhesion to the skin – The gel promoted a faster epithelial tissue regeneration and higher collagen deposition than the commercial 3 M dressing	(Zhao et al. 2020a)
HA/fibrin	None	Membrane	– The fibrin-HA biomaterial was grafted in immunodeficient mice for 8 weeks and compared to the autograft, a fibroin-agarose biomaterial and Biobrane® – All groups showed a proper clinical integration and epithelialization after eight weeks – The autografts and fibrin-HA dressing presented better skin structuration and higher expression of cytokeratins	(Sierra-Sanchez et al. 2020)
HA/PU	Propolis	Electrospun membrane	– The HA-based nanofibres were loaded with ethanolic extract of propolis to endow antibacterial properties to the dressing – The samples displayed greater antibacterial activity in *S. aureus* and *E. coli* – The membranes were biocompatible and accelerated wound closure, improving dermis	(Eskandarinia et al. 2020a)

(continued)

Table 5 (continued)

Polymeric combinations	Incorporated bioactive compounds	Types of wound dressing	Main findings	Refs.
			development and collagen deposition	
HA/alginate	None	Powder	– The HA/alginate powders presented a better water adsorption ability and shorter blood clotting time than commercial haemostatic agents – The dressings exhibited a controlled degradation profile in the presence of hyaluronidase and lysozyme – Cell proliferation and migration of fibroblasts were noticed when they were seeded in contact with dressings	(Chen et al. 2020)
HA/PU/starch	None	Electrospun membrane	– The SEM and TEM analysis evidenced the core-shell membrane – The coaxial nanofibres presented suitable mechanical properties for wound dressing applications – The in vitro cytotoxicity assays revealed that the nanofibres were biocompatible, which positively affect the wound healing	(Movahedi et al. 2020)
HA/PDA	None	Hydrogel	– The HA-based hydrogel was developed based on the Michael addition reaction between PDA and thiolated HA – The inclusion of PDA into HA hydrogel improved the cell affinity and tissue adhesion – The PDA-bearing hybrid hydrogel inhibited bacterial growth	(Yu et al. 2020)

(continued)

Table 5 (continued)

Polymeric combinations	Incorporated bioactive compounds	Types of wound dressing	Main findings	Refs.
HA	None	3D-printed hydrogel	– HA was modified with methacrylic anhydride and 3,3'-dithiobis (propionylhydrazide) aims to crosslink the 3D-printed hydrogel through UV irradiation and click reaction – The storage modulus of hydrogels was directly related to the increase on methacrylated HA concentration – The hydrogel presented a high swelling ratio and sustained degradation rate	(Si et al. 2019)
HA/PVDF	Active pharmaceutical ingredient ionic liquids (API-ILs)	Membrane	– The API-ILs were blended in a bilayer membrane made of PVDF and HA layers – Cell adhesion was noticed on the surface of the bilayer membrane – The release of API-ILs from dressings was crucial to avoid an exuberant inflammation response	(Abednejad et al. 2019b)
HA	None	Membrane	– A dressing was developed by using high- and low-molecular weight HA – The dressings reduced the inflammation biomarkers and accelerated the healing process – The samples treated with HA dressings showed a higher expression of defensin-2 (antimicrobial peptide), suggesting antibacterial functions	(D'Agostino et al. 2019)
HA	Platelet-rich plasma (PRP)	Scaffold	– The biofunctionalized HA scaffold composed of PRP and HA	(De Angelis et al. 2019)

(continued)

Table 5 (continued)

Polymeric combinations	Incorporated bioactive compounds	Types of wound dressing	Main findings	Refs.
			promoted a rapid reepithelization (96.8 ± 1.5%) after 30 days – PRP + HA scaffold revealed more substantial regenerative potential compared with other groups	
HA	ZIF-8	Film	– The HA films modified with ZIF-8 promoted an augment in Young's modulus and a decline in water contact angle values – The films promoted cell activity as well as ameliorated the antibacterial properties	(Abednejad et al. 2019a)
HA	Silver nanoparticles	Hydrogel	– The silver nanoparticles were biosynthesized by green technique (microwave-assisted) using corn silk extract – The gels had good mechanical properties with gelation temperature close to the body temperature – The hydrogels were biocompatible and presented antibacterial activity both Gram-positive and Gram-negative bacteria – The hydrogels allowed a faster wound closure and repair in in vitro model of wound healing	(Makvandi et al. 2019)
HA	Adipose-derived stem cells	Membrane	– The dressing stimulated the healing and reduced the inflammation when applied on burn wounds in a rat model – IL-1β and TGF-β1	(Alemzadeh et al. 2020)

(continued)

Table 5 (continued)

Polymeric combinations	Incorporated bioactive compounds	Types of wound dressing	Main findings	Refs.
			levels were lower in wounds treated with HA dressing with stem cells – HA membranes seeded with adipose stem cells accelerated wound healing, with increased expression levels of bFGF and decreased TGF-β1 in burns	
HA/corn starch	Propolis extract	Film	– The HA films incorporating propolis extract exhibited a higher antibacterial activity – The films did not present any cytotoxic effect in fibroblast cells – The enrichment of cornstarch wound dressings with HA and propolis extract improved the wound healing process in rats' skin excisions	(Eskandarinia et al. 2019)
HA/collagen	None	Hydrogel	– The hydrogel was produced through in situ couplings of phenol moieties of both compounds – The hydrogel possessed a porous structure suitable for water, nutrients and gaseous exchanges – The endothelial cells and fibroblasts proliferated within hydrogel' structure – The healing ration of the wounds treated with hydrogel was higher in comparison to the commercial drug – The collagen and HA combination promoted the development of the vasculature, epithelial layer and collagen fibres	(Ying et al. 2019)

(continued)

Table 5 (continued)

Polymeric combinations	Incorporated bioactive compounds	Types of wound dressing	Main findings	Refs.
HA/ Nanocrystalline cellulose	CS nanoparticles loaded with GM-CSF	Membrane	– The composite dressing showed appropriated mechanical properties, swelling capacity and controlled release of GM-CSF – The wounds covered with composite dressing exhibited a faster wound closure and reepithelialization	(Karimi Dehkordi et al. 2019)
HA/PDA	rGO	Hydrogel	– The hydrogel-based on HA-graft-dopamine and rGO by using a H_2O_2/HPR system was obtained – The hydrogels displayed similar mechanical properties to human skin, high swelling ratio and adequate degradation profile – The addition of PDA to the hydrogel composition conferred it antioxidant and haemostatic capacity and tissue adhesiveness – The NIR irradiation improved in vivo antibacterial behaviour – The hydrogel promoted the collagen deposition, granulation tissue thickness and vascularization	(Liang et al. 2019a)
HA/collagen	Heparin-binding EGF-like growth factor (HB-EGF)	Hydrogel	– HB-EGF in hydrogels containing HA and collagen was released over at least 72 h, promoting keratinocyte migration, EGFR-signalling and HGF expression in dermal fibroblasts – Hydrogels induced the epithelial tissue formation in wounds in a porcine skin organ culture model	(Thones et al. 2019)

(continued)

Table 5 (continued)

Polymeric combinations	Incorporated bioactive compounds	Types of wound dressing	Main findings	Refs.
HA/PVA	Cyclodextrin Naproxen	Electrospun membrane	– The addition of cyclo-dextrin in solutions promoted a stabilization during the electrospinning process – The nanofibrous membranes were stable in water after chemical crosslinking using EDC/NHS chemistry – The naproxen was loaded into nanofibres and presented a maximum release during the first 24 h of incubation	(Seon-Lutz et al. 2019)
HA	None	Hydrogel	– The HA-based hydrogels were prepared through physical freezing-thawing (HA1), and chemical cross-linking (HA2) – HA2 was the most promising treatment in promoting the wound healing with a least severe scar formation on rabbits – HA2 enhanced VEGF and α-SMA secretion; improved the skin regeneration and reduced the wound inflammation and scar development	(Hong et al. 2018)
HA/gelatin	None	Hydrogel	– The hydrogels displayed appropriate fluid uptake capacity and suitable water vapour transmission rate – The hydrogels were biocompatible and promoted cell proliferation and wound healing in vivo	(Wu et al. 2017)

acceleration. Furthermore, the HA2 hydrogel also improved the secretion of α-smooth muscle actin (α-SMA) and vascular endothelial growth factor (VEGF), which is beneficial for wound contraction angiogenesis mechanisms.

An injectable nanocomposite hydrogel dressing composed by HA and reduced graphene oxide (rGO) with adhesiveness, good mechanical properties, as well as antibacterial and radical scavenging abilities was produced by Liang and their collaborators (Liang et al. 2019a). To accomplish such purpose, the HA-graft-dopamine conjugate (HA-DA) was produced by a chemical EDC procedure to confer HA adhesiveness and antioxidant property. Afterwards, the HA-DA polymer chains were crosslinked with rGO@ polydopamine (PDA) by oxidative coupling of catechol groups between them and using as an initiator system H_2O_2/horseradish peroxidase (HRP). Then, the authors determined the water uptake ability of hydrogels, which achieved a maximum water absorption of ≈300% after 3 days. Unlike, the hydrogels degradation rate depended on the rGO@PDA concentration. Further, the presence of rGO into hydrogel formulation conferred to HA-DA/rGO hydrogels with photothermal property, which was crucial to impair the *E. coli* and *S. aureus* growth in vitro and in vivo infected skin wound model. Additionally, the hydrogels comparing to commercial Tegaderm® film showed higher in vivo wound healing effect regarding collagen metabolism, granulation thickness and wound closure.

As an alternative, Ying et al. explored an approach to improve HA hydrogels' mechanical strength through the enzymatic crosslinking method (Ying et al. 2019). In this work, the HA/collagen (COL) hydrogel was produced through the action of HRP. In the first instance, HA and collagen were branched with phenolic hydroxyl groups, which result in HA-Tyr and COL-P, covalently crosslinked with HRP and H_2O_2.

The swelling assays revealed that the individual HA-Tyr hydrogel (95%) was higher than individual COL-P hydrogel (30%), after 3 days of incubation, evidencing the high absorption ability of HA. In terms of enzymatic degradability, the COL-HA hydrogel showed at first 6 h an initial burst phase, when incubated with collagenase and hyaluronidase. In contrast, the hydrogels maintained about 70% of its weight when incubated without enzymes' presence.

In respect to the in vitro assays, the COL-HA hydrogel was biocompatible, allowing the vascular cells infiltration, while in vivo assays showed that the hydrogel stimulated the angiogenesis epithelium and consequently the reestablishment of skin tissue. Moreover, COL-HA hydrogel mimicked ECM for cell proliferation and differentiation, which promoted the formation of the vessels and collagen fibres (Fig. 8).

On the other hand, De Angelis et al. reported in vitro and in vivo assessment of a bio-functionalized scaffold composed of platelet-rich plasma (PRP) and HA in an observational study including 182 patients with chronic ulcers (diabetic and vascular) (De Angelis et al. 2019). Overall, the results evidenced that Group treated with PRP + HA encouraged the skin regeneration process, reducing the healing time. The HA provides a moist environment, induces the growth factors secretion, and acting as a transitory dermal substitute, while PRP promotes the cells' activity.

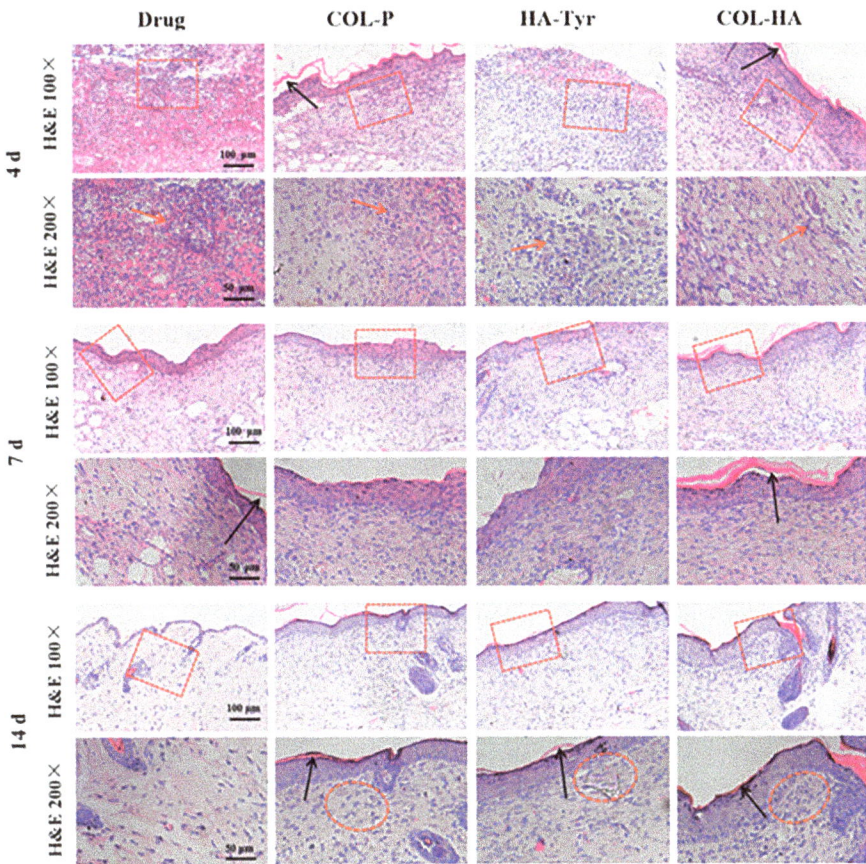

Fig. 8 Histomorphological evaluation of skin tissues stained by the haematoxylin and eosin after 4, 7 and 14 of wound induction. The groups were treated with Yunnan powder, HA-Tyr hydrogel, COL-P hydrogel and COL-HA hydrogel. The black arrow, red arrow and red round represented epidermal layer, inflammatory cells and fibroblasts and collagen fibre, respectively. Reprinted from Materials Science and Engineering C, vol. 101, Ying et al., In situ formed collagen-hyaluronic acid hydrogel as a biomimetic dressing for promoting spontaneous wound healing, 487–498, with permission from Elsevier (Ying et al. 2019)

Additionally, after 80 days of treatment, the patients treated with PRP + HA had 98.4% ± 1.3% of re-epithelization compared to 87.8% ± 4.1% in the group treated only with HA. Based on data gathered, this work suggests that the combined use with PRP and HA constitutes a promising treatment for chronic ulcer.

Si et al. used the 3D bioprinting technique to produce double-crosslinked HA-based hydrogels (Si et al. 2019). Firstly, the authors synthesized two different HA chemical derivatives to improve HA hydrogels' mechanical properties. The HA can be crosslinked via click reaction by modification with 3,3′-dithiobis (propionylhydrazide) (DTP) (HA-SH), and UV-crosslinked with methacrylic anhydride (HA-MA). Afterwards, the HA derivatives solutions were mixed and bioprinted, and then exposed to UV LED curing system. In this way, the HA-SH was crosslinked by click reaction, and simultaneously through the UV irradiation, obtaining a double-crosslinked HA-MA/SH hydrogel. The morphological characterization of hydrogels revealed that all hydrogels possessed a microporous structure with a pore size between 30 and 50μm. The swelling ratio decreased with the increase of HA-SH content (the lowest ratio was 20.62% ± 9.13%), but when the proportion between HA-SH and HA-MA is same, the water absorption performance was more controlled along 12 h of incubation. In turn, the in vitro assays demonstrated that hydrogels hyaluronidase degradation exceeded 55% in 3 days, which differ with the ratio of the HA-MA on hydrogels' composition. The biocompatibility of hydrogels was also confirmed, and the 3D bioprinting of hydrogels enables to obtain structures with pre-determined architecture, which can be tailored to different wound dressings applications.

Further, Movahedi et al. produced core-shell structured PU/starch (St) and PU/St/ HA nanofibres through coaxial electrospinning technique (Movahedi et al. 2020). The nanofibres' wettability was evaluated through the determination of water contact angles, and the results confirmed the hydrophilicity of the membranes presenting values of 61.4 ° and 42.8° for PU/St and PU/St/HA. About the biodegradation profile, the assays revealed that, after 3 weeks of incubation in PBS solution, 85–93% of the weight of electrospun membranes remained, whereas the PU/St/ HA nanofibres swelling ratio were higher (625.23%). In vitro and in vivo assays evidenced that the nanofibres were biocompatible, promoted the cell adhesion and accelerated the wound closure. Such results were attributed to HA' properties, such as (i) induction of the cellular activity, (ii) provision of a moist environment and (iii) improvement of the tissue repair (Table 5).

2.2.2   Chitosan

Chitosan (CS) is a naturally biodegradable polysaccharide, broadly used in wound healing applications. CS, derived from the alkaline N-deacetylation of chitin, is the crustacean exoskeleton's main structural component. Chitin is composed of glucosamine and N-acetyl glucosamine linked by β(1–4) glycosylic bonds.

CS, a cationic polysaccharide, displays several promising properties such as low immunogenicity, haemostasis, biocompatibility, degradability, scar prevention and antimicrobial activity with relevance for wound healing and tissue engineering (Song et al. 2018; Matica et al. 2019).

CS is biodegraded by enzymes, namely the lysozymes able to disrupt the acetylated units linkages, or by acid hydrolysis, resulting in CS oligosaccharides, able to stimulate macrophages and fibroblasts migration and proliferation (Kim et al. 2008).

Further, CS and derivatives possess antimicrobial activity through different mechanisms: i) The CS positively charged groups interacts with the negatively charged groups found on the bacterial cell wall, inducing internal osmotic imbalances and hence inhibiting microorganisms' growth; ii) CS forms a polymeric layer around bacteria, impairing the cellular transport mechanisms and nutrients absorption; and iii) CS chelate Ca^{2+} or Mg^{2+} found in the bacterial cell wall, inhibiting the bacterial growth (Simões et al. 2018).

However, CS application as wound dressings presents some limitations like the low mechanical resistance and weak water solubility. In this way, the crosslinking agents have been employed to allow the formation of the linkages between polymeric chains, adjusting the elasticity and resistance of the CS-based materials (Song et al. 2018). Furthermore, CS can also be chemically modified and combined with other polymers to overcome such handicaps (Ahsan et al. 2018). For example, methodologies like alkylation, acetylation and carboxymethylation can be adopted to tailor the CS solubility (Song et al. 2018).

CS has been widely investigated for wound healing applications owing its excellent intrinsic features. Indeed, there are already available many commercial CS-based wound dressings and works reporting its in vitro and in vivo performance (as presented in Table 6).

Intini et al. used CS as a raw material to produce skin 3D-printed scaffolds (Intini et al. 2018). The authors used an extrusion-based 3D printing technique to achieve a precise shape and spatial distribution of the 3D-CS structures. After that, co-cultures of fibroblasts and keratinocytes were seeded on 3D scaffolds, and the SEM images set the successful cellular colonization of the 3D structures. After 35 days, scaffolds were filled by cells, developing a skin-like layer composed of a co-culture of fibroblasts and keratinocytes. Moreover, in vivo assays on diabetic rat models evidenced that the CS scaffolds promoted the tissue regeneration with improved functionality compared to the wounds treated with a commercial product (wound dressing composed of a blend of carboxymethylcellulose and calcium alginate coated with a semi-permeable layer of PU).

On the other hand, Hou et al. developed cryogels with glycol chitosan (GC) and ε-polylysine (EPL) with outstanding antibacterial efficacy and haemostatic property (Hou et al. 2020). To achieve that, GC-EPL cryogels were produced by copolymerization of GC methacrylate and EPL acrylamide at sub-zero temperature ($-20\ °C$). The cryogels presented a swelling ratio around 4000%, and an interconnected porous construct with pore size between 10 and 100μm. In turn, the cryogels showed higher clotting rate and lower blood clotting index in comparison to groups treated with Combat Gauze and gelatin sponge. On the other hand, in vivo haemostatic assay also demonstrated that the GC-EPL cryogels could decrease the blood loss (90%) compared to the untreated group. The improved haemostatic properties of GC-EPL cryogels is due to the potentiated effect of CS and EPL combination that allow increasing the positive charges. Further, the cryogels showed 77.84% bacterial killing efficiency front to Methicillin-resistant *S. aureus* (MRSA) and 68.59% to *E. coli*. However, the increment in EPL concentration to 0.5% w/v promoted an augment on the killing efficiency to 99.61% and 99.84%. Finally, the in vivo assays suggested that the cryogels facilitated the normal wound healing process in MRSA infected skin injuries.

Another work performed by Chalitangkoon and their collaborators, a water-soluble derivative of CS (hydroxyethylacryl chitosan (HC)) was blended with sodium alginate (SA) to obtain films (Chalitangkoon et al. 2020). HC is a water-

Table 6 List of the most recent works available in the literature reporting the chitosan-based wound dressings

Polymeric combinations	Incorporated bioactive compounds	Types of wound dressing	Main findings	Refs.
CS	None	Hydrogel Sponge Membrane	– The CS derivative water-soluble carboxymethyl-chitosan (CMCS) was crosslinked with genipin to produce hydrogel, membrane and sponge – The sponges presented a porous structure with best water uptake, gas permeability and haemostatic efficiency – The sponges also obtained the higher performance in wound closure in vivo	(Wang et al. 2020a)
CS	*Mansoa hirsuta*	Film	– The film had an amorphous nature, thermostability and a rough structure – The incorporation of *M. hirsute* enhanced mechanical performance and films thickness – The in vivo assays revealed that the wound treatment with CS film showed an improved re-epithelization, cell proliferation, and collagen formation	(Rodrigues Pereira et al. 2020)
CS	Antimicrobial peptides	Hydrogel	– The thermoresponsive hydrogels were obtained by using β-glycerolphosphate, resulting in the hydrogel gelation after 15 min at 37 °C – The hydrogels lost 80% of its weight after 35 days of incubation and promoted a burst release of antimicrobial peptides at day 1 – The hydrogels showed excellent biocompatibility and antibacterial activity	(Rezaei et al. 2020)
CS/PLU	Curcumin	Membrane	– Membranes containing PLU and curcumin presented appropriated wettability, mechanical properties, high swelling degree and curcumin release – The membranes loaded with curcumin were effective against bacterial growth	(Enumo Jr. et al. 2020)

CS	Aluminium chloride, aluminium sulphate hydrate or iron(III) sulphate and levofloxacin	Membranes	– The strong interactions between CS and other additives resulted in dressing with increased swelling and stability – The levofloxacin incorporation avoided bacterial growth – The CS-based dressings showed biocompatibility, haemocompatibility and haemostatic competence in vitro and in vivo	(Koumentakou et al. 2020)
CS/PCL	Aloe vera	Electrospun membrane	– The weight ratio between compounds of membranes influenced the viscosity and solution conductivity – The aloe vera addition potentiated the antibacterial and biological properties of the membranes	(Yin and Xu 2020)
CS	Thyroxine	Sponge	– The thyroxine-loaded CS dressings stimulated the angiogenesis – The in vivo assays evidenced the excellent wound healing potential of the dressings	(Shahzadi et al. 2020)
CS/sodium alginate	Silver particles Para-acetylaminophenol	Film	– The silver loading increased the swelling grade and improved mechanical properties of films – The films incorporating silver had antibacterial activity against *E. coli* and *S. aureus*, without compromising the viability of Vero cells – The increase in crosslinking density and silver loading prolonged the release time of Para-acetylaminophenol from films	(Chalitangkoon et al. 2020)
CS	Iron-modified hydroxyapatite	Film	– The CS films incorporated with iron-modified hydroxyapatite had high UV protection properties – The films also were biocompatible and exhibited antimicrobial activity against both Gram-positive and Gram-negative bacteria	(Cunha et al. 2020)
CS/PVA	Cu-MOFs	Electrospun membrane	– The nanofibres displayed suitable mechanical properties, swelling ability and water vapour transmission rate – The electrospun membranes were biocompatible	(Wang et al. 2020c)

(continued)

Table 6 (continued)

Polymeric combinations	Incorporated bioactive compounds	Types of wound dressing	Main findings	Refs.
			and supported the cell adhesion – The CS/PVA nanofibres incorporating Cu-MOFs exhibited antibacterial effect and promoted the in vivo wound healing process with minimal inflammation	
CS-glucan/collagen	Aloe vera	Hollow fibres	– The dressing presented high hydrolytic stability with enhanced swelling properties – The fibres displayed excellent biocompatibility when incubated in contact with human dermal fibroblasts – The fibres-based dressing increased the wound closure after 7 days of treatment	(Abdel-Mohsen et al. 2020)
CS/oxidized konjac glucomannan	None	Hydrogel	– The hydrogel was produced through the formation of dynamic covalent Schiff-base bonds – The hydrogel was adhesive, injectable and possessed self-healing properties – In vivo assays suggested that the self-adapting hydrogel had beneficial effects on wound healing progression	(Wang et al. 2020b)
CS/alginate	*Arrabidaea chica* Verlot (*A. chica*)	Film	– The incorporation of surfactants and silicone-based compounds promoted a greater homogeneity, swelling of the polymeric matrix and more controlled *A. chica* release – The films presented a maximum mass loss of 18% after 7 days of incubation – The formulations were biocompatible and promoted cell proliferation	(Pires et al. 2020)

CS/PEG	Cephalexin and zeolitic imidazolate framework-8 (ZIF-8) nanoparticles	Film	– The ZIF-8 nanoparticles incorporation increased the tensile strength and decreased the elongation at break – The cephalexin release displayed a burst peak after 3, 8 and 10 h in acidic, neutral and alkaline media, respectively – The films were biocompatible in contact with L929 fibroblast cells	(Mazloom-Jalali et al. 2020)
CS/carragenan	None	Sponge	– The fibroblast cells remained viable, attached and proliferated when seeded on dressing – The dressing showed steady blood coagulation, implying red blood cells and platelet adhesion, which is essential to control the haemorrhage	(Biranje et al. 2020)
CS/PVA	Silver nanoparticles and *Ocimum sanctum* extract	Hydrogel	– The incorporation of silver nanoparticles in the hydrogel improved the tensile strength and elongation at break – A controlled release of *Ocimum sanctum* extract (84.3% after 16 h) was observed from hydrogels, which scavenges 63.1% of free radicals – The hydrogels inhibited the bacterial growth of Gram-negative and Gram-positive strains	(Kumar et al. 2020)
CS/PVA	None	Sponge	– The controlled freeze-drying process promoted the formation of a structure with channelled pores – The developed dressing was biocompatible, with a reasonable absorbency rate, improved mechanical integrity and low bioadhesive strength – The dressing was non-toxic and possessed good healing characteristics	(Shyna et al. 2020)
CS/PVA/PCL	Chamomile	Patch	– The CS was modified to CMCS by Michael reaction and then used to prepare an aqueous solution with PVA – The multilayer patches composed of hydrophilic	(Shokrollahi et al. 2020)

(continued)

Table 6 (continued)

Polymeric combinations	Incorporated bioactive compounds	Types of wound dressing	Main findings	Refs.
			chamomile-loaded CS/PVA layer and a hydrophobic PCL layer – The mats displayed satisfactory tensile strength and antioxidant properties – The mats incorporated with 15, 20 and 30 wt% chamomile possessed high antibacterial efficiency	
GC/ EPL	None	Cryogel	– The cryogels yielded lower amounts of blood loss – The incorporation of EPL improved the antibacterial activity, preventing the bacterial infection during wound healing – The cryogel promoted a higher healing efficiency in bacteria-infected wounds	(Hou et al. 2020)
CS	Graphene oxide and calcium silicate	Film	– The biomaterial presented good tensile strength, compatible breathability and water absorption – The dressing exhibited photothermal performance, leading to photothermal antibacterial and antitumor efficacy – The biomaterials stimulated the tissue formation on in vivo chronic wound model	(Xue et al. 2019)
CS	Quaternary ammonium chitosan nanoparticles (TMC NPs)	Sponge	– The TMC NPs incorporated into CS sponges improved the antibacterial activity – The CS sponge exhibited an outer hydrophobic surface and an inner hydrophilic surface – The hydrophobic area showed waterproof and it was able to prevent the adhesion of the external contaminants	(Xia et al. 2020)

PVA/CS	Azadirachta indica (neem)	Electrospun membrane	– The hydrophilic surface was capable of absorbing exudate and inhibiting the bacteria growth – The technique allowed the synthesis of a top layer (CS/neem) and an inner layer of PVA – The addition of CS and neem extracts improved the thermal stability, moisture management properties, and antibacterial activity	(Ali et al. 2019)
CS7oxidized dextran	None	Hydrogel	– The hydrogel composed of hydrophobically modified CS and oxidized dextran was able to coagulate whole heparinized blood – The hydrogel also showed antibacterial properties and presented wound healing functions in an infected wound model of rat skin	(Du et al. 2019)
CS/cordycepin	None	Hydrogel	– The hydrogel was produced by non-covalent bonds through one-step 'freeze-thaw' method, without using cross-linking agents – The hybrid hydrogels presented higher biocompatibility, appropriate swelling degree and mechanical properties – The in vivo assays confirmed that the hydrogels improved the re-epithelization mechanism of skin wounds and collagen deposition	(Song et al. 2019)
CS/sericin	Silver and moxifloxacin	Film	– The films displayed good swelling profile with a sustained in vitro moxifloxacin release – The films incorporated with moxifloxacin exhibited promising antibacterial activity against all tested strains – The films promoted collagen reorganization, neovascularization and mild epidermal regeneration after 7 days of treatment	(Shah et al. 2019)
CS/chondroitin sulphate	None	Polyelectrolyte complex	– The prepared dressings showed high swelling and porosity – The materials presented good blood compatibility, low blood-clotting index and excellent antibacterial properties	(Sharma et al. 2019)

(continued)

Table 6 (continued)

Polymeric combinations	Incorporated bioactive compounds	Types of wound dressing	Main findings	Refs.
CS	Silver nanoparticles	Film	– The cell density was a fourfold increase in cells seeded in contact with composite dressings – Films presented high antibacterial activity against all bacterial strains – The fibroblasts adhered and proliferated at the surface of films, expressing the characteristic proteins of ECM (tropoelastin, procollagen type I and Ki-67)	(Hernandez-Rangel et al. 2019)
CS	None	3D-printed scaffold	– 3D cell cultures of human fibroblasts and keratinocytes were obtained between 20 and 35 days of incubation – After 35 days, the best cell growth on 3D CS scaffold was obtained, where the fibroblasts and keratinocytes were seeded together – The layer of fibroblast and keratinocyte cells was formed into 3D CS printed scaffolds	(Intini et al. 2018)

soluble CS resulted from the modification of CS via Michael addition reaction with hydroxyethylacrylate. The authors also added Ag particles to HC/SA films through an in situ chemical reduction immersion method to augment the films antibacterial activity. As main results, the authors verified that Ag incorporation into HC/SA films improved the mechanical attributes and antibacterial activity, without compromising the biocompatibility.

Wang et al. produced a self-adapting hydrogel of CS and oxidized konjac glucomannan (OKGM) through the dynamic Schiff-base bond formation, allowing adaptation to irregular wounds under natural conditions (Wang et al. 2020b). In this way, CS was selected as backbone and OKGM as crosslinker, to obtain the hydrogel network. Since CS presents a limited solubility in aqueous media, a protonated tranexamic acid aqueous solution was selected to dissolve CS. Further, the aldehyde groups of OKGM can perform the reduction of silver nitrate to elemental silver to stabilize the reduced Ag. The authors then evaluated the hydrogels' adhesion ability by a lap shear test, where the higher adhesion strengths of hydrogels could be attributed to the hydrogen-bonding linkages and chemical crosslinking due to the Schiff-base reaction between amine groups of tissue proteins and the aldehyde groups of the hydrogels. It is also possible to verify that the hydrogel possessed self-adaptability, making viable its application as an injectable hydrogel. The antibacterial assay demonstrated that the in situ synthesized AgNPs avoided the *E. coli* and *S. aureus* growth. In contrast, the hydrogels did not induce cytotoxicity when they were incubated with mouse fibroblast cells, and they showed higher therapeutic effect than commercial hydrogel (AquacelAg®) when applied on the full-thickness skin injury.

Shahzadi et al. prepared CS membranes by freeze gelation method incorporating a pro-angiogenic molecule (thyroxine) (Shahzadi et al. 2020). After the CS hydrogel production, it was performed an immersion in thyroxine solution overnight to promote the total absorption of thyroxine into CS hydrogel. Afterwards, the authors characterized the hydrogels' swelling rate, and they verified that the CS-thyroxine hydrogel started swelling from 200% (after 15 min) and then increased and reached to 1300% after 8 days of incubation. Similarly, the hydrogels released 15–20% of drug within 15–20 min, and after 24 h, almost 70%. Also, the thyroxine incorporated into hydrogels was crucial to induce the blood vessels' formation and in combination with CS encouraged the healing process.

In turn, Wang et al. used a water-soluble CS derivative- CMCS and genipin (as crosslinker agent) to produce several wound dressings: hydrogel, membrane and sponge (Wang et al. 2020a). Concerning the water absorption ability, the CMCS sponges presented a higher ability due to its porous structure. Further, the expression of α-SMA, MMP-1 and TGF-β1 were determined and the results demonstrated that: i) the maximum amount of α-SMA protein was registered on CMCS sponges, ii) CMCS sponges also induced the secretion of TGF-β1 by fibroblasts and iii) CMCS sponges demonstrated the lowest expression of MMP-1 protein. From the in vivo assays, the wounds treated with CMCS sponges revealed faster-wound closure and complete re-epithelialization after 14 days.

2.2.3 Alginate

Alginate is an anionic polysaccharide obtained from the brown seaweed cell wall and excreted by some bacteria. It has low cost, excellent biodegradability, biocompatibility, gelling ability, and is ease to functionalize (Pawar and Edgar 2012).

In wound dressing applications, alginate is widely used due to its biocompatibility, chemical and physical cross-linking abilities, mild and physical gelation process, non-thrombogenic nature, and the structural similarity to the structure of ECM. In general, alginate-based wound dressings are accurately accepted due to its favourable ability to maintain a physiological moist environment, minimize bacterial colonization and facilitate the healing process (Song et al. 2018; Chandika et al. 2015).

However, alginate's main drawbacks are the weak mechanical features, low cell adhesion, and non-degradability in mammals, since they did not possess the enzyme (i.e., alginase) responsible for cleaving the polymer chains. Thus, scientists have been combined alginate with other polymers or molecules to boost their mechanical properties and degradability (Lee and Mooney 2012). The ionically cross-linked alginate gels can be disintegrated by releasing the divalent cations, due to exchange reactions with monovalent ions such as sodium ions.

Despite the existence of some alginate-based wound dressings offered by the market (e.g., Algisite Ag®, Sorbalgon Ag®, Gentell Calcium Alginate Ag®) (Sarheed et al. 2016), the researchers have been dedicating to the design of more promising alginate-based wound dressings (as presented in Table 7).

Summa et al. reported the composite polymeric material's characterization based on SA and povidone-iodine (PVPI) in animal model (Summa et al. 2018). Such combination enabled to combine the outstanding alginate' wound healing properties with the PVPI' antimicrobial (bactericidal and fungicidal) properties. So, the SA/PVPI films were prepared through the casting technique, and then its biocompatibility was demonstrated since the percentage of viable cells was 93.2%, after seeded for 24 h in contact with films. Further, the in vitro and in vivo anti-inflammatory assays showed the inhibition of the production of pro-inflammatory cytokines (IL-1β and IL-6) by the films, which was essential to regulate the inflammation phase of the healing process. Similarly, the SA/PVPI films increased the hydroxyproline levels and reduced the re-epithelialization time.

Further, Ahmed et al. prepared calcium alginate (CA) wafer dressings incorporating ciprofloxacin (CIP) (Ahmed et al. 2018). To accomplish such purpose, the CA-CIP mixtures were subjected to the lyophilization cycles to obtain wafers. The authors then determined the swelling, porosity, moisture content and water absorption of dressings, verifying that the CA-CIP wafer dressings can manage wound exudate, preventing the maceration of healthy skin cells. Further, the initial burst release of CIP (59.40 ± 0.64, 74.39 ± 3.59 and 91.43 ± 1.21% of CIP was released after 6 h, from the wafers loaded with 0.005, 0.010 and 0.025% of CIP) lead to the rapid eradication of the *E. coli*, *S. aureus* and *P. aeruginosa*. More important, the

Table 7 Description of the most recent works reporting the production and characterization of alginate-based wound dressings

Polymeric combinations	Incorporated bioactive compounds	Types of wound dressing	Main findings	Refs.
Alginate	Papain	Membrane	– The papain was immobilized at the surface of alginate membranes to improve its wound healing properties – The enzyme remained active after immobilization, while the matrix protected the enzyme from deactivation – About 64.1% of the enzyme was released from the membrane after 24 h of incubation	(Moreira Filho et al. 2020)
Sodium alginate (SA)/PEG	*Satureja cuneifolia* plant extract (SC)	3D-printed scaffolds	– The 3D scaffolds restrained *E. coli* and *S. aureus* growth – The fibroblasts remained viable on/within 3D alginate scaffolds	(Ilhan et al. 2020)
Alginate	Ibuprofen	Hydrogel	– The loading of ibuprofen into the scaffold was improved by cause the high surface area of the alginate hydrogels – The ibuprofen was sustained release over 12–24 h – The hydrogel supported the in vivo wound healing process by suppressing the inflammation and maintaining the wound hydration	(Johnson et al. 2020)
SA/pNIPAM	Diclofenac sodium Basic fibroblast growth factor (bFGF)	Hydrogel	– The thermosensitive hydrogels were produced by incorporating nanogels of pNIPAM loaded with diclofenac sodium and basic fibroblast growth	(Lin et al. 2020)

(continued)

Table 7 (continued)

Polymeric combinations	Incorporated bioactive compounds	Types of wound dressing	Main findings	Refs.
			– The hydrogels presented a desirable storage modulus, high swelling ratio, and a suitable water vapour transmission rate – The controlled release of diclofenac (92%) and bFGF (80%) promoted a better healing effect with a wound contraction of 96% after 14 days	
SA	Borosilicate bioglass (BBG)	Membrane	– The composite dressing possessed good water absorption performance – The dressings showed outstanding wound healing ability in full-thickness skin defects in rats	(Wu et al. 2020)
Alginate	Sulphanilamide	Hydrogel	– The alginate fibres crosslinked with glutaraldehyde presented higher mechanical properties and low swelling degree – The hydrogel fibres exhibited bactericidal activity towards *S. aureus* and *E. coli* – The surface of hydrogel fibres stimulated the cells adhesion and proliferation	(Sun et al. 2020)
Alginate	Vicenin-2	Film	– The film dressings were smooth, translucent, and with good flexibility – The film was able to promote the release of Vicenin-2 to the wound area in a controlled manner	(Tan et al. 2020)

(continued)

Table 7 (continued)

Polymeric combinations	Incorporated bioactive compounds	Types of wound dressing	Main findings	Refs.
SA/kappa-carrageenan	Silver nanoparticles	Bio-platform	– The polymeric bio-platforms exhibited exceptional antibacterial properties towards *S. aureus* and *E. coli* – The dressings promoted the formation of fibrous tissues, hair follicles and wound area contraction on mice with a second-degree burn	(Zia et al. 2020)
SA/PLA/PVA	None	Electrospun membrane	– In vitro experiments showed that nanofibrous membranes provide adequate support for human and rat fibroblasts growth – The electrospun membranes enhanced the wound healing in vivo – The SA/PLA/PVA dressing reduced the inflammatory response and promoted the protein deposition	(Bi et al. 2020)
SA/poloxamer 407/ pluronic F-127/PVA	Amikacin	Hydrogel	– The hydrogel membrane had good mechanical properties, outstanding swelling properties and surface porosity – The amikacin-loaded hydrogels exhibited higher zone inhibition towards *S. aureus* and *P. aeruginosa* – The hydrogel membranes promoted the formation of granulation tissue, re-epithelization and faster wound closure	(Abbasi et al. 2020)

(continued)

Table 7 (continued)

Polymeric combinations	Incorporated bioactive compounds	Types of wound dressing	Main findings	Refs.
Alginate	None	Membrane	– A clinical trial was done to compare the therapeutic properties of the alginate dressing and negative pressure wound therapy – The alginate demonstrated similar healing efficacy to that of negative pressure wound therapy	(Casanova et al. 2020)
Alginate	Vitamin D3	Hydrogel	– The hydrogels were biodegradable presenting a weight loss percentage of 89% after 14 days – The hydrogels were cytocompatible and haemocompatible – In vivo assays demonstrated that the hydrogels promoted the re-epithelialization and granular tissue formation	(Ehterami et al. 2020)
Alginate	Chlorhexidine hexametaphosphate (CHX-HMP)	Film	– The alginate film provided the release of CHX over 14 days – The dressings exerted an inhibitory effect against bacterial growth – After 7 days, the alginate film incorporated with CHX presented an enhanced antibacterial effect in comparison to the silver alginate	(Duckworth et al. 2020)
Alginate	Naringenin	Hydrogel	– The developed hydrogels presented a suitable porosity with interconnected pores and an appropriate biodegradation profile (89% after 14 days)	(Salehi et al. 2020)

(continued)

Table 7 (continued)

Polymeric combinations	Incorporated bioactive compounds	Types of wound dressing	Main findings	Refs.
			– The hydrogels were biocompatible and promoted a greater wound closure than the gauze-treated wound	
Alginate	Zinc oxide nanoparticles (ZnO)	Electrospun membrane	– The fibroblasts and keratinocytes proliferated at the surface of alginate membranes, and ZnO provided strong antibacterial properties – The mats exhibited mechanical properties and water vapour permeability values similar to those found on human skin	(Dodero et al. 2019)
Bacterial cellulose	Alginate	Membrane	– The alginate was impregnated into bacterial cellulose hydrogels in order to contribute to moist environment at the wound site – The composite dressings presented enhanced water-retention properties, antibacterial activity, and reduced adhesiveness to the wound tissue	(Sulaeva et al. 2020)
Sodium alginate/alginate	None	Self-healing hydrogel	– The self-healing hydrogel was produced by mixing adipic acid dihydrazide-modified gelatin with monoaldehyde-modified sodium alginate – The spraying of the two-precursor solution resulted in rapid filming after 2–21 s	(Du et al. 2020)

(continued)

Table 7 (continued)

Polymeric combinations	Incorporated bioactive compounds	Types of wound dressing	Main findings	Refs.
			– The antibacterial experiments showed that an effective barrier was formed during 12 h	
Alginate/pectin	Bovine serum albumin	Foam	– The composite foams were crosslinked by using calcium ions and presented suitable rheological properties – High levels of pectin induced an increase in the water absorption capacity of the foams – The pectin content also influenced the drug-release ability of foams	(Oh et al. 2020)
Alginate	PCL nanoparticles loaded with curcumin	Membrane	– The membrane exhibited a high swelling capacity and adherence to the skin – The dressing was able to regulate the loss of transepidermal water, and its transparent aspects enabled the wound monitoring – The loading of curcumin into nanoparticles facilitated the drug permeation	(Guadarrama-Acevedo et al. 2019)
TEMPO	Alginate	3D-printed scaffolds	– The alginate scaffold was obtained through the ionic crosslinking with calcium chloride – The aerogels with a remarkable water absorption ability were achieved by the freeze drying of the 3D-printed hydrogels	(Espinosa et al. 2019)
Alginate	Eudragit nanoparticles containing edaravone	Hydrogel	– The nanoparticles containing edaravone conferred the remarkable antioxidant ability, sequestrating the ROS	(Fan et al. 2019)

(continued)

Table 7 (continued)

Polymeric combinations	Incorporated bioactive compounds	Types of wound dressing	Main findings	Refs.
			– The hydrogels improved the wound healing process in diabetic mice	
Alginate/PVA	Honey	Electrospun membrane	– The increase in honey content improved the antioxidant activity of the nanofibrous membranes – The honey-loaded nanofibres also demonstrated antibacterial effects towards *S. aureus* and *E. coli*	(Tang et al. 2019)
Alginate	Ciprofloxacin	Wafers	– The alginate-based wafers possessed the ideal wound healing features (wettability and porosity) – The ciprofloxacin release showed an initial burst peak, which is crucial to inhibit the bacterial growth – The dressings showed biocompatibility with human adult keratinocytes	(Ahmed et al. 2018)
Oxidized sodium alginate and polyacrylamide (PAM)	Dopamine	Hydrogel	– The self-healing of the hydrogel was achieved by the hydrogen bonds and dynamic Schiff cross-linking – The hydrogel exhibited efficient self-healing ability, high tensile strength and ultrastretchability – The catechol groups of dopamine augmented the cell and tissue interaction with hydrogel	(Chen et al. 2018a)

(continued)

Table 7 (continued)

Polymeric combinations	Incorporated bioactive compounds	Types of wound dressing	Main findings	Refs.
Sodium alginate	Povidone-iodine	Film	– The dressing blended the wound healing properties of alginates with the bactericidal and fungicidal of povidone-iodine – The films reduced the inflammatory response in in vivo and in vitro assays – The animals treated with films showed a higher wound closure when compared to untreated animals	(Summa et al. 2018)

CIP-loaded wafers exhibited enhanced water absorption capacity, bacterial inhibition and biocompatibility than Algisite Ag® commercial dressing.

In another study, Tang et al. used SA combined with polyvinyl alcohol (PVA) to produce electrospun membranes loaded with honey, by using the electrospinning technique (Tang et al. 2019). After the preparation of PVA and SA aqueous solutions, the honey was added at different concentrations. The authors verified that the honey incorporation from 0% to 20% augmented the nanofibres' diameter from 379 ± 65 nm to 528 ± 160 nm. Furthermore, the nanofibrous membrane with 20% of honey presented the lowest fluid handling ability ($12 \pm 8\%$) and the highest weight loss ($84.82 \pm 0.42\%$) in PBS for 24 h. Also, honey promoted a high radical scavenging activity ($66 \pm 7\%$ after 9 h of interaction) and high bacterial inhibition for *E. coli* and *S. aureus*. However, the honey/SA/PVA nanofibrous membrane presented excellent biocompatibility, facilitating cell proliferation.

Du et al. prepared a self-healing hydrogel formed between adipic acid dihydrazide modified gelatin (Gel-ADH) and monoaldehyde-modified sodium alginate (SA-mCHO) (Du et al. 2020). Through these chemical modifications, the hydrogel is obtained by dynamic and Schiff base bonds between monoaldehyde-modified SA and ADH-modified Gel, resulting in rapid gelation (2 -21 s). Besides, the rapid spray filming ability offers rapidity and flexibility for covering wounds with different shapes and sizes. The hydrogels showed superior cytocompatibility, and the bacterial barrier assays evidenced that the hydrogels provide significant barriers to *S. aureus* and *Candida albicans* for 12 h.

Recently, Dodero et al. produced alginate membranes incorporating zinc oxide (ZnO) nanoparticles through the electrospinning technique, which was then exposed to a washing cross-linking process to achieve greatly stable wound dressings

(Dodero et al. 2019). In this way, the mats' crosslinking process was done using different ionic crosslinking agents (Ca^{2+}, Sr^{2+} or Ba^{2+} ions) and its influence on biological properties was evaluated. The in vitro assays demonstrated that the strontium- and barium-cross-linked membranes promoted cell viability results similar to the commercial porcine collagen membrane (control group). However, the calcium-cross-linked mats displayed good stability over 10 days of incubation in physiological conditions. In turn, the antibacterial experiments confirmed that the ZnO nanoparticles confer antibacterial activity of mats against *E. coli*, maintaining the cells' viability. Overall, the results suggested that the alginate-based mats could be successfully used to prepare surgical patches and wound healing products.

In another work, Johnson et al. used pressurized gas expanded liquid technology (PGX) to produce alginate hydrogels with a very high surface area that facilitated the ibuprofen loading (Johnson et al. 2020). As the main findings, the authors verified that the PGX processing generated highly interconnected networks of alginate with high internal surfaces areas (≈200 m^2/g), which allowed high loadings of ibuprofen (> 8 wt%). Moreover, the drug-loaded alginate scaffolds were crosslinked with calcium ions resulting in durable, bulk and structured hydrogels. In vivo results suggested that the drug-loaded alginate hydrogels accelerated the wound healing and restoration of native skin structure within 21 days. The combination of the alginate (provides to the wound site a moist environment) with the delivery of ibuprofen (regulates the inflammatory response) presented promising properties to be employed in skin injuries treatment.

2.3 Synthetic Polymers

The natural polymers have revealed excellent biological performance on the wound healing process, however, the control of their mechanical features and degradation rates is a major limitation. Unlike, synthetic polymers with predictable, reproducible and improved mechanical and physical properties can be synthesized under standardized conditions. In this way, the degradable synthetic polymers are preferred to non-degradable polymers, since they can be replaced by new skin tissue.

Among different biodegradable polymers, Poly(lactide-co-glycolide) (PLGA), Polyethylene glycol (PEG), Polyurethane (PU) and Polycaprolactone (PCL) are the most reported in wound dressing applications.

PLGA can be synthesized varying lactide/glycolide ratios to tailor the degradation rates, hydrophobic/hydrophilic balance, mechanical properties and crystallinity. This biodegradable copolymer, approved by the FDA displays biocompatibility, mechanical performance and are easily manipulated to obtain selected shapes and sizes (Song et al. 2018; Lanao et al. 2013). Zhao et al. prepared PLGA nanofibre constructs through the electrospinning technique (Zhao et al. 2019). To improve the electrospun scaffold's biological performance, the authors modified the nanofibres' surface with PDA and then loaded them with bFGF and ponericin G1 to improve tissue remodelling and antibacterial properties of the membrane. After that, the authors verified that the PLGA nanofibre scaffold functionalized with PDA presented morphological and surface character that can encounter skin regeneration requirements. Besides, the loading of bFGF and ponericin G1 on the nanofibrous constructs improved the adhesion and proliferation of cells, promoting the increasement of the expression of epidermal repair-related genes. Finally, the in vivo assays confirmed the PDA-PLGA/bFGF/ponericin G1 nanofibre scaffold's excellent potential to promote wound healing, collagen deposition and tissue vascularization.

PEG is a biocompatible, flexible, hydrophilic and non-immunogenic polymer, making it a promising material for the development of wound dressings (Song et al. 2018). Further, its mechanical, thermal properties and crystallinity can be stabilized through blending the PEG with other polymers, whereas the biological features can be improved by the functionalization with bioactive molecules, like growth factors (Shahverdi et al. 2014). Chen et al. produced a PEG-based hydrogel to be used as wound sealants (Chen et al. 2018b). According to the obtained data, the PEG-based hydrogels showed good biocompatibility and presented better results in wound closure and bleeding control compared to a commercial product (Coloskin). In general, the results demonstrated that the biodegradable adhesive PEG-based hydrogels present suitable handling procedures, worthy tissue adhesion, controlled degradability and elastomeric mechanical properties to be used as tissue adhesive sealant.

PU has been a favourable choice for medical devices, due to its toughness, durability, biocompatibility, biostability and biodegradability (Guelcher 2008). PU presents semi-permeability, protecting the wound from external environment and bacterial invasion. The PU membrane provides favourable moist environment for cell proliferation and hence healing process (Song et al. 2018). The limited adherence of PU-based wound dressings can be surpassed through the covering by collagen or collagen-based peptides, promoting the cell adhesion (El-Sayed et al. 2011). Namviriyachote et al. developed a PU foam dressing with the addition of natural polyols and incorporating silver and asiaticoside to endow antimicrobial properties (Namviriyachote et al. 2019). The incorporation of alginate and hydroxypropyl methylcellulose in PU foams improved the water absorption ability and compressive strength. In turn, the foam dressings showed satisfactory release of silver and asiaticoside. When applied in a deep partial-thickness wounds porcine model, the groups treated with PU foams incorporating antimicrobial molecules presented an improved epithelial cells and fibroblasts proliferation and consequently tissue repair process.

Finally, PCL is a saturated aliphatic biodegradable polyester employed in different biomedical applications and has been approved by FDA. PCL can result from the degradation of linear aliphatic polyester and autocatalyzed bulk hydrolysis (Labet and Thielemans 2009). It has distinctive mechanical properties, and is highly valued as non-toxic, biodegradable and bioresorbable polymer, be easily sorted out into different forms and shapes (Mir et al. 2018).

PCL is degraded by microorganisms, hydrolytic and enzymatic (under physiological conditions), or intracellular mechanisms. However, the PCL degradation rate is slow (2–4 years) in comparison to the PLGA. Due to this reason, PCL is more attractive and preferably employed for applications on long-term implants and drug delivery systems (Song et al. 2018).

In skin regeneration purposes, PCL is mainly used to produce electrospun membranes, since it is easily electrospunable and provides the structural integrity to the electrospun nanofibres. Gámez-Herrera et al. decorated the PCL-based nanofibres with electrosprayed PLGA microparticles containing the natural antibacterial compound—thymol (Gámez-Herrera et al. 2020). Through this strategy, the authors obtained dressings that inhibit bacterial growth, without compromising the human dermal fibroblasts and keratinocytes cells. Furthermore, the in vivo assays indicated that the synthesized dressings reduced the bacterial load after 7 days in infected murine excisional wounds.

3 Conclusions and Future Perspectives

The wounds treatment (acute and chronic) constitutes a significant clinical challenge worldwide. Nowadays, the clinicians protect the wounds by using dressing materials capable of assuring a moist environment and offering alleviation of pain symptoms. However, the most used wound dressings are non-degradable, impairing the healing process during its removal/replacement. In this way, the biodegradable polymers have arisen as an excellent strategy to produce biodegradable wound dressings able to act as a transitory supportive matrix as well as providing cues and signals to stimulate the functional tissue connections during tissue regeneration.

In general, natural biodegradable polymers are preferable to synthetic polymers because of their excellent intrinsic biocompatibility. Despite this, natural polymers are mechanically inferior, and the modulation of their chemical properties is not easy as for synthetic polymers.

Thus, it seems that the future for biodegradable polymers is the development of hybrid polymers, mixing different polymers, via various procedures like blending, chemical and grafting reactions. Indeed, numerous recent studies have been demonstrated the excellent effects of these polymeric combinations in wound healing applications.

Nevertheless, an emerging interest arise in the development of 3D matrices that act simultaneously as a matrix for cell proliferation and drug delivery system, incorporating and delivering bioactive agents that propel the healing process. In this way, the future passes by using more precise and advanced technologies (e.g., 3D bioprinting techniques and electrospinning) and nanotechnology to produce the nanosized particles for drug delivery that will boost the wound healing process. Such combination of advanced manufacturing techniques and nanotechnology will constitute an excellent strategy to produce more efficient biodegradable wound dressings to treat severe and devastating skin injuries that affect millions of people worldwide.

References

Abbasi AR, Sohail M, Minhas MU, Khaliq T, Kousar M, Khan S et al (2020) Bioinspired sodium alginate based thermosensitive hydrogel membranes for accelerated wound healing. Int J Biol Macromol 155:751–765

Abdel-Mohsen AM, Frankova J, Abdel-Rahman RM, Salem AA, Sahffie NM, Kubena I et al (2020) Chitosan-glucan complex hollow fibers reinforced collagen wound dressing embedded with aloe vera. II. Multifunctional properties to promote cutaneous wound healing. Int J Pharm 582:119349

Abednejad A, Ghaee A, Morais ES, Sharma M, Neves BM, Freire MG et al (2019b) Polyvinylidene fluoride-hyaluronic acid wound dressing comprised of ionic liquids for controlled drug delivery and dual therapeutic behavior. Acta Biomater 100:142–157

Abednejad A, Ghaee A, Nourmohammadi J, Mehrizi AA (2019a) Hyaluronic acid/carboxylated Zeolitic Imidazolate framework film with improved mechanical and antibacterial properties. Carbohydr Polym 222:115033

Acevedo CA, Sánchez E, Orellana N, Morales P, Olguín Y, Brown DI et al (2019) Re-epithelialization appraisal of skin wound in a porcine model using a salmon-gelatin based biomaterial as wound dressing. Pharmaceutics 11(5):196

Adeli-Sardou M, Yaghoobi MM, Torkzadeh-Mahani M, Dodel M (2019) Controlled release of lawsone from polycaprolactone/gelatin electrospun nano fibers for skin tissue regeneration. Int J Biol Macromol 124:478–491

Ahlawat J, Kumar V, Gopinath P (2019) Carica papaya loaded poly (vinyl alcohol)-gelatin nanofibrous scaffold for potential application in wound dressing. Mater Sci Eng C Mater Biol Appl 103:109834

Ahmed A, Getti G, Boateng J (2018) Ciprofloxacin-loaded calcium alginate wafers prepared by freeze-drying technique for potential healing of chronic diabetic foot ulcers. Drug Deliv Transl Res 8(6):1751–1768

Ahsan SM, Thomas M, Reddy KK, Sooraparaju SG, Asthana A, Bhatnagar I (2018) Chitosan as biomaterial in drug delivery and tissue engineering. Int J Biol Macromol 110:97–109

Akhavan-Kharazian N, Izadi-Vasafi H (2019) Preparation and characterization of chitosan/gelatin/ nanocrystalline cellulose/calcium peroxide films for potential wound dressing applications. Int J Biol Macromol 133:881–891

Alemzadeh E, Oryan A, Mohammadi AA (2020) Hyaluronic acid hydrogel loaded by adipose stem cells enhances wound healing by modulating IL-1β, TGF-β1, and bFGF in burn wound model in rat. J Biomed Mater Res B Appl Biomater 108(2):555–567

Ali A, Shahid MA, Hossain MD, Islam MN (2019) Antibacterial bi-layered polyvinyl alcohol (PVA)-chitosan blend nanofibrous mat loaded with Azadirachta indica (neem) extract. Int J Biol Macromol 138:13–20

Altman GH, Diaz F, Jakuba C, Calabro T, Horan RL, Chen J et al (2003) Silk-based biomaterials. Biomaterials 24(3):401–416

Alves P, Santos M, Mendes S, S PM, K DdS, C SDC et al (2019) Photocrosslinkable Nanofibrous asymmetric membrane designed for wound dressing. Polymers (Basel) 11(4):653

Amer S, Attia N, Nouh S, El-Kammar M, Korittum A, Abu-Ahmed H (2020) Fabrication of sliver nanoparticles/polyvinyl alcohol/gelatin ternary nanofiber mats for wound healing application. J Biomater Appl 35(2):287–298

Arthe R, Arivuoli D, Ravi V (2020) Preparation and characterization of bioactive silk fibroin/ paramylon blend films for chronic wound healing. Int J Biol Macromol 154:1324–1331

Azarniya A, Tamjid E, Eslahi N, Simchi A (2019) Modification of bacterial cellulose/keratin nanofibrous mats by a tragacanth gum-conjugated hydrogel for wound healing. Int J Biol Macromol 134:280–289

Bazmandeh AZ, Mirzaei E, Fadaie M, Shirian S, Ghasemi Y (2020) Dual spinneret electrospun nanofibrous/gel structure of chitosan-gelatin/chitosan-hyaluronic acid as a wound dressing: in-vitro and in-vivo studies. Int J Biol Macromol 162:359–373

Beishenaliev A, Lim SS, Tshai KY, Khiew PS, Moh'd Sghayyar HN, Loh HS (2019) Fabrication and preliminary in vitro evaluation of ultraviolet-crosslinked electrospun fish scale gelatin nanofibrous scaffolds. J Mater Sci Mater Med 30(6):62

Bi H, Feng T, Li B, Han Y (2020) In vitro and in vivo comparison study of electrospun PLA and PLA/PVA/SA fiber membranes for wound healing. Polymers 12(4):839

Biranje SS, Madiwale PV, Patankar KC, Chhabra R, Bangde P, Dandekar P et al (2020) Cytotoxicity and hemostatic activity of chitosan/carrageenan composite wound healing dressing for traumatic hemorrhage. Carbohydr Polym 239:116106

Casanova D, Guerreschi P, Sinna R, Bertheuil N, Philandrianos C, Chignon-Sicard B et al (2020) ALGINATE versus NPWT in the preparation of surgical excisions for an STSG: ATEC trial. Plast Reconstr Surg Glob Open 8(3):e2691

Chalitangkoon J, Wongkittisin M, Monvisade P (2020) Silver loaded hydroxyethylacryl chitosan/sodium alginate hydrogel films for controlled drug release wound dressings. Int J Biol Macromol 159:194–203

Chan TR, Stahl PJ, Li Y, Yu SM (2015) Collagen–gelatin mixtures as wound model, and substrates for VEGF-mimetic peptide binding and endothelial cell activation. Acta Biomater 15:164–172

Chandika P, Ko S-C, Jung W-K (2015) Marine-derived biological macromolecule-based biomaterials for wound healing and skin tissue regeneration. Int J Biol Macromol 77:24–35

Chattopadhyay S, Raines RT (2014) Collagen-based biomaterials for wound healing. Biopolymers 101(8):821–833

Chen T, Chen Y, Rehman HU, Chen Z, Yang Z, Wang M et al (2018a) Ultratough, self-healing, and tissue-adhesive hydrogel for wound dressing. Acs Appl Mater Inter. 10(39):33523–33531

Chen SL, Fu RH, Liao SF, Liu SP, Lin SZ, Wang YC (2018b) A PEG-based hydrogel for effective wound care management. Cell Transplant 27(2):275–284

Chen J, Gao K, Liu S, Wang S, Elango J, Bao B et al (2019) Fish collagen surgical compress repairing characteristics on wound healing process in vivo. Mar Drugs 17(1):33

Chen YW, Lu CH, Shen MH, Lin SY, Chen CH, Chuang CK et al (2020) In vitro evaluation of the hyaluronic acid/alginate composite powder for topical haemostasis and wound healing. Int Wound J 17(2):394–404

Chouhan D, Janani G, Chakraborty B, Nandi SK, Mandal BB (2018b) Functionalized PVA-silk blended nanofibrous mats promote diabetic wound healing via regulation of extracellular matrix and tissue remodelling. J Tissue Eng Regen Med 12(3):e1559–e1e70

Chouhan D, Thatikonda N, Nilebäck L, Widhe M, Hedhammar M, Mandal BB (2018a) Recombinant spider silk functionalized silkworm silk matrices as potential bioactive wound dressings and skin grafts. Acs Appl Mater Inter. 10(28):23560–23572

Craciunescu O, Gaspar A, Trif M, Moisei M, Oancea A, Moldovan L et al (2014) Preparation and characterization of a collagen-liposome-chondroitin sulfate matrix with potential application for inflammatory disorders treatment. J Nanomater 2014:1–9

Cunha CS, Castro PJ, Sousa SC, Pullar RC, Tobaldi DM, Piccirillo C et al (2020) Films of chitosan and natural modified hydroxyapatite as effective UV-protecting, biocompatible and antibacterial wound dressings. Int J Biol Macromol 159:177–185

D'Agostino A, Maritato R, La Gatta A, Fusco A, Reale S, Stellavato A et al (2019) In vitro evaluation of novel hybrid cooperative complexes in a wound healing model: a step toward improved bioreparation. Int J Mol Sci 20(19):4727

De Angelis B, D'Autilio M, Orlandi F, Pepe G, Garcovich S, Scioli MG et al (2019) Wound healing: in vitro and in vivo evaluation of a bio-functionalized scaffold based on hyaluronic acid and platelet-rich plasma in chronic ulcers. J Clin Med 8(9):1486

Diniz FR, Maia R, Rannier L, Andrade LN, M VC, da Silva CF et al (2020) Silver nanoparticles-composing alginate/Gelatine hydrogel improves wound healing in vivo. Nanomaterials (Basel) 10(2):390

Dodero A, Scarfi S, Pozzolini M, Vicini S, Alloisio M, Castellano M (2019) Alginate-based electrospun membranes containing ZnO nanoparticles as potential wound healing patches:

biological, mechanical, and physicochemical characterization. Acs Appl Mater Inter. 12 (3):3371–3381

Du Y, Li L, Peng H, Zheng H, Cao S, Lv G et al (2020) A spray-filming self-healing hydrogel fabricated from modified sodium alginate and gelatin as a bacterial barrier. Macromol Biosci 20 (2):e1900303

Du X, Liu Y, Wang X, Yan H, Wang L, Qu L et al (2019) Injectable hydrogel composed of hydrophobically modified chitosan/oxidized-dextran for wound healing. Mater Sci Eng C Mater Biol Appl 104:109930

Duan Y, Li K, Wang H, Wu T, Zhao Y, Li H et al (2020) Preparation and evaluation of curcumin grafted hyaluronic acid modified pullulan polymers as a functional wound dressing material. Carbohydr Polym 238(10):116195

Duckworth PF, Maddocks SE, Rahatekar SS, Barbour ME (2020) Alginate films augmented with chlorhexidine hexametaphosphate particles provide sustained antimicrobial properties for application in wound care. J Mater Sci Mater Med 31(3):1–9

Edwards N, Feliers D, Zhao Q, Stone R, Christy R, Cheng X (2018) An electrochemically deposited collagen wound matrix combined with adipose-derived stem cells improves cutaneous wound healing in a mouse model of type 2 diabetes. J Biomater Appl 33(4):553–565

Ehterami A, Salehi M, Farzamfar S, Samadian H, Vaez A, Sahrapeyma H et al (2020) A promising wound dressing based on alginate hydrogels containing vitamin D3 cross-linked by calcium carbonate/d-glucono-δ-lactone. Biomed Eng Lett 10:1–11

El-Aassar MR, Ibrahim OM, Fouda MMG, El-Beheri NG, Agwa MM (2020) Wound healing of nanofiber comprising Polygalacturonic/hyaluronic acid embedded silver nanoparticles: in-vitro and in-vivo studies. Carbohydr Polym 238:116175

El-Sayed S, Mahmoud K, Fatah A, Hassen A (2011) DSC, TGA and dielectric properties of carboxymethyl cellulose/polyvinyl alcohol blends. Phys B Condens Matter 406(21):4068–4076

Enumo A Jr, Argenta DF, Bazzo GC, Caon T, Stulzer HK, Parize AL (2020) Development of curcumin-loaded chitosan/pluronic membranes for wound healing applications. Int J Biol Macromol 163:167–179

Eskandarinia A, Kefayat A, Agheb M, Rafienia M, Amini Baghbadorani M, Navid S et al (2020b) A novel bilayer wound dressing composed of a dense polyurethane/Propolis membrane and a biodegradable Polycaprolactone/gelatin Nanofibrous membrane. Sci Rep 10(1):3063

Eskandarinia A, Kefayat A, Gharakhloo M, Agheb M, Khodabakhshi D, Khorshidi M et al (2020a) A propolis enriched polyurethane-hyaluronic acid nanofibrous wound dressing with remarkable antibacterial and wound healing activities. Int J Biol Macromol 149:467–476

Eskandarinia A, Kefayat A, Rafienia M, Agheb M, Navid S, Ebrahimpour K (2019) Cornstarch-based wound dressing incorporated with hyaluronic acid and propolis: in vitro and in vivo studies. Carbohydr Polym 216:25–35

Espinosa E, Filgueira D, Rodriguez A, Chinga-Carrasco G (2019) Nanocellulose-based inks-effect of alginate content on the water absorption of 3D printed constructs. Bioengineering (Basel) 6 (3):65

Fan X, Chen K, He X, Li N, Huang J, Tang K et al (2016) Nano-TiO2/collagen-chitosan porous scaffold for wound repairing. Int J Biol Macromol 91:15–22

Fan Y, Wu W, Lei Y, Gaucher C, Pei S, Zhang J et al (2019) Edaravone-loaded alginate-based nanocomposite hydrogel accelerated chronic wound healing in diabetic mice. Mar Drugs 17 (5):285

Farahani H, Barati A, Arjomandzadegan M, Vatankhah E (2020) Nanofibrous cellulose acetate/gelatin wound dressing endowed with antibacterial and healing efficacy using nanoemulsion of Zataria multiflora. Int J Biol Macromol 162:762–773

Felician FF, Xia C, Qi W, Xu H (2018) Collagen from marine biological sources and medical applications. Chem Biodivers 15(5):e1700557

Feng Y, Li X, Zhang Q, Yan S, Guo Y, Li M et al (2019) Mechanically robust and flexible silk protein/polysaccharide composite sponges for wound dressing. Carbohydr Polym 216:17–24

Frykberg RG, Banks J (2015) Challenges in the treatment of chronic wounds. Adv Wound Care 4 (9):560–582

Gámez-Herrera E, García-Salinas S, Salido S, Sancho-Albero M, Andreu V, Pérez M et al (2020) Drug-eluting wound dressings having sustained release of antimicrobial compounds. Eur J Pharm Biopharm 152:327–339

Garcia-Orue I, Santos-Vizcaino E, Etxabide A, Uranga J, Bayat A, Guerrero P et al (2019) Development of bioinspired gelatin and gelatin/chitosan bilayer Hydrofilms for wound healing. Pharmaceutics 11(7):314

Ghalei S, Nourmohammadi J, Solouk A, Mirzadeh H (2018) Enhanced cellular response elicited by addition of amniotic fluid to alginate hydrogel-electrospun silk fibroin fibers for potential wound dressing application. Colloids Surf B Biointerfaces 172:82–89

Giuri D, Barbalinardo M, Sotgiu G, Zamboni R, Nocchetti M, Donnadio A et al (2019) Nano-hybrid electrospun non-woven mats made of wool keratin and hydrotalcites as potential bio-active wound dressings. Nanoscale 11(13):6422–6430

Govindarajan D, Duraipandy N, Srivatsan KV, Lakra R, Korapatti PS, Jayavel R et al (2017) Fabrication of hybrid collagen aerogels reinforced with wheat grass bioactives as instructive scaffolds for collagen turnover and angiogenesis for wound healing applications. Acs Appl Mater Inter. 9(20):16939–16950

Graça MF, Miguel SP, Cabral CS, Correia IJ (2020) Hyaluronic acid-based wound dressings: a review. Carbohydr Polym 241:116364

Guadarrama-Acevedo MC, Mendoza-Flores RA, Del Prado-Audelo ML, Urban-Morlan Z, Giraldo-Gomez DM, Magana JJ et al (2019) Development and evaluation of alginate membranes with curcumin-loaded nanoparticles for potential wound-healing applications. Pharmaceutics 11 (8):389

Guelcher SA (2008) Biodegradable polyurethanes: synthesis and applications in regenerative medicine. Tissue Eng Part B Rev 14(1):3–17

Hadisi Z, Farokhi M, Bakhsheshi-Rad HR, Jahanshahi M, Hasanpour S, Pagan E et al (2020) Hyaluronic acid (HA)-based silk fibroin/zinc oxide Core–Shell electrospun dressing for burn wound management. Macromol Biosci 20(4):1900328

Hadisi Z, Nourmohammadi J, Nassiri SM (2018) The antibacterial and anti-inflammatory investigation of Lawsonia Inermis-gelatin-starch nano-fibrous dressing in burn wound. Int J Biol Macromol 107(Pt B):2008–2019

Hamasaki S, Tachibana A, Tada D, Yamauchi K, Tanabe T (2008) Fabrication of highly porous keratin sponges by freeze-drying in the presence of calcium alginate beads. Mater Sci Eng C 28 (8):1250–1254

Han L, Li PF, Tang PF, Wang X, Zhou T, Wang KF et al (2019) Mussel-inspired cryogels for promoting wound regeneration through photobiostimulation, modulating inflammatory responses and suppressing bacterial invasion. Nanoscale 11(34):15846–15861

Hartrianti P, Nguyen LT, Johanes J, Chou SM, Zhu P, Tan NS et al (2017) Fabrication and characterization of a novel crosslinked human keratin-alginate sponge. J Tissue Eng Regen Med 11(9):2590–2602

Hashimoto T, Kojima K, Tamada Y (2020) Higher gene expression related to wound healing by fibroblasts on silk fibroin biomaterial than on collagen. Molecules 25(8):1939

Helary C, Abed A, Mosser G, Louedec L, Letourneur D, Coradin T et al (2015) Evaluation of dense collagen matrices as medicated wound dressing for the treatment of cutaneous chronic wounds. Biomater Sci 3(2):373–382

Hernandez-Rangel A, Silva-Bermudez P, Espana-Sanchez BL, Luna-Hernandez E, Almaguer-Flores A, Ibarra C et al (2019) Fabrication and in vitro behavior of dual-function chitosan/silver nanocomposites for potential wound dressing applications. Mater Sci Eng C Mater Biol Appl 94:750–765

Hong L, Shen M, Fang J, Wang Y, Bao Z, Bu S et al (2018) Hyaluronic acid (HA)-based hydrogels for full-thickness wound repairing and skin regeneration. J Mater Sci Mater Med 29(9):150

Hou S, Liu Y, Feng F, Zhou J, Feng X, Fan Y (2020) Polysaccharide-peptide Cryogels for multidrug-resistant-bacteria infected wound healing and hemostasis. Adv Healthc Mater 9(3): e1901041

Hu W, Wang Z, Zha Y, Gu X, You W, Xiao Y et al (2020) High flexible and broad antibacterial Nanodressing induces complete skin repair with Angiogenic and follicle regeneration. Adv Healthc Mater:e2000035

Hu G, Xiao L, Tong P, Bi D, Wang H, Ma H et al (2012) Antibacterial hemostatic dressings with nanoporous bioglass containing silver. Int J Nanomedicine 7:2613

Huang S, Liu H, Liao K, Hu Q, Guo R, Deng K (2020) Functionalized GO Nanovehicles with nitric oxide release and Photothermal activity-based hydrogels for bacteria-infected wound healing. ACS Appl Mater Interfaces 12(26):28952–28964

Huang TY, Wang GS, Tseng CC, Su WT (2019) Epidermal cells differentiated from stem cells from human exfoliated deciduous teeth and seeded onto polyvinyl alcohol/silk fibroin nanofiber dressings accelerate wound repair. Mater Sci Eng C Mater Biol Appl 104:109986

Hubner P, Donati N, Quines LKM, Tessaro IC, Marcilio NR (2020) Gelatin-based films containing clinoptilolite-ag for application as wound dressing. Mater Sci Eng C Mater Biol Appl 107:110215

Ilhan E, Cesur S, Guler E, Topal F, Albayrak D, Guncu MM et al (2020) Development of Satureja cuneifolia-loaded sodium alginate/polyethylene glycol scaffolds produced by 3D-printing technology as a diabetic wound dressing material. Int J Biol Macromol 161:1040–1054

Inal M, Mulazimoglu G (2019) Production and characterization of bactericidal wound dressing material based on gelatin nanofiber. Int J Biol Macromol 137:392–404

Intini C, Elviri L, Cabral J, Mros S, Bergonzi C, Bianchera A et al (2018) 3D-printed chitosan-based scaffolds: an in vitro study of human skin cell growth and an in-vivo wound healing evaluation in experimental diabetes in rats. Carbohydr Polym 199:593–602

Jantrawut P, Bunrueangtha J, Suerthong J, Kantrong N (2019) Fabrication and characterization of low Methoxyl pectin/gelatin/Carboxymethyl cellulose absorbent hydrogel film for wound dressing applications. Materials (Basel) 12(10):1628

Jao D, Mou X, Hu X (2016) Tissue regeneration: a silk road. J Funct Biomater 7(3):22

Jin H-S, Park SY, Kim K, Lee Y-J, Nam G-W, Kang NJ et al (2017) Development of a keratinase activity assay using recombinant chicken feather keratin substrates. PLoS One 12(2):e0172712

Johnson KA, Muzzin N, Toufanian S, Slick RA, Lawlor MW, Seifried B et al (2020) Drug-impregnated, pressurized gas expanded liquid-processed alginate hydrogel scaffolds for accel-erated burn wound healing. Acta Biomater 112:101–111

Ju HW, Lee OJ, Lee JM, Moon BM, Park HJ, Park YR et al (2016) Wound healing effect of electrospun silk fibroin nanomatrix in burn-model. Int J Biol Macromol 85:29–39

Karimi Dehkordi N, Minaiyan M, Talebi A, Akbari V, Taheri A (2019) Nanocrystalline cellulose-hyaluronic acid composite enriched with GM-CSF loaded chitosan nanoparticles for enhanced wound healing. Biomed Mater 14(3):035003

Kenawy E, Omer AM, Tamer TM, Elmeligy MA, Eldin MSM (2019) Fabrication of biodegradable gelatin/chitosan/cinnamaldehyde crosslinked membranes for antibacterial wound dressing applications. Int J Biol Macromol 139:440–448

Kheradvar SA, Nourmohammadi J, Tabesh H, Bagheri B (2018) Starch nanoparticle as a vitamin E-TPGS carrier loaded in silk fibroin-poly (vinyl alcohol)-Aloe vera nanofibrous dressing. Colloids Surf B: Biointerfaces 166:9–16

Kim IY, Seo SJ, Moon HS, Yoo MK, Park IY, Kim BC et al (2008) Chitosan and its derivatives for tissue engineering applications. Biotechnol Adv 26(1):1–21

Konop M, Czuwara J, Kłodzińska E, Laskowska AK, Sulejczak D, Damps T et al (2020) Evaluation of keratin biomaterial containing silver nanoparticles as a potential wound dressing in full-thickness skin wound model in diabetic mice. J Tissue Eng Regen Med 14(2):334–346

Konop M, Czuwara J, Kłodzińska E, Laskowska AK, Zielenkiewicz U, Brzozowska I et al (2018) Development of a novel keratin dressing which accelerates full-thickness skin wound healing in diabetic mice: in vitro and in vivo studies. J Biomater Appl 33(4):527–540

Konop M, Sulejczak D, Czuwara J, Kosson P, Misicka A, Lipkowski AW et al (2017) The role of allogenic keratin-derived dressing in wound healing in a mouse model. Wound Repair Regen 25 (1):62–74

Konstantinow A, Fischer TV, Ring J (2017) Effectiveness of collagen/oxidised regenerated cellulose/silver-containing composite wound dressing for the treatment of medium-depth split-thickness skin graft donor site wounds in multi-morbid patients: a prospective, non-comparative, single-Centre study. Int Wound J 14(5):791–800

Koumentakou I, Terzopoulou Z, Michopoulou A, Kalafatakis I, Theodorakis K, Tzetzis D et al (2020) Chitosan dressings containing inorganic additives and levofloxacin as potential wound care products with enhanced hemostatic properties. Int J Biol Macromol 162:693–703

Kumar A, Behl T, Chadha S (2020) Synthesis of physically crosslinked PVA/chitosan loaded silver nanoparticles hydrogels with tunable mechanical properties and antibacterial effects. Int J Biol Macromol 149:1262–1274

Kundu B, Rajkhowa R, Kundu SC, Wang X (2013) Silk fibroin biomaterials for tissue regenerations. Adv Drug Deliv Rev 65(4):457–470

Labet M, Thielemans W (2009) Synthesis of polycaprolactone: a review. Chem Soc Rev 38 (12):3484–3504

Lanao RPF, Jonker AM, Wolke JG, Jansen JA, van Hest JC, Leeuwenburgh SC (2013) Physicochemical properties and applications of poly (lactic-co-glycolic acid) for use in bone regeneration. Tissue Eng Part B Rev 19(4):380–390

Lee KY, Mooney DJ (2012) Alginate: properties and biomedical applications. Prog Polym Sci 37 (1):106–126

Lee CH, Singla A, Lee Y (2001) Biomedical applications of collagen. Int J Pharm 221(1–2):1–22

Li H, Chen Y, Lu W, Xu Y, Guo Y, Yang G (2020) Preparation of electrospun gelatin mat with incorporated zinc oxide/graphene oxide and its antibacterial activity. Molecules 25(5):1043

Liang A, Zhang M, Luo H, Niu L, Feng Y, Li M (2020) Porous poly(Hexamethylene Biguanide) hydrochloride loaded silk fibroin sponges with antibacterial function. Materials (Basel) 13 (2):285

Liang Y, Zhao X, Hu T, Chen B, Yin Z, Ma PX et al (2019a) Adhesive hemostatic conducting injectable composite hydrogels with sustained drug release and Photothermal antibacterial activity to promote full-thickness skin regeneration during wound healing. Small 15(12): e1900046

Liang YP, Zhao X, Hu TL, Han Y, Guo BL (2019b) Mussel-inspired, antibacterial, conductive, antioxidant, injectable composite hydrogel wound dressing to promote the regeneration of infected skin. J Colloid Interf Sci 556:514–528

Lin CW, Chen YK, Lu M, Lou KL, Yu J (2018) Photo-crosslinked keratin/chitosan membranes as potential wound dressing materials. Polymers (Basel) 10(9):985

Lin X, Guan X, Wu Y, Zhuang S, Wu Y, Du L et al (2020) An alginate/poly (N-isopropylacrylamide)-based composite hydrogel dressing with stepwise delivery of drug and growth factor for wound repair. Mater Sci Eng C Mater Biol Appl 115:111123

Lin Z, Wu T, Wang W, Li B, Wang M, Chen L et al (2019) Biofunctions of antimicrobial peptide-conjugated alginate/hyaluronic acid/collagen wound dressings promote wound healing of a mixed-bacteria-infected wound. Int J Biol Macromol 140:330–342

Liu S, Liu X, Ren Y, Ph W, Pu Y, Yang R et al (2020) Mussel-inspired dual-crosslinking hyaluronic acid/ε-polylysine hydrogel with self-healing and antibacterial properties for wound healing. Acs Appl Mater Inter 12(25):27876–27888

Liu J, Yan L, Yang W, Lan Y, Zhu Q, Xu H et al (2019) Controlled-release neurotensin-loaded silk fibroin dressings improve wound healing in diabetic rat model. Bioact Mater 4:151–159

Majumder S, Ranjan Dahiya U, Yadav S, Sharma P, Ghosh D, Rao GK et al (2020) Zinc oxide nanoparticles functionalized on hydrogel grafted silk fibroin fabrics as efficient composite dressing. Biomol Ther 10(5):710

Makvandi P, Ali GW, Della Sala F, Abdel-Fattah WI, Borzacchiello A (2019) Biosynthesis and characterization of antibacterial thermosensitive hydrogels based on corn silk extract, hyaluronic acid and nanosilver for potential wound healing. Carbohydr Polym 223:115023

Marin Ş, Albu Kaya MG, Ghica MV, Dinu-Pîrvu C, Popa L, Udeanu DI et al (2018) Collagen-polyvinyl alcohol-indomethacin biohybrid matrices as wound dressings. Pharmaceutics. 10 (4):224

Matica MA, Aachmann FL, Tøndervik A, Sletta H, Ostafe V (2019) Chitosan as a wound dressing starting material: antimicrobial properties and mode of action. Int J Mol Sci 20(23):5889

Mazloom-Jalali A, Shariatinia Z, Tamai IA, Pakzad S-R, Malakootikhah J (2020) Fabrication of chitosan–polyethylene glycol nanocomposite films containing ZIF-8 nanoparticles for application as wound dressing materials. Int J Biol Macromol 153:421–432

Mehrabani MG, Karimian R, Mehramouz B, Rahimi M, Kafil HS (2018) Preparation of biocompatible and biodegradable silk fibroin/chitin/silver nanoparticles 3D scaffolds as a bandage for antimicrobial wound dressing. Int J Biol Macromol 114:961–971

Mehta MA, Shah S, Ranjan V, Sarwade P, Philipose A (2019) Comparative study of silver-sulfadiazine-impregnated collagen dressing versus conventional burn dressings in second-degree burns. J Family Med Primary Care 8(1):215

Miguel SP, Simoes D, Moreira AF, Sequeira RS, Correia IJ (2019) Production and characterization of electrospun silk fibroin based asymmetric membranes for wound dressing applications. Int J Biol Macromol 121:524–535

Mir M, Ali MN, Barakullah A, Gulzar A, Arshad M, Fatima S et al (2018) Synthetic polymeric biomaterials for wound healing: a review. Prog Biomater 7(1):1–21

Mogoşanu GD, Grumezescu AM (2014) Natural and synthetic polymers for wounds and burns dressing. Int J Pharm 463(2):127–136

Mogoşanu G, Popescu FC, Busuioc CJ, Pârvănescu H, Lascăr I (2012) Natural products locally modulators of the cellular response: therapeutic perspectives in skin burns. Romanian J Morphol Embryol 53(2):249–262

Moreira Filho RNF, Vasconcelos NF, Andrade FK, Rosa MF, Vieira RS (2020) Papain immobilized on alginate membrane for wound dressing application. Colloids Surf B Biointerfaces 194:111222

Movahedi M, Asefnejad A, Rafienia M, Khorasani MT (2020) Potential of novel electrospun core-shell structured polyurethane/starch (hyaluronic acid) nanofibers for skin tissue engineering: in vitro and in vivo evaluation. Int J Biol Macromol 146:627–637

Nagai T, Suzuki N (2000) Isolation of collagen from fish waste material—skin, bone and fins. Food Chem 68(3):277–281

Namviriyachote N, Lipipun V, Akkhawattanangkul Y, Charoonrut P, Ritthidej GC (2019) Development of polyurethane foam dressing containing silver and asiaticoside for healing of dermal wound. Asian J Pharm Sci 14(1):63–77

Navarro J, Clohessy RM, Holder RC, Gabard AR, Herendeen GJ, Christy RJ et al (2020) In vivo evaluation of three-dimensional printed, keratin-based hydrogels in a porcine thermal burn model. Tissue Eng Part A 26(5–6):265–278

Nayak KK, Gupta P (2017) Study of the keratin-based therapeutic dermal patches for the delivery of bioactive molecules for wound treatment. Mater Sci Eng C Mater Biol Appl 77:1088–1097

Oancea A, Popa E, Zărnescu O, Angelescu N (2000) The effects of cicatrizant wound compresses. Chirurgia (Bucharest) 95(3):245–252

Oh GW, Nam SY, Heo SJ, Kang DH, Jung WK (2020) Characterization of ionic cross-linked composite foams with different blend ratios of alginate/pectin on the synergistic effects for wound dressing application. Int J Biol Macromol 156:1565–1573

Oryan A, Jalili M, Kamali A, Nikahval B (2018) The concurrent use of probiotic microorganism and collagen hydrogel/scaffold enhances burn wound healing: an in vivo evaluation. Burns 44 (7):1775–1786

Pacheco MS, Kano GE, Paulo LA, Lopes PS, de Moraes MA (2020) Silk fibroin/chitosan/alginate multilayer membranes as a system for controlled drug release in wound healing. Int J Biol Macromol 152:803–811

Panico A, Paladini F, Pollini M (2019) Development of regenerative and flexible fibroin-based wound dressings. J Biomed Mater Res B Appl Biomater 107(1):7–18

Parenteau-Bareil R, Gauvin R, Berthod F (2010) Collagen-based biomaterials for tissue engineering applications. Materials. 3(3):1863–1887

Park M, Shin HK, Kim BS, Kim MJ, Kim IS, Park BY et al (2015) Effect of discarded keratin-based biocomposite hydrogels on the wound healing process in vivo. Mater Sci Eng C Mater Biol Appl 55:88–94

Pawar SN, Edgar KJ (2012) Alginate derivatization: a review of chemistry, properties and applications. Biomaterials 33(11):3279–3305

Petersen W, Rahmanian-Schwarz A, Werner J-O, Schiefer J, Rothenberger J, Hübner G et al (2016) The use of collagen-based matrices in the treatment of full-thickness wounds. Burns 42 (6):1257–1264

Pires ALR, Westin CB, Hernandez-Montelongo J, Sousa IMO, Foglio MA, Moraes AM (2020) Flexible, dense and porous chitosan and alginate membranes containing the standardized extract of Arrabidaea chica Verlot for the treatment of skin lesions. Mater Sci Eng C Mater Biol Appl 112:110869

Ramalingam R, Dhand C, Leung CM, Ezhilarasu H, Prasannan P, Ong ST et al (2019) Poly-ε-caprolactone/gelatin hybrid electrospun composite nanofibrous mats containing ultrasound assisted herbal extract: antimicrobial and cell proliferation study. Nano 9(3):462

Rehman SRU, Augustine R, Zahid AA, Ahmed R, Tariq M, Hasan A (2019) Reduced graphene oxide incorporated GelMA hydrogel promotes angiogenesis for wound healing applications. Int J Nanomedicine 14:9603–9617

Rezaei N, Hamidabadi HG, Khosravimelal S, Zahiri M, Ahovan ZA, Bojnordi MN et al (2020) Antimicrobial peptides-loaded smart chitosan hydrogel: release behavior and antibacterial potential against antibiotic resistant clinical isolates. Int J Biol Macromol 164:855–862

Ricci E, Cutting K (2016) Evaluating a native collagen matrix dressing in the treatment of chronic wounds of different aetiologies: a case series. J Wound Care 25(11):670–678

Rodrigues Pereira J, Suassuna Bezerra G, Alves Furtado A, de Carvalho TG, Costa da Silva V, Lins Bispo Monteiro A et al (2020) Chitosan film containing Mansoa hirsuta fraction for wound healing. Pharmaceutics 12(6):484

Rouse JG, Van Dyke ME (2010) A review of keratin-based biomaterials for biomedical applications. Materials 3(2):999–1014

Roy DC, Tomblyn S, Isaac KM, Kowalczewski CJ, Burmeister DM, Burnett LR et al (2016) Ciprofloxacin-loaded keratin hydrogels reduce infection and support healing in a porcine partial-thickness thermal burn. Wound Repair Regen 24(4):657–668

Sakthiguru N, Sithique MA (2020) Fabrication of bioinspired chitosan/gelatin/allantoin biocomposite film for wound dressing application. Int J Biol Macromol 152:873–883

Salehi M, Ehterami A, Farzamfar S, Vaez A, Ebrahimi-Barough S (2020) Accelerating healing of excisional wound with alginate hydrogel containing naringenin in rat model. Drug Deliv Transl Res 11(1):142–153

Samadian H, Salehi M, Farzamfar S, Vaez A, Ehterami A, Sahrapeyma H et al (2018) In vitro and in vivo evaluation of electrospun cellulose acetate/gelatin/hydroxyapatite nanocomposite mats for wound dressing applications. Artif Cells Nanomed Biotechnol 46(sup1):964–974

Samadian H, Zamiri S, Ehterami A, Farzamfar S, Vaez A, Khastar H et al (2020) Electrospun cellulose acetate/gelatin nanofibrous wound dressing containing berberine for diabetic foot ulcer healing: in vitro and in vivo studies. Sci Rep 10(1):8312

Sarheed O, Ahmed A, Shouqair D, Boateng J (2016) Antimicrobial dressings for improving wound healing. In: Alexandrescu V (ed) Wound healing-new insights into ancient challenges, pp 373–398

Schiefer JL, Rath R, Held M, Petersen W, Werner J-O, Schaller H-E et al (2016) Frequent application of the new gelatin-collagen nonwoven accelerates wound healing. Adv Skin Wound Care 29(2):73–78

Selvaraj S, Duraipandy N, Kiran MS, Fathima NN (2018) Anti-oxidant enriched hybrid nanofibers: effect on mechanical stability and biocompatibility. Int J Biol Macromol 117:209–217

Senthil R, Berly R, Bhargavi Ram T, Gobi N (2018) Electrospun poly(vinyl) alcohol/collagen nanofibrous scaffold hybridized by graphene oxide for accelerated wound healing. Int J Artif Organs 41(8):467–473

Seol YJ, Lee H, Copus JS, Kang HW, Cho DW, Atala A et al (2018) 3D bioprinted BioMask for facial skin reconstruction. Bioprinting 10:e00028

Seon-Lutz M, Couffin AC, Vignoud S, Schlatter G, Hebraud A (2019) Electrospinning in water and in situ crosslinking of hyaluronic acid /cyclodextrin nanofibers: towards wound dressing with controlled drug release. Carbohydr Polym 207:276–287

Shah A, Ali Buabeid M, Arafa EA, Hussain I, Li L, Murtaza G (2019) The wound healing and antibacterial potential of triple-component nanocomposite (chitosan-silver-sericin) films loaded with moxifloxacin. Int J Pharm 564:22–38

Shahverdi S, Hajimiri M, Esfandiari MA, Larijani B, Atyabi F, Rajabiani A et al (2014) Fabrication and structure analysis of poly (lactide-co-glycolic acid)/silk fibroin hybrid scaffold for wound dressing applications. Int J Pharm 473(1–2):345–355

Shahzadi L, Bashir M, Tehseen S, Zehra M, Mehmood A, Chaudhry AA et al (2020) Thyroxine impregnated chitosan-based dressings stimulate angiogenesis and support fast wounds healing in rats: potential clinical candidates. Int J Biol Macromol 160:296–306

Sharma S, Swetha KL, Roy A (2019) Chitosan-chondroitin sulfate based polyelectrolyte complex for effective management of chronic wounds. Int J Biol Macromol 132:97–108

Shen X-R, Chen X-L, Xie H-X, He Y, Chen W, Luo Q et al (2017) Beneficial effects of a novel shark-skin collagen dressing for the promotion of seawater immersion wound healing. Mil Med Res 4(1):1–12

Shokrollahi M, Bahrami SH, Nazarpak MH, Solouk A (2020) Multilayer nanofibrous patch comprising chamomile loaded carboxyethyl chitosan/poly (vinyl alcohol) and polycaprolactone as a potential wound dressing. Int J Biol Macromol 147:547–559

Shyna S, Shanti Krishna A, Nair PD, Thomas LV (2020) A nonadherent chitosan-polyvinyl alcohol absorbent wound dressing prepared via controlled freeze-dry technology. Int J Biol Macromol 150:129–140

Si H, Xing T, Ding Y, Zhang H, Yin R, Zhang W (2019) 3D bioprinting of the sustained drug release wound dressing with double-crosslinked hyaluronic-acid-based hydrogels. Polymers (Basel) 11(10):1584

Sierra-Sanchez A, Fernandez-Gonzalez A, Lizana-Moreno A, Espinosa-Ibanez O, Martinez-Lopez-A, Guerrero-Calvo J et al (2020) Hyaluronic acid biomaterial for human tissue-engineered skin substitutes: preclinical comparative in vivo study of wound healing. J Eur Acad Dermatol Venereol 34(10):2414–2427

Silva LP, Pirraco RP, Santos TC, Novoa-Carballal R, Cerqueira MT, Reis RL et al (2016) Neovascularization induced by the hyaluronic acid-based spongy-like hydrogels degradation products. ACS Appl Mater Interfaces 8(49):33464–33474

Silvestro I, Lopreiato M, Scotto d'Abusco A, Di Lisio V, Martinelli A, Piozzi A et al (2020) Hyaluronic acid reduces bacterial fouling and promotes fibroblasts' adhesion onto chitosan 2D-wound dressings. Int J Mol Sci 21(6):2070

Simões D, Miguel SP, Ribeiro MP, Coutinho P, Mendonça AG, Correia IJ (2018) Recent advances on antimicrobial wound dressing: a review. Eur J Pharm Biopharm 127:130–141

Singaravelu S, Ramanathan G, Raja MD, Nagiah N, Padmapriya P, Kaveri K et al (2016) Biomimetic interconnected porous keratin-fibrin-gelatin 3D sponge for tissue engineering application. Int J Biol Macromol 86:810–819

Song DW, Kim SH, Kim HH, Lee KH, Ki CS, Park YH (2016) Multi-biofunction of antimicrobial peptide-immobilized silk fibroin nanofiber membrane: implications for wound healing. Acta Biomater 39:146–155

Song R, Murphy M, Li C, Ting K, Soo C, Zheng Z (2018) Current development of biodegradable polymeric materials for biomedical applications. Drug Des Devel Ther 12:3117–3145

Song R, Zheng J, Liu Y, Tan Y, Yang Z, Song X et al (2019) A natural cordycepin/chitosan complex hydrogel with outstanding self-healable and wound healing properties. Int J Biol Macromol 134:91–99

Srivastava CM, Purwar R, Gupta AP (2019) Enhanced potential of biomimetic, silver nanoparticles functionalized Antheraea mylitta (tasar) silk fibroin nanofibrous mats for skin tissue engineering. Int J Biol Macromol 130:437–453

Stubbe B, Mignon A, Declercq H, Van Vlierberghe S, Dubruel P (2019) Development of gelatin-alginate hydrogels for burn wound treatment. Macromol Biosci 19(8):e1900123

Sulaeva I, Hettegger H, Bergen A, Rohrer C, Kostic M, Konnerth J et al (2020) Fabrication of bacterial cellulose-based wound dressings with improved performance by impregnation with alginate. Mater Sci Eng C Mater Biol Appl 110:110619

Summa M, Russo D, Penna I, Margaroli N, Bayer IS, Bandiera T et al (2018) A biocompatible sodium alginate/povidone iodine film enhances wound healing. Eur J Pharm Biopharm 122:17–24

Sun X, Ma C, Gong W, Ma Y, Ding Y, Liu L (2020) Biological properties of sulfanilamide-loaded alginate hydrogel fibers based on ionic and chemical crosslinking for wound dressings. Int J Biol Macromol 157:522–529

Taheri P, Jahanmardi R, Koosha M, Abdi S (2020) Physical, mechanical and wound healing properties of chitosan/gelatin blend films containing tannic acid and/or bacterial nanocellulose. Int J Biol Macromol 154:421–432

Tan WS, Arulselvan P, Ng S-F, Taib CNM, Sarian MN, Fakurazi S (2020) Healing effect of Vicenin-2 (VCN-2) on human dermal fibroblast (HDF) and development VCN-2 hydrocolloid film based on alginate as potential wound dressing. Biomed Res Int 2020:4730858

Tang Y, Lan X, Liang C, Zhong Z, Xie R, Zhou Y et al (2019) Honey loaded alginate/PVA nanofibrous membrane as potential bioactive wound dressing. Carbohydr Polym 219:113–120

Thones S, Rother S, Wippold T, Blaszkiewicz J, Balamurugan K, Moeller S et al (2019) Hyaluronan/collagen hydrogels containing sulfated hyaluronan improve wound healing by sustained release of heparin-binding EGF-like growth factor. Acta Biomater 86:135–147

Tort S, Acartürk F, Beşikci A (2017a) Evaluation of three-layered doxycycline-collagen loaded nanofiber wound dressing. Int J Pharm 529(1–2):642–653

Tort S, Acarturk F, Besikci A (2017b) Evaluation of three-layered doxycycline-collagen loaded nanofiber wound dressing. Int J Pharm 529(1–2):642–653

Tu H, Wu G, Yi Y, Huang M, Liu R, Shi X et al (2019) Layer-by-layer immobilization of amphoteric carboxymethyl chitosan onto biocompatible silk fibroin nanofibrous mats. Carbohydr Polym 210:9–16

Türe H (2019) Characterization of hydroxyapatite-containing alginate–gelatin composite films as a potential wound dressing. Int J Biol Macromol 123:878–888

Udhayakumar S, Shankar KG, Sowndarya S, Rose C (2017) Novel fibrous collagen-based cream accelerates fibroblast growth for wound healing applications: in vitro and in vivo evaluation. Biomater Sci 5(9):1868–1883

Ulubayram K, Aksu E, Gurhan SI, Serbetci K, Hasirci N (2002) Cytotoxicity evaluation of gelatin sponges prepared with different cross-linking agents. J Biomater Sci Polym Ed 13 (11):1203–1219

Unalan I, Endlein SJ, Slavik B, Buettner A, Goldmann WH, Detsch R et al (2019) Evaluation of electrospun poly (ε-caprolactone)/gelatin nanofiber mats containing clove essential oil for antibacterial wound dressing. Pharmaceutics. 11(11):570

Varshney N, Sahi AK, Poddar S, Mahto SK (2020) Soy protein isolate supplemented silk fibroin nanofibers for skin tissue regeneration: fabrication and characterization. Int J Biol Macromol 160:112–127

Veerasubramanian PK, Thangavel P, Kannan R, Chakraborty S, Ramachandran B, Suguna L et al (2018) An investigation of konjac glucomannan-keratin hydrogel scaffold loaded with Avena sativa extracts for diabetic wound healing. Colloids Surf B Biointerfaces 165:92–102

Venkatesan J, Anil S, Kim S-K, Shim MS (2017) Marine fish proteins and peptides for cosmeceuticals: a review. Mar Drugs 15(5):143

Voigt J, Driver VR (2012) Hyaluronic acid derivatives and their healing effect on burns, epithelial surgical wounds, and chronic wounds: a systematic review and meta-analysis of randomized controlled trials. Wound Repair Regen 20(3):317–331

Wang J, Chen YP, Zhou GS, Chen YY, Mao CB, Yang MY (2019a) Polydopamine-coated Antheraea pernyi (a. pernyi) silk fibroin films promote cell adhesion and wound healing in skin tissue repair. Acs Appl Mater Inter 11(38):34736–34743

Wang D, Li W, Wang Y, Yin H, Ding Y, Ji J et al (2019b) Fabrication of an expandable keratin sponge for improved hemostasis in a penetrating trauma. Colloids Surf B Biointerfaces 182:110367

Wang YF, Li PF, Xiang P, Lu JT, Yuan J, Shen J (2016) Electrospun polyurethane/keratin/AgNP biocomposite mats for biocompatible and antibacterial wound dressings. J Mater Chem B 4 (4):635–648

Wang Y, Xie R, Li Q, Dai F, Lan G, Shang S et al (2020b) A self-adapting hydrogel based on chitosan/oxidized konjac glucomannan/AgNPs for repairing irregular wounds. Biomater Sci 8 (7):1910–1922

Wang S, Yan F, Ren P, Li Y, Wu Q, Fang X et al (2020c) Incorporation of metal-organic frameworks into electrospun chitosan/poly (vinyl alcohol) nanofibrous membrane with enhanced antibacterial activity for wound dressing application. Int J Biol Macromol 158:9–17

Wang D, Zhang N, Meng G, He J, Wu F (2020a) The effect of form of carboxymethyl-chitosan dressings on biological properties in wound healing. Colloids Surf B Biointerfaces 194:111191

Wei LG, Chang HI, Wang YW, Hsu SH, Dai LG, Fu KY et al (2019) A gelatin/collagen/ polycaprolactone scaffold for skin regeneration. Peerj 7:e6358

Wiegand C, Buhren B, Bünemann E, Schrumpf H, Homey B, Frykberg R et al (2016) A novel native collagen dressing with advantageous properties to promote physiological wound healing. J Wound Care 25(12):713–720

Woltje M, Bobel M, Bienert M, Neuss S, Aibibu D, Cherif C (2018) Functionalized silk fibers from transgenic silkworms for wound healing applications: surface presentation of bioactive epidermal growth factor. J Biomed Mater Res A 106(10):2643–2652

Wu S, Deng L, Hsia H, Xu K, He Y, Huang Q et al (2017) Evaluation of gelatin-hyaluronic acid composite hydrogels for accelerating wound healing. J Biomater Appl 31(10):1380–1390

Wu C, Zhang Z, Zhou K, Chen W, Tao J, Li C et al (2020) Preparation and characterization of borosilicate-bioglass-incorporated sodium alginate composite wound dressing for accelerated full-thickness skin wound healing. Biomed Mater 15(5):055009

Xia G, Zhai D, Sun Y, Hou L, Guo X, Wang L et al (2020) Preparation of a novel asymmetric wettable chitosan-based sponge and its role in promoting chronic wound healing. Carbohydr Polym 227:115296

Xie H, Chen X, Shen X, He Y, Chen W, Luo Q et al (2018) Preparation of chitosan-collagen-alginate composite dressing and its promoting effects on wound healing. Int J Biol Macromol 107:93–104

Xuan C, Hao L, Liu X, Zhu Y, Yang H, Ren Y et al (2020) Wet-adhesive, haemostatic and antimicrobial bilayered composite nanosheets for sealing and healing soft-tissue bleeding wounds. Biomaterials 252:120018

Xue J, Wang X, Wang E, Li T, Chang J, Wu C (2019) Bioinspired multifunctional biomaterials with hierarchical microstructure for wound dressing. Acta Biomater 100:270–279

Yamauchi K, Maniwa M, Mori T (1998) Cultivation of fibroblast cells on keratin-coated substrata. J Biomater Sci Polym Ed 9(3):259–270

Yang W, Xu H, Lan Y, Zhu Q, Liu Y, Huang S et al (2019) Preparation and characterisation of a novel silk fibroin/hyaluronic acid/sodium alginate scaffold for skin repair. Int J Biol Macromol 130:58–67

Yao CH, Lee CY, Huang CH, Chen YS, Chen KY (2017) Novel bilayer wound dressing based on electrospun gelatin/keratin nanofibrous mats for skin wound repair. Mater Sci Eng C Mater Biol Appl 79:533–540

Ye S, Jiang L, Su C, Zhu Z, Wen Y, Shao W (2019) Development of gelatin/bacterial cellulose composite sponges as potential natural wound dressings. Int J Biol Macromol 133:148–155

Yin J, Xu L (2020) Batch preparation of electrospun polycaprolactone/chitosan/aloe vera blended nanofiber membranes for novel wound dressing. Int J Biol Macromol 160:352–363

Ying H, Zhou J, Wang M, Su D, Ma Q, Lv G et al (2019) In situ formed collagen-hyaluronic acid hydrogel as biomimetic dressing for promoting spontaneous wound healing. Mater Sci Eng C 101:487–498

Yu QH, Zhang CM, Jiang ZW, Qin SY, Zhang AQ (2020) Mussel-inspired adhesive Polydopamine-functionalized hyaluronic acid hydrogel with potential bacterial inhibition. Glob Chall 4(2):1900068

Zhai M, Xu Y, Zhou B, Jing W (2018) Keratin-chitosan/n-ZnO nanocomposite hydrogel for antimicrobial treatment of burn wound healing: characterization and biomedical application. J Photochem Photobiol B 180:253–258

Zhang X, Chen Z, Bao H, Liang J, Xu S, Cheng G et al (2019a) Fabrication and characterization of silk fibroin/curcumin sustained-release film. Materials (Basel) 12(20):3340

Zhang W, Chen LK, Chen JL, Wang LS, Gui XX, Ran JS et al (2017) Silk fibroin biomaterial shows safe and effective wound healing in animal models and a randomized controlled clinical trial. Adv Healthc Mater 6(10):1700121

Zhang J, Jeevithan E, Bao B, Wang S, Gao K, Zhang C et al (2016) Structural characterization, in-vivo acute systemic toxicity assessment and in-vitro intestinal absorption properties of tilapia (Oreochromis niloticus) skin acid and pepsin solublilized type I collagen. Process Biochem 51 (12):2017–2025

Zhang Y, Lu L, Chen Y, Wang J, Chen Y, Mao C et al (2019b) Polydopamine modification of silk fibroin membranes significantly promotes their wound healing effect. Biomater Sci 7 (12):5232–5237

Zhang X, Ookawa M, Tan Y, Ura K, Adachi S, Takagi Y (2014) Biochemical characterisation and assessment of fibril-forming ability of collagens extracted from bester sturgeon Huso huso× Acipenser ruthenus. Food Chem 160:305–312

Zhao J, Han F, Zhang W, Yang Y, You D, Li L (2019) Toward improved wound dressings: effects of polydopamine-decorated poly (lactic-co-glycolic acid) electrospinning incorporating basic fibroblast growth factor and ponericin G1. RSC Adv 9(57):33038–33051

Zhao X, Wang L, Gao J, Chen X, Wang K (2020a) Hyaluronic acid/lysozyme self-assembled coacervate to promote cutaneous wound healing. Biomater Sci 8(6):1702–1710

Zhao DY, Yu ZC, Li Y, Wang Y, Li QF, Han D (2020b) GelMA combined with sustained release of HUVECs derived exosomes for promoting cutaneous wound healing and facilitating skin regeneration. J Mol Histol 51(3):251–263

Zhou K, Krug K, Stachura J, Niewczyk P, Ross M, Tutuska J et al (2016) Silver-collagen dressing and high-voltage, pulsed-current therapy for the treatment of chronic full-thickness wounds: a case series. Ostomy Wound Manage 62(3):36–44

Zia T, Usman M, Sabir A, Shafiq M, Khan RU (2020) Development of inter-polymeric complex of anionic polysaccharides, alginate/k-carrageenan bio-platform for burn dressing. Int J Biol Macromol 157:83–95

Zou YN, Xie RQ, Hu EL, Qian P, Lu BT, Lan GQ et al (2020) Protein-reduced gold nanoparticles mixed with gentamicin sulfate and loaded into konjac/gelatin sponge heal wounds and kill drug-resistant bacteria. Int J Biol Macromol 148:921–931

Role of Biodegradable Polymer-Based Biomaterials in Advanced Wound Care

Haren Gosai, Payal Patel, Hiral Trivedi, and Usha Joshi

1   Introduction

Skin is considered as the largest organ in the body and any breakage, tearing, or defects in the skin are termed as wounds. Wounds generally develop as a result of physical/thermal damage or due to presence of any underlying pathological conditions (Percival 2002). Wounds are classified on the basis of skin damage, nature of wound repair process, number of skin layers damaged, and area of the skin disrupted (Xiao Liu and Jia 2018). They can be further categorized into either acute or chronic, based on the mechanism of the wound healing. Acute wound most commonly includes tissue injuries and healed within 8–12 weeks. However, chronic wounds have tendency to last longer and the recovery process can take up to 12 weeks or more (Bryant and Nix 2015). The presence of co-morbidities like diabetes, osteoarthritis and infection makes treatment of wounds even more precarious. In people with such co-morbidities, chronic wounds are often observed and if they left untreated it may lead to hospitalization and in some extreme cases lead to amputations (Frykberg and Banks 2015).

Wound healing is a complex, dynamic, and multi-phase physiological process which involves cellular, biochemical, and enzymatic components (Sezer and Cevher 2011). Even though, wounds have capability to heal by themselves, appropriate wound-dressing is necessary as it prevents infection and accelerate the rate of wound repair (Mir et al. 2018). Since, ancient time humans have used various natural materials such as honey, animal fats, and plant fibers for wound-dressing, as found

H. Gosai (✉) · P. Patel · H. Trivedi
Department of Bioscience, School of Science, Indrashil University, Rajpur-Kadi, Gujarat, India
e-mail: harengiri.gosai@indrashiluniversity.edu.in

U. Joshi
Department of Botany, Faculty of Science, The Maharaja Sayajirao University, Vadodara, Gujarat, India

© The Author(s), under exclusive license to Springer Nature Singapore Pte Ltd. 2021   599
P. Kumar, V. Kothari (eds.), *Wound Healing Research*,
https://doi.org/10.1007/978-981-16-2677-7_18

Table 1 Types of wound dressings

Dressing	Advantages	Limitations	Trade name
Gauze	Cost effective, readily available at clinics and pharmacy, easy to use	Dry dressing, poor barrier protection, tissue damage on replacement/removal, once saturated with exudate becomes ineffective	PETRONET®, Medifin™, Bactisafe™, CutiCell™
Foam	Available in different shapes and sizes, comfortable, easy to apply, easy to remove	Additional tape/adhesive needed if non-adherent, not useful for dry/non-draining wounds, need frequent changing to prevent excess moisture accumulation	Aquacel™, Biatain®, PolyMem®, Tegaderm™
Film	Transparent, examination without removal is possible, impermeable, reduces friction	Does not absorb moisture, sticks to the wounds, not for draining wounds	Bioclusive®, Cardinal Health, DermaView™
Alginate	Highly absorptive, non-occlusive, hemostatic properties for bleeding wounds, easily removable, can be used on infected wounds	Secondary dressing required for adherence, produces odor, not suitable for dry wounds, burns	Maxorb®, Sorbalgon®, Algicell®
Hydrofiber	Highly absorptive, no need for frequent changing, easy to remove	Non-adherent, requires secondary dressing, not suitable for dry wounds	DuoDERM™, Versiva®
Hydrocolloid	Self-adherent, comfortable, available indifferent shapes/sizes/thickness, reduces pain, moderately absorptive, thermal insulation	Sometimes difficult to remove, produces odor, may leave residue in wound bed, not suitable for heavy draining wounds	Amerex®, Comfeel®, CovaWound™
Hydrogel	Rehydrate wound, reduces pain, promotes autolytic debridement, easily removable, can be used for infection with topical medications	Non-adherent, require secondary dressing, not suitable for heavy draining wounds, may macerate skin sometimes	Biolex™, DermaGel™, Carrasyn®

in written medical records in Egypt (Mir et al. 2018; Sezer and Cevher 2011). Traditionally, dry dressings—cotton, lint, wool bandages, and gauzes are used for treatment of small wounds and cuts. These materials keep the wounded area dry and allow the blood, exudates, and tissues to scab over. However, they provide poor barrier against germ invasion, need adhesives, and can damage new tissues during replacement or removal (Lei et al. 2019). Hence, studies conducted have put forward the hypothesis that modern moist-dressings are comparatively better than dry-dressings and promote higher rate of healing. One such study conducted by Shi et al. (2020) has reported that modern moist-dressings have better biocompatibility, degradability, improve the microenvironment and relieve pain.

In the last few years, a wide range of wound-dressings have been introduced with properties suitable for treatment of different kinds of wounds (Table 1), with the choice of dressing dependent on the location, surface area/depth, type of tissue damaged, exudate, and condition of the wounded skin (Weir 2020). A perfect wound-dressing absorbs wound exudates, provides efficient barrier, maintain a moist microenvironment, thermal stability, and prevents further tissue damage (Das and Baker 2016). Though, it must be acknowledged that there is no one specific dressing that can be used to manage a variety of wounds let alone at different stages of healing process (Psimadas et al. 2012).

Biodegradable polymer-based wound-dressings have emerged as the new improved, efficient, and eco-friendly option for the management of the wounds. It has been estimated that the global market for biopolymer based wound-care is expanding and will continue to do so in the following years (Song et al. 2018). Biodegradable polymers or biopolymers as the name indicates are polymers of natural or synthetic derivation that can be broken down into simpler components. They are diverse in compositions and have tunable physical behavior. Moreover, they are renewable, cost effective, and found in wide range of variety (Smith et al. 2016). Furthermore, biopolymers are regarded as excellent source material for wound-care because of their bioactive properties that facilitates cell growth, has potential for regeneration, provides antimicrobial environment and immunomodulation (Sahana and Rekha 2018). Another property that makes bio-polymers a potential candidate for wound-care is their capacity to absorb a huge quantity of water. Use of biopolymers with drugs that are directly released into wounds have also become popular in last few years (Smith et al. 2016). Both natural and synthetic types of biopolymers are being used in wound-care but natural bio-polymers are preferred for to their biodegradability, lower antigenicity and renew-ability (Sezer and Cevher 2011).

Natural biopolymers include proteins, polysaccharides, polyesters, and proteo-glycans with alginate, cellulose, carrageen chitosan, collagen, etc. being the most extensively used biopolymers (Moohan et al. 2020). Despite having certain advan-tages, the mechanical properties and rate of degradation are not easily manipulated in natural biopolymers. In contrast, synthetic biopolymers have demonstrated to be more easily modified as they are produced under controlled conditions. Some of the synthetic biopolymers used such as saturated aliphatic polyesters (polyglycolic acid, PLA), polyanhydrides, polyurethane, and polyphosphazenes in wound management (Song et al. 2018).

Engineering of biopolymer to construct suitable biomaterial for wound manage-ment also plays a major role (Piraino and Selimović 2015) in the development process. The selection of the polymer used for developing biomaterial is a crucial factor in defining the properties of the given biomaterial. When designing biomate-rial, the most essential requirement is biocompatibility—property that ensures that a particular material can perform with a suitable host response (Piskin 1995). This biocompatibility is dependent on several biological and physicochemical properties of the biopolymer including shape-structure, biodegradability, hydrophobicity/hydrophilicity, molecular weight, solubility, and material chemistry (Kohane and Langer 2008). Some other properties of the biopolymers aside from biocompatibility that needs to be considered are a) should not generate a continued inflammatory response, (b) should have degradation rate that coincides with their function, (c) should have mechanical properties that are suitable for their appropriate use, (d) should not produce toxic degradation products that cannot be easily resorbed or excreted, and (e) should have necessary permeability and processability for intended use (Ulery et al. 2011). Advancements in technology have allowed the development of several biomaterials for wound care and management. Wound dressings composed of foam, hydrogels, hydrocolloidal, or those with incorporated biologicals with specific properties have become more readily available. Other advanced treatments like artificial skin substitutes are also being studied along with many other techniques that will find their application in the future (Dai et al. 2020).

In this book chapter, we have taken a measured look on the specific properties of several biopolymers which could be helpful to select appropriate biopolymers for wound care management. The novel biomaterials, their designing and applications have been also discussed. The chapter also focused on the recent development in the field of wound care management and the most probable direction it could gain in coming years. This chapter could be useful to researchers, students, and stakeholders for generating effective wound care management.

2 Wound Healing Process

Wound healing process developed as an evolutionary advantage and necessary factor for survival after injury (Sorg et al. 2017). The process of wound healing is complicated and involves extensive mediation between the different cellular constituents of the skin cells and the surrounding extracellular matrix (ECM), in a highly sophisticated manner (Eming et al. 2016). Since, wound healing is a frequent cause of mortality and morbidity, it poses an immense challenge to the field of clinical care. To promote healing of the injuries, to limit the scarring on patients, and restoration of tissue functions are the major goals of the wound repair system. To accomplish these objectives, there have been several advancements to develop better dressings, therapies, and techniques for wound management (Velnar et al. 2009). Thus, fully understanding the mechanism of wound healing at a cellular and biochemical level is crucial for designing better therapeutic approaches. The events taking place during the wound healing process can be arranged into three overlapping and sequential phases: the inflammatory phase, the proliferation phase, and the remodelling phase (Cañedo-Dorantes and Cañedo-Ayala 2019).

2.1 The Inflammatory Phase

The early vascular inflammatory phase begins immediately within few seconds of the injury and involves coagulation and hemostasis. The major target of this phase is to sound alarm to the body and prevent further damage. The activation of coagulation cascade is initiated upon breaking of the skin, which leads to platelet coagulation and formation of a "clot plug" to stop exsanguination and infection (Robson et al. 2001). The next secondary inflammatory phase is responsible for elimination of the pathogen and cleaning the wound. Vasodilation takes place to facilitate the migration of leukocytes to the wound area, this is followed by characteristic inflammation features—edema, dolour, and erythema (swelling, pain, and redness). The cell response starts within 24 hours and can last up to 2–4 days (Gonzalez et al. 2016). The predominant phagocytic cells (macrophage and neutrophils) along with the activated immune cells mount a host response against the invading pathogens and helps in cleaning the wound by or autolyzing the necrotic tissues and debris (Shankar et al. 2017).

2.2 The Proliferation Phase

The proliferation phase begins approximately around 48 hours after the injury and continues for around 2 weeks, thereafter. The phase occurs when homeostasis and an immune response have been achieved and is characterized by tissue damage repair (Sorg et al. 2017). The proliferative phase focuses on: (i) the proliferation of fibroblasts and their differentiation, (ii) production of collagen III and its interaction with skin cells, (iii) formation of new tissues by reepithelization, (iv) angiogenesis—formation of new vessels by endothelial proliferation, and (v) repairing of the damaged nerves and regeneration. Most of the functions involved in the proliferative phase are carried out by the predominant macrophage cells (Cañedo-Dorantes and Cañedo-Ayala 2019). The final step in the proliferative phase is the formation of granulation tissues comprising of a highly dense mixture of fibroblasts, granulocytes, macrophages, capillaries, and loose collagen bundles (Reinke and Sorg 2012).

2.3 The Remodelling Phase

The concluding phase of the wound repair process is termed as the remodelling phase and takes place from 21st day of the injury and continues for around 1 year or more. The phase is characterized by wound maturation and regaining of skin integrity. This is achieved by the turnover of collagen III into collagen I along with the decrease in the cellularity due to apoptosis (Thiruvoth et al. 2015). During this phase, the development of new epithelium and formation of scar tissues is observed. Scar demarcation can be identified as the end of the remodelling phase. However, the tissues in the wound area gain only 80% of their previously demonstrated tensile strength (Gurtner and Wong 2013). The damage sustained by the hair follicles and the sweat glands present at the wound area cannot be repaired and they are permanently lost after the wound repair (Robson et al. 2001).

3   Factors Affecting Wound Healing Process

Several factors play a significant role in impeding the healing process of the wounds. These factors may either be local—those impacting wounds only or systemic—any underlying morbidity that could interfere with the repairing mechanism (Beyene et al. 2020). Some of these factors have been reported to be interlinked and work concurrently to impair the healing process.

Hypoxia: Though most of the wounds are hypoxic due to sustained tissue damage, a certain amount of oxygen is necessary for wounds to get healed. Oxygen is required by the neutrophils and macrophages to conduct phagocytosis. It has also been found to be essential in collagen deposition for tissue repair (Harper et al. 2014).

Infection: Infection indicates the presence of pathogens in the wound area. Microorganisms that are present in environment or skin surface can enter the body through wounds and colonize or multiply there. This leads to prolongation of the inflammatory phase ultimately causing delay in wound repair (Guo and DiPietro 2010).

Necrosis: Presence of dead or necrotic tissues can delay healing. Both slough (wet tissue) and dry eschar must first be removed by either wound cleansing or phagocytic cells to continue with the healing process (Thomas Hess 2011).

Immunosuppression: People suffering from cancer, HIV, or malnutrition have been reported to have immunosuppression, which is responsible for interfering with the regular wound repair process. Additionally, drugs or therapies (radiation) taken for the treatment of the disease may aggravate inflammation and ultimately hamper the healing cascade (Harper et al. 2014).

Chronic Diseases: Diabetes mellitus, coronary artery disease, and vascular artery disease are some of the chronic diseases that have been reported to impede wound healing. Cardiorespiratory diseases alter the available oxygen supply, creating a hypoxic environment unsuitable for wound repairment. In case of diabetes, the high blood sugar level influences the working of leukocyte preventing them from carrying out their part in the healing process (Tsioufis et al. 2012).

Since numerous factors influence wound healing process, maintenance of a healthy, sterile, and stable microenvironment is a prerequisite for efficient wound repairment. An acute wound may turn into a chronic wound if required steps are ignored. To prevent this, it is critical to select the appropriate wound dressing, ensure their proper application, to carry out effective wound cleansing, and perform constant monitoring of the wound. The dressings must be changed regularly to make sure there is no infection, exudate leakage, or presence of odor. If some chronic illness is present, specific precautions pertaining to that disease must be followed to avoid any complications in wound healing (Armstrong et al. 2018).

4 Biodegradable Polymers and their Properties

As discussed earlier, complex process like wound healing requires a material that is biocompatible with host tissue. Biocompatibility can be defined as the ability of any biomaterial to function with a suitable host response in a particular wound healing process (Piskin 1995). Host tissue response to the material is controlled by many biological and physiochemical factors of biodegradable polymers. Some of the factors that can be mentioned are surface energy, material chemistry, molecular weight, solubility, mechanism of degradation and/or erosion, lubricity, hydrophilicity or hydrophobicity, and shape and structure of the biodegradable polymer can influence the material's biocompatibility (Kohane and Langer 2008). Biocompatibility of a polymer with host tissue should be constant phenomenon as the physico-chemical, mechanical, and biological properties of a biodegradable polymer can differ with time. Essentially, biodegradation of polymers changes their properties like tensile strength, color, and shape and therefore, the polymer under consideration should exhibit compatibility to the host tissue during the whole healing process. Moreover, the degradability not only depends on the environmental factors like heat, light of chemicals, but also on the chemical structure of polymer and environmental condition (Bismarck et al. 2002). The products generated in the degradation process of a polymer will differ in compatibility to the host tissue when compared to the parent material. A biopolymer can be considered ideal for healing process if its degradation products are not harmful to the body.

Several other important properties must be considered while selecting the biodegradable polymers. These are as follows:

- Time taken by biomaterial for degradation should be equal to the regeneration or healing process to ensure proper remodelling of the tissue.
- Biopolymer should constantly exhibit appropriate permeability and processability for its specific wound healing application.
- The mechanical properties and compatibility of biopolymer should remain unchanged during the entire degradation process. In addition to this, the properties should remain unaffected during the patient's day-to-day activities (Song et al. 2018).
- When the biopolymer under consideration implanted in vivo, it should not stimulate a toxic response or inflammation to the host tissue (Williams 2009).
- The material should have an acceptable shelf life.
- Degradation products should not be harmful, and easily digested and removed from the host body (Schmitt and Polistina 1963).
- In the case of scaffold-guided tissue engineering, the biodegradable polymer should be processable into a proper shape fitting the defect's site, with a proper micro-nanostructure (Puppi et al. 2010).
- The polymeric materials should be designed into a scaffold structure so as to carry mechanical properties and degradation rate which is suitable for maintaining the spaces required for cell ingrowth and matrix creation. Furthermore, it should be able to bear stresses and loading (Nair and Laurencin 2007).

Other essential qualities of polymers, like particle size, bulk density, surface area, and morphology should be considered when the polymers are used as controlled drug delivery systems because these may affect drug release from the system.

Both naturally derived and synthetic polymers find their applications in biomedical field (Table 2). Plant-based natural polymers include polysaccharides like cellulose, alginate, and dextran while animal-based polymers are collagen, silk, chitosan, etc. These biomaterials must be selected carefully because they have characteristic bioactivity and therefore, show high variation in compatibility to the host tissue. If they are not compatible, these biomaterials can provoke immunogenic response when used in the healing process. In contrast to this, synthetic polymers exhibit more predictable physical properties and are generated by chemically engineered degradation profiles, which render them biologically inert. Synthetic polymers like polyglycolide (PGL), polylactide (PLA), and polycaprolactone (PCL) are exploited highly as they are linked by ester bond which enables their hydrolytic degradation. Other hydrolytically degradable molecules include anhydride, carbonate, urea, amide, thioester, urethane, imine, and imide bonds, which have the sites for cleavage under biological conditions. Enzymatic cleavage between carbon−carbon bonds leading to the susceptibility to hydrolysis using acid and base in many molecules such as sulfonamides, phosphonates, and ethers. However, electronic biodegradable polymers utilize the above-mentioned hydrolyzable linkages to degrade in physiological, aqueous conditions (Feig et al. 2018).

5   Novel Biomaterials (Natural and Synthetic)

Human body has a complex structure and it is difficult to find the desired characteristics from the single polymeric biomaterials for the wound healing process. Recently researchers have shown the advancement in synthesis and design of biodegradable polymer and polymers are being designed for specific medical applications (Bianchera et al. 2020). This has led to integration of versatile and combinational approaches to design the biomaterial which has paved path for innovation in discovering novel biodegradable biomaterials. This area has witnessed more advances and researchers now have developed medical devices like delivery vehicles for pharmacological applications, three-dimensional (3D) porous scaffolds for tissue engineering, and temporary prostheses (Li et al. 2020). Efforts are being made to use biomaterials as bio-ink for 3D bioprinting because biodegradable materials owing biological and physicochemical properties, can replicate the properties of different tissues (Zidarič et al. 2020).

One more area that has shown progress is the generation of tissue repairing hydrogels. These hydrogels have the property of dual adhesiveness, which adheres to both tissue and implant biomaterial. Moreover, these gels show bioactivity which helps in the regeneration of tissues. For instance, Gao et al. (2019) have developed a unique bioglass (BG) (oxidized sodium alginate (OSA)) composite hydrogel with dual adhesive and bioactive properties. This has opened a vast area for potential

Table 2 Properties of biodegradable polymers and medical use

Biodegradable polymer	Source	Property	Medical use	References
Collagen	Fibroblasts	High biocompatibility and biodegradability	Nerve regeneration, drug delivery Vitreous replacement, skin replacement, skin patches, bond-filling and repair, enhanced epithelialization rate	Powell et al. (2008)
Gelatin	Skin, bones, and connective tissues of animals such as domesticated cattle, chicken, pigs, and fish	Biocompatible, biodegradable, and nonimmunogenic	Treatment of severe burn wounds	Jaipan et al. (2017)
Silk fibroin	*Nephila clavipes* and *Araneus diadematus* spiders, *Bombyx mori* domestic silkworms, and *Antheraea pernyi* and *Samia cynthiaricini* wild silkworms	Robust mechanical strength with high tensile strength, modulus, stiffness, and extensibility, enzymatic biodegradation with controllable rate, payloads stabilization capability due to hydrophobic interactions with β-sheet crystallite domains	Biological drug delivery, gene therapy, wound healing, bone regeneration	Shavandi et al. (2017)
Keratins (Ker)	Wool, hair, nails, feathers, and horns	Biodegradability, biocompatibility, and mechanical durability	Drug delivery, wound healing, tissue engineering, and cosmetic applications	Arslan et al. (2017)
Chitosan (CS)	Exoskeletons and shells of crustaceans	Biocompatibility analgesic effect, hemostatic effect, antitumor activity, and muco-adhesive properties	Antitumor drug delivery, gene delivery, protein and peptide drug delivery, antibiotic delivery, polyphenol delivery	Zhao et al. (2015)
Hyaluronic acid (HA)	Connective tissue of mammals	Non-immunogenic polysaccharide, hygroscopic nature	Promotion a scar-free wound regeneration Drug delivery agent for different routes such as nasal, oral, pulmonary,	Mele (2016)

(continued)

Table 2 (continued)

Biodegradable polymer	Source	Property	Medical use	References
			ophthalmic, topical, and parenteral, as an aid in eye surgery	
Alginate (Alg)	Brown algae and also produced by some bacteria	Biocompatibility and gelation can be done easily	Stimulate reparative wound processes	Patel et al. (2007)
BSA fibers	*Purified from blood*	Biocompatible and biodegradable	Wound applications	Dror et al. (2008)
Polyurethanes and their derivatives	Synthetic	Biocompatibility, strength, and flexibility	Wound dressing, to make surgical drapes, tubing, hospital bedding, and injection equipment implants	Davis and Mitchell (2008)
Silicone	Synthetic	Nontoxic, nonallergenic, and highly biocompatible	Skin treatment in severe burns and wounds	Momeni et al. (2009)

clinical applications. Kaygusuz et al. (2017) have invented the alginate-based wound dressings by combining it with antimicrobial properties of cerium ions and chitosan. To generate these dressings, alginate films were crosslinked with chitosan mixed cerium (III) solution. These dressings were flexible, ultraviolet protecting, and antibacterial.

In addition to this, there is a growing interest in the development of tunable and versatile electrospun biocomposite fibers as wound dressings. The main advantage of using these fibers in wound healing process is their resemblance to natural extracellular matrix which fastens the healing process. Their high surface area-to-volume ratio, tunable porosity, sufficient gas exchange, and possibility to include different active substances and living cells into the fibers further enhances the process. Both solution (blend, coaxial, and emulsion) electrospinning and green solvent-free electrospinning approaches can be utilized to produce ultrafine biocompatible fibers from different materials which can be utilized for wound (Palo et al. 2019).

6 Wound Healing Mechanism of Natural Polymers

Wound healing is a complex process expecting the use of an ideal dressing material for facilitation of the process. According to current trends, biodegradable materials are frequently exploited for the wound healing processes. There are various diverse

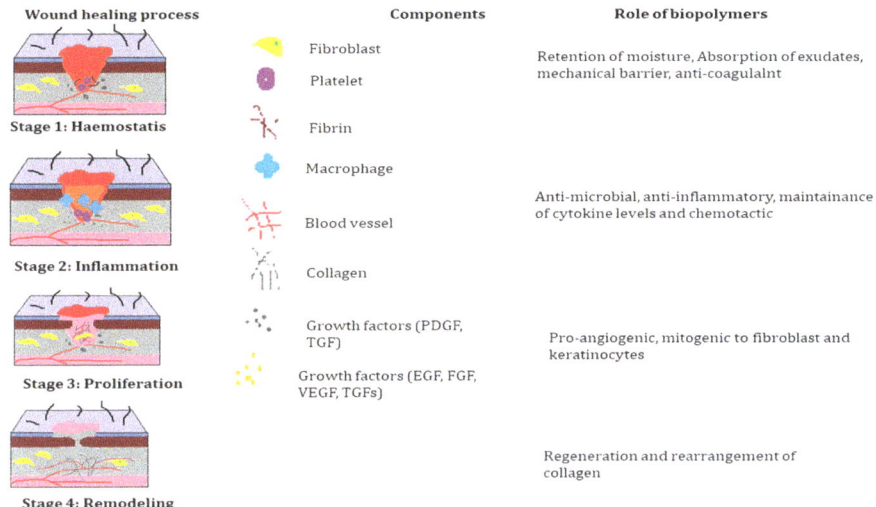

Fig. 1 Role of biopolymers in wound healing process

mechanisms developed for these materials that how they act on wound and participate in different stages of wound healing process. Each biomaterial has its own mechanism and role in wound healing process.

As discussed earlier, wound healing process consists of precisely regulated and well-balanced events like coagulation and inflammatory phase, hemostatic phase, remodeling phase, and proliferation phase. Biopolymers used for wound care or wound healing process must exhibit their impact on any wound healing process to stimulate rapid healing in less time. Biopolymers were found to exhibit a positive impact on every stage in this process as indicated in Fig. 1. Frequently used biopolymers for wound healing process are chitosan, alginate, collagen, fucoidan, etc. (Sahana and Rekha 2018).

6.1   Chitosan

As described before in this chapter, healing of wound composed of four diverse stages. Chitosan alone and chitosan-based polymer has known to have positive impact on each and every step after application as wound healing material.

Coagulation and hemostasis: Surface-induced thrombosis accompanied with blood coagulation and fast coagulation in vivo by affecting platelet activation can be stimulated by Chitosan. It also helps to block endings with pain reduction and blood clotting by acting as hemostat.

The inflammatory phase: Chitosan-based hydrogels forms a suitable microenvironment to conduct healing process by regulating associated cells and releasing factors. This hydrogel-based dressing material can restrict secretions of inflammatory mediators such as interleukin 1β, interleukin 8, prostaglandin E, and others with increasing different tissue repairing processes (Jayakumar et al. 2011). Previous work also indicated that action of leukocytes, neutrophils, and macrophages, tissue granulation stimulation for suitable inflammatory reaction are also enhanced by Chitosan-based hydrogels (Takei et al. 2012).

Proliferation: Platelets are responsible for releasing growth factors, Transforming Growth Factor (TGF)-β1 and platelet derived growth factor (PDGF)-AB, which can be stimulated by chitin and Chitosan, especially when applied in high concentrations (Okamoto et al. 2003). It also activates tumoricidal activity of macrophages for 3D growth by supplying non-protein matrix. Hydrogels based on chitosan could promote regular collagen deposition, angiogenesis, fibroblast proliferation, and amplify level of natural hyaluronic acid (HA) production at the wound site by depolymerizing itself for the liberation of *N*-acetyl-β-D-glucosamine (Jayakumar et al. 2011).

Remodeling: Dermal tissue component, the *N*-acetyl glucosamine (NAG) plays a key role in repair of scar tissues, present in chitin and chitosan (Archana et al. 2013). It had been already proved that chitosan films of low deacetylation degree were found significant in dressing superficial wounds (Borderud et al. 2015).

Chitosan alone, able to heal the acute wounds but chronic wounds have different patterned mechanism could be utilized (Ito et al. 2013). Chitosan-based hydrogels are appropriate to deliver growth factors, stem cells antimicrobial agents, and peptides to balance the biochemical events of inflammation in the chronic wound and thus enhance healing.

6.2   Collagen

Collagen is the main component of ECM and predominant inside connective tissues. The key points to maintaining balance between synthesis and breakdown of collagen which decides future of wound healing (Schwartz et al. 2002). It has an important role in an inflammatory phase as numerous proteolytic enzymes are secreted and small fragments of collagen [Arg-Gly-Asp] acting as mitogenic for fibroblast and chemotactic for macrophages thus stimulating proliferation and formation of granulation tissue (Bellis 2011). It also promotes migration as gelatin interacts with keratinocytes and epithelial–mesenchymal transmission necessary for tissue repair (Pastar et al. 2014). There are various processes utilized such as reduction in size to nanoscale, surface modification, and tagging with anti-inflammatory. Depolymerization and antimicrobial substances for the enhancement of physiological action of collagen. There are various collagen-based healing materials used such as membrane, films, powder, sponges, and hydrogel. Most frequently used are Derma Col, Fibracol, Biopad, citrix, Cellerate, Cutimed, Stimulen, Helix bioactive collagen, and Biostep (Sahana and Rekha 2018).

6.3   Alginate

At the wound bed, alginate presents a wet environment and takes in exudates. Alginates found to decrease infection, pain, and odor. It also plays role in hemostasis phase of healing process. Ion exchange occurs when dressing having alginate material comes in contact with exudates of wound. Alginate dressing materials have calcium ions and exudates or blood has sodium ions, so replacement takes place and as a result of which alginate fibers forms a gel (Jones et al. 2006). Alginates are also able to promote monocytes and thus directly affect the healing process. Alginates have low adhesive properties and so mixed with chitosan to increase adhesion, proliferation, and cell interaction (Yao et al. 2012). Several commercially available wound care products have alginate as their main ingredient such as Nu-dermetc, Kaltostat, Algicell, Calcicare, 3 M Tegaderm, Algisite, Cutimed alginate, and Dermalginate (Sahana and Rekha 2018).

6.4   Hyaluronic Acid

D-glucuronic acid and N-acetyl-d-glucosamine are repeating units joined by β-1,4 and β-1,3 glycosidic linkages alternatively to form hyaluronic acid. High molecular weight hyaluronan possesses anti-inflammatory properties while in case of low molecular weight, it has pro-inflammatory activities. The pro-angiogenic effect is carried out by the degradation product of HA. It is found to alter healing process with reduction of scar and decreased collagen synthesis when post-injury supplementation of exogenous is carried out. Several commercial wound care products having HA are Dermaplex, Regenecare, Hyalomatrix, Hyalofill, etc. It is also part of skin care protective products with cosmetics outcomes (Litwiniuk et al. 2016; Sahana and Rekha 2018).

6.5   Cellulose

Cellulose is preferred in wound care as it has properties to retain moisture and wet wounds are found to heal quicker because of sufficient delivery of growth factors and other molecules to the healing tissue (Sulaeva et al. 2015). Cellulose has a porous structure that mimics ECM of skin and helps in tissue regeneration. It can absorb exudates and result in intake of cell debris (Kucińska-Lipka et al. 2015). Cellulose-based commercially available dressings are Suprasorb, Dermafill, and Curity Exu-Dry Cellulose (Sahana and Rekha 2018).

   Each biopolymer has its own mechanism to cure wounds as indicated in Table 3. It suggests the ability of biopolymers to connect themselves with the ongoing events of wound healing and increase the healing rate by stimulating various processes.

Table 3 Biological mechanism of different biopolymers

Biopolymer	Biological mechanism	
Collagen	Chemotactic for macrophages, stimulates proliferation of fibroblast, induces release of ECM components by fibroblast	Bellis (2011); Pastar et al. (2014)
Cellulose	Maintenance of moisture, exudates absorption	Kim et al. (2018)
Alginic acid	Promoted monocytes, stimulate fibroblast proliferation and migration	Jones et al. (2006)
Hyaluronic acid	Proliferation and migration, promote fibroblasts and keratinocytes, anti-inflammatory	Huang et al. (2018)
Chitosan	Migration and proliferation, induce fibroblast and keratinocytes	Shi et al. (2018)
Fucoidan	Mitogenic to keratinocytes and fibroblast, Angiogenic	Park et al. (2017)

Table 4 Specificity of biopolymer toward types of wounds

Biopolymer	Wound types	References
Collagen	Foot ulcers, chronic wounds, large open cuts, bed-sores, surgical wounds, minor burns, heavy-to-low excretion wounds	Rehfeld et al. (2017); Bellis (2011)
Cellulose	Plastic/reconstructive surgeries, chronic wounds, burns	Sulaeva et al. (2015)
Alginate	Cavity wounds, postoperative wounds, infected wounds, heavy to moderately exhausting wound, surgical incisions or dehisced wounds, pressure ulcers, partial and full-thickness wounds, dermal wounds	Aderibigbe and Buyana (2018); Yang and Jones (2009)
Hyaluronic acid	Chronic wounds, wounds having partial and full thickness	Litwiniuk et al. (2016)
Chitosan	Acute wounds and pressure ulcers	Ahmed and Ikram (2016)

Biopolymers also possess their appropriateness toward the types of wounds as indicated in Table 4. Collagen, cellulose, alginate, HA and Chitosan have their own priority for the type of wounds which also indicate their specific mechanism of action but at the same time they can carry various disadvantages such as collagen cannot be used for patience allergic to birds, products of swine or cattle, third degree burns and dry wounds. Similarly, cellulose could not be utilized for burns of second and third degree. Alginate is also not applicable to treat dry, eschar wounds, and also for third-degree wounds (Sahana and Rekha 2018).

Thus, biopolymers have been successively utilized for wound care. Biopolymers with diversified origins are well studied because of their excellent environment-friendly properties. Still many issues related to the use of biopolymers that restrict their use but it will be solved in future and biopolymers will emerge as one of the most significant wound healing mechanisms.

7   Applications of Biodegradable Materials in Wound Healing

Healing of wound, a tightly restricted physiological progression involving different subsequent stages and any destruction in this sequence can lead toward chronic wounds and indirectly affecting the quality of life of patients. Nowadays, novel wound dressing material is highly expected to have astonishing wound repair and skin rejuvenation advantage instead of preventing wound infection (Qu et al. 2019). Traditional wound healing process involves the use of gauze made up from cotton (Sezer and Cevher 2011) having numerous drawbacks like disrupting epithelium, which was newly formed leading to eliminate and fast damaging newly formed epithelium upon removal and causing rapid drying out of the wound bed (Siritientong et al. 2014) but they became popular due to their lower cost. The standard protocol carried out today is swabbing, dressing, and cleaning (Dreifke et al. 2015). If a skin lesion is extended, allograft or autograft might be used and could give rise to immunological issues like immune rejection (Suarato et al. 2018). As far as characteristics are concerned about the ideal dressing material, it must eliminate unnecessary exudates to escape maceration of tissue, support autolytic debridement, maintains wetness, sufficient water vapor, and oxygen permissible within the wound. In favor of application/removal from patient's body, it should be adhesive and flexible.

Hence, in search of dressing materials with novel functions that could resolve the troubles of traditional dressing materials, various biomaterials have gained attention of scientific community due to their biodegradability and biocompatibility. Natural protein- as well as carbohydrate-based biomaterials are exploited for their use in wound dressing material.

Protein-based biomaterials include utilization of collagen, gelatin, silk protein, etc. Collagen is considered as protein which is abundant in animal providing strong support to tissues (An et al. 2016). Collagen offers verities of dressing material in different forms like electrospun fibers, scaffolds, and hydrogels, which could be applied on ulcers and burn wounds (Yoon et al. 2018). A multistructured nanofibrous dressing material is synthesized using poly-ε-collagen electrospun matrix having transforming growth factor-β1 (TGF-β1) and customized with polypeptide-based nanocarriers found to speed up healing and wound closure (Albright et al. 2018). Mimura et al. (2008) had suggested a significant role of gelatin in wound healing process by performing and concluded that wound healing process of cornea can be promoted through gelatin hydrogels by transplanting fibroblast precursors into corneal stroma. Silk protein synthesized by various arthropods like spiders and silkworms has also attracted the focus as in vitro study revealed that silk mats having epidermal growth factors and ciprofloxacin (Chouhan et al. 2017) increased keratinocytes proliferation and human dermal fibroblast.

Naturally, derived polysaccharides have also been found efficient in this field. Lloyd et al. (1998) stated that homoglycans are biocompatible material that occurs naturally and exploited in wound healing to modulate the cellular responses.

Hyaluronic acid (HA) has a hygroscopic nature and so it is exploited to synthesize hydrogel structure so it can promote a scar-free wound healing and keratinocyte migration and angiogenesis (Dreifke et al. 2015). Chitosan (CS) being intrinsic antifungal, antibacterial, mucoadhesive, and hemostatic makes it the most preferable dressing material. Du et al. (2009) had reported the role of chitosan membrane in prop up healing process of wound and reduced scar tissue formation in a corneal alkali burn of rabbit. There are numerous CS-based dressing materials proposed like CS-gelatin sponge (Lanzhen et al. 2007), CS-silk hydrogels (Silva et al. 2012), and CS-aloe-vera membrane (Wani et al. 2010). Alginate also offers its use in this field by removing microbial cells and necrotic tissues when utilized along with enzymatic components and antimicrobial. It can also initiate reparative wound healing process by providing polysaccharide base during its utilization (Patel et al. 2007).

7.1 Advantages of Biopolymers to Limit Microbial Formation on Wound

There are many studies that reported the formation of biofilms on chronic and acute wounds, mixed etiologies and full-thickness burns, etc. (Metcalf and Bowler 2013). Reports also suggest that half of the chronic wounds contain biofilm. If a majority of biofilm-containing wounds have delayed healing mechanisms then biofilm could be contributing many billions of dollars to the global cost for wound management process. Thus, based on the clinical and in vivo experiments, it is required to devise effective anti-biofilm strategies, to encourage wound healing in clinical practice. Various potential anti-biofilm agents have been proposed such as ethylenediaminetetraacetic acid (EDTA), xylitol, and lactoferrin, however, they do not have higher efficiency to prevent biofilm formation on wound. Recent studies suggested that biodegradable polymers such as chitosan, collagen, and polyhydroxyalkanoates are found to be effective antibiotic carriers (Pavithra and Doble 2008). Mogoşanu and Grumezescu (2014) have reported cellulose and nanocellulose prevent and control biofilm formation on wound. Ward et al. (2020) have also reported that a medium chain length polyhydroxyalkanoate prevents colonization of *Pseudomonas aeruginosa,* one of the most common bacteria that infect chronic wounds.

1. **Limitations of biopolymers as wound dressing materials**

   • Sometimes advantages may come with some restrictions also. Biopolymers/
     biomaterials are biocompatible with human tissue which is their main advan-
     tage but due to this property it cannot be exploited for extended time periods.
   • Productivity is another main concern since they are derived from the natural
     resources so it is very difficult to synthesis a sufficient quantity of products for
     practical use.

- Origin of synthesis also limits the use of these materials as products may vary slightly in physiological and biological activities even though they are produced by same animal but residing in different areas.
- These products are easily degradable under various conditions like heat and light. So they must be utilized instantly after extraction to avoid any kind of non-uniformity.
- These materials can provide an excellent growth environment for microorganisms due to which chances of contamination increase while their sterilization can also stimulate negative impacts on the structural properties.

Despite having many disadvantages, biodegradable and biocompatible polymers have gained attention from scientific community due to their diversified role in wound healing process.

2. **Future perspectives**

Many approaches for wound healing are present nowdays, still the consequences of healing during chronic conditions are still conceded. Thus, the medical and healthcare community required wound care products having tunable physicochemical and biochemical properties such as hemostasis, bioresponsive, and antimicrobial. Therefore, advancement in knowledge of biopolymer with these properties will lead to the developments in technologies of bioengineering and regenerative medicine and aid in wound healing and tissue engineering. Moreover, the addition of knowledge and understanding of biopolymer and their mechanisms will lead to the successful application of biopolymer-based scaffolds, drug, or growth factors loaded dressing, 3D artificial skin, etc. Also, advances in biopolymer engineering with nanotechnology will help to develop effective technologies such as bioprinting and 3D electrospinning. These technologies which could be alternate solutions in replacement of extracellular matrices by growth factor delivery, cell-based therapy, and tissue engineering to restore injured tissue.

8 Conclusion

This book chapter discussed recent advancements and the advantages of biopolymers looked at their applications in wound healing management. Wound healing is a very complex and dynamic process and depends on interaction between the cells, growth factors, and extracellular matrix. Various biopolymers have been extensively studied and reported as excellent biocompatible and bioactive compounds. Currently, collagen, alginate, chitosan, etc. are the biopolymers that are frequently used in the wound care industry. Advancements in the knowledge and understanding of convergent technology for biopolymers will shape the future of medical and healthcare industry to satisfy the unmet needs.

References

Aderibigbe BA, Buyana B (2018) Alginate in wound dressings. Pharmaceutics 10(2):42. https://doi.org/10.3390/pharmaceutics10020042

Ahmed S, Ikram S (2016) Chitosan based scaffolds and their applications in wound healing. Achiev Life Sci 10:27–37. https://doi.org/10.1016/j.als.2016.04.001

Albright V, Xu M, Palanisamy A et al (2018) Micelle-coated, hierarchically structured nanofibers with dual-release capability for accelerated wound healing and infection control. Adv Healthc Mater 7:1800132. https://doi.org/10.1002/adhm.201800132

An B, Lin Y-S, Brodsky B (2016) Collagen interactions: drug design and delivery. Adv Drug Deliv Rev 97:69–84. https://doi.org/10.1016/j.addr.2015.11.013

Archana D, Singh BK, Dutta J et al (2013) In vivo evaluation of chitosan–PVP–titanium dioxide nanocomposite as wound dressing material. Carbohydr Polym 95:530–539. https://doi.org/10.1016/j.carbpol.2013.03.034

Armstrong DG, Meyer AJ, Sanfey H et al (2018) Basic principles of wound management. Post TW

Arslan YE, Sezgin Arslan T, Derkus B et al (2017) Fabrication of human hair keratin/jellyfish collagen/eggshell-derived hydroxyapatite osteoinductive biocomposite scaffolds for bone tissue engineering: from waste to regenerative medicine products. Colloids Surfaces B Biointerfaces 154:160–170. https://doi.org/10.1016/j.colsurfb.2017.03.034

Bellis SL (2011) Advantages of RGD peptides for directing cell association with biomaterials. Biomaterials 32:4205–4210. https://doi.org/10.1016/j.biomaterials.2011.02.029

Beyene RT, Derryberry SL, Barbul A (2020) The effect of comorbidities on wound healing. Surg Clin North Am 100:695–705. https://doi.org/10.1016/j.suc.2020.05.002

Bianchera A, Catanzano O, Boateng J et al (2020) The place of biomaterials in wound healing. Ther Dressings Wound Heal Appl:337–366. https://doi.org/10.1002/9781119433316.ch15

Bismarck A, Aranberri-Askargorta I, Springer J et al (2002) Surface characterization of flax, hemp and cellulose fibers; surface properties and the water uptake behavior. Polym Compos 23:872–894. https://doi.org/10.1002/pc.10485

Borderud SP, Li Y, Burkhalter JE et al (2015) Reply to discrepant results for smoking and cessation among electronic cigarette users. Cancer 121:2287. https://doi.org/10.1002/cncr.29306

Bryant R, Nix D (2015) Acute and chronic wounds-E-book. Elsevier Health Sciences

Cañedo-Dorantes L, Cañedo-Ayala M (2019) Skin acute wound healing: a comprehensive review. Int J Inflam June:1–15. https://doi.org/10.1155/2019/3706315

Chouhan D, Chakraborty B, Nandi SK et al (2017) Role of non-mulberry silk fibroin in deposition and regulation of extracellular matrix towards accelerated wound healing. Acta Biomater 48:157–174. https://doi.org/10.1016/j.actbio.2016.10.019

Dai C, Shih S, Khachemoune A (2020) Skin substitutes for acute and chronic wound healing: an updated review. J Dermatolog Treat 31:639–648. https://doi.org/10.1080/09546634.2018.1530443

Das S, Baker AB (2016) Biomaterials and nanotherapeutics for enhancing skin wound healing. Front Bioeng Biotechnol 4:1–20. https://doi.org/10.3389/fbioe.2016.00082

Davis FJ, Mitchell GR (2008) Polyurethane based materials with applications in medical devices BT - bio-materials and prototyping applications in medicine. In: Bártolo P, Bidanda B (eds) . Springer US, Boston, MA, pp 27–48

Dreifke MB, Jayasuriya AA, Jayasuriya AC (2015) Current wound healing procedures and potential care. Mater Sci Eng C 48:651–662. https://doi.org/10.1016/j.msec.2014.12.068

Dror Y, Ziv T, Makarov V et al (2008) Nanofibers made of globular proteins. Biomacromolecules 9:2749–2754. https://doi.org/10.1021/bm8005243

Du Y, Zhao Y, Dai S et al (2009) Preparation of water-soluble chitosan from shrimp shell and its antibacterial activity. Innov Food Sci Emerg Technol 10:103–107. https://doi.org/10.1016/j.ifset.2008.07.004

Eming SA, Martin P, Tomic-Canic M (2016) Wound repair and regeneration: mechanisms, signaling, and translation Sabine. Sci Transl Med 6:1–36. https://doi.org/10.1126/scitranslmed.3009337.Wound

Feig VR, Tran H, Bao Z (2018) Biodegradable polymeric materials in degradable electronic devices. ACS Cent Sci 4:337–348. https://doi.org/10.1021/acscentsci.7b00595

Frykberg RG, Banks J (2015) Challenges in the treatment of chronic wounds. Adv Wound Care 4:560–582. https://doi.org/10.1089/wound.2015.0635

Gao L, Zhou Y, Peng J et al (2019) A novel dual-adhesive and bioactive hydrogel activated by bioglass for wound healing. NPG Asia Mater 11:1–11. https://doi.org/10.1038/s41427-019-0168-0

Gonzalez ACDO, Andrade ZDA, Costa TF et al (2016) Wound healing - a literature review. An Bras Dermatol 91:614–620. https://doi.org/10.1590/abd1806-4841.20164741

Guo S, DiPietro LA (2010) Critical review in oral biology & medicine: factors affecting wound healing. J Dent Res 89:219–229. https://doi.org/10.1177/0022034509359125

Gurtner GC, Wong V (2013) Wound healing: normal and abnormal. In: Grabb and Smith Plastic Surgery, pp 13–19

Harper D, Young A, McNaught CE (2014) The physiology of wound healing. Surgery 32:445–450. https://doi.org/10.1016/j.mpsur.2014.06.010

Huang J, Ren J, Chen G et al (2018) Tunable sequential drug delivery system based on chitosan/hyaluronic acid hydrogels and PLGA microspheres for management of non-healing infected wounds. Mater Sci Eng C 89:213–222. https://doi.org/10.1016/j.msec.2018.04.009

Ito T, Yoshida C, Murakami Y (2013) Design of novel sheet-shaped chitosan hydrogel for wound healing: a hybrid biomaterial consisting of both PEG-grafted chitosan and crosslinkable polymeric micelles acting as drug containers. Mater Sci Eng C 33:3697–3703. https://doi.org/10.1016/j.msec.2013.04.056

Jaipan P, Nguyen A, Narayan RJ (2017) Gelatin-based hydrogels for biomedical applications. MRS Commun 7:416–426. https://doi.org/10.1557/mrc.2017.92

Jayakumar R, Chennazhi KP, Srinivasan S et al (2011) Chitin scaffolds in tissue engineering. Int J Mol Sci 12:1876–1887. https://doi.org/10.3390/ijms12031876

Jones V, Grey JE, Harding KG (2006) Wound dressings. BMJ 332:777–780. https://doi.org/10.1136/bmj.332.7544.777

Kaygusuz H, Torlak E, Akın-Evingür G et al (2017) Antimicrobial cerium ion-chitosan crosslinked alginate biopolymer films: a novel and potential wound dressing. Int J Biol Macromol 105:1161–1165. https://doi.org/10.1016/j.ijbiomac.2017.07.144

Kim MH, Park H, Nam HC et al (2018) Injectable methylcellulose hydrogel containing silver oxide nanoparticles for burn wound healing. Carbohydr Polym 181:579–586. https://doi.org/10.1016/j.carbpol.2017.11.109

Kohane DS, Langer R (2008) Polymeric biomaterials in tissue engineering. Pediatr Res 63:487–491. https://doi.org/10.1203/01.pdr.0000305937.26105.e7

Kucińska-Lipka J, Gubanska I, Janik H (2015) Bacterial cellulose in the field of wound healing and regenerative medicine of skin: recent trends and future prospectives. Polym Bull 72:2399–2419. https://doi.org/10.1007/s00289-015-1407-3

Lanzhen HE, Liu Y, Yang D (2007) Preparation and performance of chitosan-gelatin sponge-like wound-healing dressing. Chinese J Tissue Eng Res 11:5252–5256

Lei J, Sun L, Li P et al (2019) The wound dressings and their applications in wound healing and management. Heal Sci J 13:1–8

Li Y, Xu T, Tu Z et al (2020) Bioactive antibacterial silica-based nanocomposites hydrogel scaffolds with high angiogenesis for promoting diabetic wound healing and skin repair. Theranostics 10:4929–4943. https://doi.org/10.7150/thno.41839

Litwiniuk M, Krejner A, Speyrer MS et al (2016) Hyaluronic acid in inflammation and tissue regeneration. Wounds 28:78–88

Lloyd LL, Kennedy JF, Methacanon P et al (1998) Carbohydrate polymers as wound management aids. Carbohydr Polym 37:315–322. https://doi.org/10.1016/S0144-8617(98)00077-0

Mele E (2016) Electrospinning of natural polymers for advanced wound care: towards responsive and adaptive dressings. J Mater Chem B 4:4801–4812. https://doi.org/10.1039/C6TB00804F

Metcalf DG, Bowler PG (2013) Biofilm delays wound healing: a review of the evidence. Burn Trauma 1. https://doi.org/10.4103/2321-3868.113329

Mimura T, Amano S, Yokoo S et al (2008) Tissue engineering of corneal stroma with rabbit fibroblast precursors and gelatin hydrogels. Mol Vis 14:1819

Mir M, Ali MN, Barakullah A et al (2018) Synthetic polymeric biomaterials for wound healing: a review. Prog Biomater 7:1–21. https://doi.org/10.1007/s40204-018-0083-4

Mogoşanu GD, Grumezescu AM (2014) Natural and synthetic polymers for wounds and burns dressing. Int J Pharm 463:127–136. https://doi.org/10.1016/j.ijpharm.2013.12.015

Momeni M, Hafezi F, Rahbar H et al (2009) Effects of silicone gel on burn scars. Burns 35:70–74. https://doi.org/10.1016/j.burns.2008.04.011

Moohan J, Stewart SA, Espinosa E et al (2020) Cellulose nanofibers and other biopolymers for biomedical applications. A Rev Appl Sci 10. https://doi.org/10.3390/app10010065

Nair LS, Laurencin CT (2007) Biodegradable polymers as biomaterials. Prog Polym Sci 32:762–798. https://doi.org/10.1016/j.progpolymsci.2007.05.017

Okamoto Y, Yano R, Miyatake K et al (2003) Effects of chitin and chitosan on blood coagulation. Carbohydr Polym 53:337–342. https://doi.org/10.1016/S0144-8617(03)00076-6

Palo M, Özliseli E, Sen KD et al (2019) 11 electrospun biocomposite fibers for wound healing applications. Green Electrospinning 265

Park JH, Choi SH, Park SJ et al (2017) Promoting wound healing using low molecular weight fucoidan in a full-thickness dermal excision rat model. Mar Drugs 15:1–15. https://doi.org/10.3390/md15040112

Pastar I, Stojadinovic O, Yin NC et al (2014) Epithelialization in wound healing: a comprehensive review. Adv Wound Care 3:445–464. https://doi.org/10.1089/wound.2013.0473

Patel G, Patel G, Patel R et al (2007) Sodium alginate: physiological activity, usage & potential applications. Drug Deliv Technol 7:28–37

Pavithra D, Doble M (2008) Biofilm formation, bacterial adhesion and host response on polymeric implants—issues and prevention. Biomed Mater 3:34003. https://doi.org/10.1088/1748-6041/3/3/034003

Percival NJ (2002) Classification of wounds and their management. Surg–Oxford Int Ed 20:114–117. https://doi.org/10.1383/surg.20.5.114.14626

Piraino F, Selimović Š (2015) A current view of functional biomaterials for wound care, molecular and cellular therapies. Biomed Res Int 2015:. doi:https://doi.org/10.1155/2015/403801

Piskin E (1995) Biodegradable polymers as biomaterials. J Biomater Sci Polym Ed 6:775–795. https://doi.org/10.1163/156856295X00175

Powell HM, Supp DM, Boyce ST (2008) Influence of electrospun collagen on wound contraction of engineered skin substitutes. Biomaterials 29:834–843. https://doi.org/10.1016/j.biomaterials.2007.10.036

Psimadas D, Georgoulias P, Valotassiou V et al (2012) Molecular nanomedicine towards cancer. J Pharm Sci 101:2271–2280. https://doi.org/10.1002/jps

Puppi D, Chiellini F, Piras AM et al (2010) Polymeric materials for bone and cartilage repair. Prog Polym Sci 35:403–440. https://doi.org/10.1016/j.progpolymsci.2010.01.006

Qu J, Zhao X, Liang Y et al (2019) Degradable conductive injectable hydrogels as novel antibacterial, anti-oxidant wound dressings for wound healing. Chem Eng J 362:548–560. https://doi.org/10.1016/j.cej.2019.01.028

Rehfeld A, Nylander M, Karnov K (2017) Compendium of histology: a theoretical and practical guide. Springer

Reinke JM, Sorg H (2012) Wound repair and regeneration. Eur Surg Res 49:35–43. https://doi.org/10.1159/000339613

Robson MC, Steed DL, Franz MG (2001) Wound healing: biologic features and approaches to maximize healing trajectories. Curr Probl Surg 38:A1. https://doi.org/10.1067/msg.2001.111167

Sahana TG, Rekha PD (2018) Biopolymers: applications in wound healing and skin tissue engineering. Mol Biol Rep 45:2857–2867. https://doi.org/10.1007/s11033-018-4296-3

Schmitt EE, Polistina RA (1963) Uspto, ed. United States

Schwartz AJ, Wilson DA, Keegan KG et al (2002) Factors regulating collagen synthesis and degradation during second-intention healing of wounds in the thoracic region and the distal aspect of the forelimb of horses. Am J Vet Res 63:1564–1570. https://doi.org/10.2460/ajvr.2002.63.1564

Sezer AD, Cevher E (2011) Biopolymers as wound healing materials: challenges and new strategies. In: Pignatello R (ed) Biomaterials applications for nanomedicine. InTech, Rijeka, pp 383–414

Shankar M, Ramesh B, Kumar R et al (2017) Wound healing and its importance. Der Pharmacol Sin 1:24–30

Shavandi A, Silva TH, Bekhit AA et al (2017) Keratin: dissolution, extraction and biomedical application. Biomater Sci 5:1699–1735. https://doi.org/10.1039/C7BM00411G

Shi C, Wang C, Liu H et al (2020) Selection of appropriate wound dressing for various wounds. Front Bioeng Biotechnol 8:1–17. https://doi.org/10.3389/fbioe.2020.00182

Shi L, Zhao Y, Xie Q et al (2018) Moldable Hyaluronan hydrogel enabled by dynamic metal–bisphosphonate coordination chemistry for wound healing. Adv Healthc Mater 7:1700973. https://doi.org/10.1002/adhm.201700973

Silva TH, Alves A, Popa EG et al (2012) Marine algae sulfated polysaccharides for tissue engineering and drug delivery approaches. Biomatter 2:278–289. https://doi.org/10.4161/biom.22947

Siritientong T, Angspatt A, Ratanavaraporn J et al (2014) Clinical potential of a silk sericin-releasing bioactive wound dressing for the treatment of split-thickness skin graft donor sites. Pharm Res 31:104–116. https://doi.org/10.1007/s11095-013-1136-y

Smith AM, Moxon S, Morris GA (2016) Biopolymers as wound healing materials. Elsevier Ltd

Song R, Murphy M, Li C et al (2018) Current development of biodegradable polymeric materials for biomedical applications. Drug Des Devel Ther 12:3117–3145. https://doi.org/10.2147/DDDT.S165440

Sorg H, Tilkorn DJ, Hager S et al (2017) Skin wound healing: An update on the current knowledge and concepts. Eur Surg Res 58:81–94. https://doi.org/10.1159/000454919

Suarato G, Bertorelli R, Athanassiou A (2018) Borrowing from nature: biopolymers and biocomposites as smart wound care materials. Front Bioeng Biotechnol 6:1–11. https://doi.org/10.3389/fbioe.2018.00137

Sulaeva I, Henniges U, Rosenau T et al (2015) Bacterial cellulose as a material for wound treatment: properties and modifications. A review. Biotechnol Adv 33:1547–1571. https://doi.org/10.1016/j.biotechadv.2015.07.009

Takei T, Nakahara H, Ijima H et al (2012) Synthesis of a chitosan derivative soluble at neutral pH and gellable by freeze–thawing, and its application in wound care. Acta Biomater 8:686–693. https://doi.org/10.1016/j.actbio.2011.10.005

Thiruvoth F, Mohapatra D, Sivakumar D et al (2015) Current concepts in the physiology of adult wound healing. Plast Aesthetic Res 2:250. https://doi.org/10.4103/2347-9264.158851

Thomas Hess C (2011) Checklist for factors affecting wound healing. Adv Skin Wound Care 24:192. https://doi.org/10.1097/01.ASW.0000396300.04173.ec

Tsioufis C, Bafakis I, Kasiakogias A et al (2012) The role of matrix metalloproteinases in diabetes mellitus. Curr Top Med Chem 12:1159–1165. https://doi.org/10.2174/1568026611208011159

Ulery BD, Nair LS, Laurencin CT (2011) Biomedical applications of biodegradable polymers. J Polym Sci Part B Polym Phys 49:832–864. https://doi.org/10.1002/polb.22259

Velnar T, Bailey T, Smrkolj V (2009) The wound healing process: An overview of the cellular and molecular mechanisms. J Int Med Res 37:1528–1542. https://doi.org/10.1177/147323000903700531

Wani MY, Hasan N, Malik MA (2010) Chitosan and Aloe vera: two gifts of nature. J Dispers Sci Technol 31:799–811. https://doi.org/10.1080/01932690903333606

Ward A, Dubey P, Basnett P et al (2020) Toward a closed loop, integrated biocompatible biopolymer wound dressing patch for detection and prevention of chronic wound infections. Front Bioeng Biotechnol 8. https://doi.org/10.3389/fbioe.2020.01039

Weir D (2020) Wound dressings. In: Alavi A, Maibach H (eds) Local wound Care for Dermatologists. Updates in clinical dermatology. Springer, CHAM, pp 25–34

Williams DF (2009) On the nature of biomaterials. Biomaterials 30:5897–5909. https://doi.org/10.1016/j.biomaterials.2009.07.027

Xiao Liu M, Jia G (2018) Modern wound dressing using polymers/biopolymers. J Mater Sci Eng 07:7–10. https://doi.org/10.4172/2169-0022.1000454

Yang D, Jones KS (2009) Effect of alginate on innate immune activation of macrophages. J Biomed Mater Res Part A 90A:411–418. https://doi.org/10.1002/jbm.a.32096

Yao R, Zhang R, Luan J et al (2012) Alginate and alginate/gelatin microspheres for human adipose-derived stem cell encapsulation and differentiation. Biofabrication 4:25007. https://doi.org/10.1088/1758-5082/4/2/025007

Yoon D, Yoon D, Cha H-J et al (2018) Enhancement of wound healing efficiency mediated by artificial dermis functionalized with EGF or NRG1. Biomed Mater 13:45007. https://doi.org/10.1088/1748-605X/aaac37

Zhao W, Liu W, Li J et al (2015) Preparation of animal polysaccharides nanofibers by electrospinning and their potential biomedical applications. J Biomed Mater Res Part A 103:807–818. https://doi.org/10.1002/jbm.a.35187

Zidarič T, Milojević M, Gradišnik L et al (2020) Polysaccharide-based bioink formulation for 3D bioprinting of an in vitro model of the human dermis. Nano 10:1–18. https://doi.org/10.3390/nano10040733

Atmospheric Pressure Plasma Therapy for Wound Healing and Disinfection: A Review

Alphonsa Joseph, Ramkrishna Rane, and Akshay Vaid

1 Introduction

The wound healing process, particularly in the skin, has been studied extensively for more than a century. A wound typically involves breaking the layers of the skin and damaging the underlying tissues. An array of biochemical events takes place when the layers of skin are broken to repair the damage for healing. They are blood clotting, inflammation, tissue growth, and tissue remodeling (Nguyen et al. 2009; Rieger et al. 2015). Several signal molecules like cytokines, chemokine, enzymes, and proteins also participate in the healing process. The healing process is complex in nature as it is prone to be interrupted by various factors like infection, diabetes, venous or arterial diseases, and metabolic deficiencies. The wound then becomes chronic in nature and takes longer time to heal (Enoch and Price 2004). The main features of these wounds include severe inflammation, relentless infections, and drug resilient biofilms. This is mainly due to the excessive amounts of signal molecules like pro-inflammatory cytokines, proteases, senescent keratinocytes, endothelial, fibroblasts, and macrophages (Woo et al. 2007; Stojadinovic et al. 2008). Hence, maintaining a proper balance of signal molecules and growth factors can reverse the chronicity of wounds for improved healing.

It is well known that some wounds are burdened with different levels of bacterial or microbial loads. Bryant et al. defined that the microbial load in a wound exists in five forms: contamination, colonization, critical colonization, biofilm, and infection (Bryant and Nix 2011). Hence, asepsis of microbes is primarily important. There are

A. Joseph (✉)
Institute for Plasma Research (IPR), Gandhinagar, Gujarat, India

Homi Bhabha National Institute, Mumbai, Maharashtra, India
e-mail: alphonsa@ipr.res.in

R. Rane · A. Vaid
Institute for Plasma Research (IPR), Gandhinagar, Gujarat, India

many methods used to reduce the bacterial load and they include debridement, appropriate wound cleansing, use of antibiotic medicines, application of ointments, and use of proper bandages. However, in most wounds, the effect of these traditional medical antiseptics and ointments are not effective due to the presence of some bacteria which are resistant to drugs like Methicillin-Resistant *Staphylococcus aureus* (MRSA). Therefore, there is a need for a new method for skin disinfection to attend to the quick healing of chronic wounds (Vasilets et al. 2009; Lloyd et al. 2010; Haertel et al. 2014; Weltmann and von Woedtke 2017).

While there are innumerable traditional and non-traditional treatments available for wound healing therapy, very few have shown their effectiveness in accelerating wound repair. It has been found that traditional treatments are expensive and the recovery time is very long (Xiong 2018). Moreover, modern synthetic allopathic-based medicines also have limitations like they are allergic and resistant in some patients. Some medicines are also very expensive (Majumdar and Sangole 2016). It is also reported that though several curative wound healing techniques are employed; very few satisfactory therapies for chronic wounds are available (Xu et al. 2015a, b, 2015). This has provoked the scientists and researchers to discover alternative novel technologies for wound healing and disinfection and validate its use.

Cold atmospheric pressure (CAP) plasma-based techniques are recently being considered as a potential alternative therapy for chronic wound healing and disinfection (Salehi et al. 2015). Cold atmospheric plasmas are generated at atmospheric pressure and exist in weakly ionized gaseous state. They consist of ground and excited atoms, radicals, neutral molecules, ions, negatively charged particles, and ultraviolet radiation (Lieberman and Lichtenberg 2005). As the temperature of these plasmas are low (approx. 40 °C), living tissues can be easily treated. The Reactive Nitrogen Oxygen Species (RNOS) along with electric fields generated from these plasmas have been shown to decrease bacteria in wounds to promote healing (Duchesne et al. 2018). For this reason, CAP plasma has been extensively used in the last decade in biomedical research, particularly in dermatology, wound healing, blood coagulation, treatment of cancer, disinfection of infectious matter, sterilization of medical devices, and cosmetics (Daeschlein et al. 2015; Mohd Nasir et al. 2016; Klämpfl et al. 2012; Chatraie et al. 2018; Schmidt et al. 2017; Isbary et al. 2012; Heinlin et al. 2010; Duval et al. 2013; Kang et al. 2014; Vandamme et al. 2012; Nasruddina et al. 2014; O'Connor et al. 2014). CAP plasma is also popular for sterilizing heat-sensitive surfaces and medical equipment (Kong et al. 2009). Various in vitro and in vivo investigations with CAP plasma have indicated the growth and migration of fibroblasts as well as epithelial cells to the site of inflammation. Moreover, CAP plasma has also shown an affirmative effect on the epidermal growth factors and initiation of angiogenesis for promoting chronic wound healing (Fridman et al. 2008). About less than 5% ROS and more than 80% RNS generated from plasma have been reported to penetrate tissues having thicknesses of 500µm (Woedtke et al. 2013). However, the actual penetration depths of plasma are yet not clear. Studies of CAP plasma on chronically infected wounds, pruritus, shingles, and various skin diseases have been reported to be treated successfully (Lademann et al. 2011; Ulrich et al. 2015). Hence, since CAP plasma treatments are pain free; have short treatment times, nontoxic, non-allergic, and have sterilizing properties, they are favorable for wound healing (Daeschlein et al. 2012; Lademann et al. 2013).

1.1 Plasma as a Useful Tool in Medical Field

Though, plasma is a familiar word in biology, where it is described as one of the components of blood, the term "Plasma" used in this chapter is the fourth state of matter. This state of matter is obtained by providing energy to the gaseous state of matter. Naturally, there are always few electrons and ions in any volume of neutral gas. The source of these free charged particles can be due to the interaction of cosmic rays or radioactive radiation with the gas (Conrads and Schmidt 2000). Usually, laboratory plasma is formed by supplying electrical energy to the neutral gas. On application of an electric field, free charge carriers already present in the gases accelerate and collide with gas molecules. The charged particles also collide with the surfaces of the electrodes. These collisions result in multiplication of charged particles. This generation of charged particles is eventually balanced by the loss of the charged particle so that steady-state plasma is developed (Conrads and Schmidt 2000). The plasma, thus produced is an ionized gas containing approximately the same numbers of ions and electrons. It also contains neutrals that are not ionized. Along with electrons and ions, enormous amount of reactive species are produced depending upon the type of gas used for formation of plasma. Various power sources like direct current (DC), capacitive coupled radio frequency (RF), inductively coupled RF, and microwave can be used to form plasma.

Non-thermal (cold) and thermal (hot) plasmas are the common two types of plasmas. In non-thermal plasma, the temperature of ions and neutrals is less than the electron temperature. In the case of thermal plasmas, the temperature is relatively similar for electrons and heavy particles and can typically reach temperatures of several thousand degrees kelvin (Roth 1995). These plasmas can function over a wide span of pressure regimes. Accordingly, plasmas are further categorized as low pressure and atmospheric pressure plasmas. Typically, low-pressure plasma operates in a glow discharge mode with typical pressure range of 10–100 Pa while atmospheric pressure plasma operates in an arc discharge mode, also called as hot plasma (Lieberman and Lichtenberg 2005). Even though the generation of low-pressure plasma is comparatively easy, it requires vacuum chamber and pumping systems which are costly. Secondly, continuous plasma treatment is not possible by using low-pressure plasmas. Hence, atmospheric pressure plasmas are more preferred for applications like material processing, biomedical applications where expensive pumping systems can be avoided (Becker et al. 2005).

Both thermal as well as non-thermal plasmas are useful for few biomedical applications. The thermal plasmas that are hot are mainly used for ablation and cauterization of biological tissue and in cosmetics, e.g., Argon plasma coagulation (APC), for removal of wrinkles, regeneration of skin (Canard and Védrenne 2001; Pereira-Lima et al. 2000; Foster et al. 2008). Non-thermal plasmas on the other hand do not exhibit any damaging effect as obtained from thermal plasma due to its ambient temperatures and are hence known as Cold Atmospheric Pressure (CAP) plasma. Such plasmas are widely applicable in the area of medicine. The "Plasma Medicine" term usually relates to the use of plasma for biomedical applications. Plasma medicine includes several low-temperature applications of plasmas like sterilization; wound healing; cancer treatment; dermatology, dentistry, and treatment of implants (Laroussi 2018).

1.2 Cold Atmospheric Pressure Plasma Devices

According to classical Paschen law, if one goes for atmospheric pressure, inter-electrode distance has to be reduced. In case of plasmas generated at atmospheric pressure, the gap between the electrodes is few millimeters (Eliasson and Kogelschatz 1991). Mostly, the plasmas produced at atmospheric pressure are thermal plasmas where the neutral gas temperatures are very high and are similar to the temperature of other species like electrons and ions. Now the question remains that whether there is any method to produce atmospheric pressure plasma at low temperatures? The answer is yes. This type of plasma with low temperature and generated at atmospheric pressure can be achieved by maintaining proper electrode geometry, discharge gas, and frequency of the applied voltage (Kogelschatz 2003). As far as biomedical applications are concerned, the dielectric barrier discharge (DBD) plasma and pencil-like plasma produced by Atmospheric Pressure Plasma Jet (APPJ) have been used extensively. Dielectric barrier discharges are very much suitable for obtaining the atmospheric pressure diffused plasma in large volumes while APPJ is suitable for localized treatment like skin infection or tooth cleaning.

1.2.1 Dielectric Barrier Discharge

DBD (Eliasson and Kogelschatz 1991; Kogelschatz 2003) is a kind of AC discharge that operates at atmospheric or higher pressures. This DBD discharges are also employed for the production of ozone gas using air or oxygen (Yao et al. 2015). Nowadays, Ozonizers are effective tools for ozone generation and are being used worldwide for water treatment. In these discharges, the presence of dielectric material on one or both the electrodes is important while generating the plasma. The gap between the two electrodes for the discharge is very small (~few millimeters). These discharges at atmospheric pressure are created using AC high voltage generators of 1–50 kV and frequencies of 50 kHz to 1 MHz. Figure. 1 shows various electrode configurations, which are used to generate DBD discharges.

Typically, in the DBD, a dielectric material exists on one of the metal electrodes. The ceramic materials, glass, quartz, and thin layers of polymers are preferred as dielectric materials. As the dielectric material does not permit the DC current to flow, alternating voltages (AC) are used for operating these discharges. Also, the stability

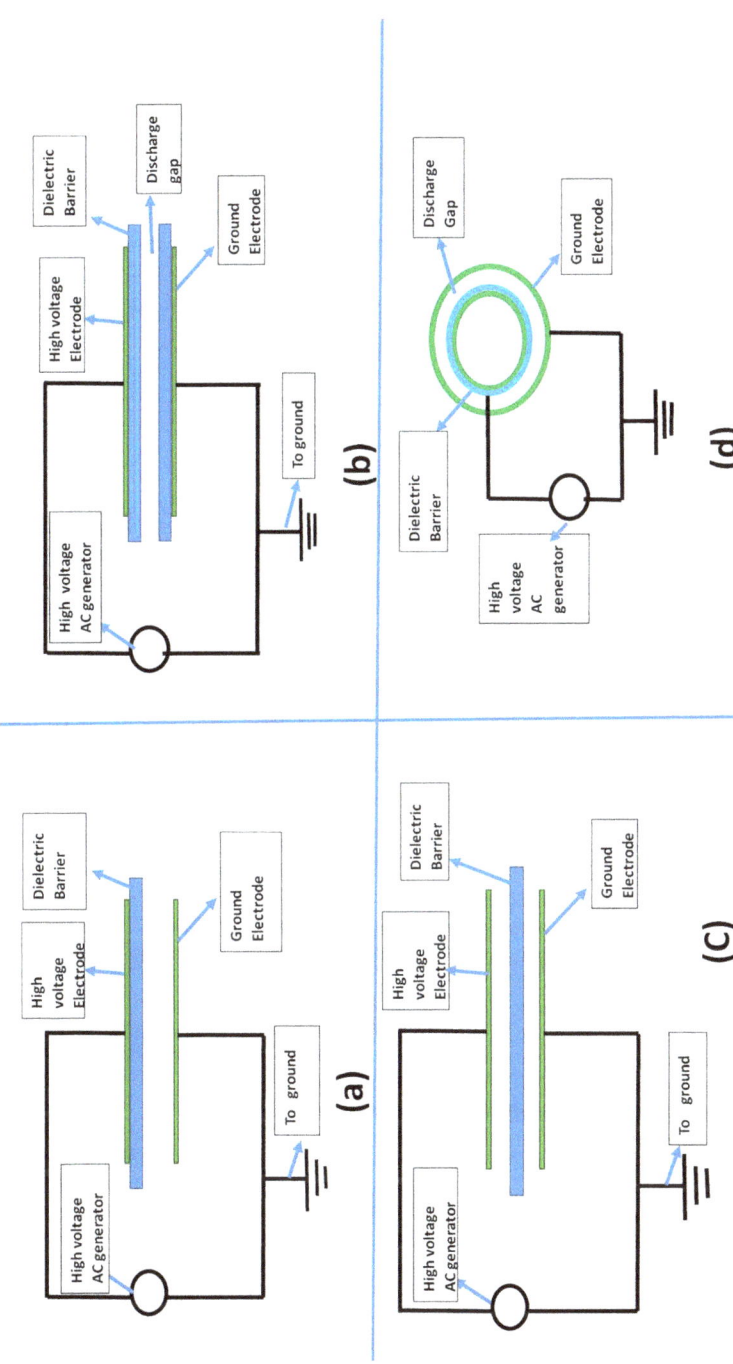

Fig. 1 Schematic of different configurations for dielectric barrier discharge systems showing (**a**) single dielectric planar, (**b**) double dielectric planar, (**c**) single dielectric with both side discharge, and (**d**) concentric cylindrical discharge

of the discharge mostly depends upon the type of discharge gas. The helium gas gives a stable glow discharge while gases like nitrogen, oxygen, and argon are unstable and change to filamentary discharges. However, it is still possible to operate them in the stable glow discharge mode by changing electrode configuration. The dielectric constant, as well as the thickness of the dielectric material, determines the displacement current passing through dielectric layer. In most applications, the dielectric material attached to the electrode limits the high current and hence arcing phenomenon in the gas space. Such type of DBD plasma is more suitable for continuous surface modifications at atmospheric pressure. As compared to other types of discharges, the DBD is the most suitable method to produce non-thermal plasma at atmospheric pressure. In addition to that scaling up to higher dimensions is comparatively easy in the DBD.

1.2.2 Atmospheric Pressure Plasma Jet

Recently, many researchers have developed an atmospheric pressure plasma jet, a kind of DBD in cylindrical electrode configuration. The discharge generated using atmospheric pressure plasma jet shows many characteristics similar to conventional glow discharge produced at low pressure. A typical schematic of an atmospheric pressure plasma jet and the APPJ device developed by the Institute for Plasma Research (IPR) in contact with human skin is shown in Fig. 2. This configuration composed of two cylindrical electrodes. The plasma-generating gas like helium, argon, oxygen, or their combinations are used. By applying high voltages between the electrodes, plasma is produced. The ionized gas that exits through a nozzle is termed as plasma jet. The plasma jet or plume coming out through the nozzle is used to treat the substrate, which is kept at few millimeters. Non-thermal plasma jets operating at atmospheric pressure are emerging as a novel research topic in plasma physics. The ongoing and future research in this field will be beneficial for material treatment as well as for medical applications. The main advantage of this type of plasma is its capability to supply significant amounts of reactive species at low gas

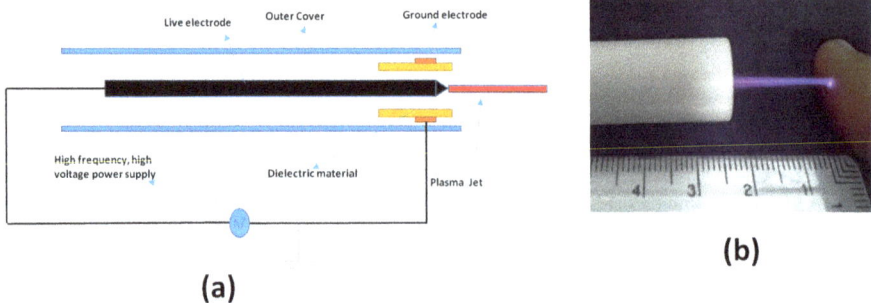

Fig. 2 Atmospheric pressure Plasma Jet developed by the Institute for Plasma Research (**a**) Schematic and (**b**) Plasma jet in contact with human skin

temperatures. The plasma plume that comes out from the nozzle interacts with its surrounding environment like ambient air. The interaction changes the properties of the discharge plasma by releasing reactive species and in turn also the treated surface. Thus, this type of plasma jet is not restricted by defined physical walls, but increases their usage for material treatment applications like human skin and teeth (Pan 2010).

In the last few decades, developments have been made in the design of an APPJ for its application in different areas. Various plasma jet arrangements exist depending upon the electrode configuration, gases employed, and power supply used to generate the plasma. Plasma jets can be transferred arc or non-transferred arc. When one of the electrodes is outside the body of jet, it is referred to as transferred arc jet. In a non-transferred arc jet, an electric field is generated between the two electrodes and the plasma jet or plume protrudes outside due to the flow of plasma generating gas. Plasma jets have been generated using noble gases like argon and helium. There exist different configurations of the plasma jets depending upon the usage of dielectric material and they can be categorized as Dielectric free electrode (DFE) (Babayan and Jeong 1998; Jeong et al. 1998), Dielectric barrier discharge (DBD) (Abuzairi et al. 2016; Lu et al. 2008), and DBD-like jets (Léveillé and Coulombe 2005; Shashurin et al. 2009). Some configurations are also of single electrode (SE) type plasma jets (Stoffels et al. 2003; Lu 2009). In case of DFE jets, in absence of dielectric material, the plasma produced touches the electrodes. However, the widely used configuration of DBD jets composed of the quartz tube, which acts as a dielectric material and the electrodes are present on the external side of this tube.

Both DBD and APPJ devices are used for biomedical applications. Recently, many plasma devices are commercially available for wound healing. Some of them are kINPen, a CE certified RF argon plasma jet, PlasmaDerm, MicroPlasma α and β device, and Plasma jet (Tigres, Plasma MEF technology) (Xiong 2018).

2 Interactions of Cold Atmospheric Plasma with a Living Cell

Atmospheric pressure plasma interacts with the living cells by various mechanisms as reported in the literature. Plasma contains charged particles, various reactive oxygen species (ROS), reactive nitrogen oxygen species (RNOS), excited molecules, and ultraviolet (UV) photons and these play an important role when plasma interacts with the biological matter. It is reported in the literature that (Moisan et al. 2001; Stoffels et al. 2008; Boudam et al. 2006):

1. Microbial DNA in the cell gets destroyed by UV radiation.
2. Intrinsic photo desorption erodes the microorganism by breaking their chemical bonds.
3. Atomic and molecular radicals etch the cell surface, thus destroying them.

Table 1 Production of ROS from plasma

ROS	Formula	Plasma reaction
Hydrogen peroxide	H_2O_2	$OH^. + OH^. \rightarrow H_2O_2$
Hydroxyl radical	$OH^.$	$e + N_2 + M \rightarrow N_2^+ + 2e + M$ $N_2^+ + H_2O \rightarrow N_2 + H_2O^+$ $H_2O^+ + H_2O \rightarrow OH^. + H_3O^+$
Superoxide	O_2^-	$e + O_2 + M \rightarrow O_2^- + M$
Singlet oxygen	$O_2\left(\Delta_g^1\right)$	$O_2 + e \rightarrow O_2\left(\Delta_g^1\right) + e$

(M- third body particle)

Table 2 Production of RNS from plasma

RNS	Formula	Plasma reaction
Nitric oxide	NO	$N_2 + e \rightarrow N + N + e$ $N + O + N_2 \rightarrow NO + N_2$
Peroxynitrite	$ONOO^-$	$NO^+ + O_2^- \rightarrow ONOO^-$
Nitrite	$NO2^-$	$NO + O_3 \leftrightarrow NO_2 + O_2$ $2\,NO_2 + H_2O \rightarrow HNO_2(aq) + HNO_3(aq)$

4. Oxidative damage is done to the membranes and DNA by ROS/RNOS species.
5. Charged particle accumulates on the cell membrane thus leading to its rupture.

Cellular proliferation, differentiation, and viability are also influenced by active species in the plasma (Lin et al. 2015). Both ROS and RNOS play a vital part as normal components of the cellular metabolism. ROS and RNOS include hydroxyl ions, atomic oxygen, singlet delta oxygen, superoxide, hydrogen peroxide, and nitric oxide. Various reactions involved in the production of ROS and RNOS are given in Tables 1 and 2, respectively (Arjunan 2011). For example, peroxidation of unsaturated fatty acids is caused by hydroxyl species. It has been reported both ROS and RNOS assist in wound healing. ROS induces oxidative stress. Wound healing is positively influenced at lower concentration of ROS. The regulation of immune deficiencies, cell proliferation, induction of phagocytosis, control on collagen synthesis, and angiogenesis are mainly affected by RNOS. It is established in the literature that nitric oxide regulates the biological functions by working as intracellular messenger (Mone et al. 2014; Stallmeyer et al. 1999; Lu 2008). Both ROS and RNOS are capable of reducing the microbiological burdens and bacterial colonization which is one of the main contributing and well-recognized factor of impairing wound healing. Moreover, a direct influence on the cellular level during wound healing is also demonstrated by the presence of these species. For this reason, atmospheric pressure plasma jet has a greater potential for the treatment of chronic and non-healing wounds.

3   Anti-Microbial Ability of APPJ for Disinfection of Wound on Skin

Researchers have shown from their works that plasma has disinfectant properties and can be used in various skin-related treatments. Scholz et al. have reported that APPJ has potential to inactivate both Gram-positive and -negative bacteria, fungi, viruses, and yeasts (Scholz et al. 2015). It is well known that excluding the gut, the next place in the body where more microorganisms are found is the skin. Among the microorganisms, bacteria are found to be the most abundant followed by fungi and viruses on a normal skin of a healthy human. Bacteria usually enter into the body through cuts, punctures, surgery, burns, or preexisting skin disorder and develop skin infection thereby hindering the process of wound healing (Robson 1997). As plasma is demonstrated to deactivate bacteria with high efficiency, the use of APPJ can help in reduction of the bacterial load in open wounds. Among the various types of bacteria, nine different species are present on a relevant wound and their strains were identified as shown in Table 3 (Daeschlein et al. 2012).

Daeschlein et al. showed that after APPJ treatment for two minutes using argon gas, highest reduction factor (RF) was found for PA followed by HS, MSSA, CA, and EF (Helmke et al. 2011). Darmawati et al. also studied microorganisms like SA, PA, Methicillin-Resistant *Staphylococcus aureus* (MRSA), and Carbapenem-Resistant *Pseudomonas aeruginosa* (CRPA), with argon-based APPJ for 2 min of treatment. They observed that the inhibition zones were highest for CRPA and lowest for SA. These and the previous results indicate that APPJ was more effective for Gram negative bacteria inactivation (Darmawati et al. 2019). It is known that the sensitivity of killing is dependent on cell wall thickness (Ermolaeva et al. 2011). APPJ treatment was more effective in deactivating Gram negative than the Gram positive bacteria because of the presence of their thin cell wall thickness. It has been reported by many researchers that the surface of the living organisms undergo a mechanical action referred to as "etching" due to the various reactive species

Table 3   Common microbial reference strains used for in vitro studies

Sr. No.	Species	Reference strain
1.	*Escherichia coli (EC)*	ATCC 25922
2.	*Pseudomonas aeruginosa* (PA)	ATCC 15442; ATCC 9027
3.	*Klebsiella group (K. pneumoniae* sp. *pneumoniae, K. oxytoca)*	ATCC 700324
4.	Methicillin-sensitive *Staphylococcus aureus* (MSSA)	ATCC 1924
	Staphylococcus aureus (SA),	ATCC 6538
5.	*Staphylococcus epidermidis (SE)*	ATCC 12228
6.	Proteus group (*P. mirabilis, P. vulgaris*)	ATCC 6380
7.	*Enterococcus faecalis (EF)*	ATCC 29212
	Enterococcus faecium (EF)	ATCC 6057
8.	*Candida albicans (CA)*	ATCC 10231
9.	Hemolytic streptococci (HS) (group A and B)	ATCC 27956

produced by plasma (Chau et al. 1996; Lerouge et al. 2000). The so-called etching causes fissure in the membranes of microorganism, which leads to their deactivation. Laroussi et al. compared the inactivation of *E. coli* (10^6 CFU/ml) on agar for 120 s using a plasma formed using helium gas alone and mixture of helium with oxygen gases. They observed that the inactivated area was more by using helium/oxygen gas mixtures compared to that obtained by pure helium (Laroussi et al. 2006). A more detailed review of the use of APPJ for microbial inactivation can be obtained from the work done by Ehlbeck et.al (2011). Hort et al. demonstrated the effect of APPJ using both air and nitrogen gases at 100 W powers on *P. aeruginosa*, *S. aureus*, and *C. albicans*. It was seen that air and nitrogen plasma significantly affected *P. aeruginosa* and *S. aureus* than *C. albicans*. Among the gases employed in APPJ, nitrogen exhibited a greater microbiocidal effect (Hort et al. 2017). Hence, for bacterial disinfection gases like argon, helium with oxygen gas mixture, and nitrogen are commonly used depending upon the type of bacteria.

Chronic wounds do not heal because of the presence of biofilms and they are observed in 90% of chronic wounds. Biofilms are a collection of different bacterial species (planktonic form of bacteria) having a protective layer adhering to the wound surface in the form of glycocalyx (Attinger and Wolcott 2012). As these biofilms are resistant to antimicrobial treatments therefore they need to be eradicated. Conventional techniques that use high heat and chemicals are not recommended as they use toxic chemicals and cause surface degradation. APPJ has also shown to have a great potential to deactivate the biofilms. Most bacteria that are present in the biofilm are *E. coli*, *P. aeruginosa*, *S. aureus*, and *Streptococcus pyogenes*. Xu et al. in their work demonstrated that atmospheric pressure plasma jet (APPJ) with helium gas completely inactivated S. *aureus* biofilms in 10 min and reported a 99% reduction in biofilm as compared to untreated samples. They attributed the biofilm inactivation to the presence of reactive species present in the plasma and plasma-induced intracellular ROS (Xu et al. 2015a, b, 2015). Theinkom et al. also studied the effect of APPJ using air against *E. faecalis* (EF) for 10 min treatment and reported a reduction in CFU of E. *faecalis* by ≥5 log. It was observed that the results obtained by chlorhexidine and UVC radiation were also comparable with the effectiveness of APPJ. On examination by spectrometric measurements, it was observed that cytoplasmic membranes were intact after this treatment (Theinkom et al. 2019). Patenall et al. studied the growth of biofilms using helium CAP plasma jet for 5 min. Biofilm grown for 8 h showed 4–5 log reduction in bacterial cells compared to 2 log reductions which were achieved on biofilms grown for 12 h. Hence, reducing the formation of biofilms by CAP plasma is time dependent (Patenall et al. 2018).

4   In Vitro Studies Using APPJ on Skin

Skin is the largest organ of the body and it covers about 2 m^2 on the body. Life would not be possible without skin. The epidermis, which is the outermost part of the skin, is considered to be the most immunologic organ as it is responsible for all types of skin diseases (Wende et al. 2010). The wound healing occurs only when the skin cells stop cell division and migrate for wound closure (Liu et al. 2017). It is reported that wound closure is induced in the infected 2-D skin model using APPJ (Hunt et al. 2000). This was achieved as there was an enhancement of proliferation rate of skin cells after APPJ treatment, which is a key requirement for healing wounds. Moreover, after plasma treatment, the overall antibacterial efficacy also increased due to extracellular trap formation (Fritsch 1998). All these factors are responsible for reduction in wound inflammation and aids wound healing.

Several researchers have indicated that the plasma cell interactions mainly depend on the plasma source, time duration as well the type of cell it is being treated (Kim et al. 2010, Kalghatgi et al. 2011, O'Connell et al. 2011). Low plasma treatment durations are beneficial as it stimulates cell viability, proliferation, differentiation, and migration, whereas longer time durations lead to cell apoptosis/necrosis (Xiong 2018). Most of the skin cells such as keratinocytes, fibroblasts, and endothelial cells are affected in a positive manner by the CAP plasma treatment as reported in the literature. It is seen that these cells migrate in wound beds resulting in timely wound healing. Duchesne et al. found that a quicker scratch closure was observed at 24 hrs. After the scratch when the treatment time was fixed for 1 min. Among the three cells studied, i.e., keratinocyte, endothelial, and fibroblasts cells, they showed 70%, 50%, and 25% smaller scratches, respectively, than that of the control sample. The short-lived ROS and RNS species were chiefly responsible for improving the scratch wound closures (Duchesne et al. 2018).

During the complex process of wound healing, both keratinocytes and fibroblasts secrete various cytokines such as IL-1 and IL-6 and growth factors like transforming growth factor α (TGF-α), which facilitate via intracellular signaling pathways. This in turn influences cellular mechanisms such as proliferation, migration, adhesion, or a contraction for wound healing. Arndt et al. studied fibroblasts cells using two types of plasma jets that are commercially available namely MicroplaSter and KINpen. Treatment on fibroblasts with the MicroplaSter and KINpen devices induced expressions of diverse wound healing relevant cytokines as well as growth factors for influencing cellular mechanisms that are important for wound healing. In their investigations, they indicated that the plasma not only activates the fibroblasts but also helps to convert into myofibroblasts which are responsible for wound contraction. Skin homeostasis and physiological tissue repair are maintained by myofibroblasts (Hinz 2016, Arndt et al. 2018).

In another study by Arndt et al. an exposure of APPJ for 2 min was done on primary human dermal fibroblasts (2F0621, 9F0438, 9F0889). A significant amount of increase in several proteins, e.g., CD 40 Ligand (CD154), GRO alpha (CXCL1), IL-1 ra (IL-1F3), IL-6, IL-8, MCP-1(CCL2), and Serpine E1 (PAI-1) were observed after treatment (Arndt et al. 2013). The same jet was also used to study Human Cytokine Arrays (array kit ARY005) for a treatment time of 2 min. After incubating for 24 hrs, it was found that on mRNA level, molecules of TGF-ß1 and TGF-ß2 were induced after the APPJ treatment. Since these transforming growth factors are responsible for cell migration and proliferation, synthesis of extracellular matrix, angiogenesis, remodeling, and the breaking strength of the repaired tissue, it can be suggested that APPJ activates and facilitates in healing wound (Amento and Beck 1991). It is also necessary that extracellular matrix (ECM) components like α-SMA are produced for wound healing. After APPJ treatment it was found that both collagen type 1 and α-SMA mRNA expressions were induced suggesting that it influenced both collagen synthesis and activation of fibroblasts. Hence, all the studies mentioned above revealed that the important genes, which are crucial for wound healing, are activated by CAP plasma treatment (Folkman 2006).

Endothelial cells play an important role in angiogenesis where they produce growth factors like fibroblast growth factor-2 (FGF2) which is involved in the formation of blood vessels. Several chemical signals secreted by the body control the process of angiogenesis. These signals along with the growth factors induce endothelial cells for the development of new blood vessels (Nugent and Iozzo 2000; Shekhter et al. 2005). Kalghatgi et al. studied endothelial cells after treating it with APPJ for 30 seconds and demonstrated that the proliferation rate was twice than that of the untreated cells after five days of plasma treatment. The high rate of proliferation was due to the fibroblast growth factor-2 and reactive oxygen species (Kalghatgi et al. 2010).

Other factors responsible for wound healing are the adhesion molecules like α2 and β1-integrin, E-cadherin, and the epidermal growth factor receptor (EGFR), which play a key role in cell migration and proliferation for cell–cell and cell–matrix interactions. Haertal et al. demonstrated the application of APPJ (kINPen) device on HaCaT-keratinocytes for 10 and 30 s and the adhesion molecules were observed to increase significantly after 10 s (Haertel et al. 2011).

Studies have also been carried out using the Tigres MEF plasma device on 3D skin models consisting of a keratinocyte containing an epidermal layer and a fibroblast/collagen dermal matrix. Various plasma parameters like process gas (air/nitrogen), input power (80–300 W), and treatment time (5 to 20s) were varied to understand the effect on 3D skin model. It was observed that less treatment durations exhibited better cell compatibility. Compared to nitrogen gas, usage of air was more damaging. Moreover, higher input power gave adverse effects on 3D skin model morphology and also on the release of inflammatory cytokines. Hence, low treatment times, lower input power, and use of nitrogen gas were recommended for wound decontamination (Wiegand et al. 2016).

5   In Vivo Studies of APPJ on Mice

Xu et al. studied the effect of argon gas APPJ on wound healing in the skin of the mice (Xu et al. 2015a, b, 2015). Two full-thickness wounds of 4 mm diameter were made on either side of the dorsal midline of each mouse. APPJ treatment was given for 10– 50 s in intervals of 10s for 14 days without any interruption on one wound of each mouse. It was found that there was an improvement and almost complete closure of the wound toward the 13th day when the treatment time was less than 40 s compared to the other wounds that was not treated by APPJ. Whereas when the treatment time was 50 s, the healing of wound was lesser even on the 14th day. The reactive species so formed were able to inactivate the bacteria around wound and promote its healing. However, as the treatment time increased, the same reactive species played a detrimental role in wound healing causing cell death by apoptosis or necrosis (Xu et al. 2015a, b, 2015). Both Dunnill et al. and Kurahashi et al. in their studies reported that ROS was not only responsible for disinfection during the inflammatory phase, but also in the regulation of tissue repair which involves migration, proliferation, and angiogenesis (Dunnill et al. 2017; Kurahashi and Fujii 2015). However, excessive ROS production creates an imbalanced redox homeostasis, which in turn hinders the healing process (Dunnill et al. 2017, Kurahashi and Fujii 2015).

Kubinova et al. and Zhang et al. analyzed the effect of APPJ when operated with air on the healing of skin wound model in rats (Kubinova et al. 2017; Zhang et al. 2019). The skin wounds were treated daily for 14 days for 1 and 2 min. It was observed that on the seventh day, there was a significant epithelization and wound contraction compared to the wounds that were not treated. Gene expression analysis indicated no enhanced inflammatory reaction which would interrupt the process of wound healing (Kubinova et al. 2017, Zhang et al. 2019). When compared to Xu et al.'s work, the APPJ treatment performed by Kubinova et.al gave better results with air as a processing gas; they attributed the reason for better results to the increased NO species produced with plasma. On evaluation of the species formed from the plasma, it was found that air plasma treatment was able to produce a large amount of NO species in the living tissue which, in turn, accelerated the healing of the wound and its closure (Hsu et al. 2006).

A study was conducted by Arndt et al. using MicroPlaSter beta device for 2 min using argon gas on mice, which was subjected to two 6-mm full-thickness wounds made on both sides of the dorsal midline (Arndt et al. 2013). The results were then compared with mice which were treated with only argon gas without creating plasma. Improved wound closure was observed with Argon plasma treatment on the third and fifth day and the wounds healed after 15 days. However, there were some differences between the two treatments. After APPJ treatment, an increased quantity of macrophages was observed in the early phases of wound healing mainly due to an elevation of the immune defense which cleans the wound area by inducing phagocytosis. (Arndt et al. 2013).

6   Clinical Studies Using APPJ

Clinical trials using atmospheric pressure plasma jet have been done for disinfection of infectious matter which is essential for healing chronic wounds. Clinical trials have also been carried out on patients, to evaluate the change in the wound dimensions in both width and length using some of the commercially available plasma jets.

Isbary et al. carried out clinical studies using argon-based MicroPlaSter alpha and beta commercially available plasma jet devices on 24 patients infected with chronic wounds. The treatment was carried out for 2 min every day. The results were compared with another study on 36 patients using APPJ for 5 min. Using standard procedures bacterial load was measured after the treatments for different times with these devices. It was observed that MicroPlaSter beta device showed a major reduction in bacterial load (23.5%, $P < 0.008$) compared to that treated by MicroPlaSter alpha device (40%, $P < 0.016$). The treatment did not have any side effects on the patients. Both the devices were found to be effective with 2 min on chronic wounds (Isbary et al. 2010).

In order to assess the wound closure dimensions after APPJ treatment using MicroPlaSter device Isbary et al. conducted another study on 70 patients with chronic wounds and the treatment time varied from 3–7 min. Wound dimensions were measured before and after the treatment and compared with wounds that were not treated. The same treatment was also performed on patients having chronic venous ulcers for 3–7 min. A significant reduction in the dimensions, i.e., width and length of the wounds were observed in patients with chronic wounds whereas there was only a reduction in ulcer width in patient having chronic venous ulcers. The studies indicate that further research with the use of APPJ for chronic venous ulcers has to be pursued (Isbary et al. 2013). Clinical studies using PlasmaDerm and KINPen have also been tried on patients and have indicated a reduction in bacterial colonization by tenfold in the wound volume than the patients who have not been given the plasma treatment (Emmert et al. 2013; Lademann et al. 2013).

7 Mechanisms Involved in Wound Healing Using Plasma

Wound healing is a complex phenomenon involving various pathways. Reactive Oxygen Species (ROS) and Reactive Nitrogen Oxygen Species (RNOS) of CAP play a vital role in healing and sterilization of the wounds. Wound healing mechanisms by atmospheric pressure plasma is proposed in the following two ways:

1. Interaction of ROS with the affected area.
2. Interaction of RNOS with the affected area.

Both these mechanisms help side by side in order to heal the wound and by assisting in one or more ways during hemostatic, inflammatory, proliferative, and maturation phases. The major species in ROS and RNOS which interact with the wound area are hydrogen peroxide (H_2O_2) and nitrogen oxide (NO). Both these mechanisms are explained as below:

Effect of ROS

1. ROS are involved in fibroblast-associated collagen production and the synthesis of growth factors (Sen et al. 2002).

2. In the initial stage of homeostasis, ROS helps in recruiting platelet, platelet activation, and mediating tissue factors such as TF-mRNA (Graves 2012).
3. ROS increases messenger ribonucleic acid (mRNA) expression of anti-inflammatory cytokines and decreases mRNA expression of pro-inflammatory cytokines (Lee et al. 2016).
4. ROS induces the expression of TGF-β (Fathollah et al. 2016).
5. ROS (H_2O_2) behaves as a second messenger for platelet-derived growth factors, vascular endothelial growth factors, and tissue growth factors (Graves 2012).

Effect of RNOS

1. Nitric oxide (NO) gas molecules help in the expansion of microvessel lumens during growth of new blood capillaries and in the opening of reserve collateral, which helps in the improvement of tissue nutrition and renewal of cell population (Vasilets et al. 2015).
2. Scar tissues soften as collagen bundles are loosened by NO resulting in forming younger tissues with the renewal of cell pool of fibroblasts (Vasilets et al. 2015).
3. NO also prevents dysplastic transformation of fibroblasts and recurrent keloid growth (Vasilets et al. 2015).
4. The NO synthase is responsible for the inflammatory and proliferative phase of the healing process and they are significantly found in bone fracture areas and also in wounds developed due to burns (Fridman and Freidman 2013).
5. Nitrogen oxide species (NOS) are linked to macrophage activation in the wound, cytokine synthesis, proliferation of fibroblasts, and epithelialization (Fridman and Freidman 2013).
6. NO helps in the delivery of immune system components to the site of infection because of increasing the microcirculation (Fridman and Freidman 2013).
7. RNOS forms nitrated fatty acids (NO_2-FAs) which are electrophile biomolecule. It induces anti-inflammatory polymorphic neutrophils, which help in wound healing (Fridman and Freidman 2013).

8 Conclusion

Cold atmospheric pressure plasma addresses several modes of wound healing action like anti-inflammatory, antimicrobial, tissue stimulation, re-epithelialization, and neovascularization. From the in vitro and in vivo studies, the antimicrobial as well as the wound healing properties of plasma have been clearly established. Furthermore, CAP plasma has also successfully demonstrated wound healing capabilities during clinical studies without any side effects. Cold atmospheric pressure based plasma devices indeed show a high potential in wound healing as it helps in inactivating the microorganisms in the first stage and in the later stage stimulates cell proliferation and migration. This makes plasma a novel and a promising alternative remedial technique for wound healing, particularly, chronic wounds.

References

Abuzairi T, Okada M, Bhattacharjeeb S, Nagatsu M (2016) Surface conductivity dependent dynamic behaviour of an ultrafine atmospheric pressure plasma jet for micro scale surface processing. Appl Surf Sci 390:489–496

Amento EP, Beck LS (1991) TGF-beta and wound healing. Ciba Found Symp 157:115–123

Arjunan KP (2011) Ph. D Thesis on Plasma produced reactive oxygen and nitrogen species in angiogenesis. Drexel University, http://hdl.handle.net/1860/3763

Arndt S, Schmidt A, Karrera S, von Woedtke T (2018) Comparing two different plasma devices kINPen and Adtec SteriPlas regarding their molecular and cellular effects on wound healing. Clinical Plasma Medicine 9:24–33

Arndt S, Unger P, Wacker E, Shimizu T, Heinlin J, Li YF, Thomas HM, Morfill GE, Zimmermann JL, Bosserhoff AK, Karrer S (2013) Cold atmospheric plasma (CAP) changes gene expression of key molecules of the wound healing machinery and improves wound healing in-vitro and in-vivo. PLoS One 8(11):e79325–e79333

Attinger C, Wolcott R (2012) Clinically addressing biofilm in chronic wounds. Adv Wound Care 1 (3):127–132

Babayan SE, Jeong JY, Tu VJ, Park J, Selwyn GS, Hicks RF (1998) Deposition of silicon dioxide films with an atmospheric-pressure plasma jet. Plasma Sources Sci Technol 7:286–288

Becker KH, Kogelschatz U, Schoenbach KH, Barker RJ (2005) Non-equilibrium air plasmas at atmospheric pressure. IOP Publishing Ltd

Boudam MK, Moisan M, Saoudi B, Popodivici C, Gherardi N, Massines F (2006) Bacterial spore inactivation by atmospheric-pressure plasmas in the presence or absence of UV photons as obtained with the same gas mixture. J Phys D Appl Phys 39:3494–3507

Bryant RA, Nix DP (2011) Acute & chronic wounds: current management concept, 4th edn. Mosby Elsevier, Missouri, pp 270–278

Canard JM, Védrenne B (2001) Clinical application of argon plasma coagulation in gastrointestinal endoscopy: has the time come to replace the laser? Endoscopy 33:353–357

Chatraie M, Torkaman G, Khani M, Salehi H, Shokr B (2018) In vivo study of non-invasive effects of non-thermal plasma in pressure ulcer treatment. Sci Rep 48(1):5621–5632

Chau TT, Kao KC, Blank G, Madrid F (1996) Microwave plasmas for low-temperature dry sterilization. Biomaterials 17:1273–1277

Conrads H, Schmidt M (2000) Plasma generation and plasma sources, plasma sources Sci. Technol 9(2000):441–445

Daeschlein G, Napp M, Lutze S, Arnold A, von Podewils S, Guembel D, Junger M (2015) Skin and wound decontamination of multidrug-resistant bacteria by cold atmospheric plasma coagulation. J Dtsch Dermatol Ges 13(2):143–150

Daeschlein G, Scholz S, Ahmed R, Majumdar A, Von Woedtke T, Haase H, Niggemeier M, Kindel E, Brandenburg R, Weltmann KD, Junger M (2012) Cold plasma is well-tolerated and does not disturb skin barrier or reduce skin moisture. J Dtsch Dermatol Ges 10:509–515

Darmawati S, Rohmani A, Nuranic LH, Prastiyanto ME, Dewi SS, Salsabila N, Wahyuningtyas ES, Murdiya F, Sikumbang IM, Rohmah RN, Fatimah YA, Widiyanto A, Ishijima T, Sugama J, Nakatani T, Nasruddin N (2019) When plasma jet is effective for chronic wound bacteria inactivation, is it also effective for wound healing? Clin Plasma Med 14:100085–100094

Duchesne C, Frescaline N, Lataillade JJ, Rousseau A (2018) Comparative study between direct and indirect treatment with cold atmospheric plasma on in-vitro and in vivo models of wound healing. Plasma Med 8(4):379–401

Dunnill C, Patton T, Brennan J, Barrett J, Dryden M, Cooke J, Leaper D, Georgopoulos NT (2017) Reactive oxygen species (ROS) and wound healing: the functional role of ROS and emerging ROS-modulating technologies for augmentation of the healing process. Int Wound J 14 (1):89–96

Duval A, Marinov I, Bousquet G, Gapihan G, Starikovskaia SM, Rousseau A, Janin A (2013) Cell death induced on cell cultures and nude mouse skin by non-thermal, nanosecond-pulsed generated plasma. PLoS One 8(12):e83001–e83012

Ehlbeck J, Schnabel U, Polak M, Winter J, Von Woedtke T, Brandenburg R, von dem Hagen T, Weltmann K-D (2011) Low temperature atmospheric pressure plasma sources for microbial decontamination. J Phys D Appl Phys 44:013002–013020

Eliasson B, Kogelschatz U (1991) Modelling and applications of silent discharge plasmas. IEEE Trans Plasma Sci 19(2):309–323

Emmert S, Brehmer F, Hänßle H, Helmke A, Mertens N, Ahmed R, Simon D, Wandke D, Maus-Friedrichs W, Däschlein G, Schön MP, Viölb W (2013) Atmospheric pressure plasma in dermatology: ulcus treatment and much more. Clin Plasma Med 1(1):24–29

Enoch S, Price P, Cellular, molecular and biochemical differences in the pathophysiology of healing between acute wounds, chronic wounds and wounds in the elderly., Article in World Wide Wounds, (2004)

Ermolaeva SA, Varfolomeev AF, Chernukha MY, Yurov DS, Vasiliev MM, Kaminskaya AA, Moisenovich MM, Romanova JM, Murashev AN, Selezneva II, Shimizu T, Sysolyatina EV, Shaginyan IA, Petrov OF, Mayevsky EI, Fortov VE, Morfill GE, Naroditsky BS, Gintsburg AL (2011) Bactericidal effects of non-thermal argon plasma in-vitro, in biofilms and in the animal model of infected wounds. J Med Microbiol 60(1):75–83

Fathollah S, Mirpour S, Mansouri P, Dehpour AR, Ghoranneviss M, Rahimi N, Naraghi ZS, Chalangari R, Chalangari KM (2016) Investigation on the effects of the atmospheric pressure plasma on wound healing in diabetic rats. Sci Rep 6:19144–19152

Folkman J (2006) Angiogenesis. Annu Rev Med 57:1–18

Foster KW, Moy RL, Fincher EL (2008) Advances in plasma skin regeneration. J Cosmet Dermatol 7(3):169–179

Fridman A, Freidman G (2013) Plasma Medicine, ISBN 978–0–470-68969-1. Plasma Medicine (1st Edition) Published by Wiley ISBN-13: 978–0–470-68969-1, ISBN: 0–470–68969-2

Fridman G, Gutsol A, Shekhter AB, Vasilets VN, Fridman A (2008) Applied plasma medicine. Plasma Process Polym 5(6):503–533

Fritsch P (1998) Dermatologie und Venerologie. Springer, Berlin

Graves DB (2012) The emerging role of reactive oxygen and nitrogen species in redox biology and some implications for plasma applications to medicine and biology. J Phys D Appl Phys 45:263001–263043

Haertel B, von Woedtke T, Weltmann KD, Lindequist U (2014) Non-thermal atmospheric pressure plasma possible application in wound healing. Biomol Ther 22(6):477–490

Haertel B, Wende K, von Woedtke T, Weltmann KD, Lindequist U (2011) Non-thermal atmospheric-pressure plasma can influence cell adhesion molecules on HaCaT-keratinocytes. Exp Dermatol 20(3):282–284

Heinlin J, Morfill G, Landthaler M, Stolz W, Isbary G, Zimmermann JL, Shimizu T, Karrer S (2010) Plasma medicine: possible applications in dermatology. J Dtsch Dermatol Ges 8 (12):968–976

Helmke A, Hoffmeister D, Berge F, Emmert S, Laspe P, Mertens N, Vioel W, Weltmann K-D (2011) Physical and microbiological characterization of Staphylococcus epidermidis inactivation by dielectric barrier discharge, Plasma Processes and Polymers b8:278–286

Hinz B (2016) The role of myofibroblasts in wound healing. Curr Res Trans Med 64:171–177

Hort K, Beier O, Wiegand C, Laaouina A, Fink S, Pfuch A, SchimanskiA GB, Hipler UC (2017) Screening test of a new pulsed plasma jet for medical application. Plasma Med 7(2):133–145

Hsu YC, Hsiao M, Wang LF, Chien YW, Lee WR (2006) Nitric oxide produced by iNOS is associated with collagen synthesis in keloid scar formation. Nitric Oxide-Biol Chem 14:327–334

Hunt TK, Hopf H, Hussain Z (2000) Physiology of wound healing. Adv Skin Wound Care 13:6–11

Isbary G, Heinlin J, Shimizu T, Zimmermann JL, Morfill G, Schmidt HU, Monetti R, Steffes B, Bunk W, Li Y, Klaempfl T, Karrer S, Landthaler M, Stolz W (2012) Successful and safe use of

2 min cold atmospheric argon plasma in chronic wounds: results of a randomized controlled trial. Br J Dermatol 167(2):404–410

Isbary G, Morfill G, Schmidt HU, Georgi M, Ramrath K, Heinlin J, Karrer S, Landthaler M, Shimizu T, Steffes B, Bunk W, Monetti R, Zimmermann JL, Pompl R, Stolz W (2010) A first prospective randomized controlled trial to decrease bacterial load using cold atmospheric argon plasma on chronic wounds in patients. Br J Dermatol 163(1):78–82

Isbary G, Stolz W, Shimizu T, Monetti R, Bunk W, Schmidt H-U, Morfill GE, Klämpfl TG, Thomas HM, Heinlin J, Karrer S, Landthaler M, Zimmermann JL (2013) Cold atmospheric argon plasma treatment may accelerate wound healing in chronic wounds: results of an open retrospective randomized controlled study in vivo. Clin Plasma Med 1(2):25–30

Jeong JY, Babayan SE, Tu VJ, Park J, Henins I, Hicks RF, Selwyn GS (1998) Etching materials with atmospheric pressure plasma jet. Plasma Sources Sci Technol 7:282–285

Kalghatgi S, Friedman G, Fridman A, Clyne AM (2010) Endothelial cell proliferation is enhanced by low dose non-thermal plasma through fibroblast growth factor-2 release. Ann Biomed Eng 38(3):748–757

Kalghatgi S, Kelly CM, Cerchar E, Torabi B, Alekseev O, Fridman A, Gary Friedman G, Azizkhan-Clifford J (2011) Effects of non-thermal plasma on mammalian cells. PLoS One 6(1):e16270–e16280

Kang SU, Cho JH, Chang JW, Shin YS, Kim KI, Park JK, Yang SS, Lee JS, Moon E, Lee K, Kim CH (2014) Non thermal plasma induces head and neck cancer cell death: the potential involvement of mitogen activated protein kinase-dependent mitochondrial reactive oxygen species. Cell Death and Disease-Nature 13(5):1056–1066

Kim SJ, Chung TH, Bae SH (2010) Induction of apoptosis in human breast cancer cells by a pulsed atmospheric pressure plasma jet. Appl Phys Lett 97:23702–23706

Klämpfl TG, Isbary G, Shimizu T, Li Y, Zimmermann JL, Stolz W, Schlegel J, Morfill GE, Schmidt HU (2012) Cold atmospheric air plasma sterilization against spores and other microorganisms of clinical interest. Appl Environ Microbiol 78(15):5077–5082

Kogelschatz U (2003) Dielectric-barrier discharges: their history, discharge physics, and industrial applications. Plasma Chem Plasma Process 23:1–46

Kong MG, Kroesen G, Morfill G, Nosenko T, Shimizu T, van Dijk J, Zimmermann JL (2009) Plasma medicine: an introductory review. New J Phys 11:115012–115047

Kubinova S, Zaviskova K, Uherkova L, Zablotskii V, Churpita O, Lunov O, Dejneka A (2017) Non-thermal air plasma promotes the healing of acute skin wounds in rats. Sci Rep 7:45183–45193

Kurahashi T, Fujii J (2015) Roles of Antioxidative enzymes in wound healing. J Develop Biol 3 (2):57–70

Lademann O, Kramer A, Richter H, Patzelt A, Meinke MC, Czaika V, Weltmann KD, Hartmann B, Koch S (2011) Skin disinfection by plasma-tissue interaction: comparison of the effectivity of tissue-tolerable plasma and a standard antiseptic. Skin Pharmacol Physiol 24:284–288

Lademann J, Ulrich C, Patzelt A, Richter H, Kluschke F, Klebes M, Lademann O, Kramer A, Weltmann KD, Lange-Asschenfeldt B (2013) Risk assessment of the application of tissue-tolerable plasma on human skin. Clin Plasma Med 1:5–10

Laroussi M (2018) Plasma medicine: a brief introduction. Plasma 1:47–60

Laroussi M, Tendero C, Lu X, Alla S, Hynes WL (2006) Inactivation of bacteria by the plasma pencil. Plasma Process Polym 3:470–473

Lee OJ, Ju HW, Khang G, Sun PP, Rivera J, Cho JH, Park SJ, Eden JG, Park CH (2016) An experimental burn wound-healing study of non-thermal atmospheric pressure micro plasma jet arrays. J Tissue Eng Regen Med 10:348–357

Lerouge S, Guignot C, Tabrizian M, Ferrier D, Yagoubi N, Yahia L (2000) Plasma-based sterilization: effect on surface and bulk properties and hydrolytic stability of reprocessed polyurethane electrophysiology catheters. J Biomed Mater Res 52:774–782

Léveillé V, Coulombe S (2005) Design and preliminary characterization of a miniature pulsed RF APGD torch with downstream injection of the source of reactive species. Plasma Sources Sci Technol 14:467–476

Lieberman MA, Lichtenberg AJ (2005) Principles of plasma discharges and materials processing, 2nd edn. Wiley, New York

Lin A, Chernets N, Han J, Alicea Y, Dobrynin D, Fridman G, Freeman TA, Fridman A, Miller V (2015) Non-equilibrium di electrical barrier discharge treatment of mesenchymal stem cells: charges and reactive oxygen species play the major role in cell death. Plasma Process Polym 12:1117–1127

Liu M, Duan XP, Li YM, Yang DP, Long YZ (2017) Electrospun nano fibers for wound healing. Materials Sci Eng C: Materials Biol Appl 76:1413–1423

Lloyd G, Friedman G, Jafri S, Schulz G, Fridman A, Harding K (2010) Gas plasma: medical uses and developments in wound care. Plasma Process Polym 7(3–4):194–211

Lu X (2008) The roles of the various plasma agents in the inactivation of bacteria. J Appl Phys 104 (5):53309–53313

Lu X (2009) An RC plasma device for sterilization of root canal of teeth. IEEE Trans Plasma Sci 37 (5):668–673

Lu XP, Jiang ZH, Xiong Q, Tang ZY, Hu XW, Pan Y (2008) An 11cm long atmospheric pressure cold plasma plume for applications of plasma medicine. Appl Phys Lett 92:081502–081504

Majumdar A, Sangole P (2016) Alternative approaches to wound healing, Chapter from: wound Healing - New insights into Ancient Challenges, ISBN: 978–953–51-2678-2

Mohd Nasir N, Lee BK, Yap SS, Thong KL, Yap SL (2016) Cold plasma inactivation of chronic wound bacteria. Arch Biochem Biophys 605(1):76–85

Moisan M, Barbeau J, Moreau S, Pelletier J, Tabrizian M, Yahia LH (2001) Low-temperature sterilization using gas plasmas: a review of the experiments and an analysis of the inactivation mechanisms. Int J Pharm 226:1–21

Mone Y, Monnin D, Kremer N (2014) The oxidative environment: a mediator of interspecies communication that drives symbiosis evolution. Proc R Soc B Biol Sci. 22 281 (1785):20133112–20133121

Nasruddina YN, Kanae M, Heni S, Rahayu E, Nur M, Ishijima T, Enomotod H, Uesugi Y, Sugama J, Nakatani T (2014) Cold plasma on full-thickness cutaneous wound accelerates healing through promoting inflammation, re-epithelialization and wound contraction. Clin Plasma Med 2(1):28–35

Nguyen DT, Orgill DP, Murphy GT (2009) The pathophysiologic basis for wound healing and cutaneous regeneration. In: Orgill DP, Blanco C (eds) Biomaterials for treating skin loss. Elsevier, pp 25–57

Nugent M, Iozzo R (2000) Fibroblast growth Factor-2. Int J Biochem Cell Biol 32:115–120

O'Connell D, Cox LJ, Hyland WB, McMahon SJ, Reuter S, Graham WG, Gans T, Currell FJ (2011) Cold atmospheric pressure plasma jet interactions with plasmid DNA. Appl Phys Lett 98 (4):43701–43703

O'Connor N, Cahill O, Daniels S, Galvin S, Humphreys H (2014) Cold atmospheric pressure plasma and decontamination. Can it contribute to preventing hospital-acquired infections? J Hosp Infect 88(2):59–65

Pan J (2010) A novel method of tooth whitening using cold plasma microjet driven by direct current in atmospheric-pressure air. IEEE Trans Plasma Sci 38(11):3143–3151

Patenall BL, Hathaway H, Sedgwick AC, Thet NT, Williams GT, Young AE, Allinson SL, Short RD, Jenkins ATA (2018) Limiting Pseudomonas *aeruginosa* biofilm formation using cold atmospheric pressure plasma. Plasma Med 8:269–277

Pereira-Lima JC, Busnello JV, Saul C (2000) High power setting argon plasma coagulation for the eradication of Barrett's esophagus. Am J Gastroenterol 95:1661–1668

Rieger S, Zhao H, Martin P, Abe K, Lisse TS (2015) The role of nuclear hormone receptors in cutaneous wound repair. Cell Biochem Funct 33(1):1–13

Robson MC (1997) Wound infection. A failure of wound healing caused by an imbalance of bacteria. Surg Clin North Am 77(3):637–650

Roth JR (1995) Industrial Plasma Engineering, Vol-1 Principles, IOP publishing Ltd.

Salehi S, Shokri A, Khani MR, Bigdeli M, Shokri B (2015) Investigating effects of atmospheric-pressure plasma on the process of wound healing. Biointerphases 10:029504–029512

Schmidt A, Bekeschus S, Wende K, Vollmar B, von Woedtke T (2017) A cold plasma jet accelerates wound healing in a murine model of full-thickness skin wounds. Exp Dermatol 26 (2):156–162

Scholz V, Pazlarova J, Souskova H, Khun J, Julak J (2015) Nonthermal plasma? A tool for decontamination and disinfection. Biotechnol Adv 33(6):1108–1119

Sen CK, Khanna S, Babior BM, Hunt TK, Ellison EC, Roy S (2002) Oxidant-induced vascular endothelial growth factor expression in human keratinocytes and cutaneous wound healing. J Biol Chem 277:33284–33290

Shashurin A, Shneider MN, Dogariu A, Miles RB, Keidar M (2009) Temporal behavior of cold atmospheric plasma jet. Appl Phys Lett 94:231504–231507

Shekhter AB, Serezhenkov V, Rudenk T, Pekshev A, Vanin A (2005) Beneficial effect of gaseous nitric oxide on the healing of skin wounds. Nitric Oxide-Biol Chem 12(4):210–219

Stallmeyer B, Kampfer H, Kolb N, Pfeilschifter J, Frank S (1999) The function of nitric oxide in wound repair: inhibition of inducible nitric oxide-synthase severely impairs wound re epithelialization. J Investig Dermatol 113(6):1090–1098

Stoffels E, Kieft IE, Sladek REJ (2003) Superficial treatment of mammalian cells using plasma needle. J Phys D Appl Phys 36:2908–2913

Stoffels E, Sakiyama Y, Graves B (2008) Cold atmospheric plasma: charged species and their interactions with cells and tissues. IEEE Trans Plasma Sci 36:1441–1457

Stojadinovic A, Carlson JW, Schultz GS, Davis TA, Elster EA (2008) Topical advances in wound care. Gynecol Oncol 111:S70–S80

Theinkom F, Singer L, Cieplik F, Cantzler S, Weilemann H, Cantzler M, Hiller K-A, Maisch IDT, Zimmermann JL (2019) Antibacterial efficacy of cold atmospheric plasma against enterococcus *faecalis* planktonic cultures and biofilms in-vitro. PLoS One 26:1–15

Ulrich C, Kluschke F, Patzelt A, Vandersee S, Czaika VA, Richter H, Bob A, Hutten J, Painsi C, Huge R, Kramer A, Assadian O, Lademann J, Lange-Asschenfeldt B (2015) Clinical use of cold atmospheric pressure argon plasma in chronic legulcers: a pilot study. J Wound Care 24 (5):196–203

Vandamme M, Robert E, Lerondel S, Sarron V, Ries D, Dozias S, Sobilo J, Gosset D, Kieda C, Legrain B, Pouvesle JM, Pape AL (2012) ROS implication in a new antitumor strategy based on non-thermal plasma. Int J Cancer 130(9):2185–2194

Vasilets VN, Gutsol A, Shekhter AB, Fridman A (2009) Plasma medicine. High Energy Chem 43:229–233

Vasilets VN, Shekhter AB, Guller AE, Pekshev AU (2015) Air plasma generated nitric oxide in treatment of skin scars and articular musculoskeletal disorders: preliminary review of observations. Clin Plasma Med 3(1):32–39

Weltmann KD, von Woedtke T (2017) Plasma medicine — current state of research and medical application. Plasma Physics and Controlled Fusion 59(1):14031–14043

Wende K, Landsberg K, Lindequist U, Weltmann KD, von Woedtke T (2010) Distinctive activity on thermal atmospheric-pressure plasma jet on eukaryotic and prokaryotic cells in a cocultivation approach of keratinocytes and microorganisms. IEEE Trans Plasma Sci 38:2479–2485

Wiegand C, Fink S, Beier O, Horn K, Pfuch A, Schimanski A, Grünler B, Hipler UC, Elsner P (2016) Dose and time-dependent cellular effects of cold atmospheric pressure plasma evaluated in 3D skin models. Skin Pharmacol Physiol 29:257–265

Woedtke TV, Emmert S, Metelmann HR, Stefan Rupf S, Weltmann KD (2013) Perspectives on cold atmospheric plasma (CAP) applications in medicine. Phy Plasmas. 27, 070501 (2020) 27:324–331

Woo K, Ayello EA, Sibbald RG (2007) The edge effect: current therapeutic options to advance the wound edge. Adv Skin Wound Care 20:99–117

Xiong Z (2018) Cold Atmospheric Pressure Plasmas (CAPs) for skin wound healing, Chapter from plasma medicine–concepts and clinical applications TY - ISBN: 978–1–78923-112-0

Xu Z, Shen J, Zhang Z, Ma J, Ma R, Zhao Y, Sun Q, Qian S, Zhang H, Ding L (2015) Inactivation effects of non-thermal atmospheric-pressure helium plasma jet on staphylococcus aureus biofilms. Plasma Process Polym 12:827–835

Xu MS, Shi XM, Cai JF, Chen SL, Li P, Yao CW (2015a) Dual effects of atmospheric pressure plasma jet on skin wound healing of mice. Wound Repair Regen 23:878–884

Xu GM, Shi XM, Cai JF, Chen SL, Li P, Yao CW, Chang ZS, Zhang GJ (2015b) Dual effects of atmospheric pressure plasma jet on skin wound healing of mice. Wound Repair Regen 23:878–884

Yao S, Wu Z, Han J, Tang X, Jiang B, Lu H, Yamamoto S, Kodama S (2015) Study of ozone generation in an atmospheric dielectric barrier discharge reactor. J Electrost 75:35–42

Zhang JP, Ling G, Chen QL, Zhang KY, Wang T, An GZ, Zhang X-F, Li HP, Ding GR (2019) Effects and mechanisms of cold atmospheric plasma on skin wound healing of rats. Contrib Plasma Physics 59:92–101

Quorum Sensing as a Therapeutic Target in the Treatment of Chronic Wound Infections

VT Anju, Madhu Dyavaiah, and Busi Siddhardha

1 Introduction

Wound is defined as a simple or a serious skin or tissue injury, which can spread to other tissues and anatomical structures such as muscles, subcutaneous tissue, nerves, tendons, and bone. Skin is the most probable organ which has a high risk for impairment, injury, scratches and burns. A wound is created when a damage occurs to the epithelium layer and connective tissues which eventually weakens the primary immune system of the human body. Thus, reconstruction of functions of epithelium or other layers of skin is most important. This process is called wound repair or healing. Wound healing follows through a cascade of overlapping phases (Negut et al. 2018). The biology of wound healing involves the coagulation or hemostasis, inflammation, migration, proliferation, re-epithelialization and finally restoration (Fig. 1) (Garraud et al. 2017). As soon as wound develops, thromboxane A2 and prostaglandin helps in the rapid vasoconstriction at the site of injury. A cascade of clotting processes initiates along with vasoconstriction. The first to arrive at the site are platelets, which favours hemostasis and produces several growth factors and cytokines (vascular endothelial growth factor, platelet-derived growth factor and fibroblast growth factor). These chemoattractants encourage the movement of inflammatory cells like monocytes, neutrophils, lymphocytes and macrophages to the site of injury. Initially, at 24 hours the neutrophils land at wound site to phagocytize bacteria and to clear microbial and other debris from the damaged

VT Anju · M. Dyavaiah
Department of Biochemistry and Molecular Biology, School of Life Sciences, Pondicherry University, Pondicherry, India

B. Siddhardha (✉)
Department of Microbiology, School of Life Sciences, Pondicherry University, Pondicherry, India
e-mail: siddhardha.mib@pondiuni.edu.in

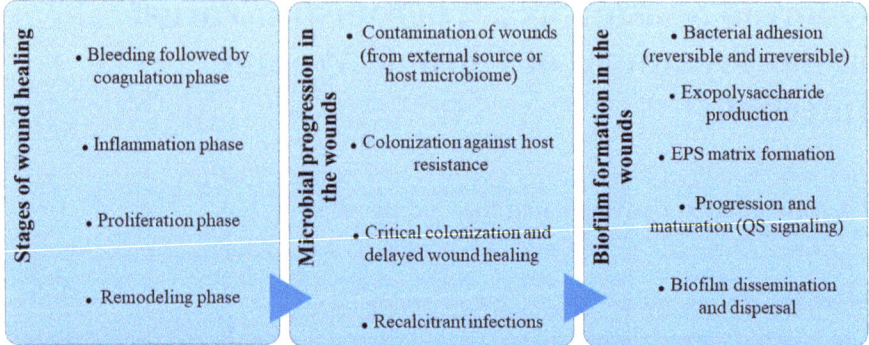

Fig. 1 Illustration of different stages of wound healing, progression of microorganisms and biofilms in the wounds

area. Reactive oxygen species produced by polymorphonuclear leukocytes mediates the elimination of external contaminants. The vital step that proceeds in 48–72 hours during the process of wound healing is the accumulation of macrophages. The phase of proliferation usually starts with migrated macrophages, which helps in the inflammatory healing through the discharge of cytokines, attraction of blast cells, and removal of debris. In the remodelling phase, fibroblast cells lay down for the components of extracellular matrix and allow re-epithelialization. Eventually, the wound is fully repaired up to 70–80% of its original tensile strength after angiogenesis and new capillary bed formation by epithelial cells. Wound contracture is completed by myofibroblasts through actin filaments (Rodrigues et al. 2019).

The processes involved in wound healing should occur at the exact time and endure for a specific period. Wound healing is delayed when these conditions fail to occur. There are several factors that interfere with different phases of wound healing and results in improper or impaired healing. Wounds that failed to accomplish rapid healing process due to the impaired tissue repair are delayed acute and chronic wounds. These wounds never follow a normal wound healing process rather enters into a stage of pathologic inflammation. Most non-healing chronic wounds are related to diabetes mellitus, blood pressure or venous stasis disease and ischemia (Guo and DiPietro 2010). Current statistics draws to the fact that more than six million of the world population is severely affected by chronic wounds. In a study conducted in India, says that 4.5 and 10.5 per 1000 population are affected by chronic and acute wounds, respectively (Shukla et al. 2005). Acute wounds are emerged due to external damage caused by bites, surgical injuries, abrasions, burns, lacerations, gunshot injuries or burns. Acute wounds heal within a stipulated time irrespective of the wound's nature (Bowler et al. 2001). Generally, chronic wounds arise as minor traumatic injuries such as insect bites, simple scratch on dry skin or penetrating injuries. These injuries are supposed to heal within few days to weeks. Patients with underlying pathologies may cause non-healing chronic wounds. These pathologies include non-diabetic neuropathies, diabetic-induced pathologies, venous insufficiency, atherosclerosis, thrombosis, arterial insufficiency, varicosis, macroangiopathy, microangiopathy, immobility and excessive pressure. Different types of chronic wounds are venous ulcers, arterial ulcers, diabetic ulcers and pressure ulcers. Different pathologies of these wounds are elongated inflammation, tenacious infections and development of antibiotic-resistant biofilms contributed by microbial species, which leads to the formation of non-healing wounds (Demidova-Rice et al. 2012).

Impaired wound healing may also have contributed by dysbiosis of host skin microbiota. Microbial species present on the skin are beneficial to the host and the dysbiosis of skin microbiota can alter their normal gene expression levels to pathogenesis. A small injury on the skin may cause the reason for colonization and proliferation of skin microbiota and other pathogenic microorganisms in the wounds, which impede the healing process (Williams et al. 2018). Some studies suggested a multifaceted impact of cutaneous microbiome on wound healing. In order to stimulate the skin health and rejuvenation, an equilibrium among different types of microorganisms is required. Most skin microbiome is composed of the species of *Streptococcus*, *Corynebacterium*, *Staphylococcus*, various anaerobes and *Pseudomonas* (Cogen et al. 2008). These microbes have contributed to both favourable and harmful effects on the host, based on their load and cutaneous environment. The ultimate role of skin commensal is to dissuade the invasion of pathogens and to stimulate the host immune system. This is interrupted often by the development of tissue injury and ultimately causes delayed wound healing (Johnson et al. 2018).

A series of local and systemic host responses are initiated at the wound surface and surrounding areas after the colonization by microorganisms. These host responses can be purulent discharge or painful erythrema. Generally, acute and chronic wounds are populated mainly by polymicrobial communities. The severity of infection is likely to be dependent on various factors like, site, type and size of wounds, incidence of exogenous contamination, burden and nature of microorganism, immune status of host and pathogenicity contributed by the microbial load in the wound (Bowler et al. 2001).

2   Chronic Wound Biofilms

The transition of wounds from acute to chronic state is greatly influenced by the establishment of several microbial populations and development of biofilms in the wound bed. It is also reported that biofilms present in 6% and 90% of total acute and chronic wounds, respectively (Attinger and Wolcott 2012). The microbial biofilms found to be residing in non-healing wounds include species of *Pseudomonas*, *Staphylococcus*, *Corynebacterium* and other varieties of organisms. These biofilms are recalcitrant to antibiotics and thus exert negative impact on wound healing (Williams et al. 2018). Biofilms are generally regarded as a population of microorganisms either of single or mixed species attached to a substrate or host surface and protected from external harsh stimuli through their exopolymeric substances (EPS) matrix. Around, 99.9% of all microorganisms found in natural and pathogenic environments exist in their biofilm state (Black and Costerton 2010) (Fig. 1).

As these biofilm exhibits extreme resistance to antibiotics and exhibits biofilm tolerance, their significant risk on public health is a challenging threat. These biofilms are responsible for several chronic and persistent infections such as cystic fibrosis, otitis media, pneumonia and so on. Biofilm communities may also form by fungi and viruses other than bacteria. Their EPS matrix contains proteins, extracellular DNA, carbohydrates and other biomolecules (Clinton and Carter 2015). Biofilms resist the action of antibiotics through decreased metabolic activity, EPS which acts as a mechanical barrier and expression of antibiotic-resistant genes through plasmids. The physiology of biofilms inhibits the penetration of various immune cells thus escape easily from host-mediated phagocytosis. Even, biofilms stimulate the chronic inflammation in the wound environment and thus deteriorate the healing cycle (Goldberg and Diegelmann 2020).

Majority of the pathogens causing biofilm infections occur as polymicrobial which increases the severity of infection and complicates the treatment. Biofilms are dynamic architectures that includes 5 stages of formation and they are reversible attachment to surface, irreversible attachment, cell proliferation, growth and differentiation and dispersion of cells. Biofilms escape from antibiotic therapy through the production of various virulence factors and toxins. A cell-to-cell communication, called quorum sensing is associated with the production of different virulence factors and the development of biofilms (Clinton and Carter 2015).

Wound infections are often caused when the epidermal barrier is removed or damaged due to some reasons. This damage invites microbial contamination and colonization in the wounds. Chronic wounds often fail to heal or delay the healing process owing to the development of chronic biofilms. In a previous report, wounds formed in patients suffering from cystic fibrosis were not healed or responded to antibiotic treatment. This was further explained that polymorphonuclear leukocytes and antibiotics failed to remove refractory biofilms of *P. aeruginosa* present in their chronic wounds (Li et al. 2020). According to the studies, only a small population of microorganisms surviving in the wounds are identified and others are yet to be revealed. If one could explore the whole microbiome involved in non-healing chronic wounds, more therapeutic strategies can be developed based on the type of host–pathogen interactions. A study by a group of researchers surveyed microbiome of 30 wounds found in humans. They could identify 12 different bacterial genera from wounds. A large population of microbes from wounds were strict and facultative anaerobes, among which many were not identified using standard culturing techniques and this provided an evidence of incredibly diverse microbial flora in chronic wounds (Dalton et al. 2011).

Wound microbiome conferring to various infections can be explored using various in vitro, in vivo and in silico approaches. The application of in vitro culture techniques enables to study the types of microbial flora or biofilms, immune responses of host, virulence factors of biofilms and to mimic the wound microbiome in laboratory for further analysis. In silico techniques including various sequencing strategies helps in the identification and detection of microbial population in genera or species wise and also target its various virulence traits. There are several established animal models to study the wound microbiome. Animal models are well suited to study single species or polymicrobial species-related infections in acute and chronic wounds (Dalton et al. 2011).

There are several strategies to target biofilms at wound sites. Among them, some are thought to be involved in the sharp removal of biofilms which enhances the susceptibility of planktonic bacteria to antimicrobial therapy. But biofilms initiate to regain its mature form within 48–72 hours. BBWC, known as the Biofilm Based Wound Care is a treatment algorithm that properly focuses on the resistance of biofilms to antibiotics and highlights the necessity to develop successive and concurrent treatment methods (Hurlow et al. 2016). Traditional and current therapeutic strategies to control the growth of biofilms in wounds are discussed below.

3   Traditional Therapeutic Strategies

As the wound biofilm shows extreme tolerance to antimicrobial agents and therapy, potential strategies or agents are required to eliminate them from wounds. The available methods mainly target three steps: either prevent the formation of biofilms or removes the biofilms or kills the bacteria involved in biofilms. The antibiofilm strategies involving prevention include: aseptic methods and barrier dressings to prevent contamination, probiotics, QS inhibitors and anti-deposition agents to prevent expression of EPS and colonization and management of wound environment through pH, moisture and bacterial entrapment to prevent infection. The removal methods include biofilm disruption by physical (ultrasound methods, surfactants, biofilm entrapment etc.) or chemical methods (enzymatic, chelation methods etc.). Administration of topical antiseptics or systemic antimicrobial agents facilitates the inhibition or killing of biofilms (Metcalf et al. 2016). Sharp debridement is one of the wound debridement techniques firstly employed to remove bacterial biofilms. In this first step, necrotic tissue, inactivated tissue, poor healing tissue and foreign bodies are removed in the clinical practice. This wound debridement helps in the prevention of initial attachment or colonization of bacterial biofilms, especially by *Staphylococcus aureus* and *P. aeruginosa* which may further cause secondary bacterial infections. Hydrosurgical debridement is developed recently to eliminate the disadvantage of other debridement techniques and this focuses on the painless removal of necrotic or damaged tissue (Schultz et al. 2018; Kim and Steinberg 2012; Caputo et al. 2008).

Negative pressure wound therapy is another extensively used method for the treatment of wounds from the past 20 years. The merits of this therapy involve improved blood flow near the wound site, enhanced growth of granulation tissue, reduced tissue oedema and effective reduction of wound bioburden (Matiasek et al. 2017; Han and Ceilley 2017). Guoqi et al. 2018 showed that negative pressure wound treatment decreased the production of virulence factors and biofilm components of *P. aeruginosa* and enhanced the wound healing in an in vivo model of rabbit with ear biofilm infection. They used different atmospheric pressure of -75, -125 and -200 mmHg which significantly reduced the motility of bacteria and improved the wound healing in rabbit ear (Guoqi et al. 2018). Ultrasound is another method employed to remove biofilms. In addition, antibiotics combined with ultrasound enabled the destruction of biofilm to a long extent. In a study, gentamycin activity on biofilms of *P. aeruginosa* and *Escherichia coli* was significantly enhanced by ultrasound. Also, ultrasound in clinical wound care achieved much attention in the last few years (LuTheryn et al. 2020).

Nanomedicine in biofilm wound care is also popular as other therapeutic strategies. Nanoparticles are nanosized particles that exhibited various bioactive properties. They are widely applied in the antibiofilm therapy against various bacteria. There are nanoparticles which designed as antimicrobial agents against bacteria found in wound surface or used for targeted drug delivery towards the wound biofilms owing to their unique physical and chemical properties. Several nanoparticles-based antimicrobial coatings or dressings are available as treatment methods for biofilms present in the wounds. For instance, silver nanoparticle dressings, a broad-spectrum dressing agent was considered to be the first choice of treatment for bacterial biofilms. Around 90% of bacteria in the wounds were killed, when 5 to 10 g/ml of silver ion was used. The ability of silver ions to inhibit the growth of bacterial biofilms and to reduce the local anti-inflammatory effect, ultimately led to the rapid healing of the wounds (Koo et al. 2017; Toy and Macera 2011).

The mechanism of antibiofilm agents on the biofilm disruption is different. Some alter the cell envelop and some dislocate the matrix. There are some categories that encourages apoptosis. D-leucine and D-tyrosine act on cell envelope by interfering with the amyloid fibres required for the tight packing of biofilm structure. Zaragozic acid is another example of biofilm inhibitor which perturbs lipid rafts and cell membrane. Norspermidine, AA-861 and parthenolide are biofilm inhibitors that aids in the dislocation of EPS. Nitric oxide is an apoptosis-inducing biofilm inhibitor, which causes the disintegration of bacterial biofilms (Oppenheimer-Shaanan et al. 2013).

One of the antibiofilm agents, N-acetyl cysteine was successfully used to resolve chronic wound biofilms of *P. aeruginosa*. In this study, an in vitro biofilm system was employed which developed using the biofilms obtained from diabetic mouse chronic wounds. The action of this agent on wound healing was through the reduction of the biofilm mass and disruption of EPS. Thus, N-acetyl cysteine was able to mitigate the wound biofilms causing chronic infections in vitro (Li et al. 2020). As EPS is the hallmark of biofilm structure, dispersion of biofilm EPS can result in the destruction of biofilms. The combination of dispersion agents with antibiotics is always thought to be a good therapy against wound biofilms. Beta amylase, an enzyme formed by oral bacteria has been reformulated as a dispersion agent as it targets exopolysaccharide bonds and disrupts biofilms. Deoxyribonuclease I, DNase and glycoside hydrolase dispersion B are few examples of EPS dispersing agents. Aminoimidazole, a synthetic dispersion agent showed potential activity against biofilms of *S. aureus* (Clinton and Carter 2015).

Altogether, based on the data available so far, the first and most important step in the treatment of biofilms in the wounds is to apply antibiofilm agents, specific biocides and antibiotics which can physically remove the biofilms and/or later inhibit them. The evidence states that the management of superficial infections initiated with the design and development of topical antimicrobial agents such as polymyxin B, bacitracin and neomycin. The topical application of antibiofilm agents combined with antibiotics provided excellent results in biofilm wound care and this therapy is known to be ubiquitous (Jones and Kennedy 2012). Now, more attention is paid to the biofilm wound care to link the gap between wound clinics and in vitro microbiology conditions. This facilitated more therapeutic advancements in the field of biofilm wound care.

4 Cell to Cell Signalling in Chronic Wounds

The production of various signalling molecules in the wound area leads to intercellular interactions among organisms which may lead to either single or multi species biofilms. The cell-to-cell communications in wound biofilms are mediated mainly by acyl-homoserine lactones or autoinducer 2 molecules. Acylated homoserine lactone molecules and furanone-based systems or autoinducer 2 are observed in cell-to-cell signalling by Gram negative and Gram positive bacteria. From the available reports, these two signalling molecules are found in the majority of the chronic wounds (Rickard et al. 2010). In general, quorum sensing (QS) is regarded as a biochemical or molecular cell-to-cell signalling or intercellular communications dependent mainly on the production of autoinducers outside the cell. The change in the host–pathogen interactions and pathogenicity occurs when the population of bacteria detects these signalling molecules and responds to it. Microorganisms exhibit antibiotic tolerance, chronic and persistent infections and enhanced pathogenicity through quorum sensing pathways. QS is also found in biofilms rendering more antimicrobial resistance to them. Other than their EPS matrix, biofilms escape from host immune response by activating QS pathways. The treatment of chronic wound infections is challenged by the biofilm-activated QS pathways (Ng and Bassler 2009; Miller and Bassler 2001; Leid 2009).

The most common pathogens that worsen the healing of wounds and transforms into chronic wounds are *P. aeruginosa* and *S. aureus*. The quorum sensing systems of *P. aeruginosa* and *S. aureus* interfere with the healing of wounds. The roles of QS systems in the pathogenicity of wounds are studied using various wound infection animal models (Nakagami et al. 2011). In some circumstances, these two pathogens interact in the site of injury and leave a negative impact on the virulence through their multispecies interactions. The evolutionary dynamics among these pathogens light into the evolutionary successions, separation of niches and co-evolution of resistance, which helped them to modify the pathogenesis and damage on host. In addition, it also stated that the pathogen–pathogen interaction can also be competition other than the cooperation for resources (Rezzoagli et al. 2020).

The QS pathway is initiated when the concentration of autoinducer molecules reaches its threshold level and interacts with its cognate receptor proteins. The interaction of protein and its ligand, in turn, controls the regulation of several virulence factors. The major QS systems of *P. aeruginosa* are N-acyl-homoserine lactone (HSL) based *RhlR-RhlI* and *LasR-LasI* pathways, and the *Pseudomonas* quinolone signal (PQS) system. The *LasI* and *RhlI* are two HSL synthases that mediate the synthesis of autoinducers, N-(3-oxo-dodecanoyl)-L-homoserine lactone and N-butanoyl-L-homoserine lactone, respectively. The QS-dependent virulence factors of *P. aeruginosa* are pyocyanin, pyoverdin, exotoxin, elastase, alkaline protease and staphylolytic proteases. Other QS and biofilm-dependent factors are rhamnolipids, alginate, exopolysaccharides, siderophores and different types of motility. These toxic factors help in the colonization, differentiation and pathogenesis of *P. aeruginosa* in wound niches (Lee and Zhang 2014). There are clinical evidence showing interference of healing process in chronic wounds by QS pathways of *P. aeruginosa* (Rickard et al. 2010). The categories at great risk to develop infections caused by QS systems of *P. aeruginosa* are patients with severe burns, complication of diabetes and chronic cutaneous wounds (Mihai et al. 2014; Weinstein and Mayhall 2003). As this pathogen is able to cause persistent infections and continuous inflammation, it is difficult to mitigate using commonly used antibiotics from chronic wounds (Chaney et al. 2017).

Staphylococcus aureus is another commonly found pathogen in chronic wounds, which intrudes the wound healing process through the QS mediated virulence and toxin production. The QS pathway of this Gram positive bacteria is called as *AGR* (accessory gene regulator) system. The autoinducer molecules are peptides (AIPs or autoinducing peptides) encoded by *agr* gene. Accessory gene regulator causes the activation of several toxins, virulence factors and biofilm-mediated infections. The recalcitrance of these bacteria towards antibiotics is due to the activation of QS systems in biofilms and their interaction with innate immune responses and that leads to elevated inflammation (Kong et al. 2006; Yarwood et al. 2004). Therefore, the role of QS inhibition in the management and treatment of chronic wound infections and its effect on wound healing are discussed below.

5   QS as Therapeutic Target to Treat Chronic Wound Infections

Thus, QS of these bacteria helps to attach, colonize and proliferate within the wound tissues through the production of cell-associated and extracellular virulence factors. The inhibition of QS pathways in wound surfaces aids in the anti-infection therapy to reduce the inflammation and delayed wound healing. Quorum quenching or inhibiting agents (QSIs) are mainly classified into two groups. They are signal supply and response inhibitors (Rutherford and Bassler 2012; LaSarre and Federle 2013). The process of quorum quenching is defined as the enzymatic cleavage of QS signals, which abolish their signalling process. AHL lactonases and acylases are two quorum quenching enzymes that chop HSL rings and amide bond of AHL respectively (Nusrat et al. 2011). There are enzymes that alter the activity of AHL such as AHL reductases and oxidases. Thus, these QSIs block cell–cell signalling among bacteria or other microbes such that only host–pathogen interaction is altered through decreased virulence factor production without impeding the growth of bacteria. Thus, later immune cells eradicate these bacteria, which are less virulent and more antibiotic susceptible (Brackman and Coenye 2014; Das and Singh 2018).

Several animal models of wound infection are employed to examine the effect and inhibition of QS in wound healing and to correlate the type of clinical infection. However, studies related to QS inhibition in the wound healing are scarce. The mechanism of action of QS inhibitor or quenching molecules is through the inhibition or reduction of virulence factor production and biofilm formation. The application of QS inhibitors along with clinical antibiotics as adjuvant or combination therapy yields promising results. Yet, their ability to repair epithelial infection is not elucidated (Kalia 2013; Bhardwaj et al. 2013). The ability of QS inhibitor to abrogate the production of various factors and its effect on injured airway epithelial cells was investigated by Ruffin and co-workers in 2016 (Ruffin et al. 2016). Different strains of *P. aeruginosa* from laboratory and clinical CF and non-CF samples were used to assess the effect of 4-hydroxy-2,5-dimethyl-3(2H)-furanone on wound healing, proliferation rates, migration rates and repair of human airway epithelial cells of primary non-cystic fibrosis. The study showed a decrease in QS through the repair of airway epithelial cells. Finally, they concluded, that the inhibitor was able to act on bacterial exoproducts and counteracted with its harmful effect on wound repair. Altogether, they suggested the use of human airway epithelial cells from cystic fibrosis patients as a clinically significant pathological model to assess chronic Pseudomonal infection in wounds (Ruffin et al. 2016).

The role of traditional folk medicine and herbal formulation in wound care was always fascinating. There are few researchers still interested in exploring the herbal medicine in wound care in the modern era. Since pre-historic era, people found traditional and complementary medicine as adorable and affordable with maximum efficiency and even after the advent of modern medicine (Dorai 2012). A group of researchers investigated the anti-infective potential of herboheal (a polyherbal formulation) on wound infective bacteria. Herboheal inhibited the production of QS-associated pigments by multidrug-resistant bacteria causing wound infections. Herboheal interfered with the QS signal response of *C. violaceum, P. aeruginosa* and *S. marcescens*. This herbal-based preparation also reduced virulence in nematode model, *Caenorhabditis elegans* infected with *P. aeruginosa*. Overall, this study validated the use of herboheal as a wound care formulation against these bacterial infections (Patel et al. 2019).

Another polyherbal preparation named as *Panchvalkal* is mentioned in Indian traditional medicine. This herbal formulation affected the QS-associated virulence factors of *P. aeruginosa* in vitro and in vivo. Around 14% of the pathogen genome was significantly affected by *Panchvalkal* formulation. This formulation was able to rescue *C. elegans* from *P. aeruginosa* mediated killing. Also, the formulation enhanced the susceptibility of *P. aeruginosa* to different antibiotics such as tetracycline and cephalexin (Joshi et al. 2019). *Panchvalkal* formulation enhanced the wound healing cycle by reducing the burden of microorganisms associated with the wounds. In a study, some bacteria associated with wound infections such as *Chromobacterium violaceum*, *S. aureus* and *Serratia marcescens* were reduced significantly by the QS modulating property of *Panchvalkal*. Thus, suggested the application of *Panchvalkal* formulation in the treatment of wound infection and delayed wound healing (Patel et al. 2018). Manuka-type honeys always gained attention in wound care due to their broad-spectrum antimicrobial properties and low capacity to develop microbial resistance. Sugars present in honey are associated with the reduction of QS and biofilms in *P. aeruginosa*. Likewise, honey is proposed as a promising agent for wound dressings to treat the wound infections caused by this bacterium (Lu et al. 2019). In a similar study, honey showed a potent effect on virulence factors (exotoxin) and QS systems (*LasR* and *RhlR*) of *P. aeruginosa* (Ahmed and Salih 2019).

Abbas and Shaldam 2017 investigated the ability of Glyceryl trinitrate to inhibit QS systems of *P. aeruginosa* and five more clinical strains of *P. aeruginosa* from burn wound infections. Glyceryl trinitrate is a FDA accepted drug with antimicrobial and wound healing activity. Also, this compound possesses antifungal activity and antibiofilm activity against *Candida albicans* and *P. aeruginosa*, respectively. This compound showed QS quenching property against *Chromobacterium violaceum* through the reduction of pigment, violacein. This indicated the effect of glyceryl trinitrate on QS inhibition. They also showed a decrease in the production of several virulence determinants including pyocyanin and protease. Computational approaches exposed the strong affinity of glyceryl trinitrate to bind with the *LasR* and *RhlR* regulators in comparison to the native ligands. Therefore, they recommended the use of this QSI as a topical or systemic agent for the treatment of burn infections caused by the drug-resistant *P. aeruginosa* and its biofilms (Abbas and Shaldam 2017).

In the below section, animal models investigated to study the impact of QS inhibition in chronic wound infections are discussed.

6   Animal Models to Study QS Interference in Chronic Wounds

There are several advantages of using animal wound healing models over human models. They enable to elucidate the mechanisms of wound regeneration and repair, to test the efficiency and safety of new therapeutic methods and allows to use animals with impaired wound healing regardless of harmful effect, which is strictly prohibited in the use of humans (Grada et al. 2018). There are several murine models used for assessing the QS regulated pathogenesis of *P. aeruginosa* and its inhibition through antibacterial, anti-persister or anti-virulence therapy. These therapies serve important roles in the drug development trials against multidrug-resistant *P. aeruginosa*, which are the need of the hour. The different murine wound models practiced so far to evaluate/assess the therapeutic efficacy of QS inhibitors and antibiotics are, role of QS systems in infection, efficacy of inhibitors of antibiotic tolerance, therapeutic efficacy of antibacterial agents against acute pneumonia and effectiveness of combined topical and systemic antibiotic agents are full-skin thickness burn injury model, abdominal burn and infection model, murine persistent/relapsing full-skin thickness burn injury model, large area burn wound model, acute lung infection and open wound infection model of *P. aeruginosa* (Maura et al. 2018).

The quantification or presence of autoinducers in the wounds is a directive indicative of QS and severity of infections. Researchers quantified autoinducer, HSL produced as a result of QS signal of *P. aeruginosa* in a rat ischemic wound infection model. The study revealed the ability to speculate the severity of infection by QS pathway and this data could be applied to suggest the exact treatment to abolish QS signalling in chronic wound infections (Nakagami et al. 2008). In a study, sodium salicylate was reported to interfere with QS regulated virulence factor production in chronic wound infections caused *P. aeruginosa* in a simulated wound fluid. The QS inhibitor decreased virulence phenotypes and expression of virulence genes contributed by Las, Rhl and Pqs systems. There were other bacterial strains isolated from wounds that also produced QS signals. The QS inhibitor was active against most of the wound strains other than *P. aeruginosa*. Thus, the study suggested the use of sodium salicylate as QS inhibitory agent against wound infections caused by a broad range of bacteria and serves as an alternative to the traditional antibiotic therapy (Gerner et al. 2020).

Green tea is one of the popular drinks worldwide owing to its exceptional benefits on human health. In addition to this, traditional Chinese medicine has tagged tea with medicinal properties and as a healthful drink. One of the tea phytochemical, polyphenols holds potential biological properties such as antimicrobial, neuroprotective, anti-inflammatory, anti-carcinogenic, antioxidant and cholesterol-reducing properties. Yin et al. 2015 investigated the ability of tea polyphenols to block the QS pathways and to interrupt QS regulated virulence factor production by *P. aeruginosa*. The prophylactic potential of these phytochemicals was evidenced with reduced bacterial pathogenesis in in vivo infection models of *C. elegans* and excision wound mice. A significant decrease in biofilm development, elastase production, total proteolytic activity and motility and enhanced wound healing were witnessed in wound infection models. The study potentiated the application of tea polyphenols as a QSI in the treatment of infections caused by *P. aeruginosa* (Yin et al. 2015).

The efficacy of QSIs which target signal biosynthesis to prevent the bacterial infections of chronic wounds was demonstrated using a murine burn wound model of *P. aeruginosa* infection. In this study, topical application of lactonases and ciprofloxacin together, reduced the systematic spread of bacteria and morality in murine burn wound model. The combined topical application of enzymes and antibiotics inhibited the population so that quorum was not formed and thus prevented its spread. Therefore, the study provided more light into the quenching of QS to treat chronic wound infections (Gupta et al. 2015). Wang and co-workers elaborated the anti-QS potential of chlorogenic acid on *P. aeruginosa* using in vitro, in silico and *C. elegans* and mouse wound infection models. Chlorogenic acid protected the nematode from death from the colonization of pathogen in the gut and its pathogenesis. Also in wound infection model, the reduced pathogenicity of *P. aeruginosa* was supported by strong hastening effect in wound healing cycle. The study concluded the possibility of using chlorogenic acid in anti-virulence therapy in chronic wound infections caused by *P. aeruginosa* (Wang et al. 2019).

QS pathway was targeted in another study which coordinately regulated the virulence determinants of the pan-resistant pathogen, *P. aeruginosa*. Halogenated anthranilic acid (primary precursor in the biosynthetic pathway involved in PQS systems) analogues were used in the above study which significantly inhibited the PQS pathway and downregulated the expression of genes associated with multiple virulence factor regulator, *MvfR*. The thermal injury mice model which received the bacterial infection could restrict the systemic dissemination of bacteria and reduced the mortality rate when administered with anthranilic acid analogues. Also, the QSI showed increased osmosensitivity of other clinically relevant pathogens. Thus these analogues could be exploited in future for the design and development of potential anti-infective agents to block the infections caused by *P. aeruginosa* and other clinically relevant pathogens (Lesic et al. 2007).

Quorum sensing agents combined with antibiotics are used as adjuvant therapy for enhanced healing of infected wounds. Thus in a study, RNAIII inhibiting peptide (RIP) along with tigecycline unveiled negative effect on methicillin-resistant *S. aureus* (MRSA)-linked murine wound infection model. RIP is thought to be interfering with the *agr* system of MRSA. FS10, a RIP-derived compound along with antibiotic reduced the formation of biofilm and production of toxins by MRSA in the murine model. The animal group which received FS10 soaked and parenteral administration of tigecycline exhibited significant healing, more epithelialization and collagen deposition. This was correlated to the reduced bacterial load in the test groups in comparison to the control groups. In the future, topical application of FS10 and parenteral administration of tigecycline as adjuvants may show positive effect on wound infections caused by MRSA through a novel therapeutic target known as QS (Simonetti et al. 2016). Likewise, another RIP derivative FS8 and tigecycline were combined and tested its anti-QS efficacy in a rat model with staphylococcal vascular graft infection to prevent prosthesis biofilm. The QS inhibitor, FS8 was recommended as an adjuvant for traditional antibiotics to remove staphylococcal biofilm infections from wounds (Simonetti et al. 2013).

A group of researchers reported that the contents present in the extract of Brazilian pepper tree could be used as an anti-virulence agent to treat MRSA skin and tissue infections, which are resistant to antibiotics. Brazilian pepper tree has already found a place in Brazilian Pharmacopoeia and it was considered always as chief in the Brazilian traditional medicine. This is owing to the medicinal properties of tree such as their anti-inflammatory and anti-septic properties in the healing of ulcers and wounds. A flavone-rich extract named 430D-F5 quenched the *agr* system of *S. aureus* and subsided the dermonecrosis in mouse skin infection model. The study explained the wide potential of non-biocide inhibitory agents in the management of skin infections (Muhs et al. 2017).

Anti-virulence strategy targeting the biosynthesis of signal involved in QS of Gram negative pathogens is always a best treatment modality. Todd et al. 2017 identified a small molecule lead, ambuic acid which significantly inhibited the production of AIP in clinically significant pathogen, MRSA. They also demonstrated the same in a mouse model of MRSA skin challenge model. The QS inhibitor was able to prevent the tissue injury induced by MRSA in the animal model. Together, the outcome of work introduced a new lead for the development of QS inhibitor which targets the biosynthesis of AIP by Gram positive pathogens. Also, they exposed the possibility to convert in translational medicine for skin infections (Todd et al. 2017).

7 Future Directions and Conclusions

There are several studies dealing with the anti-QS activity of compounds or drugs in the prevention of chronic wound biofilms and associated infections in vitro conditions. Quorum sensing is regarded as a virulence strategy produced by several multidrug-resistant pathogens and its biofilms. The inhibition or quenching of quorum sensing is considered as an effective tactic to reduce and prevent the chronic infections of wounds. Among several wound infection models available, murine models are recommended to study the impact of QS and the effect of QSIs on drug-resistant bacteria. This is also due to the close physiological similarity between humans and mice. They are important as pre-clinical experiments before the human clinical trials. To date, little information is available on different animal models of wound infection. Their application in the screening steps of QSIs is dramatically reduced due to practical and ethical concerns. Another problem with animal models relies on the difficulty to compare the outcome with the clinical infections. This is owing to the need for a high load of microorganisms to mimic infections in the in vivo model and the terrible physical and mental traumatization faced by the animals. In this context, other animal models should be recommended or developed as an alternative to murine and rat models to screen different QSIs.

Several natural quorum sensing inhibitors were tested successfully against multidrug-resistant pathogens. Nevertheless, more insights are required for the application of natural QSIs to treat wound infections, as these molecules may be less toxic for topical or systemic application. In addition, more investigations are required to extract the full potential of anti-QS agents along with the antibiotic adjuvants against drug-resistant bacteria. The translation of QSIs and antibiotics into therapeutic dressing materials would be a potential method to prevent and reduce the bacterial biofilms in chronic wounds. Integration of QSIs to other antimicrobial therapy could serve as an excellent strategy to remove antibiotic-resistant strains. Antimicrobial photodynamic therapy (aPDT) is emerged as an alternative cure to eliminate antibiotic-resistant strains. aPDT requires a photosensitizer or drug which will get activated in presence of oxygen and light and leads to the generation of reactive free radicals. Thus, antimicrobial photodynamic therapy along with photoactive QSIs could be applied to treat wound infections which prevent the generation of resistant strains in the future. Also, future studies dealing with the depth of QSIs in wound care would explore a different arsenal of therapeutic agents in comparison to the conventional antimicrobial agents.

References

Abbas HA, Shaldam MA (2017) Glyceryl trinitrate is a novel inhibitor of quorum sensing in *Pseudomonas aeruginosa*. Afr Health Sci 16:1109. https://doi.org/10.4314/ahs.v16i4.29

Ahmed AA, Salih FA (2019) Low concentrations of local honey modulate ETA expression, and quorum sensing related virulence in drug-resistant *Pseudomonas aeruginosa* recovered from infected burn wounds. Iran J Basic Med Sci 22:568–575. https://doi.org/10.22038/ijbms.2019. 33077.7902

Attinger C, Wolcott R (2012) Clinically addressing biofilm in chronic wounds. Adv Wound Care 1:127–132. https://doi.org/10.1089/wound.2011.0333

Bhardwaj AK, Vinothkumar K, Rajpara N (2013) Bacterial quorum sensing inhibitors: attractive alternatives for control of infectious pathogens showing multiple drug resistance. Recent Pat Antiinfect Drug Discov 8:68–83. https://doi.org/10.2174/1574891X11308010012

Black CE, Costerton JW (2010) Current concepts regarding the effect of wound microbial ecology and biofilms on wound healing. Surg Clin North Am 90:1147–1160. https://doi.org/10.1016/j. suc.2010.08.009

Bowler PG, Duerden BI, Armstrong DG (2001) Wound microbiology and associated approaches to wound management. Clin Microbiol Rev 14:244–269. https://doi.org/10.1128/CMR.14.2.244-269.2001

Brackman G, Coenye T (2014) Quorum sensing inhibitors as anti-biofilm agents. Curr Pharm Des 21:5–11. https://doi.org/10.2174/1381612820666140905114627

Caputo WJ, Beggs DJ, DeFede JL et al (2008) A prospective randomised controlled clinical trial comparing hydrosurgery debridement with conventional surgical debridement in lower extremity ulcers. Int Wound J 5:288–294. https://doi.org/10.1111/j.1742-481X.2007.00490.x

Chaney SB, Ganesh K, Mathew-Steiner S et al (2017) Histopathological comparisons of *Staphylococcus aureus* and *Pseudomonas aeruginosa* experimental infected porcine burn wounds. Wound Repair Regen 25:541–549. https://doi.org/10.1111/wrr.12527

Clinton A, Carter T (2015) Chronic wound biofilms: pathogenesis and potential therapies. Lab Med 46:277–284. https://doi.org/10.1309/LMBNSWKUI4JPN7SO

Cogen AL, Nizet V, Gallo RL (2008) Skin microbiota: a source of disease or defence? Br J Dermatol 158:442–455. https://doi.org/10.1111/j.1365-2133.2008.08437.x

Dalton T, Dowd SE, Wolcott RD et al (2011) An in vivo Polymicrobial biofilm wound infection model to study interspecies interactions. PLoS One 6:e27317. https://doi.org/10.1371/journal. pone.0027317

Das L, Singh Y (2018) Quorum sensing inhibition: a target for treating chronic wounds. In: Biotechnological applications of quorum sensing inhibitors. Springer Singapore, Singapore, pp 111–126

Demidova-Rice TN, Hamblin MR, Herman IM (2012) Acute and impaired wound healing. Adv Skin Wound Care 25:304–314. https://doi.org/10.1097/01.ASW.0000416006.55218.d0

Dorai AA (2012) Wound care with traditional, complementary and alternative medicine. Indian J Plast Surg 45:418–424. https://doi.org/10.4103/0970-0358.101331

Garraud O, Hozzein WN, Badr G (2017) Wound healing: time to look for intelligent, 'natural' immunological approaches? BMC Immunol 18:23. https://doi.org/10.1186/s12865-017-0207-y

Gerner E, Almqvist S, Werthén M, Trobos M (2020) Sodium salicylate interferes with quorum-sensing-regulated virulence in chronic wound isolates of *Pseudomonas aeruginosa* in simulated wound fluid. J Med Microbiol 69:767–780. https://doi.org/10.1099/jmm.0.001188

Goldberg SR, Diegelmann RF (2020) What makes wounds chronic. Surg Clin North Am 100:681–693. https://doi.org/10.1016/j.suc.2020.05.001

Grada A, Mervis J, Falanga V (2018) Research techniques made simple: animal models of wound healing. J Invest Dermatol 138(10):2095–2105.e1. https://doi.org/10.1016/j.jid.2018.08.005

Guo S, DiPietro LA (2010) Factors affecting wound healing. J Dent Res 89:219–229. https://doi.org/10.1177/0022034509359125

Guoqi W, Zhirui L, Song W et al (2018) Negative pressure wound therapy reduces the motility of *Pseudomonas aeruginosa* and enhances wound healing in a rabbit ear biofilm infection model. Antonie van Leeuwenhoek, Int J Gen Mol Microbiol 111:1557–1570. https://doi.org/10.1007/s10482-018-1045-5

Gupta P, Chhibber S, Harjai K (2015) Efficacy of purified lactonase and ciprofloxacin in preventing systemic spread of *Pseudomonas aeruginosa* in murine burn wound model. Burns 41:153–162. https://doi.org/10.1016/j.burns.2014.06.009

Han G, Ceilley R (2017) Chronic wound healing: a review of current management and treatments. Adv Ther 34:599–610. https://doi.org/10.1007/s12325-017-0478-y

Hurlow J, Blanz E, Gaddy JA (2016) Clinical investigation of biofilm in non-healing wounds by high resolution microscopy techniques. J Wound Care 25:S11–S22. https://doi.org/10.12968/jowc.2016.25.Sup9.S11

Johnson T, Gómez B, McIntyre M et al (2018) The cutaneous microbiome and wounds: new molecular targets to promote wound healing. Int J Mol Sci 19:2699. https://doi.org/10.3390/ijms19092699

Jones CE, Kennedy JP (2012) Treatment options to manage wound biofilm. Adv Wound Care 1:120–126. https://doi.org/10.1089/wound.2011.0300

Joshi C, Patel P, Palep H, Kothari V (2019) Validation of the anti-infective potential of a polyherbal '*Panchvalkal*' preparation, and elucidation of the molecular basis underlining its efficacy against *Pseudomonas aeruginosa*. BMC Complement Altern Med 19:19. https://doi.org/10.1186/s12906-019-2428-5

Kalia VC (2013) Quorum sensing inhibitors: an overview. Biotechnol Adv 31:224–245. https://doi.org/10.1016/j.biotechadv.2012.10.004

Kim PJ, Steinberg JS (2012) Wound care: biofilm and its impact on the latest treatment modalities for ulcerations of the diabetic foot. Semin Vasc Surg 25:70–74. https://doi.org/10.1053/j.semvascsurg.2012.04.008

Kong K-F, Vuong C, Otto M (2006) Staphylococcus quorum sensing in biofilm formation and infection. Int J Med Microbiol 296:133–139. https://doi.org/10.1016/j.ijmm.2006.01.042

Koo H, Allan RN, Howlin RP et al (2017) Targeting microbial biofilms: current and prospective therapeutic strategies. Nat Rev Microbiol 15:740–755. https://doi.org/10.1038/nrmicro.2017.99

LaSarre B, Federle MJ (2013) Exploiting quorum sensing to confuse bacterial pathogens. Microbiol Mol Biol Rev 77:73–111. https://doi.org/10.1128/MMBR.00046-12

Lee J, Zhang L (2014) The hierarchy quorum sensing network in *Pseudomonas aeruginosa*. Protein Cell 6:26–41. https://doi.org/10.1007/s13238-014-0100-x

Leid JG (2009) Bacterial biofilms resist key host defenses. Microbe 4:66–70

Lesic B, Lépine F, Déziel E et al (2007) Inhibitors of pathogen intercellular signals as selective anti-infective compounds. PLoS Pathog 3:e126. https://doi.org/10.1371/journal.ppat.0030126

Li X, Kim J, Wu J et al (2020) N-acetyl-cysteine and mechanisms involved in resolution of chronic wound biofilm. J Diabetes Res 2020:1–16. https://doi.org/10.1155/2020/9589507

Lu J, Cokcetin NN, Burke CM et al (2019) Honey can inhibit and eliminate biofilms produced by *Pseudomonas aeruginosa*. Sci Rep 9:18160. https://doi.org/10.1038/s41598-019-54576-2

LuTheryn G, Glynne-Jones P, Webb JS, Carugo D (2020) Ultrasound-mediated therapies for the treatment of biofilms in chronic wounds: a review of present knowledge. Microb Biotechnol 13:613–628. https://doi.org/10.1111/1751-7915.13471

Matiasek J, Domig KJ, Djedovic G et al (2017) The effect of negative pressure wound therapy with antibacterial dressings or antiseptics on an in vitro wound model. J Wound Care 26:236–242. https://doi.org/10.12968/jowc.2017.26.5.236

Maura D, Bandyopadhaya A, Rahme LG (2018) Animal models for *Pseudomonas aeruginosa* quorum sensing studies. Methods Mol Biol 1673:227–241. https://doi.org/10.1007/978-1-4939-7309-5_18

Metcalf D, Bowler P, Parsons D (2016) Wound biofilm and therapeutic strategies. In: Microbial biofilms–importance and applications InTech. https://doi.org/10.5772/63238

Mihai MM, Holban AM, Giurcă Neanu C et al (2014) Identification and phenotypic characterization of the most frequent bacterial etiologies in chronic skin ulcers. Romanian J Morphol Embryol 55:1401–1048

Miller MB, Bassler BL (2001) Quorum sensing in bacteria. Annu Rev Microbiol 55:165–199. https://doi.org/10.1146/annurev.micro.55.1.165

Muhs A, Lyles JT, Parlet CP et al (2017) Virulence inhibitors from Brazilian peppertree block quorum sensing and abate Dermonecrosis in skin infection models. Sci Rep 7:42275. https://doi.org/10.1038/srep42275

Nakagami G, Morohoshi T, Ikeda T et al (2011) Contribution of quorum sensing to the virulence of *Pseudomonas aeruginosa* in pressure ulcer infection in rats. Wound Repair Regen 19:214–222. https://doi.org/10.1111/j.1524-475X.2010.00653.x

Nakagami G, Sanada H, Sugama J et al (2008) Detection of *Pseudomonas aeruginosa* quorum sensing signals in an infected ischemic wound: an experimental study in rats. Wound Repair Regen 16:30–36. https://doi.org/10.1111/j.1524-475X.2007.00329.x

Negut I, Grumezescu V, Grumezescu A (2018) Treatment strategies for infected wounds. Molecules 23:2392. https://doi.org/10.3390/molecules23092392

Ng W-L, Bassler BL (2009) Bacterial quorum-sensing network architectures. Annu Rev Genet 43:197–222. https://doi.org/10.1146/annurev-genet-102108-134304

Nusrat H, Shankar P, Kushwah J et al (2011) Diversity and polymorphism in AHL-lactonase gene (*aiiA*) of *Bacillus*. J Microbiol Biotechnol 21:1001–1011. https://doi.org/10.4014/jmb.1105.05056

Oppenheimer-Shaanan Y, Steinberg N, Kolodkin-Gal I (2013) Small molecules are natural triggers for the disassembly of biofilms. Trends Microbiol 21:594–601. https://doi.org/10.1016/j.tim.2013.08.005

Patel P, Joshi C, Kothari V (2019) Antipathogenic potential of a Polyherbal wound-care formulation (Herboheal) against certain wound-infective gram-negative bacteria. Adv Pharmacol Sci 2019:1–17. https://doi.org/10.1155/2019/1739868

Patel P, Joshi C, Palep H, Kothari V (2018) Anti-infective potential of a quorum modulatory polyherbal extract (*Panchvalkal*) against certain pathogenic bacteria. J Ayurveda Integr Med. https://doi.org/10.1016/j.jaim.2017.10.012

Rezzoagli C, Granato ET, Kümmerli R (2020) Harnessing bacterial interactions to manage infections: a review on the opportunistic pathogen *Pseudomonas aeruginosa* as a case example. J Med Microbiol 69:147–161. https://doi.org/10.1099/jmm.0.001134

Rickard AH, Colacino KR, Manton KM et al (2010) Production of cell-cell signalling molecules by bacteria isolated from human chronic wounds. J Appl Microbiol 108:1509–1522. https://doi.org/10.1111/j.1365-2672.2009.04554.x

Rodrigues M, Kosaric N, Bonham CA, Gurtner GC (2019) Wound healing: a cellular perspective. Physiol Rev 99:665–706. https://doi.org/10.1152/physrev.00067.2017

Ruffin M, Bilodeau C, Maillé É et al (2016) Quorum-sensing inhibition abrogates the deleterious impact of *Pseudomonas aeruginosa* on airway epithelial repair. FASEB J 30:3011–3025. https://doi.org/10.1096/fj.201500166R

Rutherford ST, Bassler BL (2012) Bacterial quorum sensing: its role in virulence and possibilities for its control. Cold Spring Harb Perspect Med 2:a012427–a012427. https://doi.org/10.1101/cshperspect.a012427

Schultz GS, Woo K, Weir D, Yang Q (2018) Effectiveness of a monofilament wound debridement pad at removing biofilm and slough: ex vivo and clinical performance. J Wound Care 27:80–90. https://doi.org/10.12968/jowc.2018.27.2.80

Shukla VK, Ansari MA, Gupta SK (2005) Wound healing research: a perspective from India. Int J Low Extrem Wounds 4:7–8. https://doi.org/10.1177/1534734604273660

Simonetti O, Cirioni O, Cacciatore I et al (2016) Efficacy of the quorum sensing inhibitor FS10 alone and in combination with Tigecycline in an animal model of staphylococcal infected wound. PLoS One 11:e0151956. https://doi.org/10.1371/journal.pone.0151956

Simonetti O, Cirioni O, Mocchegiani F et al (2013) The efficacy of the quorum sensing inhibitor FS8 and tigecycline in preventing prosthesis biofilm in an animal model of staphylococcal infection. Int J Mol Sci 14:16321–16332. https://doi.org/10.3390/ijms140816321

Todd DA, Parlet CP, Crosby HA et al (2017) Signal biosynthesis inhibition with ambuic acid as a strategy to target antibiotic-resistant infections. Antimicrob Agents Chemother 61:e00263–e00217. https://doi.org/10.1128/AAC.00263-17

Toy LW, Macera L (2011) Evidence-based review of silver dressing use on chronic wounds. J Am Acad Nurse Pract 23:183–192. https://doi.org/10.1111/j.1745-7599.2011.00600.x

Wang H, Chu W, Ye C et al (2019) Chlorogenic acid attenuates virulence factors and pathogenicity of *Pseudomonas aeruginosa* by regulating quorum sensing. Appl Microbiol Biotechnol 103:903–915. https://doi.org/10.1007/s00253-018-9482-7

Weinstein RA, Mayhall CG (2003) The epidemiology of burn wound infections: then and now. Clin Infect Dis 37:543–550. https://doi.org/10.1086/376993

Williams H, Campbell L, Crompton RA et al (2018) Microbial host interactions and impaired wound healing in mice and humans: defining a role for BD14 and NOD2. J Invest Dermatol 138:2264–2274. https://doi.org/10.1016/j.jid.2018.04.014

Yarwood JM, Bartels DJ, Volper EM, Greenberg EP (2004) Quorum sensing in *Staphylococcus aureus* biofilms. J Bacteriol 186:1838–1850. https://doi.org/10.1128/JB.186.6.1838-1850.2004

Yin H, Deng Y, Wang H et al (2015) Tea polyphenols as an antivirulence compound disrupt quorum-sensing regulated pathogenicity of *Pseudomonas aeruginosa*. Sci Rep 5:16158. https://doi.org/10.1038/srep16158

Biofilm: A Challenge to Overcome in Wound Healing

Debaprasad Parai, Pia Dey, and Samir Kumar Mukherjee

1 From Planktonic to Biofilm

Nature is often unfavourable due to its unpredictability. Everyone prefers to follow the principle of 'strength through unity' for combatting it rather than individually. Microbes are also not out of this rule as they are commonly found in the form of multicellular aggregates known as biofilm. This is a cluster of microbial cells attached to various biotic and/or abiotic surfaces and interfaces by the formation of a protective extracellular polymeric substances [EPS] matrix (Costerton et al. 1999; Stoodley et al. 2002; Hall-Stoodley et al. 2004). Many bacterial species transform themselves from planktonic to biofilm, depending on the physical stress in their growing environment and to take advantage of the greater nutrients' availability from the substrata. The biofilm mode of lifestyle can also afford UV exposure, metal toxicity, acid exposure, dehydration and it could restrict the penetration of antibiotics and antimicrobial agents as well (Grossart 2010; Teschler et al. 2015; Yin et al. 2019). Hence, bacterial biofilms remain a major concern in a broad range of area like the environment, food industry and especially the biomedical sector.

2 How this Lifestyle Grows Up?

Microbial species adopt a wide variety of mechanisms through which they can contact a surface, firmly attach, promote cellular communications and initiate growth as a complex structure. In contrast to the establishment of individual microorganisms

D. Parai (✉)
ICMR-Regional Medical Research Centre, Bhubaneswar, India

P. Dey · S. K. Mukherjee
Department of Microbiology, University of Kalyani, Kalyani, West Bengal, India

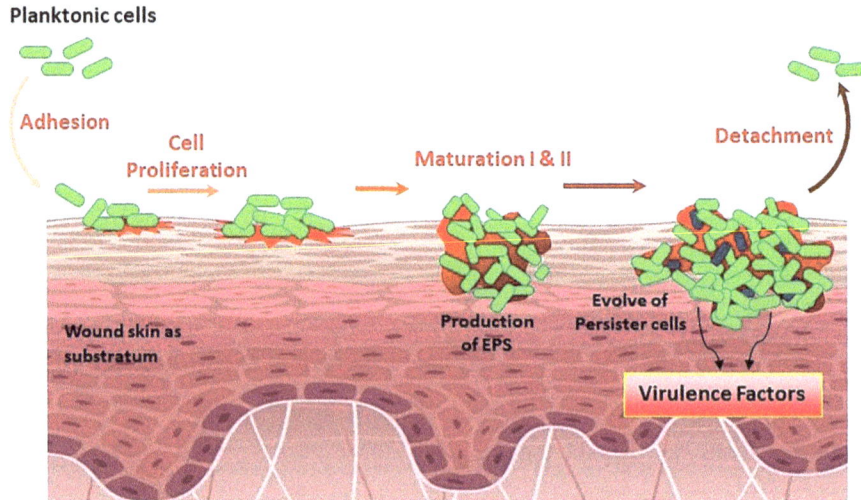

Fig. 1 Schematic diagram depicting the stages involved in biofilm development

by classical competition of natural selection, the bacterial community in a biofilm develops entirely by self-organization and cooperation (Hall-Stoodley et al. 2004; Jefferson 2004). The formation of biofilm follows a sequence of mechanical, chemical and biological processes. Microbes enter into biofilm state through five stages (Fig. 1), which is initiated by a preliminary reversible attachment, then an irreversible attachment followed by maturation stage I, maturation stage II and finally dispersion of the bacterial consortium within the developed biofilm (Sauer et al. 2002; Stoodley et al. 2002; Kaplan 2010). The initial two stages involve with the adhesion of planktonic microbial cells to the substratum by physical forces like van der Waals forces, steric interactions and electrostatic interactions (Marić and Vranes 2007; Garrett et al. 2008). It can also attach to the surfaces employing bacterial appendages such as pili or flagella, which is further modulated by different types of factors like temperature, pressure and surface functionality. Hydrophobic interaction between the surface and the bacteria plays an important role in biofilm cells aggregation where the bacteria can adhere to a hydrophobic nonpolar surface through this interaction (Garrett et al. 2008; Tribedi and Sil 2014). The extent of microbial attachment is greatly influenced by cell surface hydrophobicity which is further strengthened by the presence of extracellular filamentous appendages. Hydrophobic interactions act by increasing the non-polarity between the adhering surface and the attached microbial cells. Concisely, microbial cells initiate biofilm formation with a loose adhesion followed by specific, strong and irreversible binding to a particular substratum (Hall-Stoodley et al. 2004).

The first two stages of biofilm development are followed by maturation stages with the construction of complex architecture, channels and pores (Davies 1998). This phase witness the onset of microbial communications by the production of autoinducers that orchestrate the expression of biofilm-specific genes (Dubern and

Diggle 2008; Joo and Otto 2012; Parai et al. 2018). This particular phase involves a set of precisely synchronized events which includes the development of EPS matrix to stabilize the biofilm network along with the production of extracellular DNA (eDNA) for cellular communication and stabilization of biofilm. Experimental evidence has reported that young Pseudomonad biofilms are an easy target for DNase treatment than their mature counterparts which suggest the stabilizing role played by eDNA during the initial stages of biofilm formation when EPS components are barely present (Whitchurch et al. 2002; Gloag et al. 2013). Irreversible adhesion leads to cellular aggregation followed by the formation of microcolonies. In the maturation stage, biofilms adapt to the external environment by manipulating their integral structures, physiological aspects and basic metabolism. The final stage in the lifecycle of biofilm is the dispersion phase where the bacterial cells disseminate from the biofilm matrix and revive to their motile form from their sessile form. Various biochemical changes occur within the dispersing biofilm such as the production of different saccharolytic enzymes responsible for the degradation of stabilizing polysaccharides, thereby aiding the colonization of the bacterial cells to new surfaces (Sutherland 2001; Guilhen et al. 2017). The dispersal strategies can be broadly explained by three different mechanisms, namely seeding, erosion and sloughing. Seeding dispersal was best explained in non-mucoid *Pseudomonas aeruginosa* biofilms, where the microcolonies differentiate to form stationary bacteria of sessile phenotype, and the inner region containing liquefied microcolonies of motile cells. These planktonic motile cells then come out of that colony structures via swimming which leaves a hollow mound inside. *Staphylococcus aureus* exhibit erosion dispersal where their biofilm aggregates surrounded by EPS are found to be continuously shed from the core structure. These dispersed cells are found to have physiological similarity with the biofilm rather than their planktonic parts. Another strategy for biofilm dispersal happens in the later phase of biofilm development where a large portion of the biofilm detaches suddenly from the core aggregation. Although it is known that single cells can actively move across surfaces through gliding and twitching motility in some species, there are pieces of findings for shear-mediated motion exhibited by *P. aeruginosa* and mixed-species biofilms (Hall-Stoodley et al. 2004; Kaplan 2010).

3   Antimicrobial Resistance of Biofilm

Biofilms show increased survival and resistance to conventional antibiotics mainly by the protection conferred by EPS. Biofilm forming bacteria have a general trait of increased antibiotic resistance (Stewart and William Costerton 2001; Davies 2003). Biofilm cells are 10 to 1000 times less susceptible to a specific antibiotic as compared to their planktonic counterparts (Gilbert et al. 2002). Several mechanisms are accounted for the drug resistance mechanisms conferred by biofilm, viz. (i) formation of chemical and physical barriers against antibiotic penetration; (ii) evolution of resistant phenotypes called persister or mutator cells; (iii) antimicrobial agent inactivation by chemical modification, hydrolysis, alteration of the target sites and activation of efflux pumps; (iv) presence of eDNA to chelate cationic antibiotics; (v) expression of antibiotic resistance genes and (vi) heterogenicity in biofilm cells population which indeed trigger stress responses under unfavourable conditions like low oxygen and nutrients (Mah and O'Toole 2001; Davies 2003; Høiby et al. 2010). Even sub-inhibitory concentrations of aminoglycosides could induce biofilm formation in *Escherichia coli* and *P. aeruginosa* as part of their defensive mechanisms (Hoffman et al. 2005). Since biofilm formation is common for most bacterial pathogens found in chronic wound infection, the increasing antibiotic resistance of biofilm is a serious concern in wound healing (Donlan 2001; Metcalf and Bowler 2013).

4 Biofilm and its Clinical Relevance

The adherent behaviour of bacteria is more predominant and a clinically relevant process over the planktonic one. Biofilms have severe harmful pathogenic manifestations on human health, that contribute to almost 80% of the total chronic and recurrent microbial infection (Römling and Balsalobre 2012). This is mainly due to the increased antibiotic resistance of biofilm compared to its planktonic counterparts from the same bacterial culture (Hall-Stoodley and Stoodley 2009; Vasudevan 2014). Biofilm formation has been robustly found in diseases like otitis media, dental caries, chronic wound infection, osteomyelitis, chronic rhinosinusitis, recurrent urinary tract infection, endocarditis, cystic fibrosis-associated lung infection and many more. Some common bacteria which are frequently produced biofilm during the establishment of the disease include *S. aureus, P. aeruginosa, E. coli, Staphylococcus epidermidis, Streptococcus* sp., *Candida* sp. A complete list can be found in Table 1. Biofilms are also a major concern in clinical settings causing medical device-related infections. Surface depositions of biofilm on medical devices like intravenous catheters and cardiac pacemakers were first observed by an electron microscope in the early 1980s (Donlan and Costerton 2002). Microorganisms like *S. epidermidis* and *S. aureus* are most frequently found in association with medical devices, followed by *P. aeruginosa* and an array of other opportunistic pathogens which cause infections during an invasive medical intervention or chemotherapy in an immune-compromised individual. Biofilm formation on medical implants has even led to the characterization of a new infectious disease called chronic polymer-associated infection. Medical devices that are well documented as biofilm substratum are mainly intravenous catheters, cardiac pacemakers, joint prostheses, prosthetic heart valves, peritoneal dialysis catheters, dentures, contact lenses, cerebrospinal fluid shunts and endotracheal tubes (Donlan and Costerton 2002; Litzler et al. 2007; Veerachamy et al. 2014; Yadav et al. 2020).

Table 1 List of biofilm-associated diseases and their causative organisms

Disease name	Biofilm forming microorganism	References
Cystic fibrosis	*Staphylococcus aureus, Pseudomonas aeruginosa, Haemophilus influenzae, Streptococcus pneumoniae*	Ciofu et al. (2015)
Chronic wound infection	*S. aureus, P. aeruginosa, Enterococcus faecalis, Proteus spp.*	James et al. (2008a, b)
Chronic otitis media	*S. aureus, P. aeruginosa, Staphylococcus epidermidis, S. pneumoniae, H. influenzae, Moraxella catarrhalis*	Thornton et al. (2011)
Endocarditis	*Streptococcus* spp., *Staphylococcus* spp., *Candida* spp., *Aspergillus fumigatus*	Donlan and Costerton (2002)
Periodontitis and dental caries	*Fusobacterium nucleatum, Peptostreptococcus micros, Bacteroides intermedius, Bacteroides forsythus, Porphyromonas gingivalis, Tannerella forsythia, Treponema denticola, Bifidobacterium denticum*	Colombo and Tanner (2019)
Pleuropulmonary infections	*P. aeruginosa, S. aureus, H. influenzae, Candida albicans, S. pneumoniae, Klebsiella* sp., *Escherichia coli, Acinetobacter* sp., *A. fumigatus*	Boisvert et al. (2016)
Osteoarticular infections (osteomyelitis)	*S. aureus, Streptococcus pyogenes, H. influenzae, Enterococcus* spp.	Brady et al. (2008)
Skin infection	*S. aureus, S. epidermidis, Propionibacterium acnes*	Coenye et al. (2008)
Rhinosinusitis	*S. aureus, S. pneumoniae, H. influenzae, M. catarrhalis*	Maina et al. (2018)
Urinary tract infection	*E. coli, Klebsiella pneumoniae, Proteus mirabilis, Serratia* spp., *P. aeruginosa, S. aureus, E. faecalis*	Hatt and Rather (2008)
Medical device-related infection	*S. aureus, S. epidermidis, P. aeruginosa, K. pneumoniae, E. faecalis, C. albicans, P. mirabilis*	Jamal et al. (2018)

5   Characteristics of Chronic Wounds

The physiological process of wound healing is attributed to four overlapping phases of haemostasis, inflammation, proliferation and repair of the matrix involving epithelialization and remodelling. A chronic wound can be defined as a wound that is restrained from progressing through an inflammatory stage to its next subsequent stages, thereby disrupting the orderly events of wound healing (Attinger and Randy 2012). Depending upon their aetiologies and pathogenesis, they can be categorized into three groups: venous leg ulcers, diabetic foot ulcers and pressure ulcers. Owing to their unique origins, non-healing chronic wounds are unlikely to have a customary cure for all its types. However, non-healing chronic wounds share certain common features, such as the release of excessive levels of proinflammatory cytokines by the necrotic tissues, proteases, reactive oxygen species (ROS) and senescent cells. The situation is worsened by the collateral existence of persistent infections along with a deficiency of functional stem cells. The constant influx of immune cells are secondarily stimulated by repetitive tissue injury that ends up in a prolonged proinflammatory cytokine cascade which further elevates the level of proteases (Frykberg and Banks 2015). Inside acute wounds, proteases are meticulously regulated by their inhibitors but these protease levels surpass the inhibitor counts in case of chronic wounds. Proteases are tightly regulated by their inhibitors in acute wounds but in the case of chronic wounds, protease levels exceed their respective inhibitor counts. This causes the destruction of extracellular matrices (ECM), and the degradation of the growth factors and their receptors. The feature prevents the wound from proceeding into the proliferative phase and also attracts further inflammatory cells, thus lengthens the inflammatory stage (Mccarty and Percival 2013). Similarly, increased ROS production in chronic wounds damages the ECM proteins and causes other cell damages instead of providing defence against microorganisms (Schreml et al. 2010). Fibroblasts are essential cells in the wound healing process, which in case of venous leg ulcers and pressure ulcers become phenotypically altered and exhibit senescence with diminished ability to proliferate for the formation of granulation tissue (Stanley and Osler 2001; Clark 2008). Additionally, non-functional senescent keratinocytes at the periphery of wounds proliferate, but they cannot fully differentiate into migrating keratinocytes (Morasso and Tomic-canic 2005). These ultimately advance the epithelial build-up, often seen around the edge of the chronic wounds (Attinger and Randy 2012). Mesenchymal stem cells (MSC) are reported to have a crucial role in wound healing (Ennis et al. 2013). But deficient and defective MSCs in case of chronic wounds are found to obstruct the remodelling of microvasculature for which they get recruited into the circulation in response to injury. Hence, the key to reverse the chronicity of non-healing wounds is by exploring the underlying molecular and physiological mechanisms behind the phenomenon of non-healing and restoring the optimal balance of cytokines, growth factors, proteases and metabolically competent cells which can, in turn, improve the extent of healing.

6   Biofilm in Chronic Wounds

The assemblage of three-dimensional bacterial biofilm is the prevailing phenotype of bacteria in a chronic non-healing wound bed fused with the extracellular matrix. This sessile form of bacteria assists its inhabitation in a chronic wound more proficiently than its planktonic counterparts (Wei et al. 2019a, b). Evidence has confirmed the presence of bacterial biofilms in chronic wounds by experimentally inducing it in animal models along with the demonstration of clinical wounds (Wright et al. 1999; Thomson 2011; Metcalf et al. 2014). Various topologically distinct types of wound biofilm viz. static, colonized, slimy, pigmented found in human chronic infections are represented in Fig. 2. Bacterial biofilm levels up the severity of chronic wounds by diverse factors starting from the heterogeneous nature of biofilm consortium to interspecies relation within that mixed population. The presence of biofilm gradient and persister cell community, incompetent antimicrobial penetration along

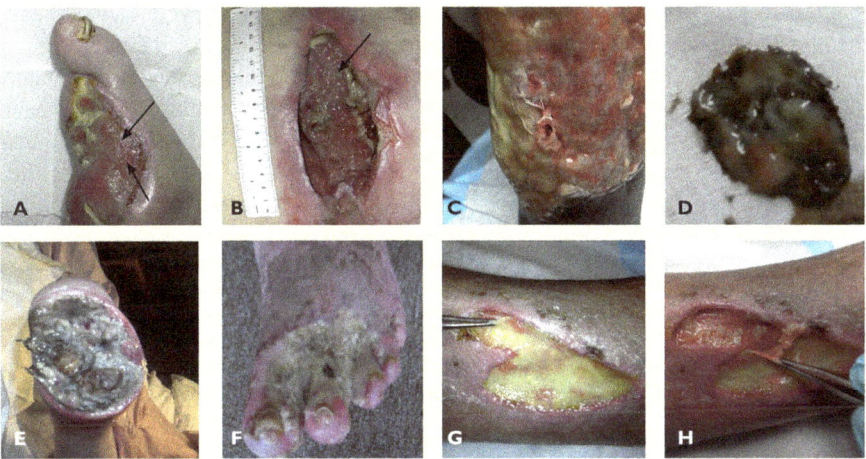

Fig. 2 Different visualization of wound biofilm. A static, non-progressing wound on a stable diabetic patient. Arrows indicate the suspected appearance of biofilm (**a**). A heavily colonized biofilm in wound bed of a stage IV pressure ulcer. Arrow indicates a possible biofilm layer with a slightly thicker and opaque appearance (**b**). An ischaemic and infected wound with suspected biofilm-forming through and over a previously-applied gauze dressing (**c**). A chronic wound bed with green pigmented biofilm (**d**). Viscous, pale, green-blue, slimy suspected biofilm covered on a forefoot amputation (**e**). Wound surface exhibited yellow suspected biofilm, possibly mixed with slough as found in a failed surgical wound (**f**). Biofilm re-formed quickly over granulation tissue in an ischaemic wound (**g**). The same suspected biofilm could be removed atraumatically using forceps to reveal the granulation tissue beneath (**h**). Images are reproduced with kind permissions from authors (Metcalf et al. (2014) and the Journal of Wound Care

with variable pathogenesis imposed by respective residing microorganisms further intensifies the chronicity. The spread of infection is not dependent on the number of species present within the biofilm, rather it can be regulated by the abundance in normal skin microflora (Brandwein et al. 2016). Though some pathogenic species (notably *S. aureus* and *P. aeruginosa*) can replace the harmless skin commensals and cause infection, which then requires probiotic therapy in the form of re-colonization with 'healthy commensal' bacteria (Krutmann 2009). Interspecies relation is another important factor that determines the net behaviour of the community whether they belong to 'functional equivalent pathogroups' where they act synergistically leading to a chronic wound (Lewis 2005). Most of the chronic wounds are persistent infections and the biofilms found here are pre-formed (Ito et al. 2009; Li et al. 2020). Inhibition of biofilm formation is always easier than to eradicate a pre-formed or mature biofilm which already consists of a special type of cells called persister cells which neither grow nor die easily. Moreover, these cells are almost resistant to conventional antibiotic dosages which further emerge the threat of antimicrobial resistant strains when tried with higher dosages for their removal in clinical settings. In vitro studies have found that a test antibiofilm compound can disperse a pre-formed biofilm only in higher concentration compared to the dose required for the inhibition of the same biofilm during formation (de la Fuente-Núñez et al. 2014;

Algburi et al. 2017). One of the major reasons behind this is the reversible features of the first two phases in a biofilm developmental pathway as discussed earlier. Once it starts to establish the complex interactions among the cell population and modulates phenotypically, it becomes ready to escape and endure regular antibiotic treatments (Keren et al. 2004; Percival et al. 2011; Olsen 2015). Clinical studies have also reported re-emergence of biofilm communities from wound bed even after their successful removal following surgical debridement of chronic wounds. Additionally, biofilms in chronic wounds may differ from the types of infection hindering the application of specific treatment approaches for chronic wounds (Wright et al. 2002).

7 Role of Biofilm Matrix in Wounds

The prolonged persistence of bacterial biofilm interferes with the potentiality of their treatment within chronic wounds. Bacterial biofilm can invade such wounds earliest by 10 hours after its inception, and persist until the wound remains open (Harrison-Balestra et al. 2003; Kim and Steinberg 2012). An array of growth associated factors like oxygen, nutrient, pH, osmolality and chemical agent concentration regulates the microenvironment of a biofilm. The main reason behind this prevalence is biofilm composition, which includes EPS along with proteins, polysaccharides, lipids, eDNA, water, ions and nutrients. This concomitantly determines the types and stages of microorganisms present within the biofilm matrix (Percival et al. 2011; Jensen et al. 2017). Biofilms evade antimicrobial challenges by multiple mechanisms as discussed earlier, of which the most important being resistance imparted by EPS. Incomplete penetration of the antimicrobials to a chronic wound completely differs from its acute state in terms of susceptibility towards antimicrobials. The inert nature of EPS retards the diffusion of antibiotics, making biofilms a predominant cause of persistent infections (Metcalf and Bowler 2013; Mendoza et al. 2019).

8 Prevalence, Detection and Management of Biofilm in Wounds

Bacterial biofilms are one of the most predominant causes behind delayed healing of approximately 60–80% of chronic non-healing skin wounds (James et al. 2008a, b; Brandenburg et al. 2018). Factors like wound size, position, moisture content, host immunity, comorbidity can influence the chronicity and healing time of wound infection. The microenvironment of chronic wounds facilitates biofilm form of bacterial growth, as it accommodates necrotic debris with low oxygen tension and dampened but sustained host immune response (Wu et al. 2019). The lethality of a chronic wound may vary from prolonged inflammation to debridement and even amputation of distal limbs in extreme cases. However, the prevalence of biofilm in the case of acute wounds is not at par with their chronic counterparts. About 60% of chronic non-healing wounds requiring debridement are connected with biofilms, while 6% of acute wounds were identified to be laden with biofilm (James et al. 2008a, b; Metcalf and Bowler 2013).

Detection of biofilm in chronic wound infection mainly focuses on two primary aspects: biofilm localization and identification of the causative organisms. Although the presence of biofilms have been reported widely in a majority of chronic non-healing wounds, yet there is a lack of precise techniques to achieve these two primary goals (Hurlow et al. 2016). Tissue biopsy from debrided wound base is crucial for precise identification of the causative organism other than inefficient superficial wound swabbing (Lipsky et al. 2012; Høiby et al. 2014). Current biofilm detection methods can be broadly categorized into morphological assay, microbiological assay and molecular assay.

8.1 Morphological Methods

With a thickness of less than 100µm, lack of distinct distinguishable feature and heterogeneous distribution throughout the wound base and eschars make it difficult to visualize wound biofilms with the naked eye (Bjarnsholt et al. 2013). Thus, more than one sample should be drawn both from upper and deep wound layers to increase the accuracy of the respective diagnostic methodology (Wu et al. 2019). Slough, debris or exudates are sometimes visually mistaken as biofilms by healthcare professionals. Hence diagnosis of biofilm rely upon high-resolution microscopy such as scanning electron microscopy (SEM) and confocal laser scanning microscopy (CLSM) since wound swabbing or histological examinations may lead to false-negative results depending on the sample collection site. Microbial electrochemical technology is another precise optical method to visualize electrochemically active biofilms which can be explored by healthcare professionals for uncomplicated identification of biofilms (Schmidt et al. 2017).

8.2 Microbiological Methods

Microbiological assays are almost carried out daily in clinical settings to identify the causative agents of acute infections. On contrary, biofilm-associated chronic wound infections are challenging because chronic wounds harbour multiple pathogens, typically 2–5 species co-existing within one ulcerative wound (Dowd et al. 2008). Most of the bacteria in chronic wounds enter a slow-growing stationary phase which is characterized as a viable but non-culturable (VBNC) state (Li et al. 2014). This might be an adaptive mechanism for surviving a stressful microenvironment consisting of antimicrobial exposure, low pH, less oxygen and limited nutrients. Culturing becomes even more difficult by the presence of multispecies within the same biofilm matrix. Moreover, VBNC bacteria require special cultural techniques, such as under temperature stress and starvation. Several other physical methods like sonication, and chemical methods like dithiothreitol treatments along with standard tissue culture methods are reported for acquiring clinical samples. But neither a single method exhibits a gold standard for the collection of chronic wound samples through microbiological assays nor any study has examined the effectiveness of these methods.

8.3 Molecular Methods

Metagenomic evaluations utilizing conserved 16S ribosomal RNA sequences were developed to overcome the limitations encountered by microbiological assays with direct identification of the pathogens involved in chronic wound associated biofilm formation. Development of several advanced techniques such as denaturing gradient gel electrophoresis (DGGE), which is a common method for semi-quantitative (around 40%) analysis of complex biofilm diversity within a chronic wound. DGGE separates DNA sequences having identical length but different sequences based on greater stability as observed in the case of G–C pairing over A–T pairing when exposed to the DNA denaturants (Kolbert and Persing 1999; Davies et al. 2004). Other sophisticated methods include partial ribosomal amplification and pyrosequencing (PRAPS), full ribosomal amplification, cloning and Sanger sequencing (FRACS) and partial ribosomal amplification, denaturing gradient gel electrophoresis, and Sanger sequencing (PRADS) applied individually or in combination to identify, characterize and quantify bacterial strains that escape conventional culturing technique (Dowd et al. 2008). Partial ribosomal amplification holds the capability of amplifying up to 600-bp only but full ribosomal amplification can amplify the entire 16S rRNA of a bacteria. Due to the diversity and variability of bacterial load within a chronic wound, different types of molecular methods are employed to determine bacterial identity, diversity and epidemiology (Dowd et al. 2008). These technologies are quite old in the study of environmental microbiota and are being used nowadays to extensively examine the complex microbiological structure in a chronic wound biofilm. Peptide nucleic acid (PNA) is a synthetic DNA analogue that displays more appropriate and easy nucleic acid binding and fluorescence in situ hybridization (FISH) is a popular method to detect specific DNA sequence on a chromosome. PNA-FISH is a modified and combined form of PNA and FISH with higher specificity and sensitivity, that is designed to target the 16S rRNA of specific bacterial species within a biofilm. PNA-FISH in combination with CLSM is a useful tool to separately identify each species clusters present in a tissue sample collected from a wound of heterogeneous biofilm (Malic et al. 2009; Kirketerp-Møller et al. 2008).

8.4   Other Methods

A set of methodologies falling under a particular category cannot meet the demand of accurately detecting biofilms in chronic wounds in a complex, intricate clinical scenario. Although molecular techniques appear to be quite promising in identifying the causative bacteria of biofilm, it fails to differentiate between true pathogens from other contaminations (Dowd et al. 2008). Henceforth, the newly emerging diagnostic methods for chronic wound biofilms come into play. This involves two techniques: wound blotting which is a newly proposed methodology to provide a non-invasive and precise evaluation of wound beds. Another one is a transcriptomics approach to differentiate true pathogens from other bacteria in the biofilm matrix. Next-generation sequencing of RNA (RNAseq) is a highly modern approach to identify the gene expression in a chronic wound biofilm and thus noted as a cutting-edge technique to evaluate a huge sample pool in a short time.

Acute infections are predominantly involved with planktonic phenotypes having more aggressive and rapidly dividing cells that invade the host tissues and stimulate a strong inflammatory response (Mendoza et al. 2019). Acute infections can be eradicated via the combined action of antimicrobials with cellular and humoral host immune responses. Chronic non-healing infections composed of highly tolerant bacterial biofilms require removal of the colonized device in most of the cases or surgical excision of infected tissue (Frykberg and Banks 2015). The immense contribution of biofilms in the chronicity of non-healing wounds has forced researchers to propose new guidelines for biofilm-based wound care (Schultz et al. 2017). This involves meticulous sharp debridement, application of broad-spectrum antibiotics with the administration of local antibiofilm agents in case of chronic and critically ischemic wounds (Wolcott and Rhoads 2008). Modern approaches like ultrasound-based debridement, nanoparticle therapies, phage therapy, quorum sensing inhibitors (QSIs), cationic antimicrobial peptides, phytochemicals, EPS degrading enzymes are now being implemented to manage the wound infections and subsequently to attenuate biofilm progression (Jones and Kennedy 2012; Wei et al. 2019a, b; Dhar and Han 2020). Among these, QSIs are a well-known class of antibiofilm agents that aim to disrupt biofilm formation and to impede virulence factor productions by blocking quorum sensing pathways (Kalia et al. 2014). The formation of an inactive complex with QS receptors leads to the cessation of the whole signalling pathway as found by the QSIs mode of action (El-Mowafy et al. 2014; Maisuria et al. 2016; Parai et al. 2020). There are facets of QSI limitations also which either completely or partially block the QSI activity, for example C-30 injections can negatively regulate the expression of some QS-mediated virulence genes such as *lasA*, *lasB*, *hcnAB*, *rhlAB*, *chiC*, *phnAB* and *phzABCDEFG* without hindering the expression of *lasI/R* and *rhlI/R* gene clusters (Hentzer et al. 2003). Fatty acids are currently reported to exhibit selective inhibition and/or cause complete disruption of biofilm formation, due to their wide diversity and ability to block biofilm formation by interfering with microbial virulence. They mainly act by enhancing the susceptibility of microorganisms towards other antimicrobials or sometimes by acting as a signal molecule, targeting several QS regulated gene functions (Kumar et al. 2020).

9 Implications of Biofilm on Wound Healing

The process of wound healing is a complex biological process that proceeds in an orderly manner involving four distinct but overlapping phases—homeostasis, inflammation, proliferation and remodelling. The phases involve explicit temporal interactions between numerous types of cells, extracellular matrix molecules and soluble factors (Velnar et al. 2009). The implications of biofilm on chronic wounds are not only confined to retard healing or as a financial burden, but it also can develop into an overt or persistent infection derived from the insertion of a medical prosthetic device. The situation can also give rise to antimicrobial resistance along with the increasing risk of tissue toxicity (Mendoza et al. 2019). Generally, wound healing involves an orderly transition that includes epithelialization, granulation tissue formation, inflammatory phase of angiogenesis, tissue reorganization and tissue regeneration (Dovi et al. 2004). Unlike normal wound healing where the inflammatory phase is generally short and lasts for a few days, chronic wounds exhibit an extended inflammatory phase, which ultimately delays the progression of wound healing. This prolonged inflammatory phase witnesses the release of a wide variety of cytokines and blood cells, especially neutrophils and macrophages, with changes in oxygen gradient and pH level at the wound site. Bacterial biofilms often avail a fringe benefit of this situation by manipulating cellular (leukocytes, keratinocytes, endothelial cells and fibroblasts) functions, inflammatory cellular response, cutaneous innate immune response and repair phase (Zhao et al. 2016; Omar et al. 2017). The diversity of microbial species in a bacterial biofilm enclosed within the protective layer of EPS ensures their uninterrupted presence in chronic wounds throughout infection. The inert nature of EPS safeguards the bacteria from antibiotic medicament, thereby imparting antibiotic resistance along with persistent infection. Antibiotics mainly target metabolically active and actively proliferating bacterial cells. As a result, bacteria within a biofilm matrix have reduced susceptibility towards antibiotic treatments as the biofilm is mainly dominated by slow-growing or starved dormant cells, which are usually exposed to nutrient limitation (Crabbé et al. 2019). Furthermore, spatial heterogeneity of bacterial cells within biofilms empowers an important survival strategy whereby some of the cells under different metabolic states, are more definite to survive in any metabolically directed attack (Mah and O'Toole 2001).

10 Future Perspective

Although non-healing wounds have various in vivo limitations that restrains their recovery or complete cure but biofilm remains to be one of the predominant causes behind its lack of rehabilitation. Biofilms on the other hand are difficult to eradicate due to various constraints regarding their isolation, identification and accurate remediation. Despite an array of potential treatments for biofilm-associated chronic wounds are present to clinicians, no reliable diagnostic methods are currently available for the detection and identification of biofilms. Sophisticated imaging techniques are available but they always require special equipment, highly trained personnel and precision of time. Sometimes it seems to be impractical in clinical settings on the regular basis owing to its huge percentage of patients. However, the current methods available such as wound blotting-based sharp debridement can precisely remove the biofilm-affected wound portion without causing much collateral damage to adjacent tissue. A comprehensive understanding of the chronic recurrent infection caused by biofilms and its drug-resistance mechanism can help to chalk out explicit therapeutic care. In a parallel approach, studying the host–microbe interaction on a molecular level can unlock the underlying intricate mechanisms which can further be used in understanding and improvising the strategies dealing with the chronicity of non-healing wounds. Rapid progress in the field of biofilm research along with innovative strategies may set the foundation for accurate diagnostics and effective treatment of chronic wounds. Although new antibiotic therapeutics are being constantly explored, novel findings and their applications are poles apart due to some unresolved constraints. The primarily high concentration of the therapeutics beyond the permissible limit, followed by cost-effectiveness and effectual storage potential are a few reasons. Therefore, recently virulence attenuation approaches are being examined that involve drugs that do not kill the bacteria but interfere with their ability to produce virulence factors. This in turn will enfeeble the biofilm, sensitizing the bacteria to antibiotics and eventually accelerating or resuming wound healing.

References

Algburi A, Comito N, Kashtanov D, Dicks LMT, Chikindas ML (2017) Control of biofilm formation: antibiotics and beyond. Appl Environ Microbiol 83(3):e02508–e02516

Attinger C, Randy W (2012) Clinically addressing biofilm in chronic wounds. Adv Wound Care 1:127–132

Bjarnsholt T, Alhede M, Alhede M et al (2013) The in vivo biofilm. Trends Microbiol 21:466–474

Boisvert AA, Cheng MP, Sheppard DC et al (2016) Microbial biofilms in pulmonary and critical care diseases. Ann Am Thorac Soc 13(9):1615–1623

Brady RA, Leid JG, Calhoun JH et al (2008) Osteomyelitis and the role of biofilms in chronic infection. FEMS Immunol Med Microbiol 52(1):13–22

Brandenburg KS, Calderon DF, Kierski PR et al (2018) Novel murine model for delayed wound healing using a biological wound dressing with *Pseudomonas aeruginosa* biofilms. Microb Pathog 122:30–38

Brandwein M, Steinberg D, Meshner S (2016) Microbial biofilms and the human skin microbiome. npj Biofilms Microbiomes 2:3

Ciofu O, Tolker-Nielsen T, Jensen PØ et al (2015) Antimicrobial resistance, respiratory tract infections and role of biofilms in lung infections in cystic fibrosis patients. Adv Drug Deliv Rev 85:7–23

Clark RAF (2008) Oxidative stress and 'senescent' fibroblasts in non-healing wounds as potential therapeutic targets. J Invest Dermatol 128:2361–2364

Coenye T, Honraet K, Rossel B et al (2008) Biofilms in skin infections: *Propionibacterium acnes* and acne vulgaris. Infect Disord Drug Targets 8(3):156–159

Colombo APV, Tanner ACR (2019) The role of bacterial biofilms in dental caries and periodontal and peri-implant diseases: a historical perspective. J Dent Res 98:373–385

Costerton JW, Stewart PS, Greenberg EP (1999) Bacterial biofilms: a common cause of persistent infections. Science 284:1318–1322

Crabbé A, Jensen PØ, Bjarnsholt T (2019) Antimicrobial tolerance and metabolic adaptations in microbial biofilms. Trends Microbiol 27:850–863

Davies DG (1998) The involvement of cell-to-cell signals in the development of a bacterial biofilm. Science 280:295–298

Davies D (2003) Understanding biofilm resistance to antibacterial agents. Nat Rev Drug Discov 2:114–122

Davies CE, Hill KE, Wilson MJ et al (2004) Use of 16S ribosomal DNA PCR and denaturing gradient gel electrophoresis for analysis of the microfloras of healing and nonhealing chronic venous leg ulcers. J Clin Microbiol 42:3549–3557

de la Fuente-Núñez C, Reffuveille F, Haney EF, Straus SK, Hancock REW (2014) Broad-spectrum anti-biofilm peptide that targets a cellular stress response. PLoS Pathog 10:e1004152

Dhar Y, Han Y (2020) Current developments in biofilm treatments: wound and implant infections. Eng Regen 1:64–75

Donlan RM (2001) Biofilms and device-associated infections. Emerg Infect Dis 7:277–281

Donlan RM, Costerton JW (2002) Biofilms: survival mechanisms of clinically relevant microorganisms. Clin Microbiol Rev 15:167–193

Dovi JV, Szpaderska AM, DiPietro LA (2004) Neutrophil function in the healing wound: adding insult to injury? Thromb Haemost 92:275–280

Dowd SE, Sun Y, Secor PR et al (2008) Survey of bacterial diversity in chronic wounds using pyrosequencing, DGGE, and full ribosome shotgun sequencing. BMC Microbiol 8:43

Dubern J-F, Diggle SP (2008) Quorum sensing by 2-alkyl-4-quinolones in *Pseudomonas aeruginosa* and other bacterial species. Mol BioSyst 4:882–888

El-Mowafy SA, Abd El Galil KH, El-Messery SM, Shaaban MI (2014) Aspirin is an efficient inhibitor of quorum sensing, virulence and toxins in *Pseudomonas aeruginosa*. Microb Pathog 74:25–32

Ennis WJ, Sui A, Bartholomew A (2013) Stem cells and healing: impact on inflammation. Adv Wound Care 2:369–678

Frykberg RG, Banks J (2015) Challenges in the treatment of chronic wounds. Adv Wound Care 4:560–582

Garrett TR, Bhakoo M, Zhang Z (2008) Bacterial adhesion and biofilms on surfaces. Prog Nat Sci 18:1049–1056

Gilbert P, Maira-Litran T, McBain AJ et al (2002) The physiology and collective recalcitrance of microbial biofilm communities. Adv Microb Physiol 46:202–256

Gloag ES, Turnbull L, Huang A et al (2013) Self-organization of bacterial biofilms is facilitated by extracellular DNA. Proc Natl Acad Sci 110:11541–11546

Grossart HP (2010) Ecological consequences of bacterioplankton lifestyles: changes in concepts are needed. Environ Microbiol Rep 2:706–714

Guilhen C, Forestier C, Balestrino D (2017) Biofilm dispersal: multiple elaborate strategies for dissemination of bacteria with unique properties. Mol Microbiol 105:188–210

Hall-Stoodley L, Costerton JW, Stoodley P (2004) Bacterial biofilms: from the natural environment to infectious diseases. Nat Rev Microbiol 2:95–108

Hall-Stoodley L, Stoodley P (2009) Evolving concepts in biofilm infections. Cell Microbiol 11:1034–1043

Harrison-Balestra C, Cazzaniga AL, Davis SC (2003) A wound-isolated *Pseudomonas aeruginosa* grows a biofilm *in vitro* within 10 hours and is visualized by light microscopy. Dermatol Surg 29:631–635

Hatt JK, Rather PN (2008) Role of bacterial biofilms in urinary tract infections. Curr Top Microbiol Immunol 322:163–192

Hentzer M, Wu H, Andersen JB et al (2003) Attenuation of *Pseudomonas aeruginosa* virulence by quorum sensing inhibitors. EMBO J 22:3803–3815

Hoffman LR, D'Argenio DA, MacCoss MJ et al (2005) Aminoglycoside antibiotics induce bacterial biofilm formation. Nature 436:1171–1175

Høiby N, Bjarnsholt T, Givskov M et al (2010) Antibiotic resistance of bacterial biofilms. Int J Antimicrob Agents 35:322–332

Høiby N, Bjarnsholt T, Moser C et al (2014) ESCMID guideline for the diagnosis and treatment of biofilm infections. Clin Microbiol Infect 21(Suppl 1):S1–S25

Hurlow J, Blanz E, Gaddy JA (2016) Clinical investigation of biofilm in non-healing wounds by high resolution microscopy techniques. J Wound Care 25(Suppl 9):S11–S22

Ito A, Taniuchi A, May T, Kawata K, Okabe S (2009) Increased antibiotic resistance of *Escherichia coli* in mature biofilms. Appl Environ Microbiol 75(12):4093–4100

Jamal M, Ahmad W, Andleeb S et al (2018) Bacterial biofilm and associated infections. J Chin Med Assoc 81(1):7–11

James GA, Swogger E, Wolcott R et al (2008a) Biofilms in chronic wounds. Wound Rep Regen 16:37–44

James GA, Swogger E, Wolcott R et al (2008b) Biofilms in chronic wounds. Wound Repair Regen 16:37–44

Jefferson KK (2004) What drives bacteria to produce biofilm? FEMS Microbiol Lett 236:163–173

Jensen LK, Johansen A, Jensen HE (2017) Porcine models of biofilm infections with focus on pathomorphology. Front Microbiol 8:1961

Jones CE, Kennedy JP (2012) Treatment options to manage wound biofilm. Adv Wound Care (New Rochelle) 1(3):120–126

Joo H-K, Otto M (2012) Molecular basis of *in vivo* biofilm formation by bacterial pathogens. Chem Bio 19:1503–1513

Kalia VC, Wood TK, Kumar P (2014) Evolution of resistance to quorum sensing inhibitors. Microb Ecol 68:13–23

Kaplan JB (2010) Biofilm dispersal: mechanisms, clinical implications, and potential therapeutic uses. J Dent Res 89:205–218

Keren I, Kaldalu N, Spoering A et al (2004) Persister cells and tolerance to antimicrobials. FEMS Microbiol Lett 230:13–18

Kim PJ, Steinberg JS (2012) Wound care: biofilm and its impact on the latest treatment modalities for ulcerations of the diabetic foot. Semin Vasc Surg 25:70–74

Kirketerp-Møller K, Jensen PØ, Fazli M et al (2008) Distribution, organization, and ecology of bacteria in chronic wounds. J Clin Microbiol 46:2717–2722

Kolbert CP, Persing DH (1999) Ribosomal DNA sequencing as a tool for identification bacterial pathogens. Curr Opin Microbiol 2:299–305

Krutmann J (2009) Pre- and probiotics for human skin. J Dermatol Sci 54:1–15

Kumar P, Lee JH, Beyenal H, Lee J (2020) Fatty acids as antibiofilm and antivirulence agents. Trends Microbiol 28(9):753–768

Lewis K (2005) Persister cells and the riddle of biofilm survival. Biochemistry (Mosc) 70:267–274

Li L, Mendis N, Trigui H et al (2014) The importance of the viable but non-culturable state in human bacterial pathogens. Front Microbiol 5:258

Li Y, Xiao P, Wang Y, Hao Y (2020) Mechanisms and control measures of mature biofilm resistance to antimicrobial agents in the clinical context. ACS Omega 5(36):22684–22690

Lipsky BA, Berendt AR, Paul B et al (2012) 2012 infectious diseases society of America clinical practice guideline for the diagnosis and treatment of diabetic foot infections. Clin Infect Dis 54: e132–e173

Litzler PY, Benard L, Barbier-Frebourg N et al (2007) Biofilm formation on pyrolytic carbon heart valves: influence of surface free energy, roughness, and bacterial species. J Thorac Cardiovasc Surg 134:1025–1032

Mah TFC, O'Toole GA (2001) Mechanisms of biofilm resistance to antimicrobial agents. Trends Microbiol 9:34–39

Maina IW, Patel NN, Cohen NA (2018) Understanding the role of biofilms and superantigens in chronic rhinosinusitis. Curr Otorhinolaryngol Rep 6(3):253–262

Maisuria VB, De Los Santos YL, Tufenkji N, Déziel E (2016) Cranberry-derived proanthocyanidins impair virulence and inhibit quorum sensing of *Pseudomonas aeruginosa*. Sci Rep 6:30169

Malic S, Hill KE, Hayes A et al (2009) Detection and identification of specific bacteria in wound biofilms using peptide nucleic acid fluorescent *in situ* hybridization (PNA FISH). Microbiology (Reading) 155:2603–2611

Marić S, Vranes J (2007) Characteristics and significance of microbial biofilm formation. Period Biol 109:115–121

Mccarty SM, Percival SL (2013) Proteases and delayed wound healing. Adv Wound Care 2:438–447

Mendoza RA, Hsieh J-C, Galiano RD (2019) The impact of biofilm formation on wound healing. In: Wound healing–current perspectives. IntechOpen, London, UK, pp 235–250

Metcalf DG, Bowler PG (2013) Biofilm delays wound healing: a review of the evidence. Burns Trauma 1:5–12

Metcalf DG, Bowler PG, Hurlow J (2014) A clinical algorithm for wound biofilm identification. J Wound Care 23:137–142

Morasso MI, Tomic-canic M (2005) Epidermal stem cells: the cradle of epidermal determination, differentiation and wound healing. Biol Cell 97:173–183

Olsen I (2015) Biofilm-specific antibiotic tolerance and resistance. Eur J Clin Microbiol Infect Dis 34:877–886

Omar A, Wright JB, Schultz G et al (2017) Microbial biofilms and chronic wounds. Microorganisms 5:9

Parai D, Banerjee M, Dey P, Mukherjee SK (2020) Reserpine attenuates biofilm formation and virulence of *Staphylococcus aureus*. Microb Pathog 138:103790

Parai D, Banerjee M, Dey P et al (2018) Effect of reserpine on *Pseudomonas aeruginosa* quorum sensing mediated virulence factors and biofilm formation. Biofouling 34:320–334

Percival SL, Hill KE, Malic S et al (2011) Antimicrobial tolerance and the significance of persister cells in recalcitrant chronic wound biofilms. Wound Repair Regen 19:1–9

Römling U, Balsalobre C (2012) Biofilm infections, their resilience to therapy and innovative treatment strategies. J Intern Med 272:541–561

Sauer K, Camper AK, Ehrlich GD et al (2002) *Pseudomonas aeruginosa* displays multiple phenotypes during development as a biofilm. J Bacteriol 184:1140–1154

Schmidt I, Gad A, Scholz G et al (2017) Gold-modified indium tin oxide as a transparent window in optoelectronic diagnostics of electrochemically active biofilms. Biosens Bioelectron 94:74–80

Schreml S, Szeimies RM, Prantl L et al (2010) Oxygen in acute and chronic wound healing. Br J Dertol 163:257–268

Schultz G, Bjarnsholt T, James GA et al (2017) Consensus guidelines for the identification and treatment of biofilms in chronic nonhealing wounds. Wound Repair Regen 25:744–757

Stanley A, Osler T (2001) Senescence and the healing rates of venous ulcers. J Vasc Surg 33:1206–1211

Stewart PS, William Costerton JW (2001) Antibiotic resistance of bacteria in biofilms. Lancet 358:135–138

Stoodley P, Sauer K, Davies DG, Costerton JW (2002) Biofilms as complex differentiated communities. Annu Rev Microbiol 56:187–209

Sutherland IW (2001) Biofilm exopolysaccharides: a strong and sticky framework. Microbiol 147:3–9

Teschler JK, Zamorano-Sánchez D, Utada AS et al (2015) Living in the matrix: assembly and control of *Vibrio cholerae* biofilms. Nat Rev Microbiol 13:255–268

Thomson CH (2011) Biofilms: do they affect wound healing? Int Wound J 8:63–67

Thornton RB, Rigby PJ, Wiertsema SP et al (2011) Multi-species bacterial biofilm and intracellular infection in otitis media. BMC Pediatr 11:94

Tribedi P, Sil AK (2014) Cell surface hydrophobicity: a key component in the degradation of polyethylene succinate by *Pseudomonas* sp. AKS2. J Appl Microbiol 116:295–303

Vasudevan R (2014) Biofilms: microbial cities of scientific significance. J Microbiol Exp 1:84–89

Veerachamy S, Yarlagadda T, Manivasagam G et al (2014) Bacterial adherence and biofilm formation on medical implants: a review. Proc Inst Mech Eng H 228:1083–1099

Velnar T, Bailey T, Smrkolj V (2009) The wound healing process: an overview of the cellular and molecular mechanisms. J Int Med Res 37:1528–1542

Wei D, Zhu X-M, Chen Y-Y et al (2019a) Chronic wound biofilms: diagnosis and therapeutic strategies. Chin Med J 132:2737–2744

Wei D, Zhu XM, Chen YY et al (2019b) Chronic wound biofilms: diagnosis and therapeutic strategies. Chin Med J 132(22):2737–2744

Whitchurch CB, Tolker-Nielsen T, Ragas PC et al (2002) Extracellular DNA required for bacterial biofilm formation. Science 295:1487

Wolcott RD, Rhoads DD (2008) A study of biofilm-based wound management in subjects with critical limb ischaemia. J Wound Care 17:145–155

Wright JB, Lam K, Buret AG et al (2002) Early healing events in a porcine model of contaminated wounds: effects of nanocrystalline silver on matrix metalloproteinases, cell apoptosis, and healing. Wound Repair Regen 10:141–151

Wright JB, Lam K, Hansen D et al (1999) Efficacy of topical silver against fungal burn wound pathogens. Am J Infect Control 27:344–350

Wu Y-K, Cheng N-C, Cheng C-M (2019) Biofilms in chronic wounds: pathogenesis and diagnosis. Trends Biotechnol 37:505–517

Yadav MK, Song J-J, Singh BP et al (2020) Microbial biofilms and human disease: a concise review. In: New and future developments in microbial biotechnology and bioengineering: microbial biofilms, 1st edn. Elsevier, pp 1–13

Yin W, Wang Y, Liu L et al (2019) Biofilms: the microbial "protective clothing" in extreme environments. Int J Mol Sci 20:3423

Zhao R, Liang H, Clarke E et al (2016) Inflammation in chronic wounds. Int J Mol Sci 17:2085

The Potential of Essential Oils as Topical Antimicrobial Agents in the Age of Artificial Intelligence

Polly Soo Xi Yap, Rabiha Seboussi, Kok Song Lai, and Swee Hua Erin Lim

1 Introduction

The ability to effectively heal wounds is the key to our survival because a slow healing wound will continuously expose the devitalized tissue to colonization and establishment of a wide variety of endogenous and potentially harmful microorganisms, which can result in life-threatening infections. In this case, surgical site infection (SSI) can result in delays in wound healing, impaired cosmetic outcome, and increased healthcare cost (Badia et al. 2017). Topical antibiotics are applied to surgical wounds with dressing after surgery to reduce the risk of SSI. A clinician may choose to prescribe a topical antibiotic on a wound after careful consideration of the underlying infective process and assessing the risks and benefits of treatment. As the action of topical antibiotics is localized, there is a reduced likelihood for undesirable effects that could affect the body systemically, exhibiting symptoms such as nausea and diarrhea. Therefore, effective wound dressing loaded with antimicrobial agents after surgery is necessary to deal with such pathological conditions, with the aim of reducing SSIs. The main function of antibiotics in wound healing is to prevent the growth of pathogens causing the infection and potentially, to promote healing (Altoe et al. 2019). However, the use of antibiotics in modern medicine has given rise to the development of antimicrobial resistance

P. S. X. Yap
Jeffrey Cheah School of Medicine and Health Sciences, Monash University Malaysia, Bandar Sunway, Selangor Darul Ehsan, Malaysia

R. Seboussi
Al Ain Men's College, Higher Colleges of Technology, Al Ain, United Arab Emirates

K. S. Lai · S. H. E. Lim (✉)
Abu Dhabi Women's College, Higher Colleges of Technology, Abu Dhabi, United Arab Emirates
e-mail: lerin@hct.ac.ae

© The Author(s), under exclusive license to Springer Nature Singapore Pte Ltd. 2021
P. Kumar, V. Kothari (eds.), *Wound Healing Research*,
https://doi.org/10.1007/978-981-16-2677-7_22

(AMR). Antibiotic- and multidrug-resistant isolates are increasingly reported in infected wounds (Godebo et al. 2013; Pirvanescu et al. 2014) and one of the common causes of community-acquired infections (van Duin and Paterson 2016). Hence, wound healing and infection without effective antimicrobials in place pose a formidable medical challenge.

The emergence of AMR is accelerated by injudicious use of antibiotics; thus there remains an urgent necessity for novel or alternative antimicrobials to fight against bacterial evolution that continues to resist new drug classes. Due to concerns with regard to AMR spread, natural products, particularly, essential oils (EOs) are viewed as a promising yet understudied group of potential antimicrobial agents that can be incorporated into wound dressings. Over the last decade, many studies have been conducted to explore the use of natural products as alternative therapies, for example, testing against resistant bacteria (Moo et al. 2020; Yang et al. 2019; Yap et al. 2013) and developing formulations for antimicrobial wound dressing (Edwards-Jones et al. 2004; Pereira Dos Santos et al. 2019). However, conventional drug screening methods are too slow as opposed to the more rapidly developing levels of resistance, notwithstanding other scientific obstacles such as molecular target validation (Hughes et al. 2011), endogenous resistance potential as well as drug permeability and intracellular accumulation (Silver 2011). Advances in genetics, genomics, and artificial intelligence (AI) represent a new paradigm for drug discovery to combat AMR. In this chapter, we aim to explore the potential of taking AI-based approaches as applied in the screening of novel antimicrobial peptides (AMPs), and applying these approaches on topical antimicrobial discovery from EOs. While majority of the essential oil screening studies remain in vitro and in vivo, computational and statistical framework involving AI, machine learning, and deep learning models is a step forward for in silico research to help guide EO drug discovery. This is in tandem with increased chances of success in the wet laboratory that can be performed after predictive models have been established in silico.

2 Practical Outlook for Essential Oils Drug Discovery in Topical Wound Care

The wide range of therapeutic properties such as bactericidal, virucidal, and fungicidal effects of EOs has garnered considerable attention of great medicinal importance (Swamy et al. 2016). Their structural diversity and complex chemical composition with about 20–60 different bioactive compounds present in many of these EOs fit the multitarget hypothesis in which they are not subject to contributing to resistance by this very attribute of possessing multiple molecular targets (Silver 2011). A literature survey was performed on both PubMed® and Google Scholar, with search keywords including "essential oil," "antibacterial activity," "induction of resistance*", and "reduce susceptibility*" (https://pubmed.ncbi.nlm.nih.gov/ and https://scholar.google.com/, accessed 2020 October 25). No evidence of resistance for major clinically important pathogens has yet to be reported on EOs. Most EOs are composed of only two to three major compounds which are present in fairly high proportions (20–70%) and other minor compounds that present in low amounts. EOs are composed of mainly compounds from the terpenes family and other aromatic and aliphatic constituents (Swamy et al. 2016). Despite exhibiting a wide range of pharmacophores that offer great potential for physiochemical properties, EOs are often excluded in drug discovery (Lipinski 2016), partially due to their volatility and hydrophobicity (Feyaerts et al. 2020). It is also important to note that the chemical constituents of plant EOs can be affected by various factors such as geographical location, climate, and stage of maturity; these factors subsequently affecting the antimicrobial properties (Pichersky et al. 2006). Additionally, difference in extraction methods is also related to difference in the stereochemical properties of the EOs (Lahlou 2004). Hence, a highly integrated interdisciplinary approach is necessary to assess the drug-likeness potential of EOs and the components in silico, to ensure fewer false positives and late-stage attritions in lead development. Although EOs were known to have possessed some undesirable characteristics that result in their exclusion from Lipinski's well-known Rule of Five (Ro5) as drug candidates with good oral bioavailability (Lipinski 2016), they have some unique properties that make them ideal for topical applications. One of the key challenges for topical antimicrobial agents is that the highest concentration may remain at the skin surface and decline in the subcutaneous fat, hence, there is a potential that the local concentrations may not remain above the minimal inhibitory concentrations for high-level resistant strains. Natural terpenes are widely recognized for use as safe transdermal permeation enhancers in the pharmaceutical industry (Fox et al. 2011; Gao and Singh 1998). Previous studies have indicated that terpenes are a favorable group of compounds for transdermal administration together with lipophilic and hydrophilic topical antimicrobials in wound healing with the purpose of maximizing therapeutic effects (AbdelSamie et al. 2016; Sims et al. 2018). It is because the principal step for the dressing of infectious wounds is to prevent infection by delivering sufficiently high concentrations of antibiotics to the wound site; at the same time being biocompatible without exerting adverse effects to the surrounding tissue.

Other than enhancing the antibiotic efficacy, real potential of EOs as antimicrobial molecules is still under-scrutinized with present-day methods. As noted above, while vast majority of the EOs major compounds have so far been identified and characterized (Dhifi et al. 2016), conventional screening of EOs for drug discovery is likely to produce a library of previously seen compounds. Thus, the AI-based screening perspective should focus on models of seeing which had not previously been seen.

3 Applications of Artificial Intelligence for Antimicrobial Resistance

The commonly used framework for predictive antimicrobial testing and drug discovery includes families of algorithms of both supervised and unsupervised learning methods, including naïve Bayes (NB), Support Vector Machines (SVM), and artificial neural networks (ANN) (Table 1).

3.1 Prediction of Antimicrobial Resistance

Currently, there are two main conventional methods to detect AMR. The first one is antimicrobial susceptibility testing (AST) and establishment of the minimal inhibitory concentration (MIC) (Vasala et al. 2020), and second one is an emerging method, whole-genome sequencing method for antimicrobial susceptibility testing (WGS-AST) (Su et al. 2019; Vasala et al. 2020). The classic AST technique only allows quantification of AMR levels, but it does not explain the resistance mechanisms at mechanistic or molecular levels. On the other hand, WGS-AST is a robust and accurate tool for AMR detection with insights into the resistance mechanisms and evolution of the pathogens (Ellington et al. 2017). This is achieved by the growing effort of various types of AMR phenotypes and genome-based databases, such as complete genome multilocus sequence typing (cgMLST) (Feijao et al. 2018), Pathosystems Resource Integration Center (PATRIC, www.patricbrc.org) (Wattam et al. 2014; Wattam et al. 2017), Comprehensive Antibiotic Resistance Database (CARD, https://card.mcmaster.ca/) (Alcock et al. 2020; Jia et al. 2017), ARG-ANNOT (Antibiotic Resistance Gene-ANNOTation) (Gupta et al. 2014), and Virulence Factor Database (VFDB, http://www.mgc.ac.cn/VFs/main.htm) (Chen et al. 2016; Liu et al. 2019). Detecting AMR phenotypes from bacterial genomic data employs basic concepts in the genome-wide association studies (GWAS), i.e., to gain knowledge of the changes of DNA sequences, or mutations that present among bacterial strains and understand how these variations impact the phenotypes. Researchers have since build phenotype identification algorithms based on variants for some clinically important bacteria, such as *Escherichia coli, Klebsiella pneumoniae,* and *Staphylococcus aureus* (Gordon et al. 2014; Stoesser et al. 2013). Davis et al. have computed a k-mer-based machine learning algorithm to classify the AMR SIR (Susceptible, Intermediate, or Resistant) of *Acinetobacter baumannii, Mycobacterium tuberculosis, S. aureus,* and *Streptococcus pneumoniae* on a wide range of antibiotics including carbapenem, beta-lactam, and co-trimoxazole, from the metadata in PATRIC (Davis et al. 2016). On the other hand, a protein sequence-based machine learning algorithm was established to classify the Gram-negative bacteria for acetyltransferase (*aac*), beta-lactamase (*bla*), and dihydrofolate reductase (*dfr*). The study considered both AMR and non-AMR bacterial protein sequences by using game theory to identify protein

Table 1 Artificial intelligence algorithms and predictive models for antimicrobial drug discovery

Type	Algorithm	Resource	Description	Website	Ref.
Unsupervised learning	Principal Component Analysis (PCA)	Scikit-learn	Cluster analysis to identify representative compounds with strong in vitro antimicrobial activity	https://scikit-learn.org/stable/	(Ragno et al. 2020)
	Composite non-negative matrix factorization (cNMF)	Ianevski et al.	Prediction of drug combination synergy	–	(Ianevski et al. 2019)
Semi-supervized	Hedge (Adaboost)	Davis et al.	Species-specific AMR phenotype and genotype prediction with RAST and PATRIC annotation services	–	(Davis et al. 2016)
Supervized learning	Support Vector Machine (SVM)	Chowdhury et al.	Game theory approach to identify AMR genes	–	(Chowdhury et al. 2019)
		Artini et al.	Binary classification model to identify chemical components involved in the inhibition of biofilm formation	–	(Artini et al. 2018)
		Patsilinakos et al.	Binary classification model to identify chemical components responsible for inhibition of biofilm formation	–	(Patsilinakos et al. 2019)
	Artificial neural networks (ANN)	Fast Artificial Neural Network (FANN) library	Prediction of antimicrobial activities based on chemical composition	http://leenissen.dk/fann/wp/	(Daynac et al. 2015)
		ALOGPS v 2.0	Interactive online prediction of logP	http://www.vcclab.org/lab/alogps/	(Tetko et al. 2001)
		DeepTox	Computational method for toxicity prediction	http://www.bioinf.jku.at/	(Mayr et al. 2016)

(continued)

Table 1 (continued)

Type	Algorithm	Resource	Description	Website	Ref.
				research/ DeepTox	
	ANN-genetic algorithm	Rajkovic et al.	Process variables optimization for EO mixture anti-microbial testing	–	(Rajkovic et al. 2015)

features for the machine learning model. Finally, a supervised method, SVM, was used to predict AMR genes (Chowdhury et al. 2019).

AI-based computational tools have also been applied to predict the antimicrobial activity of EOs or to classify the EOs based on their antimicrobial activities (Su et al. 2019). As mentioned earlier, the inherent variability in composition in EOs has been a serious concern in drug screening. To overcome this, an ANN approach has been developed to select EOs with comparable antimicrobial activities, without considering their chemical variations. Daynac et al. screened several EOs chemical compositions with corresponding AST data against *S. aureus, E. coli, Clostridium perfringens,* and *Candida albicans* and retained only compounds with known antimicrobial properties (Daynac et al. 2015). The study demonstrated that the fast artificial neural network (FANN) software was able to predict the antimicrobial activities of two or more microorganisms simultaneously while the most accurate prediction was observed for *S. aureus* (Daynac et al. 2015). Rajkovic et al. showed that model based on ANN incorporated with the genetic algorithm were able to allow optimization for the experimental variables (e.g., test duration, test compound concentrations, and mass ratio), and at the same time provided good prediction accuracy for the antifungal effects of *Thymus vulgaris L.* and *Cinnamomum cassia L.* EOs (Rajkovic et al. 2015).

A growing amount of scientific and clinical evidence has implicated that bacterial biofilm formation plays a significant role in impeding wound healing (Metcalf and Bowler 2013) and it is also one of the main resistance mechanisms that make a bacterial infection hard to eradicate (Sharma et al. 2019). In general, Gram-positive bacteria such as *S. aureus, S. epidermidis,* and *Streptococcus pyogenes* are predominant species identified in the early stage of SSIs, while Gram-negative bacteria such as *E. coli, K. pneumoniae,* and *P. aeruginosa* are mainly responsible for the chronic infections (Negut et al. 2018; Simoes et al. 2018). Many experimental data has pointed to the unique anti-biofilm potential of EOs against both Gram-positive and Gram-negative bacteria (Bilcu et al. 2014; Cabarkapa et al. 2019; Firmino et al. 2018; Kumari et al. 2017; Lagha et al. 2019). Therefore, EOs with anti-biofilm activity pose a better chance as good drug candidates for wound dressing and treating infections. Experimentally, the exact mechanism on how or which EO components influenced biofilm production remains unclear and understudied. A few studies have demonstrated how AI-based applications can solve the complex matrix of EOs compositions and experimental anti-biofilm potencies leading to identification of chemical components that are mainly responsible for the anti-biofilm activity or prediction of EOs that can inhibit bacterial growth (Artini et al. 2018; Patsilinakos et al. 2019; Ragno et al. 2020).

One of the limitations of genome-based AMR prediction is that, the predictive power is dependent on the accuracy and consistency of the culture-based AST data for model training. Thus, efforts should be put on standardizing and maintaining the quality of the data for phenotypic testing so that the prediction outcomes do not diverge over time. Additionally, it is necessary to establish a comprehensive database for multiple bacterial species, because a large training dataset is required to optimize the key parameters for specific species and infections.

3.2   AI-Based Synergy Testing

The approach of combining drugs has attracted much interest due to some promising positive outcomes, such as synergizing therapeutic efficacy, reducing side effects, and preventing the emergence of AMR. Many studies have also demonstrated the potential of natural products to act synergistically with antibiotics (Khameneh et al. 2019; Naghmouchi et al. 2012; Pizzolato-Cezar et al. 2019; Yap et al. 2013), aiming to reduce antibiotic burden and restore the activity of the antibiotic. Antimicrobial peptides (AMPs) remain at the forefront of the arena in antimicrobial discoveries (Magana et al. 2020), in which they are key components of the body's innate immune system for defense against a plethora of pathogens. AMPs and the derivatives also offer great potential in terms of chemical space and they are widely distributed in nature including unconventional sources such as unculturable soil and marine bacteria (Magana et al. 2020). However, it is time-consuming and labor-intensive to identify optimal drug-natural product combinations from a large number of possible spaces (e.g., medicinal plants, Eos, and antimicrobial peptides). Hence, AI-based technologies offer a new paradigm to speed up and optimize the drug screening process. The enormous amount of natural product-based compounds available and the diversity of plausible combinations have resulted in the accumulation of the ever-expanding and highly diversified volume of low- and high-throughput experimental data. While a growing number of reports claiming to have identified promising synergisms and the respective modes of action (Cho et al. 2020; Yang et al. 2017; Yang et al. 2020), the challenge remains to find the most optimistic combination and mechanism of action to tackle specific infections and to mitigate specific resistance mechanisms in the context of lead discovery. Hence, mining the bibliome for experimental antimicrobial combinatorial data provides scientists another elevated dimension to evaluate the most promising combination on test. Existing text mining tools have also been demonstrated to be useful for extracting and reconstructing pharmacokinetic drug–drug interactions from experimental evidences and clinical text (Iyer et al. 2014; Kolchinsky et al. 2015). Jorge et al. were inspired by these computational frameworks and an iterative workflow to curate AMP and antibiotic combinations tested against major pathogens was developed (Jorge et al. 2012; Jorge et al. 2016). As a result, the curated database (http://sing.ei.uvigo.es/antimicrobialCombination/.) encompasses 1556 combinations (345 AMPs and 282 drugs), tested against *P. aeruginosa, S. aureus, E. coli, Listeria monocytogenes,* and *C. albicans* (Jorge et al. 2016). Although this workflow has been integrated between the state-of-art text mining and expert manual curation, it offers great potential to further investigate the utilization of deep learning approaches to speed up the manual curation process, and possibly for application on EOs antimicrobial testing in literature and test on pathogens that frequently cause wound infections.

Lanevski et al. employed a novel unsupervised learning algorithm, composite non-negative matrix factorization (cNMF), and significantly reduce the number of drug combination experiments, at the same time maintain high accuracy, by using the drug combination response prediction (DECREASE) model (Ianevski et al. 2019). The output of DECREASE is expressed in the predicted combination response matrices which are compatible with SynergyFinder (Ianevski et al. 2017) and Combenefit (Di Veroli et al. 2016) for synergy scoring and determination. Intriguingly, the only input needed for the DECREASE model is a submatrix of the drug combination dose–response measurements, without the need for synergistic mechanisms, or structural and/or target information. Such mechanism-unbiased approach is therefore posing great potential in identifying new synergies.

Widespread interest of AMPs have yielded unprecedented amount of AI based AMP databases and platforms (Cardoso et al. 2019; Lv et al. 2020), and this no doubt significantly contribute to training the AI models more precisely. Nevertheless, most AI-based AMP approaches are binary classification models that do not take continuous activity information into consideration. While the evolution of resistant bacteria is growing at an alarming rate, an AI model which can understand and predict the activity of drugs continuously against specific bacteria will be valuable for future study. On the other hand, AMPs are composed of bioactive small proteins, which allow sequence-based training and prediction that offers high precision (Bhadra et al. 2018) and room for drug design and optimization (Magana et al. 2020). In comparison, EOs stand a greater chance in fragment-based drug discovery; while the experimental data and phytochemical database of EOs remain largely unstructured. Hence, a substantial amount of public AI-based EOs databases are warranted.

3.3 Drug Discovery Parameters and Filters

Ideally, compounds with a possible dermal application must be able to cross the lipids layer barrier and be soluble in both lipophilic and hydrophilic environments. The *n*-octanol/water partition coefficient (denoted as log *P*) is often used as a gold standard to evaluate the compound properties (Valko 2004). However, laboratory determination of log *P* is not only time and resource consuming; it is also complicated due to the complexity of the real interaction between human skin tissue and the test compound (Kosina et al. 2018). Hence, many nonexperimental approaches employing mainly quantitative structure–activity relationship (QSAR) techniques and multiple linear regression analysis have since been developed for the estimation of log P (Katritzky et al. 2000). Tetko et al. developed a method, ALOGPS v2.0 (http://www.vcclab.org/lab/alogps/) to assess the n-octanol/water partition coefficient on the basis of neural networks trained using 12,908 molecules available from PHYSPROP database of Syracuse Research Corporation (Tetko et al. 2001). When comparing the prediction ability of ALOGPS to other log P prediction programs, it was concluded that the prediction performance was primarily determined by the diverseness of the molecular structures subjected to train the methods rather than the design of the methods or algorithms (Tetko et al. 2001). This study also demonstrated the effectiveness of using heuristics on neural networks to obtain important results without tapping into the complex problems of solvation theory for log P determination. Combination of experimental (chromatographic) and computational (ALOGPS v2.1) methods have been applied to study the influence of log P on wound healing, and it was found that compounds with higher lipophilicity (higher log P) exerted greater wound healing activity (Bakht et al. 2014). Feyaerts et al. subsequently extrapolated the ALOGPS model alongside with structure–activity relationship analyses from public databases (such as PubChem and ChemSpider) to predict the bioavailability of EOs (Feyaerts et al. 2020). By using computational and statistical methods, the study ascertained that, contrary to the established drug discovery filters such as Ro5 and Lead Likeness, most EO components passed the stringent criteria for good drug candidates (Feyaerts et al. 2020).

Early identification of potential toxicity in drug discovery is crucial to prevent significant financial and resource loss in late-stage attritions. Many AI-based predictive algorithms have been developed for this purpose to flag potentially toxic compounds. The rise of deep learning, a machine learning method based on multilayer perceptron networks outperformed traditional machine learning approaches (e.g., naïve Bayes, SVM, and random forests) in accurately predicting toxicity in the Tox21 Data Challenge (Mayr et al. 2016). As a result, DeepTox pipeline (http://www.bioinf.jku.at/research/DeepTox/) ensemble a multitask deep neural networks model for toxicity prediction was developed. The pipeline uncovered not only previously known toxicophores but also novel, previously undiscovered toxicophores, suggesting that deep learning algorithms could potentially unfold new chemical knowledge (Mayr et al. 2016). In the context of topical applications, several classification models derived from machine learning algorithms (e.g., classification and regression tree, SVM, and ANN) have been developed to predict the skin sensitization potency (Wilm et al. 2018). However, training datasets of these models were primarily based on the rodent local lymph node assay while human experimental data remain scarce, let alone well-characterized, high-quality datasets as compiled by Hoffmann et al. (2018). For these reasons, the predictive power of skin sensitization on human health is limited by insufficient data for model training in both theoretical and experimental approaches.

4   Current Understanding and Future Perspective

AI-based technologies could significantly reduce the number of possible drug candidates, in vitro and in vivo experiments, thus accelerating the drug discovery process with less cost than conventional approaches. The practice of AI is said to depend largely (~80%) on processing and cleaning the aggregated data and relatively less on the algorithm application (~20%) (Vamathevan et al. 2019). Thus, the predictive power of any AI approach relies heavily on the availability of high-quality dataset in large volumes. Training data for the algorithm needs to be accurate, and as complete as possible in order to maximize predictability. Therefore, in order to optimize key parameters specifically for EOs exploration, many public databases on commercial and non-commercial EOs have to be built to prevent possible selection bias. AI-based methods ensemble methodological and experimental data mining also prone to a high risk of bias. Various important information about the experimental methods such as natural products extraction method, antibiotic, statistical analysis, and animal model, may collectively compromise the reproducibility of the studies. Such opinion has been formed on the basis of studies performed largely in vitro during the last two decades that may have oversight when these bacteria are investigated together with the wounds. It should be emphasized that the studies reported did not investigate specifically the antimicrobial effects of the EOs on wound healing but rather the in vitro activities.

Data in various formats, such as chemical compounds, genomic and protein sequences, AST, and literature text, have been employed to train different AI-based models. Whether the type of data is even appropriate and what data should be experimentally generated and curated are also the fundamental considerations for AI-based technologies. AI-based applications are more powerful when trained by data curated with good annotations in a systematic manner, to minimize the effect of noisy training (Vamathevan et al. 2019). As discussed above, many applications presented had been developed from largely unstructured sources with variable data quality. These methods may also be subjected to bias during selection of parameters applied for data processing and cleaning. Thus, ongoing efforts should be emphasized to generate good quality open annotated data to foster the application of AI in EOs for drug discovery. Next, a sufficiently large set of experimental data relating to pharmacological endpoint of EOs for the validation of AI models is important. As had been previously mentioned, the most important part of any AI-based approach is the quality and quantity of the underlying data used to train the models. Unlike AMPs, the limited availability of data on EOs and the compounds that make the final stage into approved drugs has limited the number of pharmacological end-points forecasted by the available machine learning tools. Looking at the increasing incidences of SSI associated with AMR, there is currently an unmet need for in silico predictive tools for topical medicine drug discovery targeting wound infections.

Among the machine learning methods, emphasis has been given to the QSAR, which applies physicochemical properties to predict the biological activity of the

AMPs or the peptide sequences (Cardoso et al. 2019). It is hypothesized that, the integration of the AI-based methods, especially the antimicrobial patterns would improve the output quality (Porto et al. 2018). This hypothesis was tested on an algorithm, named Joker, which combined the AMPs sequences with linguistic model-based antimicrobial patterns, and hit searches were improved against bacteria (Porto et al. 2018). There is also a gap between selecting drug candidates that fulfil drug discovery criteria and at the same be time effective in treating resistant bacteria in wound infections. AI-based approaches that are applied to predict AST results and drug synergism, coupled with QSAR, have the potential to improve the predictive analytics and aid decision-making along the drug discovery pipeline.

Acknowledgment The authors would like to acknowledge the HCT Interdisciplinary Research Grant (113118) from the Higher Colleges of Technology for supporting this study.

References

AbdelSamie SM, Kamel AO, Sammour OA, Ibrahim SM (2016) Terbinafine hydrochloride nanovesicular gel: in vitro characterization, ex vivo permeation and clinical investigation. Eur J Pharm Sci 88:91–100. https://doi.org/10.1016/j.ejps.2016.04.004

Alcock BP, Raphenya AR, Lau TTY, Tsang KK, Bouchard M, Edalatmand A, Huynh W, Nguyen AV, Cheng AA, Liu S, Min SY, Miroshnichenko A, Tran HK, Werfalli RE, Nasir JA, Oloni M, Speicher DJ, Florescu A, Singh B, Faltyn M, Hernandez-Koutoucheva A, Sharma AN, Bordeleau E, Pawlowski AC, Zubyk HL, Dooley D, Griffiths E, Maguire F, Winsor GL, Beiko RG, Brinkman FSL, Hsiao WWL, Domselaar GV, McArthur AG (2020) CARD 2020: antibiotic resistome surveillance with the comprehensive antibiotic resistance database. Nucleic Acids Res 48(D1):D517–D525. https://doi.org/10.1093/nar/gkz935

Altoe LS, Alves RS, Sarandy MM, Morais-Santos M, Novaes RD, Goncalves RV (2019) Does antibiotic use accelerate or retard cutaneous repair? A systematic review in animal models. PLoS One 14(10):e0223511. https://doi.org/10.1371/journal.pone.0223511

Artini M, Patsilinakos A, Papa R, Bozovic M, Sabatino M, Garzoli S, Vrenna G, Tilotta M, Pepi F, Ragno R, Selan L (2018) Antimicrobial and Antibiofilm activity and machine learning classification analysis of essential oils from different Mediterranean plants against *Pseudomonas aeruginosa*. Molecules 23(2). https://doi.org/10.3390/molecules23020482

Badia JM, Casey AL, Petrosillo N, Hudson PM, Mitchell SA, Crosby C (2017) Impact of surgical site infection on healthcare costs and patient outcomes: a systematic review in six European countries. J Hosp Infect 96(1):1–15. https://doi.org/10.1016/j.jhin.2017.03.004

Bakht MA, Alajmi MF, Alam P, Alam A, Alam P, Aljarba TM (2014) Theoretical and experimental study on lipophilicity and wound healing activity of ginger compounds. Asian Pac J Trop Biomed 4(4):329–333. https://doi.org/10.12980/APJTB.4.2014C1012

Bhadra P, Yan J, Li J, Fong S, Siu SWI (2018) AmPEP: sequence-based prediction of antimicrobial peptides using distribution patterns of amino acid properties and random forest. Sci Rep 8 (1):1697. https://doi.org/10.1038/s41598-018-19752-w

Bilcu M, Grumezescu AM, Oprea AE, Popescu RC, Mogosanu GD, Hristu R, Stanciu GA, Mihailescu DF, Lazar V, Bezirtzoglou E, Chifiriuc MC (2014) Efficiency of vanilla, patchouli and ylang ylang essential oils stabilized by iron oxide@C14 nanostructures against bacterial adherence and biofilms formed by *Staphylococcus aureus* and *Klebsiella pneumoniae* clinical strains. Molecules 19(11):17943–17956. https://doi.org/10.3390/molecules191117943

Cabarkapa I, Colovic R, Duragic O, Popovic S, Kokic B, Milanov D, Pezo L (2019) Anti-biofilm activities of essential oils rich in carvacrol and thymol against *Salmonella enteritidis*. Biofouling 35(3):361–375. https://doi.org/10.1080/08927014.2019.1610169

Cardoso MH, Orozco RQ, Rezende SB, Rodrigues G, Oshiro KGN, Candido ES, Franco OL (2019) Computer-aided Design of Antimicrobial Peptides: are we generating effective drug candidates? Front Microbiol 10:3097. https://doi.org/10.3389/fmicb.2019.03097

Chen L, Zheng D, Liu B, Yang J, Jin Q (2016) VFDB 2016: hierarchical and refined dataset for big data analysis--10 years on. Nucleic Acids Res 44(D1):D694–D697. https://doi.org/10.1093/nar/gkv1239

Cho TJ, Park SM, Yu H, Seo GH, Kim HW, Kim SA, Rhee MS (2020) Recent advances in the application of antibacterial complexes using essential oils. Molecules 25(7). https://doi.org/10.3390/molecules25071752

Chowdhury AS, Call DR, Broschat SL (2019) Antimicrobial resistance prediction for gram-negative bacteria via game theory-based feature evaluation. Sci Rep 9(1):14487. https://doi.org/10.1038/s41598-019-50686-z

Davis JJ, Boisvert S, Brettin T, Kenyon RW, Mao C, Olson R, Overbeek R, Santerre J, Shukla M, Wattam AR, Will R, Xia F, Stevens R (2016) Antimicrobial resistance prediction in PATRIC and RAST. Sci Rep 6:27930. https://doi.org/10.1038/srep27930

Daynac M, Cortes-Cabrera A, Prieto JM (2015) Application of artificial intelligence to the prediction of the antimicrobial activity of essential oils. Evid Based Complement Alternat Med 2015:561024. https://doi.org/10.1155/2015/561024

Dhifi W, Bellili S, Jazi S, Bahloul N, Mnif W (2016) Essential Oils' chemical characterization and investigation of some biological activities: A critical review. Medicines (Basel) 3(4). https://doi.org/10.3390/medicines3040025

Di Veroli GY, Fornari C, Wang D, Mollard S, Bramhall JL, Richards FM, Jodrell DI (2016) Combenefit: an interactive platform for the analysis and visualization of drug combinations. Bioinformatics 32(18):2866–2868. https://doi.org/10.1093/bioinformatics/btw230

Edwards-Jones V, Buck R, Shawcross SG, Dawson MM, Dunn K (2004) The effect of essential oils on methicillin-resistant *Staphylococcus aureus* using a dressing model. Burns 30(8):772–777. https://doi.org/10.1016/j.burns.2004.06.006

Ellington MJ, Ekelund O, Aarestrup FM, Canton R, Doumith M, Giske C, Grundman H, Hasman H, Holden MTG, Hopkins KL, Iredell J, Kahlmeter G, Koser CU, MacGowan A, Mevius D, Mulvey M, Naas T, Peto T, Rolain JM, Samuelsen O, Woodford N (2017) The role of whole genome sequencing in antimicrobial susceptibility testing of bacteria: report from the EUCAST subcommittee. Clin Microbiol Infect 23(1):2–22. https://doi.org/10.1016/j.cmi.2016.11.012

Feijao P, Yao HT, Fornika D, Gardy J, Hsiao W, Chauve C, Chindelevitch L (2018) MentaLiST–a fast MLST caller for large MLST schemes. Microb Genom 4(2). https://doi.org/10.1099/mgen.0.000146

Feyaerts AF, Luyten W, Van Dijck P (2020) Striking essential oil: tapping into a largely unexplored source for drug discovery. Sci Rep 10(1):2867. https://doi.org/10.1038/s41598-020-59332-5

Firmino DF, Cavalcante TTA, Gomes GA, Firmino NCS, Rosa LD, de Carvalho MG, Catunda FEA Jr (2018) Antibacterial and antibiofilm activities of *Cinnamomum* Sp. essential oil and Cinnamaldehyde: Antimicrobial activities. Sci World J 2018:7405736. https://doi.org/10.1155/2018/7405736

Fox LT, Gerber M, Plessis JD, Hamman JH (2011) Transdermal drug delivery enhancement by compounds of natural origin. Molecules 16(12):10507–10540. https://doi.org/10.3390/molecules161210507

Gao S, Singh J (1998) In vitro percutaneous absorption enhancement of a lipophilic drug tamoxifen by terpenes. J Control Release 51(2–3):193–199. https://doi.org/10.1016/s0168-3659(97)00168-5

Godebo G, Kibru G, Tassew H (2013) Multidrug-resistant bacterial isolates in infected wounds at Jimma University specialized hospital, Ethiopia. Ann Clin Microbiol Antimicrob 12:17. https://doi.org/10.1186/1476-0711-12-17

Gordon NC, Price JR, Cole K, Everitt R, Morgan M, Finney J, Kearns AM, Pichon B, Young B, Wilson DJ, Llewelyn MJ, Paul J, Peto TE, Crook DW, Walker AS, Golubchik T (2014) Prediction of *Staphylococcus aureus* antimicrobial resistance by whole-genome sequencing. J Clin Microbiol 52(4):1182–1191. https://doi.org/10.1128/JCM.03117-13

Gupta SK, Padmanabhan BR, Diene SM, Lopez-Rojas R, Kempf M, Landraud L, Rolain JM (2014) ARG-ANNOT, a new bioinformatic tool to discover antibiotic resistance genes in bacterial genomes. Antimicrob Agents Chemother 58(1):212–220. https://doi.org/10.1128/AAC.01310-13

Hoffmann S, Kleinstreuer N, Alepee N, Allen D, Api AM, Ashikaga T, Clouet E, Cluzel M, Desprez B, Gellatly N, Goebel C, Kern PS, Klaric M, Kuhnl J, Lalko JF, Martinozzi-Teissier S, Mewes K, Miyazawa M, Parakhia R, van Vliet E, Zang Q, Petersohn D (2018) Non-animal methods to predict skin sensitization (I): the cosmetics Europe database. Crit Rev Toxicol 48 (5):344–358. https://doi.org/10.1080/10408444.2018.1429385

Hughes JP, Rees S, Kalindjian SB, Philpott KL (2011) Principles of early drug discovery. Br J Pharmacol 162(6):1239–1249. https://doi.org/10.1111/j.1476-5381.2010.01127.x

Ianevski A, Giri AK, Gautam P, Kononov A, Potdar S, Saarela J, Wennerberg K, Aittokallio T (2019) Prediction of drug combination effects with a minimal set of experiments. Nat Mach Intell 1(12):568–577. https://doi.org/10.1038/s42256-019-0122-4

Ianevski A, He L, Aittokallio T, Tang J (2017) SynergyFinder: a web application for analyzing drug combination dose-response matrix data. Bioinformatics 33(15):2413–2415. https://doi.org/10.1093/bioinformatics/btx162

Iyer SV, Harpaz R, LePendu P, Bauer-Mehren A, Shah NH (2014) Mining clinical text for signals of adverse drug-drug interactions. J Am Med Inform Assoc 21(2):353–362. https://doi.org/10.1136/amiajnl-2013-001612

Jia B, Raphenya AR, Alcock B, Waglechner N, Guo P, Tsang KK, Lago BA, Dave BM, Pereira S, Sharma AN, Doshi S, Courtot M, Lo R, Williams LE, Frye JG, Elsayegh T, Sardar D, Westman EL, Pawlowski AC, Johnson TA, Brinkman FS, Wright GD, McArthur AG (2017) CARD 2017: expansion and model-centric curation of the comprehensive antibiotic resistance database. Nucleic Acids Res 45(D1):D566–D573. https://doi.org/10.1093/nar/gkw1004

Jorge P, Lourenco A, Pereira MO (2012) New trends in peptide-based anti-biofilm strategies: a review of recent achievements and bioinformatic approaches. Biofouling 28(10):1033–1061. https://doi.org/10.1080/08927014.2012.728210

Jorge P, Perez-Perez M, Perez Rodriguez G, Fdez-Riverola F, Pereira MO, Lourenco A (2016) Construction of antimicrobial peptide-drug combination networks from scientific literature based on a semi-automated curation workflow. Database (Oxford) 2016:baw143. https://doi.org/10.1093/database/baw143

Katritzky AR, Maran U, Lobanov VS, Karelson M (2000) Structurally diverse quantitative structure–property relationship correlations of technologically relevant physical properties. J Chem Inf Comput Sci 40(1):1–18. https://doi.org/10.1021/ci9903206

Khameneh B, Iranshahy M, Soheili V, Fazly Bazzaz BS (2019) Review on plant antimicrobials: a mechanistic viewpoint. Antimicrob Resist Infect Control 8:118. https://doi.org/10.1186/s13756-019-0559-6

Kolchinsky A, Lourenco A, Wu HY, Li L, Rocha LM (2015) Extraction of pharmacokinetic evidence of drug-drug interactions from the literature. PLoS One 10(5):e0122199. https://doi.org/10.1371/journal.pone.0122199

Kosina P, Paloncyova M, Svobodova AR, Zalesak B, Biedermann D, Ulrichova J, Vostalova J (2018) Dermal delivery of selected polyphenols from *Silybum marianum*. Theoretical and experimental study. Molecules 24(1):61. https://doi.org/10.3390/molecules24010061

Kumari P, Mishra R, Arora N, Chatrath A, Gangwar R, Roy P, Prasad R (2017) Antifungal and anti-biofilm activity of essential oil active components against *Cryptococcus neoformans* and *Cryptococcus laurentii*. Front Microbiol 8:2161. https://doi.org/10.3389/fmicb.2017.02161

Lagha R, Ben Abdallah F, Al-Sarhan BO, Al-Sodany Y (2019) Antibacterial and biofilm inhibitory activity of medicinal plant essential oils against *Escherichia coli* isolated from UTI patients. Molecules 24(6). https://doi.org/10.3390/molecules24061161

Lahlou M (2004) Methods to study the phytochemistry and bioactivity of essential oils. Phytother Res 18(6):435–448. https://doi.org/10.1002/ptr.1465

Lipinski CA (2016) Rule of five in 2015 and beyond: target and ligand structural limitations, ligand chemistry structure and drug discovery project decisions. Adv Drug Deliv Rev 101:34–41. https://doi.org/10.1016/j.addr.2016.04.029

Liu B, Zheng D, Jin Q, Chen L, Yang J (2019) VFDB 2019: a comparative pathogenomic platform with an interactive web interface. Nucleic Acids Res 47(D1):D687–D692. https://doi.org/10.1093/nar/gky1080

Lv J, Deng S, Zhang L (2020) A review of artificial intelligence applications for antimicrobial resistance. Biosafety and Health 3:1

Magana M, Pushpanathan M, Santos AL, Leanse L, Fernandez M, Ioannidis A, Giulianotti MA, Apidianakis Y, Bradfute S, Ferguson AL, Cherkasov A, Seleem MN, Pinilla C, de la Fuente-Nunez C, Lazaridis T, Dai T, Houghten RA, Hancock REW, Tegos GP (2020) The value of antimicrobial peptides in the age of resistance. Lancet Infect Dis 20(9):e216–e230. https://doi.org/10.1016/S1473-3099(20)30327-3

Mayr A, Klambauer G, Unterthiner T, Hochreiter S (2016) DeepTox: toxicity prediction using deep learning. Front Environ Sci 3(80). https://doi.org/10.3389/fenvs.2015.00080

Metcalf DG, Bowler PG (2013) Biofilm delays wound healing: a review of the evidence. Burns Trauma 1(1):5–12. https://doi.org/10.4103/2321-3868.113329

Moo CL, Yang SK, Yusoff K, Ajat M, Thomas W, Abushelaibi A, Lim SH, Lai KS (2020) Mechanisms of antimicrobial resistance (AMR) and alternative approaches to overcome AMR. Curr Drug Discov Technol 17(4):430–447. https://doi.org/10.2174/1570163816666190304122219

Naghmouchi K, Le Lay C, Baah J, Drider D (2012) Antibiotic and antimicrobial peptide combinations: synergistic inhibition of *Pseudomonas fluorescens* and antibiotic-resistant variants. Res Microbiol 163(2):101–108. https://doi.org/10.1016/j.resmic.2011.11.002

Negut I, Grumezescu V, Grumezescu AM (2018) Treatment strategies for infected wounds. Molecules 23(9):2392. https://doi.org/10.3390/molecules23092392

Patsilinakos A, Artini M, Papa R, Sabatino M, Bozovic M, Garzoli S, Vrenna G, Buzzi R, Manfredini S, Selan L, Ragno R (2019) Machine learning analyses on data including essential oil chemical composition and in vitro experimental antibiofilm activities against *Staphylococcus* species. Molecules 24(5):890. https://doi.org/10.3390/molecules24050890

Pereira Dos Santos E, Nicacio PHM, Coelho Barbosa F, Nunes da Silva H, Andrade ALS, Lia Fook MV, de Lima Silva SM, Farias Leite I (2019) Chitosan/essential oils formulations for potential use as wound dressing: physical and antimicrobial properties. Materials (Basel) 12(14):2223. https://doi.org/10.3390/ma12142223

Pichersky E, Noel JP, Dudareva N (2006) Biosynthesis of plant volatiles: nature's diversity and ingenuity. Science 311(5762):808–811. https://doi.org/10.1126/science.1118510

Pirvanescu H, Balasoiu M, Ciurea ME, Balasoiu AT, Manescu R (2014) Wound infections with multi-drug resistant bacteria. Chirurgia (Bucur) 109(1):73–79

Pizzolato-Cezar LR, Okuda-Shinagawa NM, Machini MT (2019) Combinatory therapy antimicrobial peptide-antibiotic to minimize the ongoing rise of resistance. Front Microbiol 10:1703. https://doi.org/10.3389/fmicb.2019.01703

Porto WF, Fensterseifer ICM, Ribeiro SM, Franco OL (2018) Joker: an algorithm to insert patterns into sequences for designing antimicrobial peptides. Biochim Biophys Acta Gen Subj 1862 (9):2043–2052. https://doi.org/10.1016/j.bbagen.2018.06.011

Ragno R, Papa R, Patsilinakos A, Vrenna G, Garzoli S, Tuccio V, Fiscarelli E, Selan L, Artini M (2020) Essential oils against bacterial isolates from cystic fibrosis patients by means of antimicrobial and unsupervised machine learning approaches. Sci Rep 10(1):2653. https://doi.org/10.1038/s41598-020-59553-8

Rajkovic K, Pekmezovic M, Barac A, Nikodinovic-Runic J, Arsić Arsenijević V (2015) Inhibitory effect of thyme and cinnamon essential oils on *Aspergillus flavus*: optimization and activity prediction model development. Ind Crop Prod 65:7–13. https://doi.org/10.1016/j.indcrop.2014.11.039

Sharma D, Misba L, Khan AU (2019) Antibiotics versus biofilm: an emerging battleground in microbial communities. Antimicrob Resist Infect Control 8:76. https://doi.org/10.1186/s13756-019-0533-3

Silver LL (2011) Challenges of antibacterial discovery. Clin Microbiol Rev 24(1):71–109. https://doi.org/10.1128/CMR.00030-10

Simoes D, Miguel SP, Ribeiro MP, Coutinho P, Mendonca AG, Correia IJ (2018) Recent advances on antimicrobial wound dressing: a review. Eur J Pharm Biopharm 127:130–141. https://doi.org/10.1016/j.ejpb.2018.02.022

Sims KR, Liu Y, Hwang G, Jung HI, Koo H, Benoit DSW (2018) Enhanced design and formulation of nanoparticles for anti-biofilm drug delivery. Nanoscale 11(1):219–236. https://doi.org/10.1039/c8nr05784b

Stoesser N, Batty EM, Eyre DW, Morgan M, Wyllie DH, Del Ojo Elias C, Johnson JR, Walker AS, Peto TE, Crook DW (2013) Predicting antimicrobial susceptibilities for *Escherichia coli* and *Klebsiella pneumoniae* isolates using whole genomic sequence data. J Antimicrob Chemother 68(10):2234–2244. https://doi.org/10.1093/jac/dkt180

Su M, Satola SW, Read TD (2019) Genome-based prediction of bacterial antibiotic resistance. J Clin Microbiol 57(3):e01405. https://doi.org/10.1128/JCM.01405-18

Swamy MK, Akhtar MS, Sinniah UR (2016) Antimicrobial properties of plant essential oils against human pathogens and their mode of action: an updated review. Evid Based Complement Alternat Med 2016:3012462. https://doi.org/10.1155/2016/3012462

Tetko IV, Tanchuk VY, Villa AE (2001) Prediction of n-octanol/water partition coefficients from PHYSPROP database using artificial neural networks and E-state indices. J Chem Inf Comput Sci 41(5):1407–1421. https://doi.org/10.1021/ci010368v

Valko K (2004) Application of high-performance liquid chromatography based measurements of lipophilicity to model biological distribution. J Chromatogr A 1037(1–2):299–310. https://doi.org/10.1016/j.chroma.2003.10.084

Vamathevan J, Clark D, Czodrowski P, Dunham I, Ferran E, Lee G, Li B, Madabhushi A, Shah P, Spitzer M, Zhao S (2019) Applications of machine learning in drug discovery and development. Nat Rev Drug Discov 18(6):463–477. https://doi.org/10.1038/s41573-019-0024-5

van Duin D, Paterson DL (2016) Multidrug-resistant bacteria in the community: trends and lessons learned. Infect Dis Clin N Am 30(2):377–390. https://doi.org/10.1016/j.idc.2016.02.004

Vasala A, Hytonen VP, Laitinen OH (2020) Modern tools for rapid diagnostics of antimicrobial resistance. Front Cell Infect Microbiol 10:308. https://doi.org/10.3389/fcimb.2020.00308

Wattam AR, Abraham D, Dalay O, Disz TL, Driscoll T, Gabbard JL, Gillespie JJ, Gough R, Hix D, Kenyon R, Machi D, Mao C, Nordberg EK, Olson R, Overbeek R, Pusch GD, Shukla M, Schulman J, Stevens RL, Sullivan DE, Vonstein V, Warren A, Will R, Wilson MJ, Yoo HS, Zhang C, Zhang Y, Sobral BW (2014) PATRIC, the bacterial bioinformatics database and analysis resource. Nucleic Acids Res 42(Database issue):D581–D591. https://doi.org/10.1093/nar/gkt1099

Wattam AR, Davis JJ, Assaf R, Boisvert S, Brettin T, Bun C, Conrad N, Dietrich EM, Disz T, Gabbard JL, Gerdes S, Henry CS, Kenyon RW, Machi D, Mao C, Nordberg EK, Olson GJ, Murphy-Olson DE, Olson R, Overbeek R, Parrello B, Pusch GD, Shukla M, Vonstein V, Warren A, Xia F, Yoo H, Stevens RL (2017) Improvements to PATRIC, the all-bacterial bioinformatics database and analysis resource center. Nucleic Acids Res 45(D1):D535–D542. https://doi.org/10.1093/nar/gkw1017

Wilm A, Kuhnl J, Kirchmair J (2018) Computational approaches for skin sensitization prediction. Crit Rev Toxicol 48(9):738–760. https://doi.org/10.1080/10408444.2018.1528207

Yang SK, Yusoff K, Ajat M, Thomas W, Abushelaibi A, Akseer R, Lim SE, Lai KS (2019) Disruption of KPC-producing *Klebsiella pneumoniae* membrane via induction of oxidative stress by cinnamon bark (*Cinnamomum verum* J. Presl) essential oil. PLoS One 14(4): e0214326. https://doi.org/10.1371/journal.pone.0214326

Yang SK, Yusoff K, Mai CW, Lim WM, Yap WS, Lim SE, Lai KS (2017) Additivity vs synergism: investigation of the additive interaction of cinnamon bark oil and Meropenem in combinatory therapy. Molecules 22(11):1733. https://doi.org/10.3390/molecules22111733

Yang SK, Yusoff K, Thomas W, Akseer R, Alhosani MS, Abushelaibi A, Lim SH, Lai KS (2020) Lavender essential oil induces oxidative stress which modifies the bacterial membrane permeability of carbapenemase producing *Klebsiella pneumoniae*. Sci Rep 10(1):819. https://doi.org/10.1038/s41598-019-55601-0

Yap PS, Lim SH, Hu CP, Yiap BC (2013) Combination of essential oils and antibiotics reduce antibiotic resistance in plasmid-conferred multidrug resistant bacteria. Phytomedicine 20 (8–9):710–713. https://doi.org/10.1016/j.phymed.2013.02.013

Lightning Source UK Ltd.
Milton Keynes UK
UKHW020612250722
406326UK00002B/3